AIR DISTRIBUTION IN ROOMS

Ventilation for Health and Sustainable Environment

Volume I

(in two volumes)

Elsevier Science Internet Homepage
http://www.elsevier.nl (Europe)
http://www.elsevier.com (America)
http://www.elsevier.co.jp (Asia)

Consult the Elsevier homepage for full catalogue information on all books, journals and electronic products and services.

Elsevier Titles of Related Interest

BANSAL, HAUSER & MINKE
Passive Building Design.
ISBN: 0-444-81745-X

CHENG & SHEU
Urban Disaster Mitigation
The Role of Engineering and Technology.
ISBN: 0-08-041920-8

GOUMANS, SENDEN & VAN DER SLOOT
Waste Materials in Construction.
ISBN: 0-444-82771-4

MAKELAINEN
ICSAS '99, Int Conf on Light-Weight Steel and
Aluminium Structures.
ISBN: 0-08-043014-7

MARONI, SEIFERT & LINDVALL
Indoor Air Quality.
ISBN: 0-444-81642-9

MORAWSKA, BOFINGER & MARONI
Indoor Air: An Integrated Approach.
ISBN 0-08-041917-8

SAMSON *ET AL*
Health Implications of Fungi in Indoor
Environments.
ISBN: 0-444-81866-9

SAYIGH, SALA & GALLO
Architecture - Comfort and Energy.
ISBN 0-08-043004-X

TANABE
Comparative Performance of Seismic Design Codes
for Concrete Structures.
ISBN 0-08-043021-X

WILKENING
Radon in the Environment.
ISBN: 0-444-88163-8

Related Journals
Free specimen copy gladly sent on request: Elsevier Science Ltd, The Boulevard, Langford Lane, Kidlington, Oxford, OX5 1GB, UK

Applied Thermal Engineering
Atmospheric Environment
Automation in Construction
Bioresource Technology
Building and Environment
Construction and Building Materials
Energy and Buildings
Engineering Structures

Environmental Science and Policy
Fire Safety Journal
Journal of Wind Engineering and Industrial
 Aerodynamics
Renewable Energy
Solar Energy
Solar Energy Materials and Solar Cells

To Contact the Publisher
Elsevier Science welcomes enquiries concerning publishing proposals: books, journal special issues, conference proceedings etc. All formats and media can be considered. Should you have a publishing proposal you wish to discuss, please contact, without obligation, the publisher responsible for Elsevier's civil and structural engineering publishing programme:

Mr Ian Salusbury
Elsevier Science Ltd
The Boulevard, Langford Lane   Phone: +44 (0) 1865 843425
Kidlington, Oxford   Fax: +44 (0) 1865 843920
OX5 1GB, UK   E.mail: i.salusbury@elsevier.co.uk

General enquiries, including placing orders, should be directed to Elsevier's Regional Sales Offices – please access the Elsevier homepage for full contact details (homepage details at the top of this page).

ELSEVIER SCIENCE Ltd
The Boulevard, Langford Lane
Kidlington, Oxford OX5 1GB

The papers presented in these proceedings have been reproduced directly from the author's 'camera ready' manuscripts. As such the presentation and reproduction quality may vary from paper to paper.

AIR DISTRIBUTION IN ROOMS

Ventilation for Health and Sustainable Environment

Proceedings of the 7th International Conference
on Air Distribution in Rooms

9 - 12 July 2000
Reading, UK

Volume I

Keynote Papers
Indoor Environment
Predictive Methods
Air Distribution

Editor
Hazim B. Awbi

2000

ELSEVIER

AMSTERDAM · LAUSANNE · NEW YORK · OXFORD · SHANNON · SINGAPORE · TOKYO

ELSEVIER SCIENCE Ltd
The Boulevard, Langford Lane
Kidlington, Oxford OX5 1GB, UK

First edition 2000

Library of Congress Cataloging in Publication Data
A catalog record from the Library of Congress has been applied for.

British Library Cataloguing in Publication Data
A catalogue record from the British Library has been applied for.

ISBN: 0 080 43017 1

♾ The paper used in this publication meets the requirements of ANSI/NISO Z39.48-1992 (Permanence of Paper).
Printed in The Netherlands.

PREFACE

The air distribution in occupied spaces is a major issue of public concern. It is widely recognized that the quality of air and the nature of airflow can affect the health of occupants and the energy consumed in buildings and transport vehicles. ROOMVENT is the principal international conference in the field of air distribution. It was first initiated in 1987 by SCANVAC, the Scandinavian Federation of Heating, Ventilating and Sanitary Engineering Associations in Denmark, Finland, Iceland, Norway and Sweden. The first conference was held in Stockholm in 1987. The second ROOMVENT conference was held in Oslo in 1990 and since then, the meeting has been held biennially in Aalborg, Denmark, Krakow, Poland, Yokohama, Japan and, for the second time, in Stockholm in 1998. It is a privilege to host the seventh conference, ROOMVENT 2000, in Reading, United Kingdom.

The aim of the Conference has not changed since it was first held in Stockholm, which is to bring together researchers from universities and research institutes, engineers from industry and government officials and policy makers, with the goal of experiencing the latest techniques for measuring and analyzing indoor air flow, the visualization of indoor air flow patterns, the evaluation of ventilation parameters and the most recent developments in computer simulation techniques of room airflow. It is hoped that the theme of ROOMVENT 2000 *"Ventilation for Health and Sustainable Environment"* will set the scene for room air distribution research and development for the new millennium.

These Proceedings are in two volumes and contain 198 papers that have been reviewed by members of the International Scientific Committee and other experts. These papers have been written by researchers in the field of air distribution from many countries and collectively they represent the most recent advances in the field worldwide.

I am very grateful to the valuable help and enthusiasm received from members of the Programme and Organizing Committee, the work of the members of the International Scientific Committee and the generosity of the Co-sponsoring Organizations.

Hazim Awbi
Chairman, ROOMVENT 2000
July 2000

PROGRAMME AND ORGANIZING COMMITTEE

Dr. Hazim Awbi (Chairman, University of Reading)
Professor Farshad Alamdari (Building Research Establishment)
Dr. Geoffrey Brundrett (Chartered Institution of Building Services Engineers)
Dr. Gavin Davies (Ove Arup & Partners)
Dr. Edward Finch (University of Reading)
Dr Martin Liddament (Air Infiltration and Ventilation Centre)
Mrs Sheila Rogers (University of Reading)
Professor Ali Sayigh (World Renewable Energy Network)

INTERNATIONAL SCIENTIFIC COMMITTEE

Professor Francis Allard,
University of La Rochelle, France
Professor Farshad Alamdari,
Building Research Establishment, UK
Dr. Hazim Awbi,
University of Reading, UK
Professor James Axley,
Yale University, USA
Professor Richard Aynsley,
James Cook University, Australia
Professor Laszlo Banhidi,
Hungary
Dr. Dominique Blay,
Université de Poitiers, France
Dr. Qingyan Chen,
Massachuttes Institute of Technology, USA
Viktor Dorer,
EMPA, Switzerland
Professor P. Ole Fanger,
Technical University of Denmark, Denmark
Dr. Helmut Feustel,
Kessler & Luch GmbH, Germany
Professor Klaus Fitzner,
Technical University of Berlin, Germany
Professor Gian Vincenzo Fracastoro,
Polytechnic of Turin, Italy
Professor David Grimsrud,
University of Minnesota, USA
Professor Fariborz Haghighat,
Concordia University, Canada
Professor M.A. Haque,
National University of Singapore, Singapore
Professor Michael Holmes,
Ove Arup & Partners, UK
Professor Shinsuke Kato,
University of Tokyo, Japan

Professor Allan Kirkpatrick,
Colorado State University, USA
Professor Jean Lebrun,
Universiti de Liege, Belgium
Professor Kyung-Hoi Lee,
Yonsei University, Korea
Dr. Martin Liddament,
Air Infiltration and Ventilation Centre, UK
Dr. Yuguo Li,
CSIRO, Australia
Professor Paul Linden,
University of California, USA
Professor Tor-Goran Malmstrom,
KTH, Sweden
Dr. Alfred Moser,
ETH-Zentrum, Switzerland
Dr. Elisabeth Mundt,
KTH, Sweden
Professor Shuzo Murakami,
University of Tokyo, Japan
Professor Peter V. Nielsen,
Aalborg University, Denmark
Dr. Raimo Niemela,
Finnish Institute of Occupational Health, Finland
Professor Zbigniew Popiolek,
Silesian Technical University, Poland
Professor Ulrich Renz,
RWTH Aachen, Germany
Professor Claude-Alain Roulet,
Swiss Federal Institute of Technology,
Switzerland
Professor Mats Sandberg,
KTH, Sweden
Professor Olli Seppanen,
Helsinki University of Technology, Finland

Dr. C.Y. Shaw,
National Research Council, Canada
Dr. Eimund Skaret,
Norwegian Building Research Institute, Norway
Dr. Per Olaf Tjelflaat,
NTNU, Norway

Professor Hiroshi Yoshino,
Tohoku University, Japan
Professor Rong-yi Zhao,
Tsinghua University, China

CO-SPONSORING ORGANIZATIONS

The University of Reading
Air Infiltration and Ventilation Centre
American Society of Heating, Refrigeration and Air-Conditioning Engineers, Inc.
Society of Heating, and Sanitary Engineers, Japan
Building Research Establishment Ltd.
Building Services Research and Information Association
The Chartered Institution of Building Services Engineers
Dantec Electronics Ltd.
Faversham House Group Ltd.
Ove Arup & Partners
SCANVAC (Scandinavian Federation of Heating, Ventilating and Sanitary Engineering Associations in Denmark, Finland, Iceland, Norway and Sweden)
World Renewable Energy Network

CONTENTS

VOLUME I

PREDICTIVE METHODS:

AIR DISTRIBUTION:

VOLUME II

VENTILATION STRATEGIES:

APPLICATIONS:

2000 Elsevier Science Ltd. All rights reserved.
Air Distribution in Rooms, (ROOMVENT 2000)
Editor: H.B. Awbi

PROVIDE GOOD AIR QUALITY FOR PEOPLE AND IMPROVE THEIR PRODUCTIVITY

P. Ole Fanger, D.Sc.
Director, International Centre for Indoor Environment and Energy,
Technical University of Denmark, DK-2800 Lyngby, Denmark
www.et.dtu.dk

ABSTRACT

Three recent independent studies have documented that the quality of indoor air has a significant and positive influence on the productivity of office workers. A combined analysis of the results of the three studies shows a significant relationship between productivity and perceived indoor air quality. The impact on productivity justifies a much higher indoor air quality than the minimum levels prescribed in present standards and guidelines. One way of providing air of high quality for people to breathe, without involving excessive ventilation rates and energy use, is to provide "personalized air" to each individual.The application of this concept is discussed.

KEYWORDS Indoor air quality, Productivity, Ventilation, Personalized air

INTRODUCTION

In 1936, Yagou introduced a new philosophy for ventilation, the aim being to provide an indoor air quality that is perceived acceptable by people. This philosophy has since then dominated the thinking in ventilation standards for nonindustrial buildings. It is still the idea behind recent standards and guidelines such as the ASHRAE Standard 62 (ASHRAE, 1999) and the recent CR 1752 (CEN, 1998). But indoor air quality has an impact on humans beyond perception. Three recent studies have now documented that indoor air quality has a significant impact on productivity in offices and on SBS symptoms. These studies will be reviewed in the first section of this paper.

High air quality in a space can be achieved by decreasing the pollution sources, by increasing the ventilation rate, or by cleaning the air. But what really counts is the quality of the air that the occupants breathe. One option is to supply air of high quality direct to the breathing zone of each individual. The establishment of such "personalized air" will be discussed in the second section of this paper.

PRODUCTIVITY AND INDOOR AIR QUALITY

Three recent independent studies document that the quality of indoor air has a significant and positive influence on the productivity of office workers. In one study, a well-controlled normal office (field lab) was used in which two different air qualities were established by including or excluding an extra pollution source, invisible to the occupants (Wargocki et al., 1999). The two cases corresponded to a low-polluting and a non-low-polluting building as specified in the new European guidelines for the design of indoor environments (CEN, 1998).The same subjects worked for 4-1/2 hours on simulated office work in each of the two air qualities. The ventilation rate and all other environmental factors were the same under the two conditions. The productivity of the subjects was found to be 6.5% higher (P<0.003) in good air quality (Fig. 1) and they also made fewer errors and experienced fewer SBS symptoms. This study performed in Denmark has later been repeated in Sweden with similar results (Lagercrantz, 2000). A third study was performed in the Danish field lab with the same pollution sources present at three different ventilation rates: 3, 10 and 30 l/s·person. The productivity increased significantly by increased ventilation (Figure 2). The three studies involving seven experimental conditions and 90 subjects have been analysed as a whole, relating productivity to perceived air quality (Wargocki et al., 2000). The results are presented in Figure 3 and show a significant influence of perceived air quality on productivity in offices. An improvement of perceived air quality by 1 decipol increased productivity by 0.5%. The results of three blind studies document that improved air quality increases productivity significantly.

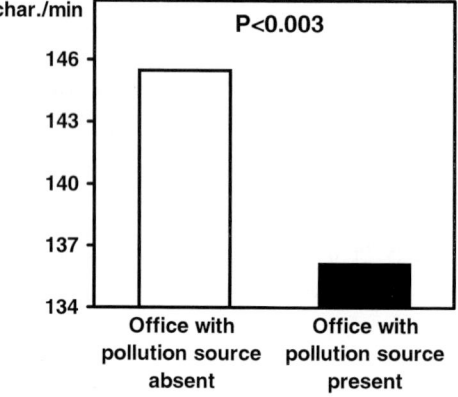

Fig. 1. Impact of indoor air pollution on productivity, i.e. number of characters typed on a PC (Wargocki et al., 1999a)

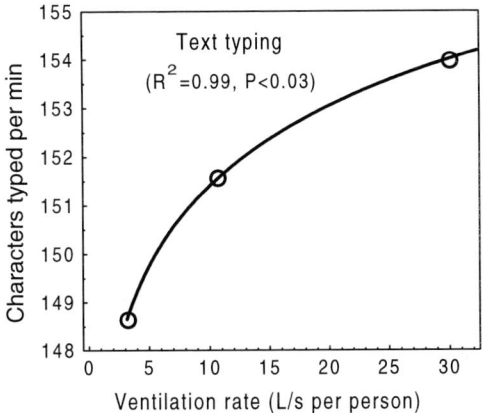

Fig. 2. Impact of ventilation rate on productivity (Wargocki, 1999b)

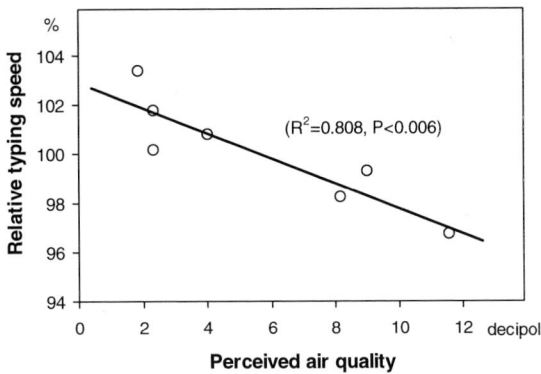

Fig. 3. Relation between perceived air quality and productivity (Wargocki, 2000)

PERSONALIZED AIR

In many ventilated rooms the outdoor air supplied is of the order of magnitude of 10 l/s·person. Of this air, only 0.1 l/s·person, or 1%, is inhaled. The rest, i.e. 99% of the supplied air, is not used. What a huge waste! And the 1% of the ventilation air being inhaled by human occupants is not even clean. It is polluted in the space by bioeffluents, emissions from building materials and sometimes even by environmental tobacco smoke before it is inhaled.

The idea of mixing ventilation is to provide the same quality of air in the entire volume of the space. This means that occupants will find the same quality of air for breathing whether they are sitting at their desk, standing on the desk or lying on the floor.

Displacement ventilation systems do acknowledge the air quality in the breathing zone but the ventilation effectiveness is usually only moderately better than with mixing ventilation. What I foresee in the future are systems that supply rather small quantities of clean air close to the breathing zone of each individual. The idea would be to serve to each occupant, clean air that is unpolluted by the pollution sources in the space. We would hesitate to drink water from a swimming pool polluted by human bioeffluents. Still we accept consuming indoor air that has previously been in the lungs of other persons and is polluted by human bioeffluents and other contaminants generated in the space. Why not serve small quantities of high-quality air direct to each individual rather than serving plenty of mediocre air throughout the space? Such "personalized air" (PA) should be provided so that the person inhales clean air from the core of the jet where the air is unmixed with polluted room air (Fig. 4). In an office the PA may, for instance, come from an outlet next to the PC on the desk. It is essential that the air is served "gently", i.e. has a low velocity and turbulence which do not cause draught (Fanger et al., 1988). By means of personalized air it is possible to provide breathing air of optimal quality. The air will be perceived as fresh and pleasant with a positive effect on human productivity as indicated in Fig. 3.

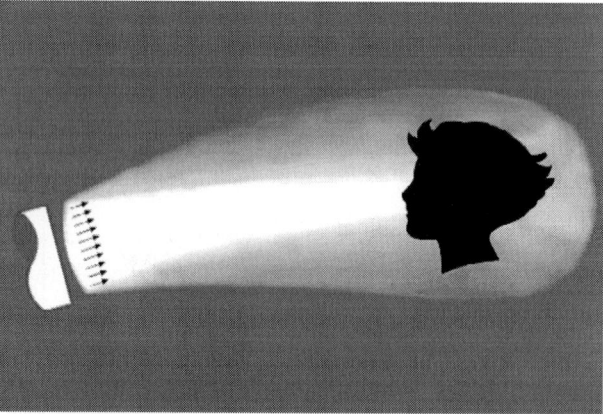

Fig. 4. The principle of personalized air (PA): small amounts of clean air
supplied directly and gently to a person's breathing zone (Fanger, 2000)

The challenge for HVAC engineering in the future will be to develop conditioning and cleaning processes so that air is perceived optimally and to develop appropriate methods for transporting this air to the breathing zone of each individual without mixing with room air.

CONCLUSIONS

- Three different studies have documented a positive effect of perceived indoor air quality on productivity in offices.

- Personalized air supplied to the breathing zone of each individual is a promising concept, allowing a quality of the air for breathing that is optimal for human perception and productivity. Further work on studying and developing this concept is recommended.

ACKNOWLEDGEMENT

The International Centre for Indoor Environment and Energy was established in 1998, based on a 10-year contract between the Danish Technical Research Council (STVF) and the Technical University of Denmark (DTU). The support of STVF and DTU is gratefully acknowledged.

REFERENCES

ASHRAE Standard 62-1999 (1999) *Ventilation for Acceptable Indoor Air Quality*, Atlanta, GA, American Society of Heating, Refrigerating and Air-Conditioning Engineers.

CEN (1998) *Ventilation for Buildings: Design Criteria for the Indoor Environment*, Brussels, European Committee for Standardization (CR 1752).

Fanger, P.O., Melikov, A.K., Hanzawa, H., Ring, J. (1988) "Air turbulence and sensation of draught", *Energy and Buildings*, **12**, 21-40.

Fanger, P.O. (2000) "IAQ in the 21st century: search for excellence", *Indoor Air*, **2**

Lagercrantz, L., Wistrand, M., Willén, U., Wargocki, P., Witterseh, T., Sundell, J. (2000) "Negative impact of air pollution on productivity: previous Danish findings repeated in new Swedish test room", Proc. of Healthy Buildings 2000 (accepted).

Wargocki, P., Wyon, D.P., Baik, Y.K., Clausen, G., Fanger, P.O. (1999a) "Perceived air quality, sick building syndrome (SBS) symptoms and productivity in an office with two different pollution loads", *Indoor Air*, **9**(3),165-179.

Wargocki, P., Fanger, P.O. (1999b) "Impact of ventilation rates on SBS symptoms and productivity in offices", Proc. of DKV-Jahrestagung, Berlin, Vol. IV, pp. 218-224.

Wargocki, P., Wyon, D.P., Fanger, P.O. (2000) "Productivity is affected by the air quality in offices", Proc. of Healthy Buildings 2000 (accepted).

Air Distribution in Rooms, (ROOMVENT 2000)
Editor: H.B. Awbi

THE QUALITY OF BUILDINGS IS EQUAL TO THE INDOOR CLIMATE QUALITY

Ulf Rengholt
CURAB Consultants, Stockholm, Sweden

ABSTRACT

Most of the worldwide industries have learned that the key to development and profitability is "consumer satisfaction". They have acknowledged the superiority of the customer and developed sophisticated methods to identify customer needs and expectations. The ventilation technology has focused more on technology, standards and regulations than on the satisfaction of building occupants. This has taken the ventilation technology far away from the dynamics a consumer-focused view can generate. In the situation where indoor climate quality becomes the main quality element of buildings, it is important to develop a better understanding of occupants' need and expectations and focus more deeply on occupant satisfaction issues.

KEYWORDS

Occupants satisfaction, facilities management, indoor climate quality, quality development, health and wellbeing

THE KING AND THE QUEEN–EMPERORS OF THE MARKET ECONOMY

The King is King and the Queen is Queen. They are consumers of products and services or occupants of buildings on the worldwide markets. What they like or dislike decides the fate of worldwide industries. They are the emperors of to-days markets. Their wishes form the target of future technologies.

Many years ago the leading worldwide industries learned that the key to development and profitability was "consumer satisfaction". In the beginning this had to do with the transformation of the Japanese nation. During the early 50-ies under the influence of the late J Edwards Deming[1], Japan changed from an underdeveloped post-war nation to one of the worlds leading industrial societies. The real base of this

---

[1] See Bibliography

extraordinary strong and fast change was just the understanding that producing products and services with a high degree of consumer satisfaction is the key to profitability and prosperity.

Since then worldwide industries in an ever-increasing scale has adopted the customer view to the development of products and services. By the help of sophisticated methods, customer needs and expectations has been identified and used as development targets.
The quality concepts, initially developed by J Edwards Deming, Crosby, Juran and others, are followed by new quality management and development tools. Among them can be mentioned Peters value-adding concepts, Norman's service-concepts postulating that cooperation between customers and service-suppliers can generate new values and new profitable business.
To summarise, quality strategies, value-adding, new service-concepts etc. have shown to be basic tools to create worldwide attractive products, gain good customer relations and profitable business. The superiority of the customer and consumer satisfaction has been acknowledged. The lodestar is customer satisfaction.

THE VENTILATION KING

What strategies will be used by the ventilation technology in the situation where the customer dominance continuously increases? What is done to satisfy the customers of the building process– the building occupants?
There are not too many signs of occupant–focused development views in ventilation technology. The meaning of customer satisfaction is usually decided by standardisation committees or national authorities regarding the customer as clusters and not as individuals. The customer (the occupant) is regarded as a consumer of fans, ducts and air-distributors – things completely uninteresting for the ordinary occupant. Quality assurance of indoor climate is still a word without connection to occupant's realities. However, the real interest of the occupant – indoor health and wellbeing – is more seldom addressed. Ventilation technology is still focused on pure technology.
The name of the Ventilation King of today is "Rules and Standards" and not consumer satisfaction. This has taken the ventilation technology far away from the dynamics a user-focused view can generate.

THE QUALITY OF BUILDINGS

The quality of buildings is equal to the indoor climate quality. Even if this statement may seem exaggerated at present, it will nevertheless be true in the near future. The process that continuously deepens the importance of IAQ and increases occupants' interest in indoor health and comfort issues has been in progress for many years.
The result is clear–the indoor climate quality will be seen as the main quality element of a building. The question is–when will the ventilation technology adopt the ongoing development and what ways will be used?

FROM COMFORT TO CHALLENGES

The ventilation technology seems to function within something that could be called "**comfort zone**". This zone provides routines and well-known tasks in a familiar environment, controlled by standards and rules. It is easy to go on doing the same thing and not pay enough attention to changing circumstances. The danger of resting in this zone is obvious.

The **learning zone** is a rather well known concept. Learning means that new research; new views and changes must be identified and applied. Learning should be an ongoing process that never stops.

Applied to indoor climate, the learning process must involve the evaluation and understanding of occupant's opinions. Here are some examples exposing common opinions among building occupants.

- Occupants rate buildings as most comfortable when conditions are stable and reasonably predictable for most of the time within acceptable but not necessarily ideal comfort thresholds.
- Occupants like operable windows, comfortable thermal conditions especially in hot summer periods, shallow plan depths, acoustic separation, good views, and usable controls that are easy for them to understand.
- Occupants want to be able to alter conditions quickly in response to unpredictable events (like glare, draughts, or noises outside). If conflicting or unsatisfactory conditions occur, occupants want to decide for themselves how to resolve the conflicts by overriding default settings rather than having conditions chosen for them.
- Occupants ask for facilities management teams, which deal with complaints sensitively and rapidly.

Occupants who perceive that they are comfortable also tend to say that they are healthy and productive at work. The meaning of this is that occupant rate the wholeness – the interaction between all relevant influences. Occupant satisfaction is closely related to the interaction between the occupants, building and building operation. Occupant satisfaction cannot be gained unless building operation and facilities management will be allowed to play their important roles.

This kind of knowledge should be identified and evaluated in the learning process. However, learning in itself may not be enough to clear a path for business development. To cultivate new markets something more must be added – risking mistakes to enter new business. In other words – entering the challenge zone.

The **challenge zone** is the area where new ideas and new business opportunities are applied. Doing things in a new way, or work together with professionals from new fields are important procedures in this zone. Putting business into the challenge zone means risking mistakes to gain new opportunities.
Translated to indoor climate business, the challenge zone implies that ventilation itself has no value. The only value there is, is user health and satisfaction. The challenge zone has been waiting for many years on ventilation technology.

THE PARADIGM SHIFT

The growing importance of customer needs and expectations appear in many ways. Talking about customers in the form of building occupants, it seems natural to assume that they perceive the indoor environment more in its entirety than in detail. The entirety has to do not only with the quality of the building itself but also with the quality of building operation and management. As a matter of fact – practice show that occupant satisfaction is more connected with how design and management factors interact to create a total building system than with design and technical features. In other words, building operation and management develops to be a powerful quality tool concerning building- and indoor climate quality.

Building operation will play a much greater role than has been acknowledged. A paradigm shift is going on. Focus of the built environment is changing from building production to building operation. Facilities Management develops to be a future key to occupant satisfaction on the whole. The interaction between buildings and occupants is the factor that decides profitability and development. The building itself is going to loose its power in favour of the power of operation quality.

THE FUTURE OF ROOM-VENT

Now, someone might say that this has nothing to do with Room-Vent. We are concerned with room air distribution and connected technology. We do not run business. Our task is to proceed with important R&D for the benefit of suitable air distribution in rooms. Our problems have to do with getting financial support for future R&D.

This is a limited view. In the new century, Room-Vent could turn into a less important role than of today. Room air-distribution technology is connected to both thermal comfort and IAQ, thus forming an important part of occupant satisfaction with the indoor climate. Turning to user-focused ventilation technology, Room-Vent could be the predecessor of a new era where consumer views, paradigm shift, and other important issues are acknowledged.

Broadening the mind and scope of Room-Vent and focusing on general issues of occupants satisfaction more than on pure technological issues can be a relevant way to proceed into the new century. However, working in the comfort zone, trusting on the obsolete power of rigid rules and standards, will sooner or later appear as an outdated strategy.

Lack of new ideas is equal to lack of profitable development. Being aware of future market needs in favour of occupant satisfaction, Room-Vent can take a scientific lead to speed up necessary changes. Turning technology-focused views into user focused issues clearly connected to comfortable, healthy and productive indoor climate is an important task. Changing the ventilation dogma "ventilation rate according to standard" from a goal to a mean is a thrilling task.

BIBLIOGRAPHY

Deming, J. Edwards (1982). *Out of the Crisis, Quality, Productivity and Competitive Position.* Cambridge University Press, UK.

Crosby, Philip B. (1985). *Quality Without Tears.* New American Library, New York.

Juran J.M. (1988). *Juran on Planning for Quality.* The Free Press, New York.

Senge P.M. (1992). *The Fifth Discipline: The Art and Practice of the Learning Organization.* Doubleday, London.

Townsend, Gebhardt J.E. (1992). *Quality in Action.* John Wiley & Sons Inc, New York.
Norman, Richard (1983). *Service Management.* Liber Forlag, Stockholm.

Naisbitt, John (1982). *Megatrends.* New York.

Peters, Thomas J. and Waterman Jr., Robert H.(1982). *In Search of Excellence.* USA.

ISO 9000-2. Quality management and quality assurance standards - Part 2: Generic guidelines for the application of ISO 9001, ISO 9002 and ISO 9003.

IS0 9000-4. Quality management and quality assurance standards - Part 4: Guide to dependability programme management.

IS0 9004-3. Quality management and quality system elements - Part 3: Guidelines for processed materials.

IS0 9004-4. Quality management and quality system elements - Part 4: Guidelines for quality improvement.

ISO IOO13. Guidelines for developing quality manuals.

Indoor climate and ventilation for tomorrows needs – Quality based principles and guidelines for the ventilation of indoor environments and workplaces (Scanvac Guidelines), Swedish Indoor Climate Institute. (1990) 112 94 Stockholm.

Air Distribution in Rooms, (ROOMVENT 2000)
Editor: H.B. Awbi

OPTIMISED INDOOR AIR QUALITY AND ENERGY SAVING STRATEGIES FOR QUALITY BUILDINGS IN MEDITERRANEAN CLIMATES: COMPATIBLE OBJECTIVES WITH COMFORT AND ENVIRONMENT ISSUES?

Luis Malheiro da Silva

LM - Projecto e Gestão de Instalações Especiais, Lda.
Estrada da Torre 83
1750-294 LISBOA, Portugal
E-mail: lms@mail.telepac.pt

ABSTRACT

The sick building syndrome is a growing concern in terms of health and is an important factor in occupancy behaviour. The building sector takes up more than 35 % of the global primary energy use in developed countries. Those two main concerns in building services design are closely related, demanding an accurate balance between strategies focused on energy consumption reduction and indoor air quality improvement, whereas controlled natural ventilation potential must be considered as an important contribution.

Designers must be aware of these facts and integrate, in the final solution for buildings, the global environment concerns and comfort standards leading all together to the quality (**Q**) building label.
The aim of this paper is to establish, in a consistent way, that those main aspects are compatible between them and can play an important role in comfort and environment issues.
This approach has been used in the design of several buildings in EXPO'98 and, from one of them, field measurements are reported, confirming design estimations.

KEYWORDS

Quality (Q) Buildings, Energy Savings, Indoor Air Quality, Natural Ventilation, Air Conditioning, Thermal Comfort

INTRODUCTION

Q building should be defined, in our opinion, as a building where it is possible to match the following main objectives simultaneously:

1. Global energy efficiency (passive and active aspects) leading to energy consumption levels and related pollutant emissions below reference values for a specific use.
2. Optimised Indoor Air Quality levels (CO_2, SO_2, NOx, O_3, etc. below reference values).
3. Comfort performance according to category A, CR 1752.
4. Automatic adaptability of the active systems to current variations in occupancy and running periods.
5. Controlled first investment and running costs.

Obviously only a team work of all the designers involved (Architects and Structural Engineers, M & E Engineers, Building Management System [*BMS*] Engineers, etc.) may lead to a final result in the scope of the **Q** building concept.

SUMMARY OF STRATEGIES TO BE CONSIDERED IN Q BUILDINGS DESIGN

In order to be considered as a **Q** building the following important passive features are recommended:

- Solar shading devices according to the geographic location and façade orientation
- Use of structural mass to increase thermal inertia (air exhaust through ducts embedded in the massive slabs, etc.)
- Use of buffer spaces as a thermal regulator (atrium, etc.)
- Low "U" values used for envelope materials
- Low-E glass
- Natural cross ventilation

The main active recommended strategies are:

- Displacement ventilation or radiation based system (cooling ceilings, etc.)
- Full use of free cooling, when weather conditions allow
- Full fresh air (100 %) during Summer or midseason operation
- Motorised operable windows
- Artificial lighting based on the desk and uplighting concepts and general use of electronic ballast technology
- Pre-cooling, night purge during Summer and pre-heating during Winter
- Solar radiation, daylight and glare control *via* motorised louvers using a solar tracking device, if needed
- Use of refrigerant gases with ODP = 0 (connection to the DH & DC if available)
- Building Management System
- Variable speed fluid distribution pumps

THERMAL ENERGY CONSUMPTION ESTIMATIONS

In the same architectural base we consider two options:

- Standard (S) building as one with an envelope conforming to the existing regulations and air mixing HVAC system.

- Quality (**Q**) building as one with improved envelope quality, displacement ventilation HVAC system, use of structural inertia from slabs and exhaust ventilation through the atrium.

For both cases thermal energy estimations were performed using two different tools (commercial software and spreadsheet respectively).

In both cases, hourly-based thermal energy calculations were obtained for a typical day of each month. While the commercial software (TRACE ULTRA 600) uses CLTD-CLF and UATD algorithms for cooling and heating capacity calculations, the spreadsheet model uses outside air temperature correction in order to obtain a constant air supply temperature of 20 °C, using official weather information.

Thermal energy estimations for standard and Q buildings are presented in Fig. 1.

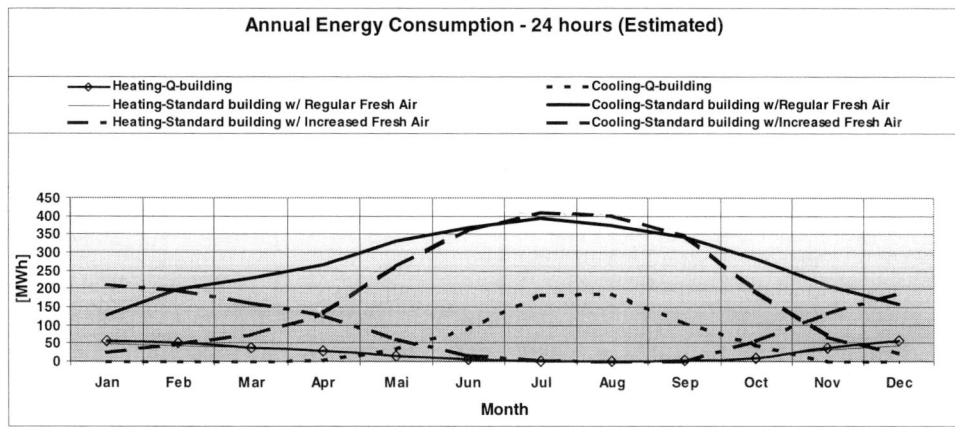

Figure 1 – Energy use estimations

Annual energy consumption estimations are presented in Table 1 below.

TABLE 1 – Energy estimations comparison between **Q** and S buildings

System	Heating	Cooling
	(M W h)	
Q building	298	648
S building w/ Regular Fresh Air	259	3276
S building w/Increased Fresh Air	1155	2348

COMFORT SIMULATIONS

For the **Q** building solution, in order to confirm that comfort conditions are obtained *CFD* software is used to perform comfort assessment (TAS-AMBIENS) and considering the volume control geometry and boundary conditions defined in Fig. 2 and Table 2.

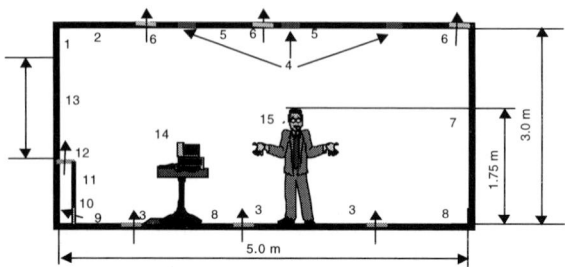

Figure 2 – Control volume and boundary conditions reference numbers

TABLE 2 – Boundary condition values for heating and cooling modes

Ref[a]. (#)	Element	Surface Temp. Heating (ºC)	Surface Temp. Cooling (ºC)	R. Humidity Heating (%)	R. Humidity Cooling (%)	Air Velocity (m/s)
1	wall	18	28	-	-	-
2	ceiling	20	25	-	-	-
3	air diffuser	21	21	35	50	0.1
4	light fixture	38	38	-	-	-
5	ceiling	20	25	-	-	-
6	return grill	free	free	-	-	free
7	partition	20	25	-	-	-
8	floor	20	25	-	-	-
9	floor	20	26	-	-	-
10	return grill	20	27	-	-	-
11	partition	16	27	-	-	-
12	air diffuser	16	28	-	-	-
13	glass	13	27	-	-	-
14	computer	-	-	-	-	-
15	occupant	-	-	-	-	-

In Figs. 3 and 4 the Percentage of People Dissatisfied *(PPD)* values obtained for 1.2 Met and 1.0/0.7 clo, Winter and Summer conditions, respectively, are presented.

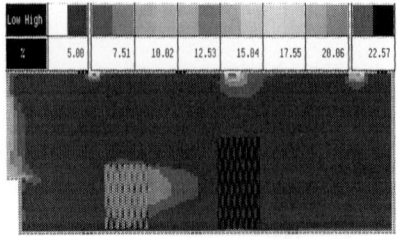

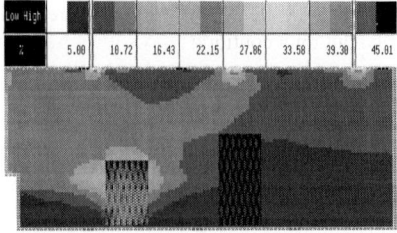

Figure 3 – *PPD* (Winter condition)   Figure 4 – *PPD* (Summer condition)

FIELD MEASUREMENTS

Taking into account the diversity of topics assessed, namely energy use, thermal comfort and indoor air quality, different field measurements were carried out for the **Q** building option which was, in fact, built (see Fig. 5), according to the above description and assumptions.

Figure 5 – Q building in EXPO'98 – Lisbon (~9100 m^2)

MEASUREMENT TECHNIQUES

Thermal Cooling and Heating Energy

Thermal energy consumption was measured by the District Heating and District Cooling (*DH & DC*) service provider, which is based on enthalpy meters. A complete year (1999) was recorded on a monthly basis.

Indoor Air Temperature and Humidity

Two types of measurements were made. One was carried out by the Building Management System (*BMS*) for the hotter months of 1999 (July and August) and the other using data loggers. In the first case four measurements per day were recorded, corresponding to 9:30, 12:00, 15:00 and 17:30 hours for all the office floors and different building wings. In the second case, ranging from 2000-03-06 to 2000-03-13, 15 minutes time interval between two consecutive measurements were carried out on a continuos basis for different representative building orientations.

Indoor Air Quality

Two techniques were used when measuring *IAQ* parameters and toxic substance concentrations. The CO_2, H_2S, NO_2, SO_2 measurement was carried out using continuous monitoring electronic equipment with interchangeable sensors for the substances other than CO_2. The equipment is provided with built in air temperature and relative humidity sensors. The contaminant gases were measured using reagent probes and aspiration pump.

The CO_2 decay concentration time was evaluated after a period of 1-hour complete fresh air shut off. The measurements of contaminant substances were made both at the air exhaust grill of the selected typical office space zone and in the office space itself.

The time sampling for the different measurements is shown in the results presentation section along with the related equipment accuracy.

RESULTS PRESENTATION

Thermal energy use

TABLE 3 – Thermal energy measurements for the whole **Q** building

Month	Mesured Heating kW h/m²·year	Cooling
Jan	8.6	1.2
Feb	6.2	1.6
Mar	4.6	4.0
Apr	2.1	5.4
Mai	0.8	7.5
Jun	0.0	13.1
Jul	0.2	14.8
Aug	0.4	14.5
Sep	0.9	12.9
Oct	3.2	8.0
Nov	6.6	3.6
Dec	6.5	1.1
Energy [kW h/m2.year]	40.0	87.7

Indoor Air Quality

TABLE 4 – Toxic substances concentration (2000-03-09)

	Cl_2	NO_x	O_3	Formol	H_2S	HF	H_2S	NO_2	SO_2
Time (hh.mm)	12:50	13:00	13:05	13:20	13:25	12:12	11:46	12:50	13:03
Concentration [p.p.m.]	0.0	0.0	0.0	0.0	0.0	0.0	0.0	0.0	0.0
Threshold Weighted Average (TWA)[ppm]	0.5	25NO\3NO3	0.1	10	10	3	7	400	500
Accuracy [p.p.m.]	0.3	2	0.01	0.5	0.3	0.2	9	6	3

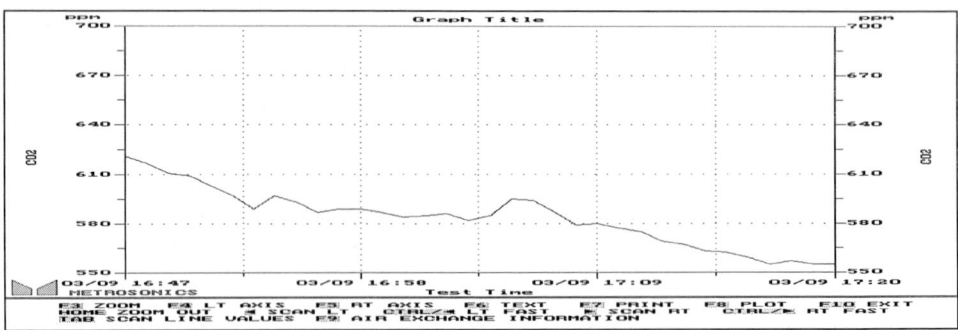

Fig. 6 – CO_2 decay with time

Thermal comfort

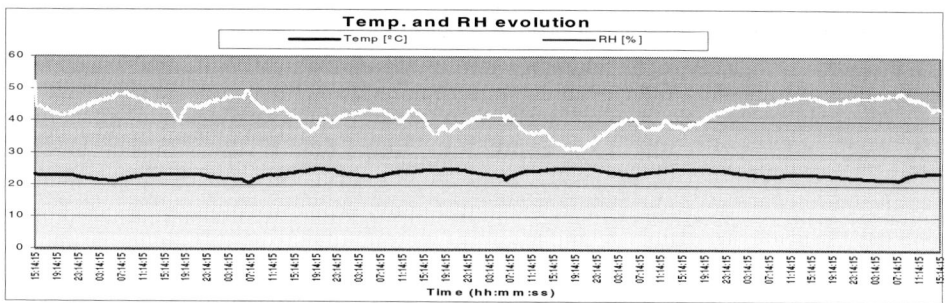

Figure 7 – Typical east oriented space

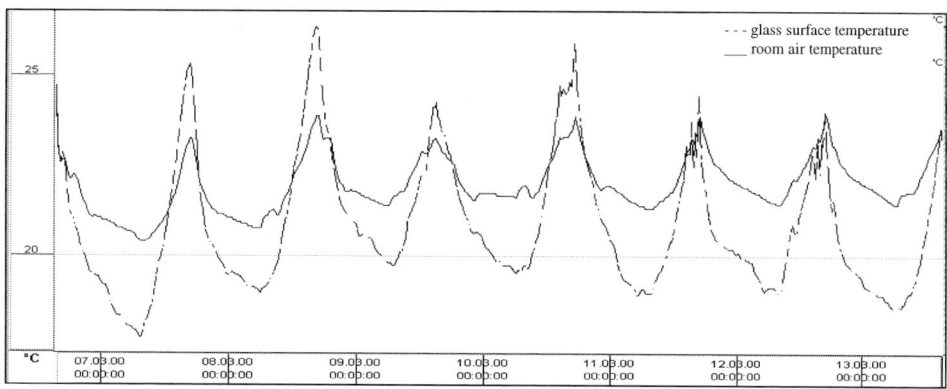

Figure 8 – Typical west oriented space air and internal glass surface temperatures

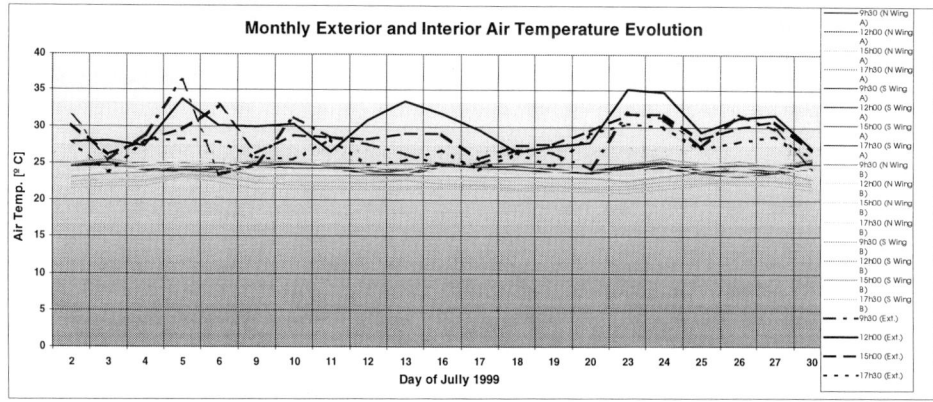

Figure 9 – Monthly evolution of daily air temperature for different hours for North, East, South and West orientations, in a typical floor

NATURAL VENTILATION

The schematic of the natural cross ventilation adopted in the building in the figure below is presented. The air is admitted through the façades and vented to the Atrium using automatically operated windows controlled by the *BMS*.

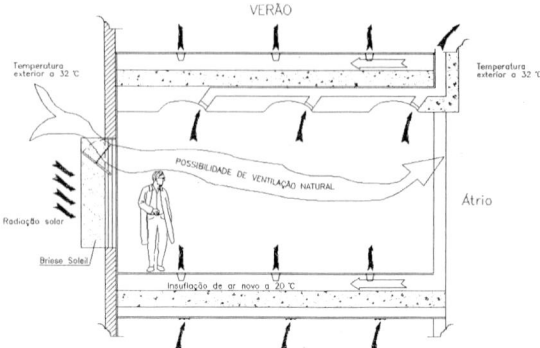

Potential usable time

According to the available weather data (Portuguese *Instituto de Meteorologia*) the calculated number of annual usable hours for the range of outside air temperature between 17 and 23 °C and mean wind speed of approximately 16 km/h are 1449 and 798 hours for 24 hours use and daily use, respectively.

Electrical energy reduction

Taking into account the installed electrical power for air handling units and associated exhaust fans the reduction on electrical energy use could be approximately 13 kWh/m^2.year and 7 kWh/m^2.year for 24 and 10 hours operation, respectively.

DISCUSSION

Putting together the results from Tables 1 and 3 and the measured values showed in 5.2, we present Table 5 below:

TABLE 5 - Energy consumption

Building type	HVAC System	Fresh Air	Heating	Cooling
Estimated Values		(L/sm^2)	(kWh/m^2.year)	(kWh/m^2.year)
Q building	Displacement	4.4	33	71
Standard building	Air Mixture	1.0	28	360
Standard building	Air Mixture w/ increased fresh air supply	4.4	127	258
Measured Values		(L/sm^2)	(kWh/m^2.year)	(kWh/m^2.year)
Q building	Displacement	4.4	40	88

from which we may notice that:

- The measured energy consumption values are about 20% higher than the estimated ones for the displacement ventilation system, which can be explained by the heavy occupation and building running procedures.
- The energy consumption according to the measurements, carried out in the **Q** building, results in 40 and 88 kWh/m^2.year for heating and cooling, respectively, (the relevant total values for energy consumption are 364 MWh for heating and 798 MWh for cooling).
- The **Q** building has a measured energy consumption corresponding approximately to 34% for cooling conditions and 32% for heating of the estimated energy consumption of the **S** building with increased fresh air volume (the same fresh air amount considered in the **Q** building solution). When compared with the **S** building, with normal fresh air volume (9000 L/s for an occupation of 900 persons), those values are respectively 24% for cooling and 41% increase for heating, respectively.

Ventilation efficiency depends mostly on the amount of handled fresh air supply. In order to achieve a better Indoor Air Quality through the dilution of indoor gas contaminants, the option for **S** building equipped with a Mixture type HVAC system leads inevitably to a more intense energy use.

The CO_2 decay with time, after a period of complete shut off, measured on the site (see Fig. 6), shows that the ventilation system of the real **Q** building tested allows a reduction of 80 ppm in a 0.5-hour period. That is, from 620 ppm to 540 ppm with an outside average CO_2 concentration of approximately 454 ppm.

The CO_2 concentration during normal operation is below the maximum values allowed for this type of utilisation and none of the toxic gases, listed in Table 4, was detected inside the building.

From the results presented we may confirm:

- The good energy performance measured for the **Q** building.
- The good *Indoor Air Quality* for the **Q** building, namely in terms of CO_2 and Toxic Gases concentrations.
- Differences in the energy consumption estimations between **S** and **Q** buildings are important contributions to energy conservation and pollutant emissions reduction.
- That comfort assessment, based in the *CFD* simulations presented in Figs. 3 and 4, on the one hand, and the measured values of air temperature and humidity presented in Figs. 7, 8 and 9, on the other hand, shows good comfort levels. *PPD* less then 15%, may easily be obtained in the **Q** building.

CONCLUSIONS

From the above considerations we may conclude:

- Good energy efficiency is compatible and convergent with *IAQ* improvement, where free-cooling is a major strategy for energy conservation
- Good energy efficiency and *IAQ* are compatible with standard comfort conditions and environmental issues

It can therefore be stated that **Q** buildings in Mediterranean climates could achieve optimum energy, comfort and indoor air quality performances in accordance with the most demanding international criteria.

BIBLIOGRAPHY

Commission of the European Communities (1992). *Indoor Air Quality & its Impact on Man.* Report N°. 11 – Guideline for Ventilation Requirements in Buildings. EUR 14449 EN.

BREAM (1993). Building Research Established Environmental Assessment Method. Building Research Establishment, UK.

ASHRAE Applications Handbook (1995). *Control of Gaseous Indoor Air Contaminants.* American Society of Heating, Refrigeration and Air-Conditioning Engineers, Atlanta, USA.

ASHRAE Fundamentals Handbook (1995). *Air Contaminants.* American Society of Heating, Refrigeration and Air-Conditioning Engineers, Atlanta, USA.

Industrial Ventilation (1992). *A Manual of Recommended Practice.*

TRACE-ULTRA 600 Manual – Thermal Load Analysis and Energy Simulation. TRANE company, USA.

TAS-AMBIENS Software Manual. Environmental Design Solutions, Ltd., UK.

Standards CR 1752 and ISO 7730.

ACKNOWLEDGEMENTS

IPS - EST – Instituto Politécnico de Setúbal - Escola Superior de Engenharia.

Carlos Soares, Dipl. Mechanical Engineer, Head of Air Conditioning Department at LM

Air Distribution in Rooms, (ROOMVENT 2000)
Editor: H.B. Awbi

SUSTAINABILITY, ENERGY AND COST
A TIGHT ROPE ACT

John Berry

Ove Arup and Partners, London
Consulting Engineers

Buildings account for more than half of the energy consumed in the industrialised world today. Little has stemmed this demand during the last twenty years and every indication is that it is more likely to increase rather than decrease. The increase is likely to come from the growth in consumption of the third world as they progress toward industrialization and percapita incomes on par with the industrialized world.

At present fuel is relatively cheap and plentiful so why should we be concerned? Surely the scientist has the answer in solar power, wind power, wave power or cold fusion to ensure that we can carry on as before? Obviously there are attractions in pollution free sources such as solar, wind and wave but there is a price to pay for all sources in one way or another. The notion of great wind farms, unless offshore, is unappealing to those living within sight or sound of those giant rotating blades. Also the impact on our global environment from tampering with nature in any significant manner should concern us. Small scale operations or experimental set ups are fine but how confident can we be that our climate will be immune from the heavy hand of man? Past attempts at major river diversions to irrigate land for crop growing for instance, give little hope for the future. One has only to remember the disastrous impact of the Ukrainian experiment, which effectively reduced fertile land to a desert, to urge caution. So why should we be concerned? Yes, energy is plentiful and relatively cheap but two factors dispel the comfortable feeling that all is well; will it be freely available in the future and what is the impact on our environment of proliferation. Fears of an energy shortage following the 1970's oil crisis undoubtedly prompted the major thrust for conservation; cost was somehow secondary. Perhaps for the first time the ability to pay was not in itself a guarantee of supply. At the same time the realisation that the burning of fossil fuels could damage the environment was gaining credibility. Excessive carbon dioxide emissions have been shown to contribute to global warming; the green house effect. Alarm was also being expressed at the depletion of the ozone layer which protects the planet from harmful solar radiation. It is generally accepted that CFC's which are used in many building insulation products as well as the large refrigeration cooling systems found in many buildings contribute to this phenomena.

The link between energy use and environmental damage is proven beyond reasonable doubt and it is this relationship which is causing unease. This new thrust for conservation policies is widespread and for once is not just linked to the standard cost effectiveness arguments. People now realise that it is impossible to put a price on environmental damage. Even for those who believe that science has the answer to all our problems, which indeed it may, the conservation ethic, as a temporary respite at least, is strong. But it goes further than this in that conservation of our natural resources and limiting potential damage to our environment must be beneficial. How can the subtleties and intangible nature of the situation be expressed in any meaningful way? Of course it could be defined in purely scientific terms but that would by definition only apply to a limited part of the argument. The general argument that we are slowly but surely undermining our planets ability to regenerate and support itself requires a far broader statement. Necessity being the mother of invention; the Green Movement was born. The word 'green' has no absolute meaning it is an expression of mans unease with the present and a feeling that we must do better in the future. Green is simply the concept of nature and something which the populace can clearly identify with. It is so widely known and understood that its success is self evident; Green peace and the Audubon Society are two examples which feature highly in the public awareness. The same cannot be said for energy conservation or the harmful effects of CO_2; so in a word it is successful. In most peoples eyes it is seen as representing a positive attitude to our environment and definitely one for the better. It knows no boundaries. It can be global or local. Now a common base exists on which to move forward and explore the new language of green thinking. Sun, light, wind, mass, volume and materials can positively influence our designs. Aspects of green thinking which perhaps above all others are potentially the most influential are harmony and balance. It is all to easy to lose sight of the implicit simplicity which should intuitively flow from a harmonious design. If it doesn't materialize then question the process!

Each building will have its own unique set of goals and objectives and these need to be determined at the beginning of a project. It should not be forgotten that the primary function of any building is that of providing shelter and comfort for its occupant, irrespective of task. A fine balance exists between energy conservation, ecology and health and these need to be addressed in a pragmatic way and not just on fashionable trends. Nevertheless accepted wisdom does need to be questioned; an open mind is essential. One of the more deeply rooted current wisdoms is the perceived need for air conditioning. Not only is this questionable in the temperate European climate, but in a good portion of the USA come to that. The equally dogmatic stance that the cocooned air conditioned environment, free from external stimulus is somehow the role model against which others should be judged is also questionable. Public opinion and user studies do not support such a universal theory. When asked which feature they most valued in a building eighty percent voted for opening windows. That is not to say that eighty percent were arguing that environmental control should be by opening the window alone they were simply expressing the desire to be able to do so. The notion of air conditioning and opening windows goes together about as well as electricity and water. This is not an argument against air conditioning but an argument for a more open minded policy toward a viable alternative. There is talk of following the lead in Holland and Switzerland and implementing legislation to limit the use of air conditioning. Legislation is not the answer. Building design, economics and user demands will force the issue.

Now, two decades later one can point to a number of outstanding buildings that reveal an ambitious architectural and engineering concept in the bold new tradition of the environmental aesthetic. They show by example that it is possible to address the social issues of the day in a balanced and harmonious manner. At one extreme there is the zero energy concept and environmental evangelism of Brenda and Robert Vales house and at the other the simple energy reduction targets of our earlier local authority housing to provide affordable heating for the socially deprived. Both are equally valid. What is apparent in the design sense is how architecture and engineering come closer together than in the conventional design situation. The sequential process of architecture then engineering has no place; the simultaneous solution of ideas is the new order. Low energy and environmental awareness are mutually compatible. It is difficult, though not impossible, to envisage an environmental design which pays little concern to energy consumption.

It is often said that designing passive buildings is more difficult and demanding than designing traditional active systems. To a large degree this is true and most environmental designs can be thought of as active design with a passive outcome, whereas the traditional approach often has the roles reversed with an active outcome from a woefully passive design input. As the passive components are the architecture the engineer enjoys a more prominent role in the overall design than is usual and is involved from the outset. This is on important step not only from the view of recognition but to be genuinely able to contribute to the building as a whole.

Without doubt the façade and building envelope play a leading part in the green building concept. It is here that the natural forces of sun, wind and light are moderated and filtered before entering the occupied rooms. It is the thermal and visual barrier between the calm interior and the natural and some times hostile exterior. The most significant contribution to the performance of the building as a whole occurs at this interface and many buildings fail as a result of inadequate façade design; more often than not the air conditioned variety. How often do we see the glazed mirror glass stump appear with scant regard for its surroundings. It is any wonder that the internal systems cannot cope; even air conditioning has its limits? Of course some external environments are hostile and a sealed building therefore is essential. Even so the green attitude of minimizing dependence on complex and sophisticated environmental control systems is still valid. Installed capacity can be halved by a thoughtful approach to the overall design. Also new system opportunities can be explored which are far removed from the conventioned low temperature solutions traditionally adopted.

So where does this leave us? The environmental lobby is well established and will not go away! Unease with the current dichotomy of plentiful energy, low cost and environmental damage is apparent! Society demands we should do better! And we can do better. It is crossing the boundary between engineering and architecture to produce a lasting environmental aesthetic which holds the key to the future. Architectural Engineering in the truest sense where the success of the energy conservation strategy depends as much on the engineering of the passive elements as the active systems.

INDOOR ENVIRONMENT

2000 Elsevier Science Ltd. All rights reserved.
Air Distribution in Rooms, (ROOMVENT 2000)
Editor: H.B. Awbi

NUMERICAL CALCULATION OF ANGLE FACTORS BETWEEN HUMAN BODY AND RECTANGULAR PLANES

Yoshiichi Ozeki,
Masaaki Konishi
Asahi Glass Co., Ltd. Research Center,
Kanagawa, Japan

Chie Narita,
Dept. of Human Env. Eng., Ochanomizu Univ., Tokyo, Japan
Shin-ichi Tanabe
Dept. of Architecture, Waseda Univ.,Tokyo, Japan

ABSTRACT

Angle factors between a human body and rectangular planes are calculated by a numerical model. Present method, which is applied for predicting the thermal radiation field in a space, is based on a numerical integration method proposed in the previous authors' paper. To confirm the validity of the calculated results, predicted angle factors for both standing and seated persons are compared with those by the experiments. It was found that predicted figures are matched quite well with those by the subjective experiments except those between the human body and the front floor. Influence of body posture on the angle factors for floor, namely between standing straight and stoop posture, was investigated and discussed.

KEYWORDS

numerical simulation, angle factor, effective radiation area, human body,
standing and seated posture, thermal comfort, infrared radiation

INTRODUCTION

Non-uniform indoor climate is often observed in a large enclosure such as an atrium and even in a narrow space such as a passenger compartment in a vehicle. In these indoor spaces, present thermal indices like SET(Gagge et al. 1971) or PMV(Fanger 1970) are still usable but should not stand alone if there is a strong radiant asymmetry. New methods for predicting thermal comfort in non-uniform spaces are highly required. In the previous paper (Ozeki et al. 1998), a numerical simulation method was proposed for predicting the effective radiation area and the projected area of a human body for any kind of posture. The validity of this method was confirmed by comparison with Fanger's and Underwood's projected area factors obtained by a photographic method for both standing and seated posture. Distribution and intensity of solar radiation to the human body surface could be predicted with enough accuracy. In this paper, angle factors between the human body and its surroundings for the prediction of thermal radiation exchange are numerically calculated, based on the numerical integration method proposed by the authors (Ozeki et al. 1996). Predicted angle factors are compared with Fanger's and Horikoshi's subjective experimental results obtained by a photographic method for both standing and seated posture(Fanger et al. 1970; Horikoshi et al. 1990). Distribution and intensity of the thermal radiation exchange between the human body and its surroundings could be predicted with enough accuracy.

REVIEW OF PREVIOUS STUDIES

Photographic methods have been applied to calculate the angle factors between the human body and its surroundings(Fanger et al. 1970; Horikoshi et al. 1990; Jones et al. 1998). Fanger et al.(1970) measured 78 types of the projected area factors for 10 male and female subjects with and without clothing for standing and seated posture by using a photographic method. They developed the basic relationships for calculating the angle factors. Horikoshi et al.(1990) measured angle factors between standing and seated posture and rectangular planes for 3 male subjects with or without clothing by using a photographic method. They used orthographic lens for measurements. Kalisperis et al.(1991) developed angle factor tables for a variety of inclined surfaces using Fanger's data. Jones et al.(1998) developed projected area data for the whole body and for individual body segments. Conventional photographic methods have a limitation in practice because of consuming very long time for measurements.

Yamazaki et al.(1983) proposed a numerical surface model of the human body by measuring the body surface of a male subject. They calculated angle factors between the human body and points on its surroundings. Miyazaki et al.(1995) verified the angle factors between a human body model, consisted of several cylindrical parts, and rectangular planes by Monte Carlo Method. However, their angle factors did not coincide to Fanger's experimental results within a sufficient accuracy. Tsuchikawa et al.(1996) proposed a numerical calculation method for evaluating angle factors between a numerical surface model of the human body and his environment, based on the contour integration method. However, concrete studies have not been conducted. Suzuki et al.(1999) proposed a numerical surface model by measuring the body surfaces of 2 standing male subjects. They compared angle factors between a numerical surface model and his environment obtained by the contour integration method and the photographic method. Nucara et al.(1999) proposed a simple algorithm for the automatic calculation of angle factors between people and composite plane surfaces. However, it is based on experimental data provided by Fanger(1970). Few studies have been conducted on calculation of the angle factors between a human body and its surroundings for any posture by numerical simulation methods with significant accuracy.

NUMERICAL SIMULATION METHOD
Human body model
The configuration of a human body including the grid system affects the characteristics of thermal radiation exchange and solar heat gain. Several human body models have been proposed for the following purposes; (1) to calculate the effective radiation area or angle factors between a human body and its surroundings (Yamazaki et al. 1983; Miyazaki et al. 1995; Suzuki et al. 1999), (2) to simulate the heat transfer characteristics around the human body by a combined numerical simulation of air flow with thermal radiation and moisture transport(Murakami et al. 1997), or (3) to simulate the temperature controlling system in a human body (Yokoyama et al. 1997). In this paper, a human body model shown in Figure 1, which represents the uneven shape such as ears, nose, mouth, fingers of hands and toes in detail, is considered to be suitable to predict heat transfer characteristics. Body shape is obtained by a commercial available software and then divided into surface elements. Height of this model and surface area of whole body are given in Table 1. Height and surface area of the present model are close to those of Fanger's subjects (1970). Human body surface is divided into 4396 quadrilateral surface elements for both standing and seated posture, which enables to conduct a combined numerical simulation of air flow with solar heat gain (Ozeki et al. 1997) and thermal radiation exchange on walls (Ozeki et al. 1998).

Angle factor between a human body and its surroundings
Angle factor F_{p-A2} between a human body surface and its surroundings is derived in Equation (1) with effective radiation area A_{eff}, angle factor F_{ij} between the i-th differential human body surface and the j-th differential wall surface, and area A_i of the i-th differential human body surface(Fanger et al. 1970).

$$F_{p-A2} = \sum_i A_i \sum_j F_{ij} / A_{eff} \qquad (1)$$

Angle factor F_{ij} between the i-th differential human body surface and the j-th differential wall surface is calculated by a numerical integration method for buildings proposed by the authors incorporating the interception of other surfaces(Ozeki et al. 1996). As the procedure for calculating angle factors can deal with any indoor geometry, the present procedure can also be applied to the human body in any posture.

Effective radiation area of a human body
The effective radiation area of a human body is defined as the surface area of a human body which

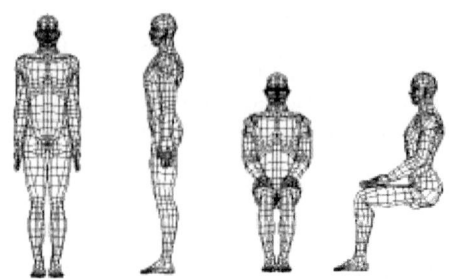

Fig.1 Human body model (standing and seated postures)

Table 1 Height and total surface area of the human body

	Present	Fanger	Horikoshi	Miyazaki	Murakami
Height(m)	1.75	1.72	1.70	1.71	1.65
Total surface area (m²)	1.72	1.74	1.69	1.58	1.69

*Fanger : mean of 10 male and female subjects
*Horikoshi : mean of 3 male subjects

directly contributes to the radiation exchange between the body and its surroundings. In case of a large sphere with a radius r_m, the effective radiation area of human body A_{eff} is derived in Equation (2) with the angle factor F_{A2-p} between sphere and human body as shown in Figure 2(Fanger et al. 1970).

$$A_{eff} = 4\pi r_m^2 F_{A2-p} \tag{2}$$

By calculating the angle factor F_{A2-p} with projected area of a human body A_p on a plane perpendicular to the direction of the differential surface element dA_2 on the sphere as shown in Figure 3, the effective radiation area can be derived from surface integration of projected area with spherical coordinate system.

$$A_{eff} = \frac{4}{\pi} \int_{\alpha=0}^{\alpha=\pi} \int_{\beta=0}^{\beta=\frac{\pi}{2}} A_p \cos\beta \ d\beta \ d\alpha \tag{3}$$

To calculate the effective radiation area of a human body in Equation (3), the projected area A_p of a human body irradiated by the parallel rays must be calculated. This projected area is equal to the surface area of the human body where parallel rays reach directly and projected on a plane perpendicular to the parallel rays as shown in Figure 3. This area is calculated by the solar heat gain simulation(Ozeki et al. 1997).

EFFECTIVE RADIATION AREA FOR BOTH STANDING AND SEATED POSTURE

To calculate the angle factors between a human body model and its surroundings, the effective radiation area of a human body must be evaluated. In this paper, the effective radiation area is calculated with Equation (3) by a numerical integration method. Ninety one integration points are set for the numerical integration, namely 13 different angles in azimuth α and 7 different angles in altitude β. Calculated effective radiation areas A_{eff} and effective radiation area factors f_{eff} are shown in Table 2. Effective radiation area and effective radiation area factor for a standing posture are predicted rather larger than those for seated. It means a seated posture has about 5% decrease of effective radiation area in radiation exchange between a human body and its surroundings than a standing posture. Predicted results for both standing and seated posture meet quite well with those of the subjective experiments obtained by Fanger within 2% accuracy, although configurations of the present human body and Fanger's are not the same.

ANGLE FACTORS BETWEEN A HUMAN BODY AND RECTANGULAR PLANES WITHIN 7 M DISTANCE
Methods

To investigate the accuracy of the calculation method of angle factors for buildings(Ozeki et al. 1996) when applied to the complex human body surface and its surroundings, angle factors between the human body and rectangular planes are calculated. They are compared with the experimental results by Fanger(1970). Three types of rectangular planes for standing and seated posture (plane A,B,C) are set as shown in Figure 4. The distance between the center of human body and each rectangular plane is set to be 7 m as in Fanger's subjective experiments. Length and width of rectangular planes are set as follows;

  a/c=0.2,0.4,0.6,0.8,1.0,1.5,2.0,3.0,5.0,
  b/c=0.2,0.4,0.6,0.8,1.0,1.2,1.4,1.6,1.8,2.0,2.5,3.0,4.0,5.0,6.0,7.0,8.0,9.0,10.0

In calculating angle factors between the human body and rectangles in Equation (1), rectangles are divided

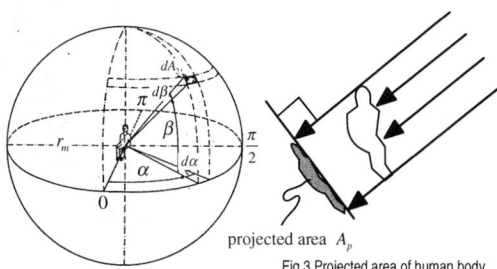

Fig.2 Notation pertinent to calculation of the effective radiation area

Fig.3 Projected area of human body

projected area A_p

Table 2 Effective radiation area and effective radiation area factor

(a) Standing posture

	Present	Fanger	Horikoshi	Miyazaki
A_{eff} (m²)	1.276	1.262	1.312	1.317
f_{eff} (-)	0.744	0.725)0.013	0.803)0.005	0.834

(b) Seated posture

	Present	Fanger	Horikoshi	Miyazaki
A_{eff} (m²)	1.176	1.211	1.214	1.224
f_{eff} (-)	0.691	0.696)0.017	0.740)0.012	0.775

*Figures by Fanger and Horikoshi were obtained from the experiments with nude subjects

into small squares with the length of 0.7 m, and the interception of other differential surfaces is taken into account (Note 1).

Results and discussion

The diagrams of predicted angle factors in each rectangular plane are shown in Figures 5-1 and 5-2. Comparison is shown for the Fanger's subjective experimental results(Fanger et al. 1970). The regression coefficient and the coefficient of determination in each rectangle are shown in Table 3.

In each rectangle, predicted angle factors for both standing and seated posture have the tendency of monotonic increase according to the rise of b/c. In the area with small b/c value, predicted angle factors

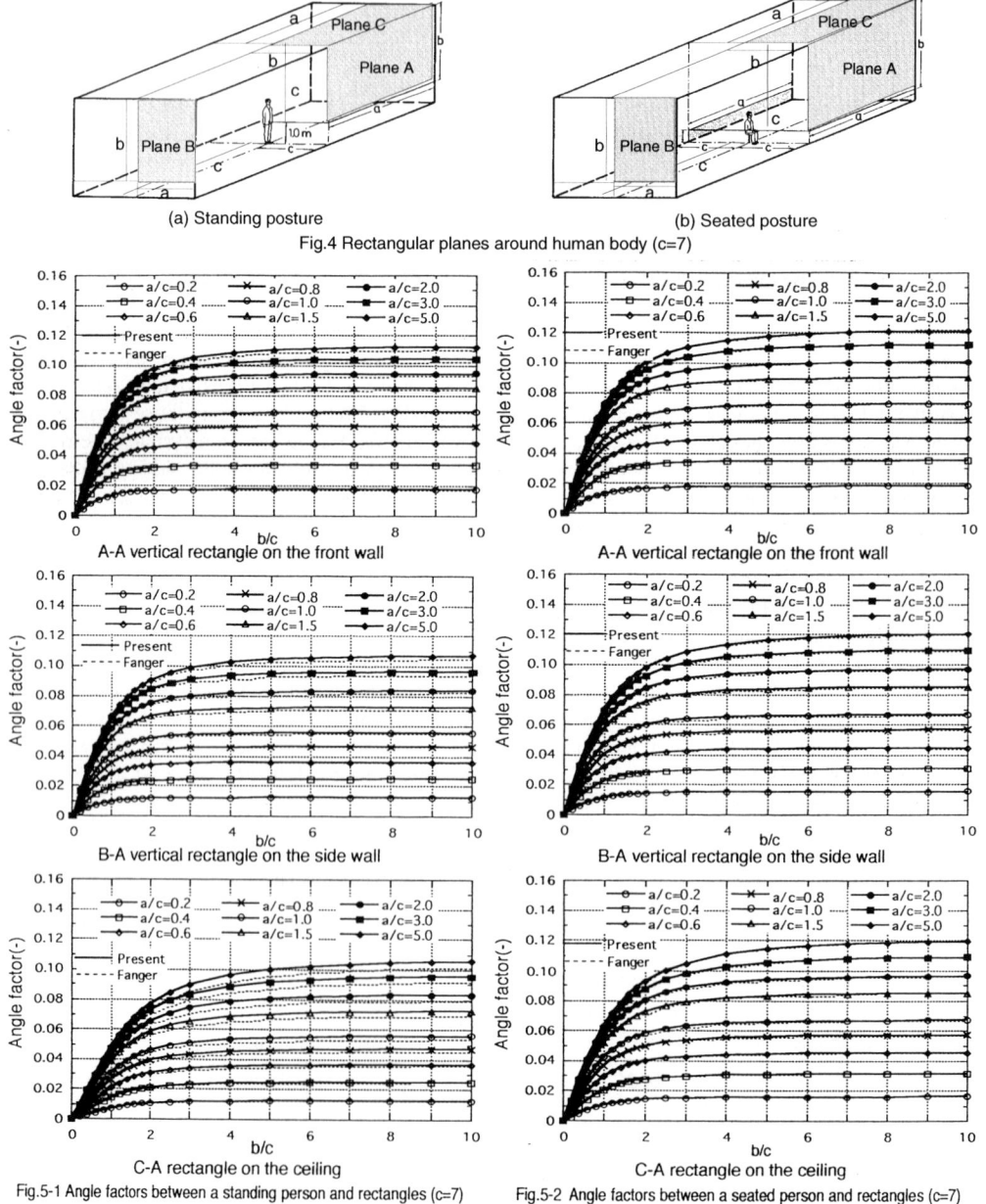

(a) Standing posture   (b) Seated posture

Fig.4 Rectangular planes around human body (c=7)

Fig.5-1 Angle factors between a standing person and rectangles (c=7)   Fig.5-2 Angle factors between a seated person and rectangles (c=7)

increase rapidly. On the other hand, in the area with large b/c value, predicted angle factors are almost constant.

Comparing predicted results and Fanger's experiments in the standing posture, the present results meet quite well with the Fanger's experimental results in the area with a/c less than 1.0. Maximum 3

Table 3 Comparison of angle factors between predicted results and measurements by Fanger

	standing posture			seated posture		
	front wall	side wall	ceiling	front wall	side wall	ceiling
Regression coeffficient	0.992	0.989	0.963	1.006	0.993	0.993
Coefficient of determination	0.999	0.994	0.999	0.999	0.999	0.999

% difference is observed at b/c=10 and a/c=5.0 in each rectangular plane. Comparing predicted results and Fanger's experiments for the seated posture, the present results also meet quite well with Fanger's experimental results. Maximum difference between the present results and the experiments is no larger than 4% in regression coefficients and the coefficients of determination in Table 3, suggesting that the present model is able to predict angle factors with enough accuracy. The present method is a useful tool for predicting them.

In the calculation of angle factors between a human body surface and a rectangular plane, Fanger introduced the parallel ray method. On the other hand, solid angle method was utilized in the present calculation. As angle factors introduced by the two methods correspond quite well, no significant difference in either method is found in the evaluation of angle factors in the case where the distance between the center of the person and rectangular planes is 7 m (Note 2). This tendency corresponds to Horikoshi's subjective experimental results(Horikoshi et al. 1990).

ANGLE FACTORS BETWEEN A HUMAN BODY AND RECTANGULAR PLANES WITHIN 2 M DISTANCE
Methods

Angle factors between the human body and its surroundings within 7 m introduced by the present method and those by Fanger's meet quite well each other. On the other hand, Horikoshi points out solid angle method should be applied when calculating angle factors between the human body and its surroundings within 1 m distance. Significant error may be occurred with parallel ray methods in this case. To verify that the present method can predict angle factors in these cases with enough accuracy, predicted angle factors are compared with Horikoshi's subjective experimental results. The distance between the center of human body and its surroundings is set no longer than 2 m as in Horikoshi's experimental conditions. Three types of rectangular planes for a standing posture (plane A,B,E) and a seated posture (plane C,D,F) are set as shown in Figure 6, respectively.

Results of calculated angle factors for vertical rectangles and discussion

Angle factors calculated for standing and seated posture are shown in Figure 6. Comparison is shown for both Fanger's and Horikoshi's subjective experimental results.

For a standing posture, calculated angle factors between the human body and rectangles (Figure 6(A),(B)) in the condition of c=2 correspond well with Fanger's experimental results within 6% accuracy, and calculated angle factors in the condition of c=1 meet well with Horikoshi's within 4% accuracy (Note 4). No significant difference is found in parallel ray and solid angle methods in calculating angle factors between the center of human body and its surroundings with the distance of 2 m. Significant difference is found in parallel ray and solid angle methods in calculating angle factors between the center of human body and its surroundings when the distance is 1 m, and angle factors predicted by solid angle methods (calculation and Horikoshi's subjective experiment) are meet well. For a seated posture, calculated angle factors between the human body and rectangles (Figure 6(C),(D)) have the same tendency as for a standing posture. It is proved that the present model can predict angle factors with enough accuracy within 2 m distance.

Results of calculated angle factors for the floor and discussion

In a standing posture, angle factors between the human body and the rectangle on the front floor are predicted larger than those on the rear floor because of feet as shown in Figure 6-(E). In a seated posture, angle factors between the human body and the rectangle on the front floor are predicted much larger than those on the rear floor as shown in Figure 6-(F). Angle factors between the human body and the rectangle on the front floor go up sharply till b/c=1.0 where feet are placed, and after that angle factors show a slight rise. Angle factors between the human body and the rectangle on the rear floor correspond well with Horikoshi's subjective experiments for both standing and seated posture. However, significant difference is found between

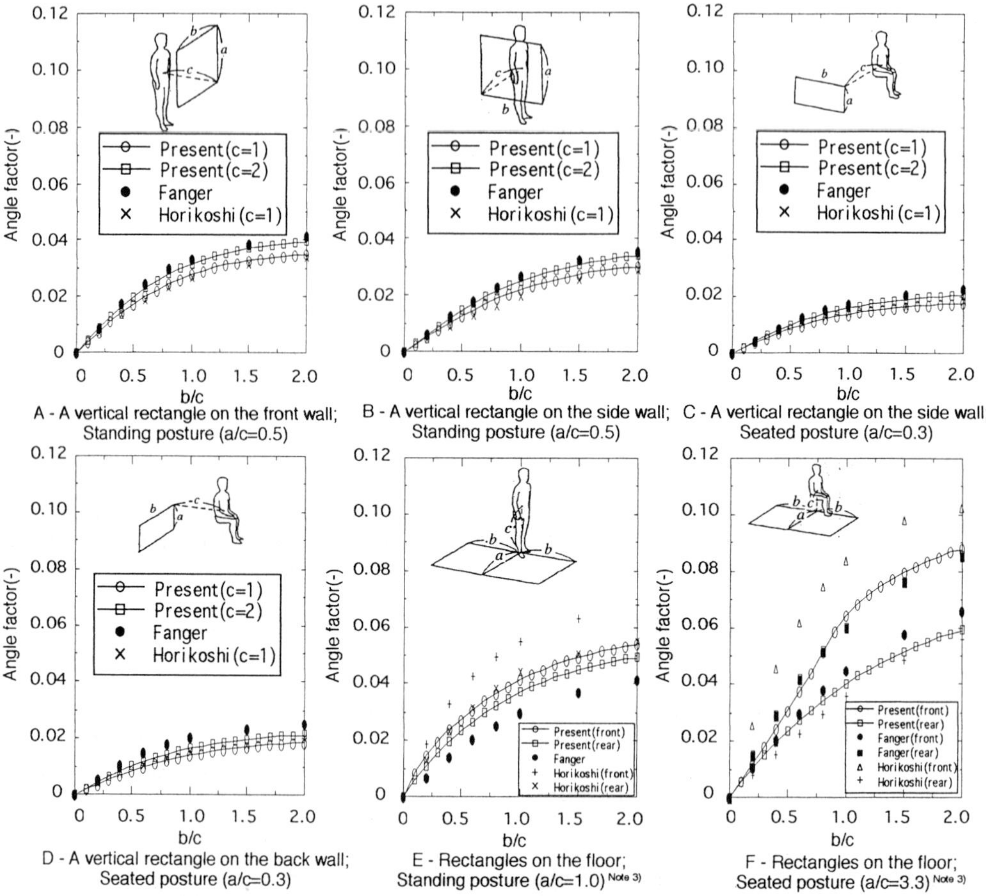

Fig.6 Angle factors between human body and rectangles (standing and seated posture)

A - A vertical rectangle on the front wall; Standing posture (a/c=0.5)

B - A vertical rectangle on the side wall; Standing posture (a/c=0.5)

C - A vertical rectangle on the side wall; Seated posture (a/c=0.3)

D - A vertical rectangle on the back wall; Seated posture (a/c=0.3)

E - Rectangles on the floor; Standing posture (a/c=1.0) Note 3)

F - Rectangles on the floor; Seated posture (a/c=3.3) Note 3)

the present and the experimental angle factors concerning the front floor. Further investigation is required to find the reason of discrepancy.

Differences in posture

We study how the differences in standing posture affect prediction results. Conducted is a stoop posture as shown in Figure 7. The effective radiation area of a stoop posture is 1.260 m². Predicted angle factors on the front and rear floors are shown in Figure 8, respectively. No significant difference is found in a tendency of angle factors. Comparing with angle factors shown in Figure 6-(E), maximum 3% difference in angle factors is observed on the front floor and maximum 7% difference on the rear floor. It is confirmed the differences between the present stoop posture and the standard standing posture have little influence on angle factors on both front and rear floors.

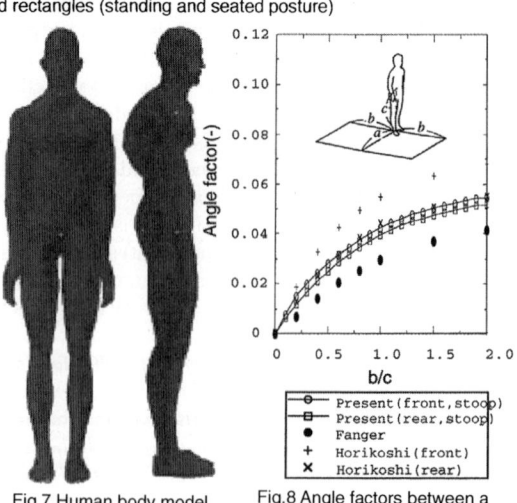

Fig.7 Human body model (A stoop posture)

Fig.8 Angle factors between a stoop posture and rectangles on the floor (a/c=1.0)

CONCLUSIONS

Angle factors between the human body and its surroundings for both standing and seated posture are calculated based on the numerical integration method proposed by the authors. The results are compared with the subjective experimental results. Following conclusions are obtained;

1)Effective radiation area and effective radiation area factors for both standing and seated posture are calculated for predicting angle factors of the whole body. Comparing with subjective experiments by Fanger, the results matched well within 2% difference.

2)Angle factors between the human body and representative rectangular planes are predicted in the case where distance between the center of persons and rectangles is 7 m as in Fanger's experiments. Predicted results matched quite well with those of Fanger's within 4% difference.

3)Angle factors between the human body and representative rectangular planes are predicted in the case where distance between the center of persons and rectangles is no longer than 2 m as in Horikoshi's experiments. Predicted results correspond quite well with those of Horikoshi's within 4% difference except angle factors on the floor.

4)Influence of body posture on the angle factors for floor, namely between standing straight and stoop posture, was investigated and discussed Differences in configuration of the human body such as a present stoop posture in a standing person have little influence on angle factors with maximum 7% difference.

Angle factors of the whole body in any posture could be predicted with enough accuracy for the practical use by the present method. This model is able to predict thermal radiation exchanges between human body surfaces and their surroundings. The present method is expected to be useful tool for predicting them.

NOTES

1) In calculating angle factors between the human body and rectangles, whether the other elements consisting of the whole human body surface intercept or not is decided by the interception of all elements consisting of the whole human body surface and lines which connect between the center of current surface element on the human body and divided surface element on rectangles. In case of crossing the line and other element on the human body, angle factor between the human body surface and a divided element on the rectangles is 0, that is, completely intercepted. In case of not crossing, angle factor between the human body surface and a divided element on the rectangles is generally calculated, that is, completely not intercepted. In case of crossing the line and other element on the human body at the side of other element, angle factor between the human body surface and a divided element on the rectangles is generally calculated and multiplied by half, that is, 50% intercepted.

2) Significant difference in parallel ray method and solid angle method must be appeared in angle factors when surroundings are close to the human body. Solid angle method is more appropriate than parallel ray one under this condition. However, in case of evaluating the solar heat gain, not only solid angle method but also parallel ray method is suitable.

3) Angle factor of the floor touching the sole of foot is set 1.0 as in Horikoshi's subjective experiments.

4) Regression coefficient of regression equation through the origin is 1.060 and the coefficient of determination is 0.991, where is obtained by comparing predicted angle factors and Fanger's experiments in the case where distance between the center of person and rectangles is 2 m. On the other hand, regression coefficient is 0.960 and the coefficient of determination is 0.990 in the case where distance between the center of person and rectangles is 1 m.

Jones, B.W., E.A. McCullough, and S. Hong. 1998. **Detailed projected area data for the human body.** ASHRAE Transactions 104(2):1327-1339.

Kalisperis, L.N., M. Steinman, and L.H. Summers. 1991. **Angle factor graphs for a person to inclined surfaces.** ASHRAE Transactions 97(2): 809-839.

Yamazaki H., M. Manabe, and S. Karashima. 1983. **Shape factors of human body viewed from walls.** Annual Meeting of AIJ, pp.157-158 (in Japanese).

Miyazaki Y., M. Saito, and Y. Seshimo. 1995. **A study of evaluation of non-uniform environments by human body model.** J. Human and Living Environment, Vol.2, No.1, pp.92-100 (in Japanese).

Tsuchikawa T., and T. Horikoshi. 1996. **Thermal radiation exchange between a man and his environment by a numerical surface model of the human body (Part 1).** Annual Meeting of AIJ, pp.355-356 (in Japanese).

Suzuki K., and N. Kakitsuba. 1999. **Development of a human body model based on the human body area and the configuration factors.** Journal of Archit. Plann. Environ. Eng. AIJ, No.515, pp.49-55 (in Japanese).

Nucara, A., M. Pietrafesa, G.Rizzo, and G. Rodono. 1999. **Human body view factors for composite plane surfaces.** Indoor Air 99, pp.650-655.

Murakami S., S. Kato, and J. Zeng. 1997. **Flow and temperature fields around human body with various room air distributions, CFD study on computational thermal manikin-Part 1.** ASHRAE Transactions 103(1): 3-15.

Yokoyama S., N. Kakuta, T. Tomigashi, Y. Hamada, M. Nakamura, and K. Ochifuji. 1997. **Development of human thermal model in steady state expressing local characteristic of each segment and its applications.** Annual Meeting of SHASE, pp.513-516 (in Japanese).

Ozeki Y., Y. Sonda, T. Hiramatsu, T. Saito, and S. Ohgaki. 1997. **Study on solar heat gain simulation for coupled analysis of radiative and convective heat transfer under complicated geometry with fine mesh.** Transactions of SHASE, No.66, pp.1-11 (in Japanese).

Ozeki Y., S. Kato, and S.Murakami. 1998. **CFD analysis on flow and temperature fields in experimental real scale atrium (Part 1).** Transactions of SHASE, No.68, pp.65-75 (in Japanese).

REFERENCES

Gagge, A.P., J.A.J. Stolwijk, and Y. Nishi. 1971. **An effective temperature scale based on a simple model of human physiological regulatory response.** ASHRAE Transactions 77(1): 247-262.

Fanger, P.O. 1970. **Thermal comfort.** Copenhagen: Danish Technical Press.

Ozeki Y., M. Konishi, C. Narita, and S. Tanabe. 1998. **Effective radiation area of human body calculated by a numerical simulation.** Roomvent'98, Vol.2, pp.173-180.

Ozeki Y., T. Saito, and S. Ohgaki. 1996. **Study on numerical prediction method of radiation exchange under complicated geometry with fine mesh.** Transactions of SHASE (The Society of Heating Air-Conditioning and Sanitary Engineers of Japan), No.62, pp.101-110 (in Japanese).

Fanger P.O., O. Angelius, and P.K. Jensen. 1970. **Radiation data for the human body.** ASHRAE Transactions 76(2): 338-373.

Horikoshi T., T. Tsuchikawa, Y. Kobayashi, E. Miwa, Y. Kurazumi, and K. Hirayama. 1990. **The effective radiation area and angle factor between man and a rectangular plane near him.** ASHRAE Transactions 96(1): 60-66.

NOMENCLATURE

A_{eff} : effective radiation area of a human body (m²)
A_i : area of i-th differential surface element (m²)
A_p : projected area of a human body (m²)
F_{A2-p} : angle factor between sphere and human body (-)
F_{ij} : angle factor between i-th differential surface element on the human body and j-th differential surface element on rectangles (-)
F_{p-A2} : angle factor between human body and a sphere (-)
f_{eff} : ratio of the effective radiation area A_{eff} to the total body surface area A_{Du} ($f_{eff} = A_{eff} / A_{Du}$) (-)
dA_2 : differential surface element on the sphere (m²)
r_m : a radius of sphere (m)
α : azimuth angle (°)
β : altitude angle (°)

Air Distribution in Rooms, (ROOMVENT 2000)
Editor: H.B. Awbi

SIMPLIFIED HUMAN BODY MODEL FOR EVALUATING THERMAL RADIANT ENVIRONMENT IN A RADIANT COOLED SPACE

Toshiyuki Miyanaga, Wataru Urabe and Yukio Nakano

Customer Systems Department, Central Research Institute of Electric Power Industry
2-11-1, Iwado-kita, Komae-shi, 201-8511 Tokyo, JAPAN

ABSTRACT

The simplified model of human body for evaluating a radiant cooled space was developed. Firstly, the model was constructed by combining some kinds of cylindrical and rectangular parts to simulate a shape of the real subject accurately. The effective body surface area which contributes to a radiation and convection heat transfer, the effective radiation area and projection area factor of the model were examined respectively. As a result, these values agreed well with those of the real subject. The geometrical validity of the model was verified. Next, a skin temperature and thermal resistance of the clothing at some typical positions on the body were defined respectively on the basis of measured values of the real subject. They were added to the model as thermal boundary conditions and the thermal human body model was constructed. Finally, the model was applied to an analysis of thermal environment in a radiant cooled space. Analyzed values of a mean surface temperature on the model agreed well with those on the real subject in several different thermal radiant conditions. It was found that the model was applicable to simulate a thermal response of the real subject adequately. The local thermal radiant environment around the model was evaluated quantitatively by visualizing analyzed values of a heat loss by radiation and plane radiant temperature on the model surface.

KEYWORDS

Radiant cooling, Thermal radiant environment, Human body, Effective radiation area, Projection area factor, Skin temperature, Thermal properties

INTRODUCTION

A radiant cooling system is an air-conditioning method that removes thermal load from a room mainly by radiation heat exchange between a cooled radiant panel and objects. Various objects and heat sources in addition to occupants are generally included in a living space. A complicated thermal radiant environment will appear in it. Analysis of the thermal radiant environment is required to systematically design the radiant cooled space for obtaining satisfactory thermal sensation. The radiant environment of a human body mainly depends on surrounding geometrical and their thermal conditions. Therefore, a human body should be numerically modeled to provide the adequate body shape and thermal properties.

Various body shaped models have been proposed for analyzing the surrounding radiant environment of a human body. Some of them are expressed as a rectangular prism extremely simplified, Nakamura (1987) and Horikoshi (1990). Some were so complicated CAD models very similar to a real body shape, Ozeki (1998) and Suzuki (1999). It is desirable to use models similar to a real body for the analysis. Computational memories required for the analysis, however, increase exponentially if so complicated models are used. This causes a serious problem in practical application.

Physiological models have been developed to simulate the thermal mechanism in a human body, Wissler (1964), Stolwijk (1966) and Gagge (1971) *etc.* It is pointed out that most of those shapes are very different from a real body shape. Important thermal informations as a skin temperature and so on can be estimated numerically by using physiological models.

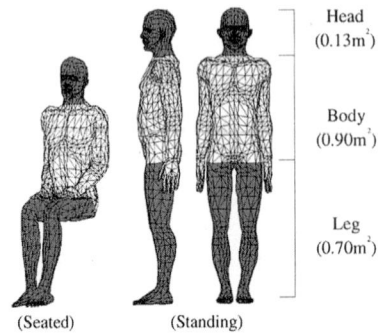

Head
(0.13m²)

Body
(0.90m²)

Leg
(0.70m²)

(Seated)   (Standing)

Figure 2 Complicated human body model
4392 small elements are included.

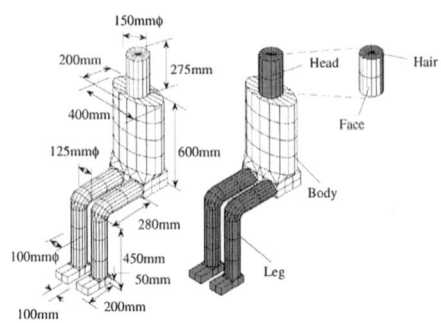

Figure 3 Simplified human body model
617 small elements are included.

EFFECTIVE RADIATION AREA

The validity of a shape of the simplified model is evaluated by using the effective radiation area A_{eff} , Fanger (1970), expressed as the following equation.

$$A_{eff} = \frac{4}{\pi} \int_{\alpha=0}^{\alpha=\pi} \int_{\beta=0}^{\beta=\frac{\pi}{2}} A_p \cos\beta \, d\alpha \, d\beta \tag{1}$$

Equation (1) is transformed into equation (2) for numerical calculation.

$$A_{eff} = \frac{4}{\pi} \sum_{j=1}^{13} \left[\sum_{i=1}^{25} A_p(\alpha_i, \beta_j) \cos\beta_j \cdot \Delta\alpha \cdot \Delta\beta \right] \qquad \alpha_i = \frac{\pi}{24}(i-1) \quad \beta_j = \frac{\pi}{24}(j-1) \tag{2}$$

Projections of both models seen from various combinations of azimuth and altitude are produced by the method illustrated in Figure 4. For example, all projections of the simplified model are shown in Figure 5. Then, the projected area A_p of each projection was calculated by counting pixels included in each projection and analyzing them. Calculated results of A_{eff} are tabulated in Table 2. A_{eff} of the simplified model is 6% larger than that of the complicated.

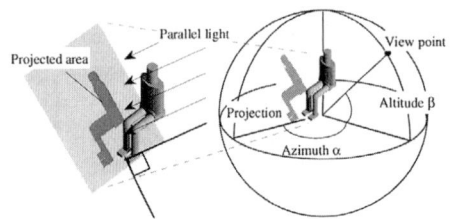

Figure 4 Production of projection of model

Table 2 Body surface area and effective radiation area
(The number in the parenthesis indicates the value
subtracted the contact area from the area of each part.)

		Complicated model	Simplified model
Surface area (m²)	Head	0.13	0.14
	Body	0.90 (0.75)	0.74
	Leg	0.70 (0.65)	0.66
	Total	1.73 (1.53)	1.52
Effective radiation area (m²)		1.22	1.29

PROJECTION AREA FACTOR

The validity of a shape of the simplified model is evaluated by using the projection area factor f_p ($=A_p/A_{eff}$), Fanger (1970). Errors of f_p at each azimuth and altitude of the simplified model to that of the complicated are shown in Figure 6. Relative large errors occurred when azimuth was 90°and 150°. The largest cause for the error is that arms were substantially omitted in the simplified model. Except previous areas, errors were within 8%. It is considered that the shape of the simplified model well simulates that of the real human body.

Small elements included in the simplified model are about one-seventh as many as those of the complicated. The above contributes much to reduction of the calculation time of view factors between small elements and saving of computational memories required for the analysis of thermal radiant environment.

THERMAL PROPERTIES OF SIMPLIFIED MODEL
MEASURING SKIN TEMPERATURE OF REAL SUBJECT

The skin temperature of the real subject was measured in the experimental room illustrated in Figure 7. The radiant panel cooled by water was equipped at the ceiling. In addition, the partition type radiant panel was equipped around the real subject. The high emmisivity paint ($\varepsilon = 0.95$) was painted on each wall surface. And the entire room was well insulated by glass wool. The indoor air temperature, water vapor pressure, temperature of each radiant panel and the outdoor air temperature were always controlled according to the target value by the

Complicated processes are required for obtaining those imformations. Physiological models have not been examined whether they are suitable to a radiant cooled space or not.

In this paper, the simplified body model which well simulates the shape and thermal property of a real body is proposed. The shape of the model is defined by simplifying the shape of the complicated model constructed on the basis of measured values of real subjects. The geometrical validity of the simplified model is examined. Next, a skin temperature and thermal resistance of the clothing at some typical positions on the body are defined respectively on the basis of measured values of real subjects. They are added to the model as thermal boundary conditions and the thermal human body model is constructed. The model is applied to an analysis of thermal environment in a radiant cooled space. The validity of the model is examined by comparing analyzed results with measured values according to a thermal response of the real subject. In addition, some practical utilities of the model are illustrated.

SHAPE OF SIMPLIFIED HUMAN BODY MODEL
MANUFACTURE OF MODEL
COMPLICATED HUMAN BODY MODEL

The complicated human body model is manufactured on the basis of a real body shape. Body shapes of five real subjects tabulated in Table 1 were measured accurately. All real subjects were male. Figure 1 shows Measured parts and mean values of measured dimension sizes. Figure 2 shows the complicated model manufactured on the basis of measured values. The body surface was divided into a head, body and leg part for comparison with the simplified model as described later. The body surface area of the model was agreed well with the value calculated by *Dubois'* equation, Dubois (1916). The complicated model simulates well a body shape of the real subject.

Table 1 Outline of real subjects

	Sex	Age	Height (cm)	Weight (kg)	Body surface area (m²)
Mean value(S.D.)	Male	30 (3.7)	173 (7.0)	66 (5.8)	1.79 (0.09)
1	Male	34	178	68	1.85
2	Male	24	165	71	1.78
3	Male	30	182	67	1.87
4	Male	30	171	68	1.79
5	Male	32	168	56	1.63

DuBois' equation: $71.84W^{0.425}H^{0.725}$, W: weight (kg),
H: height (cm)

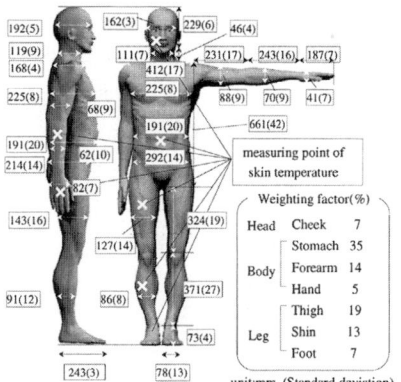

Figure 1   Real subject and mean values of
measured parts

SIMPLIFIED HUMAN BODY MODEL

Only seated posture is treated in this paper. Basically, the simplified model is expressed by combing cylindrical and rectangular parts. The shape and dimension size of each part were defined respectively to coincide several projection areas of the simplified model with the complicated (Figure 3). Arms of the model were substantially omitted for saving computational memories. 617 small elements are included in the model.

EVALUATION OF MODEL SHAPE
SURFACE AREA

Surface areas of each part of both models are tabulated in Table 2. In a head and leg part, surface areas of both models were agreed well. In a body part, however, some error was occurred. The largest cause for the error in this case is that arms were substantially omitted in the simplified model. Some areas of arms usually contact the body when a person takes seated posture. The contact area of the complicated model is about 0.2m². Dry heat transfer will not occur in the contact area. The substantial area of the body (0.75m²) can be obtained by subtracting the contact area (0.20m²) from the whole area of the body (0.90m²). The substantial area of the body was agreed well with that of the simplified model. In this case, it is considered that the surface area of the simplified model was valid.

specially designed air handling unit. Experimental conditions were tabulated in Table 3. Case 1-3 were symmetric thermal radiant environments. Case 4-6 were asymmetric environments (equal radiant cooled environments). The indoor water vapor pressure and air velocity around the real subject were always kept at 1.50KPa and 0.10m/sec respectively.

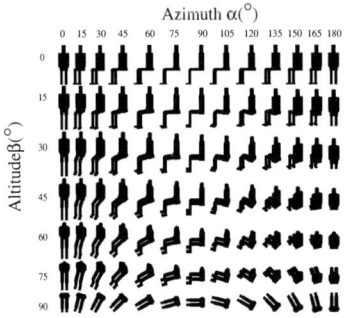

Figure 5　Projections of simplified model

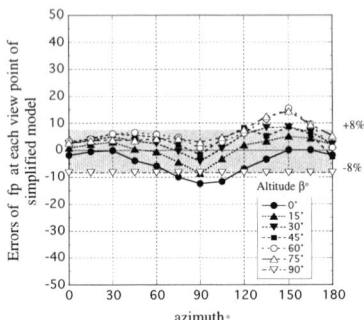

Figure 6　Errors of f_p at each view point of simplified model

Real subjects were the same as those in Table 1. A total of 30 real subjects participated in the experiment. They wore a typical summer clothing detailed in Table 4. All participants sat quietly for one and half hour. They were regarded to get accustomed to the indoor thermal environment within 30 minutes. After that, skin temperatures at eight points illustrated in Figure 1 were recorded at regular 30 seconds intervals. The thermography of the real subject was taken by a thermal infrared camera at the same time.

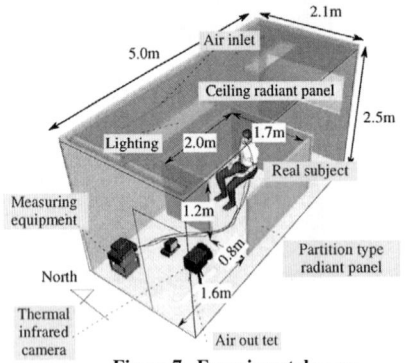

Figure 7　Experimental room

Table 3　Experimental condition for measuring skin temperature

		Surface temperature ofradiant panel (°C)		Air temperature (°C)
		Ceiling	Partition	
Symmetric radiant condition	Case 1	28	28	28
	Case 2	26	26	26
	Case 3	24	24	24
Asymmetric radiant condition	Case 4	18	28	28
	Case 5	18	26	26
	Case 6	18	24	24

DEFINITION OF SKIN TEMPERATURE

The mean skin temperature of each part of the real subject was estimated by using the experimental data and *Hardy and Dubois'* weighting factors, Hardy&Dubois (1938) . Figure 8 represents graphically relations between the mean skin temperature of each part and the indoor air temperature. There was little difference between relations of that in symmetric (Case 1-3) and asymmetric cases (Case 4-6). There may be some problems in measuring the skin temperature and the lighting and so on. It is necessary to examine the above in future.

Each mean skin temperature was estimated to be almost in proportion to the indoor air temperature. The approximate linear equation of each mean skin temperature was shown in Figure 8. Each skin temperature of the simplified model was defined by previous equations for convenient in this paper. In addition, the surface temperature of hair part of the model was defined to be constant (30°C) on basis of analyzed results of thermography.

THERMAL PROPERTIES OF CLOTHING

The clothing was considered at the body and leg part of the simplified model. It was expressed as the layer of fiber in combination with air. The thickness of the clothing was neglected. The heat transfer perpendicular to the layer was considered. Thermal properties of the clothing are tabulated in Table 4. In this case, the thermal resistance of air was considered to be 0.07m²/°C/W.

Table 4 Thermal properties of clothing

	Clothing (Material)	Heat conductivity (W/m²/°C)	Thickness *10⁻³(m)	All over heat transfer coefficient (W/m²/°C)
Body	Shirt (Polyester)	0.034	0.4	9.78
	Underwear (Cotton)	0.039	0.8	
Leg	Trouser (Polyester)	0.034	0.8	10.69

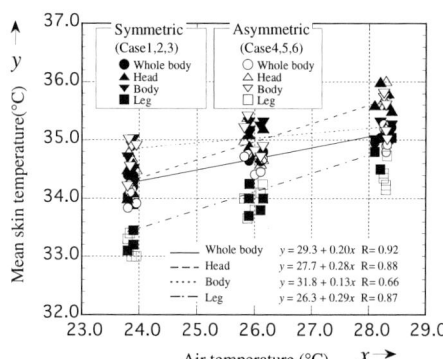

Figure 8 Relations between mean skintemperature and indoor air temperature

HEAT BALANCE AT CLOTHING SURFACE

The heat transfer by radiation, convection and conduction is considered on the clothing surface. The steady-state heat balance equation at a small element i included in the model is expressed as follows, Miyanaga (1998).

$$\sum_{k=1}^{N} A_k \varepsilon_k \sigma (T_k + 273.15)^4 G_{ki} - A_i \varepsilon_i \sigma (T_{cl,i} + 273.15)^4 + A_i h_i (T_a - T_{cl,i}) = K_i A_i (T_{cl,i} - T_{skin,i}) \qquad (3)$$

The sole part of the model was considered to be perfectly insulated. The convective heat transfer coefficient was quoted from the literature. The emmisivity of every small element was considered to be 0.95.

VERIFICATION OF VALIDITY OF SIMPLIFIED MODEL
MEAN SURFACE TEMPERATURE

The model was applied to an analysis of thermal environment and the validity of it was examined. Figure 9 shows the analysis model which simulated Figure 7. 2185 small elements are included in the analysis model. Measuring equipments were neglected. The thermal environment in each case tabulated in Table 5 were analyzed. Analysis conditions were the same as those in Table 5 and experimental conditions. The clothing surface temperature of the simplified model was unknown. A set of non linear equations are obtained for unknown temperatures. These equations are numerically solved by the Newton-Raphson method. The mean surface temperature of the simplified model was estimated by the following equation.

$$T_{cl,mean} = \sum_{i=1}^{617} (T_{cl,i} \times A_i) / \sum_{i=1}^{617} A_i \qquad (4)$$

Figure 10 shows the comparison result of the mean surface temperature between the simplified model and real subject. The mean surface temperature of the real subject was estimated by analyzing the thermography. The difference of both was less than 0.5°C. The simplified model simulated that mean surface temperatures of the real subject in asymmetric radiant conditions (Case 4-6) were 0.1-0.3°C lower than those in symmetric conditions (Case 1-3). The validity of the simplified model was verified.

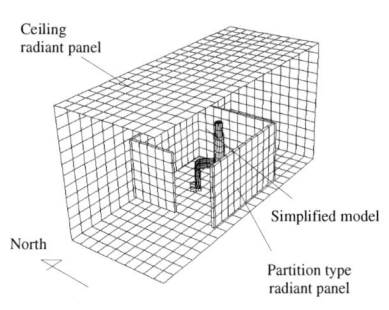

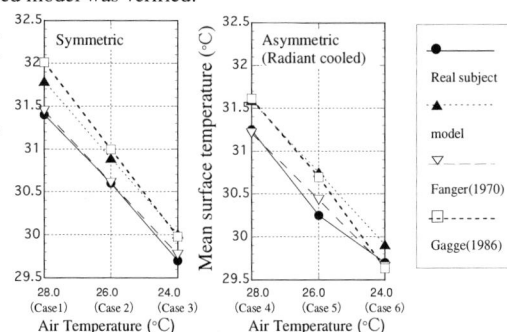

Figure 9 Analysis model
West and south walls are omitted.

Figure 10 Comparison of mean surface temperature between simplified model and real subject

PLANE RADIANT TEMPERATURE AND HEAT LOSS BY RADIATION

One example of practical informations obtained by the simplified model is shown. Figure 11 illustrates the distribution of the plane radiant temperature T*PRT* and heat loss by radiation on the model surface (Case 1 and 4). The plane radiant temperature of small element i was calculated by the following equation.

$$T_{PRT,i} = \left[\left\{ \sum_{k=1}^{N} A_k \varepsilon_k (T_k + 273.15)^4 G_{k_i} \right\} / A_i \varepsilon_i \right]^{0.25} - 273.15 \qquad (5)$$

The heat loss by radiation was calculated by equation (3). The distribution of T*PRT* was uniform in Case 1. On the other hand, T*PRT* at the top of head, shoulder and thigh was lower than that at the other part in Case 4. The heat loss by radiation increased from 31W (Case 1) to 38W (Case 4). The radiation cooling effect can be estimated quantitatively by using the simplified model.

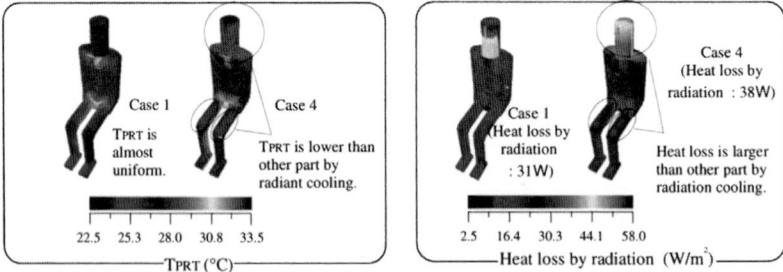

Figure 11 Visualized distribution of T*PRT* and heat loss by radiation : Case1 and Case 4

CONCLUSION

The simplified model of human body for evaluating a radiant cooled space was developed. The model was constructed by combining some kinds of cylindrical and rectangular parts to simulate a shape of the real subject accurately. The geometrical validity of the model was verified. Then, a skin temperature and thermal resistance of the clothing at some typical positions on the body were defined respectively on the basis of measured values of the real subject. They were added to the model as thermal boundary conditions and the thermal human body model was constructed. The model was applied to an analysis of thermal environment in a radiant cooled space and the practical validity of the model was verified.

References

Dubois, D. and E.F.Dubois (1916), A formula to estimate approximate surface area, if height and weight are known,*Archives of Inter. Medicine*, 17, pp.863-871.
Fanger, P.O.(1970) , *Thermal Comfort*, Danish Technical Press COPENHAGEN, pp.160-163.
Gagge, A.P., J.A.J. Stolwijk and Y. Nishi (1971), An Effective Temperature Scale Based on a Simple Model of Human Physiological Regulatory Response,*ASHRAE Trans.*, 77(1), pp.247-262.
Gagge, A.P., A.P. Fobelets and L.G. Berglund (1986), A Standard Predictive Index of Human Response to The Thermal Environment,*ASHRAE Trans.* 91-B, pp.709-731.
Hardy, J.D. and D. Dubois (1938), The Technis of Measuring Radiation and Convection, *Journal of Nutrition* Vol.15 No.5, pp.461-475.
Horikoshi, T., T. Tuchikawa, Y. Kurazumi, K. Hirayama and Y. Kobayashi (1990), Indication of Asymmetric and Uneven Thermal Radiation Environment related to Thermal Comfort and Discomfort, *J. Archit. Plann. Environ. Eng.*, AIJ, No.413,pp.21-28.
International Standard ISO 7730 (1984)
Miyanaga, T. and Y. Nakano (1998),Analysis of Thermal Sensation in a Radiant Cooled Room by Modified PMV Index, *Proceedings of ROOMVENT*, Vol.2, pp.125-31.
Nakamura, Y.(1987), Expression Method of The Radiant Field on a Human Body in Buildings and Urban Space, *J. Archit. Plann. Environ. Eng.*, AIJ, No.376, pp.29-35.
Ozeki,Y., T. Saito and S. Ogaki (1998), Effective Radiation Area of Human Body Calculated by a Numerical Simulation, *Proceedings of ROOMVENT*, Vol.2, pp.173-180.
Stolwijk, J.A.J. and J.D. Hardy (1966), Temperature Regulation in Man -A Theoretical Study-,*Pfuger Archives*, Vol.291, pp.129-162.
Suzuki, K. and N. Kakitsuba (1999), Development of a Human Body Model based on the Human Body Area and The Configuration Factors, *J. Archit. Plann. Environ. Eng.*, AIJ, No.515, pp.49-55.
Wissler, E.H. (1964), A Mathematical Model of the Human Thermal System,*Bulltein of Mathematical Biophysics* 26(2), pp.147-166.

Nomenclature

A : area (m^2) , $Aeff$: effective radiation area (m^2) , Ap : projection area (m^2) ,fp : projection area factor (=$Ap/Aeff$), G : Gebhart's absorption factor, h : convective heat transfer coefficient (W/m^2/°C), K : over-all heat transfer coefficient (W/m^2/°C) , N : total number of small element , Ta : air temperature (°C) , Tcl : surface temperature of human body (°C) , $Tskin$: skin temperature (°C), $TPRT$: plane radiant temperature (°C), Tp : surface temperature of ceiling panel (°C),α : azimuth(°), β : altitude(°),ε : emmisivity,σ : Stefan-Bolzmann constant 5.67*10^8 W/m^2/K^4 (0 °C=273.15 K) *Subscript i, j, k* : integer

Air Distribution in Rooms, (ROOMVENT 2000)
Editor: H.B. Awbi

HUMAN THERMAL SENSATION TO AIR MOVEMENT FREQUENCY

Yizai Xia, Rongyi Zhao and Weiquan Xu

Department of Thermal Engineering, Tsinghua University, Beijing, 100084, P.R.CHINA

ABSTRACT

Frequency of air movement is an important factor to influence human thermal sensation. Experiments and simulation work were done to study its effects in warm isothermal conditions. 106 subjects gave out their preferred frequencies in the temperature range of 26 $^{\circ}$C and 30.5 $^{\circ}$C and two relative humidity conditions. The results show that operative temperature and relative humidity do not have significant effects on preferred frequency in the experimental conditions. The average preferred frequency is about 0.4 Hz, and more than 80 percent of subjects chose frequencies between 0.3 Hz and 0.5 Hz. With the rising of the frequency, the sensed air velocity seems to be lowered, and the preferred velocity goes up. A TS model was developed to explain the experimental results. TS model deems that the responses of human cold cutaneous thermoreceptor (TR) have a positive dependence on the heat flow towards the skin surface at the depth of TRs when they are stimulated.

KEYWORDS

frequency, air movement, human thermal sensation, thermal comfort, cutaneous thermoreceptor, psychosensory intensity

INTRODUCTION

According to the theory of Hinze (1975), the essential characteristic features of turbulent air movement are irregularity and disorderliness, involving the impermanence of the various frequencies and also of the various periodicity and scales. Therefore, intensity and frequency are two indispensable features to describe a turbulent flow and to influence human thermal sensations in addition to mean air velocity.

Several experimental studies have examined air movement frequency and human thermal sensations. Fanger and Pedersen (1977) ever made an experiment in which subjects were exposed to a cyclic changed flow and found that the frequency in the range of 0.3—0.5 Hz cause larger uncomfortable feeling. Tanabe et al. (1994) observed human thermal sensation under seven kinds of fluctuated flow. They concluded that the air movement in a sine wave pattern makes person cooler than constant one. Air movements whose frequencies of 0.0167, 0.033 and 0.1 Hz have no different effects on human thermal sensation. After comparing the constant mode and fluctuating mode at the same mean wind speed, Arens et al. (1998) concluded that constant mode with power spectrum peaking between 0.7—1.0 Hz cools better than the fluctuation one whose peak frequency is between 0.2—0.4 Hz.

Ring et al. (1993) developed psychosensory intensity (PSI) model to simulate human cutaneous thermoreceptor (TR) responses to the sinusoidally changed stimuli of different frequencies. They concluded that a cutaneous thermoreceptor response is proportional to the time rate of change of the temperature at the depth of the TR, and the thermal sensation is proportional to the dynamic response integrated over a period of 20 s. The result displayed a response curves with peaks at about 1 Hz.

The purpose of this paper is to investigate human thermal responses under different frequencies of cyclically changed air movement. Compared with the previous experiments, this experimental method is different in that subjects can choose their preferred frequencies in a relative large range. In addition, The experimental results were explained by a newly developed model.

EXPERIMENTAL METHOD

Facilities

The experiments were conducted in the climate chamber at Tsinghua University, measuring 3.4m by 4.8m by 3m. A cyclical air movement was provided through an air box, at the outlet of which stands a series of swing blinds. Air movement with different frequency and velocity could be gotten by regulating the oscillating frequency of the blinds and the fan speed with convenient knobs. The jet was basically re-circulated isothermal air immediately taken from the chamber.

The mean radiation temperature is very close to the air temperature in the chamber. Semiconductor thermometer and dry- and wet-bulb thermometers were used to measure the air temperature and the relative humidity around the subjects. Air velocity is measured by a hot wire anemometer of the constant temperature type.

Physical Conditions Tested

All subjects were college students whose ages ranged from 18 to 24 years old. Wearing shoes, socks, underwear, trousers and light long-sleeved shirt, the subjects were instructed to enter the climate chamber. This clothing ensemble was about 0.6 clo. Adding the thermal resistance of the metal folding chair referring to the research results of. McCullough and Olesen (1994), the total clothing insulation was about 0.7 clo.

Experiments were performed at the air temperature set points of 26°C, 27.5°C, 29°C and 30.5°C, and two relative humidity of 35%, and 65%. Each condition of the lower relative humidity contains 40 sample cases, and each one of the higher relative humidity contains 22 cases. Several additional experiments were planed in case of instrument malfunction to bring the sample size up to total 106 persons. The experimental conditions are shown in Table 1.

Experimental Procedure

The experiment lasted approximately 2 hours and included one relative humidity, two air temperature conditions. In the first half an hour, the subjects were ushered into the chamber, changed clothes and adjusted to the thermal condition in the chamber. In the meanwhile, they were explained the experimental procedures and asked to fill out the background survey questionnaire. Then they were seated in front of the air box for half an hour in each temperature condition. They could adjust the air velocity and frequency to maintain comfort while filling out the questionnaire, and were allowed to do reading or writing work. At the interval of two temperature conditions, the subject could leave the chair and had a rest, but they were not allowed to leave the chamber. In the last stage of the experiment, subjects were asked to feel flows with four typical frequencies, ten minutes for each one. They gave out four corresponding preferred air velocities on the basis of feeling the same intensity of the flows.

TABLE 1
EXPERIMENTAL CONDITIONS

Operative temperature (To)	26°C, 27.5°C, 29°C, 30.5°C
Mean radiant temperature (Tmrt)	= air temperature
Relative humidity (Rh)	35%, 65%
Clothing insulation (Icl)	0.7 clo
Metabolism	sedentary (1 met)
Air velocity (v)	preferred
Air frequency (f)	preferred

RESULT

Four typical samples of air velocity at different frequencies are shown in Figure 1. By analyzing the "energy spectrum" of these samples, the peak frequencies are gotten and used to indicate the frequency of the air movement, for example, 0.16 Hz, 0.31 Hz, 0.50Hz, or 0.64Hz as shown in Figure 1.

During the experiment, subjects are asked to give out the range of frequency, which they deemed acceptable, as well as a preferred one. An analysis of variance (ANOVA) shows that the preferred frequency and the acceptable range do not change significantly with the shift of air temperature and relative humidity. In the experimental physical conditions, the average preferred frequency is about 0.4 Hz, and the acceptable range is from 0.2 Hz to 0.65 Hz.

Figure 2 indicates the distribution of preferred frequencies in the relative humidity of 35%. The distributions are similar in different temperatures. More than 80 percent of subjects chose frequencies between 0.3 Hz and 0.5 Hz. There are about 95 percent of subjects to prefer frequencies below 0.7 Hz. The alike results are obtained in the relative humidity of 65%.

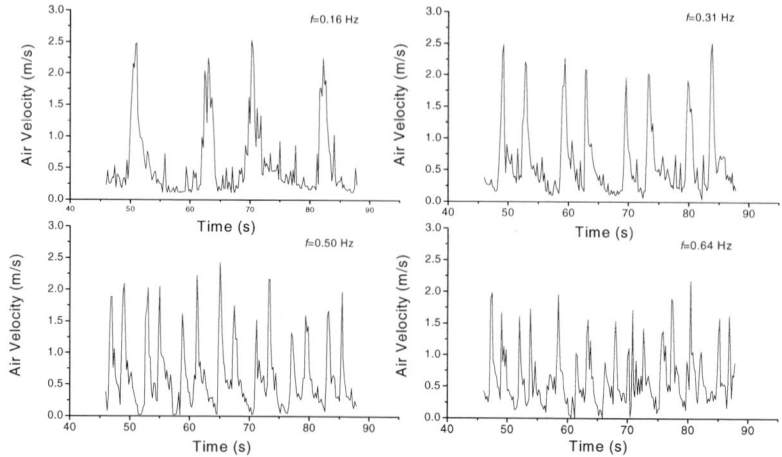

Figure 1: Four samples of air velocity under different frequencies

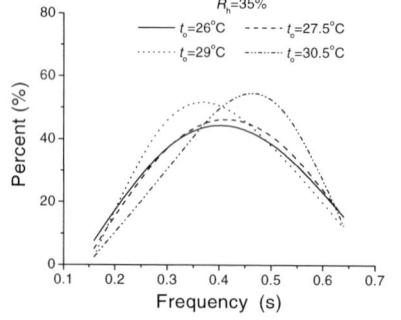

Figure 2 : Distribution of preferred frequency at relative humidity of 65%

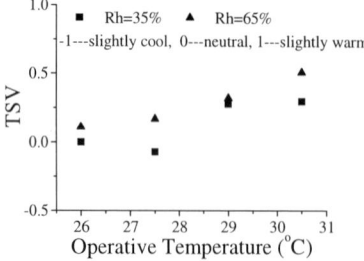

Figure 3: Mean preferred velocities in four typical frequency conditions

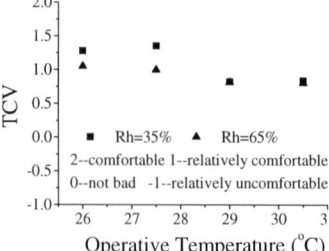

Figure 4: Mean thermal sensation votes (TSV) and thermal comfort votes(TCV) in experimental conditions

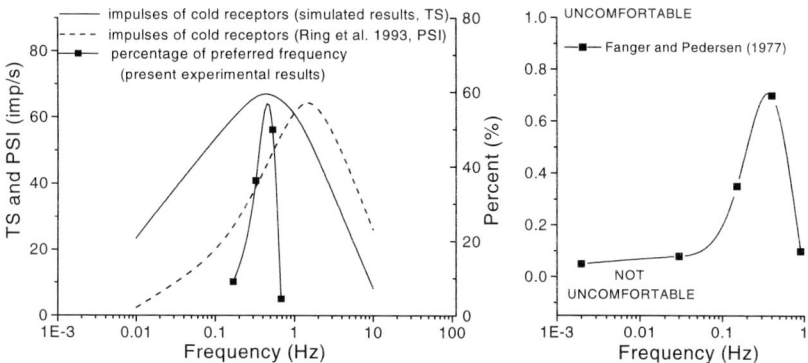

Figure 5: Comparison of simulated results (TS model and Ring's PSI model), present experimental results and the results of Fanger and Pedersen's

Under each of four typical frequencies as shown in Figure 1, subjects were asked to adjust air velocity, to which they were exposed, to choose a preferred one. The four preferred velocities should give subjects the same sensed velocities of flows. Generally, subjects reported that the sensed air velocity seemed to be lowed as the frequency was raised. Figure3 shows the mean preferred velocities in the four operative temperature conditions under the relative humidity of 65%. With the rising of the frequency, the preferred velocities also go up. For instance, in the operative temperature of 29 $^{\circ}$C, the mean velocity ascends from 1.18 m/s at 0.18 Hz to 1.27 m/s at 0.58 Hz. The same pattern occurs in the lower relative humidity.

Figure 4 show the mean thermal sensation votes (TSV) and thermal comfort votes (TCV) in experimental conditions. The average values of TSV are all in the range of –0.5 and +0.5, and TCV values are above 0.5, which indicates that most subjects can achieve thermal comfort under the experimental conditions when they can adjust the air velocity and change frequency as they want. Additionally, there shows a consistent increase in TSV votes with ambient temperature in Figure 4. This may indicate that the subjects either did not attempt or failed to restore "complete" neutrality by adjusting the air movement when the temperature was raised up.

DISCUSSION

A simulation program was developed to calculate the human skin temperature and the impulses of the thermal receptors when the skin was exposed to a dynamic flow. A new simulation model TS was rendered which accounted that the responses of the TRs are proportional to the temperature gradient with depth but not with time as presented by Ring et al. (1993). It can be supposed that the skin is exposed to a flow whose velocity changes in a sinusoidal pattern, then the skin temperatures on and below the surface will change in the same frequency as the above flow. According to the theory of Hensel (1981), the cold TRs are stimulated when the temperature where the cold TRs locate goes down. The new model deems that the responses of cold TRs have a positive dependence on the heat flow towards the skin surface when they are stimulated. The dynamic psychosensory intensity interrelated with the thermal sensation is the integration of such heat flow over a time, for example 20 s. The more the heat flow, the stronger the intensity, and the cooler the sensation. The details about the program and model will be discussed in the later paper.

Figure 5 shows the calculated psychosensory intensity with TS model and PSI model as well as the experimental results of the percentage of subjects' choosing a particular frequency. It is clearly shown that TS model matches better with the experimental results than PSI model, because most subjects choose the corresponding frequency that has the peak psychosensory intensity according to TS model.

The experimental results of Fanger and Pedersen are also shown in Fig.5. They found airflow with 0.3—0.5 Hz causes more uncomfortable draft feelings when subjects were in the cool-to-neutral state. The result of this study, that air motion with frequency of 0.3—0.5 Hz is most preferred when subjects are in the neutral-to-warm condition, is concordant with this finding. Therefore, a conclusion can be drawn that the flow with frequency of 0.3—0.5 Hz has the most influence on human cold receptor.

CONCLUSION

Experiments and simulation work were done to study the effects of frequency of air movement on human thermal sensation in warm conditions. 106 subjects gave out their preferred frequencies in the temperature range of 26 °C and 30.5 °C and two relative humidity conditions. The conclusions can be drawn as follows:

- Most subjects could achieve thermal comfort under the experimental conditions after adjusting the air velocity and frequency, as they liked.
- In the temperature range of 26 °C and 30.5 °C, and the relative humidity between 35% and 65%, operative temperature and relative humidity do not have significant effects on preferred frequency.
- With the rising of the frequency, the sensed air velocity seems to be lowered, and the preferred velocity goes up.
- TS model was developed to explain the experimental results. This model deems that the responses of cold TRs have a positive dependence on the heat flow towards the skin surface at the depth of TRs when they are stimulated.
- This study together with Fanger and Pedersen's experiment prove that the flow with frequency of 0.3—0.5 Hz has the most influence on human cold receptor.

References

Arens E., et al. (1998). A study of occupant cooling by personally controlled air movement. *Energy and Buildings* **27**, 45—59.

Fanger P.O. and Pedersen C.J.K. (1977). Discomfort due to air velocities in spaces. *Proc. Of Meeting of Commission B1, B2, E1 of Int. Instit. Refig* **4**, 289—296.

Hensel H. (1981). *Thermoreception and temperature regulation*, Academic press, London, UK.

Hinze J.O. (1975). *Turbulence*, McGraw-Hill, Inc. New York, USA.

McCullough F. and Olesen B.W. (1994). Thermal insulation provided by chairs. *ASHARE Transaction* **100:1**, 795-802.

Ring J. W. et al. (1993). Human thermal sensation: frequency response to sinusoidal stimuli at the surface of the skin. *Energy and buildings* **20**, 159-165.

Tanabe S. and Kimura K. (1994). Effects of air temperature, humidity, and air movement on thermal comfort under hot and humid conditions. *ASHRAE Transactions* **100:2**, 953—969.

© 2000 Elsevier Science Ltd. All rights reserved.
Air Distribution in Rooms, (ROOMVENT 2000)
Editor: H.B. Awbi

STUDY ON PREDICTION OF THERMAL COMFORT IN UNSTEADY AND INHOMOGENEOUS THERMAL ENVIRONMENT

Yasuhiro Nakamura[1], Minoru Mizuno[2] and Daisuke Ida[3]

[1] Department of Kansei Design Engineering, Yamaguchi University,
Yamaguchi, Ube, Tokiwadai 2-16-1, JAPAN
[2] Department of Environment and Energy System Engineering, Osaka University,
Osaka, Suita, Yamadaoka 2-1, JAPAN
[3] Daikin Industries, Ltd.
Osaka, Sakai, Kanaoka-chou 1304, JAPAN

ABSTRACT

The purpose of this study is to predict the thermal sensation under unsteady and unhomogeneous thermal environment. We carried out subject tests in the outdoors to investigate the characteristic features of thermal sensation under unsteady and unhomogeneous thermal environment. We also tried to predict the thermal sensation under unsteady and unhomogeneous thermal environment using SET* and the modified Fu's model which were developed based on transient heat balance equations for human body. The main results are as follows. The degree of the change of thermal sensation during the test period of forty minutes reaches five categories in the outdoors. The prediction by SET* has almost the same accuracy as prediction using heat loss calculated by the modified Fu's model. It is also found that the prediction method using the heat loss and the mean skin temperature calculated by the modified Fu's model has the highest accuracy of prediction of thermal sensation vote.

KEYWORDS

Thermal Sensation, Thermal Comfort, Unsteady and Unhomogeneous Thermal Environment, SET*, Fu's Model, Subject Test, Numerical Simmulation, Outdoor Thermal Environment, Mean Skin Temperature

INTRODUCTION

It is an important subject to predict thermal comfort in the unsteady and unhomogeneous thermal environment which is often formed in a dome stadium with a movable roof and in an atrium whose roof is covered by glass and is strongly affected by the solar radiation. We carried out subject tests in the outdoors because the outdoors is the extreme situation of the unsteady and unhomogeneous thermal environment. While PMV [Fanger (1970)] is an excellent index to predict the thermal comfort in steady and homogeneous thermal environment, it is not applicable to unsteady thermal environment because it is developed based on a steady heat balance equation for the human body. In this study we try to predict the thermal sensation under unsteady and unhomogeneous thermal environment by SET* [ASHRAE (1989), Gagge et al. (1986)] and the modified Fu's model because these models are developed based on unsteady heat balance equations. The accuracy of prediction of TSV (thermal sensation vote) by these models is verified by the comparison of the predicted values with TSV obtained from the subject tests.

EXPERIMENTAL METHOD

Subject tests were carried out in the outdoors for ten days from late autumn to early winter in 1998 according to the time schedule shown in Figure 1. The total number of subjects was forty and they were all healthy males of about twenty-three years old. They are requested to vote their general thermal sensation, general thermal comfort, thermal sensation of each determined part of the body, etc. every ten minutes during the test period of forty minutes. They are also requested to input their thermal sensation to their personal computers at any time when they feel rapid change of thermal sensation in addition to the beforehand determined times. The categories of each sensation are shown in Table 1. Each subject wore a 0.83 clo of uniform consisting of a T-shirt, a shirt, trunks, a sweat shirt, sweat pants, a pair of socks and sneakers.

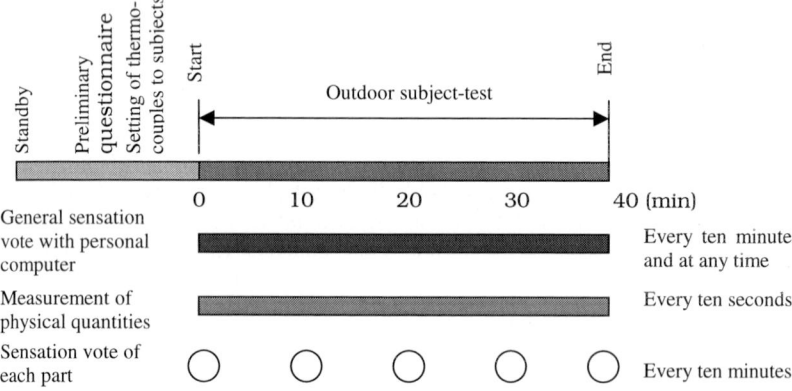

Figure 1: Time schedule of the outdoor subject-test

TABLE 1
CATEGORIES OF SENSATION VOTE

Thermal comfort		Thermal sensation				Sensation of air current and solar radiation	
1	Comfortable	5	Unbearably hot	−1	Slightly cool	1	Not feel
2	Slightly comfortable	4	Very hot	−2	Cool	2	Slightly feel
3	Neutral	3	Hot	−3	Cold	3	Feel
4	Slightly uncomfortable	2	Warm	−4	Very cold	4	Very feel
5	Uncomfortable	1	Slightly warm	−5	Unbearably cold		
6	Very uncomfortable	0	Neutral				

The physical quantities shown in Table 2 were measured during each subject test. The ranges of the change of physical quantities through the whole tests were as follows: 12.1 to 17.7 °C in the air temperature, 0.1 to 7.7 m/s in the wind velocity, 6.2 to 74.5 °C in the mean radiant temperature and 26.3 to 65.4 % in the relative humidity.

TABLE 2
MEASURED ITEMS OF PHYSICAL QUANTITIES

Measured item	Measured point or direction	Measuring instruments
Air temperature	200,300,600,1000,1500 and 1700mm above the ground	Thermocouples
Skin temperature	20 points on the body	Thermocouples
Wind velocity	200,300,600,1000,1500 and 1700mm above the ground	Multi-points anemometer
Flux of solar radiation	Front, back, right, left, up and down sides	Total hemispherical pyrheliometer
Flux of thermal radiation	Front, back, right, left, up and down sides	Total hemispherical radiometer
Ground-surface temperature	4 points around subject	Thermocouples
Wet-bulb temperature	600mm above the ground	Assman psychrometer

CHARACTERISTIC FEATURES OF OUTDOOR THERMAL ENVIRONMENT

One of the characteristic features of the outdoor thermal environment is that the air temperature, the wind velocity and the flux of the solar radiation are always fluctuating over wide ranges as shown in Figure2. Therefore the thermal sensation of the subjects is always changing and the degree of the change reaches five categories. It is essentially different from the indoor thermal environment which is usually steady and is kept in comfortable condition using air conditioning system. It is also shown that the change of the thermal sensation vote corresponds well to the fluctuation of the wind velocity and the flux of the solar radiation. The thermal sensation vote shifts to warmer categories when the flux of the solar radiation is large and shifts to colder categories when the wind velocity is large.

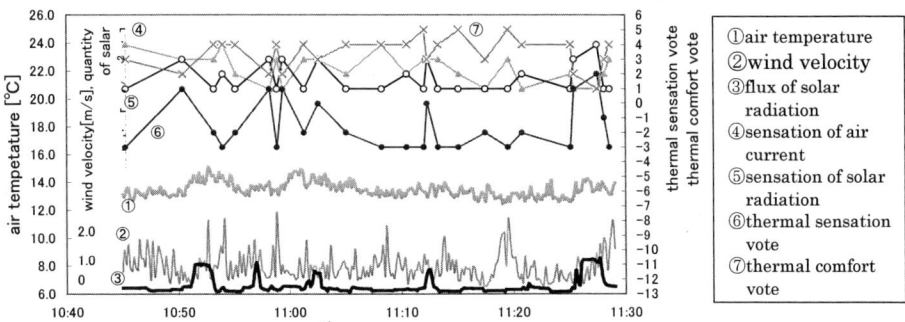

Figure 2: Change of outdoor thermal environment and thermal sensation vote

PREDICTION OF THERMAL SENSATION VOTE

In this section we discuss the predicted results of TSV under the unsteady thermal environment in the outdoors using the modified Fu's model and SET*.

Modified Fu's Model

Fu's model [Fu (1995)] is a mathematical transient model of the human thermal regulatory system utilizing finite-element techniques and is developed for the clothed human and in three dimensions. Therefore, it is applicable in a wide range of real conditions. To apply this excellent model to our subjects in the outdoors where the wind velocity always changes rapidly and the flux of the solar radiation depends strongly on the direction, we modified Fu's model about the following four points:

(1) to set the time step to 5sec as applicable to the rapid change of the thermal environment,
(2) to input thermal environmental conditions concerning the wind velocity and the flux of the solar radiation every time step,
(3) to use convective heat transfer coefficient calculated by the equation in ASHRAE Fundamentals Handbook [ASHRAE (1989)] to take into consideration the dependence of convective heat transfer coefficient on the wind velocity,
(4) to set individual boundary value of the air temperature, the wind velocity and the mean radiant temperature surrounding the body for the every surface elements of the body to take into consideration inequality of thermal environment.

Skin temperatures predicted by the modified Fu's model agree considerably well with measured skin temperatures as shown in Figure 3. Similar results are obtained for other subjects, too. The measured skin temperature of the back of the left hand is higher than the right hand because each subject is sitting to the south and the only left hand is exposed to the direct sunlight in the morning. The modified Fu's model can well simulate the temperature difference between the right hand and the left hand. Figure 4 shows an example of the change of the mean skin temperatures relative to its time average values during the subject-test period. Predicted value by the modified Fu's model coincides well with measured value. However, it is rather difficult to predict the absolute value of the mean skin temperature of an individual subject with a high accuracy because the mean skin temperatures are different among subjects even though they are exposed to the same thermal environment.

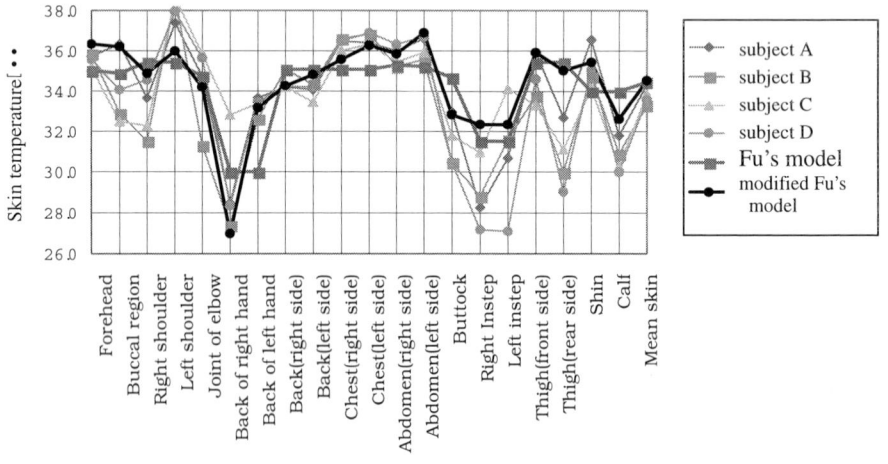

Figure3: Prediction of skin temperature by the modified Fu's model

Prediction of TSV

An adequate correlation is estimated from Figure 5 showing the correlation between TSV and the heat loss from the body. As the similar good correlation is observed at other subject tests, we tried to predict

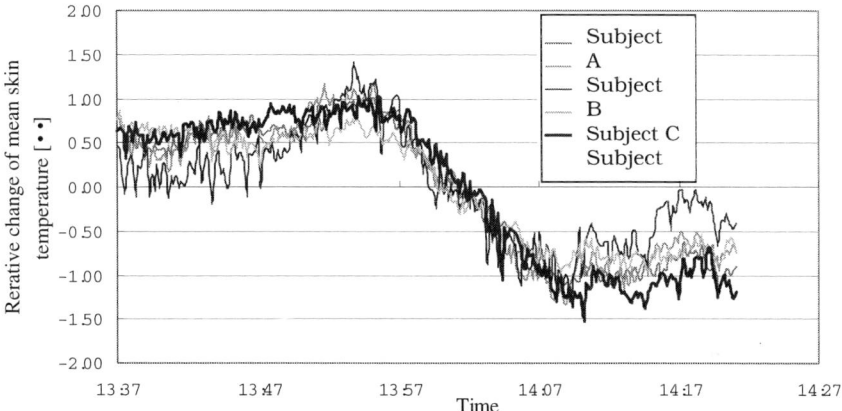

Figure4: Relative change of mean skin temperature

TSV by two methods. The method ① uses only the heat loss from the body to predict TSV and the method ② uses the heat loss and the mean skin temperature. Predicted results of TSV by two methods are shown in Figure 6. TSV is calculated using regression equations correlating the thermal sensation votes of forty subjects with heat loss calculated by Fu's model or with heat loss and the mean skin temperature. Though TSV changes frequently and rapidly and the degree of its change reaches five categories, the predicted values by two methods coincide well with TSV. The accuracy of prediction of TSV by the two methods is shown in Table 3. For example, the rate that predicted values coincide with TSV within the accuracy of ±0.5 categories is 26.7% in the method ②. The accuracy of prediction is a little higher in the method ② than in the method ①. Predicted results of TSV by the mean skin temperature alone (method ③) and SET*(method ④) are also shown in Table 3. The accuracy of prediction by the mean skin temperature alone is not good but the prediction by SET* has almost the same accuracy as method ①. It is found that the prediction method ② shows the highest accuracy of prediction among the above four prediction methods.

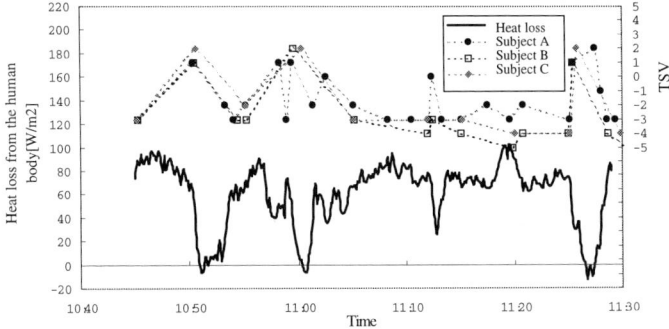

Figure 5: Correlation between TSV and heat loss

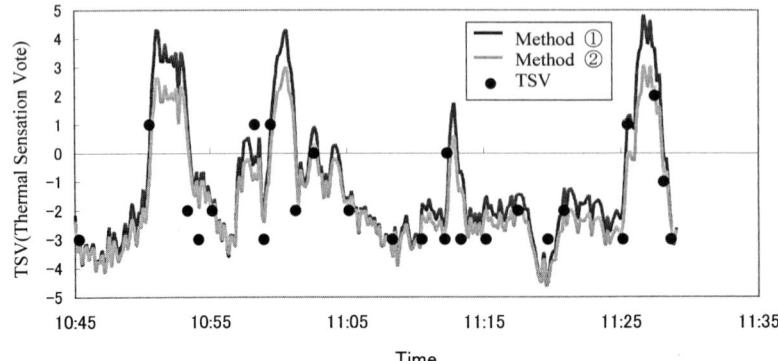

Figure 6: Prediction of TSV by method ① and method ②.

TABLE 3
ACCURACY OF PREDICTION OF THERMAL SENSATION [%]

Accuracy level within which predicted values coincide with TSV	Method ①	Method ②	Method ③	Method ④
	Heat loss	Heat loss and mean skin temperature	Mean skin temperature	SET*
±0.5 categories	22.5	26.7	18.3	24.3
±1.0 categories	43.9	49.2	38.8	43.7
±1.5 categories	65.3	68.9	57.7	62.3

CONCLUSINONS

(1) One of the characteristic features of the outdoor thermal environment is that the TSV changes frequently and rapidly according to the change of the weather and the degree of the change of TSV reaches five categories.
(2) The modified Fu's model is available to predict the difference of skin temperature caused by inequality of solar radiation and to predict the change of the mean skin temperature relative to its time average during the subject-test period.
(3) According to the results obtained by forty subject tests, the prediction method using the heat loss from the body and the mean skin temperature calculated by the modified Fu's model shows the highest prediction accuracy for TSV in unsteady and unhomogeneous thermal environment.

REFERENCES

ASHRAE (1989). Physiological Principles, Comfort, and Health (Chapter 8). *1989 Fundamentals Handbook*, 8.1-8.15

Fanger P.O.(1970). *Thermal Comfort*, McGraw-Hill, New York, USA

Fu G.(1995). A Transient, 3-D Mathematical Thermal Model for The Clothed Human. *A Dissertation for Doctor of Philosophy, Department of Mechanical Engineering, College of Engineering, Kansas State University*, UMI Number 9544220, p.270

Gagge A.P., Fobelets and Berglund L.G.(1986). A Standard Predictive Index of Human Response to The Thermal Environment. *ASHRAE Transaction*, PO-86-14, No.1, 709-731

Air Distribution in Rooms, (ROOMVENT 2000)
Editor: H.B. Awbi

PREDICTED COMFORT ENVELOPES FOR OFFICE BUILDINGS WITH PASSIVE DOWNDRAUGHT EVAPORATIVE COOLING

D. Martinez , D. Fiala, M. J. Cook and K. J. Lomas

Institute of Energy and Sustainable Development, Scraptoft Campus, De Montfort University, Scraptoft, Leicester, LE7 9SU, UK

ABSTRACT

Passive Downdraught Evaporative Cooling is a low energy strategy for maintaining thermal comfort in buildings located in hot dry climates. The thermal performance of such buildings can be predicted using simulation models. The temperatures predicted can be compared with standard comfort envelopes which show the range of temperatures and relative humidities within which occupants will be thermally comfortable. Standard comfort envelopes do not account for the adaptive opportunity which is important in free-floating buildings. Such adaptation could include changing clothing levels, local air speed (e.g. by using of fans) and solar radiation (by use of shading devices). A state of the art dynamic heat transfer and thermal comfort simulation model has been used to develop new comfort envelopes based on the adaptive behaviour theory. These "adaptive comfort envelopes" indicate occupants will be satisfied over a wider range of environmental conditions. The work indicates that increasing local air speeds is particularly effective in improving comfort when occupants are exposed to solar radiation.

KEYWORDS

Comfort envelope, adaptive behaviour, thermal sensation, PDEC, low energy buildings

INTRODUCTION

In hot dry climates, thermal mass together with night venting may keep office building comfortable without resorting to air conditioning. Active or passive evaporative cooling can assist by further reducing air temperatures. In a recent European project, Robinson et al. (1999), Bowman et al. (2000) Passive Downdraught Evaporative Cooling (PDEC) was explored as a novel way of cooling office buildings. PDEC combines the benefits of a natural ventilation strategy with a passive cooling technique. This is achieved by injecting microscopic drops of water into a hot, dry airstream using micronisers. Due to the evaporation of the water drops by the air, the relative humidity and density of the air increase and the dry-bulb temperature is reduced. Thus PDEC environments are characterised by high relative humidities, increased air speeds and, due to the free running nature of passively controlled buildings, a range of temperature fluctuations. In order to determine whether a building with PDEC will be comfortable, four issues need to be considered: (i) the local climate conditions, (ii) the thermal behaviour of the building and system, (iii) the resultant effect on indoor climate conditions, and (iv) the behaviour of the occupants.

In a former study, Fiala et. al. (1999), comfort envelopes were developed. However, these envelopes did not incorporate the adaptive behaviour of occupants in response to changes in environmental conditions that occur e.g. free-running buildings, Humphreys (1978). In this paper the comfort envelopes have been extended to account for adaptive changes in clothing, local air speed and diffuse solar radiation. The work is based on analysis of occupants' physiological and comfort responses, as predicted for combinations of relative humidity, dry-bulb temperature and solar radiation. This analysis was undertaken using a detailed dynamic model of human heat transfer and thermal comfort developed by Fiala (1998). The model predicts skin and core temperatures, sweat rates and the overall Dynamic Thermal Sensation (DTS) for any type of time-varying, asymmetric environmental conditions, clothing and activity levels. DTS is predicted according to the 7-point-ASHRAE scale running from –3 (cold) to +3 (hot) with 0 representing thermal neutrality.

EXISTING COMFORT ENVELOPES FOR PDEC BUILDINGS

In the former study, Fiala et. al. (1999), the effect of PDEC-environments on predicted thermal, regulatory and perceptual responses of occupants was studied for summer conditions in office buildings. The subjects were assumed to be wearing typical summer clothing (0.55 clo) and be engaged in typical office activities (1.2 met). Based on that analysis, zones of comfort were defined for PDEC office buildings, considering two levels of air velocity (0.3 m/s and 0.8m/s). Also included was the impact of varying solar radiation on subjects' responses (0 W/m^2 and 25 W/m^2). The analysis showed that even relative humidities of 80% were predicted to be thermally acceptable for the PDEC-building occupants. Thus the comfort envelopes (dashed lines in Figure 2), established a new upper limit for relative humidity, which allows PDEC systems to operate above the traditional 60% limit, ASHRAE 55 (1992). However, the envelopes did not indicate the tolerance to higher temperatures which adaptive behaviour might enable.

STRATEGY FOR EXTENDING THE EXISTING COMFORT ENVELOPES

In order to quantify the impact of thermal adaptation in the existing comfort envelopes, the three most influential adaptive reactions were investigated: (i) changes in clothing insulation (ii) variation in local air speed (using fans) and (iii) the manipulation the amount of diffuse solar radiation (using blinds). In principle, these adaptive actions should enable occupants to feel comfortable in a wider range of environmental conditions.

Different summer clothing ensembles for men and women, and for different ranges of operative temperatures, (light ensemble for 20°C $<T_o<$25°C and a very-light ensemble for 25°C $<T_o<$32°C) have been considered. These differences considered both the use of lighter garment fabrics and the selection of different garment items i.e. long/short sleeve, lighter shoes, dress instead of a skirt and blouse, etc. An occidental office dress-code, however, has been respected in all cases. The clothing was modelled in detail by applying individual items of an ensemble to the corresponding body elements of the multi-segmental model. The Fiala model was fed with the measured overall values of clothing insulation I_{cl} [clo], clothing area factor f_{cl} [-], and the evaporative resistance of the fabric R_{Ef} [m^{2}kPa/W], obtained from literature, McCullough et al. (1985) and (1989).

The effect of the chair was also considered. It was modelled as an item, which covers a part of the posterior leg segments. Back contact was considered but without any force. The individual clothing ensembles considered in the study were as follows:

- *Women-light outfit*: bra, pantyhose, panties, skirt knee length, blouse long sleeve, open lady shoes and chair, I_{cl}=0.63clo, i_{cl} =0.34 [1] ,
- *Women-very-light outfit*: bra, panties, long thin dress with short sleeves, open ladies shoes and chair, I_{cl}=0.42clo, i_{cl} =0.28,
- *Men-light outfit*: briefs, socks, light long trousers, long sleeve shirt, street shoes and chair, I_{cl}=0.69clo, i_{cl} =0.30,
- *Men-very-light outfit*: briefs, socks, light long trousers, short sleeve shirt, sandals and chair, I_{cl}=0.43clo, i_{cl} =0.29,

Air movement affects the evaporation of moisture from the skin and thus of the comfort perception. If occupants start feeling warmer, thermal neutrality might be maintained by switching on ceiling, desktop or floor mounted fans thus increasing the local air speed. When a PDEC system is operating (warm conditions assumed), the air speed in occupied spaces would be about 0.3 m/s whereas the air speed value may reach about 1.5 m/s when fans are switched on. For "cool" conditions, (PDEC system off), the typical air speed may be just 0.2 m/s which ensures the required ventilation levels.

The "adaptive opportunity" also includes the interaction of occupants with the building fabrics. One of the most influential effects is the use of shading devices. Operating shading devices such as blinds or louvres, will modify the diffuse solar radiation on the body. Likewise, occupants, if permitted to by the management regime may move to work in more shaded areas (away from windows). Direct solar radiation should be eliminated by the building's own external shading devices. The investigated values were zero and 50 W/m^2 (the latter ensures high natural lighting level).

PROCESS OF DEVELOPMENT

The upper and lower limits of the new extended comfort envelopes were obtained from the former study, Fiala (1999) i.e. relative humidity of 80% and a moisture content of 0.0045 kg/kg, respectively. The right and left boundaries of the new comfort zones, were defined by the operative temperature (T_o) and relative humidity at which 10% of the people will be thermally uncomfortable, ASHRAE 55,(1992)

A file matrix with combinations of operative temperatures and relative humidities was created, along with the different clothing ensembles, air speeds and solar radiation values. These were used as the input data to the simulation program. Indoor operative temperatures were investigated for a range between 20°C<T_o<32°C in successive steps of ΔT_o = 0.5K. The relative humidities investigated ranged between 10%<RH<90% in steps of 2%. The diffuse solar intensities subjected to analysis were 0 W/m^2 and 50 W/m^2. The considered air velocities were 0.2 m/s (for the range 20°C<T_o<25°C) and 0.3, 0.6, 0.9, 1.2 and 1.5 m/s (for the range 25°C<T_o<32°C). Clothing levels were studied separately for men and women, resulting in values of I_{cl}=0.69clo/0.43clo (for men light/very light ensembles) and I_{cl}=0.63clo/0.42clo (for women light/very light ensembles). A metabolic rate of 1.2 met, was used as the office activity level, ISO 7730 (1994). So, a total of nearly 26,000 different combinations of boundary conditions emerged.

The simulation series were conducted as individual two-hour-exposures to the steady environmental and personal conditions. Simulation results for which 8%<PPD<12% applies were filtered out for further data processing. This data is plotted onto a psychrometric chart, Figures 1a and 1b. The required comfort limits of PPD=10% were obtained by linear regression through the filtered data for both the right-hand side boundary, i.e. DTS>0, and the left-hand side boundary, i.e. DTS<0. A good general correlation (0.85<$|r|$<0.93) for the linear regressions was achieved.

---

[1] The overall values of I_{cl} and i_{cl}, were calculated by the Fiala model from the individual clothing items.

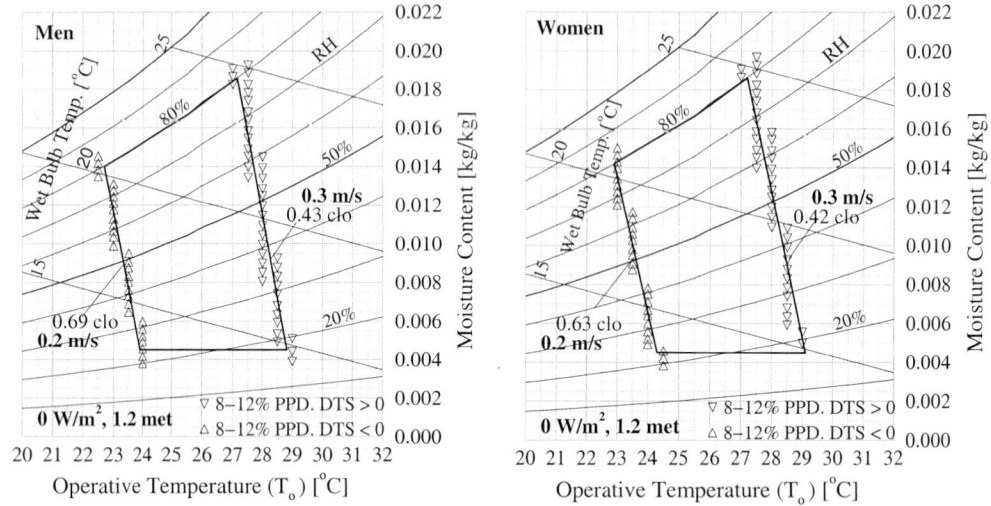

Figure 1a and 1b: Predicted comfort envelopes for men (left) and women (right) at 0 W/m² solar radiation and the corresponding minimum air speeds.

The analysis of the comfort boundaries indicated that there were no significance differences between men and women (see Figure 1a and 1b). For this reason only the results obtained for one sex (men) were considered in the further analysis.

RESULTS

In absence of solar radiation, Figure 2, the new comfort envelope ranges between 22.7°C (80 % RH) and 28.8°C (~17%RH) for minimum air speeds. This represents an enlargement of the tolerated temperature range of about 2 K when compared with the former comfort envelope (dashed lines in Figure 2) which was developed using a constant clothing level. A further extension of about 0.7 K towards warmer temperatures was achieved by increasing the air speed from 0.3 m/s to 0.6 m/s. However, further increases in the air speed produced successively less increase in tolerance to high temperatures. This is because at increased air velocities, the air temperature perceived as comfortable approaches the temperature of the body surface.

To counter the thermal effect of diffuse solar radiation of intensity 50W/m², the operative temperature must be reduced. This is indicated by the shift of the comfort envelope by about 2 K, indicating a strong effect of solar radiation on thermal comfort. It can be seen in Figure 3 that the effect of air speed on comfort is more pronounced when solar radiation enters the space than when it is excluded. This is because elevated air speeds are capable of removing more heat from the irradiated body surface. So, an increase of air speed from 0.3 m/s to 1.5 m/s extends the acceptable comfort conditions in the presence of solar radiation by 2.4 K, but only by 1.7 K in the absence of solar radiation. In both cases the variation of RH with temperature was found to be linear at a rate of about 2.5×10^{-2} K/RH% which agrees well with published data obtained from comprehensive experiments, Rohles et. al. (1971).

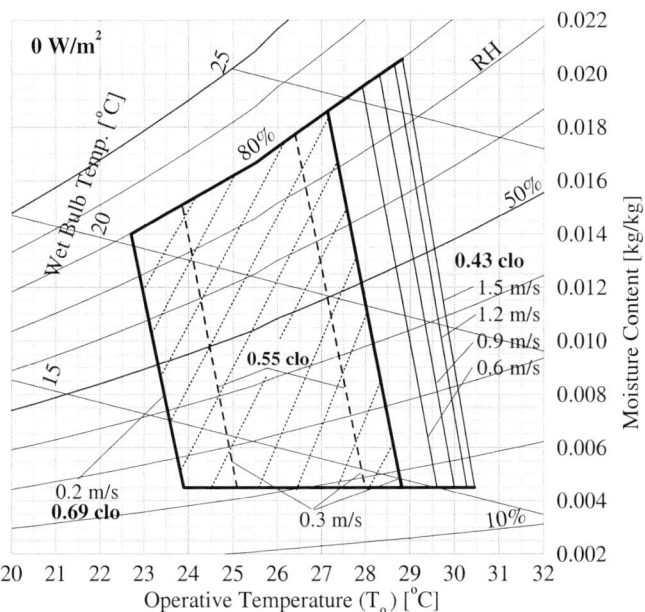

Figure 2: Predicted comfort envelopes for 0 W/m² of solar radiation and different air speeds. The former comfort envelope for 0.55clo and 0.3m/s is also included (dashed lines).

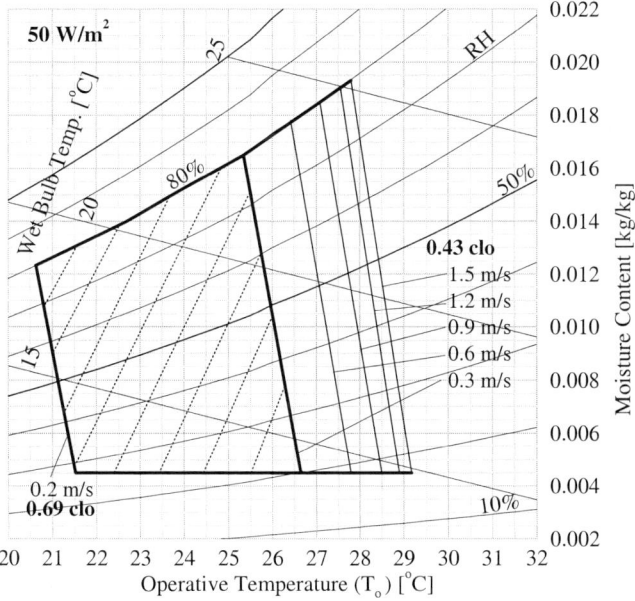

Figure 3: Predicted comfort envelopes for 50 W/m² of diffuse solar radiation and different air speeds.

CONCLUSIONS

In this study, extended comfort envelopes have been derived for hot summer conditions in office buildings. The adaptive behaviour of occupants has been investigated considering changes in the clothing level, air speed and solar radiation. As a results, the new comfort envelopes extend from 22.7°C (80 % RH) to 28.8°C (~17% RH) in absence of solar radiation. This is about 2 K wider than that obtained in the former study, Fiala et al. (1999). In the presence of diffuse solar radiation of 50 W/m^2 the comfort envelopes shift towards cooler air temperatures by about 2 K.

Increasing air speed leads to an acceptance of warmer conditions but this effect becomes less efficient as the air speed continues to rise. It was also found that increasing air speed is more effective in the presence of solar radiation.

These extended comfort envelopes were developed for PDEC buildings. However, they also may be used to examine summer comfort conditions in other types of office buildings in which thermal adaptation is possible.

REFERENCES

ASHRAE Standard 55 (1992). Thermal environmental conditions for human occupancy. ASHRAE, Atlanta, USA.

Bowman N. T., Eppel H., Lomas K. L., Robinson D. and Cook M. J. (2000). Passive Downdraught Evaporative Cooling – I: Review and precedents. *Building services Engineering Research and Technology* (**in press**).

Fiala D. (1998). *Dynamic simulation of human heat transfer and thermal comfort. Ph.D. thesis*, De Montfort University, Leicester, UK.

Fiala D, K J Lomas, D Martinez, and M J Cook. (1999). Dynamic thermal sensation in PDEC buildings. *Proceedings PLEA* (Brisbane) **99:1**, 243-248.

Humphreys M. A. (1978). Outdoor temperatures and comfort indoors. *Building research and practice* **6**, 92-105.

ISO 7730 (1994). Moderate Thermal environments. Determination of the PMV and PPD Indices. International Organisation for Standardisation, Geneva.

McCullough E.A., Jones B. W. Huck J. (1985). A comprehensive data base for estimating clothing insulation. *ASHRAE Trans* **91**, 29-47.

McCullough E. A., Jones B. W. Tamura T. (1989). A data base for determining the evaporative resistance of clothing. *ASHRAE Trans* **95**, 316-328.

Rholes F. H. and Nevins R.G. (1971). The nature of thermal comfort for sedentary man. *ASHRAE Trans* **77:1**, 239-246.

Robinson D., M. J. Cook, K.J. Lomas, and N. T. Bowman (1999) The design and control of buildings with passive downdraught evaporative cooling. *Proceedings PLEA* (Brisbane) **99:1**, 453-458.

Air Distribution in Rooms, (ROOMVENT 2000)
Editor: H.B. Awbi

SIMULATION OF PARTICLE DEPOSITION NEAR CEILING INDUCTION OUTLETS

H. Timmer and M. Zeller

Lehrstuhl für Wärmeübertragung und Klimatechnik,
Rheinisch-Westfälische Technische Hochschule Aachen,
Eilfschornsteinstraße 18, 52056 Aachen, Germany.

ABSTRACT

If air enters a room through ceiling fixtures, dust tends to deposit in the nearby region of ventilation outlets where it forms characteristic structures well known in mechanically-ventilated office rooms. Experimental investigation proves that the dust source is to be found within the room itself and is not located in the ventilation system. The objective of this paper is to analyse particle movement in close proximity to a linear diffuser by numerical simulation.

In the direction of the flow entering the room particles hit the ceiling predominantly behind webs in zones of high turbulent kinetic energy. With suitable assumptions it is found that with an increase of particle mass these depositions decrease while more particles hit the ceiling on the rear side of the outlet, i.e. where room air is induced by the momentum flux of the entering flow. The number of particles touching the ceiling increases with outlet velocity, mass flow, turbulence intensity and in a certain range with particle mass. The results suggest that turbulence plays a major role in the deposition of particles on ceilings.

KEYWORDS

Particle, deposition, dust, ceiling, boundary layer, Lagrange, low-Reynolds.

INTRODUCTION

According to Recknagel at al. (1999, p. 7) a cubic metre of air in a representative European city contains approximately 1 million solid particles. Although their shape is not ideally spherical, 98 % of the total number of particles show a characteristic diameter smaller than 1 µm, whereas 97 % of the particle mass load consists of particles larger than 1 µm.

Additional contamination sources found inside buildings such as cigarette smoke, skin particles or textile fibres tend to form particles larger than 1 μm. Every minute a single person emits millions of skin squamae (Scheer et al. 1998, p. 1). Therefore it can be assumed that the dust load inside a room also mainly consists of particles in that order of magnitude.

Experiments have shown that the particles depositing on ceilings stem from the room itself and do not enter the room through the induction outlet (Fichter et al. 1996, p. 36), which can be assured by sufficiently filtering the entering air. Raster-electron-microscope-analysis show that the particles form agglomerates with a characteristic length of about 100 μm (Fichter et al. 1996, p. 37).

Viewing the ceiling from below, Figure 1 shows a photograph of real depositions found at an induction outlet in an office room. The outlet is comprised of apertures that are separated by webs. Along the apertures air enters the room at an acute angle against the ceiling in alternating directions as indicated by the arrows.

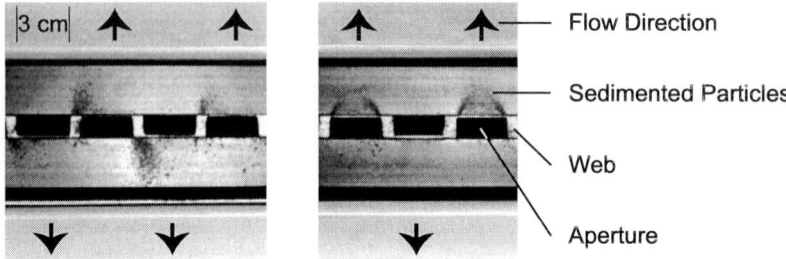

Figure 1: Characteristic depositions near ceiling induction outlets

In the literature experiments were performed with pressurised paint being blown into a full-scale office room (for instance by Finke et al. 1996). The authors found similar deposition patterns as depicted in Figure 1. Unfortunately, the distribution of particle size of vaporised paint is unknown, although an approximate medium diameter of 10 μm can be assumed (Finke et al. 1996, p. 122; Owen et al. 1992, p. 2156). All authors found a decrease of particle deposition at the ceiling as velocity and mass flow through the induction outlet increase (Rauer 1996, p. 48; Lengweiler et al. 1998, p. 321 f). They also agree that a larger angle between the ceiling and the entering airflow yields fewer particle contacts (Rauer 1996, p. 53). Furthermore, the paint deposition increases with the turbulence in the affected region (Rauer 1996, p. 73; Lengweiler et al. 1998, p. 321).

In the present paper, the experimental results of Rauer (1996) are used to allow a comparison with numerical calculations.

METHOD

Geometry

Rauer (1996, p. 26) used a commercial linear induction outlet built into a scaled-down model of a room as illustrated in Figure 2. Air enters the room through the induction outlet at the top and exits the room via the exhaust at the left bottom. Above the exhaust, a negligible flow of pressurised air vaporises paint. The plate in the middle of the room redirects the airflow and the paint upwards to the ceiling. Due to memory restrictions on the computers used for simulation, only the smaller region indicated by the dotted line in Figure 2 was calculated. The relative pressure at the boundaries to the room in the smaller region has to be given as a boundary condition.

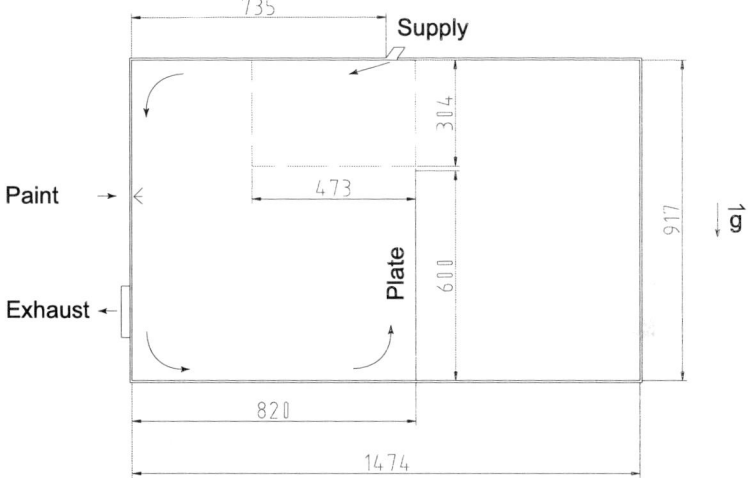

Figure 2: Geometry of the experimental room model [mm]

The induction outlet is composed of 28 elements each consisting of a web of 6 mm and an aperture length of 32 mm where air enters the room. Due to the symmetry of the problem in sufficient distance from the walls only one element is calculated. Instead of walls, cyclical boundary conditions are set at both ends of the element.

Calculation

The simulations are performed three-dimensionally with the CFD-code Fluent/4.48™. Influences such as thermo- and electrophoresis, particle shape and the properties of the ceiling are not yet taken into account. Instead, the flow is isothermal, particles are considered ideally spherical and each particle that hits the ceiling is assumed to adhere. The isothermal calculation also allows the use of the standard k-ε turbulence model, which is only problematic when buoyancy effects occur (Vogl 1997, p. 20 f).

In a first step, the whole room is calculated with a coarse grid to investigate the overall airflow and pressure distribution. The obtained relative pressure is used as a boundary condition for the required finer grid mentioned earlier that only covers a smaller region around the outlet. Thus a sufficient resolution of the boundary layer can be achieved which is necessary to reasonably calculate particle movement in close proximity to the ceiling. The two-layer model of Fluent/4.48™ is used for the finer grid and the wall function approach is used with the coarser grid.

The Lagrangian particle tracking method implemented in Fluent/4.48™ solves the force balance on a particle with a given diameter and density. If not otherwise specified, a particle density of 1,000 kg/m³ is assumed. The influence of turbulence on the particle is accounted for by the Continuous Random Walk model using a stochastic approach. Therefore, a particle with identical properties and the same starting location can move along different tracks, which would not be possible without considering turbulence. For each deposition pattern such as shown in Figure 4, 45,000 particles with a starting location 370 mm below the middle of the outlet were calculated. When a particle hits the ceiling, a user defined subroutine writes the coordinates of impact into a file, which can be read by a spreadsheet application. A two-dimensional point plot of the coordinates visualises the points of impact.

RESULTS

Calculations with the coarse grid in the complete domain show that the streamlines of the flow are circular inside the room. The air entering through the induction outlet moves to the left and downwards at the left wall, is redirected to the right at the floor and moves upward again at the plate. Trajectory calculations show that particles which do not touch any wall, floor or ceiling reach this upward flow independently of their starting location. Therefore, it is acceptable to investigate only the smaller region around the supply outlet since particles can only reach the ceiling via the upward flow. This is necessary to allow a higher grid resolution at the ceiling required by the two-layer model. It should be noted that this approach is essential for realistically calculating particle trajectories in the boundary layer; the alternative wall function does not suffice.

Calculations match the experiments of Rauer (1996) best at a particle size of 10 μm. As an example, the experimental and calculated deposition profiles for a volume flow of 120 m³/(m$_{OutletLength}$ h), which equals an outlet velocity of 4.7 m/s given the geometry of the outlet and an outlet angle of 40° against the ceiling, are presented in Figures 3 and 4. They show a perpendicular view upwards against the ceiling.

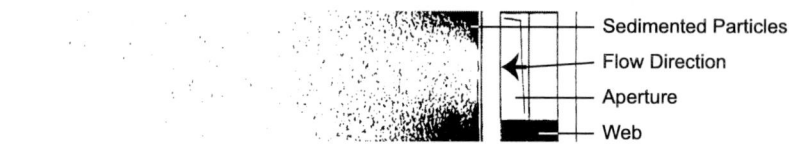

Figure 3: Experimental deposition patterns by Rauer (1996, p. A3-I-1)

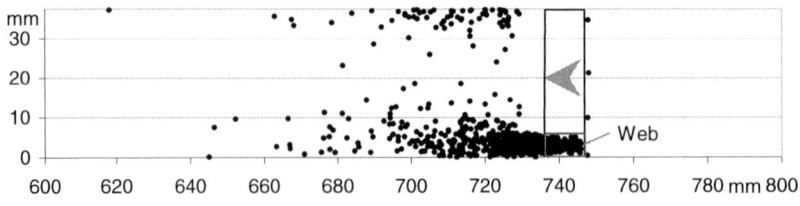

Figure 4: Calculated deposition patterns

The variation of parameters such as particle diameter, density, outlet velocity, mass flow, turbulence intensity and outlet angle influences the number of particles touching the ceiling and the structure of the deposition profiles since particles deposit in three regions: first on the web between two supply apertures, second in front of the outlet, i.e. on the ceiling in the direction of the entering flow, and third behind the outlet, i.e. where room air is induced due to the relatively high supply velocity.

The number of particles hitting the ceiling increases with the particle diameter between 0.1 and 40 μm up to 5 %. If the diameter grows to 80 μm, the number decreases. A growth of density from 800 to 1,400 kg/m^3 yields an increase of deposited particles from 1 to 2.4 %. With a higher mass, particles tend to deposit behind the outlet whereas particles with a lower mass tend to deposit in front of the outlet.

Raising the volume flow from 60 to 120 m^3/(m$_{OutletLength}$ h) or the velocity from 2.4 to 4.7 m/s leads to an increase of deposited particles from 0.8 % to 1.4 % or from 0.1 % to 2.5 % respectively. The same correlation can be found for turbulence: amplifying the turbulence intensity at the outlet from 2 % to 80 % increases the percentage of particles touching the ceiling from 0.5 % to 3.3 %. In fact, the three parameters cannot be changed individually because they are dependent on the geometry of the induction outlet.

Furthermore, it can be shown that a larger angle between the jet and the ceiling reduces the deposition of particles.

DISCUSSION

The results agree well with those of the experiments of Rauer (1996) and studies performed by the other authors mentioned above (Fichter et al. 1996, Lengweiler et al. 1998, Finke et al. 1996).

The explanation of particle deposition must take turbulence into account. Particle trajectories calculated with the mean velocities and thereby ignoring turbulence never end at the ceiling. When taking turbulence effects into account, it is striking that particles prefer regions of high turbulent kinetic energy for deposition: Figure 5 illustrates that the turbulent kinetic energy reaches its highest values in the region behind the webs, which is exactly where particles tend to deposit (Figures 3 and 4).

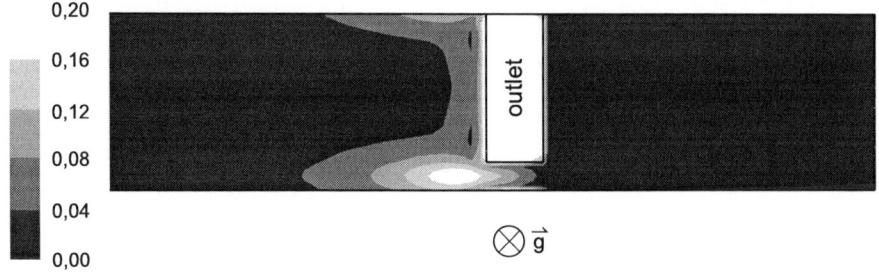

Figure 5: Calculated turbulent kinetic energy [W] directly (≈0.5 mm) below the ceiling

When analysing particle trajectories, one finds out that the trajectories of particles that hit the ceiling behind the outlet are steeper than the trajectories of particles depositing in front of the outlet. It seems that turbulence is mainly responsible for deposition in front of the outlet whereas momentum plays a significant role on the other side of the outlet.

OUTLOOK

The present paper demonstrates that numerical simulation can be used to describe the problem of particle deposition near linear induction outlets. The simulated deposition patterns are similar to those found in experiments. A sufficiently high resolution of the boundary layer is crucial for realistic trajectory calculations. Furthermore, there is evidence that turbulence plays a major role in the process of particle deposition.

Further theoretical and experimental research is being conducted on particle deposition in order to attain a deeper understanding of the phenomenon.

REFERENCES

Recknagel H., Sprenger E. and Schramek E.-R. (1999). *Taschenbuch für Heizung + Klima- technik*, München, Germany.

Scheer F. A. and Fitzner K. (1998). Dispersion and Sedimentation of Airborne Particles and Germs in Laminar and Turbulent Airflow. *RoomVent '98*, Vol. 1, 325-332.

Fichter R.-H., Knorr T. and Roth H. W. (1996). Präsentation zweier Testverfahren zur Simulation der Deckenverschmutzung durch Luftauslässe. *CCI*, Vol. 33, Nr. 12, 36-38.

Finke U. and Fitzner K. (1996). Beurteilung der Deckenverschmutzung durch Schlitzdurchlässe. *DKV Tagungsbericht*, Vol. 23, Nr. 4, 119-128.

Owen M. K. and Ensor D. S. (1992). Airborne Particle Sizes and Sources Found in Indoor Air. *Atmospheric Environment*, Vol. 26A, Nr. 12, 2149-2162.

Rauer P. (1996). *Untersuchung der Deckenverschmutzung im Bereich induktiver Decken- Zuluftdurchlässe*, Diploma Thesis, FH Köln, Germany.

Vogl N. A. (1997). *Numerische Simulation von auftriebsbehafteten Raumluftströmungen*, Aachen, Germany.

Lengweiler P., Nielsen P. V., Moser A., Heiselberg P. and Takai H. (1998). Deposition and Resuspension of Particles: Which Parameters are Important. *Roomvent '98*, Vol. 1, 317- 323.

Air Distribution in Rooms, (ROOMVENT 2000)
Editor: H.B. Awbi

DISTRIBUTION OF AEROSOLS IN TURBULENT AIRFLOWS

F. Bitter[1], K. Fitzner[1], F.A. Scheer[2]

[1] Hermann-Rietschel-Institute of Heating and Air-conditioning,
Technical University of Berlin, Germany
[2] Engineering Company Scheer, Stuttgart, Germany

ABSTRACT

To investigate the dispersion of aerosols in an airflow, a test facility was build at the Hermann-Rietschel-Institute. The distribution of aerosols across the cross section of the test duct was measured at several distances from the emission source. The turbulence intensity of the airflow in the duct was varied by installing different turbulence generating grids (1% to 20%) at the top of the test duct. Far enough downstream the source, the concentration profiles have a Gaussian distribution.
Analogous to the potential core length in a velocity profile of a round free jet, the core length of the aerosol concentration can be defined as the distance from the emission source to the cross section where the Gaussian distribution is fully developed. Up to this distance the maximum concentration in the centre of the jet remains constant, and beyond the core length it starts to decrease with increasing distance. The core length of the particle concentration is shorter than the potential core length.
The expansion of the jet depends on the Reynolds number of the jet. With low Re the core length increases due to a laminar initial flow at the nozzle. The mixing with the surrounding air in the channel is less compared to turbulent flows. With increasing distance from the source the airflow becomes turbulent.
The measurements show that the dispersion of the aerosol jet occurs faster when the turbulence intensity of the airflow increases. This results in an increase of the standard deviation of the profile.

KEYWORDS

turbulence intensity, aerosol distribution, free jet, potential core length, dispersion of germs

INTRODUCTION

The pollution load, especially the concentration of particles and germs, in clean rooms and operation theatres should be kept at a very low level. Therefore often air-conditioning systems with laminar flow ceilings are installed. The clean supply air is brought in above the working place in clean rooms or the operating table in operation theatres in form of a displacement flow. Besides the supply of clean air, the main task of the ventilation is to remove the internal pollution load caused by the persons and machines. The distribution of particles and germs throughout the entire room should be avoided and the pollution should be carried off near the source.

The influence of the air flow patterns on the sedimentation and dispersion of germs and particles was investigated by Scheer (1998). An additional result of the investigations of Scheer shows a dependence of the dispersion of aerosols on the turbulence intensity of the surrounding airflow. Therefore further investigations were made to determine this dependence. The results of this investigations are presented in this paper.

METHODS

To investigate the influence of the turbulence intensity on the dispersion of aerosols in an air flow a test facility was designed. Scheer investigated the dispersion and the deposition of aerosols and germs in a displacement flow. He determined a dependence of the dispersion of aerosols on the turbulence intensity of the air flow in the duct. The turbulence of the air flow was generated by different grids downstream of the emission source. So at the initial phase of the jet the airflow is laminar, independent on the installed grid. Therefore further investigations were made to determine the influence of the turbulence intensity with a modified test duct. In these investigations the turbulence generating grids were installed upstream of the aerosol probe.

The schematic structure of the modified test facility is shown in Figure 1.

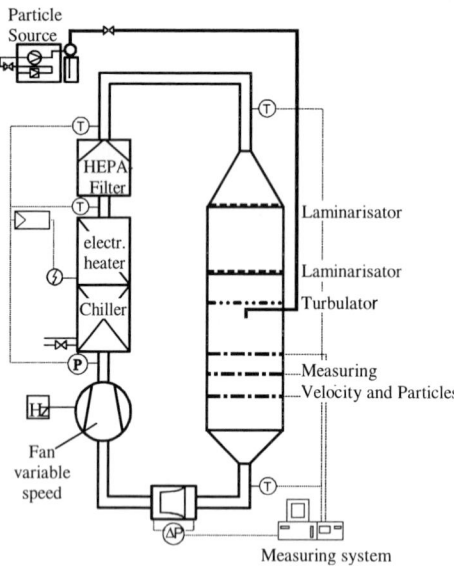

Figure 1: Modified Test Facility

The test facility consists of a closed duct system in which the test duct is integrated. The test duct is installed vertically. The air flows descending from the top to the bottom of the section. Laminarisators (finely woven fabrics) create a one directional and laminar flow. An uniform grid generates the turbulence.

The aerosols are injected into the duct by a probe in direction of the airflow. They are generated by atomization of saline water. The aerosols are transported by a carrier airflow of about 240 l/h to the probe.

A measurement bar with a moveable tube can be installed at different distances from the emission source. The tube is connected to an optical particle counter. An internal pump sucks 28,3 l/min of the air out of the test duct through the counter. At a time interval of 1 min the particles crossing the monitor are counted and divided into classes of different sizes from 0.3μm to 10μm. The tube can be positioned at every place on the bar and the bar can be moved in direction of the depth of the channel. The concentration of the aerosols can be measured at every position of the cross section of the channel.

The temperature is adjusted by an electrical heater and a water cooler. During the test isothermal conditions are reached in the test duct, so no buoyancy occurs. A HEPA-filter cleans the supply air of the test duct to ensure that no other particles are in the supply air which could lead to incorrect results. The speed of the fan is adjusted to achieve a velocity of the air flow in the duct of about 0.25 m/s.

RESULTS

Distribution of Aerosols

The initial profile of the distribution of aerosols at the outlet of the probe develops into a three dimensional Gaussian distribution with increasing distance from the source. Up to the distance where the Gaussian distribution is fully developed the maximum concentration of aerosols in the centre of the jet remains at the initial value. With increasing distance, the maximum concentration decreases and the distribution becomes wider. That is indicated by the increasing standard deviation of the Gaussian distribution. The distribution of aerosols across the cross section is illustrated in Figure 2 and the influence of the distance from the source on the distribution is shown in Figure 3.

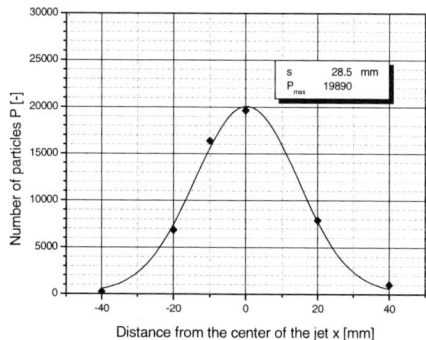 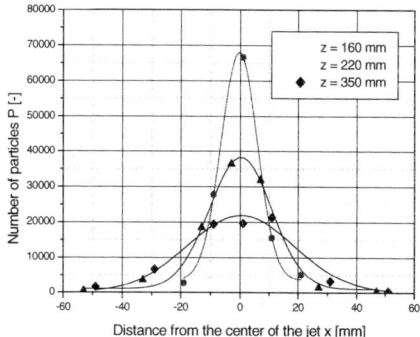

Figure 2: Distribution of the aerosols (size 0.3 µm – 1.5 µm) at a distance z = 220 mm from the emission source and a turbulence intensity of Tu = 9 %

Figure 3: Distribution of aerosols at different distances from the emission source (Tu = 5 %)

The profile of the distribution is identical for every direction on the measurement plane. Therefore it is sufficient to measure the concentration of aerosols in one direction of the plane. To get correct results the measurement bar must be placed in the centre of the aerosol jet.

Number of Aerosols in a Measurement Plane

The quantity of generated aerosols depends on the salinity of the water in the atomizer and differs between the measurements. For comparison of the results the measured aerosols must be related to the same amount of generated aerosols. The number of aerosols in a cross section must be identical for all distances from the particle source. It could be determined by the volume below the three dimensional Gaussian function.

The concentration at a distance x from the centre of the jet is given by the Gaussian function

$$c(x) = \frac{A}{s \cdot \sqrt{\pi/2}} \cdot e^{-2\frac{x^2}{s^2}}, \tag{1}$$

where A is the area below the Gaussian curve in direction x and $s=2\sigma$ is the standard deviation of the distribution. The rotation of the distribution curve in direction of the investigated plane around the normal to the cross section yield to the spatial distribution. The volume of a rotational body is given by

$$V = \pi \cdot \int_{c_1}^{c_2} [x(c)]^2 \ dc. \tag{2}$$

The Gaussian function of the concentration in Eqn. 1 must be solved on the x values. The integral in Eqn. 2 is to be solved between the limits $c_1 = 0$ (no aerosols) and $c_2 = c_{max}$, the maximum concentration of aerosols in the centre of the jet. The result is shown in Eqn. 3 and represents the number of aerosols in a cross section of the test duct at the distance z of the emission source.

$$V = \int_i c_i \ dF = \pi \cdot \int_{c_1}^{c_2} [x(c)]^2 \ dc = \tfrac{1}{2} \cdot \pi \cdot s^2 \cdot h \tag{3}$$

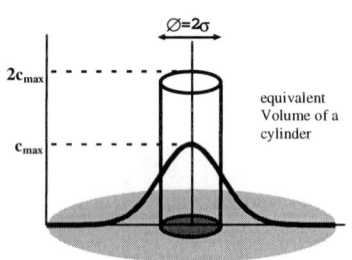

The volume below a three dimensional Gaussian function corresponds to the volume of a cylinder with the diameter of the standard deviation $s = 2\sigma$ and the height $h = 2c_{max}$, the doubled value of the maximum concentration. This is illustrated in Figure 4.

Figure 4: volume below a three dimensional Gaussian function

Comparison of the Decay of Concentration and the Decay of Velocity in a round Free Jet

The measurements show that the distribution of the aerosols can be compared to the velocity reduction in a round free jet. Figure 5 shows the velocity reduction of a free jet.

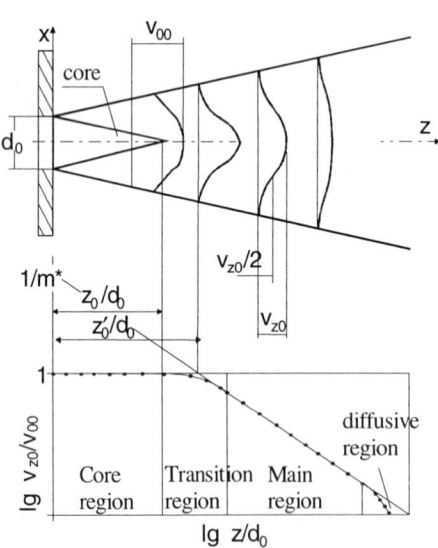

Figure 5: velocity profiles of a round free jet depending on the distance from the nozzle, Scheer (1998).

In the core region the uniform profile of the velocity changes to a Gaussian distribution. The distance from the nozzle at which the velocity in the centre of the jet starts to decrease, is called the potential core length. In the transition region the maximum starts to decrease with the distance z and then in the main region the maximum velocity starts to decrease reciprocally proportional to the distance from the nozzle. The width of the jet increases with the distance from the nozzle. The volume flow of the jet increases due to the injection of surrounding air.

The qualitative course of the function in Figure 5 (the maximum velocity depending on the distance from the nozzle) was determined on measurements with turbulent jets with high Reynolds numbers. Investigations from Regenscheidt (1976) and Schädlich (1993) have shown that the course of the curve changes at low Re numbers. Schädlich measured the decay of the velocity at different diameters of the nozzle and at different outlet velocities. Figure 6 shows the measurements from Schädlich by a jet emerging out of a round nozzle with a diameter of 4 mm (analogous to the diameter of the aerosol probe) and two different velocities. The measurement at high Re shows the expected curve. At low Reynolds numbers the potential core length increases. This could be explained by the laminar flow at the nozzle. The mixing with the surrounding air is reduced in this initial laminar section. At a certain distance from the nozzle the flow turns turbulent and the velocity in the centre of the jet starts to decrease. The reduction of the velocity occurs faster at high Re numbers because of the lower momentum of the jet. In a distance far from the nozzle the velocity curve of the jet at low Re approximates itself to the curve at high Re.

The relative maximum concentration in the centre of the aerosol jet depending on the distance from the emission source is shown in figure 7 analogous to the velocity decrease in a logarithm scale. The diagram shows the measurements at different turbulence intensities (Tu=1%, 5%, 9% and 20%).

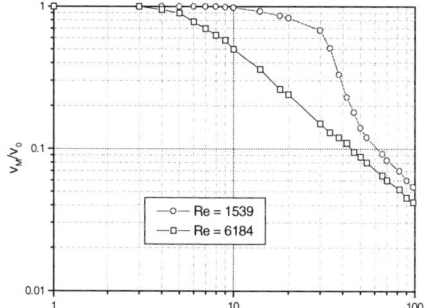

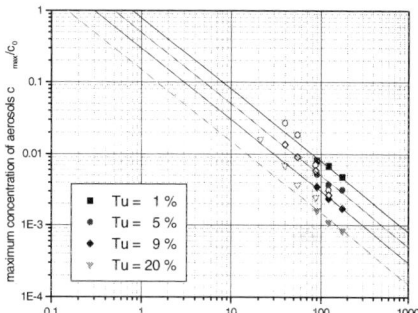

Figure 6: velocity profile of a round free jet at different turbulence intensities of the jet

Figure 7: measured concentrations at different distances and turbulence intensities of the air flow in the test section

The hollow measurement points are the new measurements with the modified test duct, the full measurement points are measurements made by Scheer (1998). The distance from the source is related to the diameter of the probe (4 mm). Analogous to the velocities the concentration decreases in the main section reciprocally proportional to the distance

$$\frac{c_{\max}(z)}{c_0} \sim \left(\frac{z}{d_0} \right)^{-1}, \tag{4}$$

where $c_{\max}(z)$ is the maximum concentration in the centre of the aerosol, z is the distance from the emission source, c_0 is the concentration at the exit of the probe and d_0 is the diameter of the probe. Similar to the potential core length of a round free jet a core length of the concentration could be defined. The core length is the maximum distance from the aerosol source at which the maximum concentration is still as high as the initial concentration. The fictitious core length z_0' found by the approximation of the measurement data with the function in Eqn. 4 is greater than the real core length z_0 (refer to the transition region in Figure 5). The regression lines are just sufficient to the measurement points at far distances from the aerosol source. This could be a result of the low turbulence (Re = 1360) of the aerosol jet (compare Figure 6). Near the outlet of the probe the jet is laminar and the mixing with the surrounding air is reduced. After the initial laminar section the flow turns turbulent. Due to the lower momentum of the jet the dispersion occurs faster than in jets at higher turbulence intensities. At far distances the dispersion of the decay of the concentration approximates itself to the expected decay at high Re numbers. Due to the constant volume of the spatial distribution in a cross section, the distribution gets wider with the decay of the maximum concentration in the centre of the jet and with increasing distance. A measure for the width is the standard deviation of the Gaussian distribution. This is shown in Figure 8.

The standard deviation is proportional to the square root of the distance

$$s \sim \sqrt{z} \qquad (5)$$

The fictitious core length z_0' depends on the turbulence intensity of the air flow in the test section. It could be determined in Figure 7 as the distance from the source where the regression line of the maximum concentration is identical to the initial concentration ($c_{max}/c_0 = 1$). In Figure 8 the fictitious core length is the length where the standard deviation is $\sqrt{2} \cdot d_0$.

Figure 9 shows the reduction of the fictitious core length with increasing turbulence intensity. The function was determined in analogy to the potential core length of round free jets. Due to the lack of measurements at different turbulence intensities this function should be seen as a first approximation.

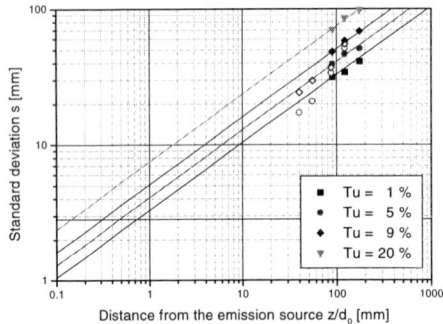

Figure 8: Standard deviation of the measurements at different distances

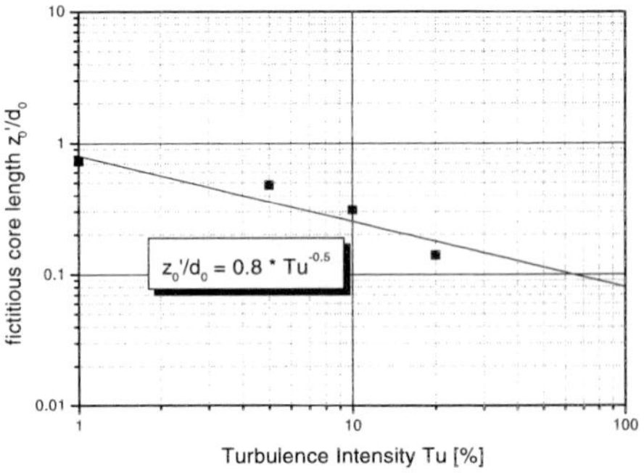

Figure 9: fictitious core length and turbulence intensity

SUMMARY

The measurements were made at low mean flow velocities in the channel of about 0.25 m/s and aerosols were injected at low turbulence intensities. This corresponds to the conditions that are often present in clean rooms and operating theatres. It can be seen, that a low turbulence intensity of the supply air flow keeps the dispersion of particles from sources inside the room within narrow limits.

Another aspect at which the dispersion of aerosols is to be considered is the atomization in humidifiers. Atomized water droplets are injected at high pressure into the airflow. The distance to the following components of an air-conditioning system and the length of the humidifier are influenced by the dispersion of the droplets. To investigate the behaviour in this case measurements at higher flow velocities of about 1 to 2 m/s and with an aerosol jet with high turbulence should be done.

REFERENCES

Scheer F.A. (1998), *Einfluß der Turbulenz einer Verdrängungsströmung in Operationsräumen auf Transport und Sedimentation von Mikroorganismen*, Dissertation, TU Berlin, Germany

Regenscheidt B. (1976). Einfluß der Reynoldszahl auf die Geschwindigkeitsabnahme turbulenter Freistrahlen, *HLH* **27:4**, 122-126.

Schädlich S. (1993), *Der Einfluß verschiedener Luftdurchlaßgeometrien auf das Freistrahlverhalten*, Dissertation, Universität Essen GHS, Germany

© 2000 Elsevier Science Ltd. All rights reserved.
Air Distribution in Rooms, (ROOMVENT 2000)
Editor: H.B. Awbi

INDOOR PARTICLE POLLUTION : EFFECT OF WALL TEXTURES ON PARTICLE DEPOSITION

M. ABADIE, K. LIMAM and F. ALLARD

LEPTAB, University of La Rochelle, Av. M. Crépeau,
17042 La Rochelle Cedex 01, France

ABSTRACT

Owing to people spend between 80 and 90% of their time indoor, prediction of indoor particle pollution levels becomes a subject of great interest for the evaluation of health risks and comfort in building. The usual approach is to consider the studied room as a perfect mixed zone and to evaluate deposition velocities (or constants of deposition) of particles. The aim of this study is the experimental determination of deposition constants for several wall textures in order to predict indoor particles concentrations. Experiments consist in injecting spherical particles (0.7, 1.0 and 5.0 μm in diameter) in a cubic box. Internal faces of the scale model are covered by the texture to be tested (wood, wallpaper, carpet and roughcast). The air is perfectly mixed by a fan and particles concentration is monitored with an optical particle counter. Global constants of deposition are determined by regression fitting of the exponential decay curves. Deposition constants for each orientation are then deduced from the Crump & Seinfeld theory (1981). Results for particles with a 5.0 micrometers diameter permit to establish a classification of the tested wall textures. For example, concentration decays for carpet are twice those for smooth textures (wood and wallpapers). In order to test values of the deduced deposition constants for particular orientation, experiments have been made covering the internal faces with several textures. Results show a good agreement between calculated and experimental values (with a relative error close to 11%). This decomposition method has then been extended to smaller particles.

KEYWORDS

Particle, pollutant, deposition, wall texture, indoor pollution, scale model.

INTRODUCTION

Prediction of indoor particle pollution levels becomes an important subject in the recent years. Models are needed for the evaluation of health risks as well as human comfort. The simplest one considers the studied room as a perfectly mixed zone in the core region (Figure 1). The mass balance of such a monozonal enclosure can be expressed by Eqn. 1 (Nazaroff & al. (1993)). Taking into account that deposition and resuspension can't easily be evaluated independently, however, in practice, it's used to consider the coefficient λ_d which takes into account their global effect ($\lambda_{de}C_i + \lambda_R D_i \approx \lambda_d C_i$), the solution is given by Eqn. 2. As a consequence, the use of this model is primarily based on the knowledge of the deposition constant, λ_d. In order to determine this decay constant in situ, measurements were carried out by several authors : Offermann & al. (1985), Roed & al. (1991), Byrne (1995) ... The major disadvantage of such measurements in real buildings is that they are not easily exploitable to geometrical and air flows configurations different from those where measurements have been done. So laboratory experiments on scale model where boundary conditions can be carefully checked were required: Okuyama & al. (1986), Holub (1988) ... All these studies are limited because of their weak panels of surface textures. This is why we proposed to determine the constant of deposition for various configurations of surface textures usually encountered in buildings.

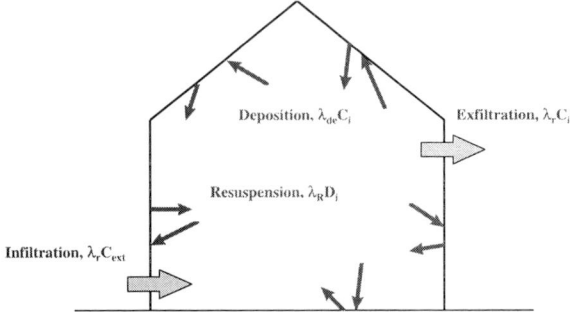

Figure 1 : Processes involved in indoor - outdoor ratios

$$\frac{dC_i}{dt} = \lambda_r C_o - \lambda_r C_i - \lambda_{de}C_i - \lambda_R D_i \tag{1}$$

$$C_i(t) = C_o \frac{\lambda_r}{\lambda_r + \lambda_d}(1 - \exp(-(\lambda_r + \lambda_d) \times t)) + C_i(0) \times \exp(-(\lambda_r + \lambda_d) \times t) \tag{2}$$

EXPERIMENTS

Experimental Device

The test chamber is a confined cubic enclosure (0.6 x 0.6 x 0.6 m) whose horizontal and vertical walls can be covered by various textures : wallpaper, carpet, wood ... (Figure 2). A three-bladed propeller suspended from the centre of the scale model ceiling and blowing towards the ceiling is employed as a turbulent source. The particle injection tube is placed at 15 cm of the agitator axis and a sample tube connected to an Optical Particle Counter (Model 227 A Met One) is positioned in the centre of the volume. Three sizes of particle are studied : 0.7 µm, 1.0 µm and 5.0 µm in diameter. These particles are usually met inside the buildings and represent a potential hazard for health. Two modes of injection are necessary. The 5.0 µm particles is come as dry powder so it's possible to introduce them into the studied volume by means of a syringe. This method is simple to implement and remains acceptable for our study because the parameter to be measured is independent of the initial concentration of particle in air. The 0.7 and 1.0 µm particles are obtained diluting concentrated solution of polystyrene latex microspheres with distilled water. Particles are separated from the solution by means of a particle generator (Hiac/Royco model 256). The injection is carried out continuously until obtaining the desired initial concentration. Injecting particles against the ascending airflow permits to obtain a rapid homogeneous concentration and to limit initial deposition on surfaces. However, deposition during injection cannot be avoided but it's supposed to have no great consequence on results.

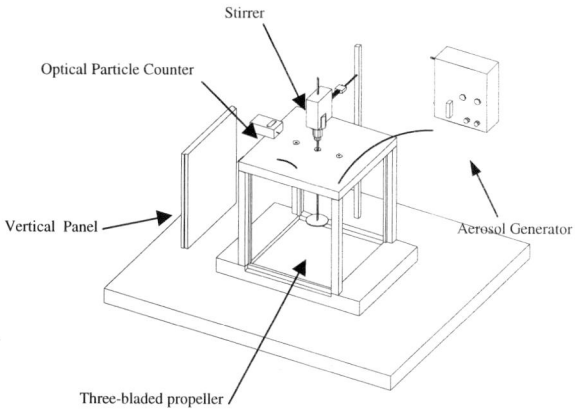

Figure 2 : Experimental facility

Measurement of the global deposition constant

Considering that there is no air renewal (case of our enclosure), equation 2 becomes:

$$C_i(t) = C_i(0) \times \exp(-\lambda_d \times t) \tag{3}$$

The value of the deposition constant is obtained by linear regression of measured particle concentration in air (only values obtained with a correlation coefficient higher than 95% were retained). No direct measurements of deposed particles on surfaces are carried out.

Determination of the deposition constant for a particular orientation

Crump & Seinfeld (1981) considered the problem of prediction of the parietal deposition of an aerosol in a closed vessel, of unspecified geometry, under turbulent agitation. Applied to a cubic enclosure, the evolution of the global and elementary deposition constants is calculated (Figure 3). For each experimentation, the values of the global constants of deposition were measured, the contribution of each surface is then deduced from the Crump & Seinfeld theory (1981).

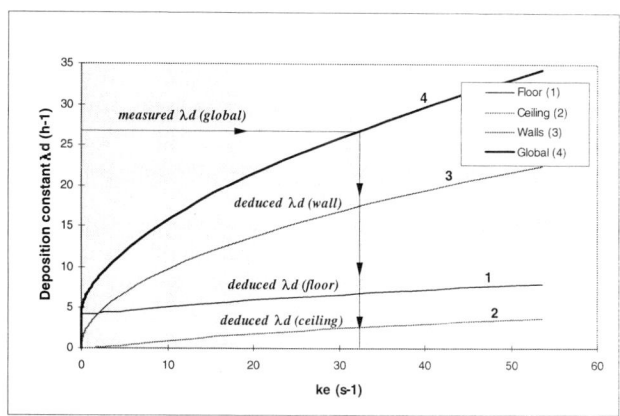

Figure 3 : Determination of deposition constants for walls, floor and ceiling

RESULTS AND DISCUSSION

Study n°1

This first study consisted in the development and the validation of an experimental protocol in order to determine the deposition constants for various walls orientations. Measurements presented here are related to the 5.0 µm particles. The deposition of these particles was studied for three textures. These textures were selected for their occurrences in the building as well as their apparent differences in behaviour with respect to particle deposition : wood (rough surface of agglomerate), roughcast and carpet (with cut velvet). The first three experiments (table 1) are intended for the study of a single wall texture. These measurements allow the establishment of a wall textures classification with respect to their potential of collecting particles, by ascending order : wood, roughcast and carpet. By application of the decomposition induced by the Crump & Seinfeld theory (1981), the deposition constants for each texture are deduced for the three orientations (table 2). The gravitational sedimentation for this particle size is significant compared to Brownian and turbulent diffusion, the deposition is greater on the floor than on the other walls. It's striking to notice that, in the case of the wood texture, the deposition on the floor represents the major part of the global deposition (70%) whereas it varies around 60% for the two other textures. From these results we can calculate the global deposition constants for other configurations using the three tested textures. The relevance of the method is shown by table 4. The average of relative error is close to 11%. The present method of decomposition of the global deposition constant seems reliable.

TABLE 1

DEPOSITION CONSTANTS MEASURED FOR SEVERAL CONFIGURATIONS OF TEXTURES

	Textures			λ_d (h^{-1})
n°	Floor	Walls	Ceiling	5.0 µm
1	Wood	Wood	Wood	11.40 ± 0.74
2	Carpet	Carpet	Carpet	20.88 ± 2.74
3	Roughcast	Roughcast	Roughcast	18.00 ± 0.59
4	Carpet	Roughcast	Wood	20.16 ± 1.53
5	Wood	3 Carpet + 1 Wood	Wood	18.54 ± 0.69
6	Wood	Wood	Carpet	13.68 ± 0.64

TABLE 2

DEPOSITION CONSTANTS FOR THREE TEXTURES AND THREE ORIENTATIONS

		λ_d (h^{-1})		
n°	Textures	Floor	Walls	Ceiling
1	Wood	4.57 ± 0.08	1.63 ± 0.14	0.32 ± 0.08
2	Carpet	5.90 ± 0.42	3.33 ± 0.48	1.66 ± 0.42
3	Roughcast	5.47 ± 0.09	2.84 ± 0.10	1.22 ± 0.09

TABLE 3

COMPARISON BETWEEN CALCULATED AND MEASURED RESULTS

n°	Calculated λ_d (h^{-1})	Measured λ_d (h^{-1})	Relative Error
4	17.57 ± 0.92	20.16 ± 1.53	15 %
5	16.52 ± 1.74	18.36 ± 0.69	11 %
6	12.74 ± 1.07	13.68 ± 0.64	7 %

Note: all these values are obtained for an equivalent surface of texture.

Study n°2

This second series of experiments is the extension of the principle developed in the first one to two other spherical particles sizes (0.7 and 1.0 µm). The studied textures were : smooth wallpaper (equivalent to wood), rough wallpaper (roughness of 2mm) and carpet. Table 4 presents the measured global constants of deposition values . It's worth seen that the deposition is an increasing function of the particle size. Following the same reasoning as previously, the constants of deposition for each orientation are deduced (table 5). Figure 4 presents the contribution of each orientations for the three particle sizes. The deposition of the 0.7 and 1.0 µm particles looks similar. For these particle sizes, the Brownian and turbulent diffusion are preponderant compared to gravitational sedimentation, thus there's no privileged direction of deposition but it's important to take the wall texture into consideration. On the one hand, for the carpet, the contribution of each orientation is the same (near to 30%) but on the other hand, for the two other wall textures, this contribution is identical to that obtained in the first study. This result can be explained by the difference of form and nature of the tested textures. The carpet is a wall texture which has a strong heterogeneity of surface made up of synthetic fibres, it's a texture of flexible nature and potentially charged of electrostatic charge. A particle which enters in collision with such a surface has a strong probability of remaining stuck. On the contrary, smooth wallpaper is a hard surface with no relief, thus particles are more susceptible to rebound. For the 5 µm particles, gravitational sedimentation is dominating, results are similar to those of the study n°1.

TABLE 4

MEASURED DEPOSITION CONSTANTS VALUES

Textures	λ_d (h^{-1})		
	0.7 µm	1.0 µm	5.0 µm
Smooth Wallpaper	0.42 ©	0.58 ± 0.15	12.30 ± 1.20
Rough Wallpaper	0.75 ± 0.12	0.86 ± 0.08	12.61 ± 0.70
Carpet	3.41 ©	3.57 ©	20.88 ± 2.74

© one measurement available

TABLE 5

DEPOSITION CONSTANTS FOR EACH ORIENTATION (FLOOR, WALL AND CEILING)

Textures	λ_d (h^{-1})								
	0.7 µm			1.0 µm			5.0 µm		
	F	W	C	F	W	C	F	W	C
Smooth Wallpaper	0.13 ± 0.00	0.06 ± 0.00	0.03 ± 0.00	0.21 ± 0.02	0.08 ± 0.03	0.02 ± 0.02	4.68 ± 0.15	1.80 ± 0.23	0.45 ± 0.15
Rough Wallpaper	0.18 ± 0.02	0.13 ± 0.02	0.08 ± 0.02	0.25 ± 0.01	0.13 ± 0.01	0.06 ± 0.01	4.71 ± 0.09	1.85 ± 0.13	0.48 ± 0.09
Carpet	0.62 ± 0.00	0.57 ± 0.00	0.52 ± 0.00	0.69 ± 0.00	0.60 ± 0.00	0.50 ± 0.00	5.90 ± 0.42	3.33 ± 0.48	1.66 ± 0.42

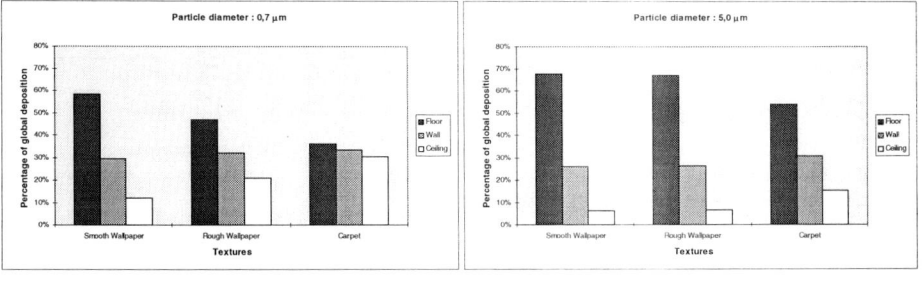

Figure 4 : Influence of orientation (0.7 and 5.0 µm particles)

CONCLUSION

The transport model of particles introduced here can be used to predict the migration and the particle deposition in a ventilated enclosure. The originally use of the Crump & Seinfeld theory (1981) presented here makes possible the constants of deposition determination for each orientation. More than this, the present study puts into relief the importance of taking particle sizes, wall textures and orientations into consideration in the prediction of indoor particle pollution. From these results, interactions between particle and wall texture will be introduced in the laboratory's numerical code (Sandu (1999)), which is a turbulent transport model of particle pollutants in cavity.

This study is part of the PhD dissertation " Contribution to the study of particle pollution: role of the walls, role of the ventilation " cofinanced by the Agency of the Environment and the Energy Management (ADEME) and the Poitou – Charentes Region.

NOMENCLATURE

C_o : Outdoor particle concentration (part./m^3),
C_i : Indoor particle concentration (part./m^3),
D_i : Deposited particles concentration (part./m^3),
k_e : Turbulence parameter (s^{-1}),
L : Characteristic length of the enclosure (cm^3),

t : Time (sec.),
λ_d : Approached constant of deposition (h^{-1}),
λ_{de} : Real constant of deposition (h^{-1}),
λ_r : Air change rate (h^{-1}),
λ_R : Resuspension parameter (h^{-1}).

REFERENCES

Sandu, A. (1999) Contribution à l'étude numérique du transport turbulent de polluants particulaires en cavités, *Thèse de doctorat des Universités de La Rochelle (France) et de Bucarest (Roumanie)*.

Byrne, M.A. (1995) An Experimental Study of the Deposition of Aerosol on Indoor Surfaces. *Thesis for the degree of Doctor of Philosophy of the University College of London.*

Crump, J.G. and Seinfeld, J.H. (1981) Turbulent Deposition and Gravitational Sedimentation of an Aerosol in a Vessel of Arbitrary Shape. *J. Aerosol Science* **12:5**, pp. 405-415.

Holub, R.F., Raes F., Van Dingenen, R. and Vanmarcke H. (1988) Deposition of aerosols and unattached radon daughters in different chambers : theory and experiment. *Radiation Protection Dosimetry* **24**, pp. 217-220.

Nazaroff W.W., Gadgil A.J. and Weschler C.J. (1993) Critique of the use of Deposition Velocity in Modelling Indoor Air Quality. *Modelling of Indoor Air Quality and Exposure, ASTM STP 1205, Niren L. Nadga, Ed., American Society for Testing and Material*, Philadelphia, pp. 81-104.

Offermann, F.J., Sextro, R.G., Fisk, W.J.,Grimsrud, D.T., Nazaroff, W.W., Nero, A.V., Revzan, K.L. and Yater J. (1985) Control of respirable particles in indoor air with portable air cleaners. *Atmospheric Environment* **19:11**, pp. 1761-1771.

Okuyanna, K., Kousaka, Y., Yamamoto, S. and Hosokaya, T. (1986) Particle loss of aerosols in a stirred tank. *Journal of Chemical Engineering of Japan* **10:2**, pp. 142-147.

Roed, J., Goddard A.J.H., Mac Curtain, J.A., Byrne, M.A. and Lange C. (1991) Reduction of dose from radioactive matter ingressed into buildings. *Proceedings of the CEC International Seminar on Intervention Levels and Countermeasures for Nuclear Accidents*, Caderache, France.

Air Distribution in Rooms, (ROOMVENT 2000)
Editor: H.B. Awbi

CFD BASED AIRFLOW MODELLING TO INVESTIGATE THE EFFECTIVENESS OF CONTROL METHODS INTENDED TO PREVENT THE TRANSMISSION OF AIRBORNE ORGANISMS

M. J. Seymour[1], A.Alani[1&2], A.Manning[3] & J.Jiang[3]

[1]Flomerics Limited, 81, Bridge Road, Hampton Court, Surrey, KT8 9HH, UK
[2]Department of Mechanical Engineering, Brunel University,
Uxbridge, Middlesex, UB8 3PH
[3]Flomerics Incorporated,

ABSTRACT

The airborne transmission of disease is a constant threat and while diseases such as Tuberculosis were considered all but extinct in the western world, the resurgence of it demonstrates that the spread of these diseases has to be taken very seriously.

This paper describes the method of application of Computational Fluid Dynamics (CFD), more appropriately called Airflow Modelling for the Building Services Industry, to the airflow and heat transfer in a Hospital Isolation Room Application. In particular it addresses how it can determine the ability of the ventilation system to limit the time during which carers, or other people present in the room, may be at risk to the airborne organisms constantly being produced by a patient coughing, sneezing or simply talking.

Research has shown that ventilation rate is no guarantee of control of these airborne organisms. Another means of minimising the risk from airborne bacteria is to apply ultraviolet germicidal irradiation (UVGI). UVGI holds promise of greatly lowering the concentration of airborne bacteria and thus controlling the spread of airborne infection among occupants.

This paper describes the techniques developed to allow the airflow simulation to be extended to simulate the motion of the droplets carrying the bacteria, their path through the room and indeed any exposure they may have to UVGI. The work will be highlighted by a series of case studies demonstrating the effect of change in the ventilation design and the effect of UVGI on the probability of survival of the bacteria.

KEYWORDS

Airflow, Modelling, CFD, Particles, Droplets, UVGI, Bacteria

INTRODUCTION

Airflow modelling in the form of Computational Fluid Dynamics has now been established for some time as another method of analysing the ventilation performance in the built environment. Based on the iterative solution of the fundamental conservation equations in the form of the Navier-Stokes equation, Airflow Modelling provides the ability to predict such parameters as the pressure, velocity and temperature at often hundreds of thousands of points in space. As experimental facilities for full and part scale modelling become fewer and further between, and computers and software become faster and more powerful, more and more designs are taking advantage of the progress and using airflow modelling to evaluate the likely success of the proposed design. This is not only of benefit in the design of conventional occupied environments where there can be capital and running cost savings by identifying improved designs (not to mention the increase in productivity), but has been shown to be of critical importance in design for contamination control. In many contamination control applications, the particles are of the sub-micron level and at this size, there is little slip relative to the air in which they are situated. As a result, the particles can be considered as completely airborne and treated as if they are gaseous.

The resurgence of Tuberculosis (TB) an Multi-Drug-Resistant forms (MDR-TB) represents an application where the assumptions of this gaseous modelling approach are insufficient for a true analysis of the ventilation performance. Patients in hospital isolation rooms constantly produce transmissible airborne organisms by coughing, sneezing or talking, which, if not under control, results in spreading airborne infection. TB infection, for example, occurs after inhalation of a sufficient number of tubercle bacilli expelled during a cough by a patient (see Federal Register, 1993). The contagion depends on the rate at which bacilli are discharged, i.e., the number of the bacilli released from the infectious source. It also depends on the virulence of the bacilli as well as external factors, such as the ventilation flow rate. In order to prevent the transmission of airborne infection, isolation rooms are usually equipped with high efficiency ventilation systems operating at high supply flow rate to remove the airborne bacteria from the rooms. However, unexpected stagnant regions, or areas of poor mixing, mean that the ventilation rate is no guarantee of good control to the spreading of airborne infection. The risk from such an airborne spread is particularly important in places where people are ill since their immunity to infection is somewhat reduced. Hospitals are however not the only place where there is such a risk, others might include shelters for the homeless.

These bacteria when expelled normally reside in small colonies on droplets that are larger than the sub-micron scale (although as water evaporates they become progressively smaller), but, more importantly these bacteria can be killed by exposure to ultra violet germicidal irradiation, or UVGI for short. To understand the effect of such exposure it is essential to know the duration and intensity of exposure for each droplet. Knowing this the probability of the bacteria remaining alive can be assessed and thus an assessment of the risk can be made.

While traditionally protection has been provided by using by ventilation systems to continually change the air, and a knowledge of the dilution of the number of bacteria is sufficient to assess the risk, it is essential to predict the path and time for droplets spent in an ultra-violet beam. Now it should be recognised that a simple streamline for a droplet is insufficient to assess the ventilation performance of the room since each droplet will take a different path due to the way in which it interacts with the turbulence in the room. Further, bacteria may be projected by the cough or the sneeze to almost any point in the room, so it is necessary to take a statistical approach to the ventilation effectiveness super-imposed on the sound methodology of Airflow Modelling.

METHODOLOGY

The methodology is based on the assumption that the number of droplets is small in relation to the total airflow and so they do not affect the air movement. On this basis, the trajectory of the droplets can be calculated after solution of the airflow and assuming the turbulence is homogeneous, a Monte-Carlo approach used to statistically calculate the expected time for particles in the UV beam.

Droplet Trajectories

The methodology for predicting turbulent particle dispersion used in this study was originally laid out by Gosman, 1981and validated by Ormancey, 1984, Shuen, 1983, Chen, 1984. Experimental validation data was obtained from Snyder, 1971. Turbulence was incorporated into the Stochastic model via the k-ε turbulence model as given in Alani, 1998.

The particle trajectories are obtained by integrating the equation of motion in 3 co-ordinates. Assuming that body forces are negligible with the exception to drag and gravity, these equations can be expressed for the x direction as:

$$m_p \frac{du_p}{dt} = \frac{1}{2} C_D A_P \rho (u - u_p) \sqrt{(u - u_p)^2 + (v - v_p)^2 + (w - w_p)^2} + m_p g_x \qquad (1)$$

$$\frac{dx_p}{dt} = u_p \qquad (2)$$

and similarly for the y and z directions

Where

u, v, w:	instantaneous velocities of air in x, y and z directions
u_p, v_p, w_p:	particle velocity in x, y and z direction
x_p, y_p, z_p:	particle moving in x, y and z direction
g_x, g_y, g_z:	gravity in x, y and z direction
A_p	cross-section area of the particle
m_p	mass of the particle
ρ	density of the particle
C_D	drag coefficient
dt	time interval

The drag coefficient for a spherical particle, taken from Wallis [16], is:

$$C_D = \frac{24}{Re}\left(1 + \frac{13}{16}Re\right)^{0.5} \qquad \text{for Re} \leq 1000 \qquad (3)$$

and $C_D = 0.44$ for Re > 1000 (4)

The Reynolds number of the particle is based on the relative velocity between particle and air.

In laminar flow, particles released from a point source with the same weight would initially follow the air stream in the same path and then fall under the effect of gravity. Unlike in laminar flow, the random nature of turbulence indicates that the particles released from the same point source will be randomly effected by turbulent eddies. As a result, they will be diffused away from the stream line at different fluctuating levels. In order to model the turbulent diffusion, the instantaneous fluid velocities in the 3 Cartesian directions, u, v and w are decomposed into the mean velocity component and the turbulent fluctuating component as:

$$u = \bar{u} + u', \ v = \bar{v} + v', \ w = \bar{w} + w'.$$

Where $\bar{u}$ and u' are the mean and fluctuating components in x-direction. The same applied for y-, z-directions. The stochastic approach prescribes the use of a random number generator algorithm which, in this case, is taken from Press, 1992 to model the fluctuating velocity. It is achieved through using a random sampling of a Gaussian distribution with a mean of 0 and a standard deviation of unity. Assuming homogeneous turbulence, the instantaneous velocities of air are then calculated from kinetic energy of turbulence:

$$u = \bar{u} + N\alpha, \ v = \bar{v} + N\alpha, \ w = \bar{w} + N\alpha \qquad (5)$$

where N is the pseudo -random number, ranging from 0 to1, with

$$\alpha = \left(\frac{2k}{3}\right)^{0.5} \qquad (6)$$

k is the turbulent kinetic energy.

The mean velocities which is the direct output of CFD determines the convection of the particles along the stream line, while the turbulent fluctuating velocity, $N\alpha$, contributes to the turbulent diffusion of the particle. In fact the time that the particle interacts with the any given turbulent eddy depends on speed of the particle and the life of the Eddy, so the turbulent contribution has to be recalculated accordingly.

Model for Impingement of Particles on Solid Surfaces

The program can either consider particles to bounce or stick when the hit a solid surface. For this application, since the particle is a droplet, when it is calculated to have hit a solid surface, the droplet is defined to stick to the surface, and is effectively eliminated from the calculation. The droplet is no longer considered to receive any further UV dose. There is no research to suggest that the particles would re-aerosolize, or detach from the surfaces. Further, the risk of infection from surface contact has not been considered, again due to lack of available literature.

Model for Killing Bacteria

As the droplets travel around the room they may pass through the UV beam. The beam itself has to be limited to high level for health and safety reasons, so the dose will depend on the actual path each droplet takes. The percentage survival (figure 1) is dependent on exposure to UV dose, defined as:

Dose = Exposed time * UV Irradiance $\qquad (7)$

This is then used to generate the probability of survival.

The percentage of survival can be written as a function of the dose as follows:

$$\% \, Survival \ = 100 \times e^{-kIt} \qquad (8)$$

Where I = UV irradiance, $\mu W/cm^2$
  t = time of UV exposure
  k = the microbe susceptibility factor, $cm^2/\mu W.s$ (k = 0.00384)

In practice the dose varies as the droplet moves trough space and so this becomes an integral of I with respect to time that can be calculated numerically.

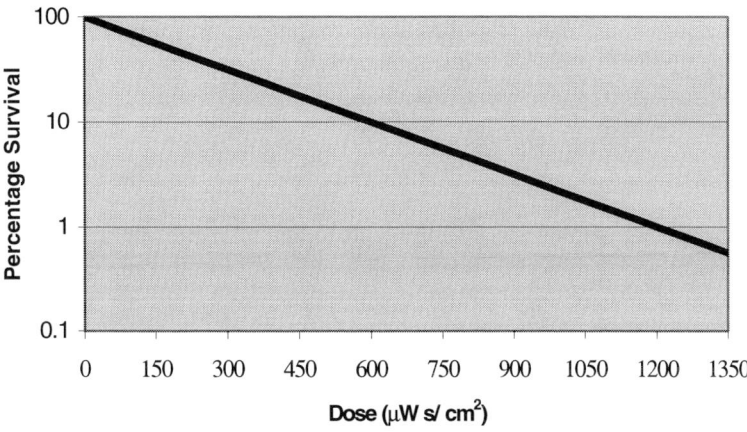

Figure 1. Survival Fraction vs. Dosage for M. tuberculosis (ASHRAE Transactions 1999, V.105 Pt. 1)

APPLICATION AND CONCLUSIONS

The application of this technique, to the design of isolation rooms itself, presents significant scope for a paper and cannot be fully documented here. The National Institutes of Health has undertaken a substantial work programme on this subject and this work has been submitted for presentation as a paper at the ASHRAE 2000 meeting (Memarzadeh, 2000).

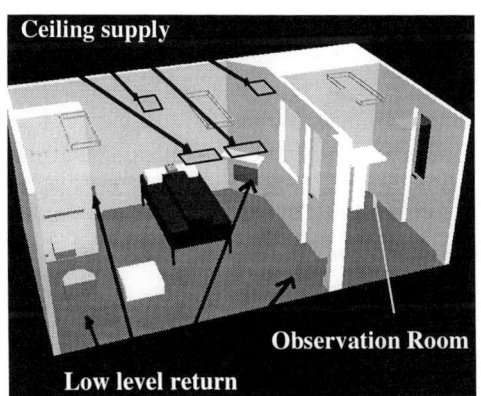

Figure 2. Isolation Room Geometry

The room, figure 2,show represents a typical isolation room with wash / toilet areas and an observation room. It is classically ventilated by laminar flow air terminals in the ceiling providing downward flow to return air grilles located at low level on the side walls. The data here show that UVGI can be used to supplement the protection provided by the ventilation system. UVGI is provided by a fitting at high level providing a shallow beam covering the whole area of the isolation room itself, figure 3.

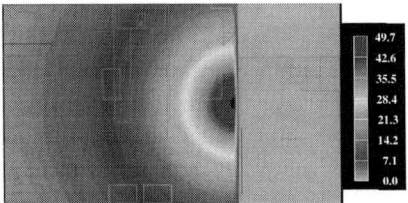

Figure 3. UVGI Intensities ($\mu W/cm^2$)

Results are presented for four different ventilation strategies. The design as shown in Figure 1, and with an additional return placed on the wall above the head of the bed. Each configuration was run at 2 ventilation rates. For simplicity of presentation the effect of UVGI is interpreted as bacteria killed after receiving a total dose of $500\mu W/cm^2$. For each simulation the number of droplets that have received a dose of $>500\mu W/cm^2$, ventilated from the room, and still alive after 300 seconds. The methodology clearly predicts that UVGI has a significant benefit with between 5 and 15% receiving a dose greater than $500\mu W/cm^2$. As a proportion of those remaining unventilated this is an important contribution of between 35% and 60%.

TABLE 1

DROPLET STATUS AFTER 300 SECONDS

	State after 300 secs	Number of Particles	Mean Life Time/s	Occupied Zone Time/s	Mean Dose $/(\mu Wsec/cm^2)$
Traditional Low Level Exhaust 14.2 ACH	Killed	383	92	34	500
	Ventilated	1749	123	61	24
	Still alive	568	300	135	36
Traditional Low Level Exhaust 18.4 ACH	Killed	260	129	60	500
	Ventilated	1982	119	56	78
	Still alive	458	300	136	102
Local Exhaust 11.4 ACH	Killed	157	120	68	500
	Ventilated	2399	71	41	77
	Still alive	144	300	183	64
Local Exhaust 17.0 ACH	Killed	142	97	51	500
	Ventilated	2458	60	35	65
	Still alive	100	300	167	60

REFERENCES

Alani A, Dixon-Hardy D, Seymour M.J, (1998) "Contaminants Transport Modelling", EngD in Environmental Technology Conference.
Chen P-P, Crowe C.T, (1984) "On the Monte-Carlo Method for Modelling Particle Dispersion in Turbulence Gas-Solid Flows", ASME-FED 10, 37 – 42.
Federal Register, (1993) "Draft Guidelines for Preventing the transmission of Tuberculosis In Health –Care Facilities, Second Edition; Notice of Comments Period", Vol.58, No.195.
Gosman D, Ioannides E, (1981) "Aspects of Computer Simulation of Liquid-Fuelled Combustors", AIAA 19th Aerospace Science Meeting 81-0323, 1 – 10.
Memarzadeh F, Jiang J, (2000) "New Research Identifies A Methodology For Minimizing Risk From Airborne Organisms In Hospital Isolation Rooms", T.C. 9.8, ASHRAE 2000
Ormancey A, Martinon J, (1984) "Prediction of Particle Dispersion in Turbulent Flow", PhysicoChemical Hydrodynamics 5, 229 – 224.
Press W.H, Teukolsky S.A, Vetterling W.T, Flannary B.P, (1992) "Numerical Recipes in FORTRAN", Second Edition, ISBN 0 521 43064 X, Cambridge University Press, Cambridge.
Shuen J-S., Chen L-D, Faeth G.M, (1983) "Evaluation of a stochastic Model of Particle Dispersion in a Turbulent Round Jet", AIChE Journal 29, 167 – 170.
Snyder W.H, Lumley J.L, (1971) "Some Measurement of Particle Velocity Auto-correlation Functions in Turbulent Flow", J. Fluid Mechanics 48, 41 – 71.

Air Distribution in Rooms, (ROOMVENT 2000)
Editor: H.B. Awbi

COMPUTATIONAL INVESTIGATION OF VENTILATION STRATEGIES TO REDUCE EXPOSURE TO NO$_2$ AND CO FROM GAS COOKING

D. I. Ross

Building Research Establishment Ltd. (BRE), Hertfordshire, WD2 7JR, UK

ABSTRACT

Gas cooking in the home can release high levels of nitrogen dioxide (NO$_2$) and carbon monoxide (CO). This study investigated the effect of various ventilation strategies to reduce personal exposure to these pollutants. It considered the effectiveness of windows, a kitchen extract fan and trickle ventilators for different dwellings, occupant behaviour, environmental conditions etc. Strategy selection was based on the need to minimise both personal exposure and energy loss. These strategies were simulated using BRE's BREEZE multi-zonal computer code. The results showed that it is best to: (a) use a window/ windows (where energy-loss acceptable) or a kitchen fan, and, (b) open all internal doors. However, whilst opening kitchen doors may be the best option for NO$_2$ and CO, it may not be appropriate for other combustion products, such as moisture and odours.

KEYWORDS

Gas cooking, residences, nitrogen dioxide, carbon monoxide, multi-zonal, modeling, ventilation

INTRODUCTION

During gas cooking, NO$_2$ and CO are emitted into the room air. Studies suggest that in many UK homes that use gas for cooking, short-term levels of one or both of these pollutants regularly exceed World Health Organisation health-based air quality guidelines, e.g. Ross (1996, 1999). Therefore there is concern that many people are being exposed to harmful levels of pollution from gas cooking.

Ventilation is the most common planned approach to reducing levels of pollution in buildings. Approved Document F to the Building Regulations (England and Wales) provides guidance on ventilation in the home to restrict the accumulation of pollutants originating within the building, where such pollutants would otherwise become a hazard to the health of the people within.

The purpose of this study was to investigate the effect of using various ventilation strategies, some based on the guidance in Approved Document F to the Building Regulations, to reduce personal exposure to NO$_2$ and CO from gas cooking. These strategies were simulated using BRE's BREEZE computer program, which can model the air and pollutant movement in a building.

BREEZE

BREEZE is BRE's multi-zonal computer program to evaluate ventilation rates and airflows in buildings. The building is taken to consist of a number of inter-connected zones (typically a zone is a room) with air moving from zones at higher pressure to those of lower pressure. The pressure differences are set up both by the actions of wind on the external surface of the building and by the temperature difference between air inside and outside, in addition to mechanical ventilation devices. BREEZE includes a contaminant analysis routine. It can allow for external sources, sources within rooms, pollutants released from surfaces and adsorption/desorption at surfaces can be modelled.

BREEZE also includes a 'consecutive analysis' routine: most input parameters (e.g. internal and external temperatures, wind speed and direction, ventilation opening areas) can be varied over time within a single run. Thus the effect of gas cooking on indoor pollutant levels can be modeled over an extended period of time, allowing for occupants' cooking and ventilation use patterns and variable environmental conditions.

For this study, BREEZE was set up to provide: (a) hourly averaged data for NO_2 and CO in each room in each house, and, (b) hourly data of the space heating gain and the air exchange rate for the whole house.

PARAMETERS FOR MODEL

Ventilation Strategies

Table 1 provides details of the ventilation devices used and their patterns of use. Eight basic conditions were selected, based on the guidance on ventilation provisions given in Approved Document F and others thought to be of interest. Case 1 is the base condition in which no ventilators are used.

TABLE 1
DESCRIPTION OF THE EIGHT VENTILATION STRATEGIES

Number	Ventilator(s)	Time used	Open area or fan extract rate
1	None	Not used	None
2	Kitchen window	During cooking	1/20th of floor area
3	Kitchen window	Constantly	1/20th of floor area
4	Kitchen window	During cooking	Fully open
5	Kitchen and living room windows	During cooking	1/20th of floor area
6	Kitchen fan	During cooking	60 l/s extract
7	Kitchen fan	During cooking and one hour afterwards	60 l/s extract
8	All trickle ventilators	Constantly	8000 mm^2 for habitable rooms 4000 mm^2 for other rooms

Dwellings

Both a two-storey terraced house (treated as an infinite row of houses) and a detached house were modeled. They had identical floor plans and are shown in Figure 1. Each floor was 2.3 m high and all doors were 0.8 m wide and 2.05 m high. The internal doors had a gap of 5 mm below and 1.5 mm around the other three sides. The window areas in the kitchen and living room were 0.89 m^2 and 1.50 m^2 respectively (no other windows were used). BRE's database of air leakage rates in UK dwellings gives a mean value of 13.1 air changes per hour at 50 Pa. This leakage was spread uniformly over all external walls and the roof.

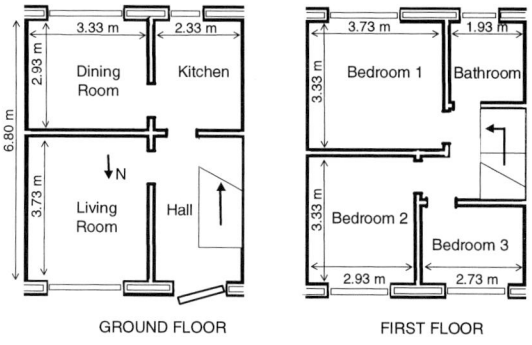

GROUND FLOOR   FIRST FLOOR

Figure 1: Plan of two-storey house

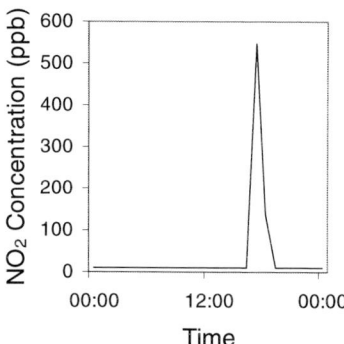

Figure 2: NO$_2$ concentration in the kitchen

Internal door positions

Three internal door positions were considered: (i) all internal doors closed, (ii) both kitchen doors fully open and all others closed, and, (iii) all internal doors open. A combination of one of the eight 'ventilation options' and one of the three 'door positions' is referred to as a 'ventilation strategy'.

Temperature

A difference between the internal and external temperatures creates a flow of air inside the house. The results would be expected to show a seasonal effect. With the large number of runs required, it was too time-consuming to model the entire year. A simpler alternative was to model a week in summer and a week in winter. The external temperature data were obtained from a set of Test Reference Years - TRYs (1985). Hourly weather data for London (Kew) were used for the first week in January and the first week in July. The internal temperatures were: (a) 18.5°C for the living rooms, and, (b) 16.5-17.0°C for all other rooms (range to provide mixing between adjacent rooms). During cooking, the kitchen temperature was raised to 19°C.

Wind

The air flow inside the house is also dependent on the pressure differences set up by the action of the wind on the external surfaces. BRE (1982) suggests that a typical wind speed for the UK is 4 ms^{-1}. To minimise the number of variables, a constant value of 4 ms^{-1} was used. Sensitivity analysis was performed for a wind speed of 0 ms^{-1}. Twelve wind directions were modeled: 0° to 330° in 30° incremental steps. The wind pressure coefficients were obtained from the BREEZE Cp database.

Cooking pattern

To best compare the effect of different ventilation strategies, a simple cooking pattern was used. The cooker was on from 5 - 6 p.m. each day. Sensitivity analysis assessed the effect of the time period chosen.

Source emissions and sorption

Based on a literature review, gas cooking emission rates of 0.1 g.hr^{-1} and 0.5 g.hr^{-1} were used for NO$_2$ and CO respectively. From the BRE indoor air quality database, the outside levels of NO$_2$ and CO were assumed to be 21.2 μg.m^{-3} and 0.40 mg.m^{-3}. NO$_2$ is removed from the indoor air through sorption by indoor surfaces. A sorption rate of 0.84 hr^{-1} was estimated using indoor and outdoor levels of NO$_2$ for homes with no known sources (using BRE's indoor air quality database). Sorption is negligible for CO.

RESULTS

Room concentrations

Figure 2 shows an example of hourly averaged NO_2 concentrations in the kitchen for a day. A daily peak occurred due to cooking. Figures 3a-c show the variation of maximum hourly-averaged NO_2 concentration with wind direction for the kitchen, living rooms (maximum of the dining room and living room) and bedrooms (maximum of all three bedrooms) respectively for the terraced home with all internal doors closed and the wind blowing towards the north. The highest level in the kitchen occurred for the wind blowing from the north as the wind and stack forces act in reverse directions through the kitchen. Similarly the highest level for the living rooms occurred with the wind blowing from the north. This was due to the wind force being dominant, driving highly polluted air from the kitchen to the dining room. Finally the highest level in the bedrooms occurred with the wind blowing from the south. In this case the wind and stack forces combine to drive the NO_2 from the kitchen into the hallway and up the stairwell.

Many cases were modeled. Overall, all the results showed wind dependency. The terraced results were symmetrical about the vertical axis, as the dwelling was treated as part of an infinite row of homes, whereas this symmetry was not evident for the detached home. As more internal doors were opened, the results became less wind dependent due to the greater mixing of air with neighbouring rooms. The effect of using additional ventilators was to increase ventilation and reduce concentrations.

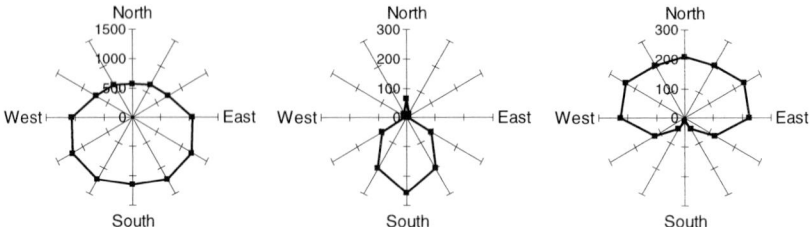

Figure 3: NO_2 concentration in: (a) the kitchen, (b) living rooms, and, (c) bedrooms

Selection by energy criteria

A simple energy criterion was used. Each ventilation strategy had to incur no net energy loss (i.e. space heating gains from cooking exceeded space heating losses from additional ventilators). Based on thermal data, the space heating gain from cooking an evening meal was 1.45 kWh. Results for each strategy were averaged over all wind directions. Ventilation options 3-5 were not acceptable during the winter period.

Selection by personal exposure

We initially considered the exposure of a person: (a) in the kitchen, and, (b) in the room of maximum exposure (apart from the kitchen). The maximum value was determined for each wind direction (typically during cooking) and the mean taken. However as this mean value was always greatest in the kitchen, and the purpose was to limit the maximum exposure, the study focused on exposure in the kitchen.

The strategies were ranked by dwelling type, season and pollutant. Tables 2-5 show the best six strategies for each parameter. The ranking is similar, if not the same, for each pollutant. Overall the results suggest it is best to:

- use a window/windows (where energy-loss acceptable) or a kitchen fan;
- open all internal doors (in some cases, it is acceptable to open only the kitchen doors).

TABLE 2
BEST STRATEGIES FOR TERRACED HOME IN SUMMER

Vent No.	Door Pos. No.	NO$_2$ Conc. (ppb)	Vent No.	Door Pos. No.	CO Conc. (ppb)
5	3	56	5	3	0.78
4	3	146	7	3	1.62
6	3	150	6	3	1.66
7	3	150	3	3	1.66
2	3	153	4	3	1.68
3	3	155	2	3	1.81

TABLE 3
BEST STRATEGIES FOR TERRACED HOME IN WINTER

Vent No.	Door Pos. No.	NO$_2$ Conc. (ppb)	Vent No.	Door Pos. No.	NO$_2$ Conc. (ppb)
2	3	116	2	3	1.25
2	2	134	2	2	1.40
6	3	148	6	3	1.59
7	3	148	7	3	1.59
6	2	173	6	2	1.80
7	2	173	7	2	1.80

TABLE 4
BEST STRATEGIES FOR DETACHED HOME IN SUMMER

Vent No.	Door Pos. No.	NO$_2$ Conc. (ppb)	Vent No.	Door Pos. No.	CO Conc. (ppb)
5	3	29	5	3	0.46
5	2	104	5	2	1.12
4	3	122	4	3	1.31
2	3	127	2	3	1.36
3	3	128	3	3	1.36
6 or 7	3	143	3	2	1.52

TABLE 5
BEST STRATEGIES FOR DETACHED HOME IN WINTER

Vent No.	Door Pos. No.	NO$_2$ Conc. (ppb)	Vent No.	Door Pos. No.	NO$_2$ Conc. (ppb)
2	3	104	2	3	1.13
2	2	117	2	2	1.24
6	3	143	6	3	1.52
7	3	143	7	3	1.52
6	2	166	2	1	1.67
7	2	166	6	2	1.72

Sensitivity analysis

Wind speed and direction

A zero wind speed was modeled for selected strategies. This is a 'worst case scenario' for ventilation as it removes the wind-driven force (in practice there would always be some air movement, for example due to local turbulence forces). The greatest increases in pollutant levels were during the summer due to the smaller stack effect during this period. Overall the greatest increase occurred for cross ventilation. The effect of trickle ventilators had a notable wind-dependence, particularly with internal doors closed.

The main analysis considered exposure averaged over 12 wind directions. It was useful to also assess the impact of wind direction. For each ventilation strategy, the ratio was calculated between the maximum and minimum exposures, across the range of wind directions. This ratio was typically greatest for window airing. For the kitchen in the terraced house, wind direction had the maximum impact for cross ventilation (ventilation strategy 5, door position 3) where the ratio approached 6. For the kitchen in the detached house, the greatest ratios (~4) occurred with the kitchen window open to 1/20th of the floor area and internal doors closed. For other rooms, the greatest ratios typically occurred during window airing and with internal doors closed. Strategies using the fan showed the minimum variation with wind direction.

Temperature difference during cooking

It was assumed that cooking occurs from 5-6 p.m. each day. However, in practice, cooking occurs at differing times during which the environmental conditions and, consequently, airflow rates may be different. This analysis focused on the effect of the external temperature during cooking on exposure in the home - the effect of variable wind conditions was considered above. Additional runs were performed for each season at three external temperatures; weekly mean and one standard deviation either side of this mean (summer:$18.0\pm4.3°C$, winter:$2.8\pm3.1°C$). The greatest effect of choice of external temperature occurred for the window opening cases. The least impact occurred for kitchen fan use.

Thermal comfort

The airflow created by the use of a ventilator device may affect the thermal comfort of the occupants. The worst cases, with high flows of cold air, were removed by the energy criterion. The highest remaining air flow rate of 28 air changes per hour occurred during the summer for cross ventilation. This would result in a flow rate of 1.5 $m.s^{-1}$ close to the window, which is quite high. If necessary, the window open area could be reduced, with a consequent reduction in the ventilation rate and an increase of pollutant levels.

DISCUSSION

This report describes a computational investigation to select the best ventilation strategies to reduce occupants' short-term exposure to NO_2 and CO produced during gas cooking. The study suggests the best approach is to open the internal doors to dilute the pollutants rather than venting them directly to outside. The best ventilation devices are either windows or extract fans. Other considerations are discussed here.

Only two cooking pollutants were studied; others include water vapour and odours. If internal doors were open, the spread of water vapour could result in condensation problems elsewhere in the home. Similarly the spread of odours may be unacceptable. Further work should consider these and other pollutants.

A sensitivity analysis showed the significant impact of wind and temperature on the effectiveness of window airing and this should be considered further. Other ventilation devices should also be considered, e.g. an extract cooker hood and a passive stack ventilator (PSV), although the effect of a PSV may be small compared to a cooker hood for reducing short term exposure.

Finally, this study assumes that each pollutant creates a similar level of risk. Further work should employ a more quantitative approach, comparing results to known health-based guidelines. It has also been assumed that all occupants are similarly susceptible to each pollutant. If the occupants in differing locations have different sensitivities, the results would need to be weighted accordingly.

ACKNOWLEDGEMENTS

This work was supported by the UK Department of the Environment, Transport and the Regions, which has given permission for it to be published.

REFERENCES

Ross D. (1999). *Continuous monitoring of nitrogen dioxide and carbon monoxide levels in UK homes.* Indoor Air '99, Scotland, Vol. 3, 147-152

Ross D. (1996). *Continuous monitoring of NO_2, CO, temperature and humidity in UK homes.* Indoor Air '96, Japan, Vol. 1, 513-518.

TRY. (1985). Test reference years - 'TRY'. Commission of the European communities, Directorate General XII for Science, Research and Development.

BRE (1982). *Principles of natural ventilation.* BRE Digest 210, CRC, UK.

Air Distribution in Rooms, (ROOMVENT 2000)
Editor: H.B. Awbi

CHARACTERIZATION OF VOCS, OZONE, AND PM $_{10}$ EMISSIONS FROM OFFICE PRINTERS IN AN ENVIRONMENTAL CHAMBER

Sanches Lam & S.C Lee

Department of Civil and Structural Engineering,
The Hong Kong Polytechnic University, Hong Kong

ABSTRACT

A stainless steel flow-through environmental chamber was used to characterize the Indoor Air Quality (IAQ) emissions from office equipment. Two types of office printers (laser-jet, and ink-jet printers) were investigated. Volatile organic compounds (VOCs), total VOC (TVOC), ozone (O_3), and respirable particles (PM_{10}) were measured. The highest emission rates of VOCs compounds were toluene, ethylbenzene, m,p-xylene, and styrene. Results showed that emissions of ozone and VOCs from laser printers were significantly higher than that from ink-jet printers. The emission rates of TVOC varied from 0.2 µg/copy(ink-jet printer) to 7.0 µg/copy(laser-jet printer).

KEY WORDS

Volatile organic compounds, Total volatile organic compounds, Ozone, Indoor air quality, Office printer, Environmental chamber

INTRODUCTION

Office environments have changed rapidly with the advent of electronic technologies; with laser-jet printers and ink-jet printers becoming commonplace. The consequence of the extensive use of modern office equipment is that office workers are exposed to an office climate giving rise to adverse health effects such as headache; mucous irritation and dryness in the eyes, nose and throat. Kreiss (1989) have also reported complaints about poor indoor air quality (IAQ) in the mid-1970s. Users of laser printers and copiers need to aware of ozone gas and its potential harmful effects. Because such indoor emissions from office equipment can be an irritant to some people, various regulatory agencies and standard setting organizations--such as Occupational Safety and Health Administration (OSHA), American Conference for Governmental Industrial Hygienists (ACGIH), the Health and Safety Executive in the United

Kingdom, and Canadian Federal Government--have established limits for the amount of indoor emissions to which employees may be exposed.

This study is aimed to identify, characterize and evaluate pollution prevention opportunities to reduce air emissions from office equipment. The objective of this study were **1)** To characterize the IAQ emissions from various types of office printers by using a dynamic environmental chamber; **2)** To determine emission rates of ozone, particulate, TVOC as well as individual VOC from different office printers; **3)** To develop pollution prevention solutions to reduce indoor air emissions from office equipment.

METHODS

Two laser printers (Printers **A** and **B**), two ink-jet printers (Printers **C** and **D**), and one all-in-one (colour printer, fax, copier, and scanner) office machine (Printer **E**) were chosen for screening. The toner recommended by the individual manufacturers was used and a new one was placed for each test. A standard printing file has been prepared for all tests. Selways et al, (1980) has found that the amount of ozone produced per copy is increased with equipment age and greatly affected by the maintenance cycle period. Therefore, all the office printers investigated here were within one year old and between 1,000 – 2,000 copies after the pervious service. The paper used under all tests were free from recycled content in order to avoid any unnecessary emissions.

All tests were conducted in a stainless steel flow-through dynamic chamber. The size of chamber is 1200(H)×1100(W)×1800(L)mm which was designed from a typical office room size in Hong Kong providing adequate space air movement and well-mixing condition. In order to minimise any chemical reaction, pollutants were transferred by Teflon tubing and all electrical and plumbing feedthroughs were sealed with inert materials. Relative humidity in the chamber was maintained within 55 ± 5% and the temperature at 23.0°C ± 0.5°C.. To ensure a well-defined test and comparable results, a set of standard guide by the American Society for Testing and Materials(ASTM), 1997 was followed. Background levels in the chamber air was measured before each test.

The tested office equipment was placed in the center of the chamber on the floor. Measurement were taken with the equipment idling. With the air exchange flow rate up to 5 ACH and an application of a mixing fan, the chamber air concentration was evaluated to be in equilibrium within 15 minutes. The equilibrium of ozone, TVOC concentrations and temperature in the chamber were monitored during continuous printing at maximum print rate for 60 pages. An integrated reading was taken every minute until 30 minutes after the machine was turned off. VOCs samples were also collected into a canister for 30 minutes from the start of an operation and were analysed using a cryogenic preconcentrator with gas chromatograph / mass spectrometry (HP 5973) by USEPA TO-14 method within 24 hours. Samples used to measure air concentrations of target VOCs using a multipoint calibration. Ozone was monitored continuously by Thermo Environmental Instruments ozone analyzer (model 49). DustTrak aerosol monitor(model 8520) with 100mm cellulose ester filter was used for the PM_{10} measurement. TVOC was measured with Mini RAE photoionization detector (ppb level).

RESULTS AND DISCUSSION

Average TVOC, ozone, and PM_{10} emissions from various types of office equipment are listed in Table 1. The emission rates of TVOC from laser printers were the highest and found to be about 6 times that of ink-jet printers. Laser printers use heat and pressure to fix an image onto the paper surface. The generated heat encouraged evaporation of VOCs compounds which are very sensitive to temperature. Ink-jet printers differ somewhat in how the toner is delivered in that a photoconductive drum is not necessary. There is a tendency for laser printers to emit larger amounts of organic volatile than ink-jet type. Wolkoff et al. (1993) has concluded in his experiment that emissions rates from laser printers ranged from 2.0-6.5µg/sheet.

TABLE 1

AVERAGE TVOC, OZONE, PM_{10} EMISSION FROM STANDARD COPIES (µG/COPY) FROM TWO LASER PRINTERS (A,B), TWO INK-JET PRINTERS (C,D) AND ONE ALL-IN-ONE OFFICE MACHINE(E)

Equipment	TVOC	Ozone	PM_{10}
A	5.7	1.2	2.6
B	7.0	1	2.4
C	1.2	0.05	0.9
D	0.7	0.05	1.3
E	0.2	0.05	1.8

TABLE 2

AVERAGE LEVELS OF VOCs FROM OFFICE EQUIPMENT A, B, D AND E

Target Compound	Idle				Operation			
	A	B	D	E	A	B	D	E
Toluene	14.7	13.8	6.22	7.89	15.3	16.3	6.43	8.17
Tetrachloroethene	0.16	0.15	0.23	0.52	0.14	0.13	0.19	0.43
Ethylbenzene	1.37	2.07	1.20	1.50	1.99	3.00	1.26	1.63
M,P-Xylene	1.16	1.22	0.86	0.90	1.56	1.68	0.92	0.87
Styrene	2.71	3.98	1.14	1.23	3.19	5.27	1.43	1.85
O-Xylene	0.86	0.99	0.69	0.58	1.96	2.29	0.68	0.58

Figures above are in ppbv

As shown in Table 2, selected VOCs were identified and quantified during both the idle and operation modes. The emissions of idle mode from all equipment were relative lower comparing with the emissions from operation. Such emissions could be either from construction materials (e.g., plastic casings) or components (e.g., cricuit boards). The greatest relative difference out of toluene, ethylbenzene, m-p-xylene, and styrene between idle and operation modes was found to be 47% for Ethylbenzene. Ethylbenzene occurs naturally in carbon. According to the toner composition data sheets given by the manufacturers, about 15% of the laser printer toners are made of carbon black, auxilliary pigment and additives. The toners used in laser printers contain a wide variety of chemicals in additional to fine, black carbon particles. A sustainable amount of VOCs were from such chemical in toner (Wokloff et al., 1993). The average temperature for laser printers were generally higher than ink-jet type during operation. Laser printer toner powders were heated to the temperature of the "fusion" roller of the laser printers examined.

Elevated temperatures used in fusion can be expected to increase the voltilization of VOCs present in the toner.

Ozone results are tabulated in Table 3. The ozone emissions rates were affected by temperature, and power consumption (Watt) (Hansen et al., 1986).

TABLE 3
RELATIONSHIP BETWEEN POWER CONSUMPTION, TEMPERATURE, AND OZONE EMISSIONS

Machine	Average Temperature ($^{\circ}$C)	Power (W) V x I	Equilibrium Ozone Concentration (ppm)	Average Ozone Emissions Rate (μg/copy)
A	30	374	10	1.2
B	32	408	9	1
C	23	12	5	0.1
D	24	12	6	0.1
E	24	15	6	0.1

Both of the average temperature and electric power were recorded to be higher from Printers A and B than that of Printers C and D (Table 3). For laser printers, electrically charged corona wires are used to add a uniform primary charge across the surface of the photosensitive drum. These wires are used to apply charges to paper surface and to electrostatically clean the cartridge drums. Therefore the ozone emission is determined by the amount of voltage over the corona wires. The results reported here for ozone can be compared to those done by other researchers (Allen et al., 1978; Selway et al., 1980; Hansen and Andersen, 1986). The electric power played an important role in the free-radical chain reaction during the formation of ozone process in a printing operation. During a printing operation, an oxygen molecule, O_2, was broken into two oxygen atoms, O. Such high energy levels encouraged the free-radical chain reaction and hence the formation of ozone. Therefore, the amount of ozone produced really depending on the amount of voltage and charge acrossing the toner drum. The reason for ink-jet printers (Printers C, D and E) having very low ozone emission rates is because of the different mechanism applied from laser printer. Ink-jet printers do not use a true photo-imaging process; that is, a photosensitive drum is not used to impart the image. Instead, the image is formed when a nozzle "sprays" the ink toward the paper character by character.

Particles problem was caused by the low transfer efficiency of the toner. Typically, about only 75 percents of the toner is transferred to the photoconductive drum. Toner particles that do not adhere to the drum become available for emission to the indoor air. The size of the toner particles in the carbon black itself is below 10μm in diameter (Hansen et al., 1986). The concentrations of particles are shown in Figure 1. The average particulate concentrations of Printer A and B were 65 μg/m^3 which were much higher than that for Printer C, D and E 20 μg/m^3, 38 μg/m^3, and 41 μg/m^3 respectively. Eggert et al., (1990) measured emissions of particulates from 20 different laser printers and found the average emission rate to be 61 μg/min. However, the maximum particulate levels of Printer A and B exceeded the National Ambient Air Quality Standard (NAAQS) which listed to be 75 μg/m^3.

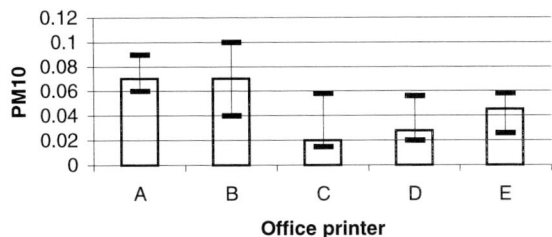

Figure 1: Minimum , average, and maximum PM$_{10}$ levels of printers A-E

POLLUTION CONTROL OPPORTUNITIES

Pollution prevention opportunities for office equipment can be applied to the machine design, and toner materials used as mentioned above. A change of fusing process may also provide some opportunities for pollution prevention. Reducing fuser temperature (by changes in pressure) may result in lower VOC emissions. The VOC emissions from ink-jet printers occurred from the volatilization and /or aerosolization of toner and toner solvents. Reformulation of toners using lower-volatility solvents can result in lower emissions. In order to improve the toner transfer efficiency to minimize the amount of toner particles available for emission into the indoor air, changes in toner particle size and a regular maintenance cycle can have a serious impact (Selway et al., 1980). A recent development by Canon using direct contact charging rollers replacing corona wires have shown to be effective to prevent the formation of electrical arcs and ozone. Such new design has resulted in lowering ozone emissions.

CONCLUSION

VOCs emissions from laser printers (Printers A & B) were higher than that from ink jet printers (Printers C, D & E), because of the higher temperature generated from fusing process and hence encouraged further evaporation of the organic compounds. Reformulation of toners using lower-volatility solvents/black carbon can result in lower VOC emissions. Also, changes in toner particle size may have an impact on fusing process. Ozone generation was only found in laser printers (Printers A and B). The ozone generated by laser printers is a by-product of the electrophotographic process, and was generated when the corona wires placed charges onto photoconductive materials. Installation of a new direct contact electrical charge rollers by many manufactures has resulted that the measured ozone concentration were significantly lower. The root cause of the particles is due to the poor transfer efficiency between the charged toner drum and paper surface. As the photoconductive surface of the drum deteriorates, the toner transfer efficiency decreases increasing the potential for indoor air emissions. A regular maintenance check is necessary to restore the transfer efficiency to its original state and reversing the trend of increasing potential for particulate pollution. Evaluation of emissions control should not only concentrate on one strategy, but also focus on source control, ventilation, and air cleaning or a combination of these. Identifying specific constituents of concern can direct efforts to reformulate the source material (e.g., toner, photoconductive surface in the case of laser printer) or make alterations in the process that will reduce the potential emission from office equipment.

94

ACKONWLEDGEMENT

The authors would like to thank the support from Dr. John Chang of U.S EPA for his kindly advise on environmental chamber design and indoor emissions characterisation.

REFERENCES

Allen,R.J., Wadden,R.A. and Ross, E.D. (1978) Characterization of potential indoor sources of ozone. American Industrial Hygiene Association Journal, **39**, 466-471.

Eggert, T.A. Grove , and I. Drabaek. (1990). Emission of Ozone and Dust from Laserprinters. Presentation of a New Emission Sopurce Test Method, Proceedings of 1990 EPA/AWMA International Symposium on Measurement of Toxic and Related Air Pollutants. Raleigh,NC, EPA-600/9-90-026 (NTIS PB91-120279).

Hansen,T, and Andersen, B (1986). Ozone and other air pollutants from photocopying machine. American Industrial Hygiene Association Journal, **47**, 659-665.

Kreiss,K. (1989). The epidemiology of building-related complaints and illness. Occupational Medicine: State of the Art Reviews, **4**, 575-592.

Selway, M.D., Allen, R.J and Wadden, R.A (1980). Ozone production from photocopying machines. American Industrial Hygiene Association Journal, **41**,455-459.

The American Society for testing and Materials (1997). Standard Guide for Small-Scale Environmental Chamber Determinations of Organic Emissions From Indoor Materials/Products (D 5116 –97).

Wolkoff,P., Wikins, C.K., Clausen, P.A, and Larsen, K.. (1993). Comparison of volatile organic compounds from office copiers and printers: Methods, Emission Rates, and Modelled Concentrations. Indoor Air **3**: 113-123.

Air Distribution in Rooms, (ROOMVENT 2000)
Editor: H.B. Awbi

EXPERIMENTAL AND NUMERICAL PREDICTION OF INDOOR AIR QUALITY

C. Teodosiu[1], S. Laporthe[1], G. Rusaouen[1] and J. Virgone[1]

[1]Centre de Thermique de Lyon (CETHIL), UPRES A CNRS 5008
Equipe Thermique du Bâtiment, Institut National des Sciences Appliquées
(INSA), Bât. 307, 20 avenue A. Einstein 69621 Villeurbanne Cedex – France

ABSTRACT

This paper is based on a dual approach (experimental and numerical) in order to predict the indoor air quality for small ventilated enclosures. The experimental part employs a ventilated test room and a tracer gas technique (constant method as gas injection) to estimate the diffusion of a pollutant. The gas used is the sulphur hexafluoride (F_6S). The numerical approach is a CFD simulation, adding a convection – diffusion equation (to determine the local mass fraction of the pollutant) to the equations normally used to solve a turbulent flow. As the injected quantity of the gas is extremely low, the convergence of the pollutant transport equation can be only reached after an important CPU. Therefore we propose a strategy for the initialization of the problem in terms of tracer gas concentrations using the experimental data. This allows us to obtain important savings in time simulation. The experimental and numerical concentration tracer gas fields are compared for isothermal and non-isothermal conditions. Further, in order to characterise in a more general manner the ventilation efficiency and the indoor air quality of the two cases studied, we evaluate the ventilation efficiency index based on the experimental and CFD values. Moreover, we add to these results the values obtained by an improved zonal model. The results achieved reveal a good agreement between the experiment and CFD calculation. On the contrary, there are important discrepancies between the ventilation efficiency index values predicted by the zonal model and the values based on the experimentation.

KEYWORDS

Experimental cell, tracer gas, numerical simulation, indoor air quality, ventilation efficiency.

INTRODUCTION

Nowadays, it is known that we spend most of our time (90%) in the enclosed spaces so the indoor air quality becomes more and more an important parameter, especially for our health. On the other hand, it is obvious that one parameter which mainly determines the level of air quality in the indoor climates is the ventilation. This underlines the importance of the studies dealing with the efficiency of the ventilation systems. Frequently, in this field of research, both experimental and numerical works are employed and our paper is based, too, on this dual approach. Using the tracer gas technique in a ventilated test room and, on the other hand, a CFD numerical simulation, we are able to have a complete description of a pollutant concentration field for the studied configurations. This allows us to quantify a ventilation efficiency parameter which express the potential of the tested ventilation system to evacuate pollutants.

EXPERIMENTAL SET-UP

The experimental set-up is the test room "MINIBAT" (CETHIL - INSA de Lyon, France). Figure 1 illustrates a simplified scheme of the experimental system. In fact, this installation includes two identical rooms (cell 1 and cell 2, $3.10 \times 3.10 \times 2.50$ m^3 each). A glass wall separates the cell 1 from an enclosed space (climatic chamber) whose temperature is controlled by the means of an air-treatment system. The temperature in this climatic chamber can vary between -10 and $30°C$. The thermal guard is maintained at a uniform temperature of $20°C$ in order to represent adjacent spaces.

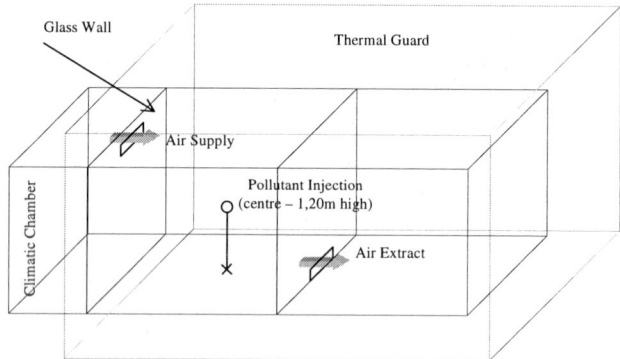

Figure 1: The experimental cell MINIBAT

Our study used exclusively the cell 1 of MINIBAT. The ventilation system of this room has a fixed supply and a mobile extract. The results presented in this work were obtained for the configuration exemplified in the figure 1. Measurements were carried out using sensors (thermocouples) to determine wall surface and air temperatures in the enclosure. The experimental methodology allowed us too, to measure the air velocity field (using hot-wire probes), as well as tracer gas concentrations. In addition, near the centre of each zone, relative air hygrometry and operative temperature are measured. The air field parameters (mean velocity, temperature) and the gas concentrations were obtained in a vertical median plane, the measurements points representing a 10 cm × 10 cm mesh grid.

We used the tracer-gas technique for studying the indoor air quality. Certainly, this technique is the most widely used to predict the movement and quality of air within an enclosed space. Furthermore, procedures using tracer gases are the only ones that can be used experimentally to:
- characterise the ventilation systems (with regard to fluxes, flows, etc.)
- quantify the quality of the air in a space (by determining its age and studying the migration of pollutants).

The tracer gas used in this study is the sulphur hexafluoride (F_6S). We have preferred F_6S as tracer gas Laporthe (to be published) mainly because of the sensibility of the Brüel and Kjaer measuring system – Brüel & Kjaer (1991). There are various ways of injecting a tracer gas into an experimental cell

Hanrion (1994), and we used in this study the constant method. This method consists of introducing the tracer gas continuously at a constant rate, throughout the measurement period; the mixing of the air in the test room should be sufficient to homogenise the concentration of the tracer gas. Moreover, the gas was injected (0.7 ml/s) into the centre of the experimental cell at a height of 1.20 m. The injection procedure used a small ball in which a large number of small holes had been made in order to assure as much as possible an uniform gas introduction.

NUMERICAL MODEL

Tracer gas concentration field prediction

The calculations were carried out by the means of a CFD code, Fluent (version 5.0), Fluent (1998). In conjunction with the equations that governed a turbulent non isotherm flow (conservation of mass, momentum and energy as well as the transport equations of turbulent kinetic energy and its rate of dissipation in the case of a two-equation turbulence model), an equation which represents the conservation of species concentration is solved. This partial differential equation has the same form as the other equations describing the conservation of a variable within the computational field. Hence, the convection-diffusion equation for predicting the local mass fraction of the pollutant, $m_{i'}$ (here sulfur hexafluoride, F_6S) takes the following form – equation (1):

$$\frac{\partial}{\partial x_i}\left(\rho u_i m_{i'}\right) + \frac{\partial}{\partial x_i} J_{i',i} = S_{i'} \tag{1}$$

In equation (1), the first term in the left represents the convective term (where x_i – the x, y and z directions, respectively; ρ - the fluid density and u_i –the velocity components) and the diffusion flux of species i', $J_{i',i}$, appears as a sum of two factors: $J_{i',i} = -\left(\rho D_{i',m} + \frac{\mu_t}{S_{ct}}\right)\frac{\partial m_{i'}}{\partial x_i}$; the first one represents the mass diffusion flux due to concentration gradients, with $D_{i',m}$ the diffusion coefficient of the species i' in the mixture; the latter one denotes the mass diffusion in turbulent flows or turbulent diffusion term added to the laminar diffusion term, with S_{ct} the turbulent Schmidt number and μ_t the turbulent viscosity. Finally, $S_{i'}$ represents the term source and its value is equal to the injection flow rate of sulfur hexafluoride that takes place in the middle of the test room - see the precedent section.

Consequently, the fluid is taken into account as a mixture of two species: air plus tracer gas (with air as bulk species). The properties for the mixture were defined as follows - for more details see Fluent (1998):
- density: ideal gas law for an incompressible flow
- viscosity and thermal conductivity: ideal-gas-mixing-law, the solver computes the values of these properties based on kinetic theory
- specific heat capacity: mass fraction average of the pure species heat capacities
- mass diffusion coefficient: constant value for the F_6S mass diffusion in the mixture (1.05×10^{-5} m^2/s).
Concerning the physical properties for the mixture constituents, constant values of specific heat capacity, thermal conductivity and viscosity were imposed for the air, as well as for the tracer gas.

Main physical and numerical hypothesis

In order to predict the flow in the test room presented at the previous section, the following physical assumptions were taken into account:
- fluid movement: 3D, incompressible, turbulent, (non) isotherm, steady
- turbulence model: a revised k-ε model, k-ε 'realizable' – Fluent (1998)
- near wall approach: two-layer.

98

Regarding the main numerical aspects of our simulations, briefly, they were as follows:
- computational domain discretisation: tetrahedral control volumes
- diffusion terms: second-order central-differenced
- convective terms: second-order upwind scheme.

Initialisation of the calculations

Firstly, the simulations were carried out initializing the concentration tracer gas field with zero values everywhere in the computational domain. In this model, we had only the term source imposed in the equation (1). Unfortunately, judging the convergence of our computations, we noticed that the evolution of the tracer gas transport equation residual (the imbalance summed over all the computational domain, for an exact definition of residuals – see Fluent (1998)) had not yet been 'stabilized'. This fact is illustrated clearly in the figure 2 where the progress of this residual is presented in function of the iterations completed.

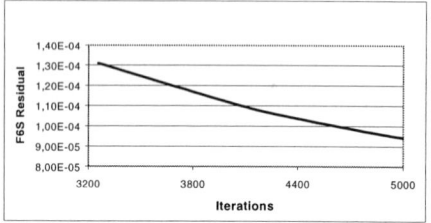

Figure 2: F_6S Residual Evolution
(zero initializing values)

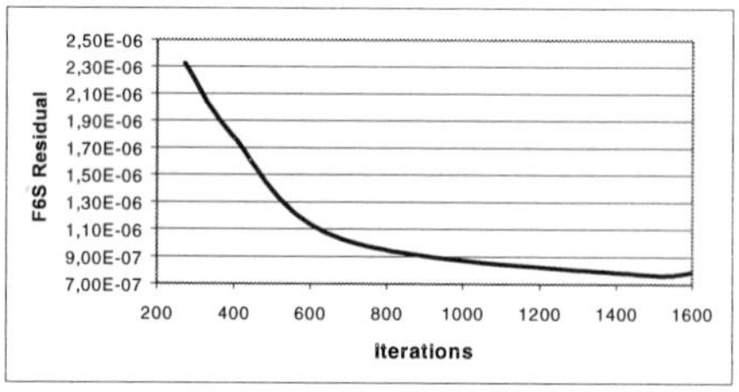

Figure 3: F_6S Residual Evolution
(mean experimental data as initializing values)

Subsequently, we tried to find out a method for accelerating the convergence of the equation (1). Therefore, we initialized the entire computational domain in term of pollutant (F_6S) mass fraction using the mean value based on the measurements carried out in the vertical median plane of the experimental cell. The residual evolution obtained in this situation is shown in the figure 3. We observe that the convergence is reached after a reduced number of iterations. This allowed us to achieve important savings in time simulations. Therefore, this approach was employed further in our simulations.

RESULTS

In our experimental – numerical comparisons, the concentration field for the tracer gas was of primary interest. Hence, we present such comparisons obtained in a vertical plane normal to the centre line of the air terminal devices. Moreover, we selected three sections in this plane in order to facilitate the comparisons. These sections are positioned at 1, 1.8 and 2.67 m, respectively, from the test room air supply.

Figures 4 and 5 show the measured and computed results for an isothermal case and for an air supply temperature of 11.2 °C at 1 air change per hour. Regrettably, there is a lack of experimental data in certain points for the last section (figure 5, x = 2.67 m) - in the case of the cold jet. These problems occurred because of the tracer gas measuring system.

The evaluation of the results given in the figures 4 and 5 shows a good agreement between the measured and simulated values, even that some discrepancies exist in the upper part of the enclosure. Most important differences (10-20%) occur in a few points between 0.75 and 1.25 m.

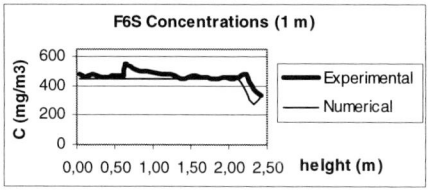
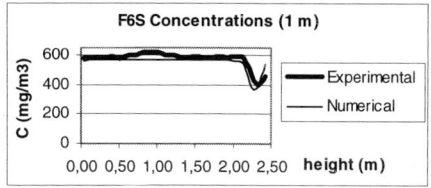
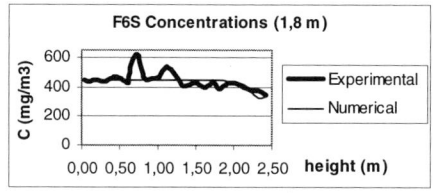
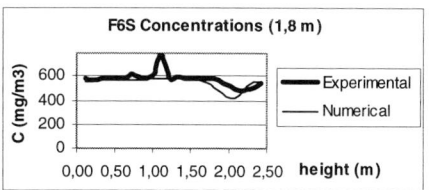
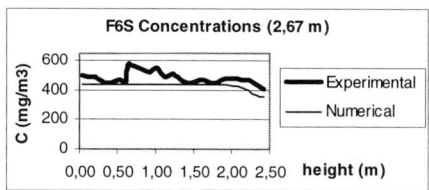
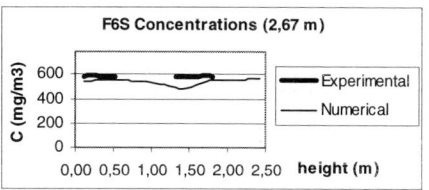

Figure 4: F$_6$S Concentration vertical distribution. Isothermal case: Re = 7100 and Ar = 0.0005

Figure 5: F$_6$S Concentration vertical distribution. Cold jet: Re = 7200 and Ar = 0.0134

This can be partially explained as an effect due to the measured values since variations from 400 mg/m^3 to 650 mg/m^3 or more, within 10 cm., are not plausible in the middle of the room, especially in our configurations which ensure an uniform air movement in the occupied zone.

In order to characterise in a more general way the ventilation efficiency and the indoor air quality of the configurations tested, we add to the results already presented, the values of an index (ε_C, equation 2) that represents the capacity of a ventilation system to evacuate pollutants, Sandberg & Sjöberg (1983).

$$\varepsilon_C = \frac{C_e(\infty)}{C(\infty)}$$ (2)

$C_e(\infty)$ – pollutant concentration in the extracted air
$C(\infty)$ – mean pollutant concentration in the room

The above equation is applied in permanent mode (meaning constant outlet concentration). In the table 1, we show the values of this index, values obtained based on the experimental data (as a mean value all over the experimental mesh), as well as values achieved using the CFD model proposed in this paper. We add to these comparisons the ventilation efficiency index values obtained by the means of an improved zonal model Bouia (1993) applied in the study Castanet (1998) for the same configurations. In building physics, the zonal model approach, that consists to split the room into different zones characterising the main driving flows, was initially used to predict the thermal behaviour of rooms but we can see its application frequently extended these days to ventilation systems and their efficiency.

For this reason, a direct comparison between simplified methods such the zonal models and CFD computations in the field of ventilation efficiency is quite interesting.

Based on the results given in the table 1, we notice that the zonal model predicts more or less reasonable the ε_C value for the non isothermal situation as the cold jet penetrates in the occupied zone and that compensates somehow the diffusion of the gas in the enclosure which is not taken into account in this simplified model. On the contrary, in an isothermal situation (and for a hot jet too, see Castanet (1998)), the result given by the zonal model is far from the experimental value (48%). In the same time, the CFD computation carried out lead to a good agreement with the value based on the measurements (a difference of 17,5% for the isothermal case and 6,7% for the non-isothermal situation). This underlines the importance and the accuracy of the CFD technique for the studies dealing with the pollutant diffusion predictions in rooms.

TABLE 1
EXPERIMENTAL AND NUMERICAL ε_C VALUES

ε_C – isothermal case (Re = 7100)			ε_C – non isothermal case (Re = 7200 and Ar = 0.0134)		
Experimental	Zonal model	CFD	Experimental	Zonal model	CFD
1.23	0.64	1.02	1.04	0.84	0.97

CONCLUSION

This study emphases the important interdependence existing between the numerical model and the measurements in the domain of indoor air quality: the CFD simulations represent surely one way to correctly express the pollutant dispersion in the enclosures but these calculations must be based on several measured values in order to reach a more quicker convergence - especially while the pollutant quantities are not so important in the room. On the other hand, the comparison based on the results given by a zonal model, in order to describe in a general way the indoor air quality, illustrated the limitations of such an approach concerning the diffusion of a pollutant in enclosures.

REFERENCES

Bouia H. (1993). *Modélisation simplifiée d'écoulements de convection mixte internes: application aux échanges thermo-aérauliques dans les locaux.* Ph.D. thesis. Université de Poitiers, France

Bruel and Kjaer (1991). *Moniteur Multigaz Type 1302 Instructions.*

Bruel and Kjaer (1991). *Multipoint Sampler and Doser Type 1303 Instructions Manual.*

Castanet S. (1998). *Contribution à l'étude de la ventilation et de la qualité de l'air interieur des locaux.* Ph.D. thesis. Insa de Lyon, France

Fluent User's Guide, Version 5.0. (1998). Fluent Inc., Lebanon – NH, USA

Hanrion, M.L. (1994). *L'Analyse de la Qualité de l'Air par la Méthode des Gaz Traceurs.* Promoclim, **25 : 6**, 359-369.

Laporthe, S., Virgone, J., Castanet, S. (to be published). *A Comparative Study of two Tracer Gases : SF$_6$ and N$_2$O.* Building and Environment.

Sandberg M. and Sjöberg M. (1983). *The Use of Moments for Assessing Air Quality in Ventilated Rooms.* Building and Environment **18 : 4**, 181-197.

Air Distribution in Rooms, (ROOMVENT 2000)
Editor: H.B. Awbi

INDOOR AIR QUALITY ASSESSMENT BY A "BREATHING" THERMAL MANIKIN

A. Melikov, J. Kaczmarczyk and L. Cygan

International Centre for Indoor Environment and Energy
Technical University of Denmark,
Building 402, 2800 Lyngby, Denmark

ABSTRACT

This paper presents a method for assessing air quality by means of a "breathing" thermal manikin developed to simulate more closely the human breathing cycle. The manikin is equipped with an artificial lung that simulates and controls the breathing cycle and the amount of respiration air as well as temperature, humidity and gas concentration of the exhaled air. The concentration of gases in the inhaled air as well as its temperature can be measured. The tests showed that the method developed is sensitive enough to assess the impact of room air distribution on the quality of air inhaled by occupants. The results show that proper simulation of humidity, temperature and gas concentration of the exhaled air as well as all three breathing modes (inhalation, exhalation and break) is important for accurate analyses of inhaled air quality.

KEYWORDS

Indoor environment, air quality assessment, breathing thermal manikin, artificial lung.

INTRODUCTION

In a calm, comfortable environment, upward free convection movement (thermal boundary layer) exists around the human body due to the temperature difference between the room air and the surface of the body (clothing temperature and skin temperature of bare body parts). The air movement around the human body is an important factor for man's thermal comfort and perceived air quality. The skin temperature of the human body will change due to convective heat transfer caused by the air movement around the body and thus will affect man's thermal sensation. The free convective flow transports air, which might be contaminated from the lower part of the space, upward to the breathing zone. It also carries the bioeffluents and vapour emitted from the human body. Furthermore, occupants' breathing generates an air movement due to exhalation. The interaction between the ventilation airflow in the room with the free convective flow around the body and the airflow of exhalation, and also the interaction between the free convective flow and the flow of exhalation are of primary importance for occupants' thermal comfort and perceived air quality.

Indoor air quality evaluations, based on chemical measurements, are usually performed in the breathing zone at some distance away from the person, often outside the thermal boundary layer. Alternatively, the person is temporarily moved while measuring the pollutants' concentration at the point of interest. The value obtained in this way is the concentration at the breathing zone, but it may differ from the actual concentration to which a person is exposed. Even when the human body is simulated by a thermal manikin, which creates an upward convective flow around the body, the difference may be large due to absence of the airflow of human respiration. A thermal manikin equipped with artificial lungs has previously been used to simulate a breathing human (Hyldgaard 1994, Brohus and Nielsen 1996). The breathing cycle consisting of inhalation, exhalation and pause has been simplified to only inhaling and exhaling. The temperature and the concentration of CO_2 (replaced by N_2O) in the air exhaled by the manikin were kept at levels close to those of an average human, while humidity was not considered. The concentration of contaminants (tracer gas) in the inhaled air was measured. Recent studies (Fang et al. 1998) however show that the temperature and humidity of the inhaled air affect perceived air quality. The air inhaled by people is perceived better when the air temperature and humidity are low. It is therefore important to measure the temperature and humidity of the inhaled air in addition to the concentration measurements.

This paper presents a method for air quality assessment by a "breathing" thermal manikin developed to simulate more closely a human being. The manikin is equipped with an artificial lung that simulates and controls the breathing cycle and the amount of respiration air as well as temperature, humidity and gas concentration of the exhaled air. The concentration of gases in the inhaled air as well as its temperature can be measured. The method was tested and some results are discussed in this paper.

THE "BREATHING" THERMAL MANIKIN

An existing thermal manikin was modified so as to simulate the breathing function of a person. For this purpose, holes were drilled in the nose and in the mouth. The manikin consists of 16 body segments that are heated and individually controlled in order to maintain a surface temperature equal to the skin temperature of an average person in thermal comfort at a given actual activity level.

Artificial lungs were designed in order to simulate realistically the breathing function of a human being and to assess the quality of the inhaled air. The artificial lungs developed consist of four systems: air transporting system; air humidifying system; system of dosing tracer gas; and system of controlling temperature of exhaled air. A simplified presentation of the four systems and their interaction is shown in Figure 1.

The air transporting system, which consists of two pumps and two valves, controls the frequency of breathing including duration of exhalation, inhalation and the break. It also controls the amount of air being inhaled and exhaled, e.g. the simulated pulmonary ventilation. The pulmonary ventilation can be set up to 13 L/min. The frequency of breathing is regulated by the settings on two connected digital timers. These parameters can be changed according to the simulated activity level.

The air to be exhaled flows through the air humidifying system that consists of a small pump and humidifier. The pump drives water through the humidifier where it is warmed up and evaporated. The system is calibrated by adjusting the amount of water dosed by the pump.

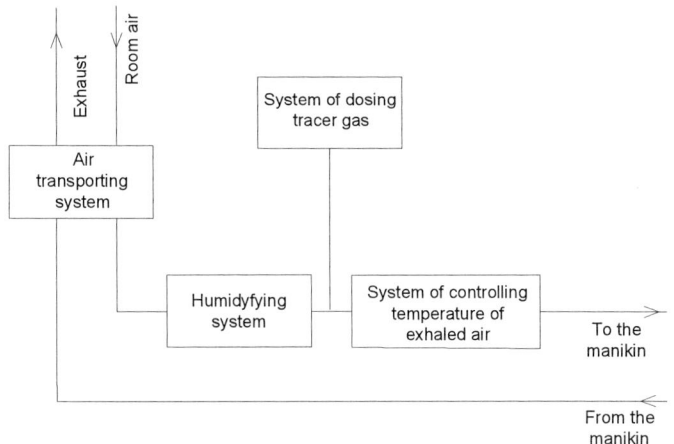

Figure 1: Diagram showing the function of the artificial lungs

The warm and humid air is then transported to the tracer gas dosing system where it is mixed with a tracer gas. Valves for pressure and flow control are used to dose the tracer gas released from a bottle. In the process of breathing, human beings produce carbon dioxide as a product of combustion in the body. The amount of this gas depends on the activity level and on the weight of the body. In the experiment, instead of CO_2, a mixture of CO_2 and N_2O was used. Both gases have the same physical features (same density) and do not react with each other. The mixture contains 90% CO_2 and 10% N_2O. The concentration of the mixture in the exhaled air from the manikin is equal to the concentration of CO_2 in the air exhaled by a seated person.

The warm, humid air treated with the appropriate gas content passes through a flexible pipe connecting the artificial lungs with the manikin and is then exhaled. The temperature of the exhaled air is maintained at a constant level by means of aluminium tape covering the pipe. The tape is heated by a silicon wire. A thermocouple mounted at the end of the flexible pipe is used to measure and control the exhaled air temperature. The system is calibrated before use.

TESTING OF THE METHOD

Experiments were designed to investigate the performance of the artificial lungs and to identify the importance of the breathing and the temperature, humidity and tracer gas in the exhaled air on the concentration and temperature of the inhaled air under different airflow conditions. The airflow conditions comprised combinations of airflows generated by the natural convective flow around the thermal manikin's body (manikin switched on/off), airflow generated by the breathing process (on/off) and airflow generated by personalized ventilation systems as described in the following.

Experimental Conditions

The breathing thermal manikin was placed behind a desk in a climate chamber, to simulate an occupant performing office work. The temperature in the chamber was kept at 26°C and the relative humidity at 40%. The climate chamber has dimensions: 5x6x2.5 m^3. It is possible to maintain temperature and relative humidity of the air inside the chamber with a high accuracy. The velocity generated by the ventilation system of the chamber is lower than 0.06 m/s (Kjerluf-Jensen et al. 1975).

Experiments were performed with and without ventilation air movement at the breathing zone. Two personalized ventilation systems were used to generate the air movement in separate experiments. The air delivered by the systems had a temperature of 20°C and a flow rate of 15 L/s and was supplied in two different ways: from a displacement diffuser and from an under-desk grill. The front panel of the diffuser is designed as a perforated plate with a cross-section of 0.54 x 0.25 m^2. The diffuser was mounted horizontally at the back of the desk, facing the seated manikin, at a distance of 0.7 m. The opening area of the grill is 0.40 x 0.06 m^2. It has deflecting blades mounted horizontally and vertically for regulating the direction of the supplied air. The grill was placed centrally under the desk with the grill opening located on the front edge of the desk.

The artificial lungs were adjusted to simulate breathing of an average sedentary person performing light physical work: breathing frequency of 10 times per minute, volume 6 L/min, breathing cycle of 2.5 s inhalation, 2.5 s exhalation and 1.0 s break, and exhaled air with a temperature of 34°C, relative humidity of 95% and tracer gas content of 3.6%. The air was exhaled from the mouth and inhaled through the nose. The manikin was dressed with underwear, T-shirt, long-sleeved shirt, pants, stockings and shoes. The total whole-body insulation value of clothing was 0.109m^{2}K/W (0.71clo).

Instrumentation and Measuring Procedure

Two tracer gases were used in the experiments: dinitrogen oxide (N_2O) and sulphur hexafloride (SF_6). The exhaled air was marked with N_2O. SF_6 was used to mark the air in the test chamber. The air delivered by personalized ventilation did not contain any tracer gas. The concentration of SF_6 and N_2O, in the inhalation air and in the air in the exhaust duct of the chamber was measured continuously by a Multi-gas Monitor Type 1302 Innova A/S connected to the Multipoint Sampler and Doser 1303. The instrument is based on the photo-acoustic infrared detection method of measurement. The concentration measured during the last 30 min of each experiment was averaged and analysed.

The temperature of the inhaled air was measured in the manikin's nose with a Thermobead sensor Series B07. The time constant of the sensor is 0.18 s in still air. The sensor is connected by a specially developed transducer to an A/D card mounted in the PC. Thus the analogue output signal from the sensor is transformed into digital, saved as a computer file and later analysed. The temperature of inhaled air was measured for 3 min at the end of each experiment with a sampling frequency of 10 Hz. All measurements were performed under steady-state room conditions. However the breathing consists of three repeating modes, inhalation, exhalation and break, which is transient in itself.

Results

The uncertainty of the N_2O measurement was assessed to be 7%, and for the SF_6 measurement to be 5% by repeated tests. The analyses of the results were based on these values. Figure 2 shows comparison of N_2O concentration measured in the inhaled air during experiments with a personalized system with a grill placed under the desk under two conditions: exhaled air treated with only N_2O and exhaled air treated with both N_2O and RH. The results in the figure clearly show that the amount of N_2O in the inhaled air is significantly higher when the exhaled air is humidified than when it is dry and treated only with N_2O. The same relation was found for the other ventilation system and also in the experiments without personalized ventilation. The content of N_2O makes the exhaled air heavier (the density is higher) in comparison with the humidified air. The air with greater density, exhaled through the mouth emerges more horizontally than the lighter humidified air, which has a stronger tendency to rise. The presence of humidity makes the exhaled air lighter and therefore it can penetrate

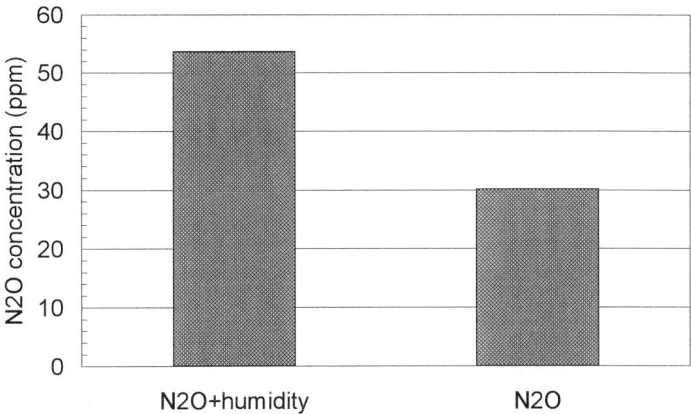

Figure 2: N_2O concentration in the inhaled air measured under different treatments of the exhaled air.

into the nostrils to grater extent than when only N_2O is added. The results showed also that the humidity had an impact on the concentration of SF_6 in the inhaled air when a personalized ventilation system was used.

In all experiments with and without personalized ventilation, a short break in breathing before each inhalation was identified as important for removing the exhaled air from the breathing zone. A pause of 1 s decreased the amount of exhaled air in inhalation by about 50% in compared to the situation where only exhalation and inhalation were simulated.

Less of the room air (14% in the case of an under-desk grill) marked with SF_6 was inhaled when the body of the manikin was the heated than when heating was turned off. Fresh air was delivered at stomach height and was directed towards the manikin. It was then entrained by the convective flow around the body and was transported up to the breathing zone. On the way to the breathing zone the air was warmed up. The measured temperature of the inhaled air was higher by 0.6°C when the heating was on.

A significant impact of the air distribution around the manikin on the quality and temperature of inhaled air was identified. Forty per cent more fresh air reached the manikin, when supplied from an under-desk grill rather than from a displacement diffuser. The low temperature and velocity of the air supplied from the displacement diffuser (due to its greater cross-section) caused the air to flow very close to the desk surface and to fall down over its edge. Only a small part of this air was delivered to the breathing zone.

The temperature of inhaled air measured in the nose was lower than that of the surrounding air when personalized ventilation with under-desk grill was used. According to recent studies (Fang et al.1998) the air delivered with these outlets would be considered as fresher than from a displacement diffuser.

The results of the tests showed that the method developed for indoor air quality assessment by means of a breathing thermal manikin is sensitive enough to identify differences in performance of air distribution systems. The amount of inhaled room air (SF_6) decreased when personalized ventilation was used. However, the amount of exhaled air in the inhaled air was measured higher with the personalized ventilation systems than in still environment. This indicates that the personalized ventilation systems used in the present experiments were not optimal. The breathing thermal manikin can be used to optimize personalized ventilation systems. Figure 3 shows the relative N_2O

concentration in inhaled air, measured without and with personalised ventilation (the concentration measured without personalized ventilation is used as a reference).

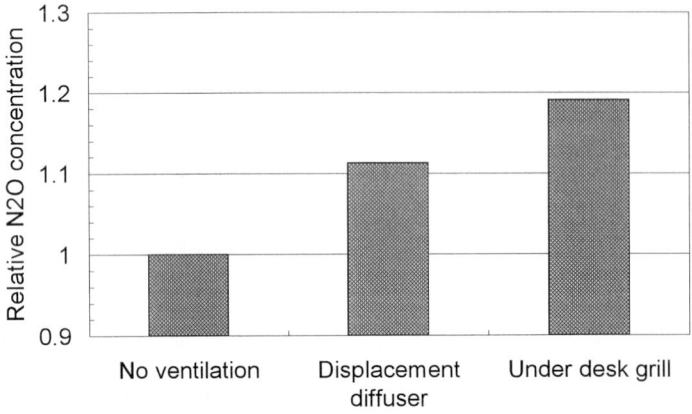

Figure 3: Relative N_2O concentration for different types of personalised ventilation.

CONCLUSIONS

A method for quality assessment of the inhaled air was developed and tested. The method includes artificial lungs simulating the human breathing cycle with realistic temperature, humidity and gas content in the exhaled air, as well as measurement of concentration and temperature of the inhaled air.

The experiments showed that proper simulation of all three breathing modes (inhalation, exhalation and break) as well as humidity of the exhaled air is important for analysing the quality of the inhaled air. Subsequent testing indicated that the method developed was sensitive enough to perform comparisons of ventilation systems as regards the quality of the inhaled air, comprising temperature and tracer gas concentration. Thus the breathing thermal manikin can be used to perform assessment of both thermal environment and air quality.

REFERENCES

Brohus, H. and Nielsen, P. V. (1996). Personal exposure in displacement ventilated rooms. *Indoor Air:* **6**, 157-167.

Fang, L., Clausen, G., Fanger, P.O. (1998). Impact of temperature and humidity on perception of indoor air quality during immediate and longer whole-body exposures. *Proceedings of Indoor Air,* **8**, 276-284.

Hyldgaard, C.E. (1994). Humans as a source of pollution. *ROOMVENT'94,* 413-433.

Kjerluf-Jensen, P., Fanger, P.O., Nishi, Y., Gagge, A.P. (1975). A new type test chamber in Copenhagen and New Haven for common investigation of man's thermal comfort and physiological reactions. *ASHRAE Journal,* January 1975, 65-68.

Air Distribution in Rooms, (ROOMVENT 2000)
Editor: H.B. Awbi

THERMAL CLIMATE ASSESSMENT IN OFFICE ENVIRONMENT – CFD CALCULATIONS AND THERMAL MANIKIN MEASUREMENTS

H. Nilsson[1], I. Holmér[1], S. Holmberg[2] and M. Sandberg[2]

[1] The Climate Group, National Institute for Working Life, 112 79 Stockholm, Sweden
Tel. +46-8-7309100, Fax. +46-87301967, E-mail: hakan@niwl.se
[2] Centre for Built Environment (HIG), 801 02 Gävle, Sweden

ABSTRACT

With increasing demand for acceptable office environment it is necessary, already in the construction phase, to estimate what effect different environmental factors have on the occupants. Thermal sensation is affected by many factors in the office environment especially thermal factors and effects from the air movements caused by different ventilation principles. In order to investigate whether CFD (Computational Fluid Dynamics) calculations and measurements with a thermal manikin are able to predict the perceived climate in an office, numerical calculations as well as full scale measurements were carried out. The heat loss and temperature of the manikin influence the air movements around the body. Thermal interaction with window, furniture, ventilation and chair are influencing the manikin. When this information is linked together with models for human thermal sensation, valuable knowledge about the thermal status of an office room is obtained. Calculations of this type enable engineers to make better predictions and early decisions in the design and construction process. It also opens possibilities to interpolate results from a reduced number of full scale tests. Results are analysed and discussed in this paper.

KEYWORDS

CFD, measurements, displacement ventilation, thermal manikin, office environment, thermal climate assessment, comfort diagram

INTRODUCTION

This investigation is a part of a larger Swedish project (The Healthy Building) that investigates how the use of modern ventilation tools can improve the thermal climate in an office room. The present investigation has been performed in a real office unit at the National Institute of Working Life (NIWL), and can therefore to some extent be regarded as a field investigation. The standard concept of displacement ventilation has been evaluated by numerical methods. Reference measurements and the use of a thermal manikin have enabled validation of simulated airflow patterns and comfort criteria. In order to investigate whether CFD calculations and measurements with a thermal manikin can predict the perceived climate in a office room numerical calculations as well as full scale measurements were carried out. This study is one step further in making computer simulations that predicts the effects of local climate disturbances well correlated with the thermal sensation experienced by subjects.

THE MANIKIN METHOD

Measurements with a thermal manikin yield a more complete, integrated and detailed information about thermal effects (Lund-Madsen and Olesen 1986, Wyon, et al. 1989). The manikin called AIMAN, a man-sized sitting thermal manikin, has been constructed of plastic foam The surface of the manikin is covered with resistance wires embedded in a hard plastic shell. The manikin surface is divided into 18 (max 33) independently controlled segments (Figure 1, 2). Once heated the manikin responds to a step change and equilibrates at the new power consumption within 20 minutes. The results are presented in a comfort diagram as variation of equivalent temperature (Nilsson, 1999, ASHRAE, 1989).

Measurement Procedure
The manikin technique has been validated in tests with subjects (Holmér, 1995). The heat losses of the manikin as well as the subjective and physiological reactions of a panel consisting of 10 subjects were obtained for altogether 20 different sets of climatic conditions. Measurements of local climate disturbances with the man–sized thermal manikin are well correlated with the thermal sensation experienced by subjects exposed to the same conditions. The manikin method represents a quick, accurate and reproducible technique for relevant, reliable and cost–effective assessment of many complex indoor climate details and their integrated effect on humans.

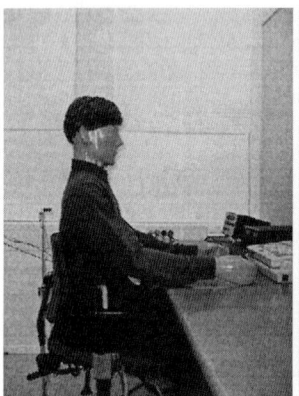

Figure 1. Picture showing manikin and computer in the office room to the left. In the middle picture the white diffuser at 0.2 m height and the grey squared mixing grill to the right side at 2.9 m and the white round exhaust at 2.9 m to the left. The right picture shows the air supply to the room from the corridor side.

The manikin was positioned in the environment that should be assessed, an office room at the National Institute for Working Life. Heat flow from the different segments of the manikin surface was measured and controlled by a computerised system. Data for the actual conditions were recorded when the manikin had reached heat equilibrium with the environment. The office rooms at the institute normally have mixing ventilation but this room has been equipped with displacement ventilation for this study. The following four cases were studied:

MIX: Mixing ventilation with 135 l/s air flow inlet and outlet 2.9 m above the floor.
DIS23: Displacement ventilation with 23 l/s air flow from the supply.
DIS40: Displacement ventilation with 40 l/s air flow from the supply.
DIS55: Displacement ventilation with 55 l/s air flow from the supply.

The diffuser choice was made with the ABB company program WinDon - Version 1.031. The program recommended a Floormaster FMC-603 with an air flow of 48 l/s at an room air temp of 24 °C and an inlet temperature of 19.4 °C. Supply air was distributed from a cooler located in the ceiling of the corridor (Figure 1). Temperature and air speed measurements were made with two Brüel & Kjær Indoor Climate

Analyzer 1213. The radiators were turned off, and windows and radiators where covered with 50 mm foam plastic.

Calculation Procedure

The numerical methods used for air flow and heat transfer predictions are described in the manuals of the CFX-4.2 suit of software (AEA Technology, Harwell, OXON, UK). A virtual manikin has been formed with a set of blocks of cubic shape and the same size and number of zones and areas as the real manikin (Figure 2). The real manikin as well as the block manikin has a free surface area of 1.6 m^2, giving the same constant heat flux to the surrounding air.

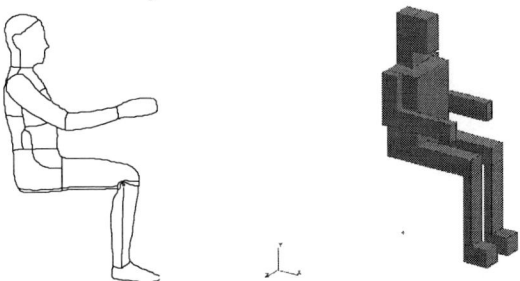

Figure 2. The geometry of the real manikin (left) and the virtual manikin (right) with 18 zones at the same locations. Co-ordinate representation (**x, y, z**).

The virtual manikin was positioned in geometry of the numerical test room (2.9, 3.0, 3.8 m) that simulates a real office room is shown in (Figure 1). Adiabatic ceiling, wall and floor conditions where assumed. Ventilation air was supplied from a simulated FMC-603 diffuser with a supply area of 0.15 m^2. The exhaust outlet was on the same wall, and had an area of 0.04 m^2. The dissipation length scale at the inlet was set to the height of the inlet 0.3 m for displacement and 0.1 m for mixing ventilation. The outlet was positioned at the back wall, se Figure 1 (middle). Boundary conditions for the CFD calculations are shown in Table 1 and 2. The incompressible buoyant flow field was calculated with three-dimensional CFD. The turbulence model used was the low-Reynolds-number k-ε model (Launder and Sharma, 1974). The body fitted grid used had a global edge length of 0.1 m, generating 32686 cells. Calculations was made with 5000 iterations.

RESULTS

Four different cases have been studied both with the manikin measurements and numerical simulations. The results from the numerical simulations are shown below (Figure 3, 4). The heat loss from the manikin zones was measured for all cases (Table 1). Temperatures at the centre of the office room as well as at the different supply inlets and exhaust outlets were measured continuously during the measurements (Table 2).

Table 1. Heat loss data (W/m^2) from a selection of manikin segments during the exposure to the four different cases. (L = Left, R = Right).

Heat loss (W/m^2)	MIX	DIS23	DIS40	DIS55
Total	**44.9**	**50.0**	**55.2**	**57.0**
Face	77.5	85.5	90.5	98.6
L hand	85.3	81.0	91.1	95.4
R hand	87.4	89.7	99.9	103.2
L foot	44.7	53.3	57.3	59.4
R foot	42.8	53.8	56.6	57.8

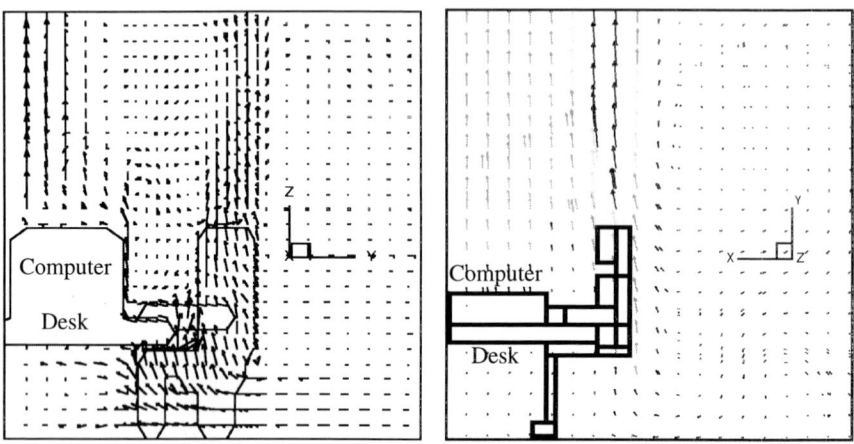

Figure 3. The work place flow structure with a person (manikin) working with a computer (70W) calculated with the uncomercial code (Holmberg et. al., 2000) with k-ε turbulence model (left) and in this paper with CFX 4.2 with low-Reynolds-number k-ε turbulence model (right). Viewed from the window side, both after 10000 iterations and without radiation calculations.

Table 2. The measured input data (°C and W/m^2) used in the CFD calculations.

Flow case	Room air at manikin (°C)	Inlet air (°C)	Outlet air (°C)	Turbulence intensity inlet	Turbulence intensity outlet
	Measured	Measured	Measured	Measured	Measured
MIX	23.6	21.2	23.8	10%	19%
DIS23	21.3	14.1	23.0	13%	33%
DIS40	20.3	15.1	22.5	11%	33%
DIS55	20.1	16.4	22.2	7%	17%

In Figure 4 below, the cooled air from the supply is distributed over the floor (left) and reaches the manikin and computer where a plume is formed by the buoyancy forces (right) and the air is elevated to the exhaust level. With higher air flows the plume effect became less pronounced due to air disturbance around the manikin.

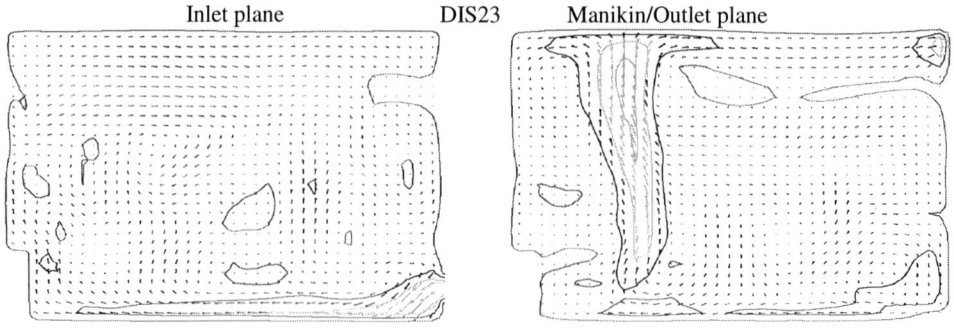

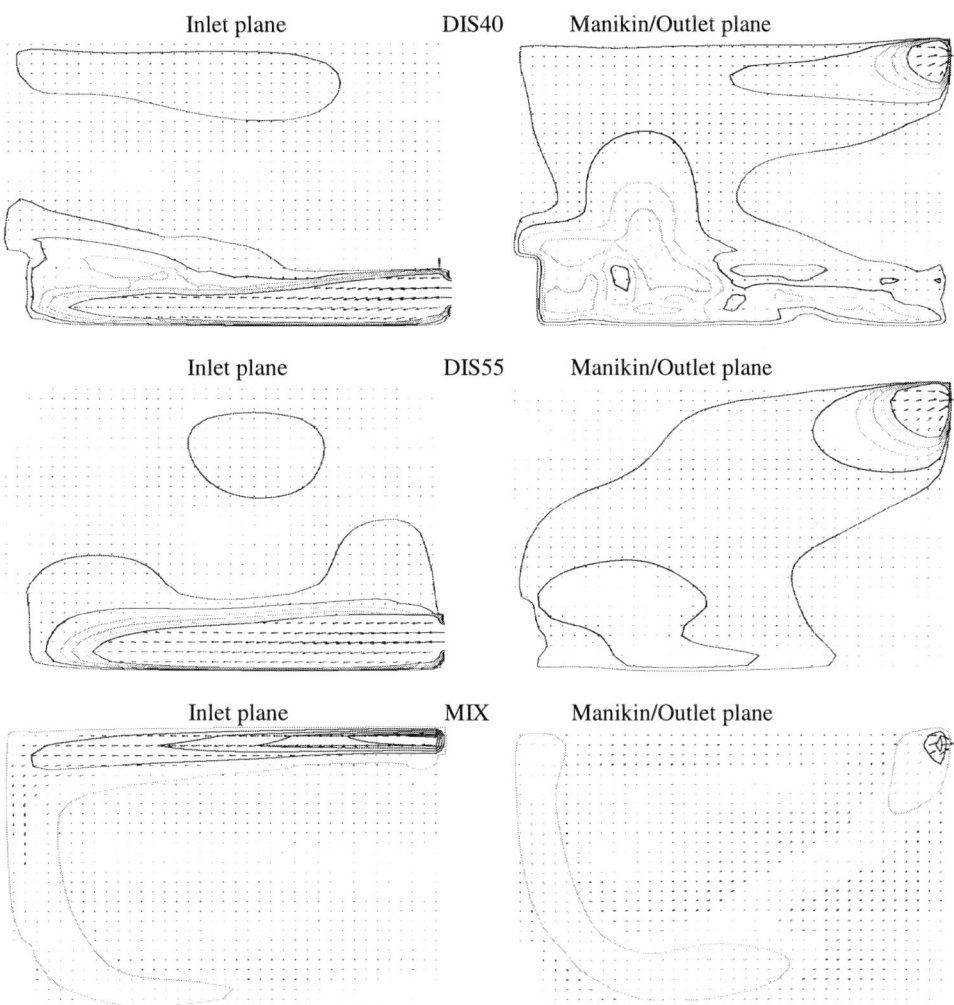

Figure 4. The flow patterns in the office room shown as velocity vectors and contours in a plane at the supply centre (left) as well as 0.1 m behind the manikin (right) for all four flow cases. With mixing ventilation air is blown right across the room at supply level and breaks down to the floor level at the opposite wall, then slowly elevate passing the manikin through the room up to the exhaust.

Table 3. Measured air velocities and temperatures compared to CFD calculated values.

Flow case	Measured	Calculation	Flow case	Measured	Calculation
MIX (°C)	23.6	22.7	DIS40 (°C)	21.9	20.6
MIX (m/s)	0.04	0.06	DIS40 (m/s)	0.06	0.06
DIS23 (°C)	22.5	21.6	DIS55 (°C)	21.3	20.1
DIS23 (m/s)	0.04	0.05	DIS55 (m/s)	0.09	0.07

112

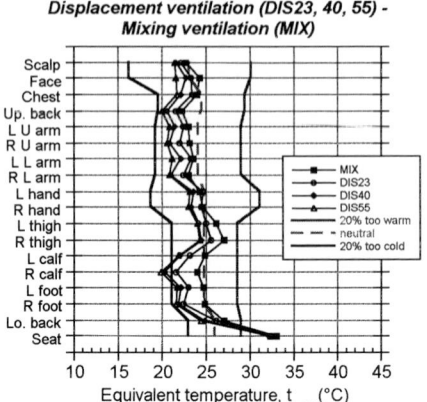

Figure 5. Comfort diagram with all four cases.

The data in figure 5 from the manikin (AIMAN) clearly shows that an increased flow rate combined with low inlet temperatures are well detected by the manikin. The values for the calves and feet are decreasing to finally reach the limit were more than 20% of the occupants would feel too cold.

DISCUSSION AND CONCLUSIONS

A mixing-ventilated office room has been equipped with displacement-ventilation and evaluated by numerical simulations and supporting measurements. Interactions between air movements and thermal comfort conditions in the office room have also been studied. The results show good agreement with the measurements made in the real environment. However, further development of the simulation of a human or manikin is needed. The manikin model should also include the same thermal behaviour as the real manikin. Future research should aim at evaluating these results in order to find more accurate methods for evaluating ventilation and office room design. CFD (Computational Fluid Dynamics) tools will be useful for this work.

ACKNOWLEDGEMENTS

This work was financially supported by the Swedish Council for Building Research (BFR). Project collaboration and support with ventilation installations has been undertaken by ABB Ventilation Products.

REFERENCES

ASHRAE. Physiological Principles, Comfort and Health. (1989). *ASHRAE Handbook*, Atlanta, 8.10-8.32.
Holmberg S, Sandberg M, Nilsson H, Holmér I (2000). Indoor air quality and climate control Parameters in Office Environment – CFD Calculations and Measurements. *Roomvent'2000*, Reading, UK.
Holmér I, Nilsson H, Bohm M, Norén O (1995). Thermal Aspects of Vehicle Comfort. Appied Human Sciences 14, 159-165.
Launder, B.E. and Sharma, B.I. (1974), Heat and Mass Transfer, Vol. 1, p. 131.
Lund-Madsen, T. and Olesen, B. W., (1986), A new method for evaluation of the thermal environment in automotive vehicles, *ASHRAE Trans 92*, part 1B, 34-38.
Nilsson H, Holmér I, Bohm M, Norén O. (1999) Definition and theoretical background of t_{eq}. Florence: *Associatione Tecnica Dell'Automobile*, 1999A4082
Wyon, D., Larsson, S., Forsgren, B. and Lundgren, I., (1989), Standard procedures for assessing vehicle climate with a thermal manikin, *SAE–Technical Paper Series*, 890049.

© 2000 Elsevier Science Ltd. All rights reserved.
Air Distribution in Rooms, (ROOMVENT 2000)
Editor: H.B. Awbi

THE AIR QUALITY AT THE BREATHING ZONE WITH DISPLACEMENT VENTILATION

H. J. Xing[1], A. Hatton[2], H.B. Awbi[3]

[1] Whitby Bird & Partners, London W1P 4DT,UK
[2] Arvin Exhaust Research and Development
Hillock Lane, Warton, Lancs. PR4 1TP, UK
[3] Department of Construction Management & Engineering, The
University of Reading, Reading RG6, 6AW, UK

ABSTRACT

This paper presents the difference in the air quality between that perceived by the occupants (breathing zone) and that in the occupied zone as a whole. An environmental chamber with displacement ventilation system has been used to carry out the measurements with the presence of a heated mannequin and other heat sources. Measurements of the age of air distribution, the air exchange index and the ventilation effectiveness were carried out at different points in the chamber for different room loads. CFD simulations were also carried out for the purpose of flow visualisation as well as the calculation of air velocity, temperature and age of air distribution. The results from the CFD simulations were compared with those from measurements.

KEYWORDS

Indoor air quality, Breathing Zone, CFD, Displacement Ventilation

NOMENCLATURE

$C_p(t)$	Concentration at a point p at time = t	$\varepsilon, \varepsilon_p$	Ventilation effectiveness
$C(0)$	Concentration at time = 0	$\bar{\tau}_p$	The local mean age of air at point p
$C_e(t)$	Concentration at the exhaust at time = t	$\langle \bar{\tau} \rangle$	The room mean-age of air
$C_s(t)$	Concentration at the supply at time = t	τ_n	Nominal time constant = room
E_p	Local air change index		volume/air flow rate

INTRODUCTION

Displacement ventilation (DV) is widely used in mainland Europe and is also gaining popularity in the UK. In DV systems, cool air supplied at low level is entrained by plumes rising from heat sources. In the case of a body plume, the air that is entrained by the plume rises to the occupant's head and subsequently breathed by the occupant, Stymne et al (1991). As a result, the air quality at the breathing zone (the nose) is expected to be better than that in other parts of the occupied zone, e.g. Brohus and Nielsen (1994) and Hatton, et al (1999). Although a considerable amount of research work has previously been done on DV systems, there has not been any major study of the air quality at the breathing zone.

This paper presents results of the local mean age of air, the air exchange index and the ventilation effectiveness in an environmental chamber using three types of DV units. Temperature distribution and air velocity were compared with the results from CFD simulations.

EXPERIMENTAL SET-UP

The experiments were carried out in the environmental chamber at the University of Reading. The chamber was ventilated by a single low level wall DV unit or two circular floor units. The chamber consisted of two compartments and was equipped with an open ventilation system which draws air from the laboratory and exhausts the contaminated air out of the building. The dimensions of the test compartment were 2.78m (length) x 2.78m (width) x 2.3 m (height). To provide a realistic office situation a light (36W fluorescent), a computer simulator (150W), a desk and a chair were placed in the chamber. Two heated plates of variable heat output giving 95W and 180W were used to represent areas of solar illuminance on a side wall.

A heated mannequin was constructed from 1mm aluminium sheet with an overall surface area of $1.60m^2$. Heating elements inside the body, head and legs of the mannequin were controlled to provide a surface temperature equal to that of a typical naked human, Olesen (1982). A polyurethane tube was attached to a copper tube (the nose) inside the head and fed through the torso and out to the gas sampler. This location represented the sampling point for the breathing zone. Throughout the tests the mannequin was unclothed to avoid possible interference of clothes with the body plume.

Different distributions of these heat sources were used to simulate an office with different heat loads. The total heat load for the experiments were varied from 140 W to 502 W, which represent typical heat loads in an office of this size.

Three types of DV units were used in the tests: DV unit 1, a flat-faced wall unit with a size 0.5m x 0.5m. DV unit 2, a semi-circular wall unit with a size of 0.5m x 0.5m high and a radius of 0.25m and two DV unit 3, which was a floor swirl unit of 0.15m in diameter.

Four-wire Platinum Resistance Thermometer has been used to measure the air temperature, the mannequin temperature and the inside and outside surface temperatures of the chamber. Other measuring devices used in the tests were an accurate Wattmeter, DANTEC omnidirectional velocity sensors and a Bruel and Kjaer SF_6 gas sampling system. 12 gas sampling tubes, velocity and temperature sensors were positioned at different points in the room. Table 1 lists the sampling points and sensor locations in the chamber and Figure 1 shows the location of measuring stands.

EXPERIMENTAL CONDITIONS AND MEASUREMENTS

Table 2 lists the 12 configurations and the room ventilation loads that were investigated with the emphasis on the air quality at the breathing zone of the mannequin. The ventilation load used here is defined as the total heat load subtracting the heat loss through the wall by conduction. Full details of the test condition for each DV unit can be found in Hatton, et al (1999).

The first priority in this work was to establish whether there is a difference between the perceived air quality and the air quality in the rest of the occupied zone. Because the local mean age of air at a point represents the time the supply air takes to reach that point, this term has been assumed throughout this paper to represent the quality of the air at that point. This is considered a plausible measure of the air quality since the residence time should be an indicator of the degree of contamination of the air at a point. The tracer decay gas technique was used to determine the age of air at a number of points in the room using SF_6.

TABLE 1

THE GAS SAMPLING POINTS AND SENSOR
LOCATIONS FOR THE TESTS

Stand No. / location	Concentration sample point No.	Velocity and temperature sample points	Height of sample point (m)
1		12, 14, 18	0.18, 0.61, 1.04
2		9	0.14
	3		0.17
		16, 13	0.35, 0.49
	5	3	0.62
		24	1.08
3		4	0.15
	10	17	0.6
	11	23	1.06
4		8	0.6
		22	1.1
5		2	0.15
		21	1.1
6		6, 1, 20, 5	0.15, 0.65, 1.1, 1.23
7	4	7	1.34
8		10, 15	0.12, 0.6
	12	19	1.1
9		11, 18	0.61, 1.04
	7		1.76
10*	3,4,5,7,8,10, 11 and 12		0.14, 0.55, 0.92, 1.18, 1.32, 1.55, 1.8
Breathing zone	9		1.21(Seated) 1.63 (Standing)
Plume of mannequin	8		1.63 (Seated) 1.8 (Standing)
Inlet	6		0.4
Exhaust	2		2.3

This is for tests using a constant emission of SF_6 gas above the heated box to measure the neutral height and the concentration distribution of gas.

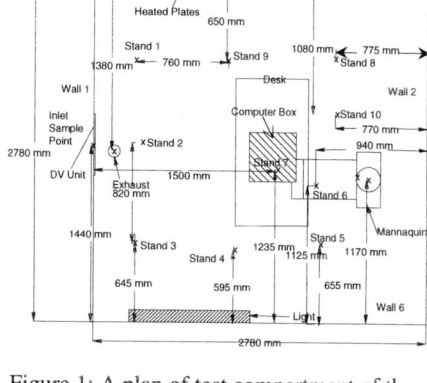

Figure 1: A plan of test compartment of the Reading Chamber and its contents

TABLE 2

CONFIGURATIONS TESTED IN THE CHAMBER

Config. No.	Man. posture	Air change rate (hour^{-1})	Ventilation load (W)			Inlet temp. (°C)
			DV1	DV2	DV3	
1	Seated	5	100	129	136	20
2	Seated	5	185	250	241	18
3	Seated	5	238	299	298	18
4	Seated	5	328	281	318	18
5	Seated	7	369	401	n/a	18
6	Standing	5	175	170	142	20
7	Standing	5	256	247	265	18
8	Standing	5	293	260	302	18
9	Standing	5	329	339	306	18
10	Standing	7	379	430	n/a	18
11	Seated	3.2	300	281	n/a	18
12	Standing	3.2	310	282	n/a	18

The local age of air at a point and the mean age of air in the room can be calculated using the following expressions:

$$\overline{\tau}_p = \frac{\int_0^\infty C_p(t)\,dt}{C(0)} \qquad (1)$$

$$\langle\overline{\tau}\rangle = \frac{\int_0^\infty t C_e(t)\,dt}{\int_0^\infty C_e(t)\,dt} \qquad (2)$$

The local mean age of air was calculated for all the sample points within the working compartment of the chamber for each test condition, and these were averaged to obtain the room mean age. The data from these tests have been analysed and are presented in the result section. The local air exchange index, (the rate of exchange of air at a particular point within the room), E_p, and the local ventilation effectiveness, ε_p, were also calculated using:

$$E_p = \frac{\tau_n}{2\bar{\tau}_p} \qquad (3)$$

$$\varepsilon_p = \frac{C_p(t)}{C_e(t)} \qquad (4)$$

Other tracer gas measurements were carried out using the steady state concentration method to measure the neutral height and the ventilation effectiveness for buoyant pollutant removal. The ventilation effectiveness, ε, was calculated from the measured concentration data using Eqn. (5).

$$\varepsilon = \frac{C_e(\infty) - C_s(\infty)}{C_p(\infty) - C_s(\infty)} \qquad (5)$$

Other measurements carried out during the tests were air temperatures and velocities.

CFD ANALYSIS

Computational fluid dynamics is a useful tool for predicting the diffusion of airborne contaminants in a room. A CFD program called VORTEX (Gan and Awbi, 1994) was used to provide a microscopic prediction of the diffusion of the tracer gas in the whole space as well as air velocity and temperature data. A nominal grid of 80x80x80 was used for the simulation. The CFD simulations were found to be particularly useful for understanding, in greater detail, the air diffusion process in the room.

The CFD code was run for all test conditions listed in Table 2. The age of air, temperature and velocity measurements were compared with the predictions.

RESULTS AND DISCUSSION

The local mean age of air at the breathing zone and in the occupied zone are plotted against the room load in Figures 2 and 3 for DV units 1, 2, and 3 with seated and standing mannequin postures.

Figure 2 shows that the local mean age of air at the breathing zone of a seated mannequin is approximately 50% lower than that in the occupied zone for the flat faced DV unit (DV 1). This varies between 35 - 50% for the three diffusers tested with the smallest difference for DV unit 3, i.e. the floor DV units. This result suggested that the mannequin entrains air from the low fresh air zone into the breathing zone and thus creates better air quality in the perceived zone than that in the rest of the occupied zone. Figure 3 shows the local mean age of air at the breathing zone and the occupied zone for a standing mannequin. The difference between the data for the occupied zone and that for the breathing zone is lower than that for a seated mannequin.

The local air exchange index for a seated mannequin in a room fitted with DV unit 1 is plotted in Fig. 4. It is clear that the local air quality at the breathing zone is better than that at all the other points except those close to the DV units.

The ventilation effectiveness profile in the room has been calculated using the data from the constant emission measurements for obtaining the neutral height, see Xing and Awbi (2000). It can be seen from Fig. 5 that the ventilation effectiveness at the breathing zone is twice that at another point of the same height in the room. This confirms that the mannequin entrains air from the lower displacement air flow zone. Although the profile with DV unit 2 was similar to that with DV unit 1, the effectiveness at a height of 0.55m was much lower than that for the case of DV unit 1. This could be due to the fact that DV unit 2 produced a greater spread of air over the floor. The ventilation effectiveness in the lower part of the chamber was much lower for DV unit 3 than for the other two DV units.

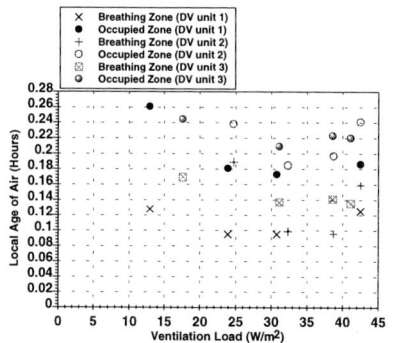

Figure 2: Local mean age of air for a seated mannequin

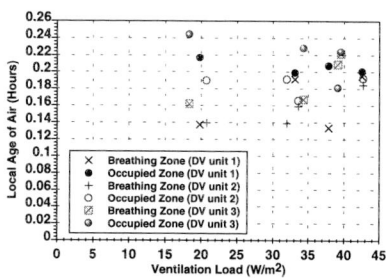

Figure 3: Local mean age of air for a standing mannequin

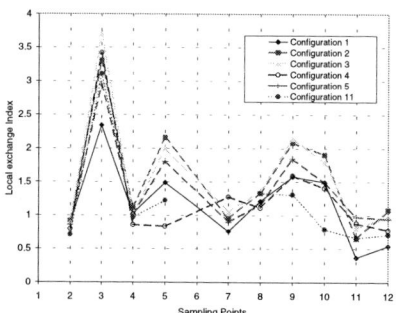

Figure 4: Local air exchange index for a seated mannequin for DV1

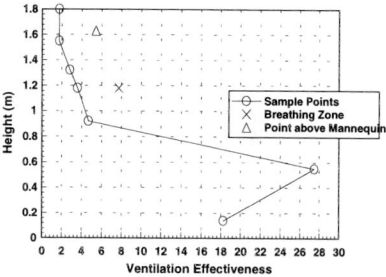

Figure 5: Ventilation effectiveness for DV1 configuration 2 with a seated mannequin

CFD simulations were carried out to compare the results with measurements. Generally, good agreement was found between the CFD prediction and experimental results. Figures 6, 7 and 8 compare the measured mean age of air, velocity and temperature with the CFD results for DV1 (configuration 4). Other comparisons can be found in Hatton, et, al (1999).

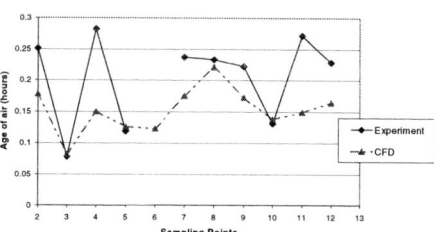

Figure 7: Comparison between measured and simulated velocity for configuration 4, DV1

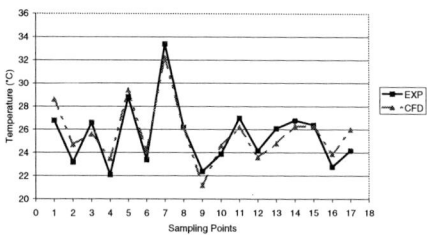

Figure 6: Comparison between measured and simulated age of air for configuration 4, DV1

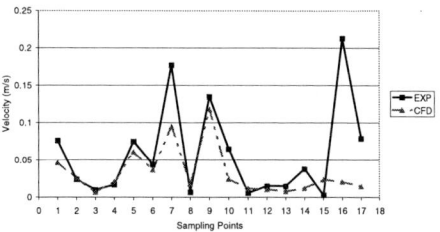

Figure 8: Comparison between measured and simulated temperature for configuration 4, DV1

CONCLUSIONS

The following conclusions can be drawn from the results presented in this paper:

- The perceived (breathing zone) air quality (represented by the mean age of air) for both seated and standing mannequin was better than the average air quality in the occupied zone for three types of DV units.
- The average perceived air quality for a seated mannequin was better than that for a standing mannequin for all three different DV units.
- The air quality in the occupied zone was found to be better for a semi-circular wall DV unit than for a flat wall DV units or floor DV units. Furthermore, the air quality was found to be better for a wall DV unit than for the floor DV units.
- The ventilation effectiveness at the breathing zone for both the seated and standing mannequin was greater than that for a point at the same height in the chamber for all the DV units.
- The local air exchange index was found to be highest close to the DV unit and at the breathing zone and was greater than at most other points in the room.
- A comparison between the CFD simulation results and the measured results showed a good agreement for most positions except those points near the DV unit.

Acknowledgement

The Engineering and Physical Sciences Research Council, UK has supported this work under Grant Reference GR/K89450. This work was also in collaboration with Halton Products Ltd and BSRIA.

References

Brohus, H. and Nielsen, PV. (1994). Contaminant Distribution around Persons in Rooms Ventilated by Displacement Ventilation. *Proc. of ROOMVENT'94*, Cracow, Poland, **1**, 293-312.

Gan, G. and Awbi, H.B. (1994). Numerical Simulation of the Indoor Environment. *Building and Environment*, **29: 4,** 449 - 459.

Hatton, A., Xing, H., and Awbi, H.B. (1999). A Study of the Air Quality in Room Ventilated with Displacement Systems. *Internal Report*, Dept. of Construction Management & Engineering, The University of Reading, UK.

International Standard 7730. ISO (1984). Moderate Thermal Environments-Determination of the PMV And PPD Indices and Specification of the Conditions For Thermal Comfort. International Standards Organisation, Geneva.

Olesen, B.W. (1982). Thermal comfort, *Bruel and Kjaer Technical Review*, **2**.

Stymne, H., Sandberg, M. and Mattsson, M. (1991). Dispersion Pattern Of Contaminants in a Displacement Ventilated Room - Implications for Demand Control, *Proc. 12th AIVC Conference*, Ottawa, 173-189.

Xing, H. Awbi, H.B. (2000). Neutral Height in a Chamber with Displacement Ventilation (to be published).

Air Distribution in Rooms, (ROOMVENT 2000)
Editor: H.B. Awbi

COMPUTATIONAL EXPOSURE AND DOSE ASSESSMENT ANALYSES FOR TRANSIENT TURBULENT FLOW AND GASEOUS POLLUTANT TRANSPORT

S. Hyun and C. Kleinstreuer

Department of Mechanical and Aerospace Engineering, North Carolina State University, Raleigh, N.C. 27695-7910 USA

ABSTRACT

A significant first step of any quantitative program for health impact assessment of air pollutants requires the accurate determination and control of pollutant gas or aerosol concentrations in a realistic test environment.

Of interest are the experimentally validated CFD simulations of transient turbulent air flow and gaseous pollutant transport in a personal exposure environment. Here, personal exposure condition is the pollutant gas concentration surrounding a subject, which is different from personal dose condition of a breathing subject. The purpose of this paper is to examine the effect of transient breathing on ambient trace gas concentrations and uptake concentrations. The system consists of a unidirectional flow chamber and a passive, spherical pollutant gas source, e.g., CO or NO_2. Personal exposure and dose assessment are determined by means of a breathing, non-isothermal manikin positioned in the chamber center and facing different directions. The transient three-dimensional turbulent thermal flow and mass transfer problem has been solved on a multi-processor SGI Workstation (Origin 2000) using CFX 5.3 (AEA Technology) with modified programs.

The modeling results indicate that measurable differences between personal exposure and inhaled dose concentrations can be found, and that advanced CFD simulations and analyses are useful tools for generating the necessary scientific understanding and quantitative information for final decision making in dosimetry-and-health-effect studies.

KEYWORDS

CFD, Transient Turbulent Flow, Exposure Concentration, Dose, Breathing Thermal Manikin

INTRODUCTION

Interest in *indoor* air quality control is rapidly increasing because most urban dwellers spend now more than 90% of their daily life in an artificial environment, e.g., offices, factories, homes, public buildings, vehicles, etc. (cf. Awbi, 1991). Although there are many studies related to outdoor air quality prediction and control, their findings cannot be directly applied to the solution of indoor air pollution problems. Toward the end of the 1970's, it was recognized that there is a potential respiratory hazard in buildings – especially livestock buildings, factories and public buildings which are developed towards larger and more mechanized systems.

Traditional methods for designing air distribution in buildings are mainly based on semi-empirical relations, which do not include the effects of room geometry. Recently, with the advanced computational fluid dynamics (CFD) simulation programs and computer aided design (CAD) tools, the air movement in ventilated enclosures has been studied (Nielsen, 1994; Andersen, 1995). Non-industrial buildings differ from industrial buildings in that indoor air pollution in the latter is directly related to the specific industrial process; thus indoor air pollution changes from factory to factory according to the type of production. The nature of chemical, biological, and physical contaminants present indoors, their sources, and their health effects have been reviewed by Maroni (1998).

Topps et al. (1998) calculated the contaminant distribution reasonably well in rooms with steady flow, with a known contaminant emission, and without people. In experimental studies contaminant distributions in ventilating rooms, especially in the breathing zone for personal exposure were investigated by Topps et al. (1998) and Mattson & Sandberg (1998). The dispersion of contaminants from human exhalation, and the interaction taking place between respiration and the convective boundary layer flow induced by the human body was investigated experimentally as well as computationally (Bjørn & Nielsen, 1996a, 1996b and 1998). Full scale breathing thermal manikins (BTM) were used to study experimentally and computationally the personal contaminant exposure due to interacting air flows between two persons (Bjørn & Nielsen, 1996b). Computational difficulties were reported and CFD simulations were limited to steady-state flow employing the standard k-ε turbulent model and box-like BTMs.

Of interest here are the experimentally validated CFD simulations of transient turbulent non-isothermal flow with gaseous pollutant transport in a personal exposure environment. The purpose of this paper is to examine the effect of transient breathing on trace gas ambient (exposure) concentrations and uptake (dose) concentrations.

THEORY

Computational Geometry
Assuming a unidirectional flow field, the exposure chamber, a manikin, and a pollutant source are modelled as shown in Figure 1 (cf. Brohus, 1997). The flow field and contaminant distribution around a person in a ventilated room depend on the system configuration – e.g., ventilating system, location of the person in the room, movement of the person, local obstacles, heat sources, as well as contaminant sources. The present system generates a uniform air flow field which is acceptable for scientific dosimetry-and-health-effect studies, neglecting local pollutant depletion and/or temporal changes in air delivery.

Governing Equations
The basic assumptions for the CFD simulations include fully turbulent non-isothermal incompressible flow. The industry standard for turbulence simulation is the κ-ε model because of its robustness at a relatively low computational cost; however, it is a poor predictor of flow separation (Durbin, 1994; 1992). Considering the relatively complex geometry of the BTM model and the need for accurate fine-

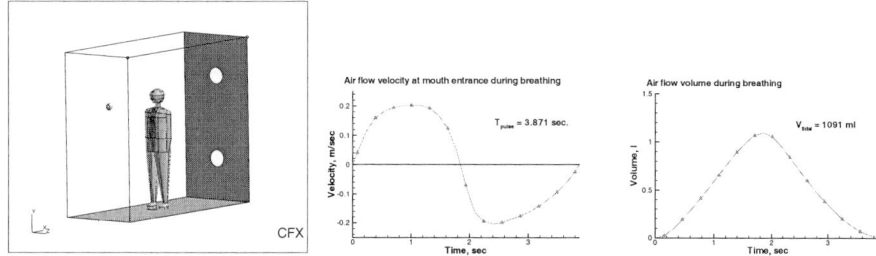

Figure 1: CFD model for a BTM and the breathing waveform

scale modelling, the ReNormalization Group (RNG) κ-ε model, which reportedly can accurately predict flow separation and reattachment (Longest et al., 2000; Durbin, 1994; Speziale et al., 1992), has been selected. The transport equations for incompressible turbulent flow employing the RNG procedure of Yakhot and Orszag (1996) are in tensor notation (Kleinstreuer, 1997):

$$\frac{\partial u_i}{\partial x_i} = 0 \qquad (1)$$

and

$$\frac{\partial u_i}{\partial t} + u_i \frac{\partial u_i}{\partial x_j} = -\frac{1}{\rho}\frac{\partial p}{\partial x_i} + \frac{\partial}{\partial x_j}\left[(v+v_t)\left(\frac{\partial u_i}{\partial x_j} + \frac{\partial u_j}{\partial x_i}\right)\right] + \frac{f_j}{\rho} \qquad (2)$$

where

$$v_l = C_\mu \frac{k^2}{\varepsilon} \quad with \quad k = \frac{1}{2}\overline{u'_i u'_i} \quad and \quad \varepsilon = v\frac{\overline{\partial u'_i}\,\partial u'_i}{\partial x_j \, \partial x_j} \qquad (2b-d)$$

Calculation of the turbulent kinetic energy, k, and dissipation rate, ε, requires two additional partial differential equations (Kleinstreuer, 1997) which contain four unknown coefficients in addition to C_μ of Eqn. 2b. Using the RNG methodology (Speziale et al., 1992; Yakhot et al., 1992), these "constants" are determined as

$$C_\mu = 0.085, \quad C_{\varepsilon 1} = 1.42 - \frac{\eta(1-\eta/\eta_\infty)}{1+\beta\eta^3}, \quad C_{\varepsilon 2} = 1.68 \; \delta_k = 0.7179, \quad and \quad \delta_\varepsilon = 0.7179 \qquad (3a-e)$$

where

$$\eta = S \cdot k / \varepsilon, \quad \eta_\infty = 4.38, \quad \beta = 0.015, \quad and \quad S = \left(2\,\overline{S}_{ij}\overline{S}_{ij}\right)^{1/2} \qquad (4a-d)$$

and $\overline{S}_{ij}$ is the mean rate-of-strain tensor. The Boussinesq model is employed for the calculation of the buoyancy force, f_j, where the fluid model density is not dependent on temperature, pressure or additional variables. The local density variation is defined as

$$\rho' = \rho\{1 - \beta(T - T_{ref})\} \qquad (5)$$

where β is the thermal expansion rate, T_{ref} is the buoyancy reference temperature, and ρ is the fluid density. For the temperature distribution calculations, the energy equation can be written as (Kleinstreuer, 1997):

$$\frac{\partial(\rho h)}{\partial t} + \frac{\partial}{\partial x_j}\left(\rho u_j h - q_j\right) = \frac{\partial p}{\partial t} + \frac{\partial}{\partial x_j}\left(u_i \tau_{ij}\right) \qquad (6)$$

The Reynolds-averaged mass transport equation can be written as (Yakhot et al., 1992):

$$\frac{\partial(\rho Y_{CO})}{\partial t} + \frac{\partial(\rho u_j Y_{CO})}{\partial x_j} - \frac{\partial}{\partial x_j}\left[\left(\rho D_{CO} + \frac{\mu_t}{Sc_t}\right)\frac{\partial Y_{CO}}{\partial x_j}\right] = S \qquad (7)$$

where the turbulent Schmidt number (Sc_t) is of the order of one and S represents the sink/source term used in the gas-uptake analysis.

Numerical Methods

The numerical solution of the Eulerian transport equations were carried out employing a user-enhanced, unstructured finite-volume based program CFX 5.3 (AEA Technology, UK), with fast turn-around times, robustness, and rapid convergence. CFX 5.3 employs an unstructured control volume mesh with triangular meshes on the surface of the geometry, which reproduces complex geometries in the computational finite volume domain. Thus, using CFX 5.3, more complex system geometries can be discretized using CAD software as a pre-processor. There is a wide range of length scales for the chamber with a BTM. For example, at the mouth of the BTM, the cell size is as small as $\Delta h = 0.002m$, while for the open chamber $\Delta h = 2.46m$.

The mesh size of the computational domain used in this study was adjusted until acceptable levels of grid independence of the solution were achieved. The resulting mesh size for the occupied chamber simulations was about 350,000 cells. The computations were performed on an SGI Origin 2000 workstation with multi-processors located at the North Carolina Supercomputing Center (RTP, NC). The computational time required for the steady-state flow simulation was approximately 12 hours.

The chamber inlet boundary condition is assumed to be steady, incompressible, three-dimensional, isotropic turbulent flow with a constant temperature. At the mouth of the pollutant inhaling BTM, a uniform velocity is applied based on the light exercise breathing pattern is applied shown in Fig. 1. The exhaled air has an assumed trace gas concentration of zero.

RESULTS

Model Validation

Brohus (1997) measured the exposure concentration of a trace gas contaminant using an inhaling BTM and calculated the trace gas concentration distribution using a simple box model. The present flow conditions are exactly the same as his experimental conditions but the BTM is more realistically generated. From experimental evaluations of the contaminant exposure concentration around a breathing thermal manikin (Brohus, 1997), it was recommended to use a finer CFD model to represent a more detailed BTM shape.

The inhaled dose concentration is non-dimensionalized via the return concentration which is an average pollution concentration at the chamber exit, i.e., $C_d^* = C_d/C_{ex}$. Figure 2 shows the inhaled non-dimensional dose concentration with and without heat generated by the BTM. The computational data points (C_d^*) agree with the measured data of Brohus (1997). With buoyancy effects, the maximum dose concentration is shown at the higher source height than that for the no-heat case, and the maximum value of the dimensionless dose concentration for the heat-on case is less than that for the heat-off case.

Case Studies

Three different cases are considered to investigate the effects of flow direction on dose concentration.

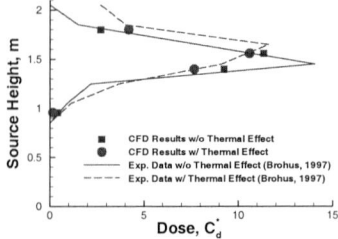

Figure 2: Personal dose of the BTM standing in unidirectional flow field

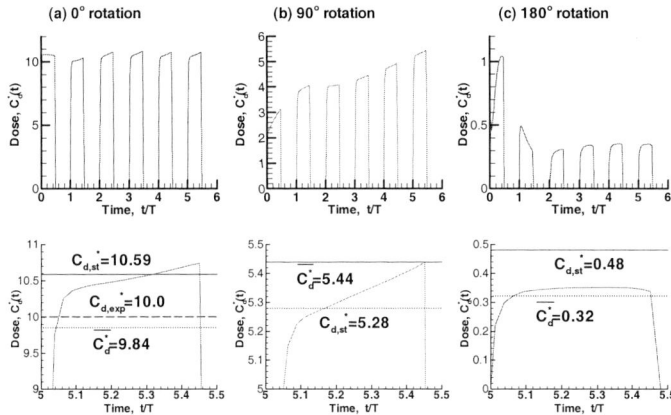

Figure 3: Transient personal dose of the BTM

Considering realistic inhalation, the transient non-dimensionalized dose concentration, $C_d^*(t)$, is shown for: (i) axial air flow moving towards the face of the BTM (Fig. 3a), (ii) axial air flow moving towards the side of the face of the BTM (Fig. 3b), and (iii) axial air flow moving towards the back of the BTM (Fig. 3c). For the 0° rotation case (Fig. 3a), the trend of $C_d^*(t)$ does not change much with time because the strong axial flow is supplied constantly to the BTM's face. At the beginning of inhalation, $C_d^*(t)$ starts with a very low value and increases because some of the exhaled air flows back; however, most of the inhaled air is supplied from the chamber inlet. Hence, the time-averaged dose concentration, $\overline{C_d^*} = \frac{1}{T}\int C_d^*(t)dt$, is less than the dose concentration for the uniform inhalation case, $C_{d,st}^*$. For the 90° rotation case (Fig. 3b), the $C_d^*(t)$ is increasing with each cycle. However, due to buoyancy, air with lower trace gas concentrations moves upward reaching the BTM's mouth and as a result, the $C_d^*(t)$ is lower than in the previous case. For the 180° rotation case (Fig. 3c), the $C_d^*(t)$ is relatively high at the beginning and converges to the quasi-steady state concentration because recirculating flow exists in front of the face. Due to the buoyant flow and wake formed behind the body, air moving upward is stronger than that for the 90° rotation case and the air with lower trace gas concentrations flows towards the BTM's mouth. The pollution concentration in the wake region decreases due to the mixing of inlet flow with the exhaled air and buoyant airflow of near-zero trace gas concentrations. The significant reduction in $C_d^*(t)$, $\overline{C_d^*}$ and $C_{d,st}^*$-values for Case (c) in the presence of a point source, is mainly due to a combination of "clean air" streams in the wake region which in part is also the breathing zone. Obviously, more simulations are needed to reach to a quasi-steady distribution for the 90° rotation case.

DISCUSSION

Previous CFD studies related to indoor air quality prediction (Bjørn & Nielsen, 1996a, 1996b and 1998; Brohus, 1997), were based on steady state simulations and required more advanced turbulence modeling as well as finer mesh generation. In this paper, the RNG κ-ε model is applied to simulate the turbulent flow and unstructured finite volume meshes are used to represent the detailed modeling of a BTM.

There is a significant difference between exposure concentration and dose concentration as well as between steady-state and transient dose concentrations. Usually, the transient dose concentration is lower than the steady-state dose concentration; however, it depends upon the surrounding flow conditions. Steady-state dose concentrations can be used as the exposure concentration, which is not affected by exhalation. If there are attached recirculation or wake regions at the face or a buoyancy influence, it is

124

necessary to simulate the system until a quasi-steady dose pattern is reached, i.e., steady-state simulations cannot be applied.. Thus, further detailed CFD modeling is necessary for indoor air quality simulations as well as for scientific dosimetry-and-health-effect studies.

REFERENCES

Andersen, M. (1995) *Particle movements in mechanically ventilated piggeries*, PhD. Thesis, The Royal Veterinary and Agricultural University, Copenhagen, Denmark.

Bjørn E. and Nielsen, P.V. (1998) CFD Simulations of Contaminant Transport between Two Breathing Persons, *Roomvent '98*, **1**, 133-140.

Bjørn E. and Nielsen, P.V. (1996a) Exposure due to Interacting Air Flows between Persons, *Roomvent '96*, **1**, 107-114.

Bjørn E. and Nielsen, P.V. (1996b) Passive Smoking in a Displacement Ventilated Room, *Roomair '96*, **1**, 887-892.

Brohus, H. (1997) *Personal Exposure to Contaminant Sources in Ventilated Rooms*, Ph. D. Thesis, Aalborg University, Denmark.

Durbin, P. A. (1994) Separated Flow Computations with the κ-ϵ-v^2 Model, *AIAA Journal*, **33:4**, 659-664.

Kleinstreuer, C. (1997) *Engineering Fluid Dynamics – An Interdisciplinary Systems Approach*, Cambridge University Press, New York, NY.

Longest, P. W., Kleinstreuer, C. and Kinsey, J. (2000) Turbulent Three-Dimensional Air Flow and Trace Gas Distribution in an Inhalation Test Chamber, *ASME Journal of Fluids Engineering* (**accepted**).

Maroni, M. (1998) Health Effects of Indoor Air Pollutions and Their Mitigation and Control, *Radiation Protection Dosimetry*, **78:1**, 27-32.

Mattson, M. and Sandberg, M. (1998) Displacement Ventilation – Influence of Physical Activity, *Radiation Protection Dosimetry*, **78:1**, 78-92.

Nielsen, P.V. (1994) *Computational fluid dynamics in ventilation*, Aalborg University, Denmark.

Speziale, C. G. and Thangam, S. (1992) Analysis of an RNG Based Turbulence Model for Separated Flows, *International Journal of Engineering Science*, **30:10**, 1379-1388.

Topps, C., Nielsen, P.V., Heiselberg, P., Sparks, L.E., Howard, E.M., and Mason, M. (1998) Experiments on Evaporative Emissions in Ventilated Rooms, *Roomvent '98*, 499-505.

Yakot, V. and Orzag, S. A. (1996) Renormalization Group Analysis of turbulence. I. Basic Theory, *Journal of Scientific Computing*, **1**, 3-51.

Yakot, V., Orszag, S. A., Thangam, S., Gatski, T. B. and Spezial, C. G. (1992) Development of Turbulence Models for Shear Flows by a Double Expansion Technique, *Physics of Fluids*, **A.4:7**, 1510-1520.

Air Distribution in Rooms, (ROOMVENT 2000)
Editor: H.B. Awbi

HORIZONTAL COLD AIR JETS INDUCED BY A FAN COIL HVAC SYSTEM AND INDOOR COMFORT

E.Rutman[1] , C.Inard[2] , A.Bailly[1] and F.Allard[2]

[1] CIAT, Avenue Jean Falconnier, 01350 Culoz-France
[2] LEPTAB, Université de la Rochelle, Avenue Michel Crepeau, 17042 La Rochelle Cedex 1 France

ABSTRACT

We present an experimental study on a real HVAC system and its optimisation to reduce local discomfort in a climate chamber. First, we analyse discomfort due to thermal effect, acoustic effect and air diffusion effect in accordance to the air flow characteristics. Lastly a general approach of the global comfort is presented.

KEYWORDS

Thermal Comfort, acoustic comfort, air quality diffusion, experiment, HVAC system

INTRODUCTION

The occupants of offices are submitted to various local discomforts, Fanger (1970) due to temperature, draft and noise. In order to improve the comfort, we can use different kinds of HVAC systems. In summer a cold air jet is supplied in the office, like the resulting air speed, temperature distribution and noise distribution within the occupied zone is as uniform as possible to minimise local discomfort. The purpose of this experimental work is to rely the cold air jet characteristics and a global approach of comfort, which takes into account thermal and acoustic comfort including air diffusion.

EXPERIMENTAL APPARATUS AND CONDITIONS

The measurements were carried out in the CIAT's laboratory located in Culoz, France. Figure 1 presents the longitudinal section of our experimental test chamber. The volume is equal to 5.2x4.05x2.5 m^3 and bounded on five sides by air volume regulated at a constant temperature level. The sixth side is submitted to the influence of a climatic housing, where we can simulate external air temperature variation. For these experiments, the value of the external temperature was set at + 30 °C. Furthermore, we simulate the presence of real internal heat loads, including computer and human bodies. These heat loads are balanced by a fan coil unit (1.2 m long and 0.6 m length) mounted on the ceiling. The dimensions of the fan coil unit supply grille are 1.19 x 0.091 m². The air velocity, the relative intensity of velocity turbulence and temperature in the occupied zone of the room are measured with a thermoanemometric sensor, (type DANTEC 54T21) placed at 76 different locations in the occupancy zone. Vertical test point located at 0.1m for foot, 0.6 m for hip, 1.1 m for head of a sitting occupant and 1.7 m for head of a standing occupant, ASHRAE (1992). This probe has been calibrated by DANTEC. The inner wall surface temperatures are measured with thermocouples type K calibrated at the CIAT laboratory. The acoustical level is measured with a microphone sensor type Bruel & Kjaer 4189 and a frequency analyser type Kontron FFT-AD3524. To calibrate the microphone, we used a pistonphone type Bruel & Kjaer 4231. For acoustic measurements, we placed our probe at 36 different locations in the occupancy zone.

Figure 1 : Longitudinal section of the experimental set-up.

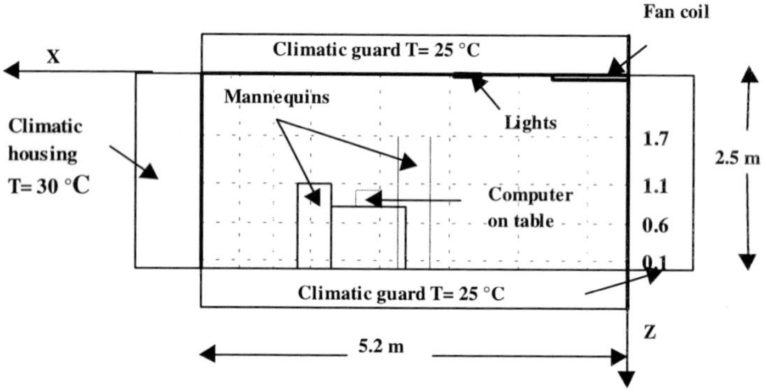

We carried out 6 tests with various supply conditions corresponding to two mechanical configurations of the fan coil. With the first one, the air jet has one supply direction, for the second one, we have three directions for the airflow. Secondly we modify the airflow rate Q_0. The cold water temperature has been set in order to get an operative temperature in the occupancy zone equal to 24.5°C ± 1.5°C. Table 1 gives the parameters used for the experiments.

TABLE 1
EXPERIMENTAL CONDITIONS

Test N°	Number of Supply direction	Q_0 m³/h	U_0 m/s	T_0 °C	ΔT_0 °C	Ar_0 x10³	Re_0
1	1	732	5.5	14.1	9.1	0.91	30517
2	1	537	4.5	11.0	12.5	1.87	24969
3	1	366	3.0	12.5	12.1	4.05	16646
4	3	722	5.6	11.8	11.8	1.14	31072
5	3	532	4.2	12.1	12.7	2.17	23304
6	3	366	2.9	11.5	13.5	4.85	16091

The Archimedes number, $Ar_0 = g\beta\Delta T_0 A_0^{0.5}/U_0^2$, and the Reynolds number $Re_0 = \rho U_0 A_0^{0.5}/\mu$ are used to characterise the supply conditions. We used the square root of the supply area (A_0) because the ratio effective length (l_0) / effective width (h_0) is equal to 13. This value is characteristic of a three dimensional jet. Measurements on the jet airflow including mean values of the inlet temperature, T_0 was measured with a Newport Omega, 0.051mm diameter, K-type thermocouple calibrate at the CIAT laboratory. For the velocity U_0, a Dantec laser Doppler anemometer unidirectional system was used.

RESULTS AND DISCUSSION

The results presented here consist in the ISO noise level, ISO (1971), the ADPI index, Koestel (1955), the floor temperature, the vertical air-temperature difference between 0.1m and 1.1 m and between 0.1 m and 1.7 m, the DR index, ISO (1995)and the PPD index, ISO(1995).

Noise level

For this kind of office we can accept a noise level ISO value between 30 dB and 35 dB, ISO (1971). Table 2 shows the number of location in the test room where the noise level comfort is satisfy. For test n° 3 and n° 6 we have always a good acoustic comfort, for test n° 2 and n° 5 we have only 50 % of location with an acceptable level and for test n° 1 and n° 4 there is no location with an acceptable level. The figure 2 shows the evolution of the noise level between the fan coil and the location in the office. In accordance with the figure 2, evolution of the ISO level with the distance, d are given by :

For tests 1 and 4 ISO = (-3.48/Log(2)) Log(d)+A_1, the correlation coefficient is 0.82   (1)
For tests 2 and 5 ISO = (-3.25/Log(2)) Log(d)+A_2, the correlation coefficient is 0.92   (2)
For tests 3 and 6 ISO = (-3.11/Log(2)) Log(d)+A_2, the correlation coefficient is 0.73   (3)

In our confined space of a 50 m^3 volume, we find a decay of 3.1 dB to 3.5 dB (see equations 1, 2 and 3) when the distance is double. In unbounded space this decay is 6 dB. Obviously we notice that the noise level increases with the airflow rate and decreases with the distance.

TABLE 2
NUMBER OF LOCATION WHERE THE NOISE LEVEL COMFORT IS SATISFY

	ISO level range		
Test n°	< 30 dB	[30 dB ; 35 dB]	< 35 dB
1	0%	0%	100%
2	0%	50%	50%
3	94%	6%	0%
4	0%	0%	100%
5	0%	50%	50%
6	94%	6%	0%

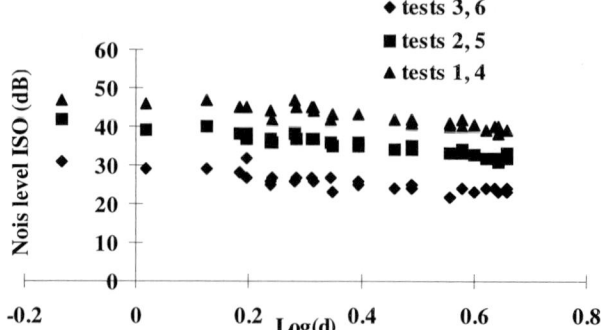

Figure 2 : Evolution of the ISO noise level with the distance between the fan coil
and the location in the office

Air Performance Index

The measured Effective Draft Temperature EDT, Koestel (1955) and ADPI index distribution in the climate chamber are shown on figure 3. The value of the ADPI fluctuates with the mechanical configuration of the fan coil. If we compare tests with a same supply flow rate, but not the same mechanical parameter, the APDI value is roughly divided by 2.

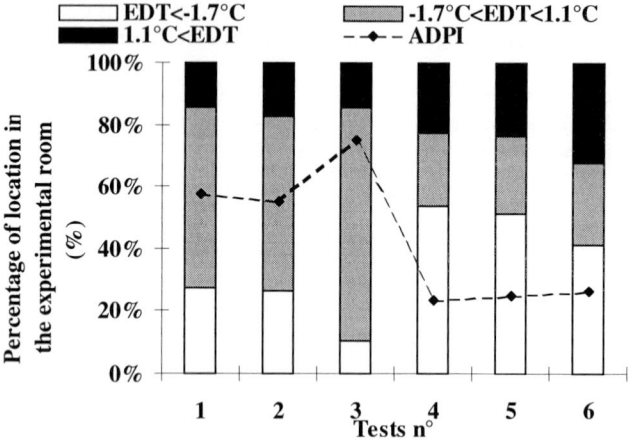

Figure 3 : EDT distribution and ADPI values

Thermal comfort

For our experiments, the percentage of location which satisfy the ISO 7730 for the floor temperature and the air-temperature difference between the head (1.7 m or 1.1 m) and the foot (0.1m) are always 100 %. Table 3 presents the number of location in the test room where we can satisfy the ISO 7730, ISO (1995) standard recommendation for the DR index and the PPD index. When the Archimedes number increases the PPD index decreases while the DR index increases. It's certainly due to the variation of the separation distance of the cold horizontal wall jet.

TABLE 3
PERCENTAGE OF LOCATION FOR THE ISO 7730 STANDARD CRITERIA

Test n°	Percentage of location for the ISO 7730 standard criteria (%)	
Test n°	PPD <= 10 %	DR <= 15 %
1	76	68
2	78	79
3	46	92
4	74	51
5	72	71
6	50	88

A global approach of comfort

Figure 4 shows evolution of average values in the occupancy zone of PPD index, DR index and ISO level, but also evolution of ADPI index. With the Archimedes number as reference, each comfort parameters has his own evolution. Thus the PPD index and ADPI index increase when the ISO level and the DR index decrease. So, we point out the difficulties to optimise the reduction of the local discomfort with occupant's preferences. In fact the best way to analyse these kinds of experiments seems to be a multicriteria analysis.

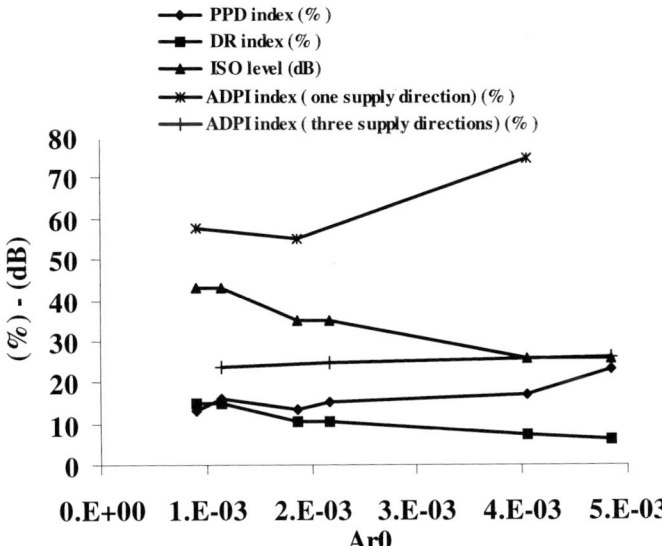

Figure 4 : Evaluation of global comfort criteria

CONCLUSION

Experimental studies of local discomfort in a test cell were studied under various conditions. A first analysis carried out on the thermal comfort and the quality of air diffusion showed the influence of the inlet Archimedes number and the separation distance of the cold horizontal wall jet. We proposed a relation between the noise level and the distance between the location and the fan coil. Lastly, the experimental studies on the comfort not seem to be related by a general law. Thus this work perspectives are the use of multicriteria analysis including the comfort indexes discuss previously.

ACKNOWLEDGEMENTS

This study was supported in part by the Research and Technology National Agency (ANRT), the Environment and Energy Management Agency (ADEME) and Electricity of France (EDF).

REFERENCES

ASHRAE (1992). Thermal environmental conditions for human occupancy. *ANSI/ASHRAE standard* **55**, Atlanta, U.S.A.

FANGER P.O. (1970). *Thermal comfort analysis and applications in environmental engineering*, Mc Graw Hill, New-York, U.S.A

KOESTEL A and TUVE G (1955). Performance and evaluation of room air distribution systems. *ASHRAE Trans.***61**, 533-550.

ISO. (1971). Assessments of noise with respect to community respons. *Technical comitee ISO/TN43* Geneva Switzerland

ISO. (1995). Moderate thermal environments. Determination of the PMV and the PPD indices and specifications of the conditions for thermal comfort. *ISO standard* Geneva Switzerland

Air Distribution in Rooms, (ROOMVENT 2000)
Editor: H.B. Awbi

EVALUATION OF THERMAL COMFORT AND LOCAL DISCOMFORT CONDITIONS BY NUMERICAL MODELLING OF HUMAN AND CLOTHING THERMAL SYSTEMS

E. Z. E. Conceição

UCEH - Universidade do Algarve - Campus de Gambelas - 8000 Faro - Portugal

ABSTRACT

The numerical program presented in this work, that simulates the human and clothing thermal responses, can be used to evaluate the global thermal comfort, in steady state and transient conditions, and the local discomfort sensations. This model is sub-divided in six parts: the human body thermal (1) and thermoregulatory (2) systems, clothing thermal system (3), the global thermal comfort in steady state (4) and in transient (5) conditions and the local discomfort (6). In this model the human body is divided in 35 cylindrical and spherical elements, being each one sub-divided in 12 slices, could be protected of the external environment through some clothing slices.

The model, after being validated, will be used to evaluate the global thermal comfort in steady state conditions and the local discomfort, for a non uniform environment, in a climatized room with a forced air system and a heated window by the solar radiation, and the global thermal comfort in transient conditions when a person enters or leaves a compartment during a Summer typical day. The non uniform air velocity and temperature field will be measured and the radiant temperature, in each element, will be calculated numerically. To the remaining variables will be used typical values.

KEYWORDS

Thermal comfort, Local discomfort, Human thermal system, Clothing thermal system, Thermoregulatory system, Thermoreceptors, Steady state and transient conditions, Energy and mass balance integral equations.

INTRODUCTION

If the environment conditions are constant the steady state thermal comfort level, that a person is subjected, evaluated by the PMV index, is calculated in accord to Fanger (1970), through the heat flux exchanged between the human body and the environment. In numerous situations the environmental conditions are not constant. In these situations the global thermal comfort level, in transient conditions (see Dear *et al.* (1993)), is evaluated through the static (constant temperature) and dynamic (temperature change) response of the warm and cold thermoreceptors located in the human body skin. Nevertheless, an occupant can express satisfaction with the environment, but feel draught sensation. This local discomfort is also evaluated by the warm and cold thermoreceptors response located in a human body skin section (see Fanger *et al.* (1988) and Ring *et al.* (1993)). To reproduce the instantaneous temperature change in the skin, in transient conditions, and calculate the temperature evolution in the warm and cold thermoreceptors, used to evaluate the thermal sensation, Dear *et al.* (1993) presented a model where the skin was divided in 36 layers.

In the model presented in this work the human body is divided in 35 cylindrical and spherical elements (see Conceição (1998)), being each one sub-divided in 4 parts: core (1 slice), muscle (2 slices), fat (2 slices) and skin (7 slices) (see figure 1). Each element could be still protected of the external environment through some clothing slices. It was verified that the skin divided in 7 slices represented a good compromise between the computational calculus time and the reproducibility of the skin thermal behavior.

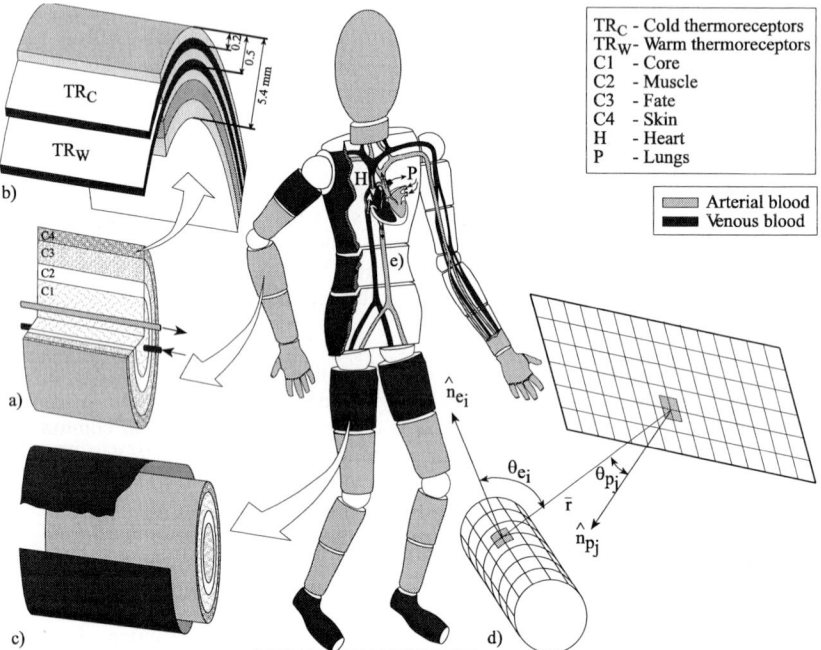

Figure 1: Scheme of the division of the human body in 35 elements and each one in 4 parts a). Division of the skin in 7 slices b) and the clothing in several slices c). Scheme used to calculate the view factors between an element and a plane surface d) and blood circulation considered in the human body e).

MATHEMATICAL MODEL

Human and Clothing Thermal Systems

The human thermal system is based on the energy and mass balance integral equations: The first one, was developed for each tissue slice of each human body element (the considered heat fluxes are the following: conduction through the tissue (1); heat generation (2); blood convection (3); heat transported from internal to external slices (4); convection between the tissue and the blood (5); heat loss by respiratory system (6); conduction (7) and evaporation (8) and radiation (9) between the skin and the clothing; heat exchange by convection (10) and evaporation (11) and radiation (12) between the body and the environment) and for the arterial and venous blood in each element (blood convection (1);

convection between the blood and the tissue (2); convection between the artery and vein (3) and heat exchange in the pulmonary system (4) and in the capillaries system (5)). In the second one, were developed, in each element, the mass balance integral equations for the blood (flow rate that inlet (1) or outlet (2) and the blood exchange in the capillaries (3)) and for the transpired water in the skin surface (transpired flow rate (1) and the released flow rate by convection to the environment (2) or diffusion to the clothing (3)). The developed termoregulatory system can be analyzed in details in Conceição (1998).

To simulate the clothing thermal system were developed energy and mass balance integral equations for each clothing slice in each element (see figure 1c): In the first kind of equations were considered the following phenomena: conduction through the clothing and immobilized air (1); heat exchange by radiation (2) and evaporation/condensation (3) between two clothing slices and between the internal clothing slices and the body skin (4) and (5); heat exchange by convection (6) and evaporation (7) and radiation (8) between the external clothing slices and the environment). In the second one, were developed the mass balance integral equations for the clothing slice (water diffusion between slices (1); adsorption (2) and desorption (3) of the water vapor) and for the adsorbed/desorbed water vapor in the clothing fibers slices (water vapor adsorption (1) and desorption (2) between the clothing fibers and the air inside the dressing).

Thermal Exchanges With the Environment

The thermal exchange between the human body and the environment are done by: convection, evaporation, pulmonary respiration and radiation (see Conceição (1998)). The environmental variables can be measured or calculated. In this work the air mean relative humidity (RH), the velocity (Vair) and temperature (Tair), in each element, can be measured experimentally and the radiant temperature (Tr), that each element is subjected, can be calculated numerically in accord the equation (1) (see Fanger (1970)). To calculate the view factor ($F_{e \to p}$) the human body is divided in 25 cylindrical elements and the room involving space in (N_S) surfaces (see figure 2). Each element (e) or plane surface (p), with inclinations, dimensions and temperatures equals to the respective body section or wall, was divided in infinitesimal areas (see figure 1d). In this calculus it was considered also the shading effect that the body elements cause in each element. To simplify it was considered that the fingers had the same radiant temperature of the respective hand and that the heat exchange between elements is negligible.

$$Tr_e = \sqrt[4]{\sum_{p=1}^{N_S} T_p^4 F_{e \to p}}, \quad \text{with } e = 1,\ldots,35. \tag{1}$$

Steady State Thermal Comfort

In this work, for steady state conditions, the global thermal comfort level was evaluated through the PMV index (see Fanger (1970)). The equation used in this calculus (see Conceição (1999)) considers only the heat fluxes exchanged between the body and the environment.

Thermal Comfort in Transient Conditions

When the skin is subjected to instantaneous temperature changes are used the static (s) and dynamic (d) response of the cold and warm thermoreceptors (R(t), see equation (2)), located below the skin surface, and an index denominated by PSI (Psycho Sensory Intensity) (see equation (3)) to evaluate the global thermal comfort or the local discomfort (see Ring et al. (1991)). The 20 seconds, in accord to the last autors, is the time that the central nervous system evaluates the thermoreceptor impulses.

$$R(t) = R_s(t) + R_d(t) = \underbrace{\phi(T(t))}_{static} + \underbrace{K \frac{dT(t)}{dt}}_{dynamic}, \qquad PSI = PSI_s + PSI_d = \underbrace{\int_0^{20} [R_s(t)]dt}_{static} + \underbrace{\int_0^{20} [R_d(t)]dt}_{dynamic} \tag{2, 3}$$

The global thermal comfort (PMV), that considers the thermal response of 35 elements (see figure 1), is evaluated by the cold and warm PSI index, the thermoreceptors numbers (N_{TR}) and the Area Summation Factor (ASF) (see equation (4)). The static part of this preliminary model was developed using the results presented in Fanger (1970). To the coefficients of the dynamic part (K) were used results presented in Dear *et al.* (1993).

$$PMV = \gamma(PSI_{Cold}^{(1)}, PSI_{Warm}^{(1)}, N_{TR}^{(1)}, ASF^{(1)}, ..., PSI_{Cold}^{(35)}, PSI_{Warm}^{(35)}, N_{TR}^{(35)}, ASF^{(35)}).$$ (4)

The local discomfort, evaluated through the percentage of dissatisfied people (PD), that considers the thermal response for only one element (j), is calculated by the equation (5). They were used experimental results present in Fanger *et al.* (1998) to develop the static and the dynamic part of this last model.

$$PD^{(j)} = \underbrace{\varphi_s(PSI_{Cold_s}^{(j)}, PSI_{Warm_s}^{(j)})}_{static\ part} + \underbrace{\varphi_d(PSI_{Cold_d}^{(j)}, PSI_{Warm_d}^{(j)})}_{dynamic\ part}.$$ (5)

Numerical Model Validation

To validate this numerical model were used experimental data presented in the bibliography and measured through an infrared "Thermo Tracer TH1100 of the NEC San-ei" system. In the first one, the data were used to validate the human body thermal behavior (see Conceição (1998)), the global thermal comfort in steady state conditions (see Conceição (1999)) and the local discomfort model, while the temperature measured by the infrared system was used to validate the clothing thermal behavior (see Conceição (1999)), the skin temperature fluctuations and the skin up-step or down-step temperature.

PRELIMINARY TESTS OF THE MODEL

After the model being validated it was used to evaluate the thermal sensation, in a non uniform environment, that an occupant is subjected, in the (P1), (P2) and (P3) positions, in a Summer typical situation (see figure 2). The people considered was 1.75 m (height), 66 Kg (weight), 1 met (activity) and 0.3 clo (T-shirt, shorts and Summer shoes).

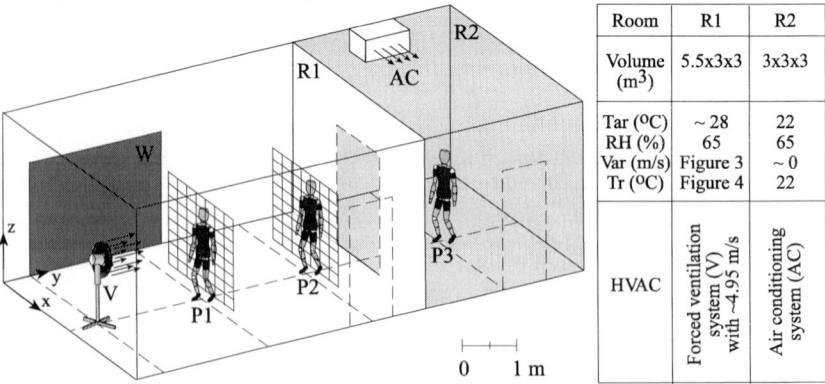

Room	R1	R2
Volume (m³)	5.5x3x3	3x3x3
Tar (°C)	~ 28	22
RH (%)	65	65
Var (m/s)	Figure 3	~ 0
Tr (°C)	Figure 4	22
HVAC	Forced ventilation system (V) with ~4.95 m/s	Air conditioning system (AC)

Figure 2: Schematic diagram of the simulated physically configuration.

The percentage of dissatisfied thermally people (global thermal comfort), in steady state conditions, calculated in (P1) and (P2) position, was, respectively, 13.9 (PMV=0.651) and 14.95 % (PMV=0.687).

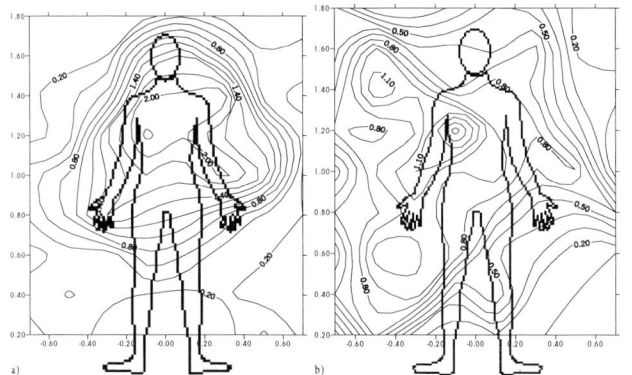 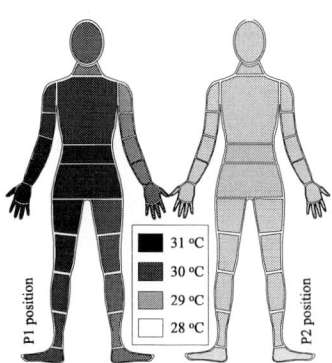

Figure 3: Air velocity field, incident of the people in the P1 a) and P2 b), determined experimentally.

Figure 4: Radiant temperature distribution determined numerically. The window (W) and the wall surfaces temperature were 45 and 28ºC.

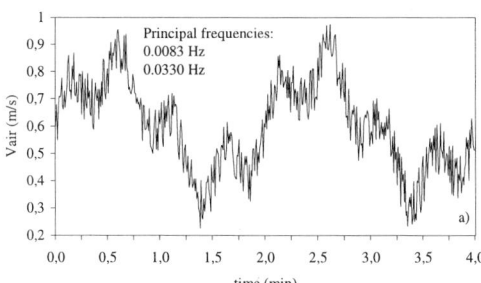 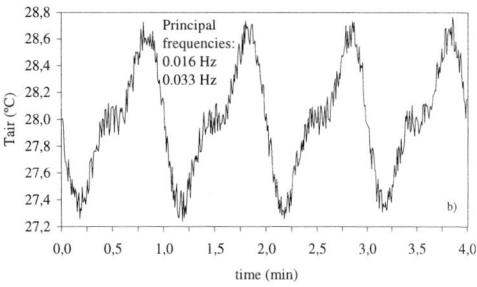

Figure 5: Evolution of air velocity (a) and temperature (b) for the head zone, in P2 position.

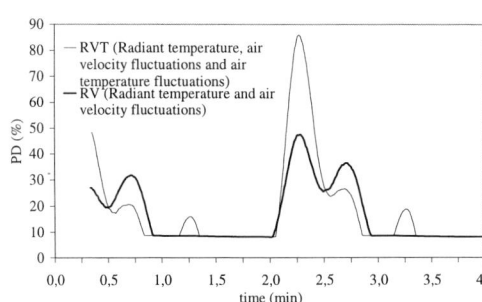 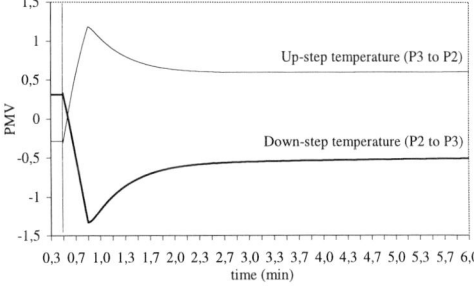

Figure 6: PD evolution, in the head zone, for the P2 position.

Figure 7: PMV evolution in up-step and down-step temperature.

To evaluate the local discomfort were created computationally, around each human body element, a pulsing air velocity (see figure 5a) and temperature (see figure 5b) signals in accord with the mean value presented in figure 3b) in the P2 position. In figure 6 is showed the evolution of predicted percentage of dissatisfied (PD), in head area, when are considered the radiant temperature and the air velocity fluctuations (RV, see figures 4 and 5a) and the radiant temperature, air velocity and temperature fluctuations (RTV, see figures 4, 5a and 5b). This figure shows that when occurs fluctuations of air temperature the thermal sensation that a person feels is modified.

Finally, the global thermal comfort evolution (PMV), in transient conditions, when the person going from P2 (room R1) to P3 (room R2) position (down-step temperature) and from P3 (room R2) to P2 (room R1) position (up-step temperature), can be analyzed in figure 7. Before entering in the second room the person stays 30 min in the first one. This figure shows that the thermal impact that an occupant is subjected is more high in down-step than in up-step. This conclusion is in accord with Dear et al. (1993).

CONCLUSIONS

In this work a model that can be used to evaluate the global thermal comfort, in steady state and in transient conditions, and local discomfort was presented. The validation tests show that the model is able to reproduce real situations. A preliminary test was done in a building compartment, with a non uniform environment, for a Summer typical situation.

The model showed that, for steady state conditions, a person is thermally comfortable in P1 and P2 positions. For transient conditions, the air velocity and temperature fluctuations influence the local discomfort and the human body thermal impact verified in down-step is more than in up-step temperature. In future researches will be done more experimental tests to study in detail the thermal comfort in a non uniform environment and the adaptation of people in climatized spaces. The influence of radiant temperature and air velocity and temperature fluctuations in the local discomfort will be also analyzed.

REFERENCES

Conceição, E. Z. E. (1998). Integral Simulation of the Human Thermal System. RoomVent´98 - 6th International Conference on Air Distribution in Rooms, KTH, Stockholm, Sweden, 14 to 17 of June.

Conceição, E. Z. E. (1999). Avaliação de Condições de Conforto Térmico: Simulação Numérica do Sistema Térmico do Corpo Humano e do Vestuário. CIAR, Lisboa, 14-16 de Outubro de 1999.

Dear, R. J.; Ring, J. W. and Fanger, P. O. (1993). Thermal Sensation Resulting from Sudden Ambient Temperature Changes. Indoor Air, N. 3, pp. 181-192.

Fanger, P. O. (1970). Thermal Comfort: Analysis and Applications in Environmental Engineering. McGraw-Hill Book Company, United States.

Fanger, P. O.; Melikov, A. K.; Hanzawa, H. and Ring, J. (1988). Air Turbulence and Sensation of Draught. Energy and Building, Vol. 12, pp. 21-39.

Ring, J. W.; Dear, R. J. and Fanger, P. O. (1991). Temperature Transients: a Model for Heat Diffusion Through the Skin, Thermoreceptor Response and Thermal Sensation. Indoor Air, N. 4, pp. 448-456.

Ring, J. W.; Dear, R. J. and Fanger, P. O. (1993). Human Thermal Sensation - Frequency Response to Sinusoidal Stimuli at the Surface of the Skin. Energy and Buildings, N. 20, pp. 159-165.

© 2000 Elsevier Science Ltd. All rights reserved.
Air Distribution in Rooms, (ROOMVENT 2000)
Editor: H.B. Awbi

VENTILATION, INDOOR AIR QUALITY AND THERMAL COMFORT

IA Raja, GS Virk, D Azzi and IW Matthews

Department of Electrical and Electronic Engineering
University of Portsmouth
Portsmouth, Hampshire PO1 3DJ
(E-mail: iftikhar.raja@port.ac.uk, Tel. 023 928 46175)

ABSTRACT

Ventilation is increasingly a subject of concern due to its relation with health and energy loss. Environmental pollution and global warming demands energy conservation and a Airtight structures are being built to reduce the energy losses due to infiltration. Reduced air change rates may lead to accumulation of CO_2 and other toxic pollutants to undesirable levels. In view of such concerns tests are carried out here to study the influence of ventilation on indoor air quality and thermal conditions in a naturally ventilated building. The results show that CO_2 accumulation of depends on the ventilation rate and the level of occupancy. CO_2 dissipation bears a high correlation with 'open window'. The study indicates that night ventilation can provide appropriate means for passive cooling in buildings and its effectiveness is strongly related to relative differences between indoor and outdoor temperatures.

KEYWORDS

Buildings, air quality, pollutants, carbon dioxide, natural ventilation, night cooling, human satisfaction,

INTRODUCTION

Ventilation in buildings is the process of changing air in the internal spaces and is required for many different reasons to meet certain requirements [ASHRAE 1991]. The need for ventilation varies with the use of building; for example in residential buildings the comfort of the occupants is highly related to ventilation where it modifies the indoor thermal environment. When related to indoor air quality ventilation, has two primary objectives, namely:

- supply of fresh air
- removal of contaminants

Ventilation, apart from controlling the indoor air quality has additional tasks such as cooling the building for human comfort, particularly in hot and warm climates. The effectiveness of night ventilation has been established in a number of studies [Awbi 1991, Hancock 1999] and passive cooling techniques are widely used. The air movement increases the dissipation of heat from the building materials significantly. The warmer air is exhausted into the low temperature atmospheric heat sink. Night ventilation can thus reduce the total cooling load of air-conditioned buildings, leading to energy saving and reduction in CO_2 emissions.

The building occupants can modify the indoor thermal conditions by manipulating various controls available to them. In naturally ventilated buildings the usual controls are openable windows and local fans. Studies have showed that control over ventilation rate (e.g. the opening of windows and doors) can provide a useful and effective tool for improving the indoor thermal conditions [Raja & Nicol 1998a]. However, its effectiveness is linked with the differences between outdoor and indoor temperatures and is more effective if the outdoor temperature is lower than the indoor temperature and when the temperature difference is high.

A study was carried out to demonstrate the influence of ventilation on indoor thermal conditions and air quality in a naturally ventilated building. The test was conducted in an office room, located in the Portland Building (which houses the Faculty of Environment) at the University of Portsmouth. Indoor thermal conditions and air quality were monitored. The number of occupants and window opening was varied in a random manner over the day, with 0, 1 and 2 occupants, and the window was closed and opened again for 30mm and 60mm. During the occupancy period, the occupants were asked to fill in questionnaires about the suitability of their environment at half hourly intervals. The indoor thermal condition and air quality was monitored by recording temperature and CO_2 concentration. The outdoor weather data for the test period was obtained from the University's weather station, operated by BMS Group in the Department of Electrical and Electronic Engineering. The analysis of indoor air quality and effect of ventilation on indoor thermal conditions is presented in this paper.

EXPERIMENTAL SETUP

The experiment is conducted in an office in Portland Building at University of Portsmouth. The building is naturally ventilated via stack-effect towers and a central glassed atrium. The room used for the test is a small office located on the second floor in the south wing of the building. It is connected internally to a corridor with a single door. The externally exposed wall is on the north. The size of the room is 3.6m x 3.1m x 2.7m. The room is ventilated through a single north facing top mounted window. The size of the window is 0.91m x 1.3m x0.2m. The ventilation is achieved via towers. These towers use the buoyancy of heated air to draw air through the central corridor of the wing. Air enters and leaves the room through the north-facing window and through infiltration around the door.

The room was monitored continuously for eight days during the summer of 1999. Thermal and environmental parameters measured are listed in Table 1. The table also contains the range and accuracy of the sensors used for monitoring. These parameters were recorded using TESTO 400, '2 Channel Multi-Function Instruments' at intervals of five minutes. The occupancy and opening of the window were varied according to a Pseudo Random Multilevel Sequence (PRMLS). The occupancy was varied from 0 to 1 and 2 persons and the window was either closed, open 30mm or open 60mm. The subjects were asked to change the window position when entering or leaving the room according to a set schedule. The subjects were university academic and research staff.

Table 1: Parameters measured and sensors used to monitor the indoor air quality.

Parameter	Sensors/Devices	Range	Accuracy
Air Temperature	Testo Comfort level Probe	0 - 50°C	±0.3°C
Air Movement	Testo Robust Hot Ball Velocity probe	0 – 5 m/s	±0.2 m/s
CO_2	Testo Ambient CO_2 Probe	0 – 10,000 ppm	±50 ppm
Mean Radiant Temperature	Tinytalk II	-50 to +300°C	±0.7°C

In addition the occupants of the room were asked to complete a questionnaire giving their feelings about the temperature in the room at half hourly intervals. They were also asked questions about their clothing and mean activity level as well as their sensation of the air quality. The analysis of subjective data is presented elsewhere [Raja et al, 2000].

VENTILATION AND AIR QUALITY

Effect of Window Open

Ventilation is increasingly a subject of concern. It bears a strong relation with health and influences energy loss. Also environmental pollution and global warming demands energy conservation and reduction in energy losses due to infiltration. Airtight structures are being strongly encouraged. Such building may lead to accumulation of CO_2 and other toxic pollutants to reach undesirable levels and become health hazards. In fact, ventilation in buildings should conform to the 'Build tight – ventilate right' philosophy. This will help to avoid uncontrolled air leakage by paying careful attention to the building 'envelope' - whilst providing controllable and adequate background ventilation in winter and 'rapid' ventilation to avoid overheating in summer and to reduce the concentration of toxic pollutants.

During the present study the CO_2 level was measured by varying occupancy from no one present to one and two persons present over a minimum period of 24 hours. The ventilation rates were varied by changing 'window open' from a closed position to level-1 (30mm opening) and level-2 (60mm opening). Using hourly values, the CO_2 concentration is plotted against input variable 'window open' for none, one and two persons in the room, as shown in Figure 1.

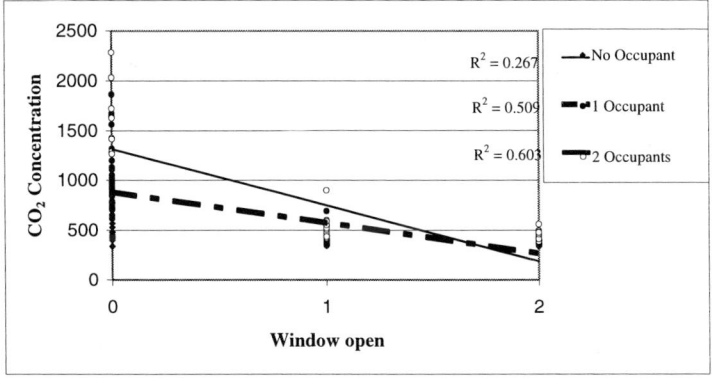

Figure 1: Dissipation of CO_2 by ventilation through 'window open'.

Effect of Occupancy

The occupants breathing raises the concentration of CO_2 and humidity in the indoor air above the level outdoors. For healthy indoor environment this needs to be replaced by fresh air at a rate equivalent to its production. This is usually achieved by ventilating the space and through air infiltration.

The effect of occupancy on indoor air quality is demonstrated in Figure 2. The figure is constructed with the data for one and two occupants for an occupancy period of two hours. Prior to occupancy the window was kept open for at least one hour while there was no body in the room. Then with occupancy, the window was closed for the first one hour and opened for the next one hour. The CO_2 concentration was recorded at 5 minutes interval. Five tests were carried out with one person and two tests with two persons in room. The values averaged over a quarter of an hour were plotted in Figure 2.

140

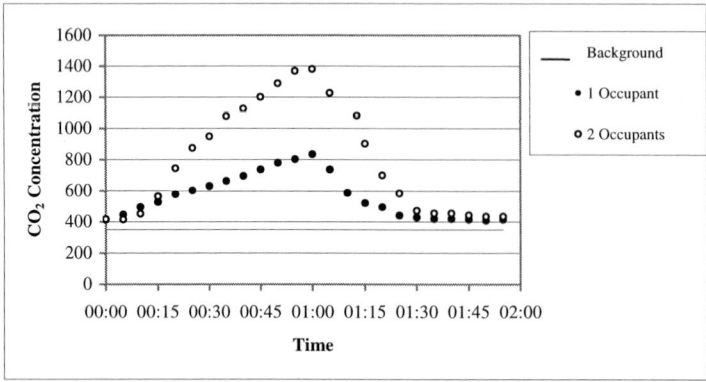

Figure 2: Effect of occupancy, accumulation and dissipation of CO_2

VENTILATION AND NIGHT COOLING

Night Ventilation Effect

Night ventilation significantly decreases the indoor air and structural temperatures. The decrease depends on the external temperature and ventilation rate. This reduces the need for additional cooling to some extent during daytime. It was originally not aimed to study the effect of night cooling on indoor thermal conditions. Therefore, the present experiment was not designed to measure certain parameters, like temperature of the structure, required for such studies. However, from the measured data, a profound effect of night cooling is noticed. Eight graphs of indoor and outdoor temperatures, one for each day, against time of the day are plotted in Figure 3.

DISCUSSION AND CONCLUSION

Figure 1 demonstrates that the dilution of indoor CO_2 concentration is strongly correlated with 'window open' i.e. the ventilation rate. The correlation coefficients are 0.51 for 1 occupant and 0.60 for 2 persons in the room. Carbon dioxide builds up during the first hour when the widow is closed, as shown in Figure 2. At the end of one hour the CO_2 concentration increases to about 1400 ppm when two persons are present in the room. It is almost double to that when there is one. By allowing fresh air through the window the indoor CO_2 concentration drops sharply during the first half an hour of 'open window' and reaches a lowest value close to the background CO_2 level. The background CO_2 was assumed to be constant at 350 ppm over the test period. The accumulation of the indoor CO_2 or pollution depends on the occupancy and ventilation rate.

The graphs in Figure 3, demonstrate the influence of outdoor temperature on indoor thermal conditions. For example, looking at the plot for 19th August 1999, the window was opened at 22:00 hours when the indoor temperature was 25°C, about 10°C higher than the corresponding outdoor temperature. The temperature drops sharply lowering the indoor temperature by 5°C during the first half an hour. The difference between indoor and outdoor temperature is narrowed to about 3°C during the night. The time 05:30 marks the start of the day, when both indoor and outdoor temperatures start rising.

Hancock studied the effect of passive night cooling of natural ventilated school buildings in Pakistan and reported a difference of 1°C between indoor and outdoor temperatures [Hancock 1999]. During the test, cross ventilation was allowed and the buildings were uninhabited at night. In the present study however, the door was kept closed restricting the ventilation through the window and occupancy was varied from none to 1 and 2 persons. This may be a possible reason that the difference between indoor and outdoor temperatures is slightly higher than that reported by Hancock. The high rate of night ventilation provides a useful level of cooling in most conventional buildings. To provide a degree of local control over the ventilation, night cooling ventilators are available which are closed automatically once a target temperature is reached. This prevents over-cooling or in extreme weather conditions, prevents rain ingress.

The results of an earlier study of thermal comfort of occupants in an office building [Raja et. al. 1998a] as shown in the Table 2, suggests that on average, the occupants close to a window are more comfortable than those further away. At 7-point scale, given in Table 3, occupants away from the window are reported warmer at 5.5 on the scale, about one level higher than those close to window (4.5 on the scale).

Table 2: Use of control by the occupants in building.

Subject code	Seating Position	Outdoor T (C)	Thermal Sensation	Door (%)	Window (%)
1.01	Away	21.0	4.75	00.0	46.0
1.02	Away	21.5	4.88	00.0	31.0
1.03	Near	20.7	4.46	00.0	79.8
1.04	Near	19.8	4.40	00.0	83.1
1.05	Near	19.9	4.19	100	74.1
1.06	Near	19.8	5.09	100	89.2
1.07	Away	20.2	6.74	99.3	96.3
1.09	Near	21.5	4.29	33.3	100
1.10	Away	19.9	4.94	12.5	56.3
1.12	Near	20.3	4.88	2.4	94.7
1.13	Near	19.9	3.73	96.8	69.9

Based on day time records (10:00am to 06:00pm) of the present study, Table 3 shows that with window closed the thermal sensation of occupants shifts toward warmer side of the scale (25% slightly warm and 5 % warm) as compared to 5% voting slightly warm when window was opened. In naturally ventilated buildings the comfort of occupants is related to the use of ventilation means. As indicated in Table 4, the occupants' discomfort decreases with the increase in the use of windows, i.e. frequency of opening windows, [Raja et. al. 1998b]. In building 2, the discomfort was reduced to only 11% with the highest use of window (81%). Low frequency of window use was compensated by the use of fan in building 14. The proportion of people recording discomfort is strongly correlated with the number of people using fans or windows and particularly when both are used together; it gives a correlation coefficient of 0.80 [Nicol & Raja 1998c]. This implies that the local controls are used in response to uncomfortable conditions.

Table 3: Percentage of votes for a comfort level.   Table 4:Frequency of the use of available controls and discomfort .

Thermal Sensation	Window Open	Window Closed	Building	Discomfort (%)	Doors	Windows	Fans
1 - Cold	0	0	1	30	0.59	0.75	0.22
2 - Cool	12	5	2	11	-	0.81	0.17
3 - Slightly Cool	28	15	4	35	-	0.70	0.67
4 - Neutral	56	50	6	10	0.67	0.40	0.18
5 -Slightly Warm	5	25	7	51	0.60	0.66	0.61
6 - Warm	0	5	8	33	0.75	0.71	0.40
7 - Hot	0	0	9	21	0.76	0.79	0.17
			11	16	0.60	0.71	0.08
			13	23	0.75	0.54	0.23
			14	19	0.76	0.11	0.80

REFERENCES

Awbi, H. B. (1991), *Ventilation of Buildings*, Chapman and Hall.

HVAC Applications ASHRAE, (1991), Chapters 1-28, Comfort Air Conditioning.

Raja, I A and Nicol J. F. (1998a), Natural ventilated buildings: Use of control for changing indoor climate, WREC V (391-394), Pergamon Press, Oxford, UK.

Raja, I. A. and Nicol J. F. (1998b), Significance of controls for achieving thermal comforts in natural ventilated buildings, EPIC 1998, Lyon, (63-68).

Raja I. A; Virk G. S; Azzi D. and Matthews I. W. (2000), Effect of ventilation on indoor thermal comfort, Proc. World Renewable Energy Congress VI, Brighton, UK.
Nicol J. F. and Raja I. A. (1998), Report to the Engineering and Physical Sciences Research Council (Unpublished).
Hancock M. (1999), Improving Thermal Comfort by Passive Thermal Design, Final Report - 1999, School of Architecture, Oxford Brookes University, Oxford UK.

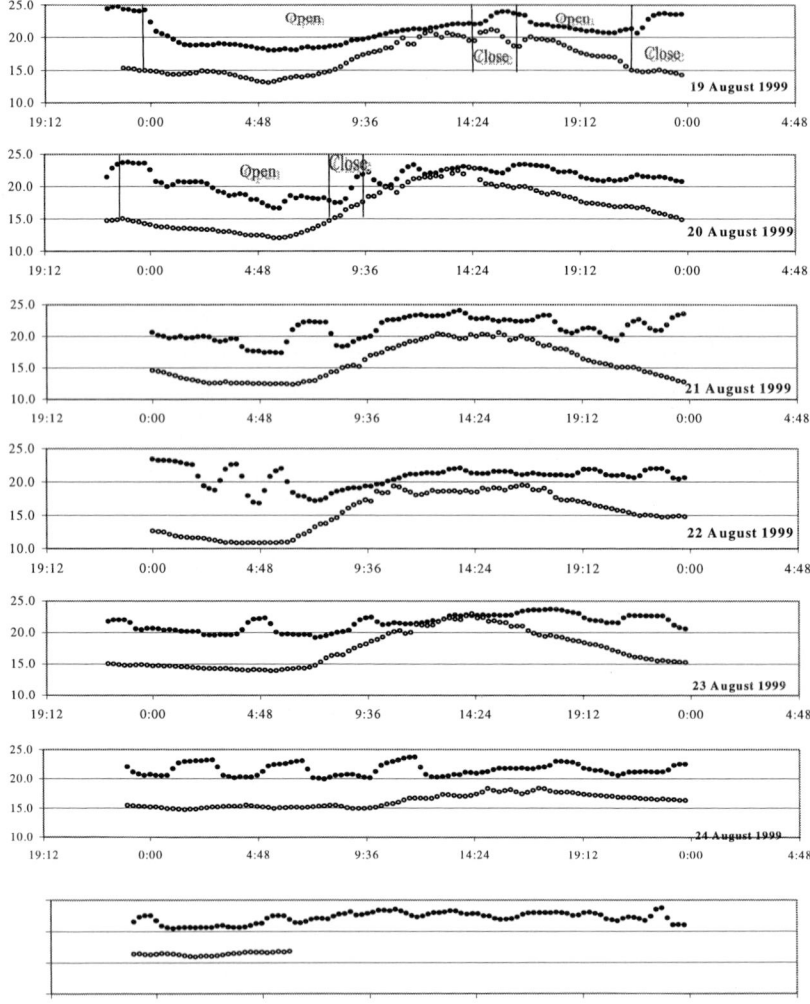

Time of the Day

Figure 3: Effect of night cooling on indoor thermal conditions
(outdoor temperature 'o', and indoor temperature '•')

Air Distribution in Rooms, (ROOMVENT 2000)
Editor: H.B. Awbi

AN ASSESSMENT AND REMEDIATION OF MOLD IN INDOOR ENVIRONMENTS

Rishi Kumar

Global Educational & Consulting Services
Mississauga, ON L5R 2A1 Canada

ABSTRACT

The health implications of indoor air contamination due to mold have become an issue of increasing concern in recent years. The role of mold is not well defined in "sick building syndrome", but it is this syndrome that most frequently prompts building investigations. Contamination of indoor air by mold in buildings may cause or exacerbate allergic type symptoms (such as wheezing, chest tightness and shortness of breath), especially in persons who have a history of hypersensitivity diseases (such as asthma, hypersensitivity pneumonitis and severe sinusitis). The purpose of this paper is to address potential health effects of mold contamination and to assist professionals who may be asked to assess a building with potential mold problems. The assessment strategy outlined in the paper covers investigation of indoor mold contamination/evaluation, remediation and preventive maintenance. It is expected that if followed, the building maintenance personnel and building managers should be able to address structural and potential environmental problems related to mold contamination in the building in a timely fashion and minimize the impact of exposure to moldy materials and contamination by building occupants.

KEY WORDS

Indoor Air Quality, Impact of Mold, Sick Building, IAQ Assessment, Healthy Buildings

INTRODUCTION

Molds belong to the fungus family. They are very common and can be found anywhere, especially in areas where it is wet or damp. Different molds like to grow on different materials. For example, "Stachybotrys atra" prefers water-saturated materials containing cellulose. This includes wallpaper, ceiling tiles, carpets, especially those with jute backings, insulated material, wood-derived building material and drywall. The presence of mold in general has been associated with respiratory symptoms and allergic reactions in susceptible persons; it may also exacerbate asthma in such persons. Mold can produce several toxic chemicals such as trichothecene mycotoxins. These mycotoxins are known to be toxic to both humans and farm animals exposed to significant quantities. There have been only a few documented cases of health problems from indoor exposure to mold. Over 50% of indoor environment problems that contribute to sick building syndrome (SBS) are caused by contaminants in the air, such as microbials and/or inhalables (smaller than 10 microns in size), respirable particles (less than 2.5

microns in size) and other specific pollutants generated internally. The remaining problems are the result of poor design, operation and maintenance of the heating, ventilating and air-conditioning (HVAC) system.

ASSESSMENT PROCESS

Reports about potential mold contamination in building should be assessed to determine whether conditions exist which favor the growth of mold. The criteria for conducting an assessment can be grouped into the following six parts:

Building History: The intent is to gain a first hand visual appreciation for the building design and floor plan. A critical inspection of the ventilation system is also important in order to thoroughly characterize the building with respect to potential sources and microbiological contaminants.

Occupant Complaints: To determine the magnitude of the problem, specifically, if the problem is widespread throughout the building or among a certain group of employees, discussions with employees, managers, union representatives, building maintenance staff and joint health and safety committee, are needed.

Complaint Area Survey: Mold requires water and nutrients for growth and proliferation. They are most often found in buildings in which there is excess moisture, often in the presence of water-damaged material. Ceiling tiles, gypsum wallboard, cardboard paper and other cellulosic surfaces should be given careful attention during the survey. There may be visible condensation on windows. Colonization of walls and other exposed surfaces may be visible. There may be a distinctive fungal odor. HVAC system should also be checked, particularly for damp filters.

Environmental Monitoring: If visible mold, water damage and occupants exhibiting related symptoms serious enough to result in lost work days are present, it is recommended that an environmental monitoring should be conducted. Mold can produce several toxic chemicals such as trichothecene mycotoxins. It is well established that mold (fungi) cause several diseases, such as systemic infections and asthma. On the other hand, mold has been identified as one of the possible causes of SBS (itching and watering eyes, nasal irritation/congestion, throat irritation, cough/wheeze, fatigue, and changed sensation of odor or taste). Environmental monitoring is also needed to confirm or rule out, a number of problem source possibilities identified during the assessment process. The ultimate goal of diagnostic indoor air mold sampling should be: (a) to determine if sufficient mold propagules, particularly those bearing irritating or immunosensitizing chemical components, are being produced and dispersed within the building to account for (or predict) symptomatology; and (b) if a connection between mold and symptoms is likely, to find and eliminate sites of mold amplification within the building. Air sampling for fungal structures is, at the most fundamental level, divided into techniques based on the culture of live propagules (Settle/Sedimentation Plates and Vacuum/Culture Pump Sampler) and techniques based on the trapping (Rotorod Sampler and Spore Traps) and visualization of living or dead materials.

Hazard Communication: Hazard communication has been defined as "the act of conveying or transmitting information between interested parties about the levels of health or environmental hazards; the significance or meaning of such hazards; or decisions, actions or policies aimed at managing or controlling such hazards". Lines of communications between building occupants, workplace health and safety officials, building managers and owners employers and union representatives should be established as soon as health complaints related to indoor air quality are received.

Destructive Testing: Destructive testing occurs when certain structures of the building have to be taken apart in an attempt to locate the source of suspected contamination. During this part, the contamination status of the building is expected to be altered by the actions taken by the assessment team, possibly through exposure of previously cryptic contamination and redistribution of such contaminants via the HVAC system or by other means.

EVALUATION CRITERIA

Assessment of a potential mold contribution to indoor air problem may follow either (1) the appearance of symptoms compatible with exposure to fungal bioaerosols, or (2) the detection of potentially problematic fungal material in a building environment. Indoor air problems may have various etiologies and should not be assumed automatically to be of microbial origin. Important factors such as non-fungal chemical contamination (such as VOCs, tobacco smoke, ozone, nitrogen dioxide, CO), potential effects of sociological issues, such as labor/management disputes on perceived environmental tolerance should be considered, and discussed further, as may be applicable before making any remedial decision.

♦ The building history suggests there is an area of readily discernible mold growth (e.g., water-damaged surfaces or stored materials, poorly maintained ducts or HVAC system, etc.); perform a visual inspection wherever this has not been done already.

♦ The building history suggests there is a hidden locus of mold growth (e.g., previously flooded wall cavity or false ceiling; previously flooded carpet backing, etc.). In these and similar cases, where symptoms warrant only, conduct an extensive physical search.

♦ The building history suggests long-time bird or bat colonization of the attic or elsewhere; assess visually while wearing a high-efficiency particulate air (HEPA) filter respirator or call an expert.

♦ Complaint area survey reveals mold or suspected mold; take scrapings or transparent tape mount for direct microscopic examination and scrapings, swabs or contact plates for culture. If a substantial area of exposed soil is found, ensure that any air sampling includes samples taken near this area to probe its significance as a source of mold propagules. Samples of soil may also be analyzed.

♦ Air sampling shows molds of indoor origin present and exceeding the benchmark levels as discussed under "Guidelines", find the sources of mold and take appropriate remedial action. If molds of indoor origin are present at negligible levels or below the benchmark level, take no further action unless indicated by other observations.

♦ In sites where no historic information is available and no water damage is recalled or seen but where assessment/investigation is still warranted by symptoms or other concerns, conduct one or more of the following tests with stringency appropriate for the degree of health concern. (a) Wall Cavities - A low stringency test is to check a few representative wall cavities behind switch faceplates. Use an alcohol-disinfected, flamed bent wire to scrap small amounts of material from back or front of wall cavity; more stringently, use a rigid endoscope (borescope) to examine cavity interior. (b) Ducts – Use existing openings or a rigid endoscope to inspect duct interiors.

♦ Water from humidifiers, cooling tower, etc. may contain profuse growth of microorganism, including molds and yeasts, as well as bacteria and protozoans. Clean humidifier water should appear clean. If it does not, the cause of turbidity may be biological or chemical. Water should be taken in a sterile container and examined for distinctive fungal filaments, yeast cells and conidia.

GUIDELINES ON MOLD

Canadian guidelines were published in "Health Canada's "Indoor Air Quality in Office Buildings: A Technical Guide: A Report of the Federal-Provincial Advisory Committee on Environmental and Occupational Health-1993": (1) If more than 50 CFU/cu. m. (colony-forming units per cubic meter-in either indoor or outdoor air; defined as a small units of biological material, such as spores, conidia hyphal fragments, capable of giving rise to individual colonies on growth medium) of a single species (other than Cladosporium or Alternaria spp.) are detected, there may be reason for concern; (2) Up to 150 CFU/cu. m. is acceptable if there is a mixture of species reflective of the outdoor air spores. Higher counts suggest dirty or inefficient air filters or other problems; (3) Up to 500 CFU/cu. m. is acceptable in summer if the species present are primarily "Cladosporium" or other tree or leaf fungi. Values higher than this may indicate failure of the filters or contamination in the building.

REMEDIATION STRATEGIES

Strategies for the remediation of indoor air quality problems caused by mold are based on the elimination of conditions that promote the amplification of these potentially hazardous organisms. Different levels (described below) of containment are necessary depending on the extent of the contamination problem. There must be a mechanism in place for ensuring an immediate response to these problems. Clean up should be conducted when the affected area is unoccupied. In all remediations, a routine follow-up inspection at 6-12 months or sooner if visible mold contamination or water damage recurs should be conducted.

Level 1: Smaller/Larger Isolated Areas (up to 30 sq. ft.) : Examples - Ceiling Tile/Drywall Panel

♦ Regular building maintenance staff can conduct cleanup. Such persons must receive training from a qualified individual on proper clean up methods, protection and potential health hazards and should be free from asthma, allergy and immune suppressive disorders. Gloves and a face respirator should be worn. A full respiratory program, in accordance with OHSA is required.

♦ Surrounding material should be covered with plastic sheets and tape before removal (for larger areas: 2-30 sq. ft.).

♦ Contaminated absorbent material should be removed in a sealed plastic bag.

♦ Surrounding areas should be cleaned with household bleach.

♦ Special containment or evacuation measures are not necessary.

Level 2: Large Scale Remediations (more than 30 sq. ft.): Example - More than one Wall Board Panel in an area which cannot be isolated from personnel

♦ Personnel trained in the handling of hazardous materials are necessary.

♦ Contaminant of the affected work area is required - (a) Complete isolation of work area from occupied spaces using plastic sheeting shield with duct tape (including openings, fixtures and HVAC components) is required; (b) An exhaust fan unit equipped with HEPA filter operating under slightly negative pressure is required; (c) Airlocks and decontamination room is needed for exit from work area.

♦ Contaminated material should be removed in double-sealed plastic bags.

♦ The work are must be cleaned by a vacuum unit equipped with HEPA filter.

♦ Cleaning crew should wear full-face respirators with HEPA cartridges or powered air-purifying respirators. After cleaning all protective clothing, headgear, foot covering and gloves should be disposed in a double-sealed plastic bag.

♦ Air monitoring should be conducted (a) during remediation to determine if spores are escaping during remediation and prior to removal of isolation barriers to assess the efficacy of the remediation; and (b) after large scale remediation, to determine its effectiveness and whether the area is safe for symptomatic persons to reoccupy.

Level 3: Remediation of HVAC Systems

♦ Personnel trained in the handling of hazardous materials are required for remediation of HVAC systems.

♦ Contaminant of the affected work area is required (similar to Level 2).

♦ Contaminated material should be removed in double-sealed plastic bags.

♦ The work are must be cleaned by a vacuum unit equipped with HEPA filter prior to the removal of isolation barriers.

♦ Cleaning crew should wear full-face respirators with HEPA cartridges or powered air-purifying respirators. After cleaning all protective clothing, headgear, foot covering and gloves should be disposed in a double-sealed plastic bag.

♦ Air monitoring should be conducted similar to Level 2.

♦ Growth supporting material should be removed from ducts with a HEPA vacuum, where practical, if not, removal of the affected component of the HVAC system is required.

♦ Contaminated material should be disinfected prior to removal. Decisions concerning the type of disinfection should be made by a qualified individual, based on the extent of the growth supporting material. Decisions as to disinfection must be based on the extent of the growth substrate in the ducts. There are numerous "biocides" such as quaternary ammonium compounds (e.g., dimethylbenzyl ammonium chloride) that are employed routinely for disinfection and cleaning surfaces, particularly in hospitals and laboratories.

♦ The causes of mold accumulation and/or growth must be identified and corrective action taken.

Many of the indoor air problems associated with mold/fungi can be remedied simply by modifying the environment where the problems occur. Fungal amplification might be eliminated under these conditions by simply raising the temperature or by dehumidifying those areas known to promote growth. Preventing condensation on walls or other surfaces can be accomplished by redirecting airflow to eliminate cold spots.

PREVENTIVE MAINTENANCE

The preventing maintenance program should be directed towards minimizing mold amplification sites by ensuring adequate drainage of sumps and drip panes, regular cleaning of dirt and slime from all constituents and replacement of filters. The frequency of conducting these procedures varies with each component, from monthly to annually. Porous lining materials should not be present in any part of the HVAC system. The building constituents include all components of the building envelope and interior. Any sources of external and internal leaks and condensation should be promptly and permanently corrected. Water-damaged insulation, ceiling and wall materials, carpets, upholstery and other porous components may need to be removed.

Several aspects of building design can be implemented to minimize the amplification of fungal contaminants in indoor environment: (a) Limiting access of the outdoor aerosol – These considerations minimize the entry of outdoor fungi into the building air, through the provision of a tight structural envelope, of particle filtration of the intake air and of climate control to minimize the need for opening windows; (b) Eliminating sites of water accumulation – Sites of unavoidable water collection in cooling and humidification systems should be constructed to be completely drained.; (c) Maintaining a sufficient humidity level – Humidity should be regulated at a sufficient level, high enough for the comfort of the occupants but not so high as to promote condensation; (d) Facilitating preventive maintenance – Sites of known and potential water accumulation should be constructed to readily allow inspection and service.

CONCLUSION

The prompt removal of contaminated material and infrastructural repair must be the primary response to mold/fungi contamination in buildings. Emphasis should be placed on preventing contamination through proper building maintenance and prompt repair of water damaged areas. Effective communication with building occupants should be an essential component of all-remedial efforts and preventive maintenance program.

REFERENCES

Burge H. (1990). Bioaerosols: Prevalence and Health Effects in the Indoor Environment. *J Allergy Clin. Immuno.* 86, 687.

Health Canada. (1993). Indoor Air Quality in Office Buildings: A Technical Guide, A Report of the Federal-Provincial Advisory Committee on Environmental and Occupational Health, 55pp

Hodgson M.J., Morey P., Leung W.Y., Morrow L., Miller D., Jarvis B.B., Robbins H., Halsey J.F. and Storey E. (1998). Building Associated Pulmonary Disease from Exposure to Stachbotrys Chartanum and Aspergillus Versicolor. *J. Occup. Environ. Med.* 40(3), 241.

Kumar R. (1998). Assessment of indoor air quality & ASHRAE Standard 62. *Symposium on Thermal & Fluids Engineering, Canadian Society of Mechanical Engineers*, Toronto, 568.

Kumar R. and Kumar A. (1997). A Guide to Indoor Air Quality Sites on the World Wide Web. *Am. Soc. Of Chem. Eng., Environ. Prog, 16-4*,W13

Kumar R., Kumar A. and Venkataraman K. (1999). A Review of Internet Sites and Literature on Indoor Mold/Fungi. *Am. Soc. Of Chem. Eng., Environ. Prog, 18-34*, F9

World Health Organization. (1988). Indoor Air Quality-Biological Contaminants. *WHO Regional Publications, European Series No. 31*, Copenhagen

© 2000 Elsevier Science Ltd. All rights reserved.
Air Distribution in Rooms, (ROOMVENT 2000)
Editor: H.B. Awbi

EMISSIONS OF VOCs FROM BUILDING MATERIALS AND THE INDOOR AIR QUALITY OF A NEW NATURALLY VENTILATED OFFICE BUILDING

C.W. F. Yu and D. R. Crump

Centre for Safety, Health and Environment
BRE, Watford WD2 7JR, UK

ABSTRACT

This study investigated the sources and concentrations of volatile organic compounds (VOCs) including formaldehyde in the air of a new office and conference centre building. The building is naturally ventilated, and was designed to demonstrate a number of innovative approaches to environmental design. Occupant surveys have shown a high level of occupant satisfaction with the indoor environment. The building and furnishing materials were, however, quite typical of current office buildings; the building therefore represented a useful opportunity to study the emission of VOCs from materials and to demonstrate methods of identifying the sources of specific compounds that are found in the air. As would be expected in a new building, a wide range of VOCs and their sources were identified, but not at hazardous concentrations. The location with the highest total VOC concentrations was measured at 1411 $\mu g/m^3$ when the building was first occupied. This declined rapidly but remained above 500 $\mu g/m^3$ during the first year of monitoring. The chemical TXIB was dominant in the VOC measurements in the building and this was found to be emitted from the carpet tiles. Most VOCs had multiple sources.

KEYWORDS

VOCs, Formaldehyde, Building materials, Emission, Sources and concentrations, Commercial buildings, Natural ventilation, Indoor air quality

INTRODUCTION

The purpose of this on-going project is to understand the relationships between the sources and airborne concentrations of volatile organic compounds (VOCs) and formaldehyde in new buildings. The work has been carried out in a new office building at BRE's Garston site. The results will be used to formulate a database of the rates of emission from the wide range of products used in the construction and furnishing of the building and this will provide guidance for specifiers who wish to select low-emitting materials for their buildings.

The building is naturally ventilated, and was designed to demonstrate a number of innovative approaches to environmental design. Occupant surveys have shown a high level of occupant satisfaction with the indoor environment. The building and furnishing materials were, however, quite typical of current office buildings; the building therefore represented a useful opportunity to study the emission of VOCs from materials and to demonstrate methods of identifying the sources of specific compounds that are found in the air.

Some air quality measurements made during the first year since the building was completed in January 1997 have already been reported (Yu et al, 1999). The present paper reports the formaldehyde and the total VOCs (TVOC) concentration and the ten major individual compounds found in the air of the ground floor and the top floor of the office building over a three-year period from January 1997 to December 1999. Diffusive samplers were used for measuring the organic compounds in the air and these were strategically placed and exposed in the open plan offices on each floor and at an outdoor location. Samples of a wide range of materials used in the construction and fitting out of the building were collected. The chemical emissions from these materials and from some consumer products used in the building were identified using environmental chamber tests. The results of the emissions work were used to identify the main sources of the organic compounds found in the air.

DESCRIPTION OF THE BUILDING

The building consists of a three-storey office block and a conference centre, which was completed in January 1997 and was occupied from April 1997. The 2000 m^2 building contains a mixture of cellular and open plan offices, as well as seminar facilities. The office block floor plates measure 30 x 13.8 m and the minimum floor-to-ceiling height is 3 m. The building has many energy saving features and is naturally ventilated, combining stack ventilation with automatic control to draw air via ducts built into the sinusoidal ceiling slabs on the first and ground floors of the office block. The internal design conditions are a minimum indoor temperature in winter of 18°C, with 25°C not to be exceeded for more than 5% of the year and 28°C for not more than 1%. The maximum design ventilation rate is 10 air changes h^{-1}.

The ventilation rate and the environmental conditions in the building were monitored as part of the Natvent[TM] project (1999) at two locations in the first floor office in August 1997 and in January 1998. The mean ventilation rates in the summer were 3.3 h^{-1} and 2.1 h^{-1} whereas in winter they were 0.78 h^{-1} and 0.74 h^{-1} respectively. The occupant control of ventilation of the top floor is by manual opening and closing of windows. This floor of the building was largely unoccupied until June 1999. The air leakage rate was measured at 50 Pa reference pressure using BREFAN (Perera & Tull, 1990); it was initially 17.34 $m^3h^{-1}m^{-2}$ of envelope area, but reduced to 15.22 $m^3h^{-1}m^{-2}$ in April 1999 after tightening work.

In the summer months during the study period, the relative humidity of the rooms was generally 55-65% and the temperature was generally 2-3°C below the target maximum of 25°C. In the winter, the relative humidity of the rooms was generally 30-45% and the temperature was generally 2-5°C above the design minimum of 18°C. An occupant evaluation survey was carried out prior to and after moving to the new office building and this found that the new building provided enhanced comfort and the level of satisfaction was higher than a range of other office buildings.

The building is carpeted throughout except the concourse area. The carpet tiles are synthetic fibre woven onto a mesh laminated to a tough, hard wearing polymeric backing. The cellular offices on the ground floor and first floor of the office building were partitioned with plasterboard and timber studs. All the interior doors and windows were timber framed; the timbers were vacuum treated and coated with lacquer. Medium density fibreboard (MDF) was used for wall panelling and for construction of window sills; these were either coated with emulsion paint, firetex paint, lacquer or vinyl lined. The undulating concrete air ducts in the sub-floor were insulated with mineral wool, with metal encased chipboard laying over the air-ducts to act as flooring. The carpet tiles were glued to the metal encased chipboard flooring. All the walls and ceilings were coated with emulsion paint. The top floor ceiling was constructed with 'Glulam' timber beam with exposed treated timber panels and joists. Pre-cast concrete curvature ceilings were installed in the ground floor and the first floor office areas. Parquet floor tiles were fixed with adhesives in the ground floor concourse area of the conference centre. Plywood was used to cover ducting in this area and for wall panelling and decking.

THE SAMPLING AND ANALYSIS OF VOCs IN THE NEW OFFICE BUILDING

Passive sampling methods (Yu et al, 1997) were used for monitoring VOCs and formaldehyde in the new office building for a three-year period from January 1997 – December 1999. The VOC passive sampler was a Perkin Elmer type adsorption tube containing Tenax TA and the formaldehyde passive sampler was a badge (GMD 570) containing a 2,4-dinitrophenylhydrazine coated filter paper. The passive samplers were strategically placed every month in the open plan offices on each floor and also in the concourse area and lecture theatre in the conference centre (Yu et al, 1999).

The VOC samplers were exposed for 28 days and were analysed by thermal desorption/gas chromatography using flame ionisation detection for quantification and ion trap detection for confirmation of the identities of compounds. The formaldehyde samplers were exposed for three days at 28-day intervals, and high performance liquid chromatography was used for the analysis.

During building construction, samples of the various types of building and furnishing materials were collected for analysis and sealed in cans or with aluminium foil. Some consumer products such as cleaning agents were also collected after the building had been occupied. These samples were then subject to emission testing using 2.4 L micro-chambers and 1 m^3 environmental test chambers, which have been described previously (Yu et al, 1997). The results were used to identify the VOCs released and compare these with the VOCs found in the different parts of the building atmosphere.

RESULTS AND DISCUSSION

Table 1 shows the concentrations of formaldehyde and some of the major VOCs and TVOC in the ground floor and top floor offices over the three-year period of monitoring. As would be expected in a new building, a wide range of VOCs and their sources were identified, but not at hazardous concentrations. The outdoor concentrations were much lower and, except for toluene, was not an important source of these compounds in the indoor air. The mean concentration for summer months (April to September) and winter months (October to March) is presented. Also included is the total number of individual compounds detected in the air by the sampling and analytical method. Figures 1 and 2 illustrate the changes in the indoor TVOC concentrations over the monitoring period.

In the ground floor office it took five months for the TVOC concentration to decline from a peak of 1225 $\mu g/m^3$ to about 400 $\mu g/m^3$. After six months, the concentration of TVOC was about 200-300 $\mu g/m^3$. The mean concentration of TVOC is generally higher in winter than summer, probably due to lower ventilation rates in winter. There are also other occasional rises in TVOC concentration as shown in Figure 1, which are probably due to differences in occupant activities. The TVOC concentration in the unoccupied top floor open plan office was consistently above the Australian indoor air quality (IAQ) guideline value (Dingle, 1993) of 500 $\mu g/m^3$ during the first year of monitoring. For the other two years it was mostly in the range 400-500 $\mu g/m^3$, which is higher than some other proposed targets for IAQ (European Concerted Action, 1992). The lower concentrations on the ground floor than the top floor probably occurred because of the higher ventilation rate on the ground floor, provided by the stack ventilation system. The top floor was mostly unoccupied during the three-year monitoring period and the windows were usually closed. With occupation, it is to be expected that concentrations would have declined more rapidly.

TABLE 1

MEAN WINTER AND SUMMER CONCENTRATIONS OF FORMALDEHYDE AND SOME VOCS ($\mu g/m^3$)

Ground Floor Open Plan Office:

	HCHO	TVOC	TXIB	Xylenes	Toluene	Texanols	Undecane	No. of VOCs
Summer 97	21	361	159	62	14	53	4	311
Winter 97	18	648	337	119	36	117	6	365
Summer 98	31	199	92	14	25	25	1	252
Winter 98	30	351	158	24	93	28	2	265
Summer 99	33	174	143	8	5	22	1	161
Winter 99	33	307	169	21	16	42	3	233

Top Floor Open Plan Office:

	HCHO	TVOC	TXIB	Xylenes	Toluene	Texanols	Undecane	No. of VOCs
Summer 97	33	628	242	25	14	29	21	345
Winter 97	19	1094	539	50	45	79	45	372
Summer 98	28	383	202	10	11	41	9	267
Winter 98	23	478	267	13	24	43	9	279
Summer 99	59	503	259	8	30	36	9	223
Winter 99*	37	409	214	12	49	22	5	259

* = October to December, HCHO = formaldehyde, TXIB = 2,2,4-trimethyl-1,3 pentanedioldiisobutyrate

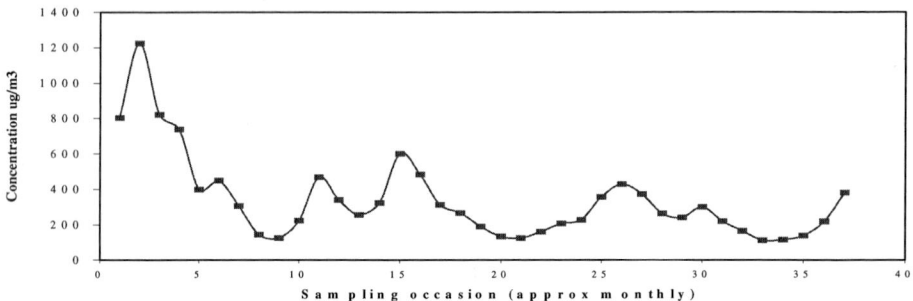

Figure 1: TVOC concentrations in the ground floor open plan office

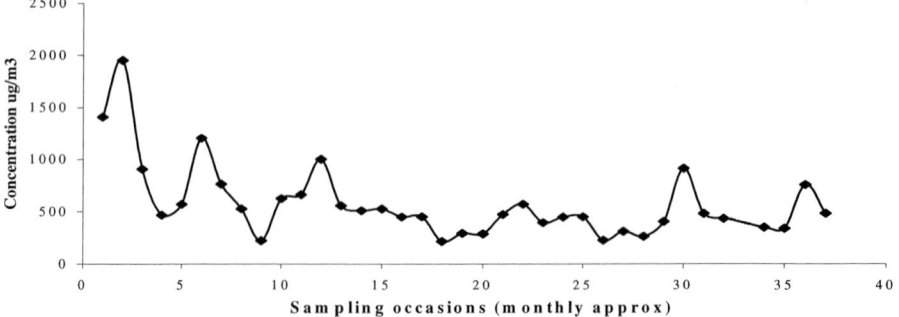

Figure 2: TVOC concentrations in the top floor open plan office

TABLE 2

THE TEN COMPOUNDS FOUND IN THE OPEN PLAN OFFICES AND IDENTIFIED EMISSION SOURCES
Upper half of table is 1st Year (7/1/97-12/1/97), lower half is 3rd Year (12/1/99-14/12/99)

Ground Floor Open Plan Office			Top Floor Open Plan Office		
	Major VOCs	Source Materials		Major VOCs	Source Materials
1	TXIB	Carpet tiles	1	TXIB	Carpet tiles
2	Xylenes	Carpet, panel, pedestal, ceramic tile & wood adhesives; carpet tiles; chairs; coatings & lacquer; parquet tiles; treated timbers; paint & coated mineral wool floor insulation	2	2-Ethyl-hexanol	Carpet tiles; ceramic adhesives; sealant; paints
			3	Xylenes	Carpet, panel, pedestal adhesives; carpet tiles; chairs; coatings & lacquer & treated timbers; paint & coated mineral wool floor insulation
3	TMBs	Carpet, panel, pedestal, ceramic tile & wood adhesives; carpet tiles; chairs; coatings & lacquers; treated timbers; parquet tiles; paint & coated mineral wool floor insulation	4	Undecane	Carpet tiles; pedestal & wood adhesives; paint, coatings; lacquers; sealant; chairs; treated timbers; coated mineral wool floor insulation; parquet tiles; cleaning products
4	Texanols	Water-based paints	5	Dodecane	Carpet tiles; pedestal & wood adhesives; paint, coatings; lacquers; sealant; chairs; treated timbers; coated mineral wool floor insulation; parquet tiles
5	Decane	Carpet tiles; pedestal & wood adhesives; paint, coatings; lacquers; sealant; chairs; treated timbers; coated mineral wool floor insulation; parquet tiles; cleaning products	6	Tridecane	Carpet tiles, treated timbers; chairs; coatings; lacquers; coated mineral wool; treated timbers
6	Toluene	Carpet, panel & ceramic adhesives; sealant; chairs; paints; coatings & lacquers; treated timbers; parquet tiles; coated mineral wool floor insulation	7	Toluene	Carpet, panel & ceramic adhesives; sealants; chairs; paints; coatings & lacquers; treated timbers; parquet tiles; coated mineral wool floor insulation
7	2-Ethyl-hexanol	Carpet tiles; ceramic adhesives; sealant; paints	8	Unknown (RT 33.11)	Not identified
8	Ethyl-benzene	Carpet, pedestal, ceramic & wood adhesives; sealant; chairs; coatings; paints; treated timbers	9	Decane	Carpet tiles; pedestal & wood adhesives; paint, coatings; lacquers; sealant; chairs; treated timbers; coated mineral wool floor insulation; parquet tiles; cleaning products
9	Butyl glycol	Water-based paints; floor polish	10	TMBs	Carpet, panel, pedestal, ceramic tile & wood adhesives; carpet tiles; chairs; coatings & lacquer; treated timbers; parquet tiles; paint; coated mineral wool floor insulation
10	Phenol	Carpet tiles; chairs; plywood			
1	TXIB	See above	1	TXIB	See above
2	Xylenes	See above	2	Pinene	Timbers; plywood; wood-based furniture; parquet adhesives; chairs; MDF panels; wood coatings; polishes & cleaning products
3	Pinene	Timbers; plywood; wood-based furniture; parquet adhesives; chairs; MDF panels; wood coatings; polishes & cleaning products			
4	Toluene	See above	3	Cyclo-hexane	Treated timbers; coatings
5	TMBs	See above	4	MEK	Carpet, parquet, pedestal & ceramic adhesives; sealant; treated timbers
6	Texanols	See above			
7	Ethyl-toluene	See above	5	Ethyl-acetate	Chipboard & furniture; timbers; plywood
8	MEK	Carpet, parquet, pedestal & ceramic adhesives; sealant; treated timbers	6	Xylenes	See above
9	Hexanoic acid	Paint, wood coatings; timbers; chipboard furniture	7	Limonene	Timbers; parquet adhesives; chairs; air freshener; cleaning products & polishes
10	Limon-ene	Timbers; parquet adhesives; chairs; polish; cleaning products; air freshener	8	2-ethyl-hexanol	Carpet tiles; sealant; paints
			9	Undecane	See above
			10	Toluene	See above

TMBs = trimethylbenzenes, 4PC = 4 phenylcyclohexene, MEK= methylethylketone, RT= retention time (minutes)

The formaldehyde concentrations were below the World Health Organisation air quality guideline of 100 µg/m^3 (WHO, 1987). In contrast to the TVOC, the summer concentrations were generally higher than those in winter. This suggests that the source of formaldehyde was stronger in summer and this offset the effect of higher ventilation. It is known that the emission of formaldehyde from products such as wood-based panels that contain urea formaldehyde adhesives increases with temperature. The sources of formaldehyde emission found were the wood-based furniture products, MDF wall panels, parquet flooring, carpet tiles and to a lesser extent, plywood panels. The major emission sources for the 10 dominant VOCs found in the air of the building is shown in Table 2.

CONCLUSIONS

The building materials were found to be a source of a wide range of VOCs. For most of the major VOCs in the air there were multiple material sources. Single sources were found for TXIB, a plasticiser in flooring products and Texanols, coalescing solvents in water borne emulsion paints. Further work is examining the influence of consumer products such as cleaning materials on the IAQ and quantifying the strength of the sources identified in the office building. The results to date indicate that, as would be expected in a new building, the building materials and furnishing products are the dominant source of VOCs in the indoor environment of a new office building.

The building did not have unusually high VOC levels and the main point of this work was not to measure levels but to identify sources. IAQ should ideally be controlled by a combination of limiting emission sources and having well planned ventilation. To reduce VOC concentrations in buildings, it is necessary to choose materials that will produce a low pollution load in the indoor air. Increasingly, information on emissions from materials is becoming available and, therefore, systematic selection of low-polluting materials should become a standard element of designing buildings for good IAQ (CEN, 1998). The availability of material emission databases, such as that being developed in this project, will enable building designers in the future to specify materials that have the least adverse impact on the building air quality.

REFERENCES

Dingle P. & Murray F. (1993). Control and regulation of indoor air: an Australian perspective. *Indoor Environment*, **2**, 217.

European Concerted Action (1992). *Guidelines for ventilation requirements in buildings.* ECA Indoor Air Quality and its Impact on Man, Report No. 11. Commission of the European Communities, Luxembourg, Report EUR 14449 EN.

European Committee for Standardisation, CEN/TC 158 (1998). *Ventilation for buildings – Design criteria for the indoor environment.* CEN Report CR 1752, December 1998. E. C. Brussels.

NatVent CD-ROM (1999), *Environmental Office, Case Study Summary.* BRE, Watford, UK.

Perera M.D.A.E.S & Tull R.G. (1990). BREFAN – a diagnostic tool to assess the envelope air leakiness of large buildings. *Proceedings of the CIB W17 meeting, 'Quality of the Air and Air Conditioning'.*

World Health Organisation (1987). *Air Quality Guidelines for Europe.* WHO, Copenhagen.

Yu C.W.F., Crump, D.R., & Squire, R.W. (1997) Sources and concentrations of formaldehyde and other volatile organic compounds in the indoor air of four newly built unoccupied test houses. *Indoor & Built Environment*, **6**, 45-55.

Yu C.W.F., Crump D.R. & Squire R.W. (1999). The impact of VOC emission on the indoor air quality of a newly built energy efficient office building at BRE. *Proceedings of Indoor Air 99*, Vol. 2, 489-494.

© 2000 Elsevier Science Ltd. All rights reserved.
Air Distribution in Rooms, (ROOMVENT 2000)
Editor: H.B. Awbi

INVESTIGATION OF CABIN AIR QUALITY IN COMMERCIAL AIRCRAFTS

Shun-Cheng Lee[1], Sanches Lam[1] and Fred Luk[2]

[1]Environmental Engineering Unit,
Hong Kong Polytechnic University, Hong Kong
[2] Cathay Pacific Airways, Hong Kong

ABSTRACT

This project covered 16 aircrafts including both smoking and non-smoking flights from June 1996 to August 1997. The parameters concerned were carbon dioxide (CO_2), humidity, temperature, carbon monoxide (CO), ozone (O_3), bacteria, fungus, and respirable suspended particulate (RSP). Compared with the Federal Aviation Administration (FAA) standard, CO_2, CO and ozone levels on all flights were within such standards. Peak levels of CO_2 and particulate were observed during both boarding and de-boarding periods. For the smoking flights, the average particulate level ($138\mu g/m^3$) was much higher comparing with the non-smoking flights ($7.6\ \mu g/m^3$). Low humidity in long haul flights caused uncomfortable conditions to the cabin crews. The average temperature was within the range of 23+/- $2^{\circ}C$.

KEY WORDS

Cathay Pacific, Aircraft; Cabin air quality; Hong Kong; Cabin crew, Humidity

INTRODUCTION

Air travel has become an essential form of transportation in modern society. The health and comfort of air travellers depends on the complex interplay of several factors such as the adequacy of the ventilation system to allow fresh air in the occupied zone; the concentration of contaminants; the temperature; and relative humidity. (O'Donnell et al, 1991). For years, flight attendants have reported various health problems - from chronic bronchitis to difficulties in pregnancy(Nagda et al, 1992). Environmental engineers have become aware of the effects of pollution in confined spaces and have revised ventilation. Spengler (1994) stated in his report that reduced amounts of outdoor air do not necessarily translate to poor air quality and increased risk of disease. Air cleaning and removal of pollutant sources mitigate some of the effects of decreasing dilution air. This was the first intensive study for aircraft air quality in Hong Kong. The detailed objectives of this paper are: **1)** To assess the indoor air quality on 16 Cathay Pacific Airways commercial aircrafts; **2)** To compare the Indoor Air Quality (IAQ) levels with existing regulations and standards including FAA, American Society of Heating, Refrigerating, and Air-Conditioning Engineers (ASHRAE), International Standard Organisation (ISO), etc; and **3)** To provide and evaluate different remedial measures for IAQ improvement in the aircraft for CPA.

METHODOLOGY

The sampling size covered three different types of aircrafts including Boeing 747-400, Airbus 330 and Airbus 340 both smoking and non-smoking flights. Details of flight information is listed in Table 1 below. The sampling strategy and methodology employed are summarized in Table 2.

TABLE 1
FLIGHT INFORMATION

Flight	Date	Duration (hour:minute)	Occupancy (%)	Flight	Date	Duration (hour:minute)	Occupancy (%)
1	Sept 2, 96	11:29	-	9	Feb 28, 97	3:20	100
2	Sept 5, 96	13:05	-	10	Mar 2, 97	3:11	69
3	Sept 19, 96	12:52	-	11[*]	May 8, 97	1:25-stopover-3:00	60
4	Sept24, 96	11:30	-	12[*]	May 13, 97	5:00	67
5	Jan 13, 97	12:35	93	13	Jun 8, 97	14:15	34
6	Jan 15, 97	11:35	88	14	Jun 12, 97	6:55-stopover-12:50	-
7	Jan 24, 97	13:15-stopover-1:20	83-stopover-33	15[*]	Jun 28, 97	1:25	91
8	Jan 28, 97	1:20-stopover-11:50	32-stopover-78	16	Jun 28, 97	1:25	-

[*]*Smoking flight, Fight 3 & 16 are Airbus 340, Flight 9 is Airbus 33*

TABLE 2

SUMMARY OF SAMPLING METHODOLOGY AND STRATEGY

Parameters	Measurement Technique	Sampling interval	Flight(s) measured
Carbon monoxide	Electro-chemical	Every5 minute	9, 10, 11, 12, 13, 14, 15, 16
Carbon dioxide	Non Dispersive Infrared	Every5 minute	All
Temperature	Thermistor	Every 5 minute	All
Relative humidity	Thin-film capacitive	Every 5 minute	All
Ozone	Passive ozone badges or Bio-check enzyme	Integrated sample taken in the test flight.	6, 8, 10, 11, 16
Respirable Suspended Particulate	Light-scattering	Every 5 minute	5, 6, 11, 12, 13, 14, 15, 16
Microbiological Organisms	Burkard air sampler with agar plates	Up to twice per flight	8, 10, 12

Air from the cabin was sampled at a specific sampling frequency (Table 2) during the entire flight and its contents of NO, SO_2, NO_2 , and CO were measured. Humidity, temperature, CO_2, and RSP were continuously monitored. The sampling and analysis of the chemical and gaseous contaminants were performed according to the standard methods acquired from American Public health association, American conference Governmental Industrial Hygienists, American Society for Testing Materials, and National Institute for Occupational Safety and Health.

RESULTS AND DISCUSSION

The average CO_2 and CO levels measured were below FAA standard of 30,000(Table 3). Overall average bacteria counts(75 CFU/m^3) were very low. The average temperature was $21.9^{\circ}C$, which was within the ASHRAE range ($23\pm2^{\circ}C$), and were stable during cruise. All of the measured RSP, CO_2, and CO concentrations were higher in smoking flights than non-smoking flights. (Fig 1)

TABLE 3

HEALTH REGULATIONS AND COMFORT STANDARDS FOR AIRCRAFT

Contaminant	Federal Aviation Administration	ASHRAE, 62-1988, 55-1989
Carbon dioxide (ppm)	30,000 continuous	1,000 continuous
Carbon monoxide (ppm)	50 continuous	
Ozone (ug/m3)	20 above 27,000ft	
Relative humidity (%)		20 minimum
Temperature (°C)		19-23, winter 23-26, summer

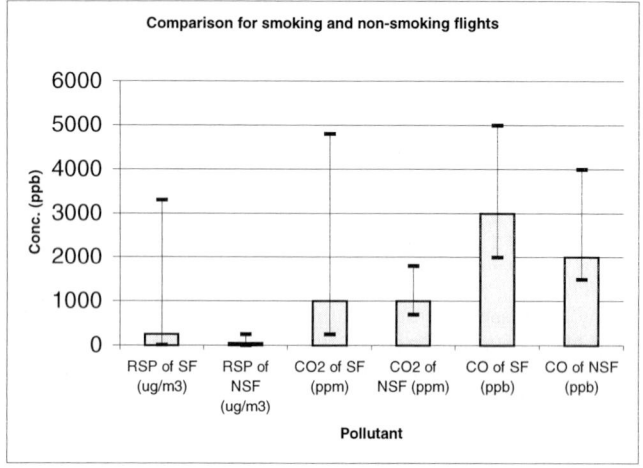

Fig 1 Comparison of smoking with non-smoking flight

The higher CO levels (3-5 ppm) in smoking flights did not exceed any relevant standards. However, the RSP levels were high for the smoking flights (71-264µg/m3). And given the low relative humidity and high ozone concentration, irritation and discomfort were likely to be important for occupants of the smoking and nearby sections. A similar temporal variation pattern of CO_2, RSP, bacteria, and fungi concentrations was dominated by the occurrence of boarding and de-boarding periods. Lower fresh air supply and the exhaust gases from the airport resulted higher CO_2 levels during boarding and de-boarding. Also, peak RSP, bacterial and fungi counts were occurred during active pre-boarding and de-boarding as passengers were retrieving luggage and leaving. There was significant difference on minimum and average humidity on long-haul, medium-haul, and short-haul flights. Both average humidity and minimum humidity dropped as flight time increased. For short-haul flight, the minimum humidity was 16.3% compared with the long-haul flights at 6.7%. (Figure 2)

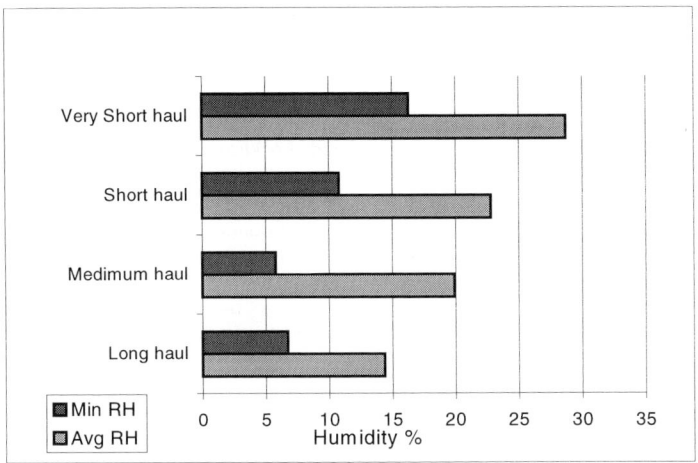

Figure 2   The impact of flight time on Humidity

The distribution of pollutants in the aircraft was not uniform. The concentration measured in any area would depend on location of the sampler in relation to the pollution source. The CO_2 levels in the rear of Economy Class was higher, suggesting a front to rear movement of the air. The humidity levels in toilet and galley area were the highest, suggesting the main sources of humidity in the aircraft were from the food preparation.

The CO_2 level was reduced by 29% due to the increase in the total cabin airflow when the ventilation was switched to high mode. But higher outside-air ventilation rates lowered humidity by 27% due to the relatively dry outside air. Temperature was not affected by the high mode ventilation.

CONCLUSION

A total of 16 flight audits on CPA aircraft were investigated in this project. In general, the aircraft air quality on CPA aircraft was satisfactory. The CO_2 levels on all flights were below the FAA standard. The relative humidity was low, with minimum 4.9% especially for long haul fights. The minimum relative humidity level recorded was below the ASHRAS standard, and was not uniform with the highest in first class and the lowest in economy class. Average particulate concentration on a smoking flight was 1,815% higher than that of non-smoking fight. CO_2 and relative humidity levels were both reduced by 29% and 27.6%, respectively, by switching ventilation from low mode to high mode. The bacteria levels on CPA flights were generally low.

ACKNOWLEDGEMENTS

This project is supported by the Environmental Office of Cathay Pacific Airways Limited and the Hong Kong Polytechnic University.

REFERENCES

Aerospace Industries Association of America, Inc. Airplane Air Conditioning System Configuration and Air Flow Data for Selected Boeing, Douglas, and Lockheed Aircraft. (unpublished document, September 17, 1985)

American Society of Heating, Refrigerating, and Air-Conditioning Engineers, Inc. ASHRAE Standard: Ventilation for Acceptable Air Quality. ASHRAE 62-1981. Atlanta, Gal: American Society of Heat, Refrigerating, and Air-Conditioning Engineers, Inc., 1981.

Malmfors, T., Thorburn D., and Westlin A. (1989). Air Quality in Passenger Cabin of DC-9 and MD-80 Aricraft. *Environmental Technology Letters*, **10**, 613-628

Nagda, N.L., Koontz, M. D., Konheim, A. G and Hammond, S. K (1992). Measurement of cabin air quality aboard commercial airliners. *Atmospheric Environment*, **12**, 2203-2210

O'Donnell,A., Donnini, G. and Nguyen, V.H (1991). Air Quality, Ventilation, Temperature and Humidity in Aircraft. *ASHRAE Journal*, April, 42-46

Spengler, J., Burge, H., Dumyahn, T., Dalhstrom, C., Muilenberg, M., Milton, D., Ludwig, and Weker, R. (1994). Harvard Aircraft Cabin Environmental Survey. Department of Environmental Health, Harvard University School of Public Health, Boston, MA, USA.

U.S. Federal Aviation Administration. FAA Aviation Forecasts, Fiscal Year~1985-1996. FAA-APO-85-2. Washington, DC.: U.S. Federal Aviation Administration, 1985

U.S. Federal Aviation Administration. Transport Category Airplanes Cabin Ozone Concentrations. Advisory Circular 120-38. Washington, D.C. U.S federal Aviation Administration, 1980.

Air Distribution in Rooms, (ROOMVENT 2000)
Editor: H.B. Awbi

EXPERIMENTAL EVALUATION OF SMOKING LOUNGES: A CASE STUDY

Fabrizio Cumo, Livio de Santoli and Gino Moncada Lo Giudice[*]

Department of Fisica Tecnica , University of Rome "La Sapienza"
Via Eudossiana, 18 – 00184 Rome – Italy

[*]CNRCSMCOA , via Montedoro 28
00186 – Roma – Italy.

ABSTRACT

The paper deals with main design criteria for smoking lounges giving a description of a measurement campaign in smoking lounges realized in the Italian House of Parliament, "Palazzo Montecitorio", in Rome. In the first part of the paper are shortly described the characteristics of the environmental tobacco smoke (E.T.S.), with a review of the available data related to sensorial responses to ETS concentrations in controlled settings. Main design criteria for smoking lounges are then pointed out, including calculation of ventilation rate, location of smoking areas, lay-out of air distribution devices, filtration issues and heat recovery related problems. The application of the above criteria complying with a CFD code, will lead to the realization of three smoking lounges in Palazzo Montecitorio. An experimental field measurement has been carried out, monitoring some ETS indicators (particulate, CO, CO_2,TVOC) and the main thermal hygrometric parameters (mean temperature, air velocity, relative humidity, radiant temperature). As a result of these measurements, important suggestions about the use of heat recovery device, the efficiency of particulate filtration system and the design ventilation rates have found confirmation.

KEYWORDS

ETS, IAQ, Smoking lounge, CFD, TVOC.

CHARACTERISTICS OF ENVIRONMENT TOBACCO SMOKE (ETS).

Environmental tobacco smoke (ETS), or second-hand smoke, is a complex mixture of both gas and particulate phase compounds composed of the aged and diluted combination of both side stream (smoke from the lighted end of a cigarette) and exhaled mainstream smoke. Most of the compounds identified in ETS have other indoor or outdoor sources, so in order to determine the contribution from smoking to these compounds in indoor air, it si necessary to measure marker compounds. The three most commonly used markers for the particulate phase of ETS (ETS-RSP), are: ultraviolet particulate matter (UVPM), fluorescent particulate matter (FPM), and solanesol. In the gas phase, nicotine concentrations do not track the concentrations of most other measurable ETS components (the most significant gas phase components are reported in table 1).The current best marker for the gas phase of ETS is 3-ethenylpyridine (3-EP); experiments performed in controlled laboratory setting suggest that odour is the primary factor associated with sensory evaluation of ETS in the air.

TABLE 1
SALES-WEIGHTED AVERAGE ETS YIELDS OF GAS-PHASE COMPONENTS

Compound	Yield (μg/cigarette)	Compound	Yield (μg/cigarette)
RSP	13,700	Solanesol	410
Scopoletin	18.2	Catechol	11.2
Carbon Monoxide	55,100	Total Hydrocarbons (FID)	27,800
TVOC (Sorbent Tube)	19,100	Isoprene	6,200
Ammonia	4,100	Acetaldeyde	2,500
Nitric oxide	1,650	Nicotine	1,590
Formaldehyde	1,330	Toluene	500
Benzene	280	m-Xylene	176

ACCEPTABILITY LEVELS OF ETS AND DILUTION VENTILATION RATES.

Some recent studies have been carried out in order to quantify the effect on smokers and non-smokers of ETS concentrations ranging from 50 to 1000 $\mu g/m^3$, considering both physical and psychological aspects. The results of this researches confirm that smokers find air quality acceptable at higher concentrations of ETS than non-smokers.

Among non-smokers in a laboratory setting, 80% acceptability of air quality is achieved when the ETS from one cigarette is diluted by 80 m3 – 120 m3 of air while smokers perceive the level of 80% of acceptability with 30 – 40 m3 of fresh air diluting each cigarette. The results of some laboratory studies on acceptability of ETS for non-smokers are shown in Figure1.

ETS concentration determined in laboratory are likely to overestimate the impact of ETS on non-smokers in real word settings; that is, 80% acceptability is likely to be achieved at higher concentration in the field than those observed in the laboratory. In table 2 are reported some measured values ($\mu g/m^3$) of ETS-RSP (mean particulate diameter of 0.141 μm with a mass median diameter of 0.21 μm; all the particles are of respirable size) in offices and restaurants. TVOC are measured by means of a FID gascromatograph (toluene as reference component). The accordance with the data appears to be satisfactory. For the evaluation of room ventilation rates it is important to underline that the dilution rates derived using statistical average smokers data agree well with those in ANSI/ASHRAE Standard 62-1989 (ASHRAE 1989).

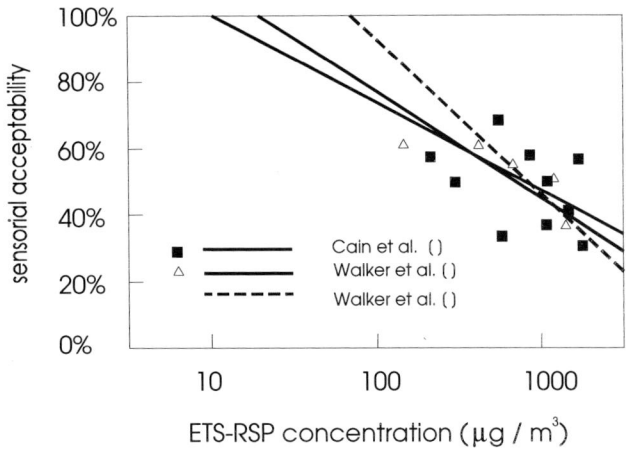

Figure 1: Acceptance data for non smokers as a function of ETS-RSP concentration

TABLE 2
TYPICAL VALUES OF ETS-RSP IN CONFINED SPACES

ETS-RSP , $\mu g/m^3$	Offices	Restaurants
Mean values with smokers	30	40
Maximum values with smokers	100	150
Mean values without smokers	10	20
Maximum values without smokers	80	100

CASE STUDY

On the base of the above illustrated criteria a pilot campaign for the realization of 35 smoking lounges in the five building propriety of the Italian House of Parliament "Camera dei Deputati" in Rome has been planned by the competent Office for the Workers Health and Safety. In the first experimental phase of the programme three smoking lounges have been realized this year in the basement and in the first floor of Palazzo Montecitorio by the Department of Fisica Tecnica of the University of Rome "La Sapienza" in cooperation with CNRCSMCOA in order to optimise the design criteria and the removal effectiveness.

In particular, in two zones the dilution air flow rate is given by an enthalpic heat-recovery device $(\eta = 70\%)$ which enables to extract the air from the smoking area and to supply the outside air nearby, almost at the same temperature (see Figure 2)

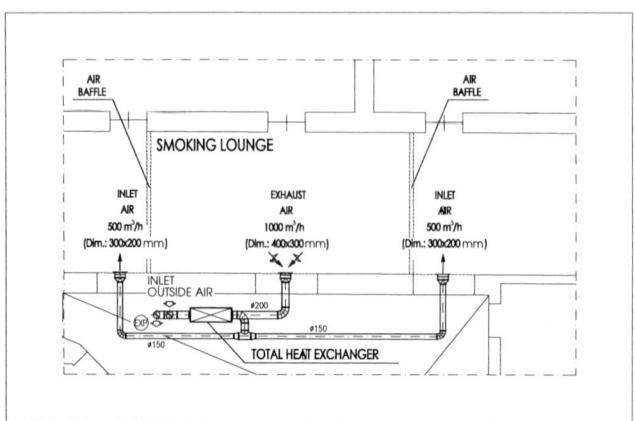

Figure 2:Schematization of the smoking lounge

In the third zone a device for the exhaust is located inside the smoking lounge and the dilution flow rate is provided by the transfer air coming from the other indoor area around. The three zones by means of air baffles are able to limit the pollutant diffusion outside the smoking lounges.

CFD SIMULATIONS

In order to optimise the main design parameters as flow-rate and air velocity, some bi-dimensional and three-dimensional numerical simulations have been carried out, using the computational fluid dynamic code Fluent[®].In Figure 3 are shown the velocity vectors of the zone schematised in figure 2 at 3 m height while in Figure 4 is shown the correspondent pollutants concentration

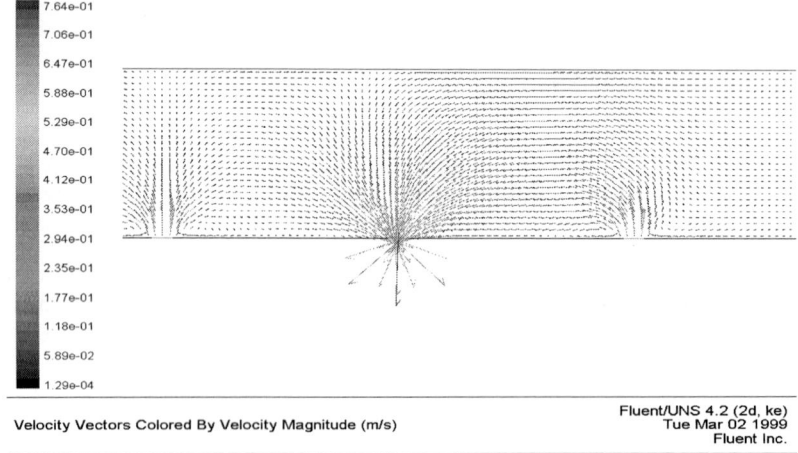

Figure 3: CFD simulation of air-flow velocity

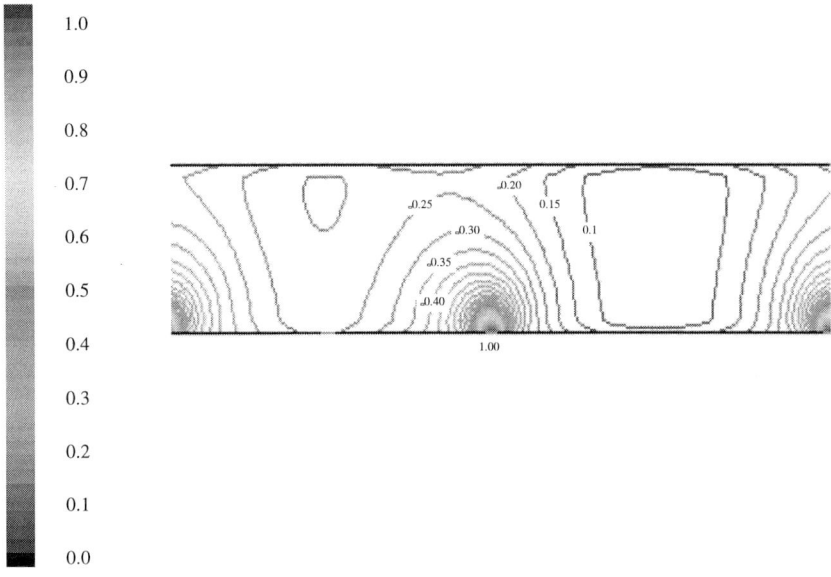

Contours of relative concentrations

Fluent/UNS 4.2 (2d, ke)
Tue Mar 02 1999
Fluent Inc.

Figure 4: CFD simulation of pollutants concentration

Figure 3 shows that the mean air velocity inside the smoking lounge is lower than 0.5 m/s with a maximum value of 0.8 m/s near the extraction of exhaust air, while in Figure 4 it is possible to notice an increase of pollutant concentration all around the extraction zone of the heat exchanger device

RESULTS OF THE EXPERIMENTAL MEASUREMENTS

The results of experimental measurements for the two smoking lounges equipped with the total heat exchanger show that the concentration values for the most significant pollutants are lower than the threshold limits, both inside and outside the smoking zone.

The on site measurement have been carried out in seven days for all the different occupational situation, using a gascromatograph (type VOCOL with a 60 m metal capillary column, diameter 0.25 mm) and a TESTO 650 CO and CO_2 analyzer.

Some experimental results regarding CO, CO_2, particulate and some VOC are shown in Table 3, where are listed concentrations for the smoking lounge, the contiguous no-smoking zone, the mean bulk pollutants concentration for the whole building and the outside air.

TABLE 3
POLLUTANTS CONCENTRATIONS IN DIFFERENT BUILDING ZONES

POLLUTANT		SMOKING LOUNGE	CONTIGUOUS NO SMOKING ZONE	BUILDING	OUTSIDE AIR
CO	(ppm)	2.4	1.9	2.0	0.0
CO_2	(ppm)	849	778	760	482
ETS-RSP	(mg/m^3)	0.318	0.212	0.057	0.034
Benzene	$(\mu g/m^3)$	23.5	18.2	10	10.7
Toluene	$(\mu g/m^3)$	80.3	68.8	48	30
Achilbenzene	$(\mu g/m^3)$	82.4	42	48.1	22.9

Some other important results have been obtained considering the pollutants concentration with the heat exchanger switched off. In such a case the benzene concentration grows of about 50%, the toluene concentration grows of about 30%,the achilbenzene concentration grows of about 20%.

The tests carried out in the smoking zone only equipped with the exhausts air device have confirmed a reduction in the performance because of increasing poor IAQ in the contiguous no-smoking zones.

CONCLUSIONS

According to the design criteria suggested by ASHRAE, it is possible to be confident that the ventilation devices employed in the three smoking lounges in this experimental phase provide a marked reduction of contaminants concentration due to the presence of ETS .Moreover the settled confined zone for smokers, coupled with an appropriate dilution flow-rate allows a reduced diffusion of ETS contaminants through the non-smokers environment. In particular the use of an high efficiency total heat exchanger yields both better thermo-hygrometric conditions and a decrease of contaminants concentration of about 50% .Measured concentration values validate computational analyses and give an important design tool in order to limit airborne pollutants due to the presence of ETS in selected zones.

REFERENCES

Nelson P., Bohanon H and Walker J.C. (1998).*Design for Smoking Areas: Part 1-Fundamentals. and Part 2- Applications.* ASHRAE Transactions 1998 V. 104, Pt 2
Cain W.S,Learder B., Isseroff R., Berglund L.G., Huey R.J., Lipsitt E.D.and Perlman D. (1983).*Ventilation requirements in buildings -Control of occupancy and tobacco smokes* .Environmental International, 15:19-28.
Leaderer B.P., Cain W.S., Isseroff R. and Berglud L. (1984).*Ventilation requirements in buildings: Part II – Particulate matter and carbon monoxide from cigarette smoking.* Atmospheric Environment, 18(1):99-106.
Leaderer B.P. and Cain W.S. (1983). *Air quality in buildings during smoking and non smoking occupancy.* ASHRAE Transactions 89:601-613.
ASHRAE Standard 62-89 (1989).*Ventilation for acceptable indoor air quality.* Atlanta
CEN PrENV 1752 (1997). *Ventilation for Buildings.*
De Santoli L. and Fracastoro G. (1998) *La qualità dell'aria negli ambienti interni.*AICARR, Milano

© 2000 Elsevier Science Ltd. All rights reserved.
Air Distribution in Rooms, (ROOMVENT 2000)
Editor: H.B. Awbi

CFD SIMULATION OF CONCENTRATION
OF GASEOUS IMPURITIES IN A TYPICAL HONG KONG
INDUSTRIAL WORKSHOP

Z Lin, T T Chow, K F Fong and J P Liu

Division of Building Science and Technology, City University of Hong Kong

ABSTRACT

One of the reasons of using displacement ventilation is that it may provide better indoor air quality in the occupied zone than mixing ventilation. It is therefore important to understand the performance on this aspect. A validated CFD model was employed to generate concentration distribution data for CO_2, radon and moisture in a typical Hong Kong industrial workshop with displacement ventilation. Analysis found that the concentration distribution affected by factors such as the source type and location and its associated plume strength and human body convection, etc. The indoor Radon comes from building materials. The indoor CO_2 and moisture are from the occupants. Because the upward free convection around a person brings the air from lower level to the breathing zone, the inhaled air is cleaner than the air at the same height. Prediction of contaminant distribution is more difficult than air temperature and flow distribution.

KEYWORDS

workshop, displacement ventilation, IAQ, gaseous impurities, CFD simulation.

INTRODUCTION

Recent evidence indicates that an increasing number of complaints in industrial indoor environments are due to worker's excessive exposure to indoor pollutants. The majority of health problems reported cannot yet be attributed to specific exposures. The problems are often called sick building syndrome. Available evidence suggests that multiple factors are involved, including indoor air quality such as microbiological and chemical exposures not adequately characterised by current assessment approaches; physical conditions such as temperature, humidity, lighting and noise; and social/psychological stressors (NIOSH, 1996). The concentration level and the thermal conditions are related to the pollutant sources and the Heating, Ventilating, and Air Conditioning (HVAC) systems used in the buildings.

To assess the performance of the mechanical ventilation system and air quality in an office building, Cheong 1996 monitored the concentrations of a number of indoor pollutants including carbon dioxide (CO_2). The pollution sources that were objectives of the research of Fleming (1986) included Radon, Nitrogen Dioxide (NO2), Formaldehyde (HCHA), Total Suspended Particles, and

Carbon Monoxide (CO). Persily (1997) pointed out that indoor CO_2 concentrations can be useful for understanding indoor air quality and ventilation. But CO_2 concentrations do not provide a comprehensive indication of indoor air quality. Also, CO_2 can also be used to estimate building air change rates. Sterling et al. (1993) used CO_2 concentrations as an index of overall ventilation adequacy. In the evaluation of indoor air quality, perhaps the most important parameter measured is the concentration of CO_2 (Stonier, 1995).

Almost all the industrial workshop buildings in Hong Kong are of high rise or multi-storey structure and adopt central air-conditioning. Many of them have only fixed windows, *i.e.* no natural ventilation. Granite and concrete are widely used as building materials in Hong Kong, and the radon emanation rate has been found to be much higher for these materials compared with those from other places (Lee and Wong, 1997). Radon level has been confirmed to be on the high side as compared to many cities in the world (HKEPD, 1994). The indoor radon build up comes mainly from radon emission from building materials in Hong Kong, which is different from the situation in the United States and Europe (Lao, 1990). Exposure to high radon levels poses a serious health risk. According to a relative risk model, the calculated number of radon-induced lung cancer death in Hong Kong in 1986 was about 13% of the total number of lung cancer death for the year.

ANSI/ASHRAE Standard 62 - 1989 recommended that humidity levels inside buildings be maintained between 30% and 60% RH for proper Indoor Air Quality. Nearly all industrial buildings in Hong Kong have mechanical cooling systems that also dehumidify the air as it is being cooled to keep humidity levels from getting too high. Many air conditioning system designers base on the design room humidity level at 50% to select the cooling coil. Historical data show that the absolute moisture content value is very high in Hong Kong and residents in the territory are most likely used to the condition.

Ventilation, or more generally HVAC, with appropriate air-handling processes, is used to create an indoor environment with acceptable air temperature, humidity, and air velocity that are important physical conditions involved in sick buildings syndrome. Displacement ventilation has been used quite commonly in Scandinavia during the past twenty years as a means of ventilation in industrial facilities to provide good IAQ and to save energy. Displacement ventilation provides a lower contaminant concentration level in the occupied zone than that in the upper zone (Heiselberg and Sandberg, 1990). Chen (1988) pointed out that both the temperature and ventilation efficiency for displacement ventilation are much higher than those for mixing ventilation, when the contaminant source is combined with the heat source. The ventilation efficiency increases as the ventilation rate increases or the cooling load decreases.

The research into displacement ventilation was mainly conducted in Scandinavian countries. Hong Kong has higher temperatures in summer than those in Scandinavian cities. Besides, the industrial workshops in Hong Kong have higher densities of occupant and lighting/equipment that produce heat. Therefore, the cooling load is much higher in Hong Kong than that in Nordic countries. In many Hong Kong industrial workshops, there are large interior zones that are mostly isolated from the external climate. Cooling is always needed in these core spaces. There is a great potential to use displacement ventilation in such spaces. Therefore, direct application of the Scandinavian results for Hong Kong design is not appropriate. A set of design guidelines for application of displacement ventilation to Hong Kong Buildings must be developed (Lin *et al.*, 1998).

CFD SIMULATION

Two main approaches are available for the study of airflow and pollutant transport in buildings: experimental investigation and computer simulation. In principle, direct measurements give the most realistic information. Because the measurements must be made at many locations, direct measurements of the distributions are very expensive and time consuming. A complete measurement may require many months of work. Moreover, to obtain conclusive results, the supply airflow and temperature from the HVAC systems and the temperatures of building enclosure should be maintained unchanged during the experiment. This is especially difficult to achieve because the outdoor conditions change over time. The temperatures of the building enclosure and the airflow and temperature from the HVAC systems will also change accordingly.

Alternatively, the airflow and pollutant transport can be determined computationally by solving a set of conservation equations describing the flow, energy, and contaminants in the system. Due to the limitations of the experimental approach and the increase in performance and affordability of high speed computers, the numerical solution of these conservation equations provides a practical option for determining the airflow and pollutant distributions in buildings. The method is the Computational Fluid Dynamics (CFD) technique, which is used to analyse indoor atmospheric problems. Due to limited computer power and capacity available at present, turbulence models have to be used in the CFD technique in order to solve flow motion. The use of turbulence models leads to uncertainties in the computed results because the models are not universal. Therefore, it is essential to validate a CFD program by experimental data (Yuan *et al.*, 1998).

A validated CFD model based on the RNG k-ε model and wall function (Lin *et al.* 1999) was used to generate data for a typical industrial workshop in Hong Kong (Figure1). The whole computational domain, the space of the room, needs to be divided into a number of finite volumes by a grid system. The flow variables, such as velocity, temperature and concentration, are solved at the centre of each finite volume. The finer grid is, the more accurate the results will be. However, fine grid will cost more computing time and capacity. A commercial CFD code (AEA Technology 1997) was used for the computations.

The internal heat sources and the ventilation rate used in the experiment are listed in Table 1. This case presents the typical situation in Hong Kong. Each occupant in the industrial workshop was simulated by a box, 0.4 m long, 0.3 m wide, and 1.2 m high. The supply air temperature was controlled at 15°C for parameter zone (3 m from the external wall) and 14°C for interior zone. The outdoor air temperature is 33°C DB and 28°C WB. Adiabatic (thermal symmetrical) condition is applied for internal partitions. CO_2 was used as an indicator for contaminants and the outdoor CO_2 concentration was 350 ppm.

Previous study found that different finish building materials (e.g. plastic wall paper) can effectively reduce the indoor radon concentration (HK Polytech 1995). For the purpose of carrying out the worst case analysis of the indoor radon concentration, it was assumed that no finishing materials are applied on the floor, walls and ceiling. The radon emission rate is 16×10^{-3} Bq/m^{2}s.

TABLE 1

MAJOR PARAMETERS USED IN THE SIMULATION

Persons [number]	Lamp [W/set]	Large machine [W]	Small machine [W/set]	Air circulation [m^3/s]	Fresh air intake [l/(person s)]
8	90	8730	3511	1.935	10

SIMULATION RESULTS

The CO_2 concentrations are found generally lower than 1310 ppm except those spots of exhaled gases as shown on Figures 2 and 3. The radon concentrations are generally lower than 56 Bq/m^3 except very narrow stagnant zones along the walls (Figures 4 and 5), which is well below 200Bq/m^3 as set in WHO's guidelines (HKEPD 1994). Relative humidity is within the range of 50% to 65%

(Figures 6 and 7) around the occupants while lower in the other regions of the industrial workshop. Shown on Figure 8 is the result for the age of air for illustrating ventilation efficiency.

DISCUSSIONS

There is a convection boundary layer around a human body. Visualisation of the boundary layer (EC report, 1997 and US EPA, 1995) showed that the expired air did not break through the convection plume around the body, but flowed upwards with the plume. Holmberg *et al.* (1987) pointed out that the free convection flow around a person may protect the breathing zone from surrounding contaminant at the head level, but it may also attract contaminant from the source below the breathing zone. Saeteri (1992) showed that CO_2 concentration in the air inhaled is lower than that at the same elevation some distance from the person because the convection flow around the human body brings fresher air from the floor level directly to the breathing zone. This has been confirmed by Murakami et al. (1997) through a detailed CFD simulation. In most offices, many contaminant sources are associated with the heat sources, such as machines, computers, and other heated equipment. A similar pattern was observed from the current simulation. The convective flow around a human body brings the air at lower zone to the breathing zone. Therefore, the occupant actually breathes air with lower contaminant concentrations than those at the nose level in the middle of the room. This is evident in Figure 3.

Stymne *et al.* (1991) showed that contaminant concentration level varies significantly in both vertical and horizontal directions, depending on the position of pollutant sources relative to the thermal plumes. The contaminant concentration in the occupied zone is high when the contaminant is combined with a weak heat source. The thermal plumes are too weak to reach the upper zone. This is especially true for the radon situation in current research since the source is the enclosure structure and is not associated with any heat source (Figure 4, 5).

Figure 8 illustrates the age of air in the industrial workshop. Clearly, the age of air in the lower part of the room is much younger than that in the upper part of the room. The age of air at the breathing level in the classroom is not older than 167 seconds. With a perfect displacement ventilation system, the ventilation effectiveness is 2. A perfect or complete displacement is impossible in practice. The corresponding age of air will be older and the ventilation effectiveness will be lower than 2. The relative short age of air is due to high ventilation rate (ACH = 40).

CONCLUSIONS

Contaminant concentration distribution depends on contaminant source type and location and its associated plume strength, etc. Low contaminant concentration may be obtained in the occupied zone when the contaminant source is associated with a heat source and the thermal plume generated by the heat source is sufficiently strong to reach the upper zone. Because the upward free convection around a person brings the air from lower level to the breathing zone, the inhaled air is cleaner than the air at the same height. Because contaminant distribution depends very much on air velocity field and, to less extent, air temperature field, prediction of contaminant distribution is more difficult than air temperature and flow distribution. The displacement ventilation performed well on the aspect of indoor concentration of carbon dioxide, radon and relative humidity for industrial workshops under Hong Kong condition, providing that the system is designed, installed, maintained and operated according to sound engineering practice.

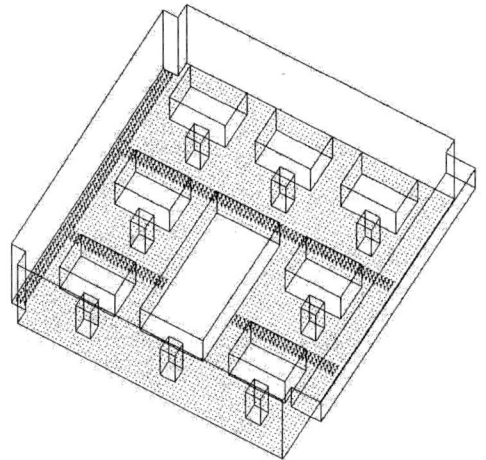

Figure 1: Typical industrial workshop

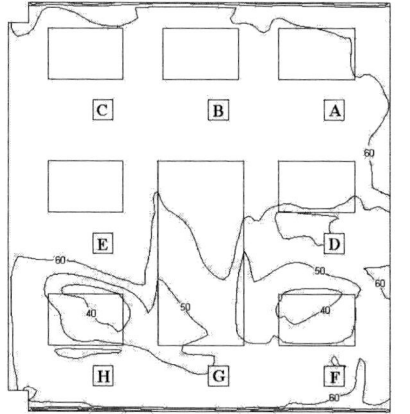

Figure 2: CO$_2$ concentration (plan view)

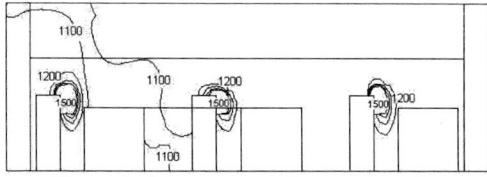

Figure 3: CO$_2$ concentration (section view)

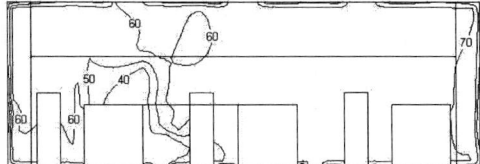

Figure 4: Radon concentration (section view)

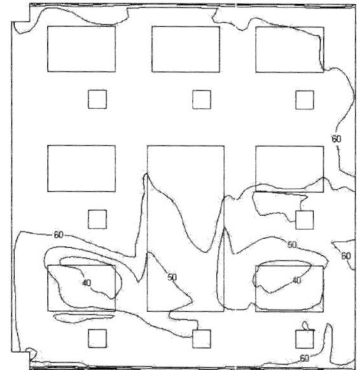

Figure 5: Radon concentration (plan view)

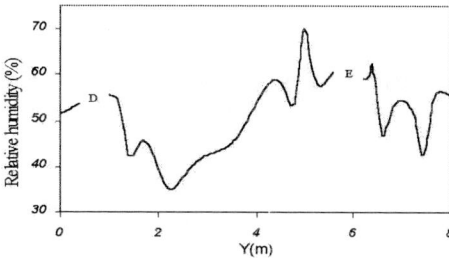

Figure 6: RH in individual offices

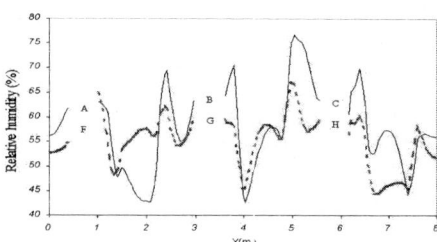

Figure 7: RH in cubicle office

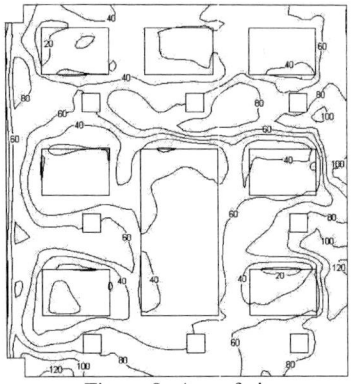

Figure 8: Age of air

REFERENCES

AEA Technology, (1997). CFX-4.2 Environment User Manual, Vol. 1 to 4.

ASHRAE. (1989). ASHRAE Standard 62-1989: Ventilation for Acceptable Indoor Air Quality, Atlanta, ASHRAE.

Chen Q. (1988) Indoor Airflow, Air Quality and Energy Consumption of Buildings, Ph.D. Thesis of Delft University of Technology, Delft.

Cheong K. W. (1996). Ventilation and air quality in an office building, *Building Services Engineering Research & Technology*, v 17, n 4, 167-176.

EC. (1997). The European data base on indoor air pollution sources in buildings, Final Report.

Fleming W. S. (1986). Indoor air quality, infiltration, and ventilation in residential buildings. Published by ASHRAE, Atlanta, GA, USA. pp. 192-207.

Heiselberg P. and Sandberg M., (1990) Convection from a slender cylinder in a ventilated room. Proceedings of *ROOMVENT '90: International Conference on Engineering Aero- and Thermodynamics of Ventilated Rooms*. Oslo.

HKEPD. (1994). Interim indoor air quality guideline for Hong Kong. EPD/TP/1/94.

HK Polytechnic University. (1995). Study of effectiveness of various finish materials in reducing the indoor radon level.

Holmberg R.B., Folkesson K., Stenberg L-G., and Jansson, G. (1987). Experimental analysis of office climate using various air distribution methods, *ROOMVENT ' 87, Stockholm*.

Lao K.Q. (1990). Controlling Indoor Radon, Measurement, Mitigation, and Prevention. *Van Nostrand Reinhold*.

Lee Y.M.R. and Wong C.K.P. (1997). "Territory-wide indoor radon follow-up survey 1995/96, Report No. EPD/TP/12/96, HKEPD.

Li X., and Jiang Y. (1996). Calculation of age-of air with velocity field. Paper presented in Post-IAQ'96 Seminar, Beijing.

Lin Z., Chow T.T., Fong K. F. and Chen Q. (1998). Review on Application of Displacement Ventilation to Hong Kong Buildings. *Proceedings of the Fifth International Conference on Tall Buildings*. Vol. I, 246-251.

Lin Z., Chow T.T., Fong K. F. and Chen Q. (1999), Validation of CFD Model for Research into Application of Displacement Ventilation to Hong Kong Buildings. The Proceedings of the 3rd International Symposium on HVAC. Shenzhen, China, 17-19 November, vol. 2, 602-613.

Miezwinki S., Nawrocki W., and Trzeciakiewicz Z. (1994). Air exchange efficiency under displacement ventilation conditions. *ROOMVENT '94*, Krakow.

Nielsen P.V. (1993). Air distribution in rooms - room air movement and ventilation effectiveness, *International Symposium on Room Convection and Ventilation Effectiveness, ISRACVE, ASHRAE*.

NIOSH. (1996). National Occupational Research Agenda, NIOSH, U.S. Department of Health and Human Services.

Persily A. K. (1997). Evaluating building IAQ and ventilation with indoor carbon dioxide. *ASHRAE Transactions*. v 103, n pt 2, 193-204.

Saeteri J. (1992). A breathing mannequin for measuring local ventilation effectiveness. ROOMVENT '92, Aalborg.

Sterling Elia, McIntyre Edward, Collett Christopher and Sterling Theodor. (1993). Field measurements for air quality in office buildings: a three-phased approach to diagnosing building performance problems, *ASTM Special Technical Publication 957*. Published by ASTM, Philadelphia, PA, USA, 46-65.

Stonier R T. (1995). CO2: powerful IAQ diagnostic tool, *Heating, Piping & Air Conditioning*, v 67, n 3, Mar. 4.

US EPA. (1995). Characterizing air emissions from indoor sources, EPA report, EPA/600/F-95/005.

Yuan X. Chen Q. and Glicksman L.R. (1998). Critical review of displacement ventilation. *ASHRAE Transactions*, v 104, n Pt 1A.

© 2000 Elsevier Science Ltd. All rights reserved.
Air Distribution in Rooms, (ROOMVENT 2000)
Editor: H.B. Awbi

SITING OF DOMESTIC CARBON MONOXIDE DETECTORS

D. I. Ross

Building Research Establishment Ltd., Hertfordshire, WD2 7JR, UK

ABSTRACT

Each year approximately 50 people die in UK homes from accidental carbon monoxide (CO) poisoning due to faulty combustion appliances. Installing CO detectors in the home can reduce this risk. This project investigated the best place to site these detectors. Both experimental investigations and computational modeling (using CFD and multi-zonal codes) were performed to investigate the distribution of CO in the home. The study showed that as the CO emission is initially hot, a buoyant plume will rise to the ceiling. Consequently, a vertical CO concentration gradient is established in the room. This gradient diminishes as the flow spreads through the home. Using data from this study, details of CO poisoning incidents and additional experiments on the audibility of the detectors, a set of siting recommendations was produced.

KEYWORDS

Carbon monoxide, laboratory study, CFD, multizone, modeling, residences

INTRODUCTION

The use of combustion appliances in the home can, under adverse conditions, generate high levels of CO. Each year this causes 50 accidental deaths and 200 serious injuries in the UK. Installing CO detectors in the home can reduce this risk. The Health and Safety Executive (HSE) and the Department of the Environment, Transport and the Regions commissioned the Building Research Establishment (BRE) to recommend where the CO detectors should be sited to achieve the best protection of the occupants. The results are to be fed into British and European standards on domestic CO detectors. This paper is a summary of the work, which is described fully by Ross et al (1999) and Ross (1999).

In providing siting recommendations, consideration needs to be given to early detection, audibility and affordability. The work programme consisted of seven tasks:
(i)   a review of CO poisoning incidents,
(ii) a literature review of the movement of CO released from a faulty domestic combustion appliance,
(iii) an experimental and computational study of the CO distribution in the room with the CO source,
(iv) an experimental and computational study of the transport of CO through the home,
(v)   research into the audibility of a CO detection system,

(vi)   recommendations on the siting of a domestic CO detection system, and,

(vii) verification of the recommendations.

DETAILS OF INCIDENCES OF CO POISONINGS

A review of CO poisoning incidents, in particular from Metra Martech (1995) and personal communication with HSE (1995-1998), identified the following relevant data.

- ➢ Most accidents were caused by fires/heaters and central heating boilers and a significant number arose from water heaters and cookers. The appliances were mainly flue-less or open-flued.
- ➢ In fatal incidents, CO was initially released into the same room as the faulty appliance.
- ➢ Approximately 40% of victims died in the lounge, 20% in the kitchen and 20% in a bedroom.

LITERATURE REVIEW OF CO MOVEMENT IN THE HOME

As the CO released will be hot and therefore buoyant, the review was extended to look at other buoyant gases, particularly smoke from fires. A number of factors were identified that may affect the distribution of CO in the home and thus the siting of CO detectors. The factors included: (a) source conditions, e.g. location, temperature, velocity and flow rate, (b) effects of surfaces, e.g. walls, ceiling, floor, partitions, objects close to walls, (c) ventilation pathways, and, (d) a heat source additional to the CO source.

DISTRIBUTION OF CO WITHIN A ROOM CONTAINING THE FAULTY APPLIANCE

A review of CO poisoning incidents suggested that CO is initially released into the room containing the CO source. To detect CO early, it is likely that a detector should be placed in that room. We therefore conducted an experimental and computational investigation of the distribution of CO in the source room.

Description of experimental studies

The experimental work was carried out in a 2.4 m high room (Figure 1) in a two-storey house at BRE. Apparatus was constructed to provide a safe and controlled method of simulating CO emissions from a combustion appliance. Thirty-six electrochemical CO sensors were used to measure the distribution of CO in the room. Early experiments showed that the results had sufficient repeatability, for a given sensor layout, that experiments could be repeated with a different sensor layout and the results combined.

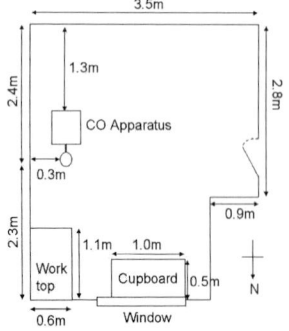

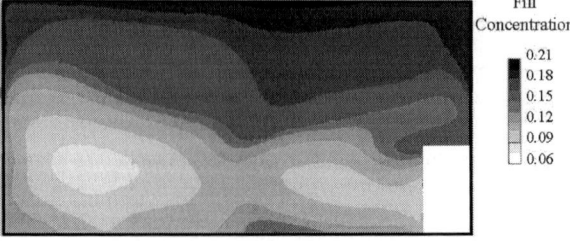

Figure 1: Plan view of test room   Figure 2: CO distribution in the test room

To simulate the CO emissions from a domestic combustion appliance, a basic experimental set-up was chosen similar to conditions that may occur in practice. The set-up was as follows:

- a 125 mm diameter circular outlet directed towards the ceiling,
- an outlet height of 860 mm,
- a source temperature of $100°C$,
- a flow-rate of 5000 cm^3s^{-1}, and,
- all doors, windows and vents in the room closed.

As there is a wide range of conditions under which CO emissions may occur, a series of sensitivity experiments was performed to determine the effects of varying these parameters. This included temperatures of 75 and $150°C$, flow-rates of 2500 and 7500 cm^3s^{-1}, different source outlet dimensions and locations and ventilation provisions open. Based on the literature review, the study investigated the effect of: (a) objects in the path of the CO flow, (b) lights switched on, (c) introducing an additional heat source and (d) having cooler external walls and windows. For verification, additional experiments were performed with actual combustion appliances; a natural gas-fired boiler and a portable LPG heater.

Description of computational studies

As a complementary approach, simulations were also performed using Computational Fluid Dynamics (CFD). A computer program called FLOVENT was selected, which has a k-ε turbulence model that has been optimised for building flows. It provided a much more detailed description of the CO flow and distribution within a room and allowed investigation of conditions that could not be set up controllably within the experimental test facility. The basic CFD model was validated by comparison with experimental data. The CFD model was then used to investigate the effects of: (a) altering the geometry of the room, (b) introducing an additional heat source to the room, (c) having cooler external walls and windows, (d) introducing a partition within the room, and, (e) introducing a sloped ceiling.

Results

The buoyant plume accelerates upwards, then slows as cooler air from the room becomes entrained into the plume, spreading and cooling the flow. Upon reaching the ceiling, it spreads in all directions across the ceiling to the surrounding walls, where it starts to descend down the walls. At this point the flow becomes more complex. Part of the flow is re-entrained into the plume, increasing the concentration in the upper part of the room, and part continues to flow down the walls. A large vertical CO concentration gradient is established, with the highest levels at the ceiling. Figure 2 shows an example CO distribution, through a vertical plane that bisects the room from north to south (relative concentration units). The area at the bottom right of the figure is a cupboard. The vertical CO concentration gradient is clearly observed. Regions of low CO concentration occur in the lower half of the room, in relatively stagnant zones.

Varying the source parameters resulted in a similar CO flow pattern and distribution. The vertical concentration gradient was reduced at lower heat input levels. Figures 3a and 3b show that opening the internal door resulted in a significant drop in CO concentration below the height of the door due to mixing with air from the adjacent hallway. With air vents open, there were reductions in CO concentration close to the vents due to airflow between the room and outside.

The CO concentration was reduced in the areas where the ceiling meets the walls and where two walls meet each other. Obstacles, e.g. furniture against a wall, impeded the flow and reduce the CO level downstream of the object. The CO concentrations at ceiling level showed no significant variation with proximity to light fittings, whether the lights were switched off or on. However, it seems prudent not to place a CO detector too close to a light fitting to avoid overheating etc.

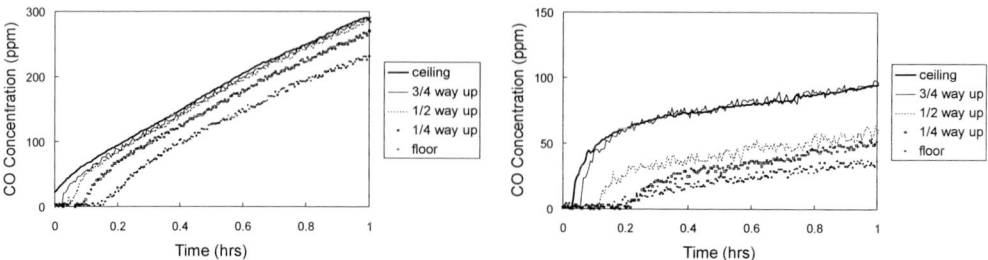

Figure 3a: CO distribution along wall   Figure 3b: CO distribution along wall with door open

Both the experimental and computational studies showed that an additional heat source could inhibit the CO plume rising directly to the ceiling. In these cases, much of the CO becomes entrained into the plume of hot air generated by the additional heat source, resulting in the CO concentration still rising to the ceiling. The overall result is that the CO concentration is still greater towards the upper part of the room.

As described, there is a flow of CO down the walls of the room. A reduction in the surface temperature (e.g. by having a cooler external wall and/or window) leads to an acceleration of the downward flow over the cooler surface as heat is transported from the air stream to the surface, reducing the temperature of the air stream. The result is a reduction in the vertical CO concentration gradient at these surfaces.

Computational studies showed that the general CO distribution was the same when the height or length of the room was doubled. In the latter case, there was a larger reduction in overall CO concentration at ceiling level from the source to the furthest wall. A partition was introduced to be typical of the situation where two small rooms have been knocked together into one room. Thus what may be left is part of the wall and a beam across the ceiling. Simulations showed that the plume first rises to the ceiling and then extends across the ceiling until it meets the partition, where it is halted briefly. As this layer cools, it pours underneath the ceiling beam. The layer is still buoyant, so it rises to the ceiling again on the other side of the beam. There is a marked reduction in CO concentration at the ceiling after passing the partition.

Finally, a computational study was carried out to investigate the effect of a sloped ceiling on CO flow. To be sure of observing any effect, a fairly large gradient of 1m rise per 4m was used. In this case, when the plume meets the ceiling, a greater proportion of the flow rises up towards the top of the sloped ceiling than flows downwards. The effect is to produce a layer of higher CO below the higher side of the ceiling.

For verification, experimental studies were performed with two flue-less combustion appliances; a natural gas-fired boiler and a LPG heater. The results were similar to those described above.

TRANSPORT OF CO THROUGH THE HOME

This stage investigated the transport of CO from the source room to other rooms in the house through both open doorways and other routes, e.g. gaps surrounding water pipes. It comprised both experimental and computational studies.

Experimental study of the transport of CO through open doorways

Experiments were performed to investigate the movement of CO, which is emitted into the room housing the combustion appliance, to other rooms in the test house through open doorways and stairways. The source conditions were similar to those described before. With the source located in a ground floor room, a large vertical CO concentration gradient was established in the source room which diminished as the CO spread through the house and mixed with surrounding air. Varying the source parameters had little effect on the results. Closing internal doors effectively reduced the volume of the home and increased levels of CO in rooms that were not closed off. Opening windows reduced the CO levels in the home especially in the rooms where the windows were open. When the source was located in a first floor room, with all the windows shut, little, if any, CO was transported to the ground floor due to the buoyancy of the gas. Again a large vertical CO concentration gradient was established in the source room which diminished throughout the upper floor. With windows open, CO was observed on the ground floor, especially from cross ventilation.

Computational study of the transport of CO through open doorways

A computational study was performed to investigate conditions that could not be investigated in a controlled manner with the test facility, in particular the effects of wind, temperature and house leakage. BRE's BREEZE multi-zonal computer code was used for this task. Initially a computer model was constructed of the experimental test house and was validated by comparison with experimental data. In summary, the results showed that wind speed and direction were typically the most important factors affecting the flow of CO through the house. They led to a wide distribution of possible CO levels in each room. In extreme cases, little CO was transported from the source room to the rest of the house.

Experimental study into other routes of transport of CO between rooms

A further series of experiments was performed to investigate other routes of transport of CO between rooms, e.g. through a broken flue that passes through the room or via gaps surrounding water pipes that pass between two rooms. The results showed that if the CO is sufficiently buoyant on entering the room, a vertical CO concentration gradient is again established. It is more complex if there is significant heat loss between rooms. In the extreme case that CO is non-buoyant, the CO distribution in the room is very dependent on source conditions and it is not possible to generalise where CO will build up fastest.

AUDIBILITY OF CO DETECTORS

Experiments were performed to study the sound transmission from a 'sounding' CO detector to other rooms. In summary, the results show that the sound level from a CO detector, which meets the British standard, BS7860 (1996), may be insufficient to wake an occupant unless located within the bedroom.

RECOMMENDATIONS FOR THE SITING OF CO DETECTORS IN THE HOME

A detailed set of recommendations were drawn up based on the needs to detect early the build-up of CO in the home and provide an audible warning that is loud enough for the occupants to hear. Affordability of the detection system was also considered. Included here is a summary of the main recommendations.

Room containing combustion appliance

In most cases, CO is initially emitted into the room housing the faulty combustion appliance. To provide early detection of CO in these cases, a detector should be placed in any room where there is a combustion appliance. If there is an appliance in more than one room and the number of detectors is limited, the detectors should be located in rooms containing a flue-less or open-flued appliance. Consideration should also be given to placing a detector in rooms most often occupied and in which the appliance is most often used. All detectors should be interlinked to give maximum protection. The main siting requirement in the room is that a detector should preferably be located on the ceiling or, if not, high up on a wall.

Other rooms

As a secondary measure, it may be desirable to have additional detectors: (a) in, or near, rooms in which occupants spend time whilst awake and may not hear the detection system, and (b) in bedrooms. As it is not possible to predict how the CO will build up in these rooms, the detector should be sited close to the typical breathing zone of the occupants. If a detector is placed in hallways outside bedrooms, it should be placed on the ceiling or high up on a wall and sited to best ensure a direct line between a detector and each bedroom to minimise sound attenuation. A cheaper alternative for these rooms may be to use a sounding device as the main role is to alert the occupants to CO build-up elsewhere. The device must be interconnected to the rest of the detection system.

TESTING CARBON MONOXIDE ALARMS IN SITU

Experiments were performed in the BRE test house to verify a number of the recommendations, by testing actual CO detectors. CO detectors were arranged in two rows, each row containing four calibrated detectors from different manufacturers with the same combination of detectors used in each row. Typically one row was located towards the top of a room, either on the ceiling itself or high up on a wall, and the second row was located on the wall just above floor level. The time taken for the detectors within each pair to respond to a build-up of CO was compared. The work was problematic, owing to the unreliability of the response from the CO detectors (at the time of the study, no CO detectors were approved to the British Standard, BS7860). However the results agreed with those previously, showing the vertical CO gradient in the source room which diminishes with increasing distance within the home.

CONCLUSIONS

This paper describes a summary of work performed to provide siting recommendations for a domestic CO detection system. It was based on experimental and computational studies and a review of existing literature and CO poisoning incidents.

ACKNOWLEDGEMENTS

This work was supported by the Health and Safety Executive and the Department of the Environment, Transport and the Regions, which have given permission for it to be published.

REFERENCES

British Standard Institution. (1996). *Specifications for carbon monoxide detectors (electrical) for domestic use.* BS7860, BSI, London, UK.

Metra Martech Ltd. (1995). *Poisoning by carbon monoxide from domestic heating appliances - Data and analysis.* Prepared for the Consumer Safety Unit, Department of Trade and Industry.

Personal communication (1995-1998). Health and Safety Executive.

Ross D. et al. (1999). *Evaluation of carbon monoxide detectors in domestic premises.* Vol. 1-5, Contract research reports 236, 237, 238, 246, 247. HSE books, Sheffield, UK.

Ross D (1999). *Carbon monoxide detectors.* Good building Guide 30, Published by CRC, London, UK.

Air Distribution in Rooms, (ROOMVENT 2000)
Editor: H.B. Awbi

PROTECTION OF NON-SMOKING PERSONS AGAINST CIGARETTE SMOKE BY AIRFLOW

H. Krühne[1] and K. Fitzner[2]

[1]Brendel Ingenieure GmbH
Trachenbergring 93, 12249 Berlin, Germany
[2]Hermann-Rietschel-Institut, TU Berlin
Marchstraße 4, 10587 Berlin, Germany

ABSTRACT

The protection of non-smoking persons against cigarette smoke is a very popular subject. In Germany the ‚pro' and ‚contra' of non-smoking regulations especially in public accessible areas like restaurants, train stations or in governmental buildings is discussed in a more and more controversial way. Especially the discussion about passive smoking and negative health effects through passive smoking lead to the demand of an effective protection of nonsmokers. But often even in ventilated rooms an effective protection of nonsmokers can not be reached, because the concept of the ventilation system and the airflow is not designed for this purpose.

The aim of the study is to show that with a normal ventilation system an effective protection of nonsmokers can be reached if some rules are observed. In the first part of this study different airflows are compared with regard to their ability to prevent a smoke transportation from smoking areas to non-smoking areas. The efficiency of the protection-effect of different airflow patterns (mixing flow, displacement flow) are discussed. In addition to the general description of the different airflow patterns the transportation of cigarette smoke as a contaminant is discussed in detail for mixing- and displacement flow.

The second part of the study offers additional possibilities in combination to the airflow to prevent a smoke transportation into non-smoking areas.

KEYWORDS

Airflow, Cigarette Smoke, Smoke Protection, Displacement Flow, Mixing Flow, Contaminant Distribution, Smoking Areas, Non-smoking Areas, Ventilation Effectiveness

INTRODUCTION

The protection of non-smoking persons against cigarette smoke is a very popular subject. In Germany the ‚pro' and ‚contra' of nonsmoking regulations especially in public accessible areas like restaurants, train stations or in governmental buildings is discussed in a more and more controversial way (Stiftung Warentest (1996). Especially the discussion about passive smoking and negative health effects through passive smoking lead to the demand of an effective protection of nonsmokers. But often even in ventilated rooms an effective protection of nonsmokers can not be reached, because the concept of the ventilation system and the airflow is not designed for this purpose.

The methods and airflow patterns described in the following chapters are not "new technology" or new scientific results but an attempt to use common and approved methods for new tasks and to provide more comfort for people in ventilated rooms. In the following sections the contaminant distribution describes especially the distribution of tobacco smoke. Also the term "smoke protection" is to be understood in the sense of "protection against smoke annoyance" and not in the sense of "prevention of negative health effects". Whether or not the smoke protection described in the following text can protect persons against the negative health effects of passive smoking is not subject of this article.

CRITERIA FOR SMOKE PROTECTION

To describe the effectiveness of a room air flow against the distribution of tobacco smoke two common values are often used to describe airflow and contaminant distribution in rooms.

1. Ventilation effectiveness
$$\varepsilon_{v,p} = \frac{c_o - c_i}{c_p - c_i} \qquad (1)$$

2. Contamination Degree
$$\mu = \frac{1}{\varepsilon_{v,p}} \qquad (2)$$

Both values give the concentration of a contaminant (for example tobacco smoke) in a room point c_P related to the exhaust concentration c_O (with c_i as the supply concentration) and describes the improvement of the air quality in a point of a room. The contamination degree μ is more useful to describe the air quality because low values of μ mean low concentrations and good air quality.

AIRFLOW AND SMOKE PROTECTION - EVALUATION OF DIFFERENT AIRFLOWS IN REGARD TO THE PROTECTIVE EFFECT AGAINST SMOKE

Normally two airflow patterns are used in ventilated spaces: The mixing flow and the displacement flow (see figure 1). The notion displacement flow (source flow) is used here to define the airflow from the "real" displacement flow in the meaning of a plug flow (figure 1 right).

The mechanics of the displacement flow (figure 1 center) are often described in literature (for example in Fitzner (1991), Krühne (1995)) and is assumed to be as known for the following chapters. The air movement of the displacement flow supports the protective effect against the contaminant distribution because the room is divided in two zones and air from the lower air zone moves upwards along the person and leads to an additional improvement of the air quality. Due to the fact that tobacco smoke is emitted from or near a person a large amount of smoke is transported upwards into the upper air layer. For the persons in the room an improvement of air quality can be reached (figure 2).

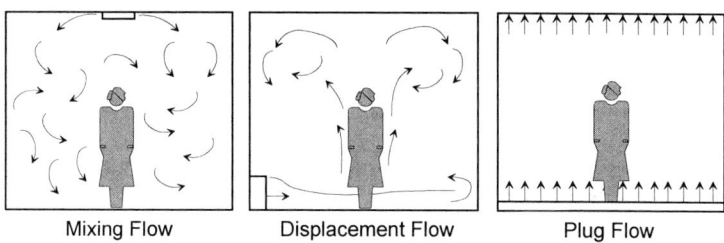

Mixing Flow   Displacement Flow   Plug Flow

Figure 1: Different airflow patterns

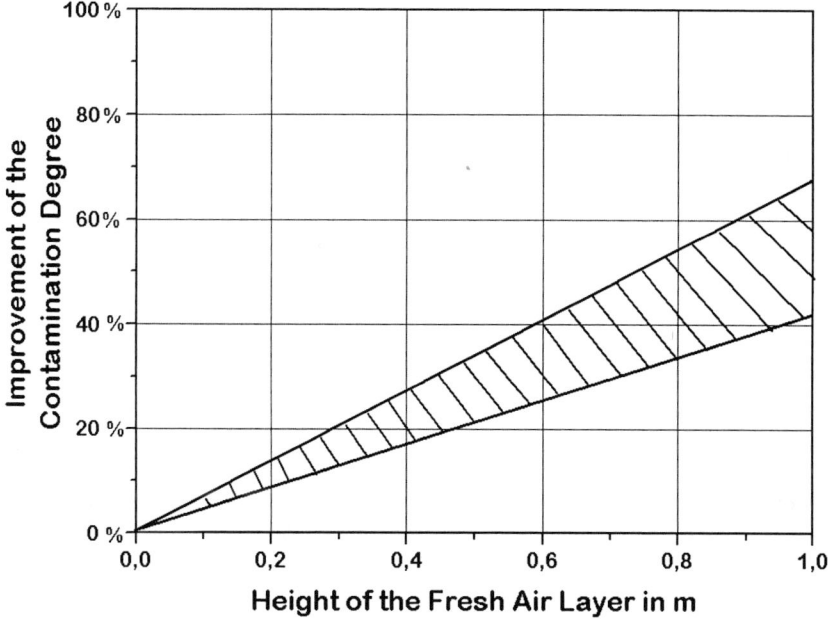

Figure 2: Improvement of the contamination degree at a person in a height of 1,1 m related to the contamination degree in the environment at the same height (Krühne and Fitzner 1994)

Together with additional steps described later a contaminant stratification like described by Brohus and Nielsen (1994) or the effects of breathing and movement (Bjørn et. al. (1997)) can be prevented. Perquisite for an improvement of the air quality is that the fresh air layer in the lower room part is higher than 0,6 m - like shown in figure 2 - then an improvement of the air quality up to 40% can be reached.

Due to the reduced horizontal contaminant distribution with displacement flow the contamination degree decreases strongly with the distance to the contaminant source (figure 3). Related to the mixing flow, were the horizontal air exchange is significantly higher than in a displacement flow, the displacement flow is much more effective to protect people against high smoke concentrations. Figure 3 shows the contamination degree with an increasing distance from the contaminant source for mixing- and displacement flow. It can be seen, that the contamination degree is significantly higher for a distance of one meter than with displacement flow because the contaminant is distributed into the room around the source and is not transported into the upper room air layer.

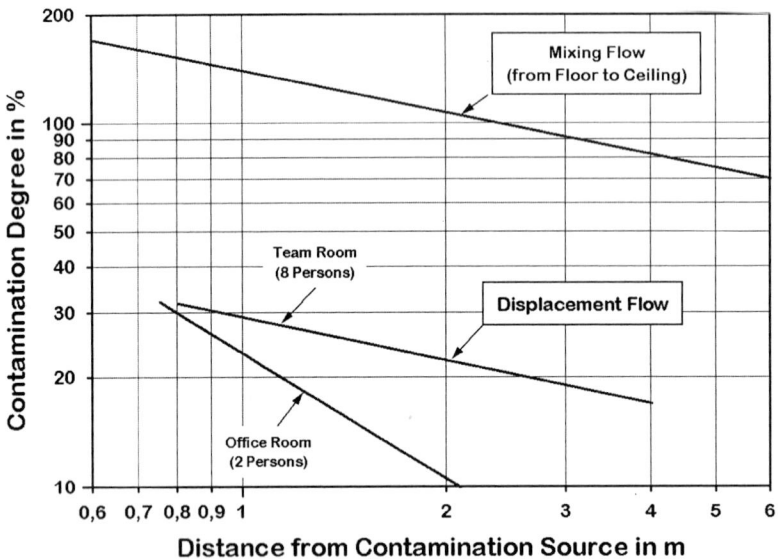

Figure 3: Decrease of contamination for mixing- and displacement flow (Fitzner (1981 and 1996))

ADDITIONAL STEPS FOR SMOKE PROTECTION AND EXAMPLES

From the airflow mechanics described above it can be seen that the displacement flow is to be preferred against the mixing flow if a smoke protection in rooms should be reached. But the smoke protection without additional supporting steps is not really sufficient. The best combination of airflow and additional steps varies with the requirements of the use of the room (e.g. restaurant, public waiting rooms, trains) and the general room conditions (geometry, floor levels, entrance situation). In general the following additional steps are possible:

With the separation of smoking and non-smoking areas (figure 4) in combination with the displacement flow a separation can be reached with relatively small distances between both areas. The protective effect is supported by the arrangement of the air inlets opposite each other in both areas.

The protective effect of displacement flow and separation can be increased if separation walls are used additionally. The separation walls must not necessarily be in the height from floor to ceiling but must provide a flow protection in the lower room level (closed to the floor) up to approx. 1,5 m (figure 5).

In rooms with areas in different floor levels the arrangement of the smoke and non-smoking areas should be as shown in figure 6 if an effective smoke protection should be reached. With the incorrect arrangement of the zones (figure 6 left) the contaminant flow is directed from the smoking to the non-smoking section and no effect can be reached. With the opposite arrangement (figure 6 right) a good smoke protection can be reached with mixing or displacement flow.

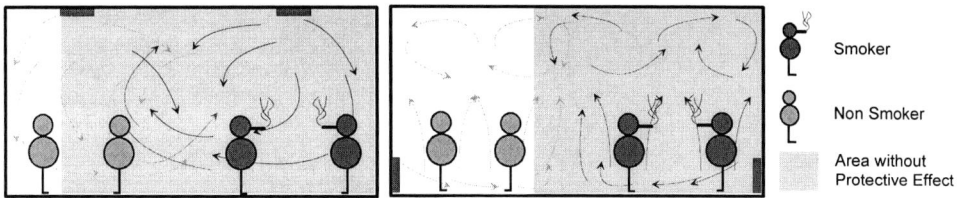

Figure 4: Local separation with mixing- and displacement flow

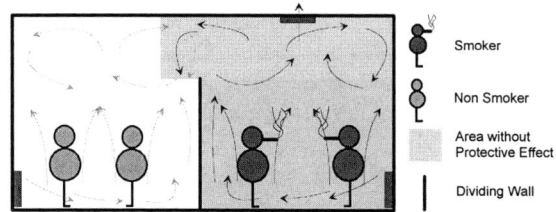

Figure 5: Influence of dividing walls

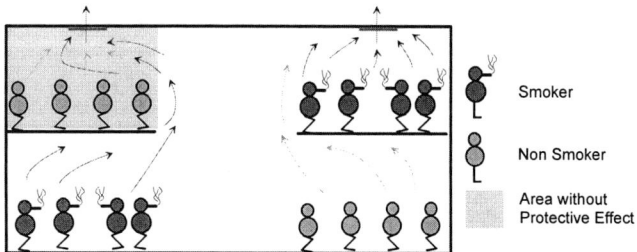

Figure 6: Location of smoking areas with separation in different levels

The use of separation walls is particularly useful if in addition to the normal air flow an overlaid cross flow takes place (figure 7). Uncontrolled overlaid cross flows are induced normally in rooms with openable windows or with doors to ambient conditions. Controlled cross flows induced by pressure differences through the ventilation system can be used to control the contaminant distribution in rooms.

The air distribution system must be designed according to the desired effect. If an air distribution in the lower part of the room is planed, the pressure difference between upper and lower room part must lead to an airflow upwards.

An additional example for overlaid cross flow is the flow in trains. Due to the pressure distribution outside of the train a directed airflow takes places against the train direction like shown in figure 8.

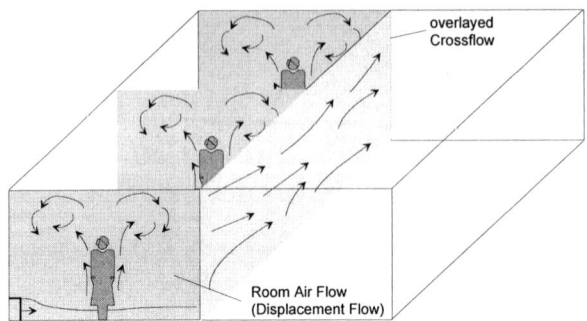

Figure 7: Room air flow and overlaid cross flow

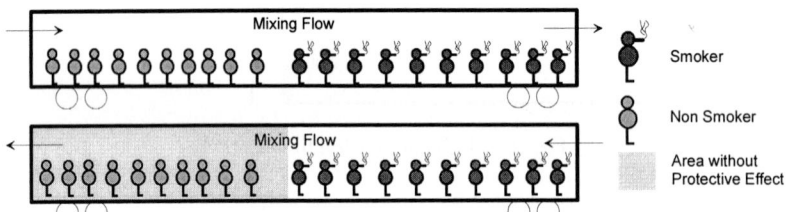

Figure 8: Airflow in trains with smoking areas

RESUME

A complete protection against tobacco smoke can be reached only in a non-smoking environment. But simple and known methods together with common airflow can provide a reasonable smoke protection for non-smoking people without expensive technology.

REFERENCES

Bjørn E. et. al. (1997). Displacement Ventilation – Effects of Movement and Exhalation. Proceedings of Healthy Buildings.

Brohus H. and Nielsen P. (1994). Contaminant Distribution Around Persons in Rooms Ventilated by Displacement Ventilation. Proceedings of Roomvent, 294-312

Fitzner K. (1996). Displacement Ventilation and Cooled Ceilings, Results of Laboratory Tests and Practical Installations. Proceedings of Indoor Air.

Fitzner K. (1991). Displacement Flow in Theory and Praxis. Clima Commerz International, 5/91.

Krühne H. (1995). Experimental and Theoretical Studies about Displacement flow. Ph.D. Thesis Berlin University of Technology.

Krühne H. and Fitzner K. (1994). Air Quality in the Breathing Zone in Rooms with Displacement Flow. Proceedings of Roomvent.

Stiftung Warentest (1996). Concepts without Consequences. Smoke Protection in Public Room (Original in German). Vol. 8 1996.

Fitzner K. (1981). Air flow experiments in full scale test rooms. ASHRAE-Transactions, Vol. 87, Pt 1.

PREDICTIVE METHODS

Air Distribution in Rooms, (ROOMVENT 2000)
Editor: H.B. Awbi

MODEL EXPERIMENTS WITH LOW REYNOLDS NUMBER EFFECTS IN A VENTILATED ROOM

Peter V. Nielsen[1], Claus Filholm[1], Claus Topp[1] and Lars Davidson[2]

[1]Department of Building Technology and Structural Engineering,
Aalborg University, Sohngaardsholmsvej 57, 9000 Aalborg, Denmark
[2]Department of Thermo and Fluid Dynamics, Chalmers University of
Technology in Göteborg, Sweden

ABSTRACT

The flow in a ventilated room will not always be a fully developed turbulent flow. Reduced air change rates owing to energy considerations and the application of natural ventilation with openings in the outer wall will give room air movements with low turbulence effects.

This paper discusses the isothermal low Reynolds number flow from a slot inlet in the end wall of the room. The experiments are made on the scale of 1 to 5. Measurements indicate a low Reynolds number effect in the wall jet flow. The virtual origin of the wall jet moves forward in front of the opening at a small Reynolds number, an effect that is also known from measurements on free jets. The growth rate of the jet, or the length scale, increases and the velocity decay factor decreases at small Reynolds numbers.

KEYWORDS

Room air flow, low Reynolds number effects, scale-model experiments, plane isothermal wall jet.

INTRODUCTION

Most of the theory behind the description of room air movement as e.g. flow elements, zonal models, throw and CFD predictions is based on an assumption of fully developed turbulent flow. This assumption represents a simplification of the theory. The flow elements can be given by equations without any adjustment for the velocity level and the turbulence model in the CFD predictions has a universal character independent of the velocity level.

The flow in a ventilated room will not always be a fully developed turbulent flow. Reduced air change rate owing to energy considerations as well as the application of natural ventilation may cause a flow with low turbulence effect in the air movement. This paper shows how it is possible to adjust the flow element of a wall jet to be valid for low turbulent flow. The measurements are also made for the validation of Large Eddy Simulation (LES) which is a promising method for the prediction of flow with low turbulence effect as shown by Davidson et al. (2000).

SCALE-MODEL EXPERIMENTS

The scale-model experiments are related to a full-scale room with the dimensions, height H, width W and length L equal to 2.5 m, 3.6 m and 4.2 m, see Topp et al. (2000). Figure 1 shows the layout of a room. The full-scale room is ventilated by a supply slot of full width with the height h_o of 2 cm located in one end wall and a return opening located in the opposite end wall.

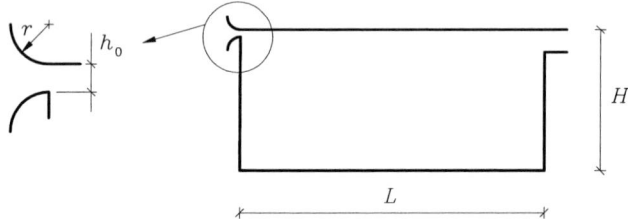

Figure1: Room geometry and supply slot.

TABLE 1

FULL SCALE AND MODEL GEOMETRY

Scale	h_o/H	W/H	L/H
1:1	0.008	1.44	1.68
1:5	0.008	1.44	1.68 - 2.88

The length of the model is different from the length of the full-scale room in some of the measurements, see Table 1. The increased length in the model makes it possible to study the penetration depth of the supply jet. It is assumed that the flow in the wall jet is a parabolic flow and therefore independent of the room length within the limits given in Table 1.

A model experiment with isothermal flow is identical with the full-scale flow if the Reynolds number Re is the same in both cases. The Reynolds number is defined as

$$Re = \frac{h_o u_o}{\nu} \tag{1}$$

where h_o is the slot height and u_o the supply velocity. ν is the kinematic viscosity of air.

Model experiments are often used to study room air movement when it is convenient to work in a small scale. Here the method is used because the reduced scale will increase the velocity level accordingly and improve the measurements at small Reynolds numbers. It is important to have the same boundary conditions in both full scale and in model scale. Figure 1 shows the contraction used in the model to obtain low turbulence in the inlet opening, $r/h_0 = 11.75$. The inlet opening in the full-scale room has a one-sided contraction and upstream installations that will generate some turbulence in the inlet flow.

WALL JET FLOW

Fully developed isothermal two-dimensional wall jet flow is given by the following equations, see e.g. Rajaratnam (1976),

$$\frac{u_x}{u_o} = K_p \sqrt{\frac{h_o}{x + x_o}} \tag{2}$$

$$\frac{\delta}{h_o} = D_p \frac{x + x_o}{h_o} \tag{3}$$

where u_x is the maximum velocity in the wall jet profile and δ is the thickness or length scale of the wall jet measured from the surface to the velocity $u_x/2$ in the profile. x_o is the distance from the supply slot to the location of the virtual origin of the jet flow. This location will in a fully developed flow often be located behind the diffuser corresponding to a positive x_o. K_p and D_p are characteristic constants for the diffuser. They are constant for the fully developed turbulent flow. The dimensionless wall jet profile (u/u_x versus y/δ) will be a universal profile for high Reynolds number flow.

Eqns. (2) and (3) are, as mentioned above, based on the assumption of fully developed flow. It is also possible to develop a corresponding set of equations for laminar flow. The behaviour of the flow in the transition regime (low turbulent regime) could be handled by expressions that connect the two sets of equations but here it is decided to express the transitional regime by the following modifications of Eqns. (2) and (3)

$$\frac{u_x}{u_o} = K_p(Re) \sqrt{\frac{h_o}{x + x_o(Re)}} \tag{4}$$

$$\frac{\delta}{h_o} = D_p(Re) \frac{x + x_o(Re)}{u_o} \tag{5}$$

$K_p(Re)$, $D_p(Re)$ and $x_o(Re)$ are considered to be functions of the Reynolds number Re and they will develop asymptotically to constant values for increasing Reynolds numbers.

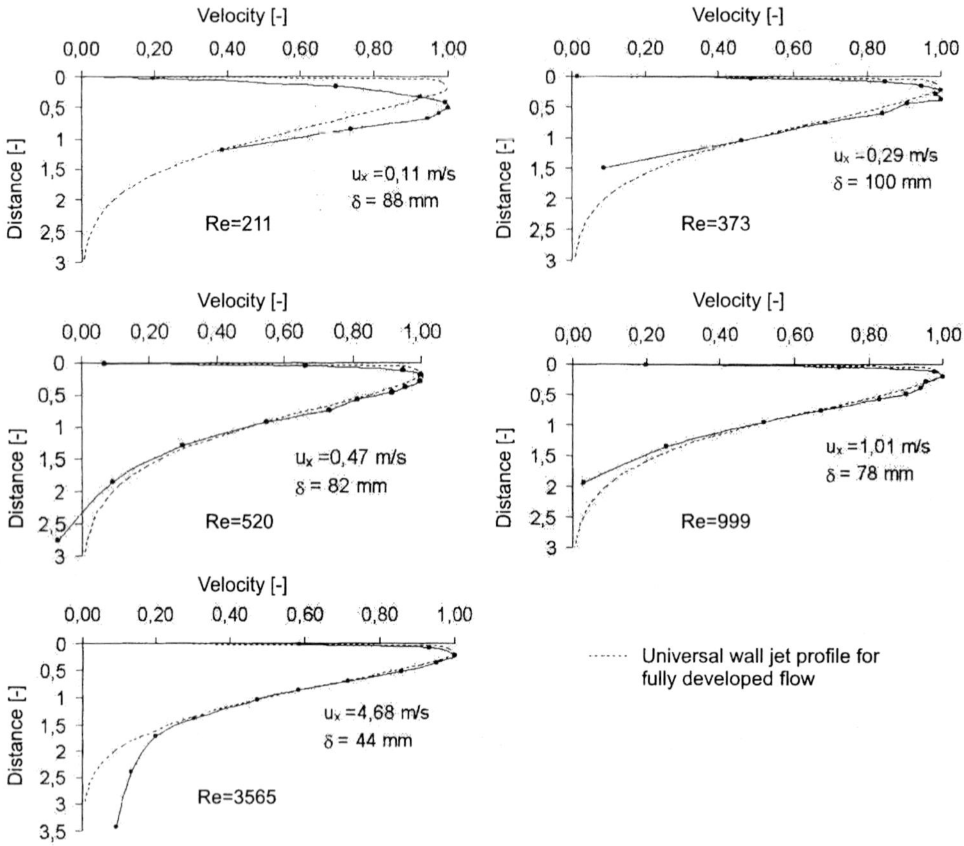

Velocity [-]

Distance [-]

$u_x = 0{,}11$ m/s
$\delta = 88$ mm
Re=211

Velocity [-]

Distance [-]

$u_x = 0{,}29$ m/s
$\delta = 100$ mm
Re=373

Velocity [-]

Distance [-]

$u_x = 0{,}47$ m/s
$\delta = 82$ mm
Re=520

Velocity [-]

Distance [-]

$u_x = 1{,}01$ m/s
$\delta = 78$ mm
Re=999

Velocity [-]

Distance [-]

$u_x = 4{,}68$ m/s
$\delta = 44$ mm
Re=3565

.... Universal wall jet profile for
fully developed flow

Figure 2: Dimensionless wall jet profile measured at different Reynolds numbers. The profile is located at $x/H = 0.8$. The universal wall jet profile for fully developed turbulent flow is given by e.g. Verhoff (1963).

Figure 2 shows the dimensionless velocity profile u/u_x versus y/δ. The profile is measured at $x/H = 0.8$ for the following Reynolds numbers 211, 373, 520, 999 and 3565. It is seen that the profile develops into the universal wall jet profile for fully developed turbulent flow. The measured profile at $Re = 211$ is similar to the laminar (Glauert) profile measured e.g. by Quintana et al. (1997). The profile has a characteristic location of the maximum velocity u/u_x at $y/\delta \sim 0.5$.

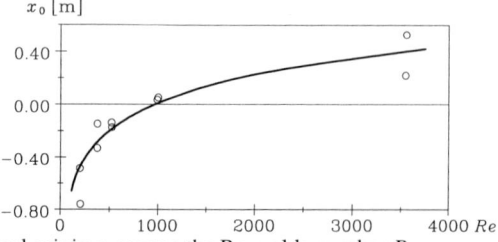

Figure 3: Distance to virtual origin x_o versus the Reynolds number Re.

A depiction of δ as a function of x for the different Reynolds numbers gives x_o and D_p. Figure 3 shows that the distance to the virtual origin x_o is negative at low Reynolds numbers corresponding to a location in front of the supply slot. This is very typical of a semilaminar flow. The jet leaves the opening as a laminar flow with a small growth in thickness. Disturbance changes the flow into a turbulent flow at some distance from the opening with a large increase in the growth rate as a consequence. The new growth rate will have a virtual origin close to the transition point from laminar to turbulent flow.

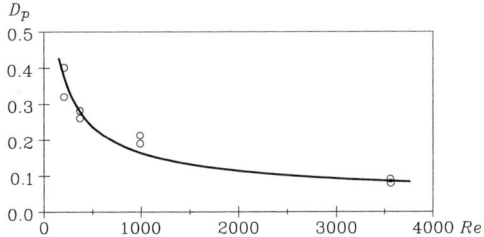

Figure 4: Growth rate D_p versus Re.

Figure 4 shows the growth rate D_p as a function of the Reynolds number. A growth rate of 0.1 is very typical of a fully developed isothermal wall jet. An increase in the growth rate at small Reynolds numbers is also typical as measured earlier by Nielsen and Möller (1988) for a ceiling-mounted slot diffuser. The K_p value for the slot is given in Figure 5. This value has also the expected asymptotic development for increasing Reynolds numbers. All the measurements in Figure 3, 4 and 5 are in good agreement with the measurements in the full-scale room, see Topp et al. (2000).

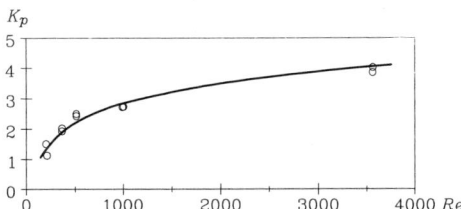

Figure 5: K_p value for the slot and the two-dimensional wall jet versus Re.

PENETRATION DEPTH OF THE WALL JET

Earlier measurements and CFD predictions show that is should be possible to measure a restricted penetration depth for the isothermal wall jet and a corresponding reattachment line in the occupied zone. The penetration depth and the location of the reattachment line l_{re} are functions of the Reynolds number at low turbulent flow and l_{re} will go to zero for $Re \rightarrow 0$, see Davidson and Nielsen (1998) and Armaly et al. (1983).

It was not possible to measure these quantities in full-scale flow, but there are some indications of a restricted penetration depth at low Reynolds numbers in the 1:5 model.

CONCLUSIONS

The flow in a ventilated room will not always be a fully developed turbulent flow but it is possible to use the equations for a fully developed flow in the transitional regime if involved "constants" are given as functions of the Reynolds number.

A detailed description of an isothermal wall jet has been developed from scale-model experiments and the results are in good agreement with full-scale experiments and earlier measurements.

ACKNOWLEDGEMENT

This research work is done in cooperation with the International Centre for Indoor Environment and Energy, the Technical University of Denmark.

REFERENCES

Armaly B. F., Durst F., Pereia J. C. F. and Schonung B. (1983). Experimental and Theoretical Investigation of Backward-Facing Step Flow. *J. of Fluid Mechanics* **127**, 473-496.

Davidson L. and Nielsen P. V. (1998). A Study of Laminar Backward-Facing Step Flow. Paper no. 83, ISSN 1395-7953 R9802, Indoor Environmental Engineering, Aalborg University.

Davidson L., Nielsen P. V. and Topp C. (2000). Low-Reynolds Number Effects in Ventilated Rooms: A Numerical Study. *Proceedings of ROOMVENT 2000*, Reading, UK.

Nielsen P. V. and Möller Å. T. A. (1988). Measurements on Buoyant Jet Flows from a Ceiling-Mounted Slot Diffuser. *Proceedings of the 3rd Seminar on "Application of fluid Mechanics in Environmental Protection 88"*, Silesian Technical University, Gliwice, Poland.

Quintana D. L., Amitay M., Ortega A. and Wygnanski I. J. (1997). Heat Transfer in the Forced Laminar Wall Jet. *Journal of Heat Transfer* **Vol. 199**, 451-459.

Rajaratnam N. (1976). Turbulent jets. Elsevier, Amsterdam, The Netherlands.

Topp L., Nielsen P. V. and Davidson L. (2000). Room Airflows with Low Reynolds Number Effects. *Proceedings of ROOMVENT 2000*, Reading, UK.

Verhoff A. (1963). The Two-Dimensional Turbulent Wall Jet With and Without an External Free Stream. Report No. 626, Princeton University, Department of Aeronautical Engineering.

Air Distribution in Rooms, (ROOMVENT 2000)
Editor: H.B. Awbi

THERMAL STRATIFICATION PRODUCED BY PLUMES AND JETS IN ENCLOSED SPACES

G.R. Hunt [1], P. Cooper [2] & P.F. Linden [3]

[1]Department of Applied Mathematics & Theoretical Physics, Silver Street,
University of Cambridge, CB3 9EW, UK.
[2] Department of Mechanical Engineering, Faculty of Engineering, University of Wollongong,
New South Wales, 2577, Australia.
[3] Department of Mechanical & Aerospace Engineering, University of California, San Diego,
9500 Gilman Drive, La Jolla, CA 92093-0411, USA.

ABSTRACT

The airflow and thermal stratification produced by a localised heat source located at floor level in a closed room is of considerable practical interest and is commonly referred to as a 'filling box'. In rooms with aspect ratios $H/R \leq 1$ (room height H to characteristic horizontal dimension R) the thermal plume spreads laterally on reaching the ceiling and a descending horizontal 'front' forms separating a stratified warm upper region and the cooler air below. The stratification is well predicted for $H/R \leq 1$ by the original filling box model of Baines & Turner (1968), although, this model represents a somewhat idealised situation as the plume is assumed to rise from a point source of buoyancy alone - in particular the momentum flux at the source is zero.

In practical situations, real sources of heating and cooling in a ventilation system often include initial fluxes of both buoyancy and momentum. New laboratory experiments were undertaken to determine the dependence of the 'front' formation and stratification on the initial momentum and buoyancy fluxes of a single source, and on the location and relative strengths of two sources from which momentum and buoyancy fluxes were supplied separately. This paper reports on the results of these experiments.

KEYWORDS

Thermal stratification, closed room, localised heat source, initial momentum flux, aspect ratio.

INTRODUCTION

The thermal stratification generated by a localised source of heat at floor level in a confined space is of considerable interest to building ventilation. Many sources of heat generation in buildings may be regarded as being localised, *e.g.* computers, occupants *etc.*, and knowledge of the developing vertical temperature profile produced by these sources is required before air quality and occupant comfort levels can be determined. In general, these sources may be classified as either *'pure'* buoyancy sources, *e.g.* an electric fire or a radiator in a hot water heating system, or as *'forced'* buoyancy sources which are characterised by non-zero initial momentum fluxes, *e.g.* in a heating system in which warm air is injected into the space.

The stratification produced by a *pure* heat source in an unventilated enclosure that initially contains cool air of a uniform temperature has been considered by Baines & Turner (1968); we refer to this paper hereafter as B&T. Using a saline plume in a water tank as a model for a heat source in air[†], B&T observed and made measurements of the developing vertical stratification. Their experiments showed that for a range of aspect ratios H/R (room height H to characteristic horizontal dimension R) the rising turbulent plume spreads laterally on reaching the ceiling to form a warm layer of air, separated by a horizontal 'initial front' (or thermal interface) from the layer of cooler air below. Turbulence in the warm upper layer decays rapidly and it soon becomes part of the non-turbulent environment. The plume now rises through this warm layer and thus arrives at the ceiling warmer than it did before. As a result, the outflow from the plume is *above* the existing warm region which is displaced downwards. This process repeats and a stable stratification develops. B&T tracked the position of the initial front and, using plume theory (Morton, Taylor & Turner 1956), showed that the relationship between its height $z = z_0$ above floor level and the time t after the plume first reached the ceiling is closely predicted by the expression

$$\frac{z_0}{H} = \left[(4\alpha/5)(18\alpha/5\pi)^{1/3} \left(t/H^{-2/3}B_0^{-1/3}R^2 \right) + 1 \right]^{-3/2} \qquad \text{for } H/R < 1, \qquad (1)$$

where B_0 denotes the initial buoyancy (or heat) flux of the source and α (≈ 0.083) is the plume entrainment coefficient. The 'filling box' behaviour, in which the initial front is well predicted by (1), was maintained only while the aspect ratio H/R was less than unity. As H/R increased, B&T observed that the initial outflow from the plume intruded vertically downwards into the cooler layer below after colliding with the side walls of the container, and that this intrusion was re-entrained by the plume. This resulted in mixing and a general overturning motion which increased in scale as H/R increased. The (stabilising) buoyant ceiling layer opposes the (destabilising) momentum flux generated by the plume and B&T suppose that if this momentum flux could be deflected downwards then a measure of the tendency towards overturning is the ratio of the inertia and buoyancy forces I/B, namely,

$$\frac{I}{B} = \frac{9\alpha}{10} \left(\frac{H}{R} \right)^2. \qquad (2)$$

In other words, overturning increases as the aspect ratio H/R increases. B&T found that the limiting value of the parameter I/B in (2) is 0.1. For $H/R < 1$ the depth of the layer formed by the initial plume outflow is small compared with H and overturning motions are negligible, *i.e.* the geometry of the space does not encourage mixing. For larger aspect ratios ($1 \lesssim H/R \lesssim 6$) an increasing pattern of overturning is observed with increasing H/R, and the enclosures geometry plays a significant role in encouraging mixing. For $H/R \gtrsim 6$ (Barnett 1991) the stratification changes dramatically from that of the filling box.

As described, B&T consider the stratification established by a *pure* source of buoyancy. In practical situations, however, real sources of heating and cooling in a ventilation system often include initial fluxes of both buoyancy and momentum. An important question then arises as to how the balance between the buoyancy and momentum of a source affects the mixing, stratification and air quality in a

---

[†] Throughout this paper the results of laboratory experiments are described as though for heat rising in air. Images of the flows observed during experiments are shown as seen in the laboratory.

room. To determine the dependence of both the 'front' formation and stratification on the initial momentum M_0 and buoyancy B_0 fluxes new experiments were performed to extend the work of B&T. First, the case of an enclosure containing a *forced* buoyancy source was considered. We then further extended the concept of a filling box containing a forced buoyancy source by considering an enclosure containing two localised sources, namely, a source of initial momentum flux only, *i.e.* a jet, and a source of initial buoyancy flux only, *i.e.* a plume. The primary motive for considering this latter configuration of sources was to determine how the stratification is modified as a result of the spatial distribution of the initial fluxes. The laboratory experiments, results and their implications to flows in rooms and buildings, are now described.

EXPERIMENTS

Two plexiglass tanks were available for the experiments; one of square cross section (tank #1, 61 cm x 61 cm) and the second of rectangular cross section (tank #2, 58.3 cm x 27.9 cm). The tank, representing a simple enclosure, was filled to a depth H with fresh water (figure 1). The aspect ratio of the enclosure was increased by increasing H. Nozzles designed to produce a turbulent jet and plume were suspended in the tank just below surface level; the large density step at the air/water interface provides a barrier to the flow and is representative of the enclosure floor. Fresh water was supplied to the jet nozzle (of outlet diameter $\phi = 1$ or 2 mm) and saline solution to the plume nozzle (nominally $\phi = 5$ mm); each supply was via a constant-head tank which ensured a constant volume flow rate Q_0. To aid visualisation, dye was added to the saline solution. The nozzles produced a turbulent flow at the point of discharge even at relatively low supply flow rates. This enabled forced plumes with relatively small jet-lengths (see (3)) compared with H to be generated. Larger jet-lengths were achieved by reducing ϕ or increasing Q_0.

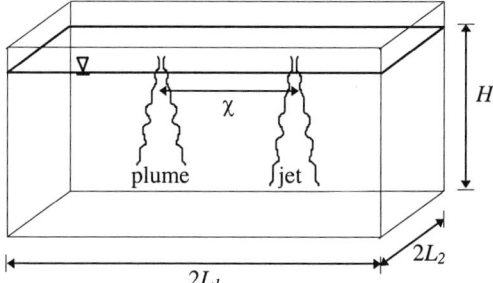

Figure 1. Schematic diagram of the apparatus showing the two-source set-up and notation.

For the single-source experiments, the nozzle was located in the central plane of the tank. For the two source experiments, the jet and plume nozzles were suspended at equal heights above the base of the tank and separated horizontally by distances of $\chi = 4$, 8, 16 or 28 cm. The tank was diffusely backlit using fluorescent lighting and the experiments filmed using a COHU camera. Images were recorded onto video tape and simultaneously via a frame grabber to computer hard disk. A mirror positioned at one end of the tank and at 45 degrees to the camera enabled the flow from both the front and side of the tank to be filmed simultaneously.

The falling turbulent saline plume is dynamically equivalent to a thermal plume rising from a heat source in air and, thus, the observed flow was simply inverted, by inverting the camera, in order to represent warm air rising in a room.

LENGTH SCALES

In an unconfined environment, turbulent flow above a buoyant source which has a non-zero initial flux of momentum is jet-like close to the source and tends towards a pure plume-like flow further above the source. A measure of the vertical height over which the flow is dominated by its initial momentum flux and is, therefore, essentially jet-like is the jet-length

$$L_j = M_0^{3/4} / B_0^{1/2} . \tag{3}$$

In a confined space, the flow may not always develop into a plume-like flow before reaching the ceiling. For $L_j \ll H$ the flow is jet-like over a relatively small vertical height and will develop into a plume-like flow well before reaching the ceiling. For $L_j \gg H$ the flow is dominated by its initial momentum flux over the entire vertical height of the enclosure and will be essentially jet-like on reaching the ceiling. Other length scales which are of significance here are the enclosure aspect ratios

$$\frac{H}{L_1} \text{ and } \frac{H}{L_2}, \tag{4}$$

(figure 1), and in the two-source case, the dimensionless horizontal source separation χ/H.

SINGLE-SOURCE EXPERIMENTS

We focused on the effect which M_0 has on the initial front formation and considered aspect ratios for which overturning and mixing is negligible in the absence of initial momentum flux. Source conditions considered ranged from an almost pure buoyancy source ($L_j/H \ll 1$) to a highly forced buoyancy source ($L_j/H \gg 1$).

For source conditions which yield $L_j \ll H$, the formation and descent of the initial front (figure 2a) was similar to the (approximately) pure buoyancy source case described by B&T. This was expected as the initial momentum flux only effects the motion over a vertical height of the order of a jet-length. Above this height the momentum flux generated by the buoyancy exceeds M_0 and thus, on reaching the ceiling the flow is plume-like, although with a greater momentum flux than for a pure buoyancy source.

On increasing L_j, the buoyant ceiling current was forced vertically downwards and intruded into the ambient layer below after colliding with the side and end walls of the enclosure. Vigorous mixing was observed and the intrusion was re-entrained by the plume, *i.e.* a pattern of overturning developed. The vertical scale of the overturning motion increased as L_j increased. Dye released in the forced plume indicated a significant radial variation in the depth of the buoyant layer during the initial development of the stratification; the layer was deepest at the perimeter of the enclosure and became progressively shallower towards the centre of the ceiling. For highly forced plumes, the radial increase in the depth of the initial outflow current was significantly greater than that produced by a pure plume of identical buoyancy flux. Entrainment into the radially spreading ceiling current produced by a pure buoyancy source is weak (for small aspect ratios) and the current tends to decrease, or remain at approximately constant depth with increasing radius. The difference in buoyant layer depths when $L_j/H = 0.14$ and $L_j/H = 1.58$ can be seen by comparing figure 2a and 2b at 180 s. The implications are thus reduced upper layer temperatures with increasing L_j/H.

B&T observed that the scale of overturning motion increased with increasing aspect ratio and thus, qualitatively, the effect of increasing M_0 is equivalent to increasing the aspect ratio. For source jet-lengths $Lj/H \ll 1$ the motion inside the enclosure was well-described by the filling box model of B&T (1). For source conditions giving $L_j/H \gg 1$ the interior was approximately well-mixed and a horizontal descending initial front was not observed.

In the square-sectioned enclosure (tank #1), the geometry may be characterised by the single aspect ratio H/L_1. The two aspect ratios H/L_1 and H/L_2 describe the rectangular-sectioned enclosure (tank #2) for which $H/L_2 < H/L_1$. In this enclosure we observed that the initial outflowing ceiling current collided with the 'front' and 'back' walls before colliding with the left- and right-hand end walls of the enclosure (figure 2a & 2b, images at $t = 15$ s). As the current spread laterally its depth increased rapidly as a result of shear induced mixing and mixing caused by collision with front and back walls. A true filling box flow was observed only when both aspect ratios were less than unity and $L_j/H < 1$.

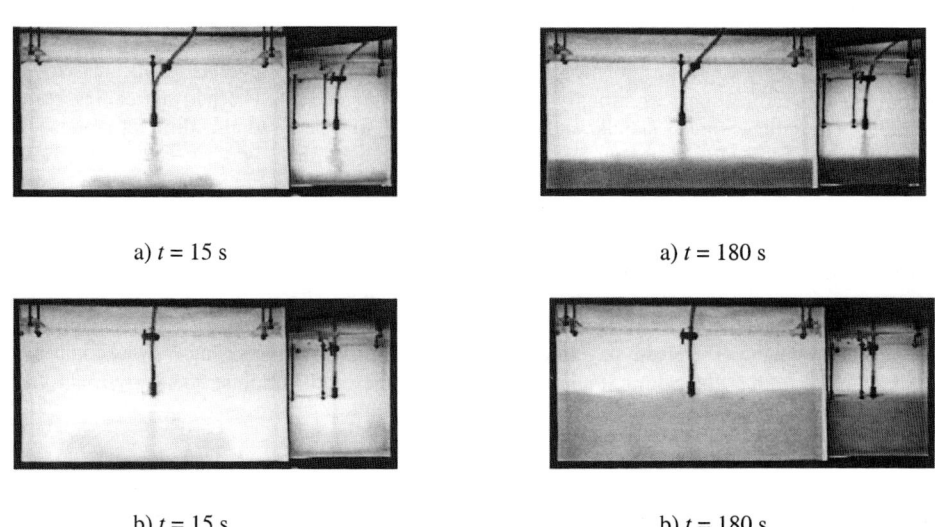

a) $t = 15$ s

a) $t = 180$ s

b) $t = 15$ s

b) $t = 180$ s

Figure 2. Forced buoyancy source (a saline plume) in an unventilated enclosure. $B_0 = 70.2 \text{ cm}^4\text{s}^{-3}$. a) Weakly forced plume with $M_0 = 44.6 \text{ cm}^4\text{s}^{-2}$ giving $L_j/H = 0.14$ (Expt 7), and b) strongly forced plume with $M_0 = 1115.6 \text{ cm}^4\text{s}^{-2}$ giving $L_j/H = 1.58$ (Expt 8). Tank #2 with $H/L_1 = 0.5$, $H/L_2 = 1.05$. Each image shows the flow viewed from the front and end face of the enclosure. Note that at $t = 180$ s the saline layer is deeper and less dense (as indicated by the reduced concentration of dye) in b) than in a).

For sufficiently highly forced sources (figure 2b), buoyant fluid mixed throughout the entire fluid column (*i.e.* from ceiling to floor level) after colliding with the front and back walls before being re-entrained into the forced plume. Dye released in the plume clearly showed the vertical extent of the recirculating flow. A similar pattern was observed when the spreading current reached the left and right hand end walls. After a short time, warm air from the forced source was mixed throughout the enclosure (figure 2b, image at $t = 180$ s).

TWO-SOURCE EXPERIMENTS

Of fundamental interest is how a developing stratification is effected by the relative strengths and spatial distribution of localised buoyancy and momentum inputs. For a range of aspect ratios these inputs are in conflict as, in general, the buoyancy input produces a stabilising stratification, whereas the momentum input induces overturning and destabilising motion. We examined how the stratification established by a single forced buoyancy source, with initial fluxes M_0 and B_0, is modified as a result of supplying these initial fluxes independently via two separate sources; namely, a source of initial momentum flux M_0 only (a jet), and a source of initial buoyancy flux B_0 only (a plume), and separated by a horizontal distance χ/H. A range of separation distances and relative jet / plume strengths were considered. Only the stratifications produced by simultaneous inputs from the two sources are described here, although, those produced by first establishing a stable stratification and then supplying initial momentum flux, and conversely, creating a turbulently well-mixed environment by an input of momentum flux and then supplying buoyancy have also been examined.

Effect of source separation χ/H (Expts 11, 12 & 13, tank #2)

For all dimensionless separation distances considered, namely, $\chi/H = 0.27$ to 1.9, the jet and plume flows did not strongly interact as they rose (although interaction was stronger as χ/H decreased), and the developing stratification was strongly horizontally inhomogeneous. The buoyant ceiling current produced by the plume spread radially and collided with the turbulent radial ceiling jet produced by the momentum source. Where the current and ceiling jet collided there was vigorous mixing and a downward flow was observed which removed buoyant fluid from the current and carried it into cooler air below. This fluid was re-entrained into both the rising jet and plume and a layer formed of intermediate temperature. At later times, fluid carried from the collision region had mixed throughout the entire volume and the temperature throughout the enclosure increased with time. After initially increasing in depth, the buoyant layer approached a steady depth, thereby indicating that the rate of removal of fluid from this layer (at the collision region) was matched by the rate of fluid supplied to the layer by the plume. The steady layer depth, which increased with increasing H, was comparable with the plume width at the ceiling. Figure 3 shows the saline stratification at 90 s for $\chi/H = 1.9$ and $\chi/H = 0.27$.

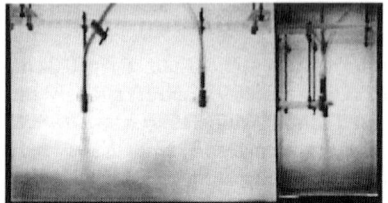 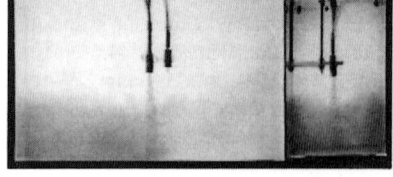

a) $\chi/H = 1.9$ (Expt 8)   b) $\chi/H = 0.27$ (Expt 13)

Figure 3. Localised buoyancy source (saline plume, LHS) and momentum source (water jet, RHS) in an unventilated enclosure. The flow is shown at 90 s. For the jet $M_0 = 1115.6$ cm^{4}s^{-2}; for the (weakly forced) plume $M_0 = 44.6$ cm^{4}s^{-2} and $B_0 = 70.2$ cm^{4}s^{-3}. Tank #2 with $H/L_1 = 0.5$, $H/L_2 = 1.05$.

Note that the flow is more symmetric (about the vertical axis midway between the sources) at the smaller separation and, in addition, the horizontal variation in density is reduced (as indicated by the concentration of the dye). This trend is expected to continue and for sufficiently small χ/H the overall motion is anticipated to be similar to that established by a single forced plume with the combined fluxes M_0 and B_0. Further experiments are planned to examine in detail the transition from two-source to single-source behaviour. Similar flows were observed in both square and rectangular section enclosures.

Effect of aspect ratio (Expts 8, 9 & 10, tank #2)

Aspect ratios of $H/L_1 = 0.5$ (figure 4a), 0.25 (figure 4b) and 0.125, with a source separation of $\chi/H = 1.9$ were considered. At each aspect ratio, the ascending flows from the sources remained separate and did not strongly interact and mix. After impacting with the ceiling, the horizontal outflows from sources collided (as described above) and buoyant fluid from the plume outflow was carried downwards and mixed throughout the enclosure. The degree of overturning decreased as the aspect ratio decreased. With an aspect ratio of $H/L_1 = 0.5$ the outflows collided roughly at the midpoint between the jet and plume centre-lines (figure 4a). As the aspect ratio decreased the collision occurred closer to the jet centre-line. The stable buoyant layer fed by the plume initially increased in depth and intruded further across the ceiling toward the origin of the turbulent ceiling jet flow. For the smaller aspect ratios considered buoyant fluid was observed to 'slide' above the opposing ceiling jet and the buoyant layer extended across the entire surface area of the ceiling (figure 4b).

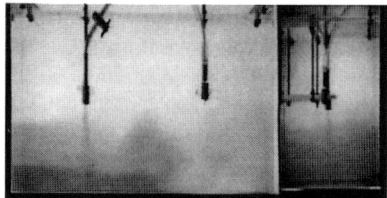 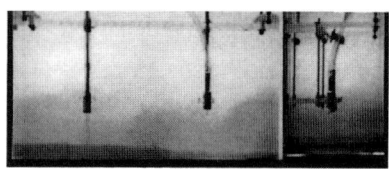

a) $H/L_1 = 0.5$ (Expt 8)   b) $H/L_1 = 0.25$ (Expt 9)

Figure 4. Localised buoyancy source (saline plume, LHS) and momentum source (water jet, RHS) in an unventilated enclosure. The flow is shown at 180 s. For the jet $M_0 = 1115.6$ cm^{4}s^{-2}; for the (weakly forced) plume $M_0 = 44.6$ cm^{4}s^{-2} and $B_0 = 70.2$ cm^{4}s^{-3}. Tank #2 with $\chi/H = 1.9$. In a) $H/L_2 = 1.05$ and in b) $H/L_2 = 0.52$.

CONCLUSIONS

We have described preliminary results of a fundamental experimental study examining the stratification established by a localised buoyant source of non-zero initial momentum flux on the floor in a closed room of aspect ratio (H/L_1) less than unity. In the absence of initial momentum flux, a stably stratified region of warm air forms at the ceiling separated by a horizontal interface from the cooler air below. Increasing the initial momentum flux leads to increased mixing and overturning motion, and as a result the warm air layer initially descends more rapidly than for the zero momentum flux case. Qualitatively, the same effects are observed on increasing the room aspect ratio. The vertical scale of the overturning motion is characterised by the dimensionless jet-length and aspect ratio of the enclosure; if two characteristic horizontal length scales are available, *e.g.* in a room of rectangular cross-section, the smaller of the aspect ratios determines the initial extent of the overturning motion. For source conditions which yield jet-lengths sufficiently larger than the room height, the heat from the source is rapidly distributed throughout the enclosure.

When buoyancy and momentum fluxes are input from separate sources, the developing thermal stratification is quite different from that established by an equivalent single source. The mixing produced by the momentum source is found to depend critically on the location and relative strengths of the sources. We are currently developing theoretical models in order to predict the stratification and flows observed in the experiments described and to provide further insight into air movement and stratification in closed spaces.

ACKNOWLEDGEMENTS

The experiments were performed at the Department of Mechanical & Aerospace Engineering at the University of California, San Diego. GRH and PC would like to thank PFL for making the visit to UCSD possible. GRH gratefully acknowledges the funding of the EPSRC.

REFERENCES

Baines, W.D. & Turner, J.S. (1968) Turbulent buoyant convection from a source in a confined region. *J. Fluid Mech.*, **37**, 51-80.

Morton, B.R., Taylor, G.I. & Turner, J.S. (1956) Turbulent gravitational convection from maintained and instantaneous sources. *Proc. Roy. Soc. A*, **234**, 1-23.

Barnett, S.J. (1991) The dynamics of buoyant releases in confined spaces. *Ph.D. thesis*. The University of Cambridge, 166 pp.

Air Distribution in Rooms, (ROOMVENT 2000)
Editor: H.B. Awbi

A STUDY ON THE CHARACTERISTICS OF AIRFLOW IN A FULL SCALE ROOM WITH A SLOT WALL INLET BENEATH THE CEILING

Zhang G.[1]; Morsing, S.[1]; Bjerg, B.[2]; Svidt, K.[3]

[1] Danish Institute of Agricultural Sciences, Department of Agricultural Engineering, Research Centre Bygholm, P.O.Box 536, DK-8700 Horsens, Denmark (guoqiang.zhang@agrsci.dk)
[2] Royal Veterinarian and Agricultural University, Department for Animal Science and Animal Health, Bülowsvej 13, 2. vej, bygn 1-03, DK-1870 Frederiksberg C, Denmark
[3] Aalborg University, Department of Building Technology and Structural Engineering, Sohngaardsholmsvej 57, DK-9000 Aalborg, Denmark

ABSTRACT

Symmetry is not a sufficient condition for the design of a ventilated room to generate two-dimensional airflow. Three-dimensional effects were observed in a symmetrically designed 3m × 5m × 8.5m test room having a 0.019m slot inlet opening height under the ceiling in the one end wall. The ceiling jet velocity profile measured in the symmetric plane agreed well with the jet models for two-dimensional flow, but large differences were found out of the symmetric plane. The velocities in the jet 4.5m downstream from the inlet wall were up to a factor of two higher on the one side than the other side. During the measurement period the side with high velocities occasionally changed without any obvious disturbance in the room. Two semi-stable statuses were observed. The measured velocities in the symmetric centre plane were only slightly effected by the switch-over and remained at the same level throughout the experiment. In both statuses the return air direction diverged 30° from the symmetric plane. Systematic validation of flow behaviour is thus necessary before the assumption of two-dimensional flow should be accepted and a study is carried out in the symmetric plane only.

KEYWORDS

Airflow Characteristics; Full-scale experiments; Slot-inlet ventilation; Attached jets

INTRODUCTION

To validate Computational Fluid Dynamics (CFD) simulations of indoor air the measured data using well-defined ventilated enclosures are recommended. The enclosures can be scale models (Nielsen, 1974; Timmons et al., 1986), or full-scale rooms (Hoff, 1995). A commonly used design is a rectangular room with a slot inlet beneath the ceiling and a slot outlet near the floor (Timmons et al., 1986; Hoff, 1995). In such room two-dimensional (2D) airflow patterns are assumed, and measurements are taken in the symmetric plane only. Few systematic experimental data are available, however, throughout well defined, full-scale rooms to check the CFD code and to analyse the boundary conditions.

An intensive joint project that the authors are involved is focusing on the application of CFD in the analysis of air motion and velocity characteristics in livestock buildings. One of inessential intentions was to validate algorithms for the simple situation of 2D flows in an empty room as a reference, and then to provide the necessary information on boundary conditions for animal housing. The hypothesis was that the use of a geometrically symmetric room was sufficient to achieve 2D airflows. In this paper results of the measurements in the empty room are presented.

MATERIAL AND METHODS

The length (L) of the experimental room was 8.5m, the height (H) 3m, and the width (W) 5m in order to represent a common piglet producing section. A slot opening with a length (w) of 5m was placed just below the ceiling. The height of the inlet slot was 0.1m at fully open. The actual opening height could be adjusted in the range form 0 to 0.1 m with a 0.24m×5m bottom-hinged flap. To simulate cool weather conditions an inlet height (h) of 0.019m was chosen.

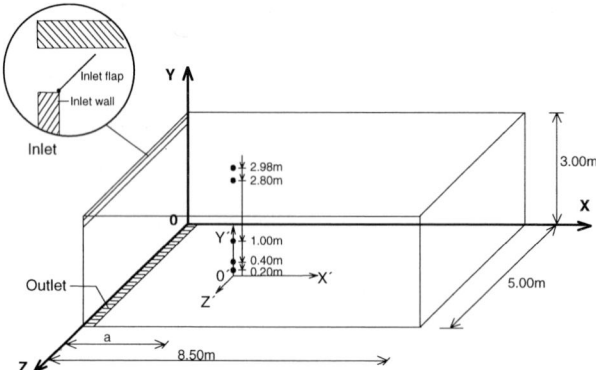

Figure 1. Room geometry and positions of inlet, outlet and sensors. a = 2.68m, 3.5m and 4.5 m.

Comparing the recommend geometry for creating 2D flows by Nielsen (1990), which was 9m (L) × 3m (H) × 3m (W) with a full width slot inlet of 0.168m (h_{max}), the differences were: (1) The W/H ratio was extended from 1 to 1.66. (2) The L/H ratio was reduced from 3 to 2.83. (3) The h/H ratio became 0.006, nearly a factor 10 smaller than Nielsen (1990) suggested.

The floor of the experimental room was raised 0.6 m above the building floor so it was possible to create a 4.8 m by 0.13 m 20% open slatted floor with exhaust to a below-floor channel along the inlet wall, figure 1. The airflow rate was measured at exhaust with an orifice system (ISO 5167-1) with estimated uncertainty of ±3%. The room was as airtight. Leakage test was performed and the leakage data was considered in the computing of inlet airflow rate.

JET MODELS

The air jet through a slot beneath the ceiling is generally referred to as an attached plan jet. The inlet the maximum velocity u_x in the jet downstream can be estimated by (Grimitlin,1970; ASHRAE, 1997)

$$\frac{u_x}{u_o} = C_p \left(\frac{h_o}{x + x_o} \right)^{1/2}$$

(1)

where u_o is the inlet velocity, C_p is the velocity decay constant for a slot inlet, h_o is the effective opening height of the inlet, x is the distance downstream from the inlet and x_o is the distance from the inlet opening to the virtual origin of the jet. For a narrow slot inlet opening without deflects x_o can be ignored. For a jet generated by a slot inlet beneath the ceiling with bottom hinged flap, C_p is approximately 3.8 (Förthmann, 1934; Nielsen, 1997). The effective opening height h_o can be estimated as the discharge coefficient, C_d, multiplied with the geometric height, h.

In the similarity region, the velocity profiles remain identical in dimensionless form. The universal velocity profile for an attached plan jet could be stated as (Verhoff, 1963):

$$\frac{u}{u_x} = 1.481 \left[\frac{y}{\delta(x + x_o)} \right]^{\frac{1}{7}} \left(1 - erf(0.68 \frac{y}{\delta(x + x_o)}) \right)$$

(2)

where y is the vertical distance from the ceiling to the point, δ is the expansion coefficient for the jet, and erf is an error function. According to flow similarity in the jet δ is considered to be a constant. It defines the slope of the line where the velocity is half of the maximum velocity. This line is often used to characterise the expansion of the jet. The product $\delta(x + x_o)$ thus expresses the vertical distance from the ceiling to the point where the velocity is half of the maximum velocity. Equation 2 includes descriptions of the boundary layer near the ceiling where the velocity increases from zero to maximum.

Another simpler way to describe an attached plan jet could be found in Becher (1972) and ASHRAE (1997), where the velocity profile is proposed as half of a free plane jet:

$$\frac{u}{u_x} = 10^{-0.3030 \left[\frac{y}{\delta(x + x_o)} \right]^2}$$

(3)

where the boundary layer near the ceiling is omitted, but the rest of the profile is similar to equation 2. In ventilated room spaces a value of δ in the range from 0.06 to 0.14 is suggested, mainly depending of room geometry, inlet design and the Reynolds number (Nielsen & Möller, 1988).

MEASUREMENTS

The inlet velocity was adjusted to 4.8 m/s in the experiments. That gave a Reynolds number of 5100 with the inlet opening h=0.019m. Velocity measurements were carried out with a multi-channel thermistor based system (Zhang & Morsing 1994). The pre-calibration in the range from 0.1 to 6 m/s provides the uncertainties within ±5% for 0.1-1 m/s and ±3% for velocity above 1 m/s.

The measurements were conducted in total 29 measurement positions in a vertical line (y-direction) on a fixture (exchangeable in two level) with the top two positions of 0.02m and 0.1m from the ceiling and 0.1m apart from each other down to the floor for the rest. The fixture was placed downstream at 2.68, 3.5, and 4.5 m from the inlet wall (x-direction) and in the planes z = 0.5, 1.5, 2.5 (symmetric plane), 3.5, and 4.5 m, resulting in 29×3×5 sensor positions in the room (figure 1). The measurement time was 30 minutes with scan rate 10 Hz and averaging every 5 minutes.

RESULTS

Prescribed velocity

In figure 2(a) the velocity profiles are shown in the symmetric plane at the three downstream positions. The dimensionless velocity u/u_x is presented as a function of the non-dimensional distance from the ceiling y/x. The velocity u_x is estimated according to equation 1. Smoke tests showed that the non-dimensional total width of the jet was in the order of 0.2. The data up to that width was thus extracted and used for non-linear parameter estimation on the basis of equation 2. The estimated expansion coefficient, δ, was 0.093 m/m ± 0.0053 (P<0.05). In figure 2(a) the estimated profile of the jet, using equation 2, is shown as a solid line. It is seen that the estimated profile fits well to the measured data. Using the estimated δ in equation 3 gives the dotted line in figure 2(a), and the deviation near the ceiling is clearly seen.

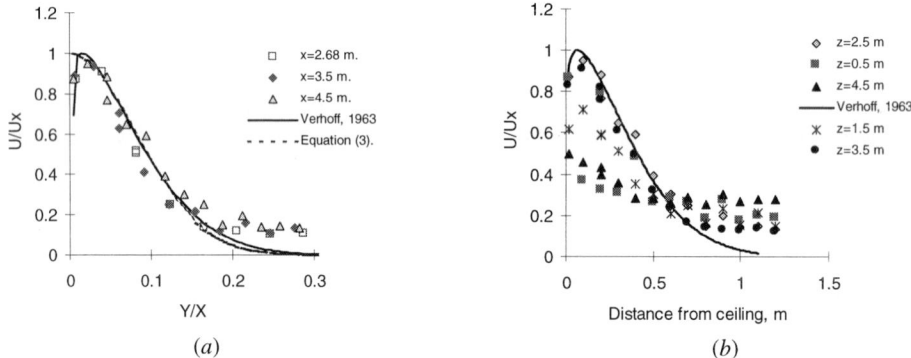

(a)   (b)

Figure 2. (a) Velocity profiles of the ceiling jet in the symmetric plane z = 2.5 m. (b) Velocity profiles of the jet in five z-planes at the distance x = 4.5 m from the inlet wall.

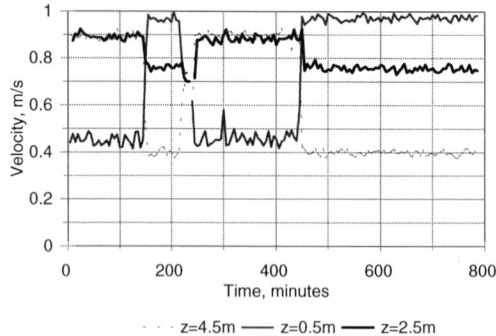

Figure 3. Velocities measured in the ceiling jet at the positions of x=4.5m; z = 0.5, 2.5, and 4.5m; and y = 2.98 m (0.2m from ceiling) showing the semi-stable airflow statuses in the room.

Two dimensional airflow

To check the two-dimensionality of the airflow in the room the data at x = 4.5 m are used. The air velocity profiles for z = 0.5, 1.5, 2.5, 3.5 and 4.5m are shown in figure 2(b). It is seen that the velocity profiles on both sides of the symmetric plane were different from the velocities measured in the symmetric plane. In general the average velocities out of the symmetric plane were lower. The velocities measured outside the symmetric plane were not in accordance with the 2D flow expectation, which assumes similar velocity profiles in all planes parallel to the symmetric plane.

To evaluate the findings, additional measurements were carried out. Six velocity sensors were placed 0.02 m and 0.2 m below the ceiling at positions z = 0.5, 2.5 and 4.5 m from the wall. Continuous measurements were performed for 8 hours in a scan rate of 5 scans/sec saving averages every 5 minutes. The results from the sensors placed 0.2 m below the ceiling are shown in figure 3. It can be seen that the velocities at one side of the symmetric plane were a factor of up to two higher than the other side. During the measurement period the side with high velocities occasionally changed without any obvious disturbance in the room. Smoke tests showed that the jet for some periods turned towards the right downstream corner, and then changed and turned to the left, showing two semi-stable statuses. The measured velocities in the symmetric centre plane were only slightly effected by the switch-over and remained at the same level throughout the experiment.

To study the impact on the return airflow pattern near the floor, a low velocity direction sensor was constructed. The sensor was placed at floor level in the centre plane (x, y, z = 6.5m, 0.2m, 2,5m). In both semi-stable conditions the return air direction diverged 30° from the symmetric plane.

DISCUSSIONS

It is surprising that the 3-D flows encountered in this study are not explicitly reported in many previous studies. Compared with the recommendation given by Nielsen (1990), the differences between the two test rooms were the ratio of inlet opening height to room height, h/H, and the ratio of room width to room height, W/H, may be the main reasons.

The ratio of inlet opening to room height used in this study was about a factor of 10 smaller than the one used by Nielsen (1990). The smaller ratio could be an important explanation for the 3-D flow. The inlet opening height and inlet air velocity chosen in this study, however, were to represent practical conditions of a weaning pig room. Increasing inlet opening height or initial air velocity may reduce the 3-D effects, but this would result in a too high ventilation rate and a too high air velocity in the occupied zone – and that would not be representative for winter ventilation of livestock buildings. Systematic validation of flow behaviour is thus necessary before the assumption of 2D flow is accepted and the measurements are carried out in the symmetric plane only.

3D CFD simulations with wall function as described by Bjerg et al. (1999) indicate that the horizontal velocity profile of a ceiling jet in a very wide room space (W/H = 18) contains a periodical variation with an interval of 3H. In the room with W/H = 1.67 the velocity profile in a horizontal plane contains a part of the periodical curve. The validation of the CFD simulation has not been performed in experiments with room width ratio larger than 1.67. In a test room with room width ratio equal to 1 and room length equal to 3H no significant 3-D effects were found, however, whether in measurements (Nielsen et al 1978) nor in CFD-calculations (Bjerg et al 1999).

CONCLUSIONS

Symmetric room geometry is not a sufficient condition for the design of a ventilated room where 2D airflow is to be generated. The velocities at one side of the symmetric plane were a factor of up to two higher than at the other side. During the measurement period the side with high velocities occasionally changed without any obvious disturbance in the room. Semi-stable flow behaviour was observed. The measured velocities in the symmetric centre plane were only slightly effected by the switch-over and remained at the same level throughout the experiment. In both conditions the return air direction diverged 30° from the symmetric plane.

The ratios of inlet opening height to room height, room width to room height, and room depth to room height may play important roles in generating 2D airflow. Fully to investigate the effects of different inlet and room geometry were beyond the scope of the present experiment, however. Systematic validation of flow behaviour is thus necessary before the assumption of 2D flow is accepted and measurements are carried out in the symmetric plane only.

Acknowledgement

The founding of the Danish Veterinary and Agricultural Research Council for the research program is greatly appreciated. Thanks are due to Professor Peter V. Nielsen, Aalborg University, Denmark, for his careful review, constructive comments and good suggestions for the manuscript.

References

ASHRAE (1997). ASHRAE handbook of fundamentals. American Society of Heating, Refrigerating and Air-Conditioning Engineers (ASHRAE), Atlanta, USA.

Becher P (1972). Varme og Ventilation (Heating and Ventilation), Vol 3, 4th Ed. Teknisk Forlag, Copenhagen.

Bjerg B; Morsing S; Svidt K; Zhang G (1999). Three-dimensional airflow in a livestock test room with two-dimensional boundary conditions. Journal of Agric. Engineering Research **74**(3), 267-274.

Förthmann, E (1934). Über turbulente Strahlausbreitung (Development of Turbulent Jets). Ing. Archiv 5.

Grimitlin M (1970). Zuluftverteilung in Räumen (Air distribution in rooms). Luft- und Kältetechnik, **5**:247-256.

Hoff S J; Janni K A; Jacobson L D (1995). Evaluating the performances of a low Reynold's number turbulence model for describing mixed-flow airspeed and temperature distributions. The Transactions of the ASAE, **38**(5):1533-1541.

Nielsen P V (1974). Flow in air conditioned rooms (model experiments and solution of the flow equation). Ph.D. Thesis, Danish Technical University, Copenhagen.

Nielsen P V; Restivo A; Whitelaw J H (1978). The velocity characteristics of ventilated rooms. ASME Journal of fluid engineering, 100:291-298.

Nielsen P V; Möller Å T A (1988). Measurement on buoyant jet flows from a ceiling-mounted slot diffuser. The Third Seminar on "Application of Fluid Mechanics in Environmental Protection – 88". Silesian Technical University, Gliwice, Poland.

Nielsen P V (1990). Specification of a two-dimensional test case. International Energy Agency, Energy Conservation in Buildings and Community Systems, Annex 20: Air flow Pattern within Buildings.

Nielsen P V (1997). The Box Method – A Practical Procedure for Introduction of an Air Terminal Device in CFD Calculation. Aalborg University, Aalborg, Denmark.

Timmons M B; Albright L D; Furry R B; Torrance K E (1986). Experimental and Numerical Study of air movement in slot-ventilated enclosure. The Transactions of the ASHRAE, 92:221-239.

Verhoff A (1963). The Two-Dimensional Turbulent Wall Jet without an External Free Stream. Report No.626, Princeton University, Princeton, USA.

Zhang G; Morsing S (1994). Multi-head thermistor anemometer. Departmental Note, 1095-2. , Danish Institute of Agricultural Sciences, Dept.Ag.Eng., Research Centre Bygholm. Horsens, Denmark.

Air Distribution in Rooms, (ROOMVENT 2000)
Editor: H.B. Awbi

SPREAD OF GRAVITY CURRENTS WITHIN
A MULTI ROOM BUILDING

C Blomqvist and M Sandberg

Royal Institute of Technology
Centre for Built Environment
Box 88, S-801 02, Gävle, Sweden
Phone +46 26 147800 Fax +46 26 147803
E-mail: blomqvist@bmg.kth.se

ABSTRACT

In dwellings ventilated by extract ventilation there are common complaints of cold draught caused by the supply air entering the room through openings close to the windows. This paper reports on studies of unconventional ways to distribute the supply air in order to minimise the risk of such problems. Experiments have been done where the supply air device is located in the hall of an apartment. The ventilation efficiency in the rooms adjacent to the hall has been studied with open and closed doors.
The behaviour of gravity currents has also been studied in scale models.
Measurements show that it is possible to ventilate rooms using only slots above and below the doors.

KEYWORDS

Extract ventilation, cold draught, gravity currents, air distribution, full scale test, scale model

INTRODUCTION

In dwellings ventilated by extract ventilation there are common complaints during wintertime of cold draught caused by the supply air entering the rooms through openings close to the windows. In order to reduce the problems unconventionally ways can be used to distribute the ventilation air. One way is to use ducts to distribute the supply air to spaces where people usually do not reside for longer periods of time e.g. the hall. However this unconventional location of the supply air devices may result in difficulties to obtain sufficiently good ventilation in some parts of the apartment, especially when the doors between the rooms are closed. When doors are open even very small temperature differences between the rooms cause large airflow rates in both directions through the doorways. Results from measurements of airflow in a doorway between two rooms of different temperature have been reported by Blomqvist, Sandberg (1998). The aim of this work is to investigate if it is possible to ventilate a room using only slots above and below the doors.
This paper also reports on a model scale study of the behaviour 2D gravity currents.

Theory

Figure 1 shows a 2-D gravity current. Entrainment of fluid (ΔQ) from the ambient occurs at the inlet and further downstream the gravity current is heated by the floor.

The velocity is governed by the local buoyancy flux $B(x)$, see e.g. Etheridge&Sandberg (1996)

$$U = kB(x)^{1/3} \tag{1}$$

Where ,k , is a constant and the buoyancy flux, B, is :

$$B(x) = \frac{Q(x)}{w} g \frac{\Delta\rho(x)}{\rho(x)} = \frac{Q(x)}{w} g'(x) \qquad [\text{m}^3/\text{s}^3]$$

The height is

$$h(x) = \frac{\dfrac{Q}{w}}{B(0)^{1/3}} \tag{2}$$

No heat transfer but entrainment of fluid

In this case the flowrate increases but at the same time the density difference decreases and the buoyancy flux becomes

$$B(x) = \frac{Q(0)}{w} g \frac{\Delta\rho(0)}{\rho(x)} \approx B(0)$$

The buoyancy flux is almost conserved and therefore the velocity and the height are according to relations (1) and (2) constant.

Heat transfer

We assume the floor, which has the temperature T_f, heats the gravity current. The heat transfer coefficient is K [watt/m·K]. For simplicity we ignore the entrainment of ambient air. We obtain the following expression for the reduction in buoyancy flux

$$B(x) = B(0)e^{-\frac{Kx}{\rho C_p \frac{Q}{w}}} \tag{3}$$

Therefore according to relation (1) the velocity will decrease with increasing distance from the supply.

Mean age of air

The local mean age of air, τ_p, in a room is one indicator of ventilation efficiency and can be determined by tracer gas measurements using the decay method, see Etheridge & Sandberg (1996). If the mixing of the air is complete τ_p is the same in the entire room and equal to the nominal time constant, $\tau_n = V/Q$.

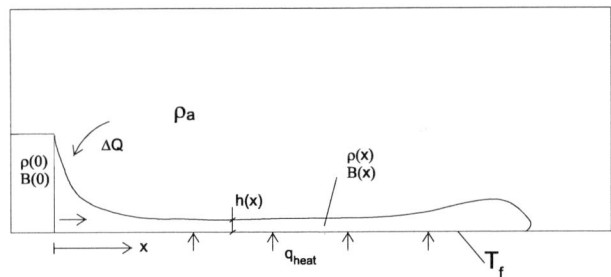

Figure 1 Example of a 2-D gravity current from a low momentum flux device

METHODS

Full scale experiments

For the experiments a test house in the laboratory hall of the department has been used. The house has been designed to look like an ordinary apartment consisting of five rooms (figure 2). The total area of the apartment is 70m² and the corresponding volume, V, is 175m³. Air has been supplied as a gravity current in the hall at floor level and the air has been distributed within the apartment through openings above and under the doors. Extract air devices are located in kitchen and bathroom and the total air flow rate, Q, has been 90 m³/h which is in accordance with the Swedish building code for a dwelling of this size. The nominal time constant for the entire test apartment will then be 1.94 h (V/Q).

The gravity current in the hall has been studied by Thörnström (1998), where the height and the temperature of the current have been determined for various temperature differences.

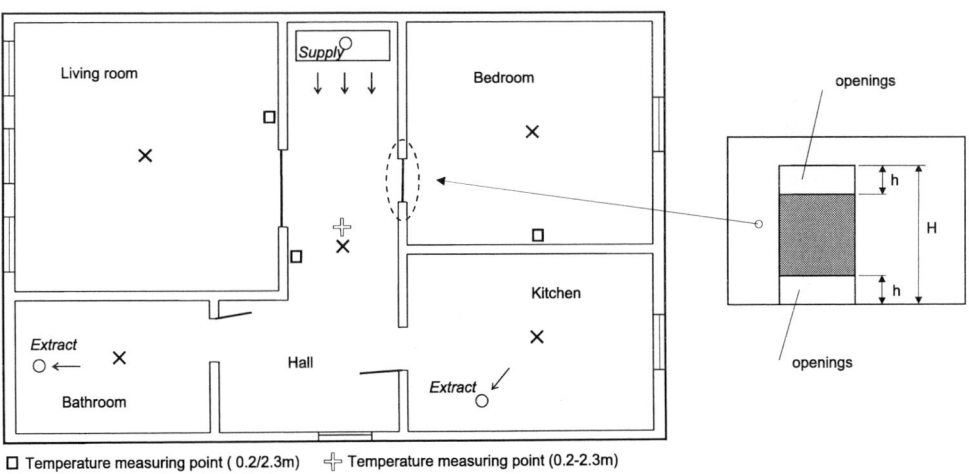

□ Temperature measuring point (0.2/2.3m) ⊹ Temperature measuring point (0.2-2.3m)
X Tracer gas measuring point (1.2m)

Figure 2. Test house

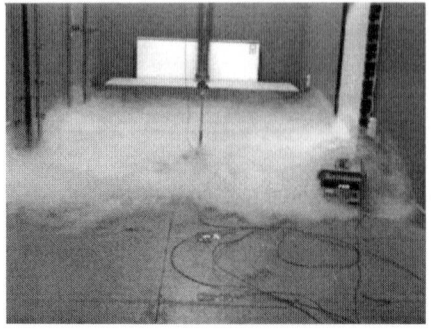

Figure 3. Gravity current from supply air device in the hall advancing towards the viewer

In the middle of each room the local mean age of the air has been determined using tracer gas (decay-method). During these measurements the temperature of the supply air has been approximately 2°C below the temperature in the test building. The height of the slots, h, above and below the doors have been varied from 0.0m (door closed) to 1.00m (door open).

In Figure 3 the supply air in the hall has been visualised by smoke.

Scale model experiments

Figure 4 shows a water model experiment to demonstrate how the supply air can be distributed within the apartment. In the hall the fluid from the supply device is transported as a 2-D gravity current continues as a radial gravity current into the adjacent rooms. Water has been used as fluid and the density difference has been achieved by using saline water as supply fluid. The fluid has been visualised by using coloured dye.

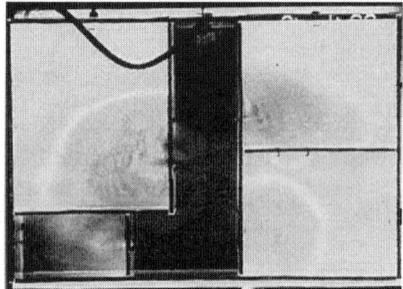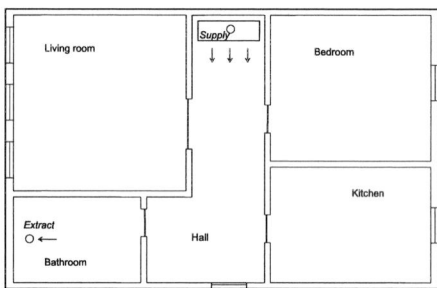

Figure 4. Scale model of the test house with a slot below each door.

To study the behaviour of gravity currents in more detail another scale model has been used. The density difference in the scale model corresponds to a temperature difference in air of 3°C. The length of the model is 2.0m. Test conditions of the experiment can be found in table 1. To make the currents visible the model has been lighted from behind and the gravity current has been videotaped.

TABLE 1

TEST CONDITIONS OF THE WATER SCALE MODEL EXPERIMENT

Height of model	2.00E-01	m	Height of supply	1.02E-01	m
Width of model	7.00E-02	m	Width of supply	7.00E-02	m
Supply flow rate	2.90E-05	m³/s	Free area	9.74E-04	m²
Supply density	1.01E+02	kg/m³	Initial velocity	2.98E-02	m/s
Ambient density	9.99E+03	kg/m³	Buoyancy flux, B(0)	4.07E-05	m³/s³
Reduced gravity, g'	9.82E-02	m/s²	B(0)$^{1/3}$	3.43E-02	m/s

RESULT

Full scale test

Study of the gravity current in the hall

A picture of the gravity current is shown in figure 3 where the incoming air has been visualised using smoke. The height of the current has been estimated to 0.07-0.10 m. Air temperatures and air velocities have been measured along the gravity current. Initial temperature difference is 3.1°C and the ambient temperature of the hall is 25.8°C. The result is shown in table 2. Based on the inlet specific buoyancy flux the theoretical velocity is 0.12 m/s. Effect of the heat transfer from the floor can be seen clearly.

TABLE 2
AIR VELOCITIES AND TEMPERATURES OF THE GRAVITY CURRENT IN THE HALL

Distance from supply [m]	Air velocity [m/s]	Air temperature [°C]
0.0	0.25	22.7
1.0	0.13	23.8
2.0	0.10	24.2
3.0	0.08	24.5
4.0	0.08	24.6
5.0	0.07	24.6

Measurements of the mean age of air in the rooms adjacent to the hall

The mean age of air in each room has been measured for various heights of the openings above and below the door. Figure 5 shows the local mean age of air in the bedroom at 1.2 m above the floor.

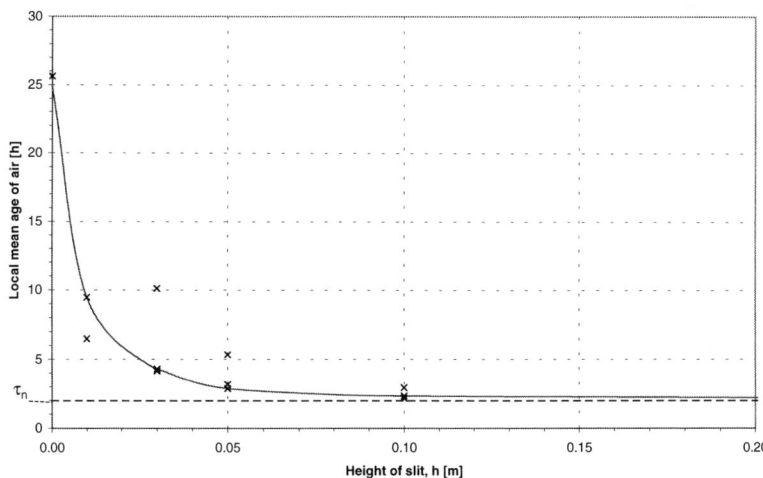

Figure 5. Local mean age of air in bedroom versus height of slot.

When the door is closed (h=0) the mean age of air is high, indicating that the room is not well ventilated. As the height of the opening is increased the mean age of air rapidly decreases. When the height of the opening is 0.10 m the mean age of air is almost the same as the nominal time constant of the apartment which means that the room is almost as well ventilated as when the door is fully open.

Scale model experiments

The 2-dimensional scale model has been videotaped and the velocity of the initial gravity current has been studied. By studying the videotape it has been found that the velocity if the gravity current is constant, (the mean value of k, in relation (1) is equal to 0.84), which is in accordance with theory.

DISCUSSION

The full scale tests show that it is possible to distribute cool air within a building using slots above and below the doors. The velocities close to the floor in the hall are small and will not cause thermal discomfort. One can see that when the openings in the doors become larger the ventilation of the room increases rapidly. When the height of the slots are greater than 0.10 m the mean age of air is almost independent of the size of the opening. This corresponds well to the observed height of the gravity current from the supply air device.

The experiments show that there are good possibilities to ventilate a room only by openings above and below the doors. If low velocity devices are used for the supply air there will be no problems of thermal comfort. However the risk of privacy problems is significant due to noise transmission through the openings. Therefore special openings must be developed.

NOTATIONS

C_p	Specific heat at constant pressure [J/kg·K]
$g'=g·\Delta\rho/\rho$	Reduced gravity [m/s²]
K	Heat transfer coefficient [W/m·K]
Q	Flow rate [m³/s]
q_{heat}	Heat flux [W]
T_f	Floor temperature [K]
V	Total volume of ventilated space [m³]
w	Width of inlet [m]
ρ_a	Density of ambient fluid [kg/m³]
$\Delta\rho$	$\rho-\rho_a$ [kg/m³]

ACKNOWLEDGEMENT

We would like to acknowledge the support by the Swedish Council for Building Research through grant no 960637-6. We also would like to thank Ragnvald Pelttari for assistance with the experiments in the water models.

REFERENCES

Etheridge, D., Sandberg, M. (1996), *Building Ventilation, Theory and Measurement*, John Wiley & Sons, Chichester, UK, p627-648

Blomqvist, C., Sandberg, M., (1998). *Transition From Bi-directional to Uni-directional Flow in a Doorway*. ROOMVENT'98, Stockholm.

Thörnström, T, (1998) *Okonventionell Tilluftsmetod* , University of Gavle (in Swedish).

Air Distribution in Rooms, (ROOMVENT 2000)
Editor: H.B. Awbi

A DYNAMIC MODEL FOR SINGLE-SIDED VENTILATION

Birgitta Nordquist and Lars Jensen

Department of Building Services, Lund University,
P.O. Box 118, S-221 00 Lund, Sweden

ABSTRACT

The aim was to develop a simple dynamic model for predicting air exchange caused by short time single-sided ventilation and necessary window opening time in classrooms. Tracer gas measurements have been made in a full-scale room. The comparison indicates that the model can be used when rough estimates of air exchange are of interest.

KEYWORDS

Window opening, single-sided ventilation, indoor climate, schools, model, measurement, tracer gas method.

INTRODUCTION

Ventilation is an important factor when a satisfactory indoor environment is to be achieved and good conditions are to be provided for people. This is especially important for people who suffer from allergic symptoms. In classrooms both insufficiency of fresh air and high room temperatures are a problem. Window opening can be a useful addition to the existing mechanical ventilation system in order to solve these problems. There should however always be an existing ventilation system where window opening can only be a complement. The disadvantages of this kind of ventilation is; that the supply air is unfiltered and unheated, heat recovery is difficult or impossible and the mechanisms are uncontrollable and variable. The disadvantage due to cold draught can be avoided by opening windows during the breaks when there is nobody in the classroom. In some conditions window opening is not advisable, for instance when there is a large amount of pollen or situated close to heavy traffic.

The main purpose of this project was to study window opening and its application in classrooms. The whole work, presented in a licentiate thesis (Nordquist, 1998), consists of three parts, the first of which will be presented in this paper. In the first part measurements were made of air exchange through an open window using tracer gas. A relationship for the time dependent airflow was suggested. The airflow decreases during the window opening period and with the help of the model an appropriate opening time can be decided.

In the second part, not presented here, computer- calculations were made of the impact of window opening on the airflows distributed by a mechanical supply and exhaust ventilation system. The total air exchange is increased in the room during the opening period. In the third part a questionnaire answered by teachers in Malmö, Norrköping and Umeå about their window opening behaviour was presented. The questions related to the frequency of window opening, the reasons for and against this. One out of two teachers open the windows during every break all through the year. The most common factors for window opening were bad air and high room temperatures and the factors for avoiding it were cold outdoor air and noise from outside.

MATERIAL AND METHODS

Dynamic Model

When air is exchanged through an opening the driving force; the temperature difference between the outdoor and the indoor air, decreases as the outdoor air enters the room. This phenomenon has been partly studied for open doors by Kiel and Wilson (1986) and Nielsen and Olsen (1992). The model of Kiel and Wilson assumes a period of constant flow until the outdoor air level reaches the middle of the opening. Then the constant flow is assumed to decrease exponentially with time until the upper level of the opening is reached. This is an approximation of the physics. The physics without mixing can however be treated correctly as will be shown in this paper, with constant flow until the lower level of the opening is reached and then a decreasing flow to zero until the upper level of the opening is reached. Single-sided ventilation has also been studied in the European PASCOOL project (Santamouris et al, 1996). Passive night cooling were studied and thermal dynamics were modelled with computer programs. This paper presents an explicit expression how to calculate air exchange for short time window opening.

The aim is that the model should give typical values of air exchange and window opening time, to be used when few parameters of a building are known. The model assumes that openings are vertical and that the mechanism for the air exchange is the temperature difference between outdoor and indoor air. Heat convection from enclosures during window opening is not taken into account, this may apply when windows are opened for short periods. If windows are opened for a long time the enclosures will be cooled, which is not advisable for energy-saving reasons during heating season. It is also assumed that the outdoor air temperature is lower than the room air temperature. Wind is not considered in the model.

Two models have been developed. The first model assumes no mixing between room air and incoming outdoor air. The second model assumes a total mixing in the room. In this model the indoor temperature is decreased during the time the window is open. It is expected that conditions would in reality be between these two extreme models. The first model shows the lowest air exchange and will be presented in this paper in the following section. The calculation is divided into two time spans. During the first interval the outdoor air is assumed to fill the room air volume below the opening and during the second phase the room air volume in front of the opening is assumed to be replaced with outdoor air, leaving a volume of unventilated air above the opening.

During the first phase the flow is constant and may be expressed as follows (Awbi, 1991)

$$q = C_d \frac{B}{3} \sqrt{g' H_0{}^3} \qquad (\text{m}^3/\text{s}) \qquad (1)$$

where q =   flow through window (m^3/s)

$C_d = 0.4 + 0.0045(T_i - T_u)$ opening orifice coefficient (-) (Kiel et al, 1986)   (2)

B = opening width (m)

$$g' = g\frac{\Delta\rho}{\rho} = g\frac{2(T_i - T_u)}{(T_i + T_u)} \qquad (\text{m/s}^2) \tag{3}$$

H_0 = opening height (m)

T_i = indoor temperature (K)

T_u = outdoor temperature (K)

The outdoor air volume V_{a1} which has entered the room through the lower part of the opening during the time t_1 (s) is

$$V_{a1} = qt_1 \qquad (\text{m}^3) \tag{4}$$

The same volume of room air is assumed to have passed trough the upper part of the opening.

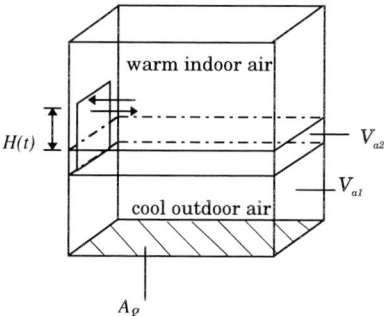

Figure 1: The outdoor air is assumed to first fill the room air volume below the opening (first phase). During the second phase the room air volume in front of the opening is assumed to be replaced with outdoor air. During the second phase the flow is assumed to decrease. The height through which the air is exchanged is a function of time as shown in the figure.

The second phase ends when the exchange stops, which is when the room volume up to the upper limit of the opening is filled with cold outdoor air. As no mixing is assumed to occur, the room air above the opening will not be ventilated. The air flow rate during the second phase is written as

$$q(t) = C_d \frac{B}{3}\sqrt{g' H(t)^3} \quad (\text{m}^3/\text{s}) \tag{5}$$

and the height $H(t)$ can be solved as

$$H(t) = \frac{1}{\left(\dfrac{at}{2} + \dfrac{1}{H_0^{0.5}}\right)^2} \qquad (\text{m}) \tag{6}$$

where t = window opening time during phase 2, $t=0$ when phase 2 starts $\qquad$ (s)

$$a = C_d \frac{B}{3A_g}\sqrt{g'} \tag{7}$$

where A_g = floor area $\quad (\text{m}^2)$

The outdoor air volume V_{a2} which has entered the room during the second phase is given by

$$V_{a2} = (H_0 - H(t)) \bullet A_g \quad (m^3) \tag{8}$$

The total amount of exchanged air V_a during both phases is calculated

$$V_a = V_{a1} + V_{a2} \quad (m^3) \tag{9}$$

Measurements

Measurements have been made in a full-scale room in a test building situated close to the department of Building Services. The floor area was 10.8 m² (3x3.6) and the room volume was 26.6 m³. A side-mounted casement window with a width of 1.05 m and a height of 1.15 m was opened by a telescopic arm for different time intervals. The room was facing south. The test cycle was as follows: The room was filled with NO_2 until a well mixed concentration of about 1000 ppm was reached. The supply was stopped and the decay of the tracer gas due to infiltration was measured for 10 minutes. During the supply and decay of the gas three fans were used to mix the air. Then the fans were turned off and the window was opened and held open for 1, 2, 3, 4 or 5 minutes. After closing the window the fans were turned on again. The decay was measured for another 10 minutes and one test was concluded.

A gas analyzer (Binos 4b), was used to determine the concentration of tracer gas sampled from 2 locations in the room. The measuring equipment was supervised by a computer. The indoor and outdoor temperature were measured to determine the temperature difference needed in the model calculations. The temperatures were measured with 9 thermocouples type T. One was placed outside and the other eight were spaced 0.3 m apart on a vertical rod placed in the middle of the test room. The indoor temperature T_i used in the model was measured just before the window was opened. T_i was calculated as the mean value of all the 8 thermocouples.

The fans were not working when the window was open in order not to disturb the air movement. In the calculation only the tracer gas concentrations before and after opening were used, as the mixing was not uniform during the time the window was open. The air exchange during the open period was assumed to be proportional to the difference in concentration right before and after the opening. The exchanged amount of air was calculated for two different cases. One case assumed no mixing between the room air and the outdoor air and the other assumed total mixing. The case with no mixing will be presented here.

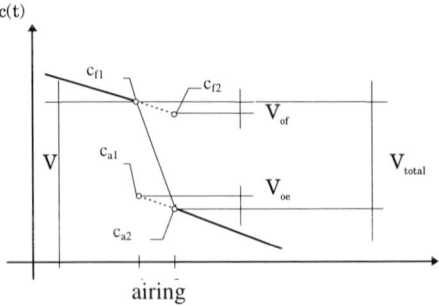

Figure 2: The gas concentrations c_{f1}, c_{f2}, c_{a1}, c_{a2} were calculated with linear regression from the measured concentrations before and after the window opening period. The dotted lines are extrapolated from the linear regression equations. The air volume V_{total} is the whole volume of air which is exchanged during the time the window is open. V_o is the volume which is exchanged due to other ventilation during airing. The difference between V_{total} and $(V_{of}+V_{oe})/2$ is the exchanged air volume V_a due to window opening.

RESULTS

Comparison between model and measurements

Measured and calculated values are presented in Figures 3 and 4. The outdoor conditions during the different measurements were fairly similar. The temperature difference was within the interval 8-12 K during 21 of the 24 measurements and the wind speed was between 2-6 m/s during 22 of 24 and the wind direction were from NW, W and SW towards SE, E, NE. The wind condition were given as hourly values/ mean values over one hour. No parametric adjustments have been made in the model.

The model presented by Kiel and Wilson (1986) has also been applied to the conditions during the measurements, assuming constant flow when the outdoor air fills the room volume below the opening.

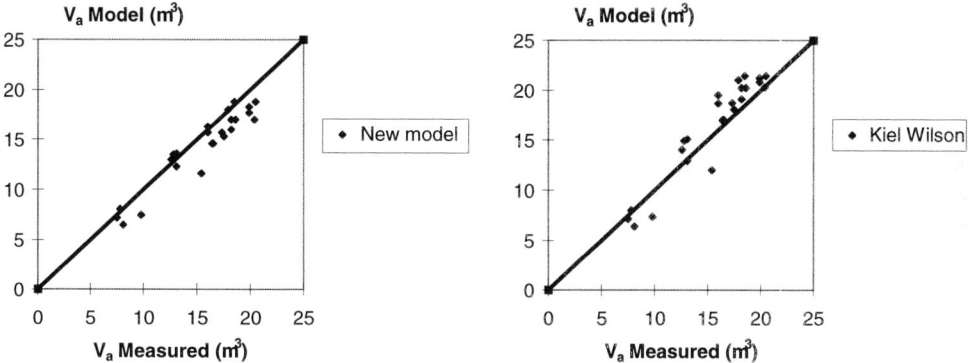

Figure 3: The calculated air volumes as a function of measured values.

Figure 3 shows that in most cases the new model slightly underestimates the measured air exchange. It also shows that the model of Kiel and Wilson and the new model give values of the same order of magnitude.

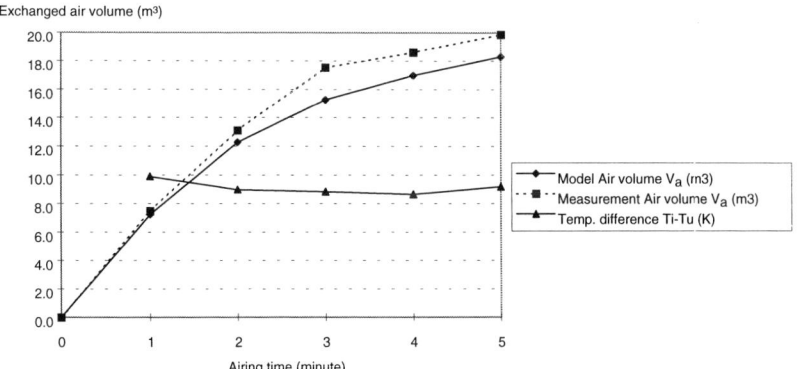

Figure 4: Calculated and measured air volumes. Window open for five different periods; 1, 2, 3, 4 and 5 minutes respectively. No mixing between room air and outdoor air is assumed.

216

Figure 4 shows the result for one of the five sets of measurements which is representative for the results. The measurement results in Figure 4 show that the exchanged air volumes are decreased during the window opening period which means that the air flow through the window opening decreases.

DISCUSSION

The model and the measurements show a congruence in window opening pattern with decreasing air flow. As regards advisable window opening periods the decrease shows that the window should be open for only shorter periods as the amount of air exchanged will be less and less the longer the window is kept open. In most cases the new model underestimates the measured air exchange by about 10-20%.

The model is, as mentioned before, designed to give relatively rough estimations. The contribution of wind to the air change is not included. If a rough value is aimed for, without considering orientation and shape of the building, it is difficult to account for changes in wind speed and wind direction. At high wind speeds it is not advisable to use the model. Heat convection from the enclosures to the air is neglected in the model. No consideration is given to the possible effect of the mechanical ventilation system. The measurements are made in a room that is relatively small compared to a classroom.

The comparison indicate that the model can be used when rough estimates of air exchange by window opening are of interest. Continued research is however needed in order to verify the model. Measurements should also be conducted in bigger rooms, such as classrooms. Additional measurements for longer periods would also be of interest. The model considers only one opening. It would also be of interest to study air exchange with several openings.

ACKNOWLEDGEMENTS

The study was funded by the Swedish Council for Building Research.

REFERENCES

Awbi H. B. (1991). *Ventilation of Buildings.* UK: London, Chapman & Hall

Kiel D. E., Wilson D. J. (1986). *Gravity driven flows through open doors.* Paper 15. Proc. of the 7th AIC Conference, Stratford-upon-Avon, United Kingdom

Nielsen A. F., Olsen E. (1992). *Luftströmmer og varmetap gjennom porter.* FORUT Teknologi AS, Norway

Nordquist B. (1998). *Vädring i skolor - ett komplement till normal ventilation?* (Window opening in schools - a complement to existing ventilation?) (in Swedish). TABK—98/1014, Dept. of Building Science, Lund Institute of Technology, Sweden

Santamouris M., Dascalaki E., Allard F. (1996). *Natural ventilation studies within the frame of PASCOOL project.* Proc. of the 17th AIVC Conf., Gothenburg, Sweden

© 2000 Elsevier Science Ltd. All rights reserved.
Air Distribution in Rooms, (ROOMVENT 2000)
Editor: H.B. Awbi

SIMULATION OF AIR FLOW DISTRIBUTION IN ROOMS BY A SYSTEMIC APPROACH

S. Soares[1], S. Domenech[2], J.C. Laborde[1], C. Laquerbe[2], L. Ricciardi[1]

[1]Institut de Protection et de Sûreté Nucléaire, Département de Protection et d'Etude des Accidents,
Service d'Etudes et de Recherches en Aérocontamination et en Confinement
CEA/Saclay – Bât. 389 – 91191 Gif-sur-Yvette Cedex, France
[2]Laboratoire de Génie Chimique de Toulouse, UMR 5503 CNRS
INP-ENSIGC – 18, chemin de la loge – 31078 Toulouse Cedex, France

ABSTRACT

In order to achieve a satisfactory level of hygiene and comfort in premises and to assess the pollutant transfers, it is necessary to control the air flow distribution. An intermediate approach between predictive numerical simulation and experimental determination of aerodynamic parameters characterizing air distribution in rooms, is the systemic approach. The paper presents the principles of this approach which is based on the residence times distribution (RTD) theory, commonly used in chemical engineering. The aim of the IDTS code recently developed is to build a model from a combination of elementary systems representing basic ideal flows (mixing flow, piston, ...). The adjustment of this model is derived from the comparison of the response to a signal injected into the model with an experimental tracer emission realized in the ventilated room inlet. The general strategy adopted, consisting of a decoupled treatment of parametric identification of structural model determination, is presented. The comparison between experimental residence times distribution on a ventilated laboratory enclosure and the simulated one shows good agreement, as well as the comparison performed with results from a Computational Fluid Dynamics (CFD) tool.

KEYWORDS

Residence times distribution, air flow, pollutant transfers, parametric and structural identifications, genetic algorithms, ventilated premises.

INTRODUCTION

The control of the airborne contamination transfers linked to the operators protection and to the facilities safety is an essential feature in the nuclear industry, particularly to prevent radiological risk. So, it is necessary to have tools in order to assess airborne contamination transfers inside a ventilated enclosure, since the ventilation constitutes a system of contamination containment.

The systemic approach concept aims at interpreting RTD in a room, whose theory is commonly used to describe flow patterns in a wide variety of applications. It appears to be of great interest, specially when the predictive method based on CFD codes cannot be used, particularly in the case of large and cluttered systems. Furthermore, this approach leads to more integrated results, giving a physical significance to the flow model within reasonable CPU times.

GENERAL PRINCIPLES

General structure of the code

The establishment of a RTD model lies on the comparison of a simulated response of a proposed model to a stimulus, with an experimental curve obtained generally through the response of the system to a tracer release, classically helium in ventilated premises. The problem can be formulated as a two level identification problem: a structural identification problem and a parametric identification one. As a consequence, in the IDTS code (Laquerbe, 1999), the modeling is split into two algorithmic loops as shown in figure 1: an inner loop performing the evolution and the choice of the model structure, and an outler loop performing the evolution and the choice of the model structures.

The parameter estimation step for a fixed model structure consists in a constrained nonlinear least square minimization problem solved by a general successive quadratic procedure. The objective function $j(p)$ is then given by the Eqn.1.

$$j(p) = \frac{1}{n_t} \sum_i \left[y_{exp}(ti) - y_{mod}(ti) \right]^2 \qquad (1)$$

n_t is the number of experimental sampling points, y_{exp} the experimental response, y_{mod} the model response, t the time and p the vector of the model parameters, constituted by the volumes and the flow rates of the unitary models.

Concerning the structural identification, an evolutionary procedure based on Genetic Algorithms (Gas) is performed to tackle the combinational aspect of the problem. The objective function considered here is given by Eqn.2.

$$j(M) = j(M, \hat{p}) + \rho.\dim(p) \qquad (2)$$

$j(M, \hat{p})$ representing the least square criterion for optimal parameters $\hat{p}$, is added to a penalty term, $\rho.\dim(p)$, related to the complexity of the model via the number of parameters included in the model $(\dim(p))$ and a weighting factor ρ.

The structures are coded into a hierarchical unique decomposition of the flow diagrams. The models are then represented through a flow system made up with ideal components (piston flow reactor, continuous stirred tank reactor) connected by flow mixers, flow splitters and recycle loops. Leaks, infiltration or short-circuits can also be introduced. The elementary building blocks are linear cascade, recycling loop and parallel distribution.

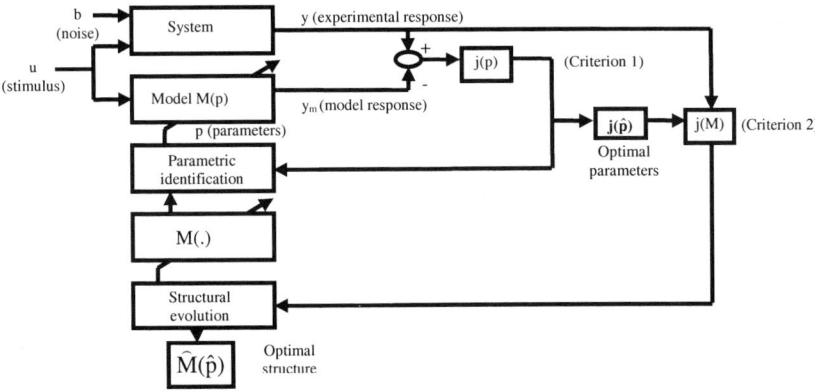

Figure 1: Structure of the general identification strategy

Genetic algorithms procedure

Theoretically developed by Holland (1975), GAs are procedures based on the mimesis of the mechanics of natural selection and genetics (see Table 1). It computes a set of individuals, called population, evolving through a set of biologically inspired operators (cross-over, mutation) constituting the reproduction scheme. In this way, new individuals are generated from parents; only the most suited elements of a population can survive and generate offspring, thus transmitting their biological heredity to new generations.

TABLE 1
ANALOGIES BETWEEN NATURAL GENETICS AND GAs

Natural genetics		Genetic Algorithms
Individual	◄──►	Coding of structures
Gene	◄──►	Elementary building block
Population	◄──►	Set of candidate structures
Generations	◄──►	Algorithm iterations

The fitness of each individuals F_i (Eqn.3) contained in a population of N_{pop} individuals is directly derived from the initial objective function $j(M_i)$.

$$1/F_i = j(M_i)/\left(\sum_{k=1}^{N_{pop}} j(M_k)/N_{pop}\right)$$ (3)

The offspring generation is obtained from the parent one in accordance to the scheme below:

    Generation of the initial population

    Estimation of the fitness of the initial population

    While the number of generations is not reached, do:

        Generate the offspring population

        → selection of individuals surviving

        → synthesis of offspring obtained through cross-over

The selection process provides a fixed percentage of individuals that survive in the new generation. The probability p_i that the individual i survives is given by Eqn.4.

$$p_i = F_i / \left(\sum_{j=1}^{N_{pop}} F_j \right)$$ (4)

After determining the surviving individuals, the population is completed with new individuals obtained through cross-over mechanisms (classical one point permutation operation). The mutation operation is then performed on the entire population with a fixed probability; it consists in the replacement of one decomposition module with another one. The optimized set of parameters are: a population of $N_{pop} = 20$ individuals, a cross-over ratio equal to 0.6, a mutation one equal to 0.1, and a number of generations equal to 50.

APPLICATION TO A VENTILATED LABORATORY ENCLOSURE

A real ventilated room, represented figure 2, has been used as an application of the IDTS computational tool. In this room, it is possible to obtain different air flows and residence times distributions by either changing the positions of blowing and exhaust openings or modifying the exhaust flow rate.

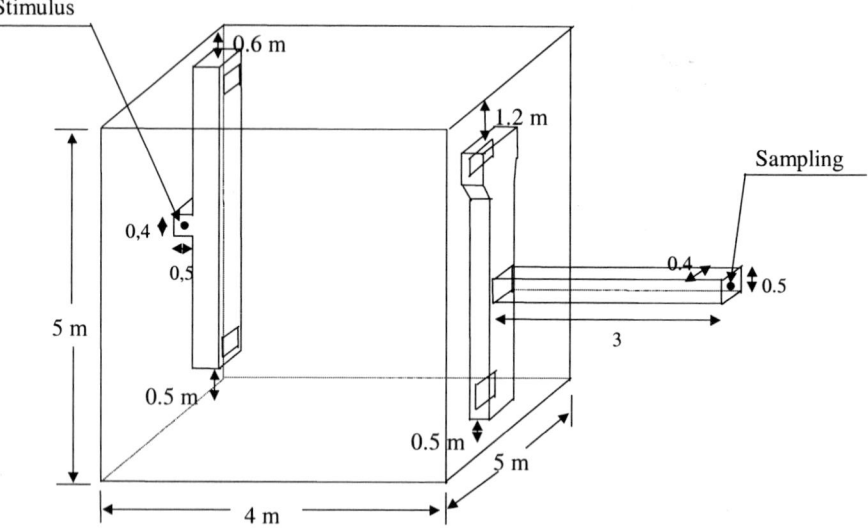

Figure 2: Ventilated room studied

The configuration studied corresponds to an exhaust flow rate equal to 1100 $m^3.h^{-1}$, and to a classical location of blowing and exhaust openings (blowing in the upper part and exhaust in the lower one). For this configuration, identifications have been realized by IDTS by using the experimental response of the system from an helium release. Among the structures proposed by the code, only one has been retained as the most physically probable; it is presented figure 3.

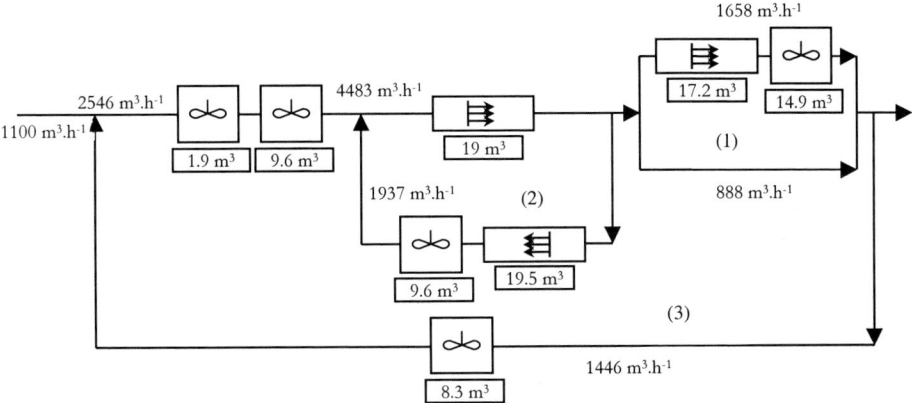

Figure 3: Best structure obtained by the code

The comparison between the experimental curve and the simulated one corresponding to the structure retained is presented figure 4; the results are quite satisfactory. The structure proposed by IDTS is analyzed by comparing the flow patterns induced by this structure and those calculated by 3D predictive numerical simulations performed by the Flovent tool. The helium transfers inside the ventilated enclosure have been calculated and are presented figure 5.

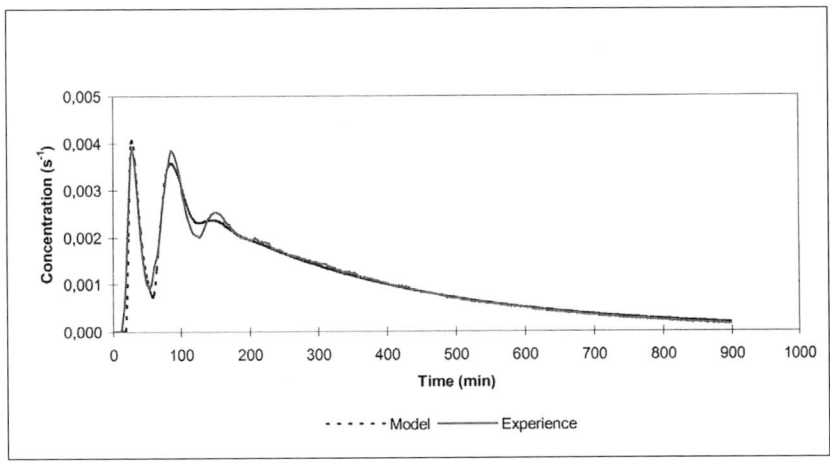

Figure 4: Comparison between experimental and calculated curves

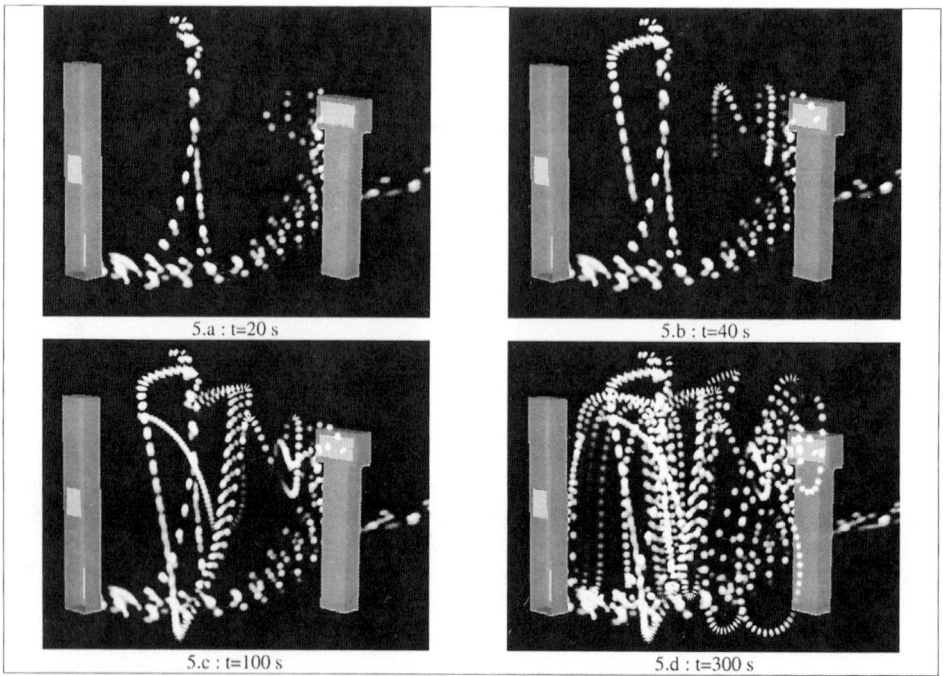

| 5.a : t=20 s | 5.b : t=40 s |
| 5.c : t=100 s | 5.d : t=300 s |

Figure 5: Visualization of the evolution of tracer released inside the ventilated enclosure by using a CFD code

Figure 5.a shows a small part of the blowing flow rate passing quickly (< 20 s) into the exhaust opening. This corresponds to the first peak on the experimental curve, and is generated by the branch n° 1 of the structure obtained by IDTS. Figure 5.b illustrates a secondary recirculation due to the exhaust opening that appears after 40 s. This flow pattern induces the second peak on the experimental curve and corresponds to the branch n° 2 of the structure. After t = 100 s (figure 5.c), the part of helium tracer contained in the main recirculation (branch n°3) is then extracted; this can explain the third peak. At least, after t = 300 s (figure 5.d), the helium tracer has filled all the volume of the cell, and its change involves an exponential decrease.

CONCLUSION

The computational package IDTS presented in this paper constitutes an useful and innovative tool to elaborate residence times distribution models in order to characterize flow patterns and associated transfers. A first application shows the good complementarity between the systemic and the CFD approaches.

References

Laquerbe C. (1999). Modelisation of the air flow in a ventilated enclosure by a systemic approach. INP Toulouse Thesis, France

Holland J. (1975). *Adaptation in natural and artificial systems*, MIT Press, Cambridge

Air Distribution in Rooms, (ROOMVENT 2000)
Editor: H.B. Awbi

MODELLING THE SPATIO-TEMPORAL TEMPERATURE DISTRIBUTION IN AN IMPERFECTLY MIXED VENTILATED ROOM

K. Janssens[1,2], D. Berckmans[1] and A. Van Brecht[1]

[1] Laboratory for Agricultural Buildings Research, Katholieke Universiteit Leuven, Belgium
[2] Fund for Scientific Research, Belgium

ABSTRACT

In this study the spatio-dynamic temperature response in a ventilated room to variations of the supply air temperature was modelled for a wide range of ventilation rates. The model structure was first formulated by applying standard heat transfer theory to zones of better mixing. Spatio-temporal temperature data were then exploited in statistical terms to estimate the physically meaningful model parameters. The dynamic model yielded an excellent fit to the experimental data and was found to characterise the spatially heterogeneous nature of the air flow pattern quite well.

KEYWORDS

imperfect mixing, ventilated room, spatio-temporal temperature distribution, air flow pattern, well mixed zone, heat transfer theory, experimental data

INTRODUCTION

Every living organism (man, animal, plant) lives in an imperfectly mixed fluid which is characterised by spatio-temporal temperature gradients under non-isothermal conditions. In many process rooms (livestock buildings, greenhouses,...) it is desirable to control these temperature gradients in order to achieve optimum process quality (production results, comfort,...) with a minimum use of energy. For this purpose, advanced model-based control theory can be applied. Before this becomes possible, it is first required to have a dynamic mathematical model which describes the spatio-temporal temperature distribution in the process room. The objective of this study was to develop such a model and to evaluate its efficacy in a laboratory test room under controlled conditions.

LABORATORY TEST ROOM

The study was carried out in a large instrumented laboratory test room (length 3 m, width 1.5 m, height 2 m, volume 9 m^3) which is schematically depicted in figure 1. It is a mechanically ventilated room with a slot inlet in the left sidewall just beneath the ceiling (1 in figure 1) and an asymmetrically positioned, circular air outlet in the right sidewall just above the floor (2 in figure 1). An envelope chamber (6 in figure 1) is built around the primary test room to minimise disturbing effects of varying laboratory conditions. The volume of the buffering interspace or buffer zone is 21 m³. The primary test room and the envelope chamber are both constructed of transparent Plexiglass through which the air flow pattern can be observed during flow visualisation experiments.

A series of five aluminium heating elements (3 in figure 1) and a shallow hot water reservoir (4 in figure 1) are placed at the floor to physically simulate the heat and moisture production of the occupant(s). A mechanical ventilation system enables an accurate control of the ventilation rate in the range 70 - 420 m³/h. A heat exchanger in the supply air duct is used to regulate the temperature of the inflowing air. Supply air temperatures in the range 10 - 30 °C can be achieved.

To measure the spatio-temporal temperature distribution in the test room, 36 thermocouples are positioned in a 3-D measuring grid (5 in figure 1). Thermocouples are further located in the air inlet and outlet, in the buffer zone and in the laboratory hall. The accuracy of the thermocouples is 0.1 °C. An intelligent data logger with programmable measurement speed is used for the data acquisition.

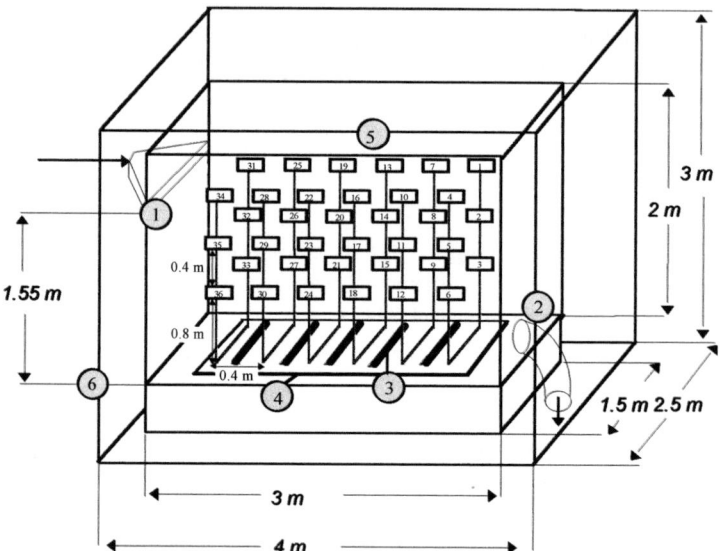

Figure 1: Schematic representation of the test room: 1. slotted air inlet; 2. air outlet; 3. heating element; 4. hot water reservoir; 5. grid of 36 thermocouples; 6. envelope chamber.

MECHANISTIC FORMULATION OF MODEL STRUCTURE

In previous research (De Moor and Berckmans, 1993) it has been demonstrated that the test room is an imperfectly mixed airspace with considerable temperature gradients. Within this imperfectly mixed airspace it is always possible to define a well mixed zone (WMZ) around a temperature sensor in which there exists a good mixing and an acceptably low temperature gradient (Berckmans et al., 1992). To describe the dynamic behaviour of temperature in the WMZ, standard heat transfer theory can be applied. In case of a constant ventilation rate this suggests a linear, first-order heat balance differential equation of the form:

$$\frac{dT_i(t).vol_1.\gamma_1.cl_1}{dt} = V_c.T_o(t).\gamma_o.cl_o - V_c.T_i(t).\gamma_1.cl_1 + Q_c + k_1.S_1.(T_{buff}(t) - T_i(t)) \qquad (1)$$

where t (s) is the time; T_i (°C) is the temperature in the WMZ; T_o (°C) is the supply air temperature; T_{buff} (°C) is the buffer zone temperature; vol_1 (m^3) is the volume of the WMZ; V_c (m^3/s) is the part of the ventilation rate entering the WMZ; Q_c (J/s) is the part of the internal heat production of the 5 heating elements entering the WMZ; γ_1 (kg/m^3) and cl_1 (J/kg.°C) are the density and the heat capacity of the air in the WMZ; γ_o (kg/m^3) and cl_o (J/kg.°C) are the density and the heat capacity of the supply air; and k_1 (J/s.m^2.°C) and S_1 (m^2) are the heat transfer coefficient and the surface area of heat exchange between the WMZ and the buffer zone. A schematic representation of the WMZ concept is given in figure 2.

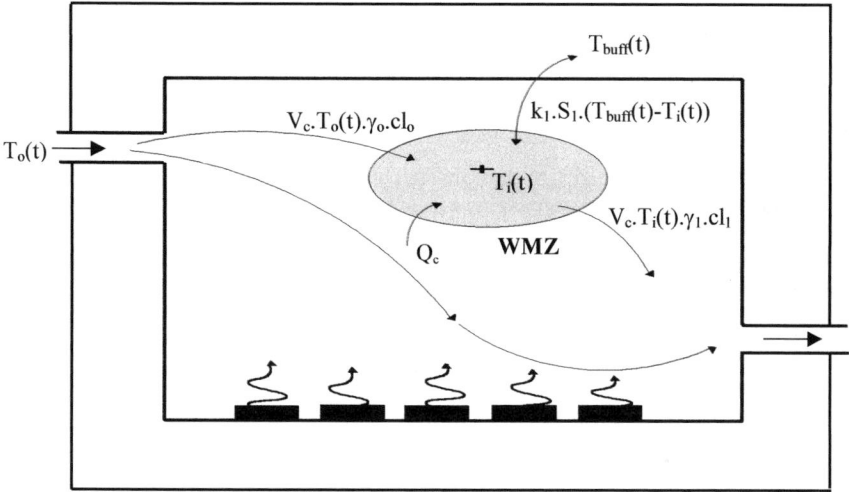

Figure 2: Schematic representation of the WMZ concept

If it is assumed that $\gamma_o \approx \gamma_1 \approx \gamma$ and $cl_o \approx cl_1 \approx cl$ and if only small temperature perturbations ($t_o(t)$, $t_{buff}(t)$ and $t_i(t)$) are considered about steady state, heat balance differential equation (1) can be written more concisely as follows (Janssens, 1999):

$$\frac{dt_i(t)}{dt} = \beta_1.t_o(t) + K_1.t_{buff}(t) - \alpha_1.t_i(t) \qquad (2)$$

where $\beta_1 = \dfrac{V_c}{vol_1}$

$K_1 = \dfrac{k_1.S_1}{vol_1.\gamma.cl}$

$\alpha_1 = \dfrac{V_c}{vol_1} + \dfrac{k_1.S_1}{vol_1.\gamma.cl}$

The WMZ concept, represented in figure 2, can be applied to each of the 36 spatially distributed sensor positions in the test room. By doing so, we obtain a set of 36 first-order differential equations of the form (2) which describe the spatio-dynamic temperature behaviour.

DATA-BASED PARAMETER ESTIMATION

To estimate the model parameters β_1, K_1 and α_1 for each of the 36 sensor positions in the test chamber, 30 identification experiments were carried out. In each experiment the supply air temperature was switched from 11.5 to 17 °C as sharply as possible, whilst maintaining a constant ventilation rate. The experiments were carried out over a range of low (130, 140, 150 and 160 m³/h) and high (200, 210, 220, 230, 240, 250, 260, 270, 280, 290 and 300 m³/h) ventilation rates with two runs at each rate. The internal heat and moisture production were maintained constant at 300 J/s and 0.5 l_{water}/h. In all experiments the supply air temperature, the buffer zone temperature, the ventilation rate and the temperature response at each of the 36 sensor positions in the test room were measured every second over a time period of 2 hours. Further the air flow pattern was visualised and filmed. In the low ventilation rate experiments (130 - 160 m³/h) the incoming air jet deflected downward to the floor resulting in an airflow pattern with stable anti-clockwise direction of rotation. In the high ventilation rate experiments (200 - 300 m³/h) the air jet remained at inlet level after entry producing an airflow pattern with stable clockwise direction of rotation. Experiments at intermediate ventilation rates (170, 180 and 190 m³/h) were not carried out, because preliminary research revealed that the airflow pattern and the spatio-dynamic temperature behaviour were unstable under these conditions.

As a typical example, figures 3(a) and 3(b) show the output of the estimated model for sensor positions 12 and 31 in comparison with the measured temperature response data for an identification experiment with ventilation rate of 290 m³/h. The model clearly yields an excellent fit to the experimental data at both sensor positions. This was the case for each of the 36 sensor positions in each of the 30 experiments (Janssens, 1999). The average model accuracy was 0.15 °C.

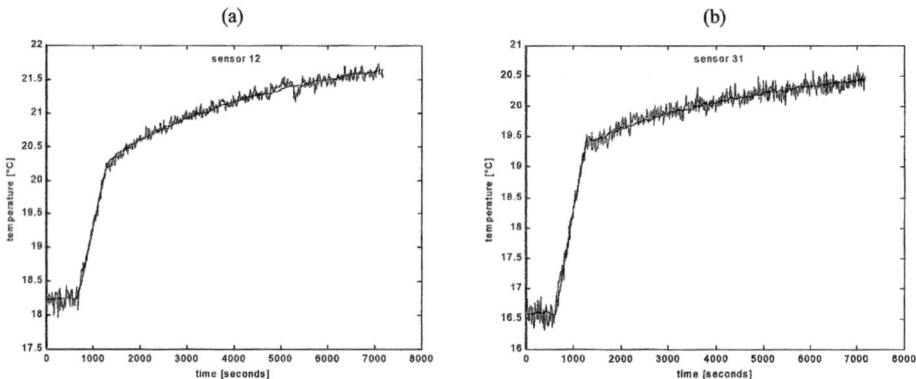

Figure 3: The output of the estimated model for sensor positions 12 (a) and 31 (b) in comparison with the measured temperature response data for an experiment with ventilation rate of 290 m³/h.

SPATIAL DISTRIBUTION OF FRESH OUTSIDE AIR

A very interesting aspect of the model lies in the physical meaning of model parameter $\beta_1 = V_c/vol_1$ (m³/s.m³) which is the local outside air change rate. By modelling the temperature responses at each of the 36 sensor positions in the 30 identification experiments, we obtained very useful information about the spatial distribution of fresh outside air in the test room for a wide of ventilation rates.

As an example, figure 4(a) shows the spatial contours of the local outside air change rate in the front and rear sensor plane of the test chamber at a high ventilation rate of 290 m³/h. The front sensor plane is the vertical xy-plane of the test room which consists of the temperature sensors 4, 5, 6, 10, 11, 12, 16, 17, 18, 22, 23, 24, 28, 29, 30, 34, 35 and 36 and which lies at a z-distance of 0.375 m from the front wall of the room. The rear sensor plane consists of the sensors 1, 2, 3, 7, 8, 9, 13, 14, 15, 19, 20, 21, 25, 26, 27, 31, 32 and 33 and lies at a z-distance of 0.375 m from the back wall. The contour plots in figure 4(a) illustrate rather well how, in this particular situation of high ventilation rate, the fresh supply air, which enters at the upper left, flows along the top of the chamber and then descends to the exit at the lower right.

Quite different behaviour occurs at low ventilation rates, as shown in figure 5(a). Here, with an airflow rate of 150 m³/h, the high local outside air change rates concentrated towards the base of the chamber show how the fresh incoming air sinks to the bottom and then moves across the chamber to the outlet at the lower right, leaving more stagnant air above it.

The visualised air flow pattern at high and low ventilation rates, as shown in figures 4(b) and 5(b), is clearly mirrored in the spatial contour plots. It is clear, therefore, that the WMZ model underlying these plots is characterising the flow behaviour quite well.

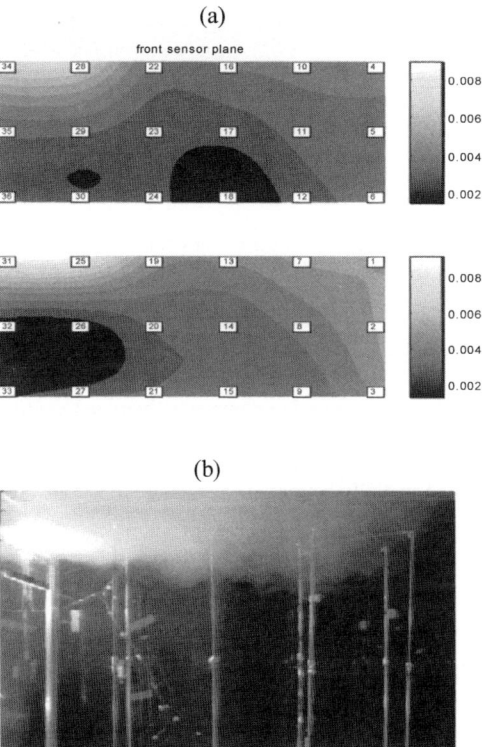

Figure 4: The spatial contours of the local outside air change rate in the front and rear sensor plane of the test chamber (a) and the visualised air flow pattern (b) at a high ventilation rate of 290 m³/h.

(a)

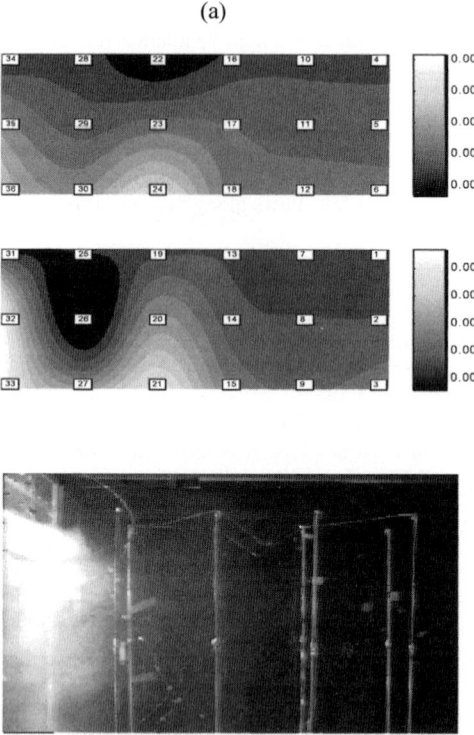

Figure 5: The spatial contours of the local outside air change rate in the front and rear sensor plane of the test chamber (a) and the visualised air flow pattern (b) at a low ventilation rate of 150 m³/h.

CONCLUSIONS

The spatio-dynamic temperature response in an imperfectly mixed ventilated test room to variations of the supply air temperature was successfully modelled for a wide range of ventilation rates. The model provided an excellent fit to the spatio-temporal temperature response data with an average accuracy of 0.15 °C and characterised the spatially heterogeneous nature of the air flow pattern quite well.

REFERENCES

Berckmans D., De Moor M. and De Moor B., 1992. New model concept to control the energy and mass transfer in a three-dimensional imperfectly mixed ventilated space. Proceedings of Roomvent '92, Aalborg, Denmark, Sept. 2-5, 1992, Vol. 2, p. 151-168.

De Moor M. and Berckmans D., 1993. Visualisation of measured three-dimensional well mixed zones of temperature in a ventilated space. Proceedings of the 14th AIVC Conference on Energy Impact of Ventilation and Air Infiltration, Copenhagen, Denmark, Sept. 21-23, 1993, p. 543-552.

Janssens K., 1999. Dynamic modelling of heat and mass transport in imperfectly mixed fluids: a comparison of two approaches. PhD thesis at the Catholic University of Leuven, Faculty of Agricultural and Applied Biological Sciences, pp. 144.

Air Distribution in Rooms, (ROOMVENT 2000)
Editor: H.B. Awbi

SOME RELATIONS REVISITED IN TRACER GAS ANALYSES USING NUMERICAL METHODS

Shia-Hui Peng

Department of Thermo and Fluid Dynamics, Chalmers University of Technology

SE-412 96 Gothenburg, SWEDEN

ABSTRACT

Tracer gas measurements have long been used to quantify the performance of ventilation systems by exploring such scales as the air exchange efficiency, the local mean age of the air, the residence time distribution and so on. The present work deals with a numerical reexamination and calibration of some relations previously derived from tracer gas analysis. The relations revisited include the approximated local air age in terms of the local, instant tracer concentration using the step-up and step-down methods, and the equations for the mean cumulative age distribution and for its frequency distribution. A mixing ventilation flow is employed in the numerical examination. The availability of these relations for practical tracer gas analyses is discussed and they are shown to be useful for efficient tracer measurements.

KEYWORDS

Tracer gas analysis, Local air age, Ventilation performance, Numerical simulation

INTRODUCTION

Tracer gas experiments have been widely employed for quantifying building ventilation performance, by which a quantitative assessment of room effectiveness can be made on the basis of a number of ventilation scales or indices. In experiments, the air flow in a ventilated space is tracked by a passive tracer gas released at the air supply opening or at an interior location. The tracer gas concentration and its time series are recorded and subsequently translated into different ventilation parameters, which are used to assess the capabilities of the ventilation system to deliver fresh air to and/or expel contaminants from the ventilated space. Different measurement methods have been employed, including the step-up method, the step-down method and the pulse method, see e.g. Etheridge and Sandberg (1996). They have been proved in both laboratory and field measurements to be very powerful approaches for analyzing ventilation performance.

Several relations were derived in a previous work based on the so-called *imaginary* tracer gas analysis using numerical and/or multi-zonal approaches (Peng *et al.*, 1997). The derivation of these relations has largely resorted to the local mass conservation, by tracking the passive tracer gas which is presumed completely to follow the ventilation flow. These relations include approximations of the local concentration in terms of the local mean age of the air, and the discrete zonal equations for the mean cumulative age distribution and its frequency distribution.

While these formulations have been approximated on a plausible basis of some *imaginary* tracer gas experiments, no comprehensive examination of them has been conducted using either experiments or numerical simulations. By means of numerical analyses, this work revisits these relations and investigates the degree to which they can reach with a certain confidence and be used as complementary formulations in practical tracer gas measurements. The relations are closely correlated with practical applications in assessing ventilation performance. It should be noted that the purpose of this work is not ultimately directed toward numerical simulations of tracer experiments. Instead, numerical methods are employed here to generate several tracer gas experiments whereby the validation of the relations inherent are calibrated. A mixing ventilation flow is used to account for such a numerical calibration.

THE RELATIONS FOR TRACER GAS ANALYSIS

The relations were approximated from the compartmental (zonal) method and formulated locally (Peng *et al.*, 1997) . They can thus be examined on any interior local cell using the CFD modeling method or at any location in the ventilated space when an experimental measurement is employed. It should be pointed out that, in the derivation, the tracer gas is assumed to be passive with no interaction with the ventilation air flow, whose concentration, C, is thus governed by the transport equation

$$\frac{\partial C}{\partial t} + \frac{\partial}{\partial x_j}\left(u_j C\right) = \frac{\partial}{\partial x_j}\left(\Gamma \frac{\partial C}{\partial x_j}\right) \tag{1}$$

where Γ is the diffusivity, in which a turbulent eddy diffusivity should be included if the flow is turbulent, and u_j are velocity components.

The *step-down analysis* is taken for a room with a volume of V and a supply air flow rate of Q. The nominal time constant is then $\tau_n = V/Q$. Before carrying out a step-down measurement, an initial concentration is set up in the room. This can be done by releasing the tracer gas at a constant rate, q, through the air supply opening for such a long period that the tracer gas concentration in the room is eventually equal everywhere, and $C_p(0) = q/Q$. The step-down measurement is then switched on by stopping the release of the tracer gas at the inlet. During the step-down procedure, the tracer gas concentration is assumed to decay exponentially with time t. This suggests that the local concentration at location P can be approximated as

$$C_p(t) \approx C_p(0)\exp\left(-E_p t\right) \tag{2}$$

where E_p is a coefficient needed to be determined.

At location P, the mean cumulative age distribution, Φ_p, can be expressed in terms of C_p as

$$\Phi_p = 1 - \frac{C_p(t)}{C_p(0)} \tag{3}$$

Using Eqn. (3), the local mean age of the air can be derived from

$$\tau_p = \int_0^\infty t\left(\frac{\partial \Phi_p}{\partial t}\right) dt \tag{4}$$

Substituting Eqn. (2) and (3) into (4) yields $E_p = 1/\tau_p$, which reasonably makes the concentration a function of the location through the local air age, τ_p. The local concentration in the step-down measurement is then approximated as

$$C_p(t) \approx C_p(0)\exp\left(-\frac{t}{\tau_p}\right) \tag{5}$$

In the *step-up measurement*, the initial tracer concentration at any location in the room is zero, i.e. $C_p(0) = 0$. During the measurement, the tracer gas keeps releasing through the air supply opening at a constant rate, q. As the local tracer gas concentration is recorded for a sufficiently long period, it will eventually become equal at all locations and $C_p(\infty) = q/Q$. Using an approximation analogous to the step-down analysis, the local concentration during the step-up procedure is formulated as

$$C_p(t) \approx C_p(\infty) \left[1 - \exp\left(-\frac{t}{\tau_p} \right) \right] \tag{6}$$

In addition, it was shown previously that the discrete zonal equations for the cumulative air-age distribution, Φ, and for its frequency distribution, ϕ, can be derived from tracer gas analyses based on the zonal model (Peng et al., 1997). Here, these equations are reformulated in differential forms for numerical use. This can readily be done using the mass transport equation for tracer gas. In the step-up method, $\Phi_p = C_p(t)/C_p(\infty)$, and $C_p(\infty) \equiv q/Q$ at an arbitrary location P. Dividing $C_p(\infty)$ on both sides of the concentration equation, (1), gives

$$\frac{\partial \Phi}{\partial t} + \frac{\partial}{\partial x_j} (u_j \Phi) = \frac{\partial}{\partial x_j} \left(\Gamma \frac{\partial \Phi}{\partial x_j} \right) \tag{7}$$

For the age frequency distribution, ϕ, we have $\phi = \frac{\partial \Phi}{\partial t}$. Differentiating Eqn. (7) with respect to t yields the governing equation for ϕ,

$$\frac{\partial \phi}{\partial t} + \frac{\partial}{\partial x_j} (u_j \phi) = \frac{\partial}{\partial x_j} \left(\Gamma \frac{\partial \phi}{\partial x_j} \right) \tag{8}$$

These equations can also be derived from similar arguments using the step-down method. They are generally available for indicating the statistical characteristics of air distribution related to the local air age for internal flow systems, although the derivation presented here is based on the tracer gas analysis. Eqn. (7) and (8) always hold in time-dependent forms since they are essentially the governing equations for time-related distribution quantities. Therefore, they should be solved transiently with proper initial values. In this work, only the derivation is given, while their use incorporated into the numerical simulations is not further explored.

The emphasis here is instead placed on the calibration of the availability of Eqn. (5) and (6) and their use in tracer gas measurements. If the two equations hold (conditionally), the local air age at an arbitrary location, P, can be determined by an instant sample of the local concentration, $C_p(t_m)$, at a measurement time t_m in the step-down or step-up procedure by rewriting them, respectively, as

$$\tau_p \approx \frac{t_m}{\ln[C_p(0)] - \ln[C_p(t_m)]}, \quad \text{with the step-down method.} \tag{9}$$

$$\tau_p \approx \frac{t_m}{\ln[C_p(\infty)] - \ln[C_p(\infty) - C_p(t_m)]}, \quad \text{with the step-up method.} \tag{10}$$

If the tracer gas is released at the inlet, $C_p(0)$ and $C_p(\infty)$ in the above expressions are equal to q/Q. When the step-up measurement is undertaken with a tracer released at an interior position, $C_p(\infty)$ in (10) varies locally. Furthermore, it should be pointed out that Eqn. (5) and (6) are not general solutions to the concentration equation under the step-up and step-down conditions. The main purpose of this work is to numerically verify to what degree these relations can be used in tracer gas analyses.

NUMERICAL CALIBRATION OF THE RELATIONS

A mixing ventilation flow is used to carry out the verification of the above relations. The air is supplied through a slot below the ceiling and exhausted from an opening on the opposite wall above the floor. The room is configured with dimensions of 9×3, and the nominal time constant is $\tau_n = 353.4$ s. The air flow is simulated using the standard $k - \varepsilon$ model, which is shown in Figure 1. Figure 1 (c) presents the

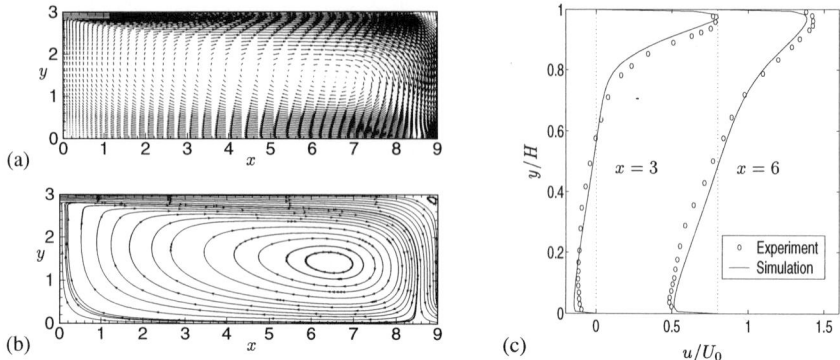

(a)

(b)

(c)

Figure 1: Simulation of the mixing ventilation flow. (a). Flow field; (b). Flow streamlines; (c). Computed velocity compared with experimental data.

velocity profiles at $x = 3$ and $x = 6$ in comparison with the experimental data. The results show that the flow computation is in reasonable agreement with the experiment.

The predicted local air age for this flow is shown in Figure 2 (a). Eight points, $pi(x, y)$ ($i = 1 - 8$), within the room and one point, pe, at the exhaust opening are used for monitoring the relations, as sketched in Figure 2 (b).

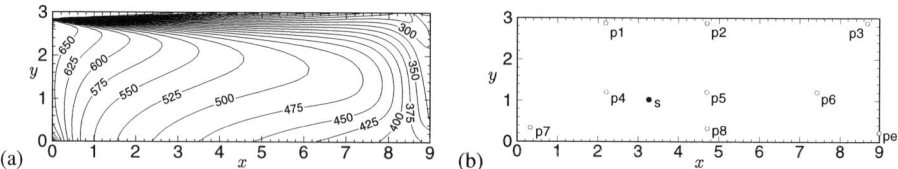

(a)

(b)

Figure 2: Simulated local air age and sketch of the monitoring points. (a). Contour lines of the simulated local air age; (b). Monitoring points used for numerical analysis.

The calibration is carried out in the following ways: computing the time-dependent concentration from its transport equation, Eqn. (1), based on the ventilation flow; comparing this simulated concentration with the one calculated from approximations (5) or (6) based on the local age. Figure 3 shows the comparison using the step-up method at the monitoring locations tabulated in Figure 2 (b). It suggests that the quality of the agreement between the two depends on the local flow features. The approximation agrees rather well with the simulation in regions with strong mixing and recirculation (at locations p4 – p8), while relatively poor agreement is found in the wall-jet where the flow is of a plug-flow type (at locations p1 – p3). In the initial stage (usually with $t/\tau_n < 1$), the simulated concentration does not even increase exponentially, as also confirmed in previous experiments (Sandberg, 1981). Without showing the results here, we pointed out that similar approximations can be reached using the step-down method, Eqn. (5), or using the step-up method by releasing the tracer in the room (at point s in Figure 2 (b)) for which $C_p(\infty)$ in Eqn.(6) should be replaced with the local, saturated tracer concentration which is often not equal to q/Q.

Figure 3 illustrates that the approximation in (5) or (6) is able to reasonably represent the exponential, transient tendency of the tracer gas concentration change after a short initial period ($t < \tau_n$) in the step-up or step-down measurement. To make these relations practically useful, they are further analyzed to

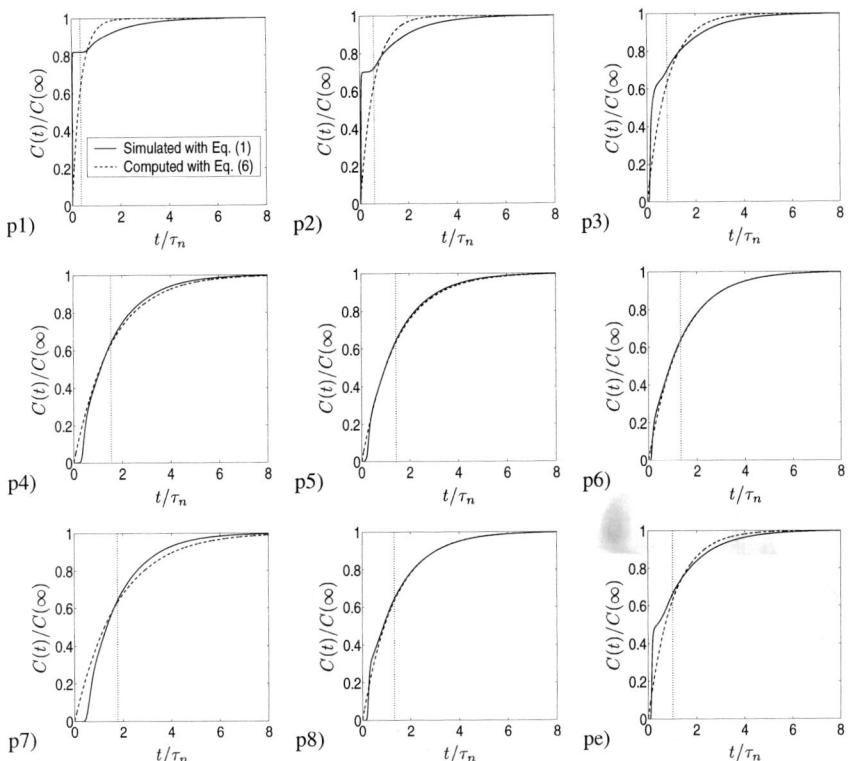

Figure 3: Local concentration approximated by Eqn. (6) using the step-up method in comparison with the numerical simulation with Eqn. (1). The vertical dotted line is the local air age, τ_p/τ_n, at the location explored.

verify whether they can be used as simple approaches to determine the local air age in terms of the local, instant tracer concentration. The local age is usually obtained by a transformation of the transient concentration series measured at the same location over a sufficiently long period. Using approximations (9) and (10), this can instead be done instantly (or in a short period in practical measurements).

The availability of Eqn. (10) (and, similarly, of Eqn. (9) with the step-down method) is calibrated through a comparison in which the local age calculated from (10) at different time instants, t_m, is compared with the numerical solution of the transport equation for τ. At $t_m = 1.5\tau_n$, Figure 4 (c), the local air age calculated from Eqn.(10) shows very good agreement with the numerical solution shown in Figure 4 (a). In general, the approximation at all *measured* time gives an age distribution in the recirculation zone rather close to the simulated one. At $t_m = \tau_n$, Figure (b), the age of the air flow near the exhaust opening is however estimated to be somewhat *younger* by the approximation than by the numerical simulation, while *older* in the region below the inlet. As t_m increases (for $t_m \geq 2.5\tau_n$), this tendency changes inversely with larger values near the exhaust and lower values below the inlet, as compared with the simulation. The results show that appropriate estimation can be achieved in the period of $1.5\tau_n \leq t_m \leq 2\tau_n$ using the approximation for regions where the flow is characterised by mixing and recirculation. Caution should however be taken for regions where the flow is of a one-way type, e.g. in regions near walls and near air supply openings.

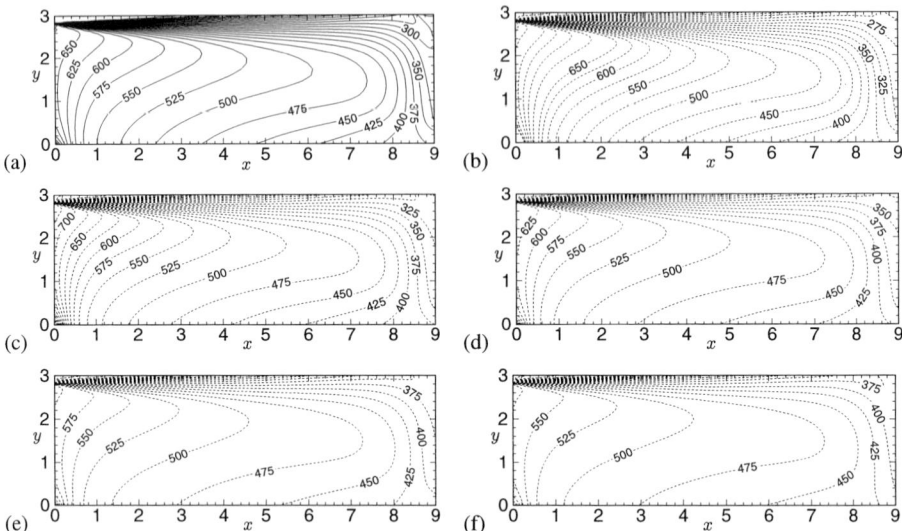

Figure 4: Comparison of the local air age: (a). Numerical simulation; (b). Calculated from Equation (10) at $t_m = \tau_n$; (c). At $t_m = 1.5\tau_n$; (d). At $t_m = 2\tau_n$; (e). At $t_m = 2.5\tau_n$; (f). At $t_m = 3\tau_n$.

CONCLUSIONS

Several relations derived previously from tracer gas analysis have been reexamined by means of numerical analyses. The governing equations for the cumulative age distribution and its frequency function are useful for numerical analysis with no need to specify a tracer gas. The approximation for the local air age formulated in terms of the instant local tracer concentration is expected to be a simple approach in realistic tracer gas measurements to efficiently determine the local air age, particularly for mixing ventilation flow as considered in this work.

The availability of the approximation was explored with a mixing ventilation flow system. It was found that the exponential growth or decay of the concentration with time holds reasonably well after a short period of about τ_n using the step-up or step-down method. This is particularly true in the recirculating flow region. The local air age approximated from these relations is in rather good agreement with the numerical solution of the local-age transport equation, provided that the tracer concentration is sampled at a time within $1.5\tau_n \le t_m \le 2\tau_n$. Erroneous estimations may however arise in regions where the flow is of a *one-way* type, e.g. in near-wall boundary layer and in the region near a displacement ventilation diffuser. More comprehensive investigation and calibration of these relations are needed by means of tracer gas experiments and in more complex ventilation configurations to achieve more general conclusions.

REFERENCES

Etheridge, D. and Sandberg, M. (1996). *Building Ventilation: Theory and Measurement*. John Wiley, Chicherster, U.K.

Peng, S.-H., Holmberg, S., and Davidson, L. (1997). On the assessment of ventilation performance with the aid of numerical simulations. *Build. and Environment*, **32**, 497–508.

Sandberg, M. (1981). What is ventilation efficiency. *Build. and Environment*, **16**, 123–135.

© 2000 Elsevier Science Ltd. All rights reserved.
Air Distribution in Rooms, (ROOMVENT 2000)
Editor: H.B. Awbi

ZONAL MODELS USING LOOP EQUATIONS AND SURFACE DRAG CELL-TO-CELL FLOW RELATIONS

James W. Axley

School of Architecture, Yale University
New Haven, CT 06520, USA

ABSTRACT

Zonal models have been proposed to bridge the gap between the whole-building macroscopic modeling methods of programs like CONTAM or COMIS and the more detailed microscopic modeling methods based on solutions of the time-smoothed Navier-Stokes equations for room airflows. This paper identifies a critical shortcoming of conventional approaches to zonal modeling by introducing alternative approaches a) to formulate the key cell-to-cell flow relations upon which zonal models are based and b) to assemble the zonal system equations. Conventional cell-to-cell flow relations based on boundary power-law formulations appear to capture gross aspects of the flow structure in rooms but fail by orders of magnitude to properly model the resistance offered to airflow. Cell-to-cell flow models based on surface drag momentum transfer may mitigate this shortcoming and appear to capture room airflow structure more accurately. Furthermore, these flow models offer a means to provide quick approximate solutions of room airflow problems (i.e., based on linear formulations of cell-to-cell flow relations) that may be acceptable for certain purposes or can be used as initial estimates for the solution of the more accurate nonlinear formulations of zonal problems.

KEYWORDS

Air Flow, Air Distribution, Modeling, Network Models, Zonal Models, Loop Equations, CONTAM

INTRODUCTION

A number of innovative mechanical and natural ventilation, cooling and heating strategies put forward in recent years depend critically on the details of airflow within rooms and their variation with time. Whole building macroscopic models, such as the CONTAM and COMIS models (Pelletret and Keilholz 1997; Walton 1997), have been developed to predict time histories of bulk airflows into, out of and between rooms in building systems but ignore the details of room airflows altogether. Microscopic models, on the other hand, provide this detail but can not yet be applied to studies of complex whole building systems, are often limited to steady conditions, and demand personnel and computational resources well beyond that of current macroscopic modeling tools. Consequently, a number of proposals have been offered to extend available macroscopic models to provide at least an approximate evaluation of the details of room airflows for whole building annual simulation studies.

These *zonal models* have quite logically been formulated by subdividing rooms into a relatively small number of discrete control volumes or *cells*, formulating heat and mass transfer relations that approximate exchanges between these cells, and assembling these relations into system equations using fundamental conservation principles. In the Systems Dynamics literature such formulations are identified, generically, as *continuity laws* and are often contrasted to the alternative *compatibility laws* that may be used to formulate macroscopic system equations (Shearer, Murphy et al. 1971). In the zonal modeling context a *compatibility* approach would use a similar subdivision of a room into cells but would assemble system equations based on so-called *path laws* or *loop equations*.

The *compatibility* approach to zonal modeling has, apparently, been largely ignored. This paper will explore this alternative approach. In addition, the formulation of the key mass transfer relations that describe cell-to-cell airflow rates in existing zonal models will be critically reviewed and an alternative approach based on momentum transfer to surrounding walls will be presented. Comparisons to existing zonal models will be made and simple applications of these alternative methods explored.

THEORY

The general approach to zonal modeling employed here follows what has become common practice in the field (Allard 1998; Wurtz, Nataf et al. 1999). A room is subdivided into a number of control volumes or *cells*; temperature and pressure fields within the room, $T(x,y,z)$ and $p(x,y,z)$, are approximated by discrete values, T_i and p_i, associated with nodes located within each cell; and system equations are formulated that relate the cell-to-cell air mass flow rate, $\dot{m}_{i,j}$, with these approximate state variables, Figure 1a. This common zonal model is analogous to a resistance network laid out on an orthogonal grid or *mesh*, Figure 1b, with node voltages V_i and currents $i_{i,j}$ corresponding to node pressures p_i and air mass flow rates $\dot{m}_{i,j}$.

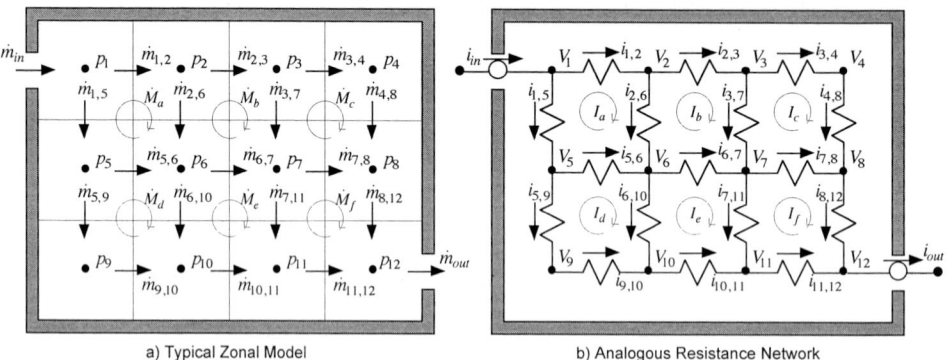

a) Typical Zonal Model   b) Analogous Resistance Network

Figure 1 A multi-cell room model and analogous electrical resistance network.

Here, within each cell, temperatures are assumed to be uniform and equal to the cell discrete value T_i and pressures are assumed to vary hydrostatically about the cell discrete value p_i as:

$$T(x,y,z)\big|_{cell\ i} \cong T_i \quad , \quad p(x,y,z)\big|_{cell\ i} \cong p_i - \rho_i g z_i \qquad (1,2)$$

where x and y are taken as horizontal coordinates, z is the vertical coordinate, z_i is the local elevation relative to the cell node, and ρ_i is the uniform air density within the cell.

System equations may then be formulated using one of two fundamental principles – *continuity* of flow at each node (i.e., mass or energy conservation in the zonal model or current continuity in the resistance network), or *compatibility* of state variable changes as one traverses any path around a continuous *loop* in the model:

$$\text{Continuity} \qquad \sum_k \dot{m}_{k,j} - \sum_l \dot{m}_{j,l} = 0 \ ; \ \left\{ \sum_k i_{k,j} - \sum_l i_{j,l} = 0 \right\} \tag{3}$$

$$\text{Compatibility} \qquad \sum_{i,j} \left(p_i - p_j \right) = 0 \ ; \ \left\{ \sum_{i,j} \left(V_i - V_j \right) = 0 \right\} \tag{4}$$

where k and l are permuted through the indices of all adjacent cells (i.e., up to six cells in the 3D case) and i,j in Equation 4 are permuted through the indices around a given loop.

System Equations: Commonly, the continuity approach has been applied in zonal modeling by formulation of *inverse* expressions that relate the cell-to-cell air mass flow rates to the linked state variables as:

$$\text{Inverse Flow Expression: } m_{i,j} = g\left(p_i, T_i, p_j, T_j \right) \ ; \ \left\{ i_{i,j} = \Delta V_{i,j} / R_{i,j} \right\} \tag{5}$$

System equations are then assembled by systematically applying the continuity equation (Equation 3) to each node using Equation 5 or semi-empirical expressions for known jets or plumes in the room. Coupled systems of nonlinear equations are thus obtained that may be solved to determine the cell state variables p_i and T_i. Unknown air mass flow rates are then recovered using Equation 5.

Alternatively, *forward* expressions may be formed that relate pressure changes to air mass flow rates and node temperatures as (with $\Delta p_{i,j} = p_i - p_j$):

$$\text{Forward Flow Expressions: } \Delta p_{i,j} = f\left(m_{i,j}, T_i, T_j \right) \ ; \ \left\{ \Delta V_{i,j} = i_{i,j} R_{i,j} \right\} \tag{6}$$

In this approach system equations are then assembled by systematically applying the compatibility equation (Equation 4) to each of a set of *independent* loops and complementing this set of equations with the node continuity equation (Equation 3) applied at *(n-1)* of the n nodes of the zonal model. Graph theory reveals that the simple *mesh loops* linking four adjacent nodes (e.g., the directed loops with hypothetical flow rates M_a, M_b, ... in Figure 1a or analogously I_a, I_b, ... in Figure 1b) yield an independent set of loop equations (Shearer, Murphy et al. 1971). Furthermore, for linear forward flow expressions the system equations may be defined in terms of these hypothetical mesh loop flow rates to minimize the number of system equations. The influence of known room jets or plumes is, in this case, used to constrain individual loop equations.

Cell-to-Cell Mass Flow Expressions: Inverse flow expressions have been based on methods used for large openings. In this approach flow resistance is assumed to occur at the boundary between cells, Figure 2a, and a power-law relation is assumed to govern the differential mass flow $d\dot{m}_{i,j}$ through differential areas of the boundary dA as:

$$d\dot{m}_{i,j} = C\rho \left(\Delta P_{i,j} \right)^n dA \tag{7}$$

C is an empirical "permeability" constant, analogous to the orifice discharge coefficient, that is assumed to have a value less than 1.0 with one recent study concluding a value of 0.83 $\text{m·s}^{-1}\text{·Pa}^{-n}$ being most reliable (Wurtz, Nataf et al. 1999). The power-law exponent n is often taken as $n = 1/2$ which corresponds to the orifice equation and the air density ρ taken as the average of adjacent cells. Finally, the driving pressure difference $\Delta P_{i,j}$ follows directly from Equation 2 above:

$$\Delta P_{i,j} = \left(p_i - \rho_i g z_i \right) - \left(p_j - \rho_j g z_j \right) \tag{8}$$

238

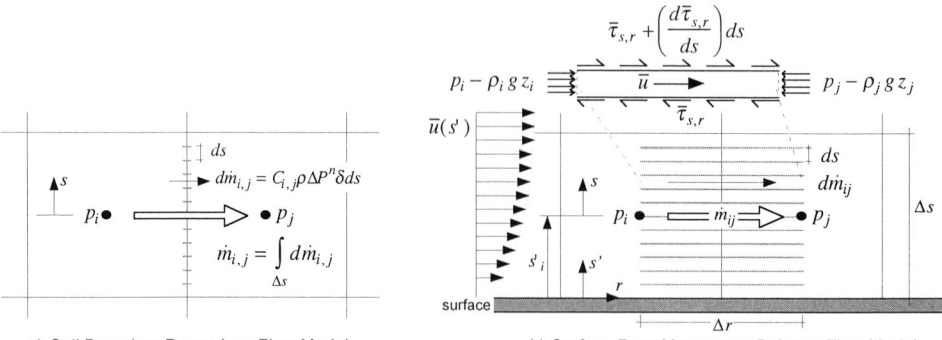

a) Cell Boundary Power-Law Flow Model   b) Surface-Drag Momentum Balance Flow Model

Figure 2 Conventional a) and proposed b) cell-to-cell flow idealizations.

Expressions for the cell-to-cell air mass flow rate $\dot{m}_{i,j}$ may then be derived by integrating Equation 7 over the cell boundary area. For isothermal flow in a 2D flow regime of arbitrary depth δ one obtains:

$$\dot{m}_{i,j} = \int d\dot{m}_{i,j} = \int_{-\Delta s/2}^{\Delta s/2} C\rho\left(\Delta P_{i,j}\right)^n \delta\, ds = C\rho\delta\,\Delta s\left(\Delta P_{i,j}\right)^n \tag{9}$$

This model implicitly assumes viscous dissipation occurs only at the boundary – a reasonable approximation for an orifice but not room airflows dominated by surface drag. Thus when applied to airflow through a room, the total pressure drop will depend linearly on the number of cells used – clearly not a reasonable outcome. Furthermore, when recast in forward form and used to form loop equations, the flow solution proves to be independent of the permeability coefficient C!

An alternative approach may be developed by considering the transfer of shear stress near wall surfaces using a momentum balance on differential flow conduits linking adjacent cells, Figure 3b. For a flow conduit oriented parallel to the nearest room surface, pressures integrated over the ends of the conduit must balance (time-smoothed) shear stresses $\bar{\tau}_{sr}$ acting over the length of the conduit:

$$\Delta P_{i,j}\, \delta\, ds' = -\frac{d\bar{\tau}_{sr}}{ds'}\, \delta\, \Delta r\, ds' \tag{10}$$

where s' and r' are local normal and tangential coordinates to the surface and Δr is the cell width parallel to the surface. The shear stress may be related to the (time-smoothed) velocity profile perpendicular to the wall $\bar{u}(s')$ using the fundamental relation for Newtonian fluids for laminar flow and, here, Prandtl's mixing length relation for turbulent flow that has proven to be effective in CFD simulations of room air flow (Chen and Xu 1998):

$$\text{Laminar Flow: } \bar{\tau}_{sr} = -\mu\frac{d\bar{u}}{ds'} \qquad , \qquad \text{Turbulent Flow: } \bar{\tau}_{sr} = -\rho\kappa^2 s'^2\left(\frac{d\bar{u}}{ds'}\right)^2 \tag{11, 12}$$

where κ is a "universal constant" with empirically determined values ranging from 0.36 to 0.40.

To effect closure, relations for the velocity profiles are needed – for this empirical equations of the following form for laminar and turbulent flow respectively may be used:

$$\text{Laminar Flow: } \bar{u} \approx \bar{u}_{\max}\sin\left(\pi s'/2S\right) \qquad , \qquad \text{Turbulent Flow: } \bar{u} \approx \bar{u}_{\max}\left(s'/S\right)^a \tag{13, 14}$$

where $\bar{u}_{\max}$ is a hypothetical asymptotic maximum approached as $s' \to S$, S is a characteristic room dimension taken as half the dimension between opposing room walls, and a is an empirically determined coefficient. Well-developed turbulent velocity profiles in tubes and ducts are well approximated with $a = 1/7$ – the value that will be used in subsequent applications.

Finally, flow relations may be derived by substituting these expressions into the momentum balance, Equation 10, and integrating over the cell height Δs. For isothermal conditions one obtains:

$$\text{2D Laminar:} \quad \Delta P_{i,j} = \frac{\mu\,\pi^2 \Delta r}{4\rho\,S^2 \delta\,\Delta s}\,\dot{m}_{i,j} \quad , \quad \text{2D Turbulent:} \quad \Delta P_{i,j} = 2k_s \frac{\kappa^2 a^3 \Delta r}{\rho\delta^2 \Delta s^3}\,\dot{m}_{i,j}^{\;2} \quad (15,\,16)$$

Here, k_s is defined uniquely for each cell position $n_s = 1,2,3,\ldots$ relative to the nearest surface with $k_s \approx 4/(4n_s - 3)$ for central cells of odd meshes or $k_s \approx 2/(2n_s - 1)$ for all other cells. The extension of this cell-to-cell airflow model to 3D flow regimes and nonisothermal conditions is straightforward. Details are available from the author and will be presented in other publications.

APPLICATIONS

Isothermal applications of zonal models should reasonably be used for validation before consideration of more challenging nonisothermal cases are considered. Two cases are considered below.

Duct Flow Modeling: The 3D variants of both the laminar and turbulent surface drag cell-to-cell flow models, Equations 15 and 16, and the power-law boundary flow model, Equation 9, with n=1/2 were applied to the problem of modeling airflow through a 6m length of a square duct 2m x 2m. Four turbulent surface-drag idealizations were considered utilizing 1x1, 2x2, 3x3, and 4x4 cell subdivisions across the section and 4 cell subdivisions along the length of the duct. These were compared to 1x1x4 cell subdivisions of both the laminar surface-drag model and the conventional power-law model as cross-sectional subdivision has no impact on the results using these flow models for this particular flow problem.

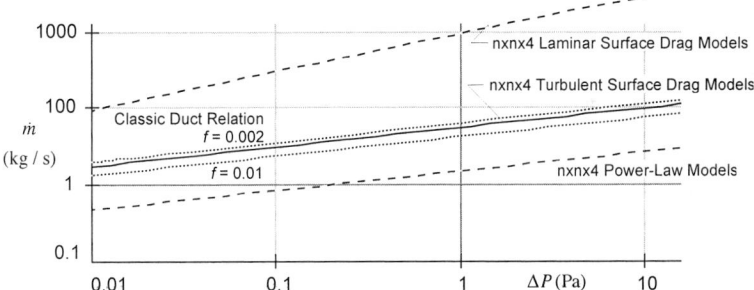

Figure 3 Comparison of modeled duct pressure-flow relations for the surface drag and power-law flow models and the classic Darcy-Weisbach equation for friction factors from $f = 0.01$ to 0.002.

The turbulent surface drag idealizations produced duct pressure-flow relations essentially identical in form to the classic Darcy-Weisbach equation, adapted to rectangular cross-sections via the hydraulic radius approximation, with effective duct friction factors of $f_{eff} = 0.0034$, 0.0034, 0.0038, and 0.0040 for the 1x1, 2x2, 3x3, and 4x4 cross-sectional subdivisions respectively, Figure 3. Likewise the laminar surface drag idealization yielded duct pressure flow relations essentially identical in form to the classic Hagen-Poiseuille solution for the laminar case. Given the appearance of the cell-subdivision length Δr in both of these surface drag models, these idealizations yielded results that were independent of the number of longitudinal cell subdivisions used. Furthermore, the turbulent surface drag idealizations provided reasonable approximations to the velocity profiles across the duct that improved with cross-sectional subdivision.

The power-law idealization, on the other hand, yields pressure-flow relations that depend on the number of longitudinal cells used and does not provide any approximation of the velocity profile. For

240

the 4-cell subdivision considered the resulting pressure-flow relation exhibited significantly greater resistance (i.e., by greater than an order of magnitude) for turbulent flow than the classic solution.

2D Forced Convection: The proposed surface drag and conventional power-law models were combined with well-established semi-empirical wall jet relations (Rajaratnam 1976) to model a 2D forced convection case studied by Chen (Chen and Xu 1998). Computed mass flow rates are shown below in Figure 4 for all idealizations. In this particular problem air is injected at the ceiling at a mass flow rate of 90 g/s (Re = 5,000) and is exhausted at the opposite floor.

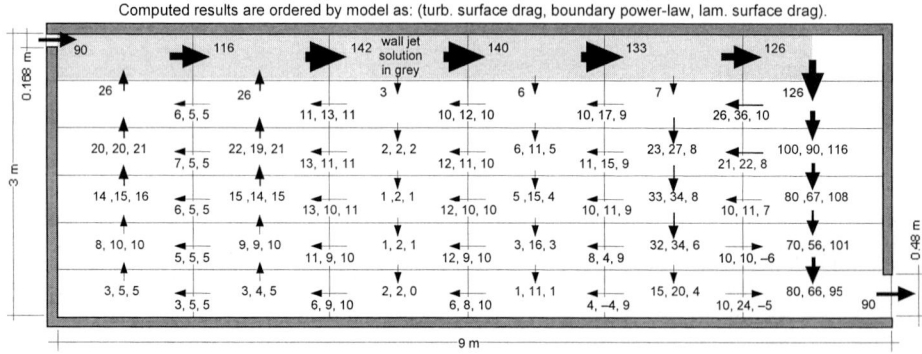

Figure 4 Comparison of computed results for a 2D forced convection problem.

Broadly speaking all cell-to-cell flow models provide similar results that compare reasonably well with the measured data and CFD computed results reported by Chen, although the turbulent surface drag model captures the nature of the flow intensification at the right wall more faithfully. Again, the power-law flow model resulted in a pressure drop from inlet to outlet that was over two orders of magnitude greater than that produced by the turbulent surface drag model. The success of the laminar surface drag model is perhaps most significant as this model produces linear system of algebraic equations, by either the continuity or compatibility approach, that may be easily solved. Furthermore, a very compact compatibility (loop) formulation of the system equations is possible using the hypothetical *mesh loop* flow rates discussed above. For quick approximate analysis the (linear) laminar surface drag solution may be acceptable. Alternatively, it may be used as a starting point for the solution of the much more difficult nonlinear analyses resulting from the use of the other models.

REFERENCES
Allard, F., Ed. (1998). Natural Ventilation in Buildings: A Design Handbook. London, James & James
Chen, Q. and W. Xu (1998). "A Zero-Equation Turbulence Model for Indoor Air Simulation." Energy and Buildings **Vol. 28**: pp. 137-144.
Pelletret, R. Y. and W. P. Keilholz (1997). COMIS 3.0 - A New Simulation Environment for Multizone Air Flow and Pollutant Transport Modeling. Building Simulation '97 - Fifth International IBPSA Conference, Prague, IBPSA.
Rajaratnam, N. (1976). Turbulent Jets. Amsterdam, Oxford, New York, Elsevier Scientific Pub. Co.
Shearer, J. L., A. T. Murphy, et al. (1971). Introduction to System Dynamics. Reading, MA, Addison-Wesley Pub. Co.
Walton, G. (1997). CONTAM96 User Manual. Gaithersburg, MD, NIST.
Wurtz, E., J.-M. Nataf, et al. (1999). "Two- and Three-dimenaional Natural and Mixed Convection Simulation using Modular Zonal Models in Buildings." International Journal of Heat and Mass Transfer(42): pp. 923-940.

Air Distribution in Rooms, (ROOMVENT 2000)
Editor: H.B. Awbi

IMPROVING SPEED AND ROBUSTNESS
OF THE COMIS SOLVER

D.M. Lorenzetti and M.D. Sohn

Indoor Environment Department
Lawrence Berkeley National Laboratory
Berkeley CA 94720, USA

ABSTRACT

The numerical investigation of airflow and chemical transport characteristics for a general class of buildings involves identifying values for model parameters, such as effective leakage areas and temperatures, for which a fair amount of uncertainty exists. A Monte Carlo simulation, with parameter values drawn from likely distributions using Latin Hypercube sampling, helps to account for these uncertainties by generating a corresponding distribution of simulated results. However, conducting large numbers of model runs can challenge a simulation program, not only by increasing the need for fast algorithms, but also by proposing specific combinations of parameter values that may define difficult numerical problems.

The paper describes several numerical approaches to improving the speed and reliability of the COMIS multizone airflow simulation program. Selecting a broad class of algorithms based on the mathematical properties of the airflow systems (symmetry and positive-definiteness), it evaluates new solution methods for possible inclusion in the COMIS code. In addition, it discusses further changes that will likely appear in future releases of the program.

KEYWORDS

Airflow network, Line search, Newton-Raphson, Multizone, Simulation, Trust region.

INTRODUCTION

The numerical investigation of a building's airflow and chemical transport characteristics involves identifying values for model parameters, such as effective leakage areas and temperatures, for which a fair amount of uncertainty and variability exist. When seeking to represent a general class of building, rather than a particular building for which design information and measured data may exist, the uncertainties become greater still. For example, as part of an effort to develop guidelines for building managers seeking to respond to indoor pollutant releases, Sohn et al. (1998) modeled a five-story office building using the multizone airflow simulation program COMIS (Feustel 1999). The model, meant to typify intermediate-size, open-style commercial spaces, identified 13 critical parameters describing the structure and its use. With the parameter values varying widely among sample spaces, clearly a single simulation cannot represent the whole range of possible behaviors.

Monte Carlo simulation provides one approach to characterizing uncertainties in the model predictions. Assigning a likely distribution to each critical parameter defines a corresponding distribution of possible building representations. Discretizing this distribution using Latin Hypercube sampling (Iman et al. 1980) yields a set of simulations, and a corresponding range of possible flow characteristics, for the building under study. Thus Sohn et al., sampling from assumed distributions of their 13 model parameters, generated 2000 specific COMIS simulations. The response of each to a pollutant release indicated, in aggregate, the uncertainty expected in a real building, about which little may be known other than that it belongs to the general class of buildings defined earlier.

Unfortunately, this technique burdens the simulation tool: first by requiring a large number of runs, and hence increasing the total execution time; second by increasing the chance that some particular combination of parameter values will define a difficult numerical problem. For example, in initial tests COMIS completed 1825 out of 2000, or about 91%, of the simulations required to characterize the five-story office.

This paper describes changes to COMIS v3.0, that begin to address these problems. Because of known nonconvergence associated with duct junctions (Feustel 1999), the investigation initially focused on methods for stabilizing the solution algorithm. However, efforts to recode the solver showed that the initialization scheme employed at each time step was largely responsible for slow and nonconverging simulations. The sections below discuss these issues, and show how relatively modest changes to the solver can improve its execution speed. At least some of these changes should appear in COMIS v3.1 (the Berkeley Labs web site http://epb1.lbl.gov/comis/ provides links to the COMIS code).

BUILDING AIRFLOW SYSTEMS

Multizone airflow models, such as COMIS, represent zones (e.g., rooms and duct junctions) as nodes of unknown pressure, connected via discrete flow paths (such as doors, cracks, and ductwork) with unknown mass flow. The governing equations describe steady-state mass conservation at the nodes, and the pressure-flow relations of the paths. COMIS adopts a nodal formulation of the problem, treating the node pressures as independent, and updating the flows to keep them in agreement with the current pressure estimates. Walton (1989) details nodal models. Wray & Yuill (1993) discuss this and other possible formulations.

The nodal formulation requires pressure-flow relations that express the flow, f_{i-j}, through the element(s) connecting node i to node j, as a function of their pressures, P_i and P_j. In a useful idealization, the flow depends only on the pressure drop between the nodes:

$$f_{i-j} = f_{i-j}\{P_i - P_j\} . \tag{1}$$

The flow element pressure drop includes wind and thermal buoyancy effects (Feustel 1999). If every pressure-flow relation follows Eqn. 1, then at least one node (typically that representing the building's surroundings) must have its pressure fixed.

The solution of the nodal equations proceeds iteratively. For each guess at the unknown pressures, the program sums the flows entering each variable-pressure node, then adjusts the pressures, seeking to enforce mass balance. Call the pressure vector at the kth iteration $P_{[k]}$, and the corresponding sums of flows $r_{[k]}$. Then $r_{[k]}$ gives a vector of residuals that the solver seeks to zero. Most nonlinear solvers use some variation on Newton-Raphson's method to zero the residuals (Dennis & Schnabel 1996). This method finds the Jacobian matrix, $J_{[k]}$, of derivatives of the residual vector, forms a local model of the residuals using those derivatives, and calculates the next set of pressures in order to zero that model:

$$P_{[k+1]} = P_{[k]} - J_{[k]}^{-1} r_{[k]} .\tag{2}$$

When the flow elements obey Eqn. 1, they yield a symmetric Jacobian. If in addition every flow element has $\partial f_{i\text{-}j}/\partial P_i \geq 0$, then the Jacobian is positive-definite (Axley 1989); note that Axley proves a stronger result than found in most of the building airflow literature. Since a symmetric positive-definite matrix cannot be singular, and may be factored without pivoting (Dennis & Schnabel 1996), flow elements of this form simplify the numerics of an equation solver considerably.

INITIALIZATION MODIFICATIONS

Each time step of a simulation requires an initial $P_{[1]}$ from which to start the iterative solution. COMIS v3.0 finds $P_{[1]}$ based on two user-specified inputs (Feustel 1997). Nominally the control flag USEOPZ determines how the program establishes initial pressures, while NOINIT determines whether or not to apply linear initialization before attempting the fully nonlinear problem. Linear initialization replaces each flow element relation with a straight-line approximation meant to represent its global behavior; see Walton (1989). With both flags set to their default values of zero, COMIS begins every time step by adjusting the zone pressures to counter thermal buoyancy, seeking a zero net pressure drop across each flow element. After setting these pressures, by default the program performs linear initialization.

As noted above, this default initialization causes about 9% of the parameterized office building simulations to fail. In all failed cases, the program completes the first time step successfully, but fails at later simulated times. A better initialization scheme, achieved by setting USEOPZ = 1, begins each new time step with $P_{[1]}$ set to the final pressures calculated at the preceding step. This allows COMIS to complete all 2000 simulations successfully. However, inspecting the source code reveals that setting USEOPZ = 1 prevents linear initialization at the first time step, and in fact prevents NOINIT from having any effect at any time step.

A second problem with USEOPZ concerns its overloaded use for controlling density updates. By default, the solver fixes zone densities at values consistent with the initial pressures, $P_{[1]}$. Setting USEOPZ = 2, or specifying LOOPRHO in the &-PR-SIMU section of the input file, nominally causes the solver to update zone densities with every pressure iteration. This forces USEOPZ to control two unrelated aspects of the program: pressure initialization and density updates. Thus setting USEOPZ = 0 or 1 effectively cancels an earlier request for density updates via the LOOPRHO option.

Inspecting the source code also shows that the linear initialization stage, if invoked, runs iteratively. That is, instead of taking a single Newton-Raphson step using the global linearization, it seeks to solve that linearized system exactly, using a series of Newton-Raphson steps. Unfortunately, not every flow element has a global linearization defined. This means the program can spend many iterations solving, or attempting to solve, a partly nonlinear system that bears only marginal resemblance to the nonlinear system of interest.

To address these problems, we updated COMIS to initialize the first time step according to a control flag STP1INIT, and to initialize subsequent time steps according to STP2INIT. Both flags may either zero the pressure drops (value 0), or zero the drops and then take a single step on the globally linearized system (value 1). In addition, setting STP2INIT = 2 initializes subsequent time steps using the pressures calculated at the previous time step. These two flags replace USEOPZ and NOINIT, respectively, in the new &-PR-CONT section of our input files. A new internal variable handles density updates as specified by the LOOPRHO option.

SOLVER MODIFICATIONS

At each time step, after any pressure initialization, the program turns to the fully nonlinear system. Close to a solution, the pure Newton-Raphson iteration defined by Eqn. 2 converges quickly, but it may diverge if the step to $P_{[k+1]}$ exceeds the range over which the derivative information in the Jacobian remains valid (Dennis & Schnabel 1996). For airflow systems containing highly nonlinear elements, such as crack elements, duct fittings, and large openings, Newton-Raphson may fail, no matter how many iterations it is allowed to take (Herrlin & Allard 1992, Feustel 1999).

For stability, COMIS uses a damping factor, μ, to damp the Newton-Raphson step:

$$P_{[k+1]} = P_{[k]} - \mu J_{[k]}^{-1} r_{[k]} . \tag{3}$$

Setting $\mu = 1$ recovers the original method, while $0 < \mu < 1$ gives the same search direction, but places $P_{[k+1]}$ at a shorter step. Recall that Newton-Raphson zeros the residual models formed using the local derivatives stored in $J_{[k]}$. Since this derivative information remains valid for short enough steps about $P_{[k]}$, and since the symmetric positive-definite Jacobian always admits inversion, then in exact arithmetic it must be possible to solve the system by repeated iterations with small enough μ.

In general, an algorithm should pick a relaxation factor small enough to avoid instability in the pressure iterates, but large enough to minimize the number of iterations required. Wray & Yuill (1993) found that a constant relaxation factor of 0.75 works well for many airflow systems; the current versions of both the COMIS and CONTAM airflow simulation programs use a variation on this idea, described by Walton (1997), to choose from among a fixed set of relaxation constants, depending on the change in the mass balances from iteration to iteration.

The mathematical literature defines several mature methods for selecting the damping factor in Eqn. 3 (Dennis & Schnabel 1996). These line search algorithms apply optimization theory to the problem of selecting μ, by forming a scalar cost function from the residual vector, and then seeking a minimum of that cost function. The sum of squares of the residuals, r-square, provides a suitable cost function. Not only does it have a global minimum at $r = 0$, the solution of the nonlinear system, but any other minima can occur only at singularities in the Jacobian. Since the positive-definite Jacobian has full rank, in exact arithmetic an airflow system solver can always take a step that reduces the cost function.

We programmed a trust region based line search algorithm, adapted from Dennis & Schnabel (1996), as a new COMIS airflow solver (SLVSEL = 6). Like all minimization methods, the algorithm reduces the cost function at every iteration. In case it tries an overly long step, and fails to decrease r-square, it tries again with smaller μ. Thus the algorithm can take multiple residual evaluations at a single iteration. A trust region algorithm updates the expected length of a successful step from one iteration to the next, expanding and contracting the trust length depending on how well the actual value of r-square matches the predicted result. Our method follows the published one fairly closely, except that in order to avoid the possibility of stagnation due to numeric effects, it imposes the constraint $\mu \geq 1E\text{-}6$.

RESULTS

Testing the trust region algorithm on a number of COMIS simulations shows that it is generally competitive with the standard solver (SLVSEL = 5), provided each solver uses the appropriate initialization scheme. For 23 COMIS input files, Table 1 records the number of iterations each solver requires to complete the first time step, initializing with either zero pressure drops (STP1INIT = 0) or linear initialization (STP1INIT = 1). The iteration count includes the one associated with linear initialization, if used. The input files labeled "a-" come from the office simulations of Sohn et al., while those labeled "b-" are standard COMIS test cases.

The results show that both solvers are sensitive to the initialization method, with the standard solver taking many more iterations on average, and the new solver taking marginally fewer iterations, when using linear initialization. The significantly greater iteration counts for the standard solver under linear initialization may indicate that one or more flow elements have unreasonable global linearizations, but this remains for later investigation. Inspecting the solution trajectories for the trust region method shows that zeroing the pressure drops, which starts the flow elements in a regime of operation where small changes in the pressures dramatically change the slope of the pressure-flow curve, causes the algorithm to take a very short step in its first iteration. The solver subsequently increases the trust length, but not as aggressively as it might. The trust region method sacrifices speed for stability, if necessary, and when the stability problems arise from e.g. a crack model operating near zero pressure drop, this may be to its detriment. Linear initialization, by jumping the solution away from these regimes of operation, avoids this behavior.

Comparing the new solver, under linear initialization, to the standard method, initialized with zero pressure drops, shows that the trust region method usually performs as well as, or slightly better than, the standard COMIS solver. However, none of the simulations shown in the table represent especially difficult problems, so one would not expect to see great differences between the two solvers. For more difficult problems, for example ones with duct fittings, we expect the trust region based solver to perform more robustly than the standard COMIS routine, due to its ability to adaptively choose finer gradations in the relaxation parameter. Again, this improved reliability will exact a toll in iteration counts on some problems.

TABLE 1

ITERATIONS REQUIRED, BY SOLVER AND INITIALIZATION SCHEME

Input file	Std solver (5)		New solver (6)		Input file	Std solver (5)		New solver (6)	
	Zero (0)	Lin (1)	Zero (0)	Lin (1)		Zero (0)	Lin (1)	Zero (0)	Lin (1)
a01	10	18	16	15	b10	4	14	5	4
a02	7	19	18	13	b11	4	14	5	4
a03	8	11	3	6	b12	3	11	3	3
b01	2	2	2	2	b13	4	13	4	4
b02	3	5	3	3	b14	5	13	5	4
b03	2	8	2	3	b15	8	15	6	5
b04	4	8	4	3	b16	8	15	6	5
b05	4	13	5	4	b17	4	9	5	3
b06	4	6	12	4	b18	4	9	5	3
b07	10	12	6	4	b19	4	9	5	3
b08	4	10	4	4	b20	3	8	4	4
b09	5	10	5	3					

CONCLUSIONS

A few relatively simple programming changes can improve the performance of the COMIS solver, and at the same time enhance the user's control of the initialization scheme. A trust region based solver performs well in the limited tests described, however, we have not challenged either solution algorithm with particularly difficult simulations. The authors invite the submission of COMIS input files known to cause the standard solver to fail, in order to further test and refine the algorithms.

The numerical experiments described here do not pursue the question of convergence difficulties associated with duct junctions, which remains open for further investigation.

REFERENCES

Axley J.W. (1989). Multi-Zone Dispersal Analysis by Element Assembly, *Building and Environment* **24:2,** 113-130.

Dennis J.E., Jr. and Schnabel R.B. (1996). *Numerical Methods for Unconstrained Optimization and Nonlinear Equations*, Society for Industrial and Applied Mathematics, Philadelphia PA, USA.

Feustel H.E. and Smith B.V. (1997). *COMIS 3.0 User's Guide*, available from the Lawrence Berkeley Laboratory, Berkeley CA, USA. Download from http://epb1.lbl.gov/comis/.

Feustel H.E. (1999). COMIS—an International Multizone Air-Flow and Contaminant Transport Model. *Energy and Buildings* **30:1,** 3-18.

Herrlin M.K. and Allard F. (1992). Solution Methods for the Air Balance in Multizone Buildings, *Energy and Buildings* **18:2,** 159-170.

Iman R.L., Davenport J.M., and Zeigler D.K. (1980). *Latin Hypercube Sampling (a Program User's Guide), Technical Report SAND79-1473*, Sandia Laboratories, Albuquerque NM, USA.

Sohn M.D., Daisey J.M., and Feustel H.E. (1998). Characterizing Indoor Airflow and Pollutant Transport Using Simulation Modeling for Prototypical Buildings. 1. Office Buildings. *Proceedings of the Eighth International Conference on Indoor Air Quality and Climate, Indoor Air 99*, Edinburgh, Scotland, **4,** 719-724.

Walton G.N. (1989). Airflow Network Models for Element-Based Building Airflow Modeling. *ASHRAE Transactions* **95:2,** 611-620.

Walton G.N. (1997). *CONTAM96 User Manual, Report NISTIR 6056*, U.S. Department of Commerce, National Institute of Standards and Technology, Gaithersburg MD, USA.

Wray C.P. and Yuill G.K. (1993). An Evaluation of Algorithms for Analyzing Smoke Control Systems. *ASHRAE Transactions* **99:1,** 160-174.

© 2000 Elsevier Science Ltd. All rights reserved.
Air Distribution in Rooms, (ROOMVENT 2000)
Editor: H.B. Awbi

STUDY OF THE EFFECT OF THERMAL STRATIFICATION ON HEAT TRANSFER IN MULTIZONE BUILDINGS

S. BURCHIU[1], A. MESLEM[2] and C. INARD[2]

[1]Université Technique de la Construction, 66 bd Pache Protopopescu, Bucarest, Roumanie
[2]LEPTAB, Université de La Rochelle, av. Michel Crépeau, 17042 La Rochelle Cedex 1, France

ABSTRACT

This study is devoted to the analysis of the influence of the thermal stratification on heat transfer through vertical large openings in multizone buildings. For that, we built a zonal model in order to describe the thermo-convective fields with an upper zone and a lower zone separated by a horizontal plane. This model has been validated by comparison with experimental results obtained under steady and unsteady states in a two zone test cell and for various thermal inputs. The results of a parametric study carried out for a multizone building (six rooms) are then presented.

KEY WORDS

Building, heat transfer, modelling, thermal stratification, multizone.

LIST OF SYMBOLS

b_0 initial width of jet (m)
Cp air specific heat (J/kg K)
g acceleration of gravity (m/s^2)
H height of the vertical opening (m)
m air mass flow rate (kg/s)
P pressure (Pa)
Pe heating system power (W)
$Q(z)$ heat flow in thermal plume (W/m)
T, Ta air temperatures (°C)
Ts surface temperature (°C)

u velocity (m/s)
W width of the vertical opening (m)
x, y, z co-ordinates (m)
zn height of neutral plane (m)
z_0 virtual origin of thermal plume (m)
ρ air density (kg/m^3)
ΔP pressure difference (Pa)
Φ_{ij} heat flow between zones i and j (W)

INTRODUCTION

In the scope of energy in buildings, the influence of the thermal stratification has a great influence on the exchanged heat flows between rooms, van der Maas and al (1992) and on the assessment of the neutral planes within large enclosures, Li and al (1998).

For that, we decided to use the concept of zonal model, Inard and al (1998) in order to be able to predict the indoor thermal stratification within buildings. The model we developed here is original because it takes into account the thermal stratification of the indoor air by using two zones in a room. These zones are split into six subfields of which five are the driving flows such as the thermal plumes (hot or cold), the thermal boundary layers (hot or cold) and the jet of the ventilation system, the last satisfying the mass balance of the zone considered. First of all, we present the different stages of the modelling.

DESCRIPTION OF THE MODEL

Here we are concerned with a problem of coupled heat transfers, where the three different modes of heat transfer appear simultaneously : convective transfers in the indoor air volume, radiative transfers between the different surfaces of the enclosed space, and conductive transfers through the walls.

Although our efforts have essentially been focused on the first one, we present the others very briefly.

Concerning the conduction, we used an analogical representation with three capacities and two resistances (3C2R) as proposed by Rumianowski and al (1989). For radiative transfers as well as for the short waves and for the long waves, we used the method of the fictitious enclosure developed by Walton (1980). The description of the convective exchanges within the building can be divided into two parts that are respectively the mass and heat exchanges through the vertical openings and the convective exchanges within a zone.

Calculation of the flows between two rooms

The main assumptions adopted are the following:
- the streamlines are horizontal and the flow is under the steady state ;
- the fluid is assumed non-viscous and incompressible ;
- the variation of the pressure is hydrostatic between the two sides of the opening ;
- every room is divided into two isothermal zones A and C.

The opening separating two rooms 1 and 2 is schematised in figure 1.

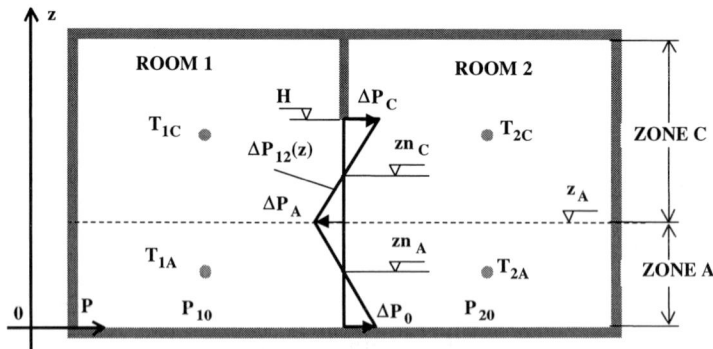

Figure 1 : Schematic diagram of a large vertical opening separating two rooms

Every room is divided into two isothermal zones (A and C) by a horizontal plane situated at a height defined arbitrarily at $z = z_A$.

In this configuration, two neutral planes can appear within the opening.

The pressure differences relative to the limits of every zone are given by :

$$\Delta P_0 = P_1(z=0) - P_2(z=0) \tag{1}$$

$$\Delta P_A = \Delta P_0 - (\rho_{1A} - \rho_{2A}) \, g \, z_A \tag{2}$$

$$\Delta P_C = \Delta P_A - (\rho_{1C} - \rho_{2C}) \, g \, (H - z_A) \tag{3}$$

The air mass flow through the opening between co-ordinates z1 and z2 and from room i toward room j is defined by :

$$m_{ij} = C_d \, W \, \rho_i \int_{z1}^{z2} u(z) \, dz \tag{4}$$

Using the Bernoulli' equation, we can calculate the velocity u(z) :

$$u(z) = \left[\frac{2}{\rho_i} \left(P_i(z) - P_j(z) \right) \right]^{1/2} \tag{5}$$

The integration of equation (5) gives :

$$m_{ij} = \frac{2\sqrt{2}}{3} C_d \, \frac{W}{g} \, \frac{\sqrt{\rho_i}}{|\rho_i - \rho_j|} \left\{ \left[\Delta P_{ij}(z_2) \right]^{3/2} - \left[\Delta P_{ij}(z_1) \right]^{3/2} \right\} \tag{6}$$

Applying equation (6) to the two zones A and C of the two rooms 1 and 2 permits to get the air mass flows through the opening.

The Cd coefficient is an empirical coefficient that takes into account the effects of viscosity of the fluid and a local contraction of the streamlines.

In the case of a large vertical opening, a recent analysis of the available data given by Santamouris and al (1995) showed a scattering of results that can reach 40% for the value of the coefficient Cd for natural convection flows. This survey shows the difficulty to get the accurate values of the Cd coefficient.

However, in order to give a meaningful value to the Cd coefficient, we used the results given by Pelletret and al (1991). Thus, in this configuration, the value of the Cd coefficient has been fixed to 0.5.

The heat flow through the large opening and for the room i toward the room j is defined by :

$$\Phi_{ij} = m_{ij} \, Cp \, Ti \tag{7}$$

The heat flow from room j to room i is defined by :

$$\Phi_{ji} = m_{ji} \, Cp \, Tj \tag{8}$$

Using equations (7) and (8), we can compute the net heat flow through the opening :

$$\Phi net_{ij} = \Phi_{ij} - \Phi_{ji} \tag{9}$$

Calculation of the convective heat flows within a room

Every room is represented by an upper and a lower zone including the driving flows such as thermal plumes, thermal boundary layers or jets. Every driving flow is described by a specific equation which is as follows :

Wall thermal plume, Inard (1988) :

$$m(z) = 9.5 \, 10^{-3} \, Q(z)^{1/3} \, (z - z_0) \tag{10}$$

Free thermal plume, Kofoed (1991) :

$$m(z) = 6.1 \ 10^{-3} \ Q(z)^{1/3} \ (z - z_0)^{5/3} \tag{11}$$

Thermal boundary layer, Allard (1987) :

$$m(z) = 0.004 \ (T_S - T_A)^{1/3} \ z \tag{12}$$

Jet, Rajaratnam (1976) :

$$m(x) = 0.25 \ (\frac{x}{b_0})^{1/2} \tag{13}$$

These laws permit to calculate the air mass flow through the borders between the zones A and C and the different driving flows. The altitude z_A is arbitrary. We carried out simulations with two values of z_A equal respectively to the mid-height of the opening and the mid-height of the cell. The results obtained are very similar. Thereafter, we fixed the value of z_A to the mid-height of the opening.

Including the air mass flows through the large opening and writing the mass balance of the zone A (or C) permits to calculate the air mass flow between zones A and C. Lastly , equations of the thermal balance of zones A and C are used in order to calculate air temperatures T_{1A} and T_{1C}.

This model has been validated by comparison with experimental results obtained under steady and unsteady states in a two zone test cell, for various thermal inputs. The detailed results are available in Burchiu (1998).

Cases treated

Although the model is able to describe unsteady state, we only present here results obtained in steady state.

We chose a configuration corresponding to a house of 96 m^2 with 5 rooms and a hall. A view of the house is given in figure 2. The outside temperature is equal to −7.2°C and the ventilation air flows given by a mechanical system are shown in figure 2. The temperature of the fresh air has been set at − 4.2°C.

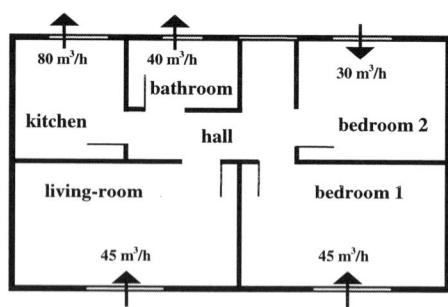

Figure 2 : Horizontal section of the house treated

Two cases for the control of the air temperatures have been considered :

- Case 1 : the set point of all the air temperatures is equal to 20±0.2°C;

- Case 2 : only the air temperature within the living room is controlled to 24.5±0.2°C, the others are not controlled.

Besides, in order to be able to quantify the influence of the thermal stratification on the heat exchanges between the rooms, we led simulations with an isothermal model, that is to say with only one air temperature calculated in each room.

The figures 3 and 4 show the results obtained that is to say the air temperature (Ta), the heating power (Pe) for each room and the net heat flow through each door. For the zonal model, the air temperature is a weighted average by the height of every zone A and C.

Considering case 1 (see figure 3), it appears that taking into account the air thermal stratification in the model has a low influence on the heating powers. Nevertheless, we can notice that these are higher for the zonal model and in the rooms where the ventilation air is exhausted that is to say the kitchen and the bathroom. It is essentially because the air is exhausted in the upper part of the room.

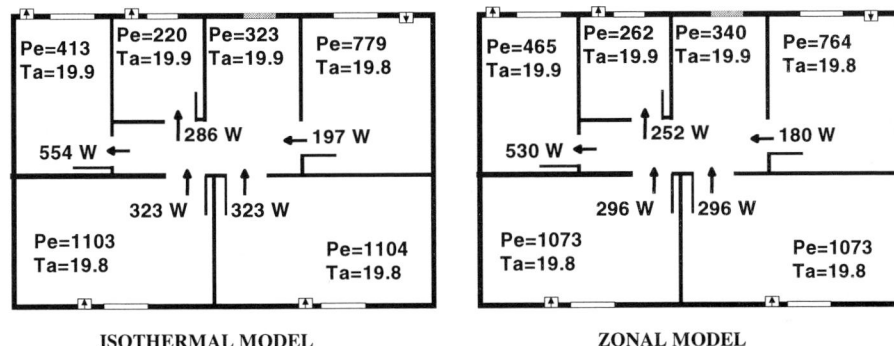

Figure 3 : Air temperature, heating power and net heat flow through the doors (case 1)

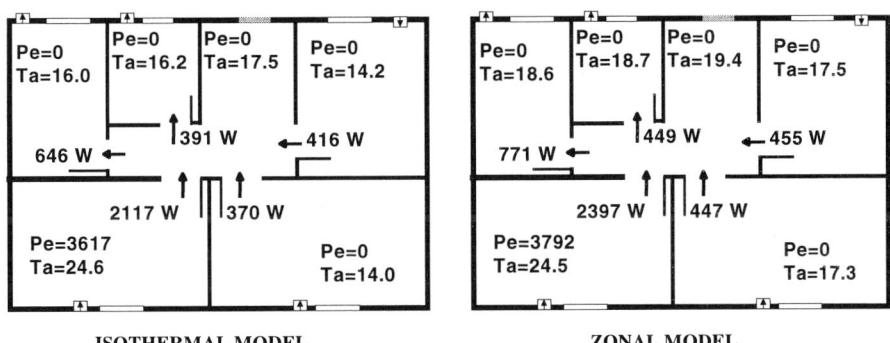

Figure 4 : Air temperature, heating power and net heat flow through the doors (case 2)

Case 2 for which the heating power is only released in the living room (see figure 4) highlights the influence of the thermal stratification on the heat transfers between rooms. Indeed, we notice that the isothermal model undervalues the heat flows through the doorways. This implies that the calculated heating power with the zonal model is higher than for the isothermal model leading in lower air temperature values for the latest. The gap between the calculated air temperatures can reach 3°C.

CONCLUSION

In this study, we presented a zonal model that allows to take into account the thermal stratification in a heated multizone building. The first results obtained showed that when the air temperatures of all the rooms have the same set point, the differences between an isothermal model and a zonal model are low. On the other hand, if the heating power is only supplied in one room, the isothermal model gives lower values of the indoor air temperatures than the zonal model. This is due to the difference in the computed heat flows through the doors.

REFERENCES

Allard F. (1987). Contribution à l'étude des transferts de chaleur dans les cavitées thermiquement entraînées à grand nombre de Rayleigh. *Thèse d'Etat*, INSA Lyon, France

Burchiu S. (1998). Etude de l'influence d'un système de chauffage sur l'état thermique et aéraulique des bâtiments multizones avec prise en compte de la stratification thermique. *Thèse de Doctorat*, INSA Lyon, France

Inard C. (1988). Contribution à l'étude du couplage thermique entre un émetteur de chaleur et un local. *Thèse de Doctorat*, INSA Lyon, France

Inard C., Meslem A. and Depecker P. (1998). Energy consumption and thermal comfort in dwelling-cells : a zonal approach. *Building and Environment* **33:5**, 279-291.

Kofoed P. (1991). Thermal plumes in ventilated rooms. *Ph. D. Thesis,* Instituttet for bygningtechnik, Aalborg, Denmark

Li Y., Delsante A., Symons J.G. and Chen L. (1998). Comparaison of zonal and CFD modelling of natural ventilation in a thermally stratified building. *Roomvent'98*, Stockholm, Sweden, **2**, 415-422.

Pelletret R., Allard F., Haghighat F. and Van Der Maas J. (1991). Modelling of large opening. *12 th AIVC Conference*, Ottawa, Canada, **1**, 99-109.

Rajaratnam N. (1976). *Turbulents jets*, Elsevier Scientific, New York, USA

Rumianowski P., Brau J. and Roux J.J. (1989). An adapted model for simulation of the interaction between a wall and the building heating system. *Thermal performance of the exterior envelopes of buildings IV Conference*, Orlando, USA, 224-233.

Santamouris M., Argiriou A., Asimakopoulos D., Klitsikas N. and Dounis A. (1995). Heat and mass transfer through large openings by natural convection. *Energy and Buildings* **23**, 1-8.

van der Maas J. (1992). Air flow through large openings in buildings. *Technical report*, International Energy Agency, Annex 20, subtask n°2.

Walton G.N. (1980). A new algorithm for radiant exchanges in room loads calculation. *ASHRAE Transactions* **86:2**, 190-208.

© 2000 Elsevier Science Ltd. All rights reserved.
Air Distribution in Rooms, (ROOMVENT 2000)
Editor: H.B. Awbi

NATURAL VENTILATION OF AN ENCLOSED ROOM BY DOORWAY EXCHANGE FLOWS

Jeremy C. Phillips[1] and Andrew W. Woods[2]

[1]Centre for Environmental and Geophysical Flows, School of Mathematics,
University of Bristol, Bristol BS8 1TW, UK.
[2]BP Institute, University of Cambridge, Cambridge CB3 0EZ, U.K.

ABSTRACT

When a single doorway between a warm enclosed room and a large cooler exterior is opened, a two-layer exchange flow is set up in the doorway, with cool air filling the room along the floor, and warm air being displaced out through the doorway. The transient filling of the room continues until the cool lower layer reaches the doorway top. This process was studied experimentally using saline solutions to create density contrasts in a laboratory tank fitted with a doorway whose length and width could be varied. The height of the dyed dense layer filling the room was tracked using image processing techniques. A mathematical model based on maximal exchange flow through the doorway shows good agreement with the experimental data, and confirms that the fractional interface height in the room scales as $\frac{A_{room}}{A_{door}} \left(\frac{H}{g'} \right)^{\frac{1}{2}}$, where A_{door} and A_{room} are the areas of the doorway and room, H is the doorway height and g' is the buoyancy contrast between the layers. The use of the models to predict room cooling times by doorway exchange flows is illustrated.

KEYWORDS

Natural Ventilation, Exchange Flow, Transient, Experiment, Mathematical Model

INTRODUCTION

With increasing energy costs and the need to maintain comfortable building environments, there has been growing interest in the natural ventilation of buildings. A particularly important class of flows result when air masses of different temperature and hence density come into contact, for example through openings between rooms which are maintained at different temperatures. Even with small temperature differences, the heat fluxes transported by exchange flows are significant compared with the heat budget of modern buildings (Dalziel and Lane-Serff, 1991). Exchange flows also provide a transport mechanism for pollutants from hotter to colder rooms within buildings and for accidental releases of dense gas within buildings.

The exchange flows studied here are displacement ventilation flows (figure 1). For an enclosed room with one high-level and one low-level opening (figure 1a), displacement ventilation corresponds to the case in which an upper warm layer of air leaves the room through the higher opening and a lower cold layer of air enters the room through the lower opening (Linden *et al.*, 1990). When a heat source is supplied within the room, a steady regime can become established, with a sharp density contrast between the two layers of fluid if the source of buoyancy is localised (figure 1a). Linden *et al.* (1990) and more recently Linden and Cooper (1996) have examined these flows in some detail, successfully testing a quantitative model for the steady state flow with a series of analogue laboratory experiments using aqueous saline solutions. For a warm enclosed room with a single low-level opening to a colder exterior, the cold incoming air remains at the base of the room and there is little mixing in the room between the incoming cold and escaping warm air (figure 1b). This case has some features in common with the displacement ventilation problem associated with two openings (figure 1a), even though the flow is through a single doorway. In this paper, we investigate both experimentally and theoretically the transient exchange flow which develops on opening a single doorway at the base of a warm room connected to a larger, cooler exterior.

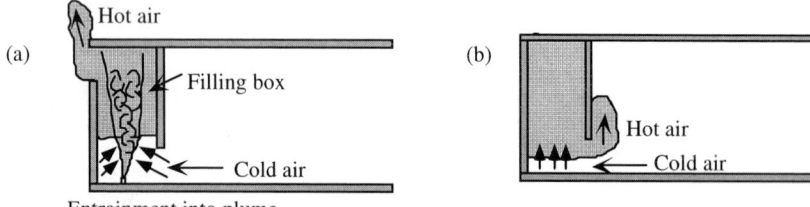

Figure 1. Displacement ventilation flows through (a) high-level and low-level openings and (b) a single opening

Our work builds on the earlier studies of Armi (1986) and Dalziel and Lane-Serff (1991), who analysed the steady-state exchange flows between effectively infinite environments on either side of a doorway. This study extends the experimental study of doorway exchange flows made by Gladstone *et al.* (1998) to include image processing methods to quantify the mixing between the counter-flowing layers, and presents a simple mathematical model of the process.

APPARATUS AND METHODOLOGY

Experiments investigating the transient filling of a room through a single doorway were conducted in a rectangular glass tank 1.2 m long x 0.37 m wide x 0.38 m deep. A 2 mm thick aluminium partition was situated on the mid-point of the tank, 0.60 m from either end, with a 20 mm wide doorway extending from the floor of the tank up to 350 mm (see figure 2). The doorway opening was controlled by a removable lock gate, which could be raised to simulate the opening of a doorway (heights 100-200 mm in 20 mm steps). Dyed saline solution (2, 3 wt%) was used in one side of the tank to simulate a reservoir of cold exterior air and clear weak saline solution (0.5 wt%) was used on the other side to simulate heated air within a finite volume room. Preliminary experiments (Gladstone *et al.*, 1998) had used the partition situated only 0.18 m from one end, to simulate a small heated room connected to a larger, cooler exterior, as illustrated in figure 1(b). However, this equipment configuration allowed only a small number of measurements to be made at the start of the experiment. For this reason, these transient experiments were conducted with the partition in located centrally in the tank.

The tank was initially filled with fresh water to a depth of 0.37 m and then the lock gate was closed and salt was added to each side of the lock to make up saline solutions. The reduced gravity (g') of the 'room' had values of 0.105 m s^{-2} and 0.174 m s^{-2} in experiments using 2 and 3 wt% saline solutions.

One side of the tank was dyed using food colouring, and the position of the height of the interface between the two saline solutions was recorded using a video camera. The video sequences were analysed using the computer image analysis package *DigImage* (Dalziell, 1993) to locate and measure the time evolution of the height of the top of the unmixed dense (dyed) saline solution in the room. Vertical transmitted light intensity profiles were measured 10 cm from the doorway on each side; the light intensity on the exterior (dyed) side of the doorway was matched with that in the room, thus locating the top of the unmixed (dyed) layer. The analysis confirmed that the mixed layer depth is small, approximately 2 cm in all experiments.

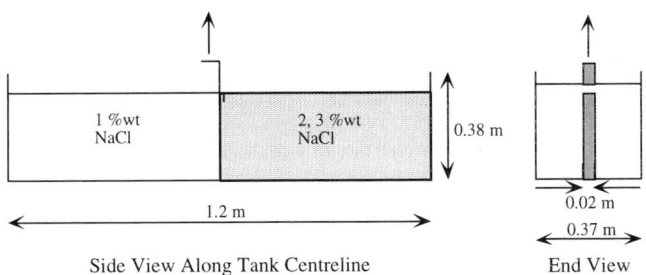

Figure 2. Diagram of the Experimental Apparatus

The experiments were designed to be analogous to real ventilation flows. For a typical doorway height (H) of 2-3 m and a temperature difference (ΔT) of 10 °C, the buoyancy contrast $g' = g(\Delta T/T_u) = g(\Delta\rho/\rho_u)$ is about 0.4 m s^{-2} (the subscript u refers to the upper layer or light fluid) The typical flow speed through a doorway scales as $(g'H)^{1/2} \sim 1$ m s^{-1}, where H is the door height and g' the reduced gravity, and so the Reynolds number of the flow $(g'H^3)^{1/2}/\nu \sim 10^5$, while the interfacial Richardson number ~ 1. The flows are therefore fully turbulent, but there is little mixing across the fluid interface (Turner, 1979, p. 92). In the experiments, we sought to reproduce this physical regime. Typical flow velocities were of order 0.1 m s^{-1}, producing Reynolds numbers of order 10^4 and Richardson numbers of order 1.

EXPERIMENTAL OBSERVATIONS

On opening the lock gate, a current of dense saline solution propagated into the room along the tank floor. At the same time, a reverse flow of the lighter weak saline solution issued from the room through the upper section of the doorway, creating the exchange flow. The temporal evolution of the interface height in both the room, h_i, and the doorway, h_d, for this experiment is shown in figure 3. No data could be obtained for the initial 15 s, during which pulses of fluid periodically passed through the doorway, rather than the continuous flow which characterised the subsequent exchange. The data suggest that after this initial transient, for $t \geq 15$ s, the interface height, h_i, increases linearly with time until $h_i \approx 0.55H$ where H is the door height. Subsequently, for $h_i > 0.55H$, the rate of ascent of the interface gradually decreased with time. After an initial phase of flow establishment, $t \leq 15$ s, the interface height at the doorway, h_d, became approximately equal to 0.55H (figure 3). This is in close agreement with the value 0.54H reported by Dalziel and Lane-Serff (1991). The interface height in the doorway remained at the constant value 0.55H until eventually the interface in the room rose above this point $h_i > 0.55H$. Subsequently, the height of the interface in the doorway increased with that in the interior of the room, so that for the remainder of the exchange flow, $h_i \approx h_d$ (figure 3).

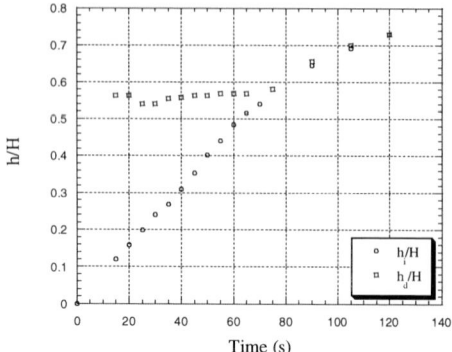

Figure 3. The evolution of interface height with time for a typical experiment

THEORETICAL ANALYSIS

In many ventilation problems, the density differences are small, so that we can use the Boussinesq approximation (Turner, 1979, p.9) in deriving the model of flow through the doorway and that in the room. If the lower layer of dense fluid has density ρ_l and flows with velocity u_l into theroom, and the lighter fluid of density ρ_u and velocity u_u flows from the room, then conservation of volume requires that

$$u_l h_d = u_u (H - h_d) \ .$$
(1)

We expect the flow to be controlled at the doorway so that the composite Froude number of the flow equals unity (Armi, 1986, Dalziel and Lane-Serff, 1991)

$$\frac{u_l^2}{h_d} + \frac{u_u^2}{H - h_d} = g'$$
(2)

where the reduced gravity $g' = \dfrac{(\rho_l - \rho_u)g}{\rho_u}$. Thus the lower layer velocity is related to the interface depth according to

$$u_l = \left[g' \frac{h_d (H - h_d)^3}{h_d^3 + (H - h_d)^3} \right]^{\frac{1}{2}} \ .$$
(3)

The rate of ascent of the density interface within the room is given by conservation of the lower layer fluid. Assuming there is no mixing of the inflowing fluid with the upper layer, conservation of mass requires that

$$A \frac{dh_i}{dt} = w h_d u_l$$
(4)

where A is the cross-sectional area of the room and w the width of the door. Therefore,

$$\frac{dh_i}{dt} = \frac{w}{A} \left[g' \frac{h_d^3 (H - h_d)^3}{h_d^3 + (H - h_d)^3} \right]^{\frac{1}{2}} \ .$$
(5)

In order to solve the system and thereby predict the time evolution of the interface height, $h_i(t)$, we must complete the model (1-5) by specifying the relationship between the depth of the interface in the room, $\hat{h}_i = h_i/H$ and that in the doorway, $\hat{h}_d = h_d/H$. During the first stage of the exchange flow ($\hat{h}_i < 0.55$), the interface height in the doorway was independent of that in the room (figure 4)

$$\hat{h}_d = 0.55 \text{ for } \hat{h}_i < 0.55 . \tag{6a}$$

Once the interface in the room had ascended above the point $\hat{h}_i = 0.55$, we observed that the doorway interface height followed that in the room (figure 4),

$$\hat{h}_d \approx \hat{h}_i \text{ for } \hat{h}_i > 0.55 . \tag{6b}$$

Equations (5) and (6) therefore suggest that

$$\hat{h}_i \approx 0.24\hat{t} \text{ for } \hat{h}_i < 0.55 \tag{7}$$

and

$$\frac{d\hat{h}_i}{d\hat{t}} \approx \left(\frac{\hat{h}_i^3 (1-\hat{h}_i)^3}{\hat{h}_i^3 + (1-\hat{h}_i)^3} \right)^{\frac{1}{2}} \text{ for } \hat{h}_i > 0.55 \tag{8}$$

where

$$\hat{t} = \frac{wH}{A} \left(\frac{g'}{H} \right)^{\frac{1}{2}} t . \tag{9}$$

COMPARISON WITH EXPERIMENTS

In figures 4a and 4b, we show experimental measurements of the dimensionless interface height, $\hat{h}_i = h_i/H$, plotted as a function of the dimensionless time $\hat{t}$ for doorway heights from 100 mm to 200 mm and buoyancy contrasts of 0.105 m s^{-2} and 0.174 m s^{-2}. The error bars represent an error of $\pm$ 2 mm in determining the height of the unmixed layer top, h_i, introduced in the estimation of the mixed layer thickness using the image processing method described above. The solid line plotted on figures 4a and 4b shows the universal model formed by coupling the solutions for the two phases of the filling process (equations 7 and 8) developed above.

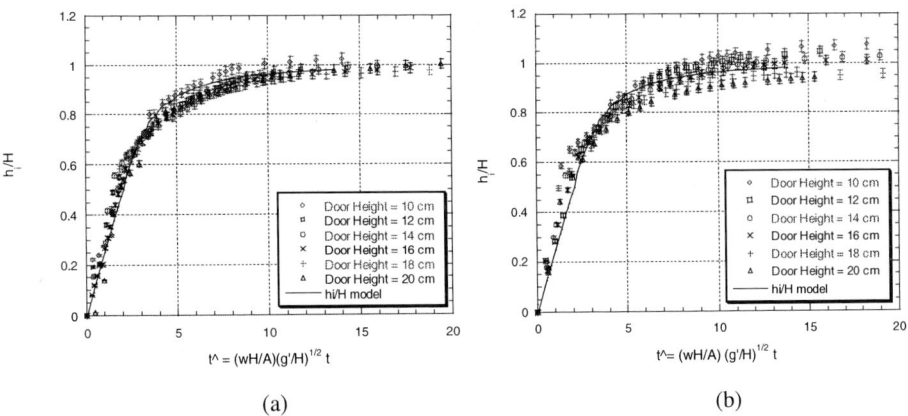

Figure 4. Evolution of dimensionless interface height with dimensionless time during transient exchange flow experiments. (a) g' = 0.105 m s^{-2} (b) g' = 0.174 m s^{-2}

At the lower buoyancy contrast used, the data appear to collapse satisfactorily within experimental error over the range of door heights. The model provides an acceptable leading order description of the filling process, although shows slight underprediction of the interface height in the first phase and slight overprediction in the second phase. The increased flux at the start of the experiment may correspond to the propagation of dense fluid into the room as a spreading gravity current, with additional hydraulic controls on this release condition not accounted for in the model. At the higher buoyancy contrast used, there is greater scatter in the experimental data, particularly during the second phase of the filling process. However, the model still provides an acceptable leading order fit to the data, particularly for intermediate door heights.

APPLICATION

The analysis and experimental study reported in this paper provides estimates for the time scale for the exchange flow to set up a steady density distribution following the opening of a doorway. The time required for the filling front in the room to extend to 90% of the doorway height is given by the expression

$$\tau_{0.9} = 5\left(\frac{A_{room}}{A_{door}}\right)\left(\frac{H}{g'}\right)^{\frac{1}{2}} .$$

(10)

We estimate filling times of order $\tau_{0.9}$ = 3-150 minutes. The larger estimate corresponds to large rooms, for example, classrooms, open offices or meeting halls, or doorways across which the temperature contrast is small. The smaller estimate corresponds to small domestic rooms or doorways across which there is a larger temperature and, for example when there is hot smoke produced by a fire within the room.

ACKNOWLEDGEMENTS

This work is sponsored by the EPSRC Built Environment Programme, UK.

REFERENCES

Armi, L. (1986). The hydraulics of two flowing layers with different densities. *J. Fluid Mech.* **163**, 27-58.

Dalziel, S. (1993). Rayleigh-Taylor instability: experiments with image processing. *Dyn. Atmos. Oceans* **20**, 127-153.

Dalziel, S. and Lane-Serff, G.F. (1991). The hydraulics of doorway exchange flows. *Building and Environment* **26**, 121-135.

Gladstone, C., Woods, A.W., Phillips, J.C. and Caulfield, C.P. (1998). Experimental study of mixing in a closed room by doorway exchange flows. *Proceedings 6th International Conference on Air Distribution in Rooms* **2**, 555-561.

Linden, P.F., Lane-Serff, G.F. and Smeed D.A. (1990). Emptying filling boxes: the fluid mechanics of natural ventilation. *J. Fluid Mech.* **212**, 309-335.

Linden, P.F. and Cooper, P. (1996). Multiple sources of buoyancy in a naturally ventilated enclosure. *J. Fluid Mech.* **311**, 177-192.

Turner, J.S. (1979). *Buoyancy effects in fluids* Cambridge University Press.

Air Distribution in Rooms, (ROOMVENT 2000)
Editor: H.B. Awbi

A HYBRID DISPLACEMENT-MIXING VENTILATION REGIME IN A NATURALLY VENTILATED ROOM

Charlotte Gladstone and Andrew W. Woods

BP Institute, Bullard Laboratories, University of Cambridge, Madingley Road, Cambridge

ABSTRACT

The evolution of the temperature profile in a warm room driven by a natural ventilation flow which develops when the room is connected to a cold exterior by two openings at different vertical heights is explored. With the openings at the top and base of the room, we find the classical displacement ventilation regime provides a leading order description of the flow. With openings at the centre and top of the room, the ventilation is hybrid, with the lower part of the room being well-mixed, and the upper part being stratified by an upward displacement ventilation flow. We present some dimensional scaling, two end-member models and a series of analogue experiments of the process. The suite of experiments involve two different configurations of openings and a variety of temperature differences between the interior and exterior temperature.

KEYWORDS

Natural ventilation, displacement, mixing, buoyancy, mathematical modelling, experiments.

INTRODUCTION

Natural ventilation of a room develops when air of one temperature inside the room exchanges through one or more openings with air of a different temperature outside. The density differences associated with the temperature difference inside and outside the room lead to a higher internal pressure at the top of the room and a lower internal pressure at the bottom of the room in comparison with the external pressure. This leads to a natural ventilation flow. Depending on the number and positions of the openings, the exchange of warm and cold air may occur through mixing ventilation or through displacement ventilation as described by Linden, Lane-Serff & Smeed (1990). In a warm room linked to a cold exterior by a single opening located near the top of the room, mixing ventilation occurs. Cold exterior fluid enters and mixes with fluid in the room as it descends to the floor. In contrast, with two openings, one located high and one low, displacement ventilation occurs. Cold exterior fluid enters the room through the lower hole displacing warm interior fluid up and out of the upper hole. Linden *et al.* (1990) considered such displacement ventilation by studying the draining of warm fluid from a room which is connected to a cold exterior through holes in the ceiling and the floor of the room. In this paper we generalise the study to consider the natural ventilation of a warm room connected to a large cold exterior by two small openings, and now account for mixing of the inflowing fluid with that originally in the room. We assume that one of the openings is located high in the room and examine the effect of positioning the second opening either in the centre or base of the room.

EXPERIMENTS

These analogue experiments are conducted using a small perspex tank with circular openings in one wall, placed inside a large reservoir which acts as the exterior. The model room has a square floor area of 311.5 cm^2 and is 28.7 cm deep. It includes three openings each with a diameter of 1.5 cm and with midpoints located at 1.05, 13.50 and 25.95 cm above the floor. These are sealed by rubber bungs. The external reservoir is filled with cold water of temperature T_{EXT}, typically 17.5°C to 20.5°C. The room is filled with warm fluid of temperature T so that the initial temperature difference between interior and exterior fluids, ΔT_0 ranges from 15°C to 50°C. An experiment is started by removing bungs from two of the three openings. Two series of experiments were conducted, both with the same values of ΔT_0. In Series 1, the uppermost and centre of the three holes are opened, i.e. those with midpoints at 25.95 cm and 13.50 cm, while the bottom hole remains closed. During Series 2, the centre hole is closed with the top and bottom holes open. Flow patterns are observed using the shadowgraph method. The evolving temperature profile within the room is monitored using seven Type K thermocouples placed at 1.9 cm, 7.4 cm, 11.4 cm, 15.0 cm, 18.4 cm, 22.2 cm and 25.6 cm above the floor of the room. These record the temperature profile every three seconds. An eighth thermouple monitors the external temperature, T_{EXT}.

Once the bungs are removed, cold exterior fluid flows into the room through the lower hole, while hot interior fluid flows out of the room through the upper hole. The room progressively becomes cooler from bottom to top with time with contrasting temperature profiles resulting depending on which configurations of openings is employed.

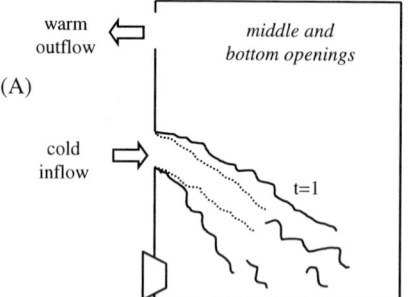
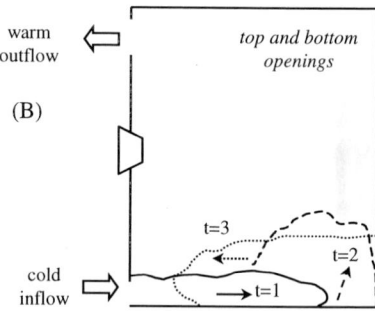

Figure 1: Schematic diagrams show the different mixing processes operating in (a) Series 1 with centre and top openings employed, and (b) Series 2 with top and bottom openings employed.

During Series 1, in which the top and centre holes are opened, cold exterior fluid enters the room and gradually descends to the floor as an inclined, entraining plume (figures 1a and 2a). This causes the lower section of the room to become well-mixed. During Series 2, with top and bottom holes opened, exterior fluid enters as an advancing gravity current very near the base of the room. This reflects back and forth off near and far walls (figures 1b and 2b). The mixing caused by the gravity current of Series 2 is confined to the beginning of an experiment and produces a thin mixed layer of 2-4cm. The ventilation process then approximates the classical displacement ventilation process, as outlined by Linden et al. (1990). The mixing driven by the descending plume in Series 1 is far more substantial, with much of the fluid in the lower half of the room becoming well mixed. These two mixing processes produce different evolving temperature profiles. In both series of experiments the bottom of the room cools first. The room then becomes progressively colder sequentially from bottom to top (figure 2). The temperature decrease is rapid at first, but becomes slower with time before reaching a constant value, which is slightly stratified with height. In Figure 2a, thermocouples below the centre hole show a similar temperature decrease indicating that this part of the room is mixed. The mixing is less pronounced in Figure 2b; indeed, over the first 500s, the temperature adjusts over a very narrow vertical range.

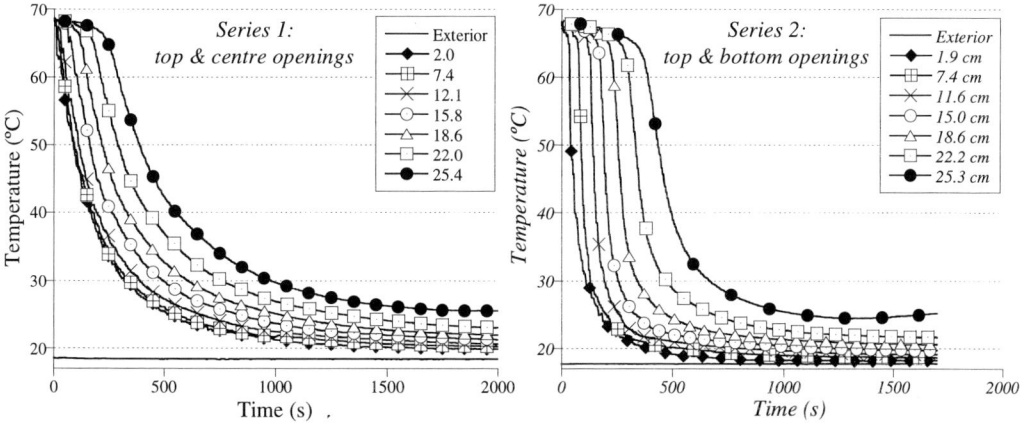

Figure 2: During both these experiments ΔT_0=50°C, and in (a) top and centre holes are opened (Series 1) and (b) top and bottom are opened (Series 2)

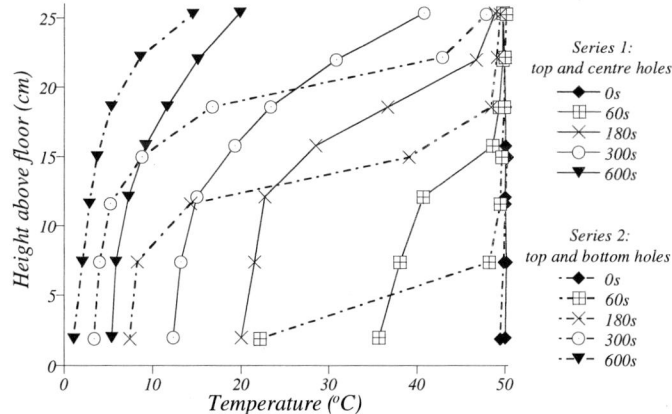

Figure 3: Two experiments, one from Series 1 and the second from Series 2, but both with $\Delta T_0 = 50$°C, are shown here. Temperature T-T_{EXT} is plotted against height at different times t.

Temperature changes in the room can be compared by plotting T-T_{EXT} as a function of height at different times, t=60s, 120s, 180s, 300s, 420s and 600s (figure 3). Experiments from Series 1 and 2 are presented, both with ΔT_0=50°C. When t=0s, T-T_{EXT} =ΔT_0 and a slight stratification develops towards the base of the room. After t=60s, the Series 1 temperature profile (top and middle openings) shows a mixed lower half of the room and an unmixed upper half (figure 3). The fluid in the lower half of the room is substantially warmer than T_{EXT} indicating significant mixing. In Series 1, the mixed layer is still present at t=180s, and by 300s the top of the room is also beginning to cool. The temperature profile in the room evolves in a very different way in the Series 2 experiment (figure 3) where top and bottom holes are opened. The lower part of the room is colder than in Series 1, indicating that there is less mixing here, while the upper part of the room remains at the initial warm temperature for longer. By reference to the idealised model displacement processes, we infer that Series 2 approximates to the simple displacement ventilation process (Linden *et al.*, 1990) with only a small amount of mixing, whereas in Series 1, the mixing ventilation process operates in the lower part of the room, while the displacement process controls the flow in the upper part of the room.

MODELLING

The volume flux through the two holes has natural scale $Q=[g\Delta Th/\Delta T_0]^{1/2}A_{HOLE}$ where g is the gravitational constant, ΔT_0 is the temperature difference between the temperature of fluid inside the room, T, and fluid outside the room, T_{EXT}, h is the distance between the midpoints of the two holes and A_{HOLE} is the area of one of the (equally-sized) openings. The timescale for the temperature profile to evolve, τ, is given in terms of the volume of the room and Q (eqn 1), $\tau = A_{ROOM}H/Q$ where A_{ROOM} and H are the cross-sectional area and height of the room respectively. In order to compare all the experiments, we introduce the dimensionless height of the room, $z^*=z/H$, the dimensionless temperature of fluid within the room, $T^*=(T-T_{EXT})/\Delta T_0$ and the dimensionless time, $t^*=t/\tau$.

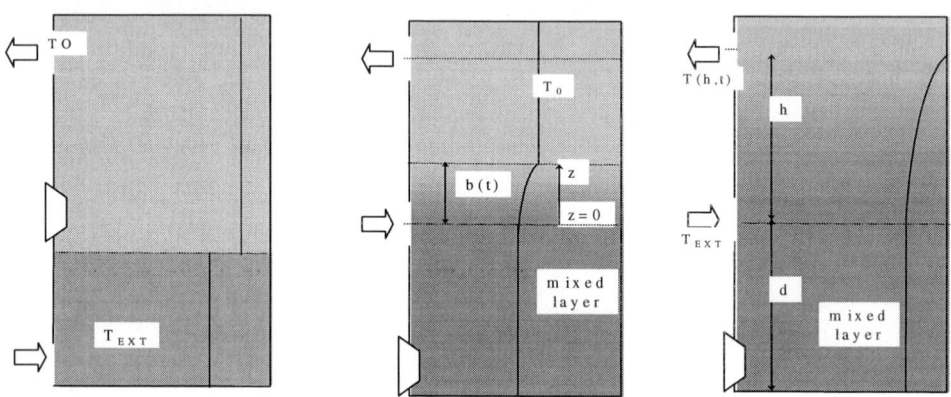

Figure 4. (a) displacement ventilation regime;
(bi) initial stages of hybrid ventilation regime; (bii) late stages of hybrid ventilation regime.

Displacement ventilation regime: This regime is essentially that described by Linden *et al.* (1990). If the interface has height h above the lower opening, and the buoyancy contrast between the interior and exterior fluid is $g'=g(T_0-T_{EXT})/T_{EXT}$, then the flow through each opening has volume flux $Q = A_{HOLE}(g'(H-h)/2)^{1/2}$. The rate of ascent of the interface between the two fluids is then given by the relation $dh/dt=Q/A_{ROOM}$, leading to the expression

$$h(t) = H - \left(H^{1/2} - \frac{A_{HOLE}}{2A_{ROOM}}\left(\frac{g'}{2}\right)^{1/2}t \right)^2$$ (1)

This relation also describes the ascent of the isotherms upwards through the system, assuming that the thin mixed zone about the interface is established in the initial stages of the process, and then remains approximately fixed as the interface ascends.

Hybrid displacement-mixing regime

In the partially mixed regime the flow is more complex, with the region below the opening being approximately isothermal, while that above the opening is stratified in temperature. We may model this flow by assuming the incoming fluid mixes the region below the inflow opening uniformly, $0<z<d$, but that above this region, $d<z<h+d$, the upward flow of warm air to the outflow vent generates a region of gradually increasing temperature, where d is the depth of fluid below the midpoint of the centre opening and h is the depth of fluid between the midpoints of the centre and upper openings. If we denote the temperature in the fluid as $T(z,t)$, and the fluid in the mixed zone to be $T_M(t)$, then the exchange flux between the two openings has the form

$$Q = A_{HOLE}\left(\frac{g}{2}\int_d^{h+d}\left(\frac{T(z,t)-T_{EXT}}{T_{EXT}}\right)dz\right)^{1/2} \tag{2}$$

The conservation of thermal energy in the lower well-mixed zone requires that $(T_M - T_{EXT})Q = -A_{ROOM}\,dT_M/dt$ where $T(d,t)=T_m(t)$. Finally, we note that $T(z,t)=T_M(\tau)$ where $z = d + \int_{t-\tau}^{t}(Q/A_{ROOM})dt$. During this time the temperature in the ascending zone has value $T(z,t) = T_{EXT} + (T_0 - T_{EXT})\exp(-(b+d-z)/d)$ for $z<b+d$, where $b(t)$ represents height of the top of the stratified zone. Above this point, the fluid in the room is unmixed and the temperature equals the original temperature of the room T_0. Noting that the conservation of mass requires that $Q=A_{ROOM}db/dt$ we see that the system may be simplified to an equation for the ascent of the initial surface of mixed fluid, which has the dimensionless form

$$\frac{db}{dt} = \frac{A_{HOLE}}{A_{ROOM}}\left(\frac{g}{2}\frac{(T_0-T_{EXT})}{T_{EXT}}[d(1-\exp(-b/d))+(h-b)]\right)^{1/2} \tag{3}$$

This relation applies until the ascending front of the mixed fluid has reached the upper outflow opening, when $b=h$. Subsequently, the spatial structure of the temperature profile in the stratified region between the two openings, $d<z<h+d$, has the form $T(z,t) = T_{EXT} + (T(h+d,t)-T_{EXT})\exp(-(h+d-z)/d)$ so that the temperature in the mixed zone $0<z<d$ has value $T_M(t) = T_{EXT} + (T(h+d,t)-T_{EXT})\exp(-h/d)$. The temperature of the mixed zone then evolves at a rate given in dimensionless units by the relation

$$\frac{dT_M}{dt} = -\frac{A_{HOLE}}{A_{ROOM}}\left(\frac{g}{2dT_{EXT}}\right)^{1/2}(T_M-T_{EXT})^{3/2}\exp(h/2d)(1-\exp(-h/d))^{1/2} \tag{4}$$

starting at the time at which the ascending mixed fluid first reaches the outflow opening. At this time the temperature of the region $0<z<d$ is $T_M(t) = T_{EXT} + (T_0-T_{EXT})\exp(-h/d)$. We take the time scale to be $\tau_s = 2(dT_{EXT}A_{ROOM}^2/A_{HOLE}^2 g(T_0-T_{EXT}))^{1/2}$, the dimensionless thickness of the ascending stratified region to be $y=b/d$, and the dimensionless temperature of the mixed zone to be $\theta = (T_M-T_{EXT})/(T_0-T_{EXT})$. Using the dimensionless time t_d we find that in the initial stage of the process $dy/dt_d = \sqrt{2}(1-\exp(-y)+Y-y)^{1/2}$ and $\theta = \exp(-y)$. Once $y=Y=h/d$, the distance between the two openings, then the temperature of the lower mixed zone evolves according to

$$d\theta/dt_d = -\sqrt{2}\theta^{3/2}\exp(Y/2)(1-\exp(-Y))^{1/2} \tag{5}$$

with solution

$$\theta(t_d) = \left(\theta^{-1/2}(t_c) + \frac{1}{\sqrt{2}} \exp(Y/2)(1 - \exp(-Y))^{1/2}(t_d - t_c) \right)^{-2} \tag{6}$$

where t_c is the time at which $y=Y$. We illustrate how the temperature profile in the fluid evolves with time according to this model in figure 5.

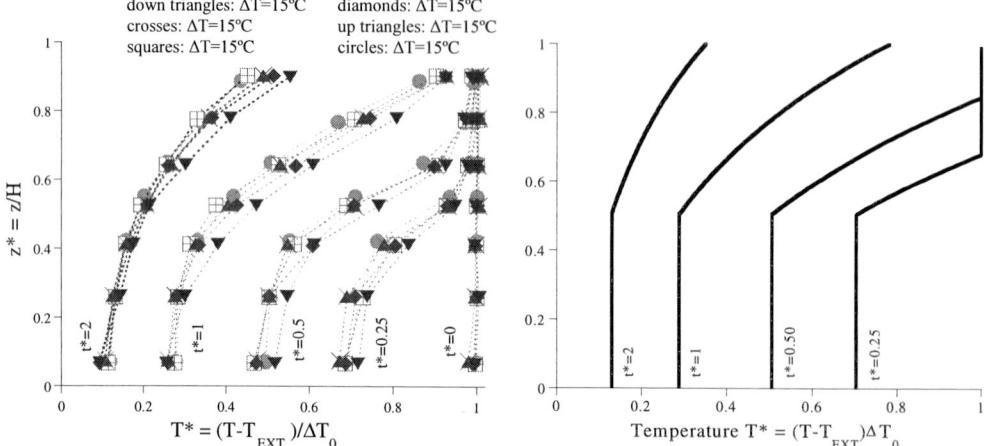

Figure 5: (A) experiments are scaled on z* and T*; (B) evolution of temperature profile as predicted by the model for the hybrid ventilation

We have described a series of experiments in which a room initially filled with relatively light fluid was allowed to exchange fluid with a relatively dense exterior through two openings at different heights in a closed room. The experiments and model identified two different regimes depending on the location of the lower opening. When it was positioned at the base of the reservoir, the flow resembled the simple displacement ventilation flow regime (see also Gladstone and Woods, RoomVent 2000) whereas when the opening was located at the midpoint of the room, the inflowing dense fluid mixed vigorously with the interior fluid leading to formation of a relatively well-mixed zone at the base of the room. However above the lower opening a steady upward displacement flow developed, and this generated a strong stratification through the upper part of the room. The dimensional scalings presented of the experimental data provide a very good leading order collapse of the experimental data, and the theoretical predictions of the rate of ascent of the mixed zone are also in good accord with the observations. The work provides new insight into the thermal structure which might develop under different natural ventilation regimes, illustrating that the location of the ventilation openings is key in terms of limiting or promoting the formation of a mixed zone.

REFERENCES

Gladstone, C and Woods, AW, (2000), The natural ventilation of a room with an areal source of heat and two openings, Proc. Roomvent 2000

Linden, P.F. (1999) The fluid mechanics of natural ventilation. *Ann. Rev. Fluid Mech.*, **31**, 210-238.

Linden, P.F., Lane-Serff, G.F. & Smeed, D.A. (1990) Emptying filling boxes: the fluid mechanics of natural ventilation. *J. Fluid Mech.*, **212**, 309-335.

Air Distribution in Rooms, (ROOMVENT 2000)
Editor: H.B. Awbi

VALIDATION OF A NEW INTEGRATED DESIGN TOOL FOR NATURALLY VENTILATED BUILDINGS

EH Mathews*, DC Arndt*

Department of Mechanical Engineering
University of Pretoria, South Africa
(*Also consultants to TEMM International)

ABSTRACT

In many cases natural ventilation is used to ensure an acceptable indoor environment. However it is difficult to design a building for acceptable ventilation rates and indoor comfort without the proper tools or guidelines. The passive building simulation tool *Building Toolbox* (Web-site: http://www.newquick.com) was extended with natural ventilation models for the design of natural ventilated buildings. The simulation tool was verified with actual measurements during three case studies to ensure its integrity and to illustrate its applicability in this field. The predicted indoor air temperatures of all the studies were within 1°C for 80% of the time.

KEYWORDS

Passive buildings, building simulation, natural ventilation, indoor comfort, thermal performance of buildings, building design tool, stack effect, wind effect.

INTRODUCTION

The principle elements, in order to provide an acceptable indoor environment in most applications, are dry bulb temperature, relative humidity, outdoor air ventilation rate and air movement (ASHRAE Fundamentals Handbook 1997 and CIBSE Guide Volume A 1986). The problem however is to predict these conditions in the design of new building facilities to ensure a healthy indoor environment.

Natural ventilation is a very common method to create acceptable conditions in building facilities but also the most difficult to design for. Most available calculation methods do not integrate the performance of the building envelope with ventilation in a dynamic fashion (Rousseau and Mathews 1994). These methods can therefore result in a very poorly designed building.

The best way to go about this will be to use a thermal simulation program, which integrates all the elements having an influence on the indoor air conditions and ventilation rates. With such a simulation tool, it would be possible to investigate the effect, the building envelope and ventilation openings will have on the outdoor air ventilation rate and the indoor air conditions.

It is always important for the user that a simulation tool predicts "real life", for confident use in practice. Therefore a verification study was performed to validate the integrity and accuracy of the models and to illustrate the value and applicability of simulation in the design of building facilities. The study was performed empirically by comparing the predicted values to measurements.

SIMULATION MODELS

An electrical analogy introduced by Van Heerden and Mathews (1996) is used to model the heat transfer processes in the building for the accurate prediction of the thermal performance of the building zones. The model makes use of a six-node electric circuit to simulate the different heat flow paths. The ventilation model makes provision for deliberate flow through purpose provided openings such as open windows and ventilators caused by natural driving forces. These ventilation rates are directly influenced by the pressure distribution on the building envelope and the characteristics of the different openings.

The pressures on the building envelope consist of wind induced pressures and pressures arising from the difference in temperature between the indoor and outdoor air. A ventilation model introduced by Mathews and Rousseau (1996) where the building may be represented by a flow network of pressure nodes coupled by non-linear flow resistances is used to predict the ventilation rates. The air flow rate through an opening can be calculated by employing the energy conservation equation in the following way:

$$Q = C_d A \left[\frac{2}{\rho} \Delta p \right]^{\frac{1}{n}} \qquad (1)$$

where Q is the flow rate (m^3/s), C_d the discharge coefficient, A the area of the opening (m^2), ρ the density of the air (kg/m^3) and Δp the pressure difference across the opening (Pa). Wind pressure is calculated as follows:

$$C_p = \frac{p_{wind} - p_{ref}}{\frac{1}{2}\rho V_h^2} \qquad (2)$$

where p_{ref} is the reference pressure (Pa), which equals the static pressure of the undisturbed airflow and may be taken as equal to the barometric pressure, and V_h the free stream velocity at roof height (m/s). The free stream velocity at roof height is given as:

$$V_h = V_{ref}\left[\frac{h}{h_{ref}}\right]^\alpha \qquad (3)$$

where h is the roof height (m), V_{ref} velocity at a specific reference height (m/s), h_{ref} the reference height (m) and α the velocity profile exponent. The orientation of the building with respect to the wind direction is taken into account by the C_p values (Rousseau and Mathews 1996). The pressure (Pa) due to thermal forces is given as:

$$p_{therm} = 0.0342\frac{yp_{ref}}{T_i T_o}(T_o - T_i) \qquad (4)$$

where y is the height (m) of the opening above ground level and T_i and T_o the indoor and outdoor temperatures ($^\circ$ C) respectively.

VALIDATION OF MODELS

The facilities of the Equine Research Centre of the faculty of Veterinary Science of the University of Pretoria at Onderstepoort were used for this study. The models were verified by comparing the simulated predictions to measured conditions.

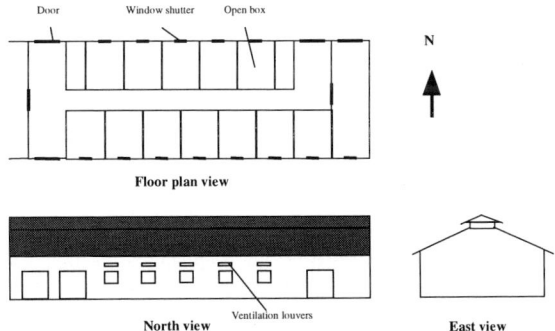

Figure 1: Schematic drawing of the stable

Building description

The stable used, provided mechanical ventilation facilities. The building had a double layer brick construction and a 27° angle pitched corrugated iron roof with a plaster board ceiling. Figure 1 displays a schematic drawing of the stable. The facility has 12 sliding window shutters and ventilation louvers for ventilation purposes.

Procedure

The following three studies were conducted for 24 hour cycles on the same stable to obtain the validity and integrity of all the models in an integrated fashion. The first study was done to verify the thermal performance of the building envelope on its own (building model). The best way to achieve this was to ventilate the building with a known rate of outside air (forced mechanical ventilation). By doing this we simultaneously pressurised the building space to eliminate any natural ventilation or infiltration. For this study all the window shutters were closed.

The second study was performed to obtain the accuracy of the building model integrated with the ventilation model (wind and thermal effects). In the third study we investigated the integration of the building, ventilation and internal heat models. An internal heat load was present in the form of horses. The window shutters were open during study 2 and 3 to allow for natural ventilation.

Measurements

Measurements were taken over a period of one week during June 1998. The measurements included indoor air temperatures, outdoor air temperatures, the radiation of the sun, wind speed and wind direction and mechanical driven air flow rates. From this data a single representative day for each study was extracted to be compared with the 24-hour day predictions provided by the simulation tool. The building was left in its passive state with the window shutters open for 24-hours to stabilise after the first study, which involved mechanical ventilation.

Results

Only the indoor air temperatures were used for evaluating the accuracy of the simulation tool. Figure 2 to Figure 4 display the predicted and actual measured indoor air temperatures for each study, as well as the outdoor air conditions for the applicable period.

It is clear from the results of case study 1 that the dynamic modelling of the building envelope is satisfying with the ventilation rate known. Good comparison between the measured and predicted indoor air temperatures in study 2 and 3 will therefore point to the accurate prediction of natural ventilation.

A moderate average wind of 1.6 m/s from a north west direction was measured during the 24 hours of case study 2. Therefore the dominant force behind the natural ventilation was wind and not temperature. This study's results reflect therefore on the accurate modelling of natural ventilation driven by wind through vertical openings.

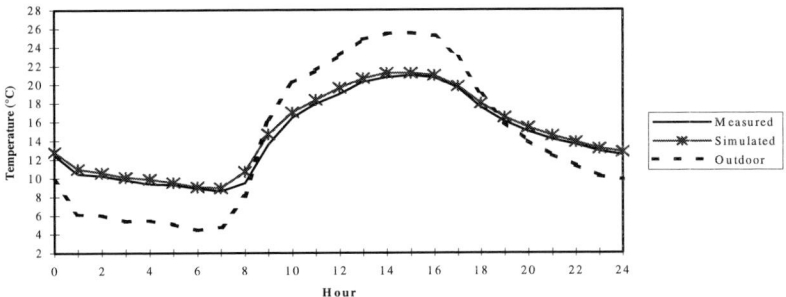

Figure 2: Indoor air, case study 1

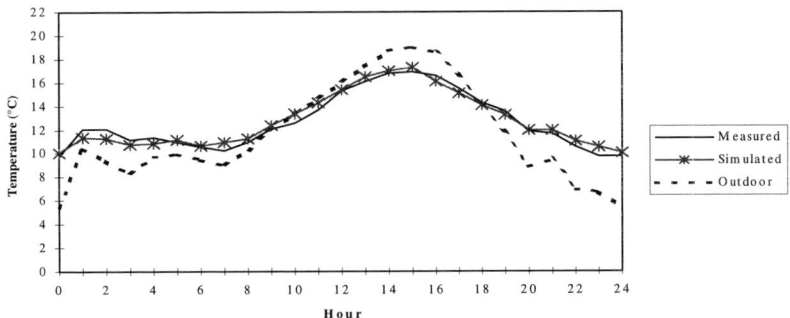

Figure 3: Indoor air, case study 2

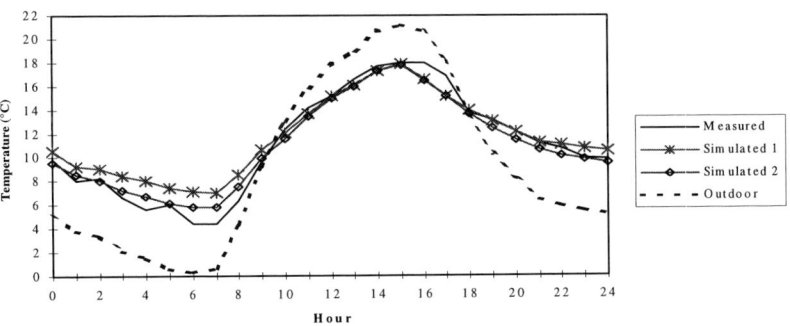

Figure 4: Indoor air, case study 3

The results of study 1, 2 and 3 are summarised in Table 1.

TABLE 1
SUMMARY OF VERIFICATION RESULTS

Study	Case study 1		Case study 2		Case study 3	
Type	Measured	Predicted	Measured	Predicted	Measured	Predicted
T_{ave} (°C)	14.4	14.8	12.6	12.7	11	11.1
Swing (°C)	12.4	12.3	7.2	7.2	13.6	12
Max. error (°C)	1.2		0.8		1.6	
Within 1°C (%)	92		100		80	
Within 2°C (%)	100		100		100	

The results obtained from the first simulation (simulation 1, figure 4) during case study 3 showed large discrepancies during the morning and late night. This result can be attributed to the inaccurate prediction of temperature induced ventilation since windless conditions existed during the time. Each window opening was treated as one opening at one specific stack height (at geometric area centre). This only made provision for one directional flow driven by the ventilation louvers at a higher vertical location.

If an opening's vertical dimension is large enough to create a temperature difference due to the stack effect, two directional flow can occur through the same opening (ASHRAE Fundamentals Handbook 1997). To make provision for this effect, each window opening was treated as two openings (simulation 2) directly above each other, each with its respective relative geometric area centre height. The results from this simulation 2 (figure 4) are a big improvement on the first one with acceptable accuracy.

No natural ventilation air changes were measured for verification purposes. But we know that the thermal performance of the building envelope is correct for a given ventilation rate. Therefore the predicted air changes calculated for natural ventilation must be correct since the comparison between the predicted temperatures is more than satisfying.

CONCLUSIONS

Three case studies were conducted to verify the simulation models' predictions. The simulated indoor air temperatures of all the case studies were within 1°C for 80% of the time. The case studies have shown that the stack effect, the driving force behind temperature induced ventilation, can be better accounted for if all the vertical openings are treated as two separate openings directly above one another. The height difference between the openings is the distance between the geometric centres of the two divided areas.

The verification results reflect on the accurate modelling of the building envelope and ventilation models (wind and temperature induced) in an integrated fashion. The results were further sufficiently accurate for the simulation models to be used with confidence in the design of forced or natural ventilated (through vertical openings) buildings.

REFERENCES

ASHRAE Fundamentals Handbook (1997). *America Society of Heating, Refrigeration and Air-conditioning Engineers*, Inc. Tullie Circle, NE Atlanta, GAZ 30329.

CIBSE Guide Volume A (1986). *Chartered Institution of Building Services Engineers*, Delta House, 222, Balham High Road, London SW12 9BS, Staples Printers.

Rousseau P.G. Mathews E.H. (1994). A new integrated design tool for naturally ventilated buildings part 1: Ventilation model, *Building and Environment*, 29, 4, pp. 461-471.

Rousseau P.G. Mathews E.H. (1996). A new integrated design tool for naturally ventilated buildings, *Energy and Buildings*, 23, pp. 231-236.

Van Heerden E. Mathews E.H. (1996). A new simulation model for passive and low energy, *In Proceedings of the 1996 PLEA Building and Urban Renewal conference*, Louvain-la-Neuve, Belgium.

© 2000 Elsevier Science Ltd. All rights reserved.
Air Distribution in Rooms, (ROOMVENT 2000)
Editor: H.B. Awbi

EFFECT OF THERMAL MASS ON THE AIRFLOW AND VENTILATION IN PASSIVE BUILDING DESIGN

Phil Jones and Kaja Kippenberg

Welsh School of Architecture
Cardiff University, Wales, UK

ABSTRACT

Air may be pre-cooled using thermal mass before it is supplied to an occupied space. One option is to pre-cool the air in a basement space and exhaust the air at high level through stacks. However, the thermal forces that determine the direction of airflow, including heat gains in the occupied space, thermal mass cooling and the external air temperature may counter each other, and result in flow reversal. This paper uses a combination of thermal dynamic and computational fluid dynamics (CFD) airflow modelling to study the ventilation and thermal performance of a hypothetical naturally ventilated 'mass cooled' auditorium.

KEYWORDS

ventilation, airflow direction, CFD, thermal mass, dynamic modelling

INTRODUCTION

Thermal mass and natural ventilation are often combined as part of a passive building design strategy. In particular thermal mass may be used to provide passive cooling in warm weather or when there are high levels of heat gains to the space. In recent years this concept has been applied to a range of building types in the UK. The thermal mass may be either directly coupled to the space, for example, as an exposed concrete ceiling found typically in offices such as Edinburgh Gate, DETR New Practice Case Study (in press), or it may be indirectly coupled, for example, in a separate adjacent space, through which the ventilation supply air is routed, as found in auditorium and theatres, e. g. Queens Building DeMontfort University and Contact Theatre in Manchester , Short et. al. (1998).

The driving forces of natural ventilation are temperature and wind effects. The temperature effects are due to the interaction of outside conditions and heating/ cooling systems, together with the internal heat gains from people, lighting and machines. The intended heat losses through the thermal capacity and transmission characteristics of the building fabric can however lead to negative buoyancy effects, which may counteract the intended ventilation strategy.

This paper examines the balance between the various factors associated with positive and negative thermal buoyancy effects. In practical ventilation design the dynamic nature of wind should also be considered in order to minimise any unwanted interaction with thermal forces. In addition to this, the air movement is influenced by resistance in the system to the airflow due to acoustic attenuators, heating coils, guidance of airflow, etc. However, for the purpose of this study these have been ignored in order to concentrate on the thermal mass effects alone.

The work is based on a series of thermal modelling studies using a dynamic building energy model and CFD airflow modelling. The paper takes an auditorium as an example case, with an underfloor thermal mass plenum, coupled to the supply airflow path.

MODELLING METHOD

In order to predict the thermal mass performance of coupled spaces both the dynamic effects associated with thermal capacity and the spatial variation of temperature and air movement must be considered. Computer models exist for both types of prediction. Previous research, Srebric et. al. (1999), has used coupled thermal dynamic and CFD modelling, with the CFD operated in steady-state mode at discrete hourly time intervals, to predict internal temperatures and air movement for a mechanically ventilated atrium. Time-varying CFD simulations are not yet practical for this type of application because of the short time steps needed (order of seconds) over relatively long periods (order of hours). One of the problems of coupling thermal dynamic and airflow models is the need to iterate between them to achieve a converged solution. The thermal dynamic model provides data for surface temperatures which take account of the time varying thermal mass effects, while the CFD model provides an instantaneous ventilation rate and pattern of air movement. Both models provide values of internal air temperature. In combining the models some ´start-up´ assumptions are needed, of either ventilation rate (if the thermal dynamic model is used first) or surface temperatures (if the CFD model is used first). The procedure is to iterate between the models until a converged solution is obtained. This can be assumed when the ventilation rate does not change for successive CFD runs. A criterion of good convergence is the agreement of the internal air temperatures of both models. Figure 1 illustrates the iterative modelling procedure. It is considered more appropriate to begin with the CFD simulation as there is likely to be more uncertainty about the ventilation rate than the surface temperatures. Initial surface temperatures may be estimated from an average of outside and inside temperatures, for the mass space and occupied spaces respectively.

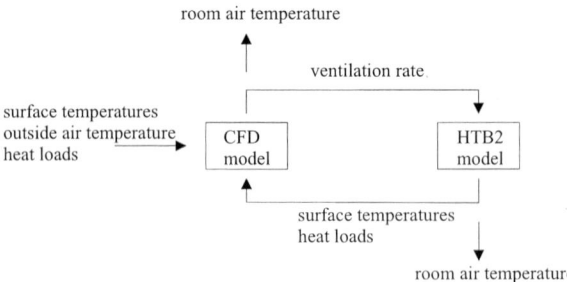

Figure 1: Diagram of iterative solution procedure

The steps in the procedure are then as follows:
1. Select the specific times for running the CFD simulations and the corresponding outside air temperatures.

2. Estimate the initial surface temperatures.
3. Run the CFD model to predict ventilation rate, airflow pattern and internal air temperatures.
4. Use the ventilation rate and direction of air flow to set up thermal dynamic model.
5. Run the dynamic model to predict the internal surface temperatures and air temperatures.
6. Update the surface temperatures in the CFD model and recalculate the ventilation rate, airflow pattern and internal air temperatures.
7. Repeat from step 4 until there is no change in the ventilation rate between successive CFD runs or until the internal air temperatures between the models agree.

The modelling procedure has been applied to a series of 3- and 2-dimensional case studies. The 3-dimensional case studies have been used to investigate the effects of external temperature, amount of exposed thermal mass, size of openings, and cooling and heating situations on the internal temperatures, ventilation rate and air movement for varied internal gains. The 2-dimensional modelling has been used to predict the impact of varying internal heat gains on ventilation rate and direction of air movement for two external temperatures in detail.

Figure 2 presents schematics of the geometry of the CFD model for both, the 2- and 3-dimensional case studies. In this study, the thermal dynamic model HTB2, Alexander (1996), and CFD model DFS-AIR were used, Jones & Waters (1993). In the dynamic model the main space is divided into two horizontal zones in a volume ratio of 4 (lower zone) to 1 (higher zone) in order to account for the temperature gradient as predicted by the CFD model.

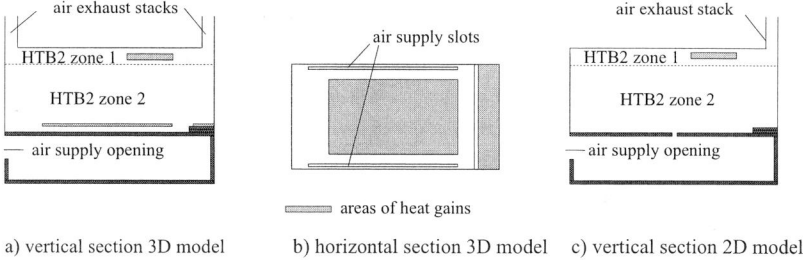

a) vertical section 3D model   b) horizontal section 3D model   c) vertical section 2D model

Figure 2: Schematics of the CFD models

Simulations were conducted for three outside air temperature conditions, corresponding to typical auditorium periods of use. This comprised a summer day and a summer evening in July, and a winter evening in January. The summer day simulation used the peak July outside air temperature, and the

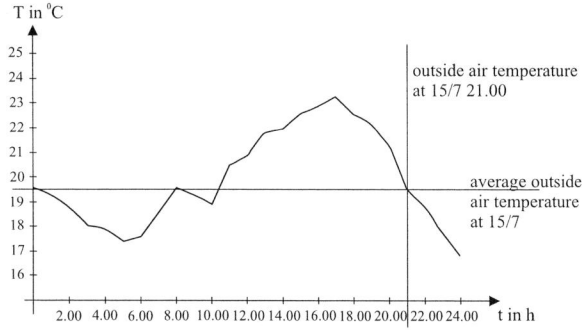

Figure 3: Diurnal outside air temperature profile at 15/7, selected for summer evening simulation

summer and winter evening simulations used the average monthly outside air temperature at 21.00 hours for July and January respectively. As an example, Figure 3 shows the diurnal outside air temperature profile for the day chosen for the summer evening simulation.

As an example, Figure 4 shows the values of internal air temperature and ventilation rate at each stage of the iteration for one summer evening simulation, case 4, described in more detail in figure 5. Convergence is achieved after three iterations.

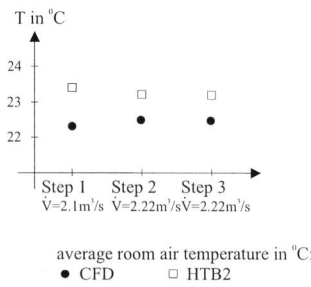

Figure 4: Iterative procedure

3-DIMENSIONAL CASE STUDIES

The modelled auditorium is 8 m high and has a floor plan area of 20 x 10 m^2, the underfloor thermal mass plenum is 4.25 m high. Thus, the volumes of the spaces are 1600m^3 and 850m^3 respectively. With a chosen construction of external cavity brick walls, carpet and ceiling tiles the auditorium has only a minor influence on the thermal mass storage performance of the model.

Figure 5 presents a selection of 3-dimensional case study results, where cases 1-8 are summer simulations and cases 9-12 winter simulations. For each case the results of the dynamic and airflow simulation is summarised. Cases 1-4 show the summer day performance for varied internal gains, ranging from 20kW to 1kW. There is a change of airflow direction from mixed to pure downward movement with decreasing heat gains. The agreement of the internal temperatures in both models is not very good for cases 1-3. This is considered to be caused by the demand of HTB2 for one distinct airflow direction, whereas the CFD simulations yield mixed directions. Cases 5 and 6 show summer evening performance for internal gains of 20kW and 1kW respectively. The direction of air movement is upward in both cases. The increased thermal mass case studies 7 and 8 correspond to cases 1 and 4, displaying summer day and evening performance for 20kW internal gains respectively. For the above mentioned reason the convergence of the internal temperatures is not satisfying for case 7, which shows mixed air flow directions. As examples for the winter simulations cases 9 and 10 (single storey mass plenum) as well as cases 11 and 12 (double storey mass plenum) show the influence of varying the supply and exhaust external opening areas - 50% and 30% of the summertime opening areas - and thermal mass on air movement and heat load. The heating is supplied at the point of entry or the air to the main space. All winter cases have rather high ventilation rates, indicating that even smaller openings could provide sufficient ventilation rates in winter.

Two further studies were carried out on the Case 1 situation, to investigate the effect of wind. Pressure gradients of 1 and 2Pa (corresponding to wind speeds of the order 2-4m/s) were applied across the supply and exhaust, with a relative negative pressure on the exhaust. The results show that the reversal in airflow at zero wind is corrected at low to moderate wind speeds.

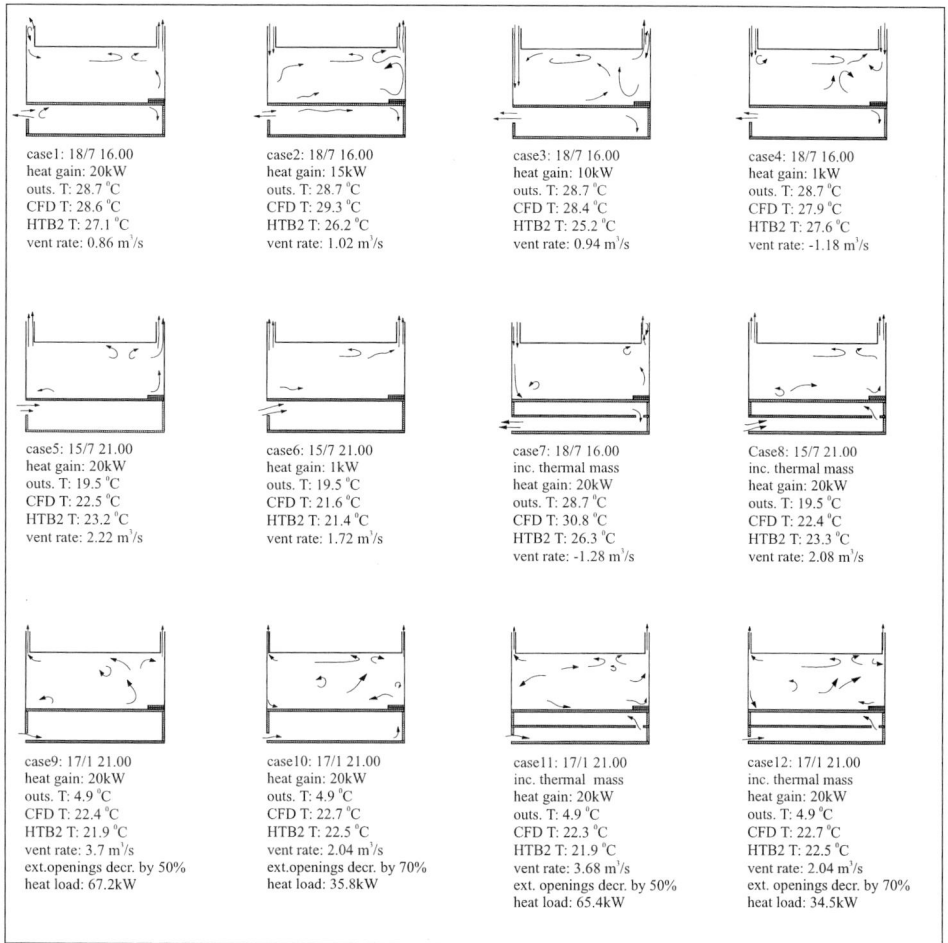

Figure 5: Selection of 3-dimensional case study results

2-DIMENSIONAL CASE STUDIES

Thus the 2-dimensional study was carried out to investigate the direction of airflow for increasing internal heat gain, and for two external air temperature conditions. The 2-dimensional case studies were carried out by CFD simulations - since HTB2 demands the direction of ventilation as an input, it was not used for this study. Figures 6 and 7 present the results of the summer day and summer evening simulations respectively. For the summer day, with a high outside air temperature, the direction of air movement is downward for low internal gains. With increasing internal heat gains, the downward ventilation rate decreases causing a rise in the internal air temperature. The direction of airflow switches from downward to upward for internal gains between 0.25 and 0.3 kW/m^2. At this point the temperature drops considerably due to the pre-cooling of supply air by the thermal mass, rising again slowly with increasing heat gains. For the summer evening case with a lower external temperature the movement of air is upward, independent of the level of heat gains.

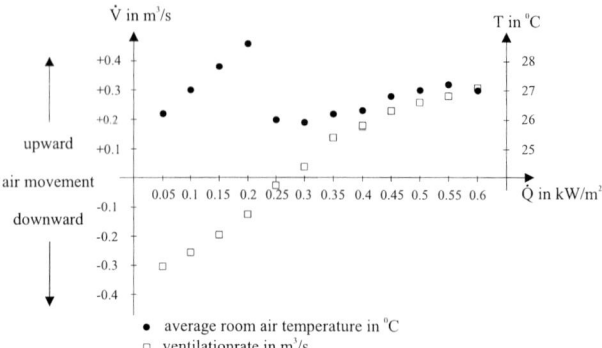

Figure 6: Diagram of ventilation rate and internal temperature versus internal heat gains for 2-dimensional summer day study

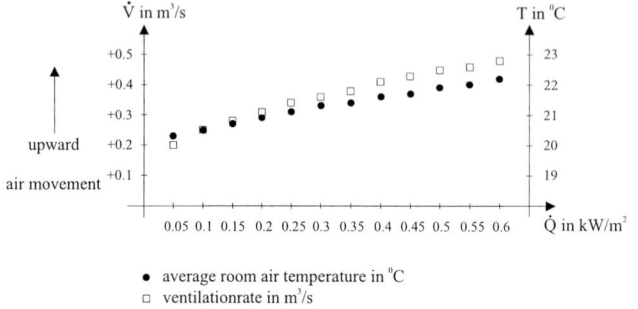

Figure 7: Diagram of ventilation rate and internal temperature versus internal heat gains for 2-dimensional summer evening study

CONCLUSIONS

The main conclusions of the research are as follows:

- For high external air temperatures there is a risk of flow reversal with air entering through the stacks and leaving via the mass plenum. The cooling of the air in the underfloor plenum creates a negative thermal mass effect. This may be countered by high internal heat gains, although these also create higher internal air temperatures, though not necessarily in the occupied zone.
- When it is cooler outside the air flows in the intended direction, with the air supplied to the occupied space about 1°C higher than the external air temperature. Increasing the thermal mass in the plenum reduces the temperature only marginally because of the reduced ventilation due to the increased negative buoyancy.
- In winter, with heat supplied to the incoming air, the flow is always upward. The openings at supply and exhaust should be controlled to avoid excessive ventilation.
- When there is wind, a pressure gradient of 1-2Pa is sufficient to ensure upward airflow for the high external air temperature situation.

- Thermal mass would be best placed above the occupied zone for warm weather cooling, where the thermal buoyancy effects would work together – of course the wind pressure gradient would need to be reversed, with devices to create a relative positive wind pressure at high level.
- The main limitation of the modelling technique is the weak convergence of internal air temperatures for mixed airflow directions. This is an area of further study.

REFERENCES

Alexander, D.K. (1996, rev. Nov. 1997). *A Model for the Thermal Environment of Buildings in Operation,* release 2.0c, Welsh School of Architecture R&D

Edinburgh Gate Building, Harlow, DETR New Practice Case Study, in press.

Jones, P. and Waters, R. (1993). The Practical Application of Indoor Airflow Modeling. *Modeling of Indoor Air Quality and Exposure, ASTM STP 1205*, Niren L. Nagda, Ed., American Society for Testing and Materials, Philadelphia, USA, 173-181.

Short, A. et. al. (1998). Design of Naturally Ventilated Theatre Spaces. *5th European Conference Solar Energy in Architecture and Urban Planning*, 69-73.

Srebric, J. et. al. (1999). A Computer Design Tool for Non-uniform Indoor Thermal Environmental Problems, *The 3rd International Symposium on Heating, Ventilation and Air Conditioning* **2,** 635-658

Air Distribution in Rooms, (ROOMVENT 2000)
Editor: H.B. Awbi

THE USE OF SOLAR AIR COLLECTORS FOR ROOM VENTILATION: A STUDY USING TWO NUMERICAL APPROACHES

A. Moret Rodrigues[1]; A. Canha da Piedade[1]; H. B. Awbi[2]

[1]*Instituto Superior Técnico/ICIST - Av. Rovisco Pais, 1049-001 Lisbon, PORTUGAL*

[2]*Department of Construction Management & Engineering, The University of Reading, UK*

ABSTRACT

Solar energy air-collectors installed on the sun-oriented building façades can be used for improving natural ventilation of adjacent rooms. The basis of the physical process is an unbalanced buoyancy force arising from the temperature difference between ambient and the air inside the room. Although difficult to control due to the variability of the climatic conditions, these devices can be used as means of reducing the need for conventional energy to provide indoor air conditions within acceptable limits required by health and comfort considerations. In this paper, two numerical approaches of different complexity are applied to study the airflow produced by a solar-air collector in a room. These are a simplified model, based on the integral equations of motion and Bernoulli theorem, which provides an insight into the main features of the flow, and a more complex computational fluid dynamics (CFD) model called VORTEX. The simplified model takes into consideration the heat and momentum transfer within the collector and room and the CFD code used has been specifically developed for studying the air movement in rooms, which solves the full elliptic Navier-Stokes equations together with the equations for the k-ε turbulence model. Results for a parametric study of a room with an air collector using both models are presented and discussed.

KEYWORDS: Solar air-collector, Integral flow equations, CFD model, Natural ventilation, Room air movement

INTRODUCTION

This paper analyses the behaviour of solar-air collectors used as passive devices contributing to comfort and health requirements in rooms. Specifically, the solar-air collector design used in this study is mainly for providing room air ventilation for the following two purposes:

- *Winter time:* to reduce the pollutant concentration of room air by preheating incoming air thereby reducing the heat losses associated with the ingress of outdoor cold air.

- *Summer time:* to increase the ventilation rates to the room through open windows by buoyancy action during the day.

This is an important application area in buildings of high occupancy density that normally require a mechanical ventilation system but where costs associated with such mechanical devices (capital and maintenance) are important barriers to their practical implementation. That is the case of school buildings (classrooms) in some countries (Moret Rodrigues *et al.*, 1996).

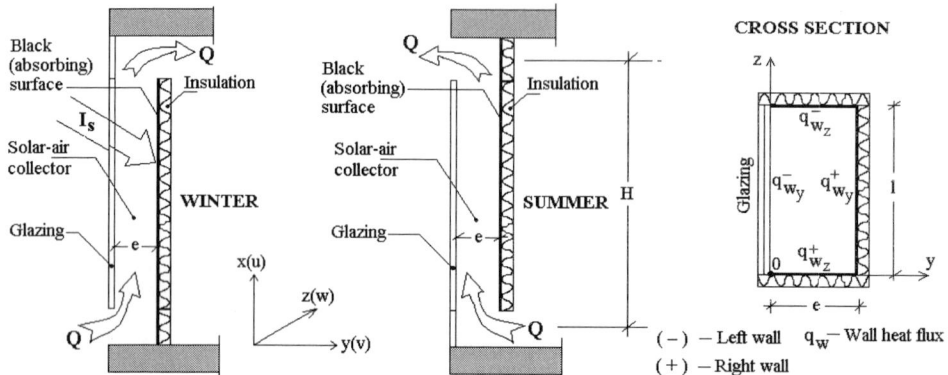

Fig. 1: Solar-air collector

The solar-air collector is basically composed of four walls forming a gap, with different functions: The external wall is a glazing, transparent to the short wave radiation (solar irradiation, I_s, Fig. 1); the internal walls are energy absorbing, by the use of a black covering sheet (absorber) on the exposed surface and made of a light insulation material. A low thermal inertia and a high level of insulation result in higher temperatures, larger heat gain to the air within the gap, and a consequent increase in the collector performance.

In winter, the outdoor air entering the collector through a lower opening is heated by contact with the absorber and the density differences produce the driving force ("stack effect") that pulls the air upwards into the room through an opening at the top of the channel (Fig. 1, winter case). Complementarily to the collector, a chimney is used in the opposite side as a path for extracting the room air, with or without the help of a fire placed at its base.

In summer, during the day, the system induces air circulation in the reverse direction to the winter case, as shown in Fig. 1 (summer case), through inlets placed on the opposite façade. At night, the system works as an extract with the fresh air entering the room through openings thus removing the heat accumulated in the enclosed space.

A simple method based on integral equations and Bernoulli theorem is developed and applied to the problem of the channel coupled with the room and chimney, in winter, and with the room and windows, in summer. This will be useful in understanding the basic parameters that govern the flow and in helping designers in their study of similar problems, mainly in the first stages of the design process. In parallel, CFD simulations are presented and results compared. This will be useful in supporting the simple model and in providing an insight of the flow within the room.

INTEGRAL MODEL

The basic model applies the steady-state, isotropic 3-D turbulent boundary layer equations of momentum, mass continuity and energy, respectively given by

$$\frac{\partial(\rho u u)}{\partial x} + \frac{\partial(\rho v u)}{\partial y} + \frac{\partial(\rho w u)}{\partial z} = \frac{\partial}{\partial y}\left(\mu_e \frac{\partial u}{\partial y}\right) + \frac{\partial}{\partial z}\left(\mu_e \frac{\partial u}{\partial z}\right) - \frac{\partial p}{\partial x} - \rho g \tag{1}$$

$$\frac{\partial(\rho u)}{\partial x} + \frac{\partial(\rho v)}{\partial y} + \frac{\partial(\rho w)}{\partial z} = 0 \tag{2}$$

$$\frac{\partial(\rho u T)}{\partial x} + \frac{\partial(\rho v T)}{\partial y} + \frac{\partial(\rho w T)}{\partial z} = \frac{\partial}{\partial y}\left(\Gamma_e \frac{\partial T}{\partial y}\right) + \frac{\partial}{\partial z}\left(\Gamma_e \frac{\partial T}{\partial z}\right) \tag{3}$$

to a nearly constant density flow in a rectangular duct (x - diffusion terms dropped for being of minor importance) – with height H and perimeter $\chi = 2(1+e)$ – subjected to a uniform heat flux (Neumann conditions) $q_w = \left[1\left(q_{w_y}^- + q_{w_y}^+\right) + e\left(q_{w_z}^- + q_{w_z}^+\right)\right]/\chi$ – on the solid boundaries. In these equations the effective viscosity is $\mu_e = \mu + \mu_t$ and the effective diffusion coefficient is $\Gamma_e = \mu_e / \sigma_e$; μ_t being the turbulent viscosity and σ_e the effective Prandtl number. Equations 1~3 are firstly integrated over the channel area $S = 1 \times e$. At the wall boundaries, the no-slip conditions $u = v = w = 0$ apply and, for the derivatives, the apparent shear stresses $\tau_{app_i} = \mu_e \partial u / \partial x_i$ and apparent heat fluxes $q_{app_i} = -c_p \Gamma_e \partial T / \partial x_i$ are equal to the wall values, τ_{w_i} and q_{w_i} respectively $(i = 1,2,3)$. From the integration of Eqn. 2 it is found that the mass flux $\dot{m} = \int_S \rho u \, dS$ is constant over the channel height. Using averaged quantities,

$u_s = \int_S u \, dS / S = Q_s / S$, $p_s = \int_S p \, dS / S$, $\rho_s = \int_S \rho \, dS / S$, Q_s being the air flow rate, and following the notation of Fig. 2, a second integration of Eqn. 1 over the channel height gives, for the winter case:

$$P_{s,2} = P_{s,1} - \frac{2\rho_e Q^2}{D_{co} S_{co}^2} \int_1^2 c_{fco,x} dx - g \int_1^2 \rho_{s,x} dx \tag{4}$$

where the fully developed flow and Boussinesq approximations (a constant axial velocity and density variations only affecting the buoyancy term, respectively) are used and the concepts of hydraulic diameter $D \equiv 4 \times S / \chi = 2 \times (1 \times e)/(1+e)$ and shear stress coefficient $c_f = 2\tau_w /(\rho_s u_s^2)$, are also introduced. Using, for the density variation, the relationship $\rho_{s,x} = \rho_e[1 - \beta(T_{s,x} - T_e)]$, where $\beta = 1/T_e$ is the coefficient of thermal expansion and $T_s = \int_S \rho u T \, dS / \dot{m}$ is the channel bulk temperature, a second integration of Eqn. 3 gives $T_{s,x} = T_{s,1} + \psi x$, with $\psi = 2q_w(1+e)/(\rho_e Q c_p)$. Thus, the buoyancy integral term of Eqn. 4 results in

$$\int_1^2 \rho_{s,x} dx = \rho_e H_{co}\left(1 - \beta \psi \frac{H_{co}}{2}\right) \tag{5}$$

For turbulent flow, $c_f = 0.079 \, Re_D^{-1/4}$ $\left(2 \times 10^3 < Re_D < 2 \times 10^4\right)$ (Bejan,1993) is adopted, with the Reynolds number given by $Re_D = QD/(Sv)$, and the correspondent integral of Eqn. 4 can easily be solved. At the inlet, the flow contraction gives rise to a minor loss that can be expressed as:

$$P_{ref,e} - P_{s,1} = \frac{\rho_e}{2}\left(\frac{Q}{C_{dco} A_{co}}\right)^2 \tag{6}$$

which is obtained by applying the Bernoulli theorem. Here, $p_{ref,e}$ is the external reference pressure, A is the inlet area of the collector and C_d is the discharge coefficient. In turn, at the outlet, and since the streamlines are approximately parallel, the static pressure across the jet can be assumed as:

$$P_{s,2} = P_3 \tag{7}$$

and assuming still room air,

$$P_3 = P_{ref,i} - \rho_i g H_{co} \tag{8}$$

where $p_{ref,i}$ is the internal reference pressure. The chimney will be simply assumed as an air extract system. This means that the chimney only contributes to the flow in terms of energy loss. At a more advanced stage of the study a fire at the base of the chimney can also be included in winter to add an additional heat source and assist the ventilation induced by the collector. Applying Eqn. 4 to this case gives (Fig. 2):

$$P_{s,5} = P_{s,4} - \frac{2\rho_i Q^2 c_{fch} H_{ch}}{D_{ch} S_{ch}^2} - \rho_i g H_{ch} \tag{9}$$

similarly,

$$P_{s,5} = P_6; \qquad P_6 = P_{ref,e} - \rho_e g H_{ch}; \qquad P_{ref,i} - P_{s,4} = \frac{\rho_i}{2}\left(\frac{Q}{C_{dch} A_{ch}}\right)^2 \tag{10}$$

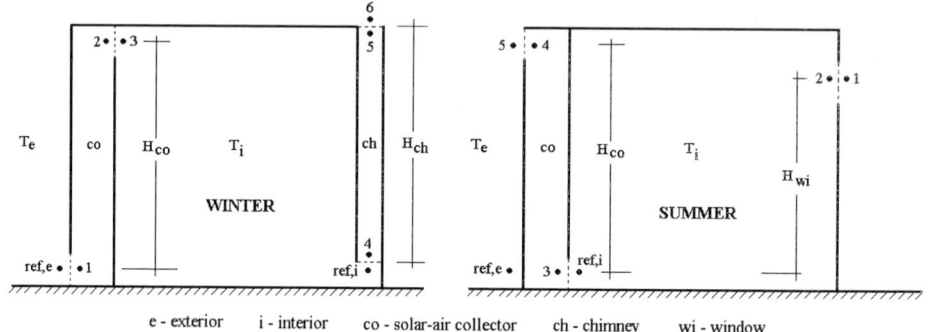

e - exterior   i - interior   co - solar-air collector   ch - chimney   wi - window

Fig. 2: Integral model

Combining Eqns. 4~10, with $\rho_i \approx \rho_e$, the flow rate in the winter case is given by the implicit equation

$$Q = \sqrt{\frac{g\beta\psi H_{co}^2}{\left(\frac{1}{C_{dco} A_{co}}\right)^2 + \left(\frac{1}{C_{dch} A_{ch}}\right)^2 + \frac{4c_{fco} H_{co}}{S_{co}^2 D_{co}} + \frac{4c_{fch} H_{ch}}{S_{ch}^2 D_{ch}}}} \tag{11}$$

The construction of the model for the summer case is identical to the winter case, with one difference and that is the chimney does not take part in the ventilation process but is replaced by one or more open windows (j=1,..n) at a height H_{wi}. In this case, it is easy to conclude that the relevant equation becomes:

$$Q = \sqrt{\frac{g\beta\psi H_{co}^2}{\left(\frac{1}{C_{co} A_{co}}\right)^2 + \left(\frac{1}{\sum_j C_{dwi_j} A_{wi_j}}\right)^2 + \frac{4c_{fco} H_{co}}{S_{co}^2 D_{co}}}} \tag{12}$$

CFD MODEL

The CFD study was carried out using a 3-D computer code - VORTEX (Awbi,1996) - which numerically solve the differential equations that govern air movement – continuity, momentum and thermal energy equations - for the combined flow in the room and other systems considered here. The program allows for the turbulent nature of the flows by solving two additional equations for the kinetic energy and energy dissipation rate which are the base of the so-called $\kappa-\varepsilon$ turbulence model. The numerical scheme uses a staggered 3-D Cartesian system where equations are discretized using the Finite Volume Method (FVM) and solved by the well known SIMPLE algorithm. To enhance the stability of the solution process under-relaxation techniques are applied to all the equations. This program has been extensively validated in solving various types of ventilation flows.

RESULTS AND DISCUSSION

The following figures present results for the cases tested, either from the integral model (INT), based on equations (11) and (12), or from the CFD model, based on the numerical solution of the flow equations. The simulations were carried out for a room size of 5m x 5m x3.2m, collector and chimney heights of H_{co}=2.9m and H_{ch}=2.7m, respectively, collector and chimney sections of S_{co}=0.2m x 0.7m and S_{ch}=0.3m x 0.3m, and opening areas of A_{co}=0.3m x 0.7m for the collector and A_{ch}=0.3m x 0.3 m for the chimney.

In Fig. 3, the variation of the air flow rate with the heat flux on the absorbing collector walls (set equal for both, i.e., $q_{w_y}^+ = q_{w_z}^- = q_{w_z}^+$) is shown for both models in the two working situations considered, i.e. summer (S) and winter (W). For the summer situation, the results shown correspond to the case of two open windows of area A_{wi}=0.70m x 0.70m each, positioned at a height H_{wi}= 2.0m on the wall opposite the collector. The corresponding curves have similar trends, with a progressive increase of airflow rate with the flux intensity, and the agreement achieved is very satisfactory considering the difference in complexity between the two models. It is important to note that despite the fact that the heating power has been set the same for the winter and summer cases, the corresponding air flow rates are different. The lower values in winter are due to the higher resistance offered to the flow by the chimney device when compared to the minor losses produced by the windows in summer.

In Figs. 4~7 some of the CFD plots are shown for both winter and summer situations. Figs. 4 and 5 are referred to the same winter simulation case: T_e=15 °C and $q_{w_y}^+ = q_{w_z}^- = q_{w_z}^+$=100 W/m². In Fig. 4 the contour lines of the air temperature for a longitudinal plane in the middle of the room are plotted. The higher values observed at the collector outlet are followed by a progressively decreasing temperature with distance from the collector, on one hand, and the distance from the ceiling, on the other. This stratification is consistent with the corresponding flow patterns shown in Fig. 5.

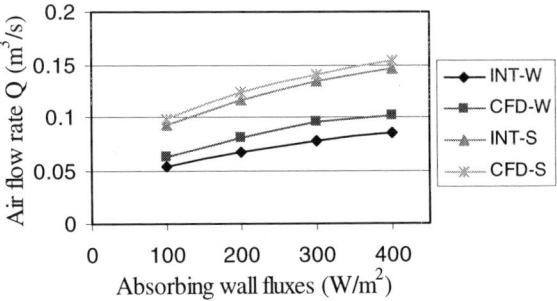

Fig. 3: Air flow rates

286

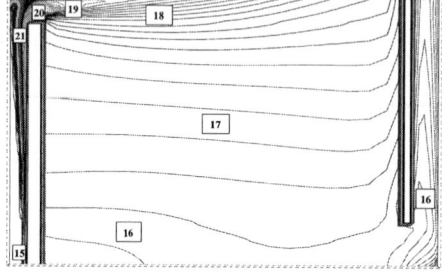

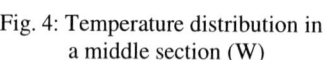

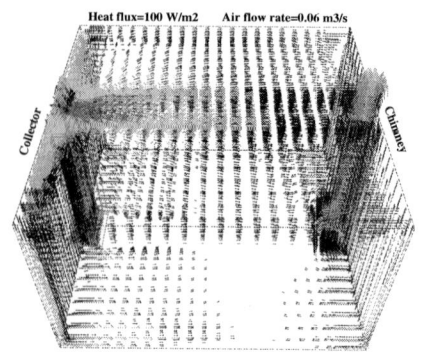

Fig. 4: Temperature distribution in   Fig. 5: Velocity distribution (W)
a middle section (W)

In the winter case, a typical situation is a warm jet leaving the channel at the top and moving towards the direction of the chimney close to the ceiling (Fig. 5). Conversely, when the flow enters through the windows (summer case), the lower velocities at inlet causes the jet to drop soon after entrance, thus resulting in a more efficient air distribution in the occupied zone (Figs. 6 and 7).

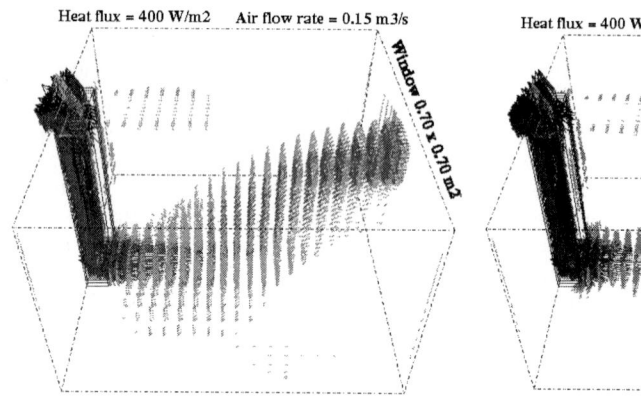

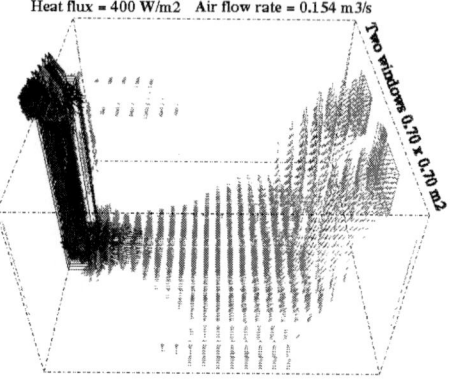

Fig. 6: Air velocities - single window (S)   Fig. 7: Air velocities - two windows (S)

To predict the distribution of CO_2 in the room the same numerical procedure described above for velocity and temperature was used but where an additional transport equation for the concentration was solved. Sixteen CO_2 sources representing seated people were uniformly distributed through the room at height of 1.35m above the floor, with each producing 4.7×10^{-3} l/s (total production $P = 75.2 \times 10^{-3}$ l/s), and an average outdoor air concentration of $\overline{C}_e = 0.4 \times 10^{-3}$ (400 ppm) was considered. A graphical representation of the concentration at this height is shown as horizontal planes in Figs. 8 and 9 and also for the sections C1 and C2 in Fig. 10. The mean concentration in the occupied zone $\overline{C}_{oc}$ (from floor to 1.8 height) and in the whole space $\overline{C}_{sp}$ are lower than the recommended limits for comfort and health (Portuguese recommendation for a period of one hour: 1.2×10^{-3}) although locally they are exceeded (Fig.10). Despite the fact that the two flow rates are nearly the same, the two windows give higher ventilation effectiveness as shown from the results of Figs. 8 ~ 10. The mean velocities for the occupied zone $\overline{V}_{oc}$ and for the whole space $\overline{V}_{sp}$ are also below the comfort limit (ASHRAE Standard 55-1992 recommends 0.25 m/s for summer).

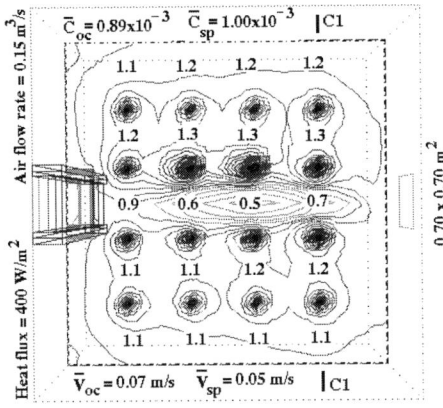

Fig. 8: CO_2 contours $(x10^{-3})$ (single window)

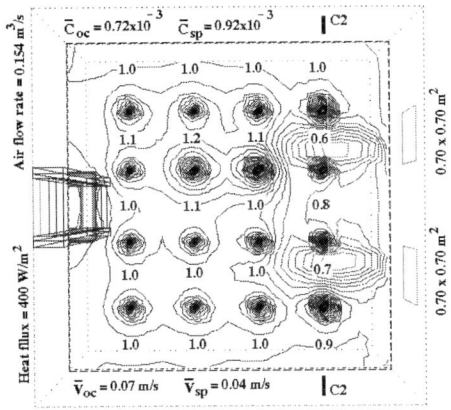

Fig. 9: CO_2 contours $(x10^{-3})$ (two windows)

Prolonging the use of the simplified method to the evaluation of the mean concentration, it is useful to compare the CFD value with that obtained through the simple equation, valid for the stationary state,

$$\left(\overline{C}_{sp} - \overline{C}_e\right) \times Q = P$$

With $P = 75.2 \times 10^{-3}$ l/s, $C_e = 0.4 \times 10^{-3}$, $Q = 0.147 \times 10^3$ l/s (INT model value – Fig. 3 - for a heat flux on the absorbing walls of 400 W/m^2) one easily conclude that the concentration value that there results, that is $\overline{C}_{sp} = 0.91 \times 10^{-3}$, agrees very well with that predicted by the CFD model (Figs. 8 and 9).

CONCLUSIONS

A simple model for room ventilation based on integral equations was developed and compared with a more complex CFD model. The agreement of results from the two models is very encouraging for so that the simple model could be used as first step in the design of these solar ventilators. The CFD simulation was undertaken in order to obtain detailed information about the flow. These results also seem feasible for the conditions being simulated in this study.

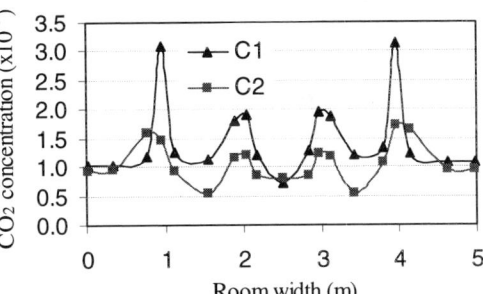

Fig. 10: CO_2 distribution (C1 and C2 in Figs. 8 and 9 respectively)

REFERENCES

ASHRAE Standard 55-1992 (1992). *Thermal Environment Conditions for Human Occupancy*, ASHRAE, Atlanta, USA.

Awbi, H.B. (1996). VORTEX-2, A Computer code for Air Flow, Heat Transfer and Concentration in Enclosures, version 2.1, U.K.

Bejan, Adrian (1985). *Heat Transfer*, John Wiley & Sons, Inc., New York.

Moret Rodrigues, A., A. Canha da Piedade, S. Domingos and S. Valente Pereira (1996). Performance of solar-air collectors for classroom ventilation: Mathematical models versus experimental results. In: Proceedings of the 4th European Conference *Solar Energy in Architecture and Urban Planning*, pp. 472-475, Berlin.

Air Distribution in Rooms, (ROOMVENT 2000)
Editor: H.B. Awbi

SOME EXAMPLES OF SOLUTION MULTIPLICITY IN NATURAL VENTILATION

Yuguo Li[1], Angelo Delsante[1], Zhengdong Chen[1] and Mats Sandberg[2]

[1] Thermal and Fluids Engineering, CSIRO Building, Construction and Engineering
PO Box 56, Highett, Victoria 3190, Australia
[2] Laboratory of Ventilation and Air Quality, Department of Built Environment
The Royal Institute of Technology, Box 88, S-801 02, Gävle, Sweden

ABSTRACT

This paper shows that under certain conditions, multiple solutions for the flow rate exist in a natural ventilation system, induced by the non-linear interaction between buoyancy and wind forces. Under certain physical simplifications, the system is governed in steady state by a non-linear algebraic equation or a system of equations. Three examples are given here: a single-zone building with two openings, a channel with two end openings, and a two-zone building with two openings in each zone. Analytical and numerical solutions are presented. It is shown that in all three cases the flow rate exhibits hysteresis. These results have significant implications for multi-zone modelling of natural ventilation and smoke spread in buildings. An experimental investigation using a small-scale water model in a water tunnel confirms that two steady-state solutions exist for a single-zone building.

KEYWORDS

Natural ventilation, multi-zone simulation, analytical methods, winds, thermal buoyancy.

INTRODUCTION

Our recent studies of natural ventilation (Li & Delsante 1998; in press) showed that there are multiple solutions of the equations governing natural ventilation when the flow rate is determined by combined buoyancy and wind forces. Some similar results were also obtained by Nitta (1996). Linden (1999) reported analytical and experimental studies of opposing wind and buoyancy forces, and also noted the phenomenon of hysteresis experimentally.

The presence of multiple solutions is significant for understanding natural ventilation control and the interpretation of results obtained from mathematical modelling. This paper provides a summary of the key findings of our work using three examples (see Figure 1). The analytical results are supported by some new experimental results from a water tunnel.

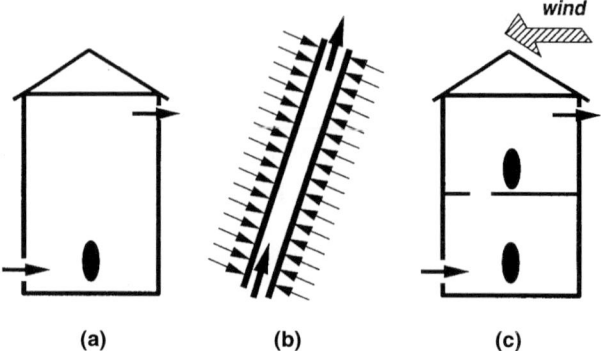

Figure 1: Three example enclosures with opposing wind. (a): single-zone building; (b): an inclined channel; (c): a two-zone building. In the channel, heat is applied to the two surfaces. In the buildings, the heat source is indicated by the solid ellipse.

A SINGLE–ZONE BUILDING

The building has two openings at different vertical levels on opposite walls, as shown in Figure 1(a). The heights of the two openings are relatively small, and the areas of the top and bottom opening are A_t and A_b respectively. There is an indoor source of heat, E. The wind force can assist or oppose the thermal buoyancy force. We assume that the indoor air is fully mixed, i.e. the air temperature is uniform, recognising that in practice the air will be stratified in some circumstances. However the fully mixed assumption is used here because it leads to relatively simple equations which nonetheless display interesting behaviour (which can then be checked by experiment), and because this assumption is used in the simpler treatments of natural ventilation.

Steady-state Solutions

Three air change parameters (α, β and γ) are used to characterise respectively the effects of the thermal buoyancy force, the envelope heat loss and the wind force:

$$\alpha = (C_d A^*)^{2/3} (Bh)^{1/3} \tag{1}$$

$$\beta = \frac{\Sigma U_j A_j}{3\rho c_p} \tag{2}$$

$$\gamma = \frac{1}{\sqrt{3}} (C_d A^*) \sqrt{2\Delta P_w} \tag{3}$$

where $A^* = A_t A_b / \sqrt{A_t^2 + A_b^2}$ is the effective area; C_d is the opening discharge coefficient (assumed to be the same for both openings); B is the buoyancy flux, given by $B = Eg/\rho c_p T_o$, where g is the acceleration of gravity, ρ and c_p are the density and heat capacity of the air respectively, and T_0 is the outdoor air temperature; ΔP_w is the wind pressure difference between the two openings; and U_j and A_j

are the U-values and area of wall element j respectively. For assisting winds, the ventilation flow rate, q, is determined by (see Li & Delsante, in press):

$$q^3 + 3\beta q^2 - 3\gamma^2 q - 2\alpha^3 - 9\gamma^2\beta = 0 \qquad (4)$$

For opposing winds, the flow rate is determined by:

$$q^3 + 3\beta q^2 = |-3\gamma^2 q + 2\alpha^3 - 9\gamma^2\beta| \qquad (5)$$

The solution for opposing winds is complex. The behaviour of the flow rate as a function of α and γ reveals that for $\beta \neq 0$ the general form of the solutions must be as shown in Figure 2, where for clarity upward flow is indicated as positive and downward flow as negative. The flow exhibits hysteresis. It can also be shown that a solution on the A–B curve is not stable (see Li & Delsante, in press).

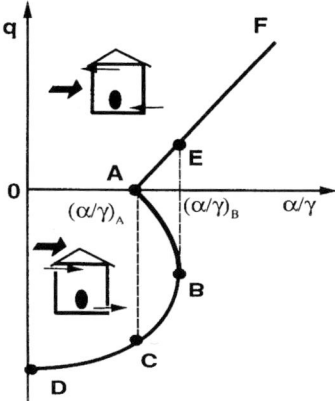

Figure 2: Analytical sketch of the opposing wind situations

A CHANNEL

Examples of airflows in a channel (Figure 1(b)) can be found in road tunnels, solar chimneys and the air gaps behind photovoltaic panels (Sandberg & Moshfegh 1998). Again, an ideal channel is considered with constant total heat flux E applied to the channel walls. We assume a linear temperature profile in the channel for both upward and downward flows. Equations for the flow rate can be derived which are the same as those for the one-zone building, but with minor changes in some parameters. Firstly, the effect of friction is now included in the definition of the effective area, which becomes

$$C_d A^* = \left(\frac{1}{(C_{d1}A_1)^2} + \frac{1}{(C_{d2}A_2)^2} + \frac{K_f}{(A_c)^2} \right)^{-\frac{1}{2}} \qquad (6)$$

where C_{d1} and C_{d2} are the discharge coefficients of the two end openings with opening areas A_1 and A_2, A_c is the cross-sectional area of the channel, and K_f is the channel friction loss coefficient, which is proportional to the length of the channel. Secondly, because of the linear temperature profile, the definition of the thermal air change parameter is also slightly changed:

292

$$\alpha = (C_d A^*)^{2/3} (\tfrac{1}{2} Bh)^{1/3}$$

(7)

With these changes, the hysteresis behaviour is the same as for the one-zone building.

A TWO–ZONE BUILDING

The two-zone building has two openings at different vertical levels in each zone, as shown in Figure 1(c). Again, we assume that the height of the vertical openings is relatively small. There is an indoor source of heat, E_i, in each zone i. Again we assume that the indoor air is fully mixed, i.e. the air temperature is uniform in each zone. Additionally, we assume that the partitions between the zones are perfectly insulated, i.e. heat loss only takes place through the external walls, roof and floor.

The flow is governed by a quartic equation (see Delsante and Li, 1999, for details). Figures 3(a)–3(d) show schematically some possible solution behaviours for the upward and downward flows as a function of a buoyancy force parameter α_D. Figure 3(a) shows the simplest case: the slope at $q = 0$ is positive for upward flow and negative for downward flow, and the flows are monotonic. Thus, for any value of α_D there is a unique solution for the flow. Note that there is no analogue for this behaviour in the single-zone case, where the slope is always positive at $q = 0$. Figure 3(b) shows similar hysteresis behaviour to that found for the single-zone case: for certain values of α_D there are three possible solutions – one upward flow (S1), and two downward flows (S2 and S3). Figure 3(c) also shows three possible solutions for certain values of α_D, but with two upward flows and one downward flow. Finally, Figure 3(d) shows five possible solutions – two upward and three downward. This set of behaviours is not exhaustive.

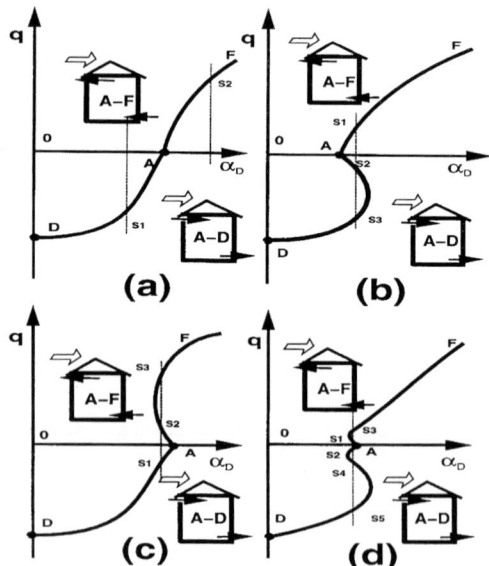

Figure 3: Some possible behaviours of the flow rate q for opposing winds in a two-zone building.

Figure 4 shows numerical solutions of the flow rate for a particular set of parameters. It is interesting to note that for small values of β/γ the solution is of the form shown in Figure 3(d), but as β/γ increases the form becomes that of Figure 3(c).

Thus, the hysteresis behaviour found in the single-zone building also exists in the two-zone building, and is indeed considerably more complex. Again we find that for opposing winds and a given set of heat gains, wind speeds and U-values, there appear to be several solutions for the flow. The full spectrum of solution behaviours has not yet been fully analysed. Furthermore, the stability of the multiple solutions has also not yet been resolved. A two-zone building presents a two-dimensional non-linear system. Further analysis of the system dynamics and identification of stable solutions will be carried out in the near future.

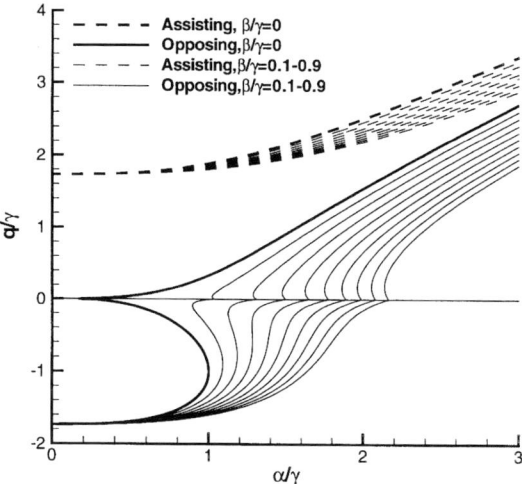

Figure 4: Scaled flow rate in a two-zone building as a function of a scaled buoyancy parameter.

SOME EXPERIMENTAL RESULTS

A small-scale water tunnel laboratory model of the single-zone building with opposing wind was constructed in order to test whether multiple solutions exist in reality. The details of this work are reported in Andersen et al., (2000). One of the key results was to experimentally demonstrate the hysteresis predicted for the single-zone building. With a constant opposing wind the buoyancy flux was increased from zero (point D on Fig. 2, but note that in the experiment there was no heat flow through external surfaces, i.e. point A lies at the origin), and steady-state flows established at various points along the D-B curve. In particular a downward flow rate was established for a buoyancy flux lying between C and B. When the buoyancy flux was increased beyond E, the flow direction reversed and became buoyancy-dominated (i.e. upward) and continued to increase as the buoyancy flux was increased to point F. Crucially, as the buoyancy flux was then progressively decreased from F, an upward flow was established for a buoyancy flux lying between E and A, which was the same flux as was used to establish a downward flow between C and B. This demonstrates the hysteresis. The experimental results are shown in Fig. 5.

294

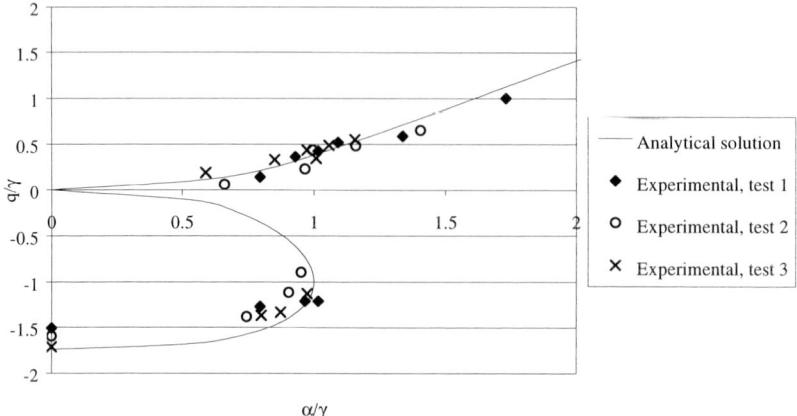

Figure 5: Experimental results from a water tunnel laboratory model of the one-zone building. Tests 1-3 refer to different wind speeds. Note that for a given wind speed, a buoyancy flux between 0.5 and 1.0 resulted in two possible flows, depending on whether the flux was increased from a low value or decreased from a high value. (Taken from Andersen *et al.*, 2000).

CONCLUSIONS

Natural ventilation flows can be quite complex even when calculated for very simple systems. For three examples it can be shown analytically that, for a certain range of parameters, multiple solutions exist in natural ventilation systems when the wind opposes thermal buoyancy. Even though the calculations assume fully mixed spaces, the existence of multiple solutions has been confirmed in small-scale laboratory tests where stratification was observed with buoyancy-dominated flows. More analytical and experimental analysis will be reported in the near future.

REFERENCES

Andersen A. A., Bjerre M., Chen Z. D., Heiselberg P. and Li Y. (2000). Experiment Modelling of Wind Opposed Buoyancy-driven Building Ventilation. Submitted to 21st AIVC Conf., The Netherlands, 26–29 Sept. 2000.

Delsante, A. and Li, Y. (1999). Natural Ventilation Induced by Combined Wind and Thermal Forces in a Two-Zone Building. Proc. IEA Annex 35 Hybvent Forum, Sydney, Australia, 28 Sept. 1999, 181-190.

Li Y. and Delsante A. (1998). On Natural Ventilation of a Building with Two Openings. Proc. 19th AIVC Conf., 'Ventilation Technologies in Urban Areas", Oslo, Norway, 28–20 Sept. 1998, 189–1996.

Li Y. and Delsante A. (in press). Natural Ventilation Induced by Combined Wind and Thermal Forces. *Building and Environment*.

Linden, P. F. (1999). The Fluid Mechanics of Natural Ventilation. *Ann. Rev. Fluid Mech.*, **31**, 201-238.

Nitta, K. (1996). Study on the Variety of Theoretical Solutions of Ventilation Network, Journal of Architecture, Planning, and Environmental Engineering (Transactions of AIJ), **480**, 31-38

Sandberg M. and Moshfegh B. (1998). Ventilation Solar Roof Air Flow and Heat Transfer Investigation. *Renewable and Sustainable Energy Reviews* **15**, 287–292.

Air Distribution in Rooms, (ROOMVENT 2000)
Editor: H.B. Awbi

THE POTENTIAL OF LARGE EDDY SIMULATION TECHNIQUES FOR MODELLING INDOOR AIR FLOWS

Shia-Hui Peng and Lars Davidson

Department of Thermo and Fluid Dynamics, Chalmers University of Technology

SE-412 96 Gothenburg, SWEDEN

ABSTRACT

Large eddy simulations (LES) were performed for flows relevant to or incurred in ventilation air motions with and without thermal effects. The emphasis was placed on the discussion of the possibility and potential of LES for modelling indoor air flows. Some prospective views were given on the capability and implementation of the LES approach. LES is a potential tool for providing detailed and accurate resolution of turbulent flow and heat transfer in analyses of indoor environment and building energy performance. In the foreseeable future, however, it will remain mainly as a complementary approach to statistical modelling and experiments owing to the limitation of computational costs.

KEYWORDS

Large eddy simulation (LES), Subgrid scale (SGS) model, Numerical simulation, Indoor air flow

INTRODUCTION

In Reynolds-Averaged Navier-Stokes (RANS) computations, the turbulence fluctuations are ruled out by a time-averaging process, leaving only the mean flow to be simulated. The statistical effect of turbulence on the mean flow is then fully modelled using the eddy viscosity or Reynolds stress models. Large eddy simulation is a modelling technique that accounts for turbulent flows, where the large, energy-carrying structures are computed directly and the effect of the small scales below the computational grid size is modelled through a subgrid-scale stress term. The large scales in LES are often distinguished from the small scales by using a spatial filtering process, which determines to what degree the large-scale eddy motions in a turbulent flow are explicitly resolved. The width of the spatial filter, which is usually related to the numerical mesh size, determines the largest size of the subgrid scales. In Direct Numerical Simulation (DNS), all the essential turbulence fluctuations are resolved from the largest to the smallest eddies, which consequently requires a sufficiently fine computational grid and is limited to low Reynolds number turbulent flows. LES is an intermediate approach between DNS and RANS. Still, the numerical grid resolution used in LES should be fine enough to resolve high wavenumber flow instabilities leading to large-scale eddy motions, whereupon the dynamically important eddies can be accurately computed and the effect of subgrid scale eddies on the large ones tends preferably to be isotropic and universal. As a result, the modelling of the effect of subgrid scales may be more amenable to successful simulations and may yield significantly smaller errors in the large-scale resolution than are incurred in RANS.

LES has been the subject of extensive studies in the past two decades. Nevertheless, studies of LES for modelling room air flows are rare. This work discusses the potential of LES techniques applied to indoor air flow modelling and briefly describes approaches to using LES. Along with some LES results, the capabilities of LES in producing flow and heat transfer quantities necessary for indoor environment analysis are discussed. It is not the intention of the present work to make an overall review of current LES approaches, but rather to outline the possibilities and potential of LES for indoor air flows in the authors' opinions. Comprehensive discussions on LES can instead be found in, e.g., Rogallo & Moin (1984) and Lesieur & Métais (1996).

MODELLING FORMULATION

LES is based on a spatial filtering operation. The filtered variable, denoted by an overbar, is defined as

$$\bar{f}(\mathbf{x}, t) = \int_D f(\mathbf{x} - \mathbf{x}', t) G(\mathbf{x}') d\mathbf{x}', \tag{1}$$

where D is the entire flow domain and G is a low-pass filter function that smoothes the fluctuations on subgrid scales. Applying this filtering operation to the flow equations yields the governing equations for the large-scale motions:

$$\frac{\partial \bar{u}_i}{\partial x_i} = 0, \quad \frac{\partial \bar{u}_i}{\partial t} + \frac{\partial}{\partial x_j} (\bar{u}_i \bar{u}_j) = -\frac{1}{\rho} \frac{\partial \bar{p}}{\partial x_i} + \nu \frac{\partial^2 \bar{u}_i}{\partial x_j \partial x_j} - \frac{\partial \tau_{ij}}{\partial x_j}, \tag{2}$$

and the filtered thermal energy equation is

$$\frac{\partial \bar{\theta}}{\partial t} + \frac{\partial}{\partial x_j} (\bar{u}_j \bar{\theta}) = \alpha \frac{\partial^2 \bar{\theta}}{\partial x_j \partial x_j} - \frac{\partial h_j}{\partial x_j}. \tag{3}$$

In Eqs (2) and (3), the SGS stress tensor, $\tau_{ij} = \overline{u_i u_j} - \bar{u}_i \bar{u}_j$, and heat fluxes, $h_j = \overline{u_j \theta} - \bar{u}_j \bar{\theta}$, are the unknowns and must be modelled. Most SGS models assume a principal alignment between the SGS stress or flux and the resolved deformation tensor, $\bar{S}_{ij} = \frac{1}{2} \left(\frac{\partial \bar{u}_i}{\partial x_j} + \frac{\partial \bar{u}_j}{\partial x_i} \right)$, or the thermal gradient, $\frac{\partial \bar{\theta}}{\partial x_j}$, using the SGS eddy viscosity, ν_t. This gives

$$\tau_{ij} = -2\nu_t \bar{S}_{ij} + \frac{\delta_{ij}}{3} \tau_{kk}, \quad h_j = -\alpha_t \frac{\partial \bar{\theta}}{\partial x_j} = -\frac{\nu_t}{Pr_t} \frac{\partial \bar{\theta}}{\partial x_j}, \tag{4}$$

where α_t is the SGS eddy diffusivity and Pr_t is the SGS Prandtl number. The modelling then becomes a task to formulate the SGS viscosity/diffusivity, which has usually been cast into a relation of $\nu_t \propto L_{sgs}^2 / \mathcal{T}_{sgs}$. The SGS length scale, L_{sgs}, is often related to the mesh size, Δ, and the time scale, $\mathcal{T}_{sgs}$, is constructed from the resolved large scales. In the famous Smagorinsky model, $\mathcal{T}_{sgs} = 1/|\bar{S}|$ with $|\bar{S}| = \sqrt{2\bar{S}_{ij}\bar{S}_{ij}}$. This suggests that $\nu_t = C\Delta^2 / \mathcal{T} = C\Delta^2 |\bar{S}|$ and thus $\alpha_t = C_t \Delta^2 / \mathcal{T} = C\Delta^2 |\bar{S}| / Pr_t$. There are also other models that derive the SGS time scale from the SGS kinetic energy, k_{sgs}, through $\mathcal{T}_{sgs} \propto L_{sgs} / \sqrt{k_{sgs}}$. This approach consequently involves an additional equation for k_{sgs} (thus termed the one-equation SGS model). The modelled k_{sgs} equation can be written as

$$\frac{\partial k_{sgs}}{\partial t} + \frac{\partial}{\partial x_j} (\bar{u}_j k_{sgs}) = \frac{\partial}{\partial x_j} \left(C_{k0} \Delta k_{sgs}^{\frac{1}{2}} \frac{\partial k_{sgs}}{\partial x_j} \right) - 2 C_{k1} \Delta k_{sgs}^{\frac{1}{2}} \bar{S}_{ij} \bar{S}_{ij} - C_{k2} \frac{k_{sgs}^{\frac{3}{2}}}{\Delta}. \tag{5}$$

Details on the implementation of such models can be found in, e.g., Davidson (1997). Many other SGS models have also been developed, see e.g. Peng (1998) for a literature trace and Rogallo & Moin (1984) and Lesieur & Métais (1996) for reviews.

For thermal flows, such as in rooms with displacement ventilation, where the thermal buoyancy plays a significant role in turbulence evolution, the assumption of local subgrid-scale equilibrium (SGS shear production *vs* SGS dissipation) inherent in the Smagorinsky model can be argued to include the buoyant production. This will in turn yield a $\mathcal{T}_{sgs}$ formulation associated explicitly with a buoyancy-related term. To avoid giving non-real solutions and to account for a part of the energy backscatter caused by thermal stratification, Peng and Davidson (1998) proposed an improved buoyant SGS model as

$$\nu_t = C\frac{\Delta^2}{\mathcal{T}} = C\Delta^2\left(|\bar{S}| - \frac{g\beta}{Pr_t|\bar{S}|}\frac{\partial\bar{\theta}}{\partial x_j}\delta_{2j}\right), \quad \alpha_t = \frac{C}{Pr_t}\Delta^2\left(|\bar{S}| - \frac{g\beta}{Pr_t|\bar{S}|}\frac{\partial\bar{\theta}}{\partial x_j}\delta_{2j}\right). \tag{6}$$

It is obvious that, for isothermal flows, this model returns to the conventional Smagorinsky model.

The SGS models often include one or more model coefficients (C's and Pr_t in the above expressions). A wide range of values has been employed in applications: the coefficients are flow-dependent. Moreover, some SGS models hold an incorrect near-wall asymptotic property, which would play a negative role in handling near-wall turbulence and would be a denominator of accuracy in simulations of, e.g., convective heat transfer over building enclosure surfaces. To overcome these and other drawbacks preserved in these base models, the Germano-Lilly dynamic procedure (Germano *et al.*, 1991, Lilly, 1992) has commonly been used to determine the model coefficient as a function of time and space. It is here termed a *dynamic procedure* rather than a dynamic model because the Germano approach provides the most powerful means to date and makes the LES more universal than ever for dealing with different flows, while the employment of this procedure needs to be based on a large eddy simulation performed using the aforementioned or other SGS base models. The philosophy of this procedure is that the small scales in the resolved eddies are assumed to be the most active part interacting with the subgrid scales. An additional test filtering operation is thus applied to the resolved scales to obtain the effect of small-scale eddies therein and, subsequently, is used to compute the SGS model coefficient. Note that the model coefficient determined by the dynamic procedure may in many cases be ill-conditioned with singularity problems and/or by having locally and temporally unphysical values. A number of methods have been proposed to remedy such problems. One usual way is to make a spatial averaging on the coefficient in the directions of flow homogeneity. For flows with no homogeneous direction, e.g. many room ventilation flows, local averaging may be used.

The numerical methods used are of significant importance for a successful LES. The governing equations can be discretised using any methods as used in RANS computations, e.g. finite difference, finite volume and finite element methods. Very commonly in LES for fundamental flows, the spectral method has been used owing to its high numerical accuracy. This method is much less flexible for geometrically complex flows, however, whereas finite volume (or finite difference) methods have so far been the most popular approaches. To resolve large-scale turbulence fluctuations, the grid resolution needs to be sufficiently fine, particularly near the wall, to account for the flow structures inherent in large-scale motions, e.g. streaks in the near-wall viscous sublayer. High-order numerical schemes, together with fine grid resolution, are always preferable in order to reach accurate LES, but are not necessary as such as are adopted in DNS. This may be justified by the numerical error associated with the schemes, which should be substantially smaller than the SGS contribution. Moreover, SGS stresses are dissipative in nature, and special care should then be taken in using dissipative, high-order upwind-biased schemes that may significantly affect the spectrum of the resolved structures. Instead, non-dissipative schemes such as central differencing are mostly used. In applications for relatively complex flows, second-order schemes have commonly been employed in both spatial and temporal discretisations.

APPLICATION EXAMPLES

Indoor air flows are often characterized by turbulence and, in many cases, in combination with a number of local flow features, such as mixing and separating, wall jet, thermal plume and stratification, vortex shedding, local laminar and transitional flows, and local relaminarisation. The modelling approach in most room air flow simulations has so far been two-equation eddy viscosity models (some have used Reynolds stress models). To deal with the local flow features as such, the RANS approach may be an awkward method. By contrast, LES is able to yield more affluent and accurate information on turbulence than RANS by resolving the flow structures and their evolution in time and space. It is thus expected to be more powerful and universal than RANS models in handling different flow phenomena such as those existing in room air motions.

Three examples are presented below to demonstrate the capability of LES. It should be pointed out that most ventilation flows are much more complex than the flows considered. Nonetheless, these flows are viewed as essential ingredients extracted from complex ventilation flows and cast in relatively simple, isolated configurations for more efficient and comprehensive studies. These flow features, alone or together, often play an essential role in the creation and control of indoor air flows, such as thermal buoyancy and stratification, natural convection boundary layer, mixing and separation. They are thus relevant to practical room air motions created by mixing, displacement and natural ventilation systems.

Figure 1 shows a LES based on Eq. (6) for the Reayleigh-Bénard (RB) convection flow ($Ra = 3.8 \times 10^5$) arising between two 6×6 planes with a heated floor and a cooled ceiling. This configuration is somewhat similar to a displacement ventilation system with cooled ceiling panels. The air motions are different, however, because this flow is driven purely by thermal buoyancy. The mean flow is statistically stationary but has very active turbulent fluctuations, which, particularly near the floor, are often necessary factors in evaluating thermal comfort. It is obvious that two-equation RANS models are incapable of handling this type of flow in which the turbulence fluctuations are ruled out by time averaging. Fig. 1 a) and b) show the time-averaged (denoted $\langle \rangle$) temperature and its fluctuation in comparison with the DNS data. Fig. 1 c) and d) illustrate the instantaneous flow field and the contour of vertical velocity, $\bar{v}$ (at the mid-depth xy-plane, only the left side is shown). Fig. 1 d) shows that the cold air (dashed lines) descends and the hot air (solid lines) rises alternatingly, causing large-scale coherent structures.

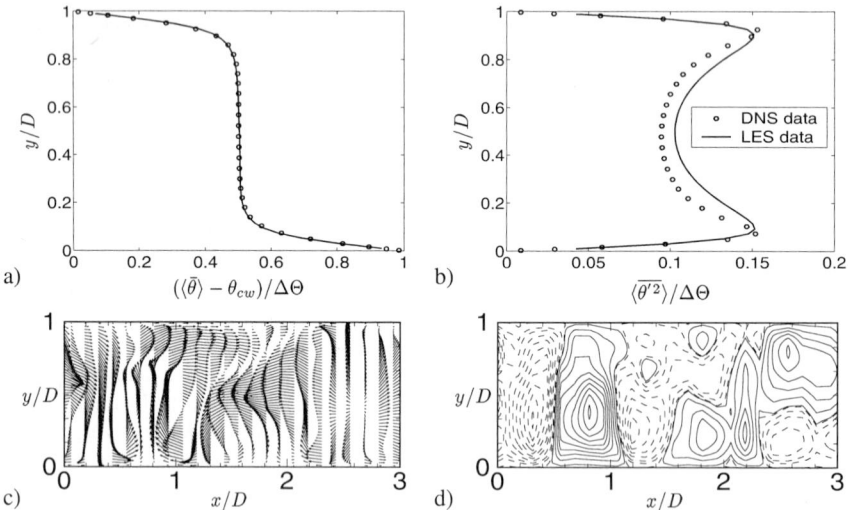

FIGURE 1: Rayleigh-Bénard convection. a). Mean temperature θ_{cw} is the temperature on the cooled ceiling and $\Delta\Theta$ the temperature difference from the heated floor); b). Temperature fluctuation; c). Instantaneous flow field; d). Contour of instantaneous vertical velocity: solid lines positive and dashed lines negative.

Figure 2 shows a LES for a buoyancy-driven flow arising in a closed square cavity ($H/W = 1$) with two opposite sidewalls differentially heated. This configuration is similar to rooms with one facet exposed to the outdoor cold/hot climate, causing natural convection flow along the room enclosure surfaces. The buoyant velocity scale, $U_0 = \sqrt{g\beta\Delta\Theta H}$, is about unity and $Ra = 1.58 \times 10^9$. The results shown are in rather good agreement with the experiment (Tian, 1997) .

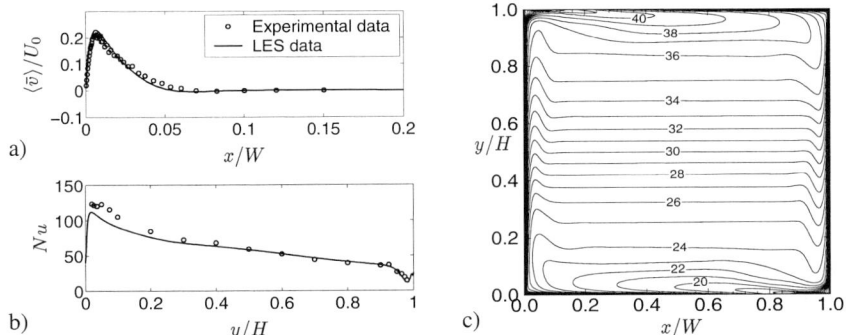

FIGURE 2: Buoyant cavity flow. a). Mean vertical velocity at $y/H = 0.5$; b). Heat transfer (Nusselt number, Nu) along the hot wall; c). Contour of mean temperature.

Figure 3 shows some LES results for a mixing ventilation flow in a room with dimensions of length $(L) \times$ height $(H) \times$ depth $(D) = 9 \times 3 \times 3$, in which the air is supplied through a slot under the ceiling and is exhausted above the floor on the opposite wall. The Reynolds number based on the inflow parameters is about 5000. A dynamic one-equation SGS model (Davidson, 1997) was used with wall functions (not along the ceiling). The LES does not show an improvement for the mean flow predictions as compared with the standard $k - \varepsilon$ model and is even worse in the near-floor region (Fig. 3 a)). This is probably due to the use of the log-law wall function, which is unsuitable for instantaneous flows, and to the relatively coarse grid used. The resolution for the turbulence quantities is reasonable, however, as shown in Fig. 3 b). Since the SGS model accounts only for the small scale eddies, the SGS eddy viscosity should be much smaller than the entire turbulent eddy viscosity obtained from a RANS model. Figure 3 c) shows that the SGS ν_t is two orders lower in magnitude than the RANS ν_t in the near-ceiling wall jet.

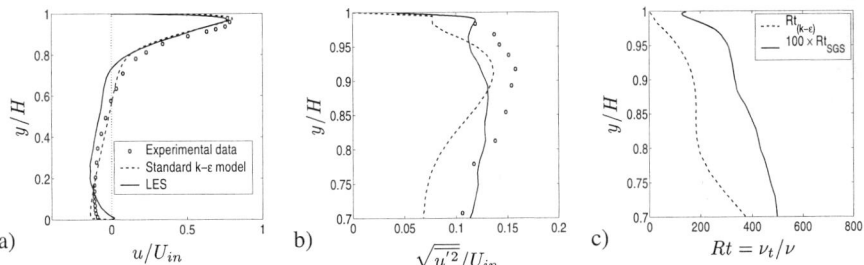

FIGURE 3: Mixing ventilation flow. a). Mean velocity at $x = H$; b). Turbulence intensity in the wall jet at $x = H$; c). Ratio of SGS/RANS eddy viscosity to molecular viscosity, $Rt = \nu_t/\nu$, in the wall jet at $x = H$.

SOME PROSPECTIVE VIEWS

Over the past twenty years the LES technique has matured considerably with significant advancements in developing new SGS models and numerical schemes. The rapid development of digital computers has also greatly facilitated the implementation of LES to complex flows of industrial and/or academic interest. At present, however, LES is viewed mainly as a complementary approach to RANS modelling and experiments and is used to improve the confidence of RANS modelling for indoor air flows.

Air movement in rooms is usually characterized by low velocities and a low turbulence level for comfort indoor environments. This makes LES possible and affordable for relatively complex room air flows. The general advantages of LES over RANS have been mentioned by many other authors elsewhere. For LES of indoor air flows, specifically, the following aspects are outlined here. **a)** Dynamic LES modelling should be more universal than any existing RANS models and usually provide more accurate predictions for indoor air flows created by and controlled with different ventilation systems; **b)** LES can provide relatively accurate measures, as a complementary approach to experiments, to develop, calibrate and validate RANS modelling of indoor air flows where no measured data are available and running measurements is too costly or may even be impossible; **c)** LES can provide quantities that are needed to evaluate indoor air quality and thermal comfort, while RANS models are incapable of providing them or inaccurate in their rendering of these quantities. For example, although little attention has been paid to thermal fluctuation for its influence on thermal comfort, it seems plausible (at least to the present authors) to argue that this quantity plays a significant role in thermal sensation of room occupants. Moreover, in studies of indoor air quality involving dispersion and deposition of indoor airborne pollutants, LES allows a much more accurate accounting of particle-turbulence interactions than does RANS upon a more powerful parameterisation of the turbulence.

In CFD simulations, higher accuracy and more profound physical insight are usually compromised by excessively costly computational resources. The same is ture for LES. While it is viewed as a potential and viable technique, LES is not at this time sufficiently ready for practical indoor air flow analyses, mainly owing to restrictions in computer power. Furthermore, more robust approaches are needed to improve the computational efficiency and accuracy. For indoor air flow analyses, special effort needs to be made on two primary aspects:

- Near-wall treatment. LES resolves instantaneous large-scale motions. The use of the log-law wall function, derived from local-equilibrium assumption for mean flows, is questionable in LES. To relax the requirement on near-wall grid refinement (in all three directions to capture structures such as low- and high-speed streaks), proper wall treatment is needed for efficient LES.

- Inflow boundary conditions. LES requires transient boundary conditions for instantaneous, filtered flow and thermal quantities. For ventilation flows, where the air is often supplied through diffusers, the inflow condition in LES must rely heavily on empirical formulation. Models/approaches to describe the inflow condition are needed in practical LES applications.

REFERENCES

Davidson, L. (1997). Large eddy simulation: A dynamic one-equation subgrid model for three-dimensional recirculating flow. In *11th Int. Symp. on Turbulent Shear Flow*, Vol. 3, pp. 26.1–26.6.

Germano, M., Piomelli, U., Moin, P., and Cabot, W. (1991). A dynamic subgrid-scale eddy viscosity model. *Phys. Fluids A* **3**, 1760–1765.

Lesieur, M. and Métais, O. (1996). New trend in large-eddy simulations of turbulence. *Annu. Rev. Fluid. Mech.* **28**, 45–82.

Lilly, D. (1992). A proposed modification of the Germano subgrid-scale closure method. *Phys. Fluids A* **4**, 633–635.

Peng, S.-H. (1998). *Modelling of Turbulent Flow and Heat Transfer for Building Ventilation*. PhD Thesis, Department of Thermo and Fluid Dynamics, Chalmers University of Technology, Gothenburg.

Peng, S.-H. and Davidson, L. (1998). Comparison of subgrid-scale models in LES for turbulent convection flow with heat transfer. In *EF Turbulent Heat Transfer-2*, Vol. II, pp.5.24–5.35.

Rogallo, R. S. and Moin, P. (1984). Numerical simulation of turbulent flows. *Ann. Revs. Fluid Mech.* **16**, 99–137.

Tian, Y. S. (1997). *Low Turbulence Natural Convection in an Air Filled Sqaure Cavity*. PhD Thesis, South Bank University, London, UK.

Air Distribution in Rooms, (ROOMVENT 2000)
Editor: H.B. Awbi

COMPARISON OF DIFFERENT SUBGRID TURBULENCE MODELS AND BOUNDARY CONDITIONS FOR LARGE-EDDY-SIMULATIONS OF ROOM AIR FLOWS

D. Müller[1], L. Davidson[2]

[1]Lehrstuhl für Wärmeübertragung und Klimatechnik
(Institute of Heat Transfer and Air-Conditioning)
Aachen University of Technology (RWTH Aachen), Germany

[2]Department of Thermo and Fluid Dynamics
Chalmers University of Technology, Sweden

ABSTRACT

The calculation results applying two equation turbulence models for the time average Navier-Stokes equations suffer sometimes from large difference to measurements of complex room air flows. Using less modelling assumptions the Large-Eddy-Simulation (LES) becomes practical to calculate room air flows because of more powerful CFD codes and computers. This paper focus on different LES-Models and its boundary conditions.

KEYWORDS

Large-Eddy-Simulation, subgrid models, isothermal room air flows, boundary conditions.

INTRODUCTION

Room air flows are complex flows with low Reynolds numbers. Different flow regimes as attached jets, jets and free convection flows exist in ventilated room. This combination of different flow types causes usually problems using time averaged Navier-Stokes equations with standard two equation turbulence models. The main reasons for this difficulties are non isotrope effects as thermal stratification and the relaminarization from turbulent to laminar flow fields. LES resolves most of the turbulent motion and needs only model assumptions for the less energy containing small scale turbulence.

The first LES model was proposed by Smagorinsky (1963). This model uses local mixing length formulation for the subgridscale stress by applying the local grid space as length scale. Some limitation of the Smagorinsky model were overcame by the first dynamic subgrid model proposed by Germano et al (1991). But this model has numerical stability problems and the remedy is to average the dynamic part of the model in some homogeneous flow direction(s) or to introduce some artificial clipping. Thus, this type of models does not seem to be applicable to real three-dimensional flows as room air flows without introducing ad hoc user modifications.

In the present study two different one-equation subgrid model are compared which eliminate the need of this type of user modifications. One model was presented by Davidson (1997) (DAV) and the idea of this model is to include all local dynamic information into the source terms of the transport equation of K_{SGS}. The second model was presented by W. Kim and S. Menon (1996) (KIM). This model is fully localised and uses no averaging in space or time for the dynamic coefficient C_τ. Similarity of the resolved and non resolved stress tensor is used to calculate the dynamic coefficient. This assumption leads to a more stable formulation for C_τ and gives the possibility to use the local value of C_τ in the momentum equation. The models will be tested for recirculating air flow in a ventilated room.

FILTERING PROCEDURE

LES uses a space averaging technique instead of the usually performed time averaging of the Navier-Stokes equations. All flow quantities have to be averaged in space by a local filter function G. Here, the a simple top-hat filter function G will be used.

$$u_i = \overline{u}_i + u_i' \;\; ; \;\; \overline{u} = \int_\Omega u(\vec{x}^*) \, G(\vec{x} - \vec{x}^*) d\vec{x}^* \;\; ; \;\; G(\vec{x}, \vec{x}^*) = \begin{cases} \frac{1}{\Delta} & \forall |\vec{x} - \vec{x}^*| \leq \frac{\Delta}{2} \\ 0 & \end{cases} \tag{1}$$

Applying this filter function G to the continuity and momentum equations gives the following equations:

$$\frac{\partial \overline{u}_i}{\partial x_i} = 0 \;\; ; \;\; \frac{\partial \overline{u}_i}{\partial t} + \frac{\partial \overline{u}_i \overline{u}_j}{\partial x_j} = -\frac{1}{\rho} \frac{\partial \overline{p}}{\partial x_i} + \nu \frac{\partial^2 \overline{u}_i}{\partial x_j \partial x_j} - \frac{\partial \tau_{ij}}{\partial x_j} \;\; ; \;\; \tau_{ij} = \overline{u_i u_j} - \overline{u}_i \overline{u}_j \tag{2}$$

Similar to the time averaging procedure we get an unknown turbulent stress tensor τ_{ij} that relates the unknown correlation of the non filtered quantities to the known filtered ones. This tensor is modelled with a Boussinesq approximation assuming a proportionality of the stress tensor and the local gradient of the filtered quantities.

$$\tau_{ij} - \frac{1}{3} \delta_{ij} \tau_{KK} = -2\nu_t \left[\overline{S}_{ij} - \frac{1}{3} \delta_{ij} \overline{S}_{KK} \right] \;\; ; \;\; \overline{S}_{ij} = \frac{1}{2} \left(\frac{\partial \overline{u}_i}{\partial x_j} + \frac{\partial \overline{u}_j}{\partial x_i} \right) \tag{3}$$

The isotropic part of the stress tensor is combined with the pressure term leading to the following formulation of the momentum equations for LES:

$$\frac{\partial \overline{u}_i}{\partial t} + \frac{\partial \overline{u}_i \overline{u}_j}{\partial x_j} = -\frac{1}{\rho} \frac{\partial \overline{P}}{\partial x_i} + \nu \frac{\partial^2 \overline{u}_i}{\partial x_j \partial x_j} + \frac{\partial}{\partial x_j} \left[\nu_t \left(\frac{\partial \overline{u}_i}{\partial x_j} + \frac{\partial \overline{u}_j}{\partial x_i} \right) \right] \tag{4}$$

LARGE EDDY SUBGRID MODELS

Smagorinsky model

The first LES subgridscale model uses a mixing length hypotheses based on the local grid dimensions. The turbulent viscosity is calculated as a product of this length scale and a turbulent velocity scale based on the filtered-field deformation tensor $\overline{S}_{ij}$.

$$l = \overline{\Delta} = \left(\Delta x \, \Delta y \, \Delta z \right)^{1/3} \;\; ; \;\; \nu_t = \left(C_S \, \overline{\Delta} \right)^2 |\overline{S}| \;\; ; \;\; |\overline{S}| = \left(2 \, \overline{S}_{ij} \overline{S}_{ij} \right)^{1/2} \tag{5}$$

The constant C_S is set to 0.18 but this value does not give reasonable results for all flow types. This model is only able to predict dissipation of resolved turbulent energy. Backscattering of turbulent energy from small scales to larger scales is not possible, because the dissipation at subgrid level is always negativ.

$$\varepsilon_{SGS} = -\left(C_S \overline{\Delta}\right)^2 |\overline{S}|^3 \leq 0 \tag{6}$$

Germano model

To calculate dynamically the value of C_S Germano (1991) applied a second test filter to the filtered Navier-Stokes equation using a larger filter length $\widehat{\Delta}$, here $\widehat{\Delta} = 2\,\overline{\Delta}$.

$$\frac{\partial \widehat{\overline{u}}_i}{\partial t} + \frac{\partial \widehat{\overline{u}}_i \widehat{\overline{u}}_j}{\partial x_j} = -\frac{1}{\rho}\frac{\partial \widehat{\overline{P}}}{\partial x_i} + \nu\left(\frac{\partial \widehat{\overline{u}}_i}{\partial x_j \partial x_j}\right) - \frac{\partial T_{ij}}{\partial x_j} \; ; \; T_{ij} = \widehat{\overline{u_i u_j}} - \widehat{\overline{u}}_i \widehat{\overline{u}}_j \tag{7}$$

The resolved stress between the two different filter level L_{ij} is determined by the resolved velocity field and expresses the difference between the test and the grid filtered stress tensor.

$$L_{ij} = \widehat{\overline{u}_i \overline{u}_j} - \widehat{\overline{u}}_i \widehat{\overline{u}}_j = T_{ij} - \widehat{\tau}_{ij} \tag{8}$$

Using the Boussinesq approximation for both filter levels gives

$$\tau_{ij} - \frac{\delta_{ij}}{3}\tau_{KK} = -2\,C_S\,\overline{\Delta}^2 |\overline{S}|\overline{S}_{ij} = -2\,C_S\,\beta_{ij} \; ; \; T_{ij} - \frac{\delta_{ij}}{3}T_{KK} = -2\,C_S\,\widehat{\Delta}^2 \left|\widehat{\overline{S}}\right|\widehat{\overline{S}}_{ij} = -2\,C_S\,\alpha_{ij} \tag{9}$$

and the constant C_S can be estimated using the non isotropic part of L_{ij}.

$$L_{ij}^\alpha = -2 C_S\,\alpha_{ij} + 2\widehat{C_S\,\beta_{ij}} \approx -2 C_S\,\alpha_{ij} + 2 C_S\,\widehat{\beta}_{ij} \; ; \; C_S(\vec{x},t) = -\frac{1}{2}\frac{L_{ij}^\alpha\left(\alpha_{ij} - \widehat{\beta}_{ij}\right)}{\left(\alpha_{mn} - \widehat{\beta}_{mn}\right)\left(\alpha_{mn} - \widehat{\beta}_{mn}\right)} \tag{10}$$

No predefined constant is necessary to close the equation set but the model suffers from its numerical stability problems and the remedy is to average the dynamic part of the model in some homogeneous flow direction(s) or to introduce some artificial clipping.

One equation model (DAV)

The idea of this one equation model is to include all local dynamic information into the source terms of the transport equation for the kinetic energy of the subgridscales K_{SGS}.

$$K_{SGS} = \frac{1}{2}\tau_{ij} = \frac{1}{2}\left(\overline{u_i u_i} - \overline{u}_i \overline{u}_i\right) \; ; \; \nu_t = C\,\overline{\Delta}\,K_{SGS}^{1/2} \tag{11}$$

Additionally a kinetic energy K for the second filter level is defined as:

$$K = \frac{1}{2}T_{ii} = \frac{1}{2}\left(\widehat{\overline{u_i u_i}} - \widehat{\overline{u}}_i \widehat{\overline{u}}_i\right) \tag{12}$$

Now, assuming similar transport of this turbulent energies on both level

$$C_{K_{SGS}} - D_{K_{SGS}} = P_{K_{SGS}} - \varepsilon_{K_{SGS}} \; ; \; C_K - D_K = P_K - \varepsilon_K \tag{13}$$

and the same formulation of the dissipation

$$\varepsilon_{K_{SGS}} = C_\varepsilon\,\frac{K_{SGG}^{3/2}}{\overline{\Delta}} \; ; \; \varepsilon_K = C_\varepsilon\,\frac{K^{3/2}}{\widehat{\Delta}} \tag{14}$$

yields to the following equation by using the second filter for the RHS of the K_{SGS} transport equation. This term should be comparable to the non filtered RHS of the K transport equation.

$$\widehat{P_{K_{SGS}}} - C_\varepsilon\,\frac{K_{SGS}^{3/2}}{\overline{\Delta}} = P_K - C_\varepsilon\,\frac{K^{3/2}}{\widehat{\Delta}} \tag{15}$$

Based on the old values for C_ε a new value is defined locally for every time step.

$$C_\varepsilon^{n+1} = \left(P_K - \widehat{P_{K_{SGS}}} + C^n_\varepsilon\,\frac{\widehat{K_{SGS}}}{\overline{\Delta}}\right)\frac{\widehat{\Delta}}{K^{3/2}} \tag{16}$$

This local value for C_ε is used in the dissipation formulation of the transport equation for K_{SGS}.

$$\frac{\partial K_{SGS}}{\partial t} + \frac{\partial \bar{u}_j K_{SGS}}{\partial x_j} = \frac{\partial}{\partial x_j}\left(\langle C\rangle_{xyz}\, \overline{\Delta} K_{SGS}^{1/2}\, \frac{\partial K_{SGS}}{\partial x_j}\right) + 2\nu_t \bar{S}_{ij}\bar{S}_{ij} - C_\varepsilon \frac{K_{SGS}^{3/2}}{\overline{\Delta}} \tag{17}$$

In diffusive term in the K_{SGS} equation above as well as in the momentum equations an average value of C is used which is computed with the requirement that the production in the whole domain remains the same.

$$\left\langle 2C\overline{\Delta} K_{SGS}^{1/2}\bar{S}_{ij}\bar{S}_{ij}\right\rangle_{xyz} = 2\langle C\rangle_{xyz}\left\langle \overline{\Delta} K_{SGS}^{1/2}\bar{S}_{ij}\bar{S}_{ij}\right\rangle_{xyz} \tag{18}$$

One equation model (KIM)

The one equation model of Menon and Kim is based on the following similarity assumption of different scales $\tau_{ij} = C_k L_{ij}$. They define a kinetic energy K_{test} of a test level that includes all energy between two length scales ($\overline{\Delta} < l < \widehat{\Delta}$).

$$K_{test} = \frac{1}{2}\left(\widehat{\overline{u_k u_k}} - \hat{\bar{u}}_k \hat{\bar{u}}_k\right) \tag{19}$$

Assuming a similar representation of this two levels we get

$$\tau_{ij} = -2C_\tau \overline{\Delta} K_{SGS}^{1/2}\bar{S}_{ij} + \frac{\delta_{ij}}{3}\tau_{kk} \quad ; \quad L_{ij} = -2C_\tau \widehat{\Delta} K_{test}^{1/2}\hat{\bar{S}}_{ij} + \frac{\delta_{ij}}{3}L_{kk} \tag{20}$$

and one can calculate the value of directly as shown.

$$C_\tau = \frac{1}{2}\frac{L_{ij}\sigma_{ij}}{\sigma_{ij}\sigma_{ij}} \quad ; \quad \sigma_{ij} = -\widehat{\Delta} K_{test}^{1/2}\hat{\bar{S}}_{ij} \tag{21}$$

Dissipation of energy at the test level takes place between the two length scales:

$$\varepsilon_{test} = (\nu + \nu_t)\left[\frac{\partial \bar{u}_i}{\partial x_j}\frac{\partial \bar{u}_i}{\partial x_j} - \frac{\partial \hat{\bar{u}}_i}{\partial x_j}\frac{\partial \hat{\bar{u}}_i}{\partial x_j}\right] \quad ; \quad \varepsilon_{test} = C_\varepsilon \frac{K_{test}^{3/2}}{\widehat{\Delta}} \Rightarrow C_\varepsilon \Rightarrow \varepsilon_{SGS} = C_\varepsilon \frac{K_{test}^{3/2}}{\overline{\Delta}} \tag{22}$$

This model gives a transport equation for K_{SGS} that is able to use the local value of C_τ in the diffusion and C_ε in the dissipation term. To prevent negative diffusion the total diffusion (laminar and turbulent) is limited to positive values.

$$\frac{\partial K_{SGS}}{\partial t} + \frac{\partial \bar{u}_j K_{SGS}}{\partial x_j} = \frac{\partial}{\partial x_j}\left[\left(\underbrace{C_\tau \overline{\Delta} K_{SGS}^{1/2}}_{>-\nu} + \nu\right)\frac{\partial K_{SGS}}{\partial x_j}\right] + 2\nu_t \bar{S}_{ij}\bar{S}_{ij} - C_\varepsilon \frac{K_{SGS}^{3/2}}{\overline{\Delta}} \tag{23}$$

BOUNDARY CONDITIONS

A common inlet boundary condition for LES is white noise. Using a random number for every time step to change the inlet boundary condition as shown in the equation below gives the desired turbulence level.

$$u_{in} = \bar{u}_{in} + \sqrt{12}\, t_{in}\bar{u}_{in}\xi \qquad -0,5 < \xi < 0,5 \tag{24}$$

But this kind of artificial turbulence has no length scale because this fluctuations have no correlation in space or time. The dissipation is very large and thus, this fluctuation will vanish fast.

$$l = \int f(r)\,dr \cong 0 \quad ; \quad \varepsilon \sim \frac{1}{l} \cong \infty \tag{25}$$

To test the influence of this boundary condition data from previous channel flow calculation is used as a second type of inlet condition. All calculations will be performed on a three dimensional grid using approximate 500.000 grid points.

TESTCASE

A simple two dimensional test case from Nielsen (1990) is used to evaluate the calculation results. The inlet Reynolds-Number is Re = 5000, the turbulence intensity is 4%.

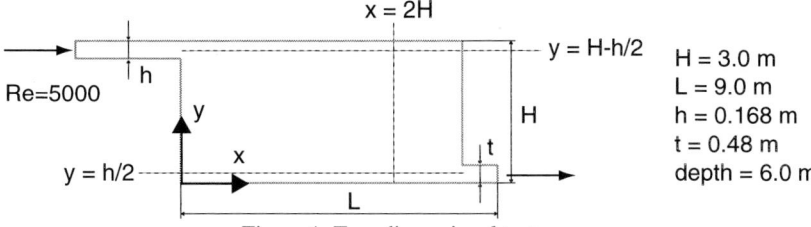

Figure 1: Two dimensional test case

RESULTS

Figure 2 shows the velocity profiles at x = 2H. All models predict the average velocity profile well. The KIM model gives a steeper gradient towards the top wall in agreement with the measurements. This result could be further improved by using channel flow data as inlet condition (KIM-CH). The turbulent fluctuation profiles indicate too low values for the KIM model with standard inlet boundary condition. The DAV models shows higher turbulent fluctuation velocity but only the KIM model with channel flow inlet data predicts the measured values.

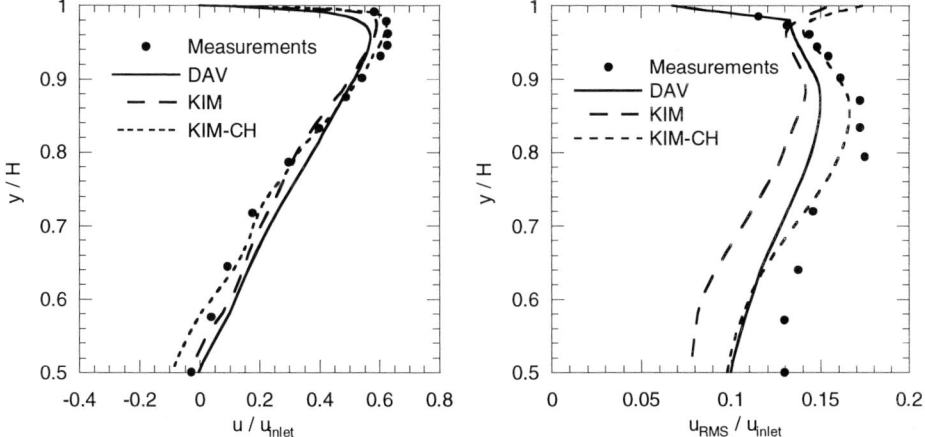

Figure 2: Velocity and turbulent fluctuation profiles at x = 2H, centre line

The profiles at y = H-h/2 are presented at Figure 3. Again, the KIM model with channel flow data shows the best agreement with the measured velocity profile. At y/H = 0 the inlet velocity is over predicted by the channel flow data because the room inlet device still shows entrance effects. The turbulent fluctuation profiles show the fast decay of standard inlet turbulence without any length scale. The calculation applying the channel flow data file preserves its turbulence level. The shape of the measured fluctuation profile is well predicted by the DAV model up to x/H = 2.5 and from the KIM model over the whole range. The level of the fluctuation velocities is to low between x/H = 2.5...3.

306

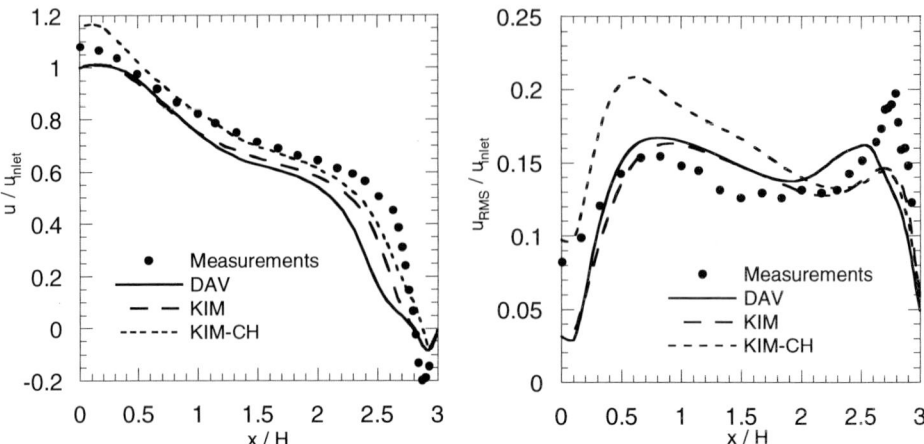

Figure 3: Velocity and turbulent fluctuation profiles at y = H-h/2, centre line

The velocity profile at y = h/2 is only well predicted by the KIM model with channel flow inlet data boundary condition. All other calculations give a to small recirculation zone of the wall jet. The calculated fluctuation velocities show a poor agreement with the measurements. The grid resolution in the lower half of the room is too low to preserve the resolved turbulent motion using a second order discretization method in space and time.

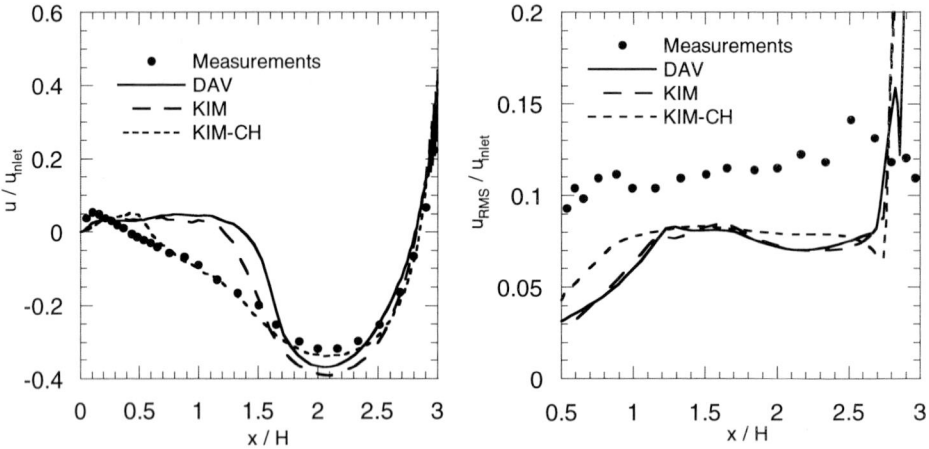

Figure 4: Velocity and turbulent fluctuation profiles at y = h/2, centre line

REFERENCE
Smagorinsky (1963) Mon. Weath. Rev. 91 (3).
Davidson, L. (1997) Large Eddy Simulation: A dynamic one-equation subgrid model for three dimensional recirculating flow, 11th Int. Symp. On Turbulent Shear Flow, Volume 3, Pages 26.1-26.6, Grenoble.
Germano, M., Piomelli, P., Moin, P., Cabot, W.H. (1991) A dynamic subgrid-scale eddy viscosity model, Physics of Fluids A, Pages 1760-1765.
Menon, S., Kim, W.-W. (1996) High Reynolds number flow simulations using the localized dynamic subgrid-scale model, 34th Aerospace Sciences Meeting, AIAA Paper 96-0425, Reno.
Nielsen, P (1990) Specification of a two-dimensional test case, ISSN 0902-7513 R9040, Aalorg University

Air Distribution in Rooms, (ROOMVENT 2000)
Editor: H.B. Awbi

LOW-REYNOLDS NUMBER EFFECTS IN VENTILATED ROOMS: A NUMERICAL STUDY

L. Davidson,[*] P.V. Nielsen and C. Topp

Dept. of Building Technology and Structural Engineering
Aalborg University, Denmark

ABSTRACT

In ventilated rooms the flow is often not fully turbulent because the Reynolds number is too low. When the flow is not fully turbulent, the traditional RANS (Reynolds Averaged Navier-Stokes) method employing $k - \varepsilon$ or $k - \omega$ is not suitable, because this method has difficulties in treating regions where the flow is laminar or not fully turbulent. There exist a number of low-Re number $k - \varepsilon$ and $k - \omega$ models, but they have all been designed to treat low-Re number effects close to walls, not regions where the flow becomes laminar far from the walls, as occurs in ventilated rooms. In the present study, we use Large Eddy Simulations (LES) which is a suitable method for simulating the flow in ventilated rooms at low Reynolds number.

EQUATIONS

With a spatial, inhomogeneous filter (denoted by a bar) applied to the incompressible Navier-Stokes equations, we obtain the momentum and continuity equations for the large scale motion

$$\frac{\partial \bar{u}_i}{\partial t} + \frac{\partial}{\partial x_j}\left(\bar{u}_i \bar{u}_j\right) = -\frac{1}{\rho}\frac{\partial \bar{p}}{\partial x_i} + \nu \frac{\partial^2 \bar{u}_i}{\partial x_j \partial x_j} - \frac{\partial \tau_{ij}}{\partial x_j}, \quad \frac{\partial \bar{u}_i}{\partial x_i} = 0 \tag{1}$$

where the subgrid stress tensor is given by $\tau_{ij} = \overline{u_i u_j} - \bar{u}_i \bar{u}_j$, and is modelled as

$$\tau_{ij} = -2\nu_{sgs}\bar{S}_{ij} = -2C_{hom}^k \Delta k_{sgs}^{\frac{1}{2}} \bar{S}_{ij}, \quad \bar{S}_{ij} = \frac{1}{2}\left(\frac{\partial \bar{u}_i}{\partial x_j} + \frac{\partial \bar{u}_j}{\partial x_i}\right). \tag{2}$$

[*]on leave from Dept. of Thermo and Fluid Dynamics, Chalmers University of Technology

The Dynamic One-Equation Model

In the present study, a one-equation dynamic subgrid model (Davidson 1997) is used. In this model, the modeled k_{sgs} equation can be written

$$\frac{\partial k_{sgs}}{\partial t} + \frac{\partial}{\partial x_j}(\bar{u}_j k_{sgs}) = \frac{\partial}{\partial x_j}\left(C_{hom}^k \Delta \, k_{sgs}^{\frac{1}{2}} \frac{\partial k_{sgs}}{\partial x_j}\right) + P_{k_{sgs}} - C_*^k \frac{k_{sgs}^{\frac{3}{2}}}{\Delta} \tag{3}$$

In the production term, the dynamic coefficient C^k is computed in a way similar to that used in the standard dynamic model (Davidson 1997). The coefficient in front of the dissipation term C_*^k is computed by assuming that the transport of the SGS kinetic energy on the grid level (k_{sgs}) is equal to the transport of the SGS kinetic energy on the test level (K). The production term in the k_{sgs} equation is computed as

$$P_{k_{sgs}} = -2C^k \Delta k_{sgs}^{\frac{1}{2}} \bar{S}_{ij} \bar{S}_{ij} \tag{4}$$

Please note that in the production term a *local* coefficient C^k is used, whereas in the momentum equations and in the diffusion term in the k_{sgs} equation a homogeneous (constant) coefficient C_{hom}^k is used. For more details, see Davidson (1997), Krajnović and Davidson (1999) and Sohankar, Davidson, and Norberg (2000).

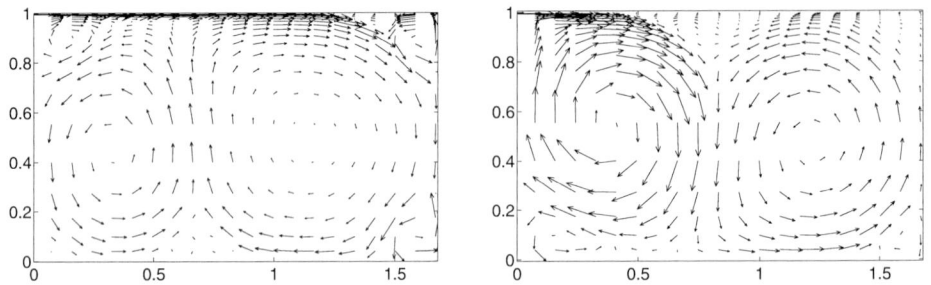

FIGURE 1: Unsteady, 2D laminar computations. Time-averaged flow field. Left: QUICK; right: central differencing.

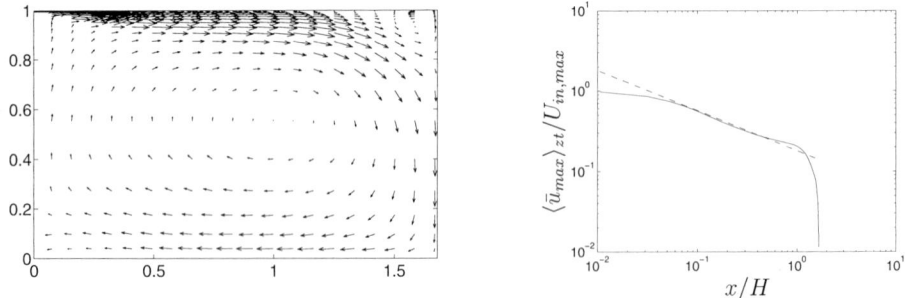

FIGURE 2: LES. Left: predicted velocity field $\langle \bar{u} \rangle_{zt}$ and $\langle \bar{v} \rangle_{zt}$. Right: peak velocity along the ceiling (solid line); the dashed line represents $\bar{u}/U_{in,max} = K_p (x/h)^{-1/2}$ with $K_p = 2.03$.

RESULTS

A finite volume code is used. For space discretization, central differencing is used for all terms. (For the 2D, unsteady, laminar computation, either the QUICK scheme or the central-differencing scheme was used for the convective terms.) Crank-Nicolson scheme is used for time discretization.

The inlet is located at the left wall, immediately below the ceiling. The outlet is located at the left wall immediately above the floor. The Reynolds number is $Re = U_{in}h/\nu = 600$ (where U_{in} denotes the bulk velocity) and $L/H = 1.68$, $W/H = 1.44$, $h/H = 0.008$. Inlet and outlet extend over the whole width of the room.

The boundary conditions are as follows. At the inlet we have laminar flow, i.e. $\bar{v} = \bar{w} = k_{sgs} = 0$ and a parabolic profile for $\bar{u}$. At all walls $\bar{u} = \bar{v} = \bar{w} = k_{sgs} = 0$. At the outlet a constant $\bar{u}$ is set from global continuity. The streamwise gradient is set to zero for the other variables. The normal gradient of $\bar{p}$ is set to zero at all boundaries.

A $80 \times 80 \times 48$ mesh is used. Constant spacing is used in the x and z directions. The cell near the floor ($y = 0$) has $\Delta y/H = 0.0105$. The cells are then stretched by 8% in the y direction up to $y/H = 0.5$ (cell 1 to 22). From $y/H = 0.5$ the cells are compressed by 8% up to $y/H = 0.992$. Ten cells (constant Δy) cover the inlet.

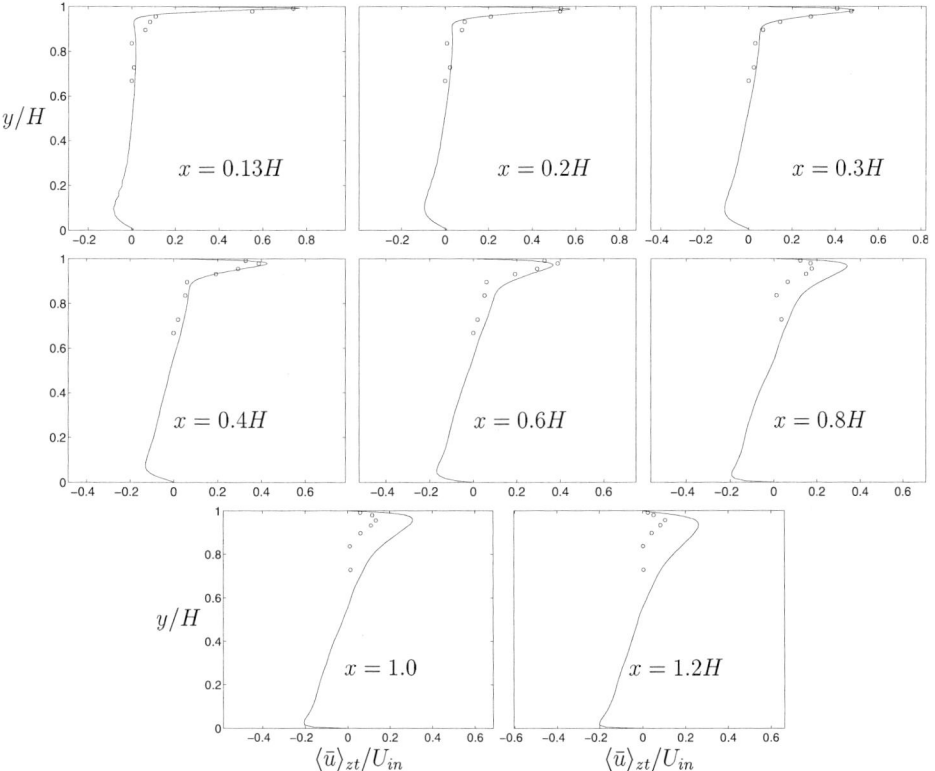

FIGURE 3: Velocity profiles. Lines: LES; markers: experiments.

310

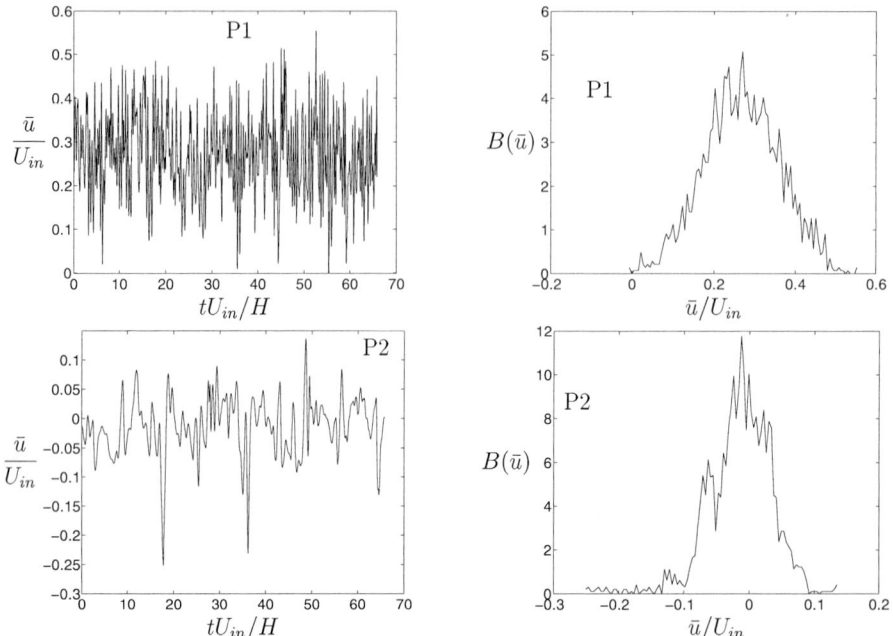

FIGURE 4: Left: time history. Right: probability function. Point P1: $x = 0.3H$, $y = 0.997H$, $z = 0.5W$; point P2: $x = 1.3H$, $y = 0.997H$, $z = 0.5W$.

Unsteady Laminar Two-Dimensional Flow

Initially we assumed that flow was fully laminar, and we computed the flow by solving the unsteady, laminar, two-dimensional equations. We used two different discretization schemes for the convective terms: the QUICK scheme and the central-differencing scheme. The time-averaged flow fields are presented in Fig. 1. When the QUICK scheme is used the wall jet stays attached to the ceiling up to $x \simeq 1.2H$. When, however, the central-differencing scheme is used, the wall jet detaches from the ceiling at $x \simeq 0.5H$ and falls down to the floor. The reason is believed to be due to the fact that the QUICK scheme is dissipative but the central-differencing scheme is not. That a scheme is dissipative means that it dampens oscillations, both numerical as well as physical ones. The central-differencing scheme does thus not dampen the oscillations in the wall jet, oscillations which may both be numerical as well as physical due to unsteadiness. The oscillations increase the entrainment into the wall jet. This is probably the main reason why the predicted wall jet falls down to the floor when central differences are used.

RANS With $k - \omega$ Model

Some attempts to compute the flow using a traditional RANS have also been carried out. The CALC-BFC code was used together with the $k - \omega$ model of Peng et al. (Peng, Davidson, and Holmberg 1997). Since the inlet conditions are laminar, it is not clear which boundary conditions should be used for k and ω. The local turbulence intensity was set to 0.01, and the inlet value for ω was varied between $\omega_{in} = \sqrt{k_{in}}/(0.1h)$ (gives $\nu_{t,in} \simeq 32\nu$) with and $\omega_{in} = k/(0.1\nu)$ (gives $\nu_{t,in} = 0.1\nu$). The variations of ω_{in} did not produce any changes in the flow pattern. It should here be mentioned that $k - \omega$ models are superior in treating flow which is close to being laminar, since the ω equation behaves well even if k goes to zero. On the contrary, the ε equation in $k - \varepsilon$

models causes problems when k goes to zero, since the ε equation includes the ratio ε/k.

The predicted flow field is very different from the laminar ones. The wall jet detaches from the ceiling almost immediately after the inlet. Both the QUICK scheme and the Hybrid scheme were tested, but they gave the same predicted flow field. The predicted flow fields differ very much from the experiments, and it seems that RANS is unable to predict this kind of low-Re number flow.

Large Eddy Simulations

Here the predictions using Large Eddy Simulations are presented. A time step of $\Delta t = 0.0026 H/U_{in}$ was used which gave a maximum convective CFL number of $CFL_{max} \simeq 0.9$. As initial flow field an instantaneous 2D laminar flow field was employed. The predicted results presented here have been averaged for a period of $T = 65 H/U_{in}$, and also averaged in the spanwise direction between $0.15 \leq z/W \leq 0.85$; this is denoted as $\langle . \rangle_{zt}$.

The predicted vector field is presented in Fig. 2, and as can be seen, the wall jet stays attached to the ceiling up to $x/H \simeq 1.4$. In Fig. 2 the peak velocity in the wall jet is also shown. The peak velocity decays as $x^{-1/2}$ and the dashed line represents the curve $\langle \bar{u} \rangle_{zt}/U_{in,max} = K_p (x/h)^{-1/2}$ with $K_p = 2.03$. The corresponding experimental value is $K_p = 2.39$ (Topp, Nielsen, and Davidson 2000). Please note that at the inlet a parabolic profile is prescribed in the predictions, and $U_{in,max}$ denotes the maximum value of the inlet profile.

In Fig. 3 the predicted $\bar{u}$ profiles are compared with experiments (Topp, Nielsen, and Davidson 2000). The agreement is, as can be seen, fairly good, at least up to $x/H = 0.6$. Downstream of this position it seems that the experimental wall jet more or less vanishes, and the experimental velocities decays much faster for increasing x than the predicted ones.

In Fig. 4 the predicted time history of $\bar{u}$ at two different points in the room are presented. Point P1 is located in the wall jet, and and the flow seems to be fully turbulent. However, when looking at the energy spectrum for that point, it is clearly seen that the flow is not fully turbulent, as the spectrum does not show any $-5/3$ region, see Fig. 5. The probability of $\bar{u}$ at Point P1 (Fig. 4) exhibits the usual form, with a well-defined peak at the mean value of $\bar{u}_{mean}/U_{in} \simeq 0.3$ (which is also seen from the time history), and its distribution is close to Gaussian. Point P2 in Fig. 4 is located in the region where the wall jet separates. Here the frequency of $\bar{u}$ is much lower. From the time history it is seen that the velocity at some instances go down to large negative values ($\bar{u}/U_{in} \simeq -0.25$). This is also sen from the probability function, where a tail of large negative $\bar{u}$ can be seen. The probability is very low, almost zero; this is because, as can be seen from the time history, it happens only once.

The spectrum in Fig. 5 for point P2 shows a region where the spectrum exhibits a $-5/3$ behavior, which should indicate that the flow were fully turbulent. However, from the time

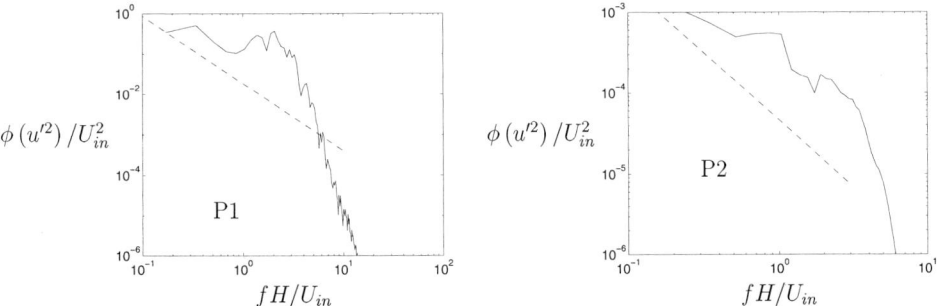

FIGURE 5: Density spectrum of $\bar{u}$ at two points P1 & P2 (see Fig. 4). The dashed lines show $\phi \propto f^{-5/3}$.

312

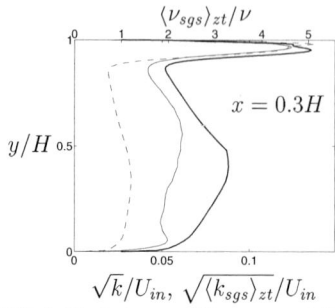
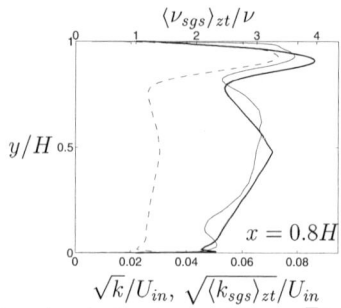

FIGURE 6: Turbulent profiles at $z = 0.5W$. Thick solid line: $\langle \nu_{sgs} \rangle_{zt}/\nu$; thin solid line: $\sqrt{k}/U_{in}$, dashed line: $\sqrt{\langle k_{sgs} \rangle_{zt}}/U_{in}$.

history in Fig. 4, it seems that the flow cannot be characterized as turbulent, because no high frequencies are present. The flow at this point, which is in the separation region, is dominated by large, unsteady structures. This can also be seen when the flow in the mid-plane ($z = W/2$), is visualized as a movie. The flow in the whole room can probably best be characterized as transitional.

In Fig. 6 profiles of the square root of the resolved turbulent, kinetic energy $k = (\langle u'^2 \rangle_{zt} + \langle v'^2 \rangle_{zt} + \langle w'^2 \rangle_{zt})/2$ are shown. As mentioned above, the flow is not believed to be fully turbulent, but it should rather be called transitional. Thus, although we call it "turbulent kinetic energy", k should here be taken as a measure of the kinetic energy of the large, unsteady structures (the streamwise fluctuation, for example, is defined as $\langle u'^2 \rangle_{zt} = \langle (\bar{u} - \langle \bar{u} \rangle_{zt})^2 \rangle_{zt})$. From Fig. 6 it can be seen that the SGS kinetic energy k_{sgs} is rather large compared to k, and in the wall jet at $x = 0.3H$ k_{sgs} is even larger. However, the SGS stresses (not shown here) are much smaller that the corresponding resolved stresses. The SGS viscosity ν_{sgs} is also shown in Fig. 6, and as can been it is everywhere smaller than 5ν. This is typical in a well-resolved LES. Recall that in RANS, the turbulent viscosity is typically two to three orders of magnitude larger than the physical viscosity.

CONCLUSIONS

In the present work the flow in a ventilated room at low Reynolds number ($Re = 600$) has been simulated both with 2D unsteady laminar approach, RANS and LES. Only LES proved to be capable of predicting this flow.

Acknowledgment
The first author was financed by a VELUX guest professor ship.

REFERENCES

Davidson, L. (1997). Large eddy simulation: A dynamic one-equation subgrid model for three-dimensional recirculating flow. In *11th Int. Symp. Turb. Shear Flow*, Vol. 3, Grenoble, pp. 26.1–26.6.

Krajnović, S. and L. Davidson (1999). Large-eddy simulation of the flow around a surface-mounted cube using a dynamic one-equation subgrid model. In S. Banerjee and J. Eaton (Eds.), *The First International Symp. on Turbulence and Shear Flow Phenomena*, New York, pp. 741–746.

Peng, S.-H., L. Davidson, and S. Holmberg (1997). A modified Low-Reynolds-Number $k - \omega$ model for recirculating flows. *ASME: Journal of Fluids Engineering 119*, 867–875.

Sohankar, A., L. Davidson, and C. Norberg (2000). Large eddy simulation of flow past a square cylinder: Comparison of different subgrid scale models (to appear). *ASME: J. of Fluids Engng 122*(1).

Topp, C., P. Nielsen, and L. Davidson (2000). Room airflows with low Reynolds number effects (to appear). In *7th Int. Conf. on Air Distributions in Rooms, ROOMVENT 2000*, Reading, U.K.

© 2000 Elsevier Science Ltd. All rights reserved.
Air Distribution in Rooms, (ROOMVENT 2000)
Editor: H.B. Awbi

NUMERICAL STUDY OF AIRFLOW STRUCTURE OF
A CROSS-VENTILATED MODEL BUILDING

T. Kurabuchi[1], M. Ohba[2], A. Arashiguchi[1] and T. Iwabuchi[1]

[1]Department of Architecture, Science University of Tokyo,
Tokyo, 162-0825, JAPAN
[2]Department of Architecture, Tokyo Institute of Polytechnics,
Kanagawa, 243-0297, JAPAN

ABSTRACT

With the purpose of evaluating validity of the application of CFD on the problems of cross-ventilation, numerical simulation was performed, using standard k- ε model and two types of modified k-ε models which improve evaluation accuracy in production term of turbulence energy, and also using LES, and the results were compared with those of the corresponding wind tunnel experiment.   As a result, it was found that the defects of the model characteristic to the standard k- ε model could be improved to a certain extent by application of the modified models.   However, from the reasons that the effects of improvement are not necessarily remarkable, it was judged that the standard k- ε model could be applicable to a certain extent in case general cross-ventilation phenomenon is to be predicted.   On the other hand, when LES is used, the factors such as wind pressure coefficient, turbulence energy, etc. are more accurately reproduced than k- ε type models.   Based on these results, the results of calculation of LES were separately assessed, and turbulence structure specific to cross-ventilation was evaluated.

KEYWORDS

Cross-Ventilation, CFD, LES, k-ε model, Wind Tunnel Experiment, Wind Pressure, Natural Ventilation

INTRODUCTION

There are now increasingly strong interests on the utilization of cross-ventilation with the purpose for environmental control in damp heat season and intermediate season. However, there is not much progress in the study to elucidate these problems because cross-ventilation is a complicated turbulence phenomenon where airflow inside and outside a building are interrelated to each other. To overcome the problems, it has been proposed to use numerical simulation, and there have been several cases where the standard k-ε model was actually applied (Frescos, 1998 and Iino et al. 1998). However, it is difficult to apply k- ε model to turbulent flow where turbulence around the flow stagnation point initially plays an important role, and the reliability is not perfectly ensured. Under such circumstances, we performed calculation using the standard k-ε model and the modified k-ε models to cope with the flow field where flow impingement exists, and also using LES. Based on the comparison with the results of the experiment newly performed, validity of the application of the models was assessed.

OUTLINE OF EXPERIMENT AND NUMERICAL SIMULATION

Building Model and Wind Tunnel Experiment

The study was performed on a building where a pair of openings was mounted at the center of the building as shown in Fig. 1. A model with height of 0.15 m was prepared and wind tunnel experiment was then performed. Fig. 2 shows the profiles of average wind velocity of approach flow and turbulence energy. For the measurement of airflow, split-film type anemometer was used. To determine the ventilation flow rate, a constant quantity of ethylene gas was continuously introduced into inner space of the model, and concentration near the opening on outflow side was measured.

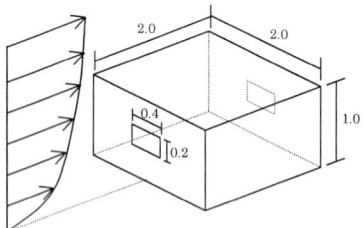

Figure 1 : Flow configuration

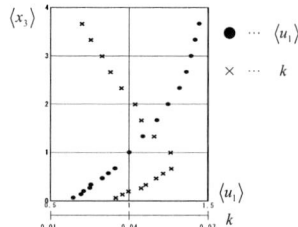

Figure 2 : Profiles of $\langle u_1 \rangle$ and k at inflow boundary

Turbulence Model and Simulation Procedure

It is known that the use of k-ε model often leads to overestimation of the turbulence production term (P_k) because of error in the evaluation of normal stress in case of the flow such as impingement of jet. As shown in Table 1, Launder and Kato (1993) demonstrated that this problem can be improved by evaluating P_k term using a product of characteristic strain parameter S and vorticity parameter ω. However, this modification is disadvantageous in that consistency of energy transfer process between mean flow and turbulent motion may be lost. Murakami et al. (1996) modified eddy viscosity v_t and proposed a model with consistency and providing similar effects. In this study, three types of k-ε model including the above standard type, LK type and MMK type were assessed. Generalized log-law proposed by Launder and Spalding (1974) was applied for handling wall boundaries. In LES, Smagorinsky model was used with setting constant of the model to 0.13. Wall shear stress was estimated by fitting instantaneous velocity component tangential to wall to the universal wall law. Separate calculation for plane channel flow was performed for generating inflow condition and the stocked data was modified so that distribution of data accurately agrees with the experimental result. After preliminary calculation, formal calculation of 200,000 steps was carried out and analyzed .

TABLE 1

APPLIED k—ε MODELS

Standard k-ε Model	$P_k = v_t S^2, v_t = C_\mu \dfrac{k^2}{\varepsilon}$	LK Model	$P_k = v_t S\omega, v_t = C_\mu \dfrac{k^2}{\varepsilon}$
$C_\mu = 0.09, S = \sqrt{\dfrac{1}{2}\left(\dfrac{\partial\langle u_i\rangle}{\partial x_j}+\dfrac{\partial\langle u_j\rangle}{\partial x_i}\right)^2}, \omega = \sqrt{\dfrac{1}{2}\left(\dfrac{\partial\langle u_i\rangle}{\partial x_j}-\dfrac{\partial\langle u_j\rangle}{\partial x_i}\right)^2}$		MMK Model	$P_k = v_t S^2, v_t = C_\mu^* \dfrac{k^2}{\varepsilon}, C_\mu^* = C_\mu \cdot MIN\left[1.0, \dfrac{\omega}{S}\right]$

COMPARISONS OF OBSERVED AND SIMULATED RESULTS

Flow Rate
In Table 2, the results of experiment on flow rate are compared with calculations.   There was no substantial difference between the models, and all of the results matched well with the experiment.

TABLE   2

COMPARISON OF OBSERVED AND PREDICTED FLOW RATE

experiment	LES	standard	LK	MMK
0.043	0.042	0.041	0.041	0.040

Airflow Patterns of Inside and Outside of the Building
Comparison of the average velocity vector is summarized in Fig. 3.   The features of velocity vector in the experimental results were recirculation at the front edge of roof surface (not identified in the experiment due to the limitation on installing probe), extensive recirculation at the lower portion on windward surface, and sudden downfall of the inflow.   In the standard k-ε model, the shape of recirculating flow in the lower portion on the windward surface was different, descending of inflow was weak, and that recirculation on the roof surface was not reproduced almost at all.   In contrast, in LK model and MMK model, the results were closer to the results of the experiment in the features that descending of inflow is slightly increased, and that the recirculating flow on roof surface is enlarged.
On the other hand, in LES, the recirculation on the lower portion on the windward surface was somewhat longer in the streamwise direction, but, the features such as shape of recirculation, reproduction of recirculation at the front edge of roof, descending of inflow into the building and aspect of wake were very close to the experimental results.

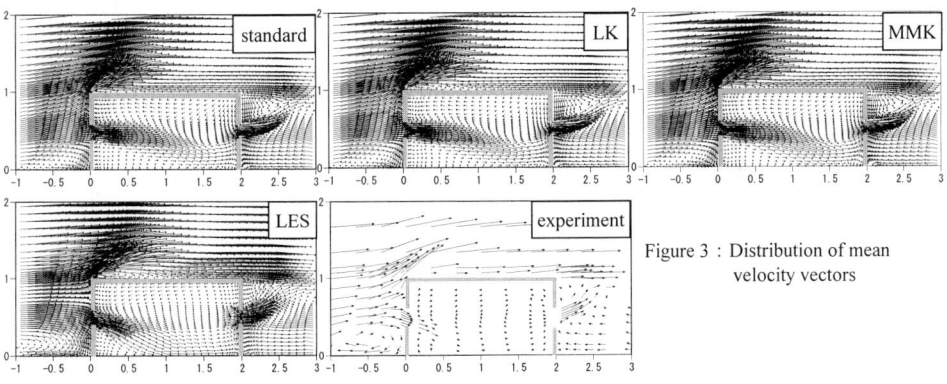

Figure 3 : Distribution of mean velocity vectors

Distribution of Turbulence Kinetic Energy
Fig. 4 shows the comparison of turbulence kinetic energy k.   In the experimental results, the value of k reached peak at the front edge of roof surface and there was also a peak in recirculating region on the lower portion on the windward surface.   Further, the airflow entering the building reached the maximum value immediately after entering.   In the standard k-ε model, the region with high k value is extensively spread from the front edge of roof to the windward surface.   In LK model, the level of k value shows extreme decrease on the windward surface of the building   Results of MMK model exhibited a distribution intermediate between the standard and LK model.   Peak associated with the recirculation on the lower portion on the windward surface was reproduced in none of k-ε models.
On the other hand, in LES, peak values on the lower portion on the windward surface or on the front edge of roof surface were reproduced.   Regarding the indoor space of the building, k assumes relatively higher value on the downstream side of the windward opening.   In the wake region, the level of k on downstream side of roof surface is somewhat higher than the experiment.

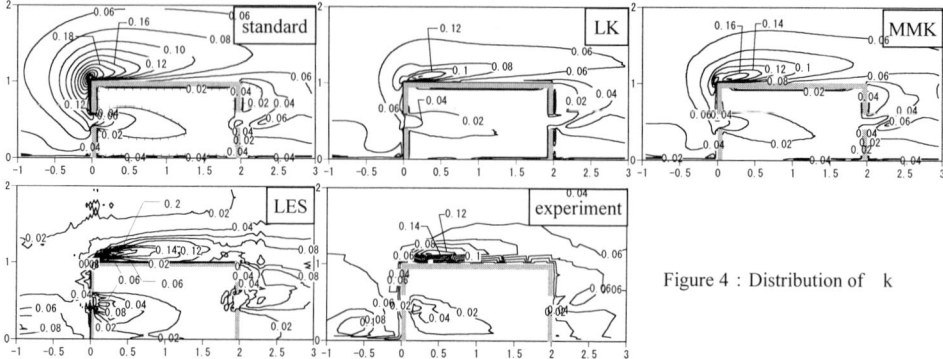

Figure 4 : Distribution of k

Distribution of Wind Pressure Coefficient

Fig. 5 summarizes the comparison between the results of experiment and those of calculation using k-ε models on wind pressure coefficient on external surface of the building in symmetrical surfaces. On the front surface of the building, overestimation of wind pressure coefficient was extremely conspicuous in the standard k-ε model, and the results became closer to the experimental results in the order of LK model and MMK model.   On the front edge of the roof surface, recirculation was not sufficiently reproduced in the standard k-ε model, and absolute value of negative pressure on this portion was extremely high.   In LK model and MMK model, extreme negative pressure did not occur as the result of the enlargement of recirculation.   Magnitude of the calculated v_t in the vicinity of windward surface is significantly small in the modified models than the standard model, and this would improve accuracy of Reynolds stress, and wind pressure coefficient consequently.

Fig. 6 represents the comparison of calculation results using LES and experiment including the case where openings are not present.   There is no sign of overestimation of wind pressure coefficient on the windward surface of the building.   In particular, in case there was the opening, wind pressure coefficient was slightly increased on the upper portion of the opening compared with the case of no opening, and the results of the experiment were reproduced very well.

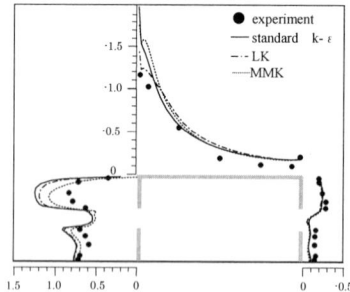

Figure 5 : Distribution of mean pressure
coefficient (experiment and k- ε)

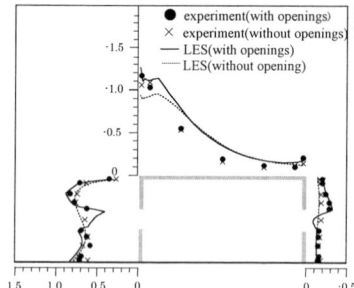

Figure 6 : Distribution of mean pressure
coefficient(experiment and LES)

ANALYSIS OF THE RESULT OF LES SIMULATION

As described above, it is evident that the results of LES match far better than k-ε type models with the experimental results, and it can be expected that LES calculation reflects actual conditions even in the case, for which it is difficult to evaluate experimentally.   In this respect, we attempted to analyze turbulence structure characteristic in cross-ventilation based on the results of LES calculation.

Momentum Balance near the windward opening

One of the features of the airflow during cross-ventilation is extreme downfall of inflow. To elucidate the cause, we evaluated momentum balance along the central axis, which passes through the center of the opening. The equation of time-averaged velocity component in vertical direction at the plane of symmetry can be approximately expressed as in the equation (1):

$$\frac{\partial \langle \overline{u_3} \rangle}{\partial t} \cong -\frac{\partial \langle \overline{p} \rangle}{\partial x_3} - \frac{\partial \langle u_1 \rangle \langle u_3 \rangle}{\partial x_1} - \frac{\partial \langle u_1'' u_3'' \rangle}{\partial x_1} - \frac{\partial \langle u_3 \rangle \langle u_3 \rangle}{\partial x_3} - \frac{\partial \langle u_3'' u_3'' \rangle}{\partial x_3} \tag{1}$$

Fig. 7 shows the average airflow field around the central axis and profile of vertical velocity component (positive in upward direction). Longitudinal profiles of each term in the equation (1) are given in Fig. 8. In the far region from the opening, vertical velocity is turned from positive to negative, and the second term on the RHS of the equation (1) is turned to positive. When the airflow approaches the opening, acceleration occurs and the second term has a high positive value. The cause to deflect the flow downward may be the negative contribution in the equation (1), which offsets the above action. According to Fig. 8, the pressure gradient in the first term of the RHS of the equation (1) opposes from the point -1.5 to around the opening except in the vicinity of opening where only momentum transport by vertical velocity component takes negative value. We may conclude the primary cause of the downfall lies in the pressure gradient due to the recirculation at the lower portion on windward surface.

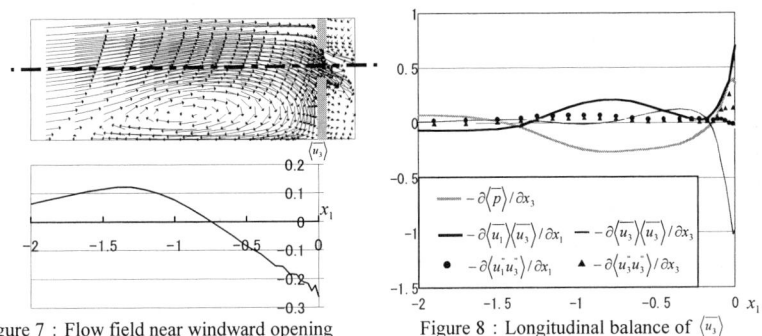

Figure 7 : Flow field near windward opening and longitudinal profile of $\langle \overline{u_3} \rangle$ at central axis of opening

Figure 8 : Longitudinal balance of $\langle \overline{u_3} \rangle$ transport equation at central axis of opening

Deformation of Virtual Flow Tube

To elucidate the mechanism of total pressure loss, the shape of a virtual flow tube passing through the windward opening was calculated by tracing trajectories of passive markers. Fig. 9 shows horizontal and vertical cross sectional views of the flow tube. The flow passing through the opening is turned to a complicated flow where acceleration and deceleration occur within short distance.

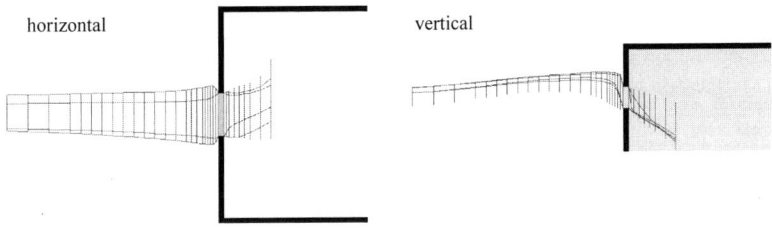

Figure 9 : Cross sectional views of virtual flow tube

318

Change of Mean Flow Energy

If energy transport equation of mean flow is applied to the flow tube, it is known that the streamwise change of total pressure corresponds to the work on the surface of the flow tube and to the production of k inside the flow tube.   Fig. 10 summarizes the changes of static, dynamic and total pressure along the central axis of the opening.   The total pressure shows extensive change on upstream side of the opening due to the negligence of streamline direction.   The result of integral averaging along the flow tube is shown in Fig. 11.   Energy change occurs between static and dynamic pressure due to acceleration and deceleration, but almost no change occurs in total pressure.   When reaching the opening, static pressure is decreased at first, and after passing through the opening, dynamic and static pressure are decreased at the same time.   The peak of k value is observed immediately after entering the opening, and this is estimated as total pressure loss due to production of k in this region

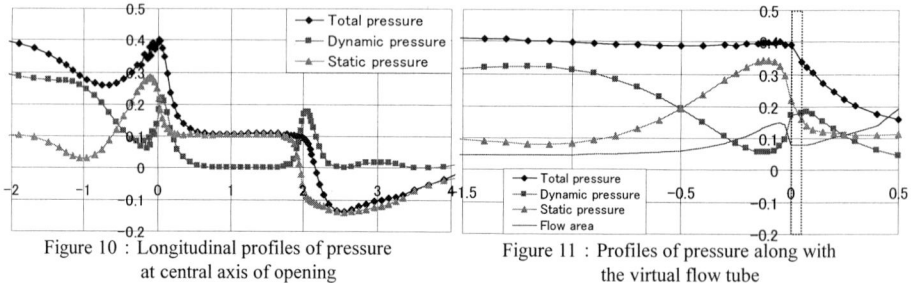

Figure 10 : Longitudinal profiles of pressure at central axis of opening

Figure 11 : Profiles of pressure along with the virtual flow tube

CONCLUDING REMARKS

The results of the present study suggest that LK and MMK model slightly improve the problems with the standard k-ε model, however, it is considered that the use of the standard k-ε model can be justified as far as the general cross-ventilation phenomenon is concerned.   On the other hand, excellent agreement was observed in the application of LES.   Based on these results, the calculation results were analyzed, and it was elucidated that the downfall of the inflow is caused by pressure gradient due to recirculation on the lower portion on the windward surface.   Further, based on the results of calculation of a virtual flow tube, the changes along the flow of total pressure loss were evaluated.

Acknowledgement

The authors extend sincere gratitude to TOSTEM Foundation for Construction Materials Industry Promotion for assistance and support to cover research cost of the present study.

Nomenclature

x_i : spatial coordinate ($i=1,2,3$: streamwise, spanwise, vertical)

u_i : velocity component
p : pressure
$\langle f \rangle$: time average of f

$\overline{f}$: filtered value of f
$f'' : = \overline{f} - \langle \overline{f} \rangle$

k : turbulent kinetic energy
ε : dissipation rate of k
v_t : eddy viscosity

References

Freskos G.O. (1998). Influence of Various Factors on the Predistions Furnished by CFD in Cross-Ventilation Simulations, *Proceeding of the 6th International Conference on Air Distribution in Rooms*, **vol.1**, 483-490

Iino Y., Kurabuchi T.,Kobayashi N. and Arashiguchi A. (1998). Study on Airflow Characteristics in and around Building Induced by Cross Ventilation Using Wind Tunnel Experiment and CFD Simulation, *Proceeding of the 6th International Conference on Air Distribution in Rooms*, **vol.2**, 307-314

Launder B.E. and Spalding D.B. (1974). The Numerical Computation of Turbulent Flows, *Comput. Methods APPL. Mech. Eng.*, **3**, 269-289.

Launder B.E. and Kato M. (1993). The Modeling Flow-Induced Oscillations in Turbulent Flow around a Square Cylinder, *ASME Fluid Eng. Conf. Jun.*, **157**, 189-199.

Murakami S., Mochida A., Kondo K., Ishida Y. and Tsuchiya M. (1996). Development of new k-ε model for flow and pressure fields around bluff body, *CWE96*, Colorado, USA.

Air Distribution in Rooms, (ROOMVENT 2000)
Editor: H.B. Awbi

FINITE ELEMENT CALCULATION OF NATURAL VENTILATION

R. Gritzki[1], H. Müller[2], W. Richter[1] and M. Rösler[1]

[1] Dresden University of Technology,
Institute for Thermodynamics and Technical Installation of Buildings,
Dresden, Germany
[2] AEROTEC Engineering GmbH,
Hamburg, Germany

ABSTRACT

The intention of this paper is not to compare discretization schemes but to show some advantages of a stabilized finite element method for modelling natural ventilation. Based on the finite element theory we present a formulation of boundary conditions that can be used for most ventilation openings in buildings. Stationary as well as transient situations can be considered without modelling of the outdoor space. Mathematical background and implementation details are discussed. Results are presented for ventilation of a living room at typical outdoor conditions.

KEYWORDS

CFD, Finite Elements, Boundary Conditions, Natural Ventilation

MOTIVATION

Using CFD for analysing flow and temperature distribution in rooms and buildings one has often to deal with boundary conditions at large openings. In case of windows and similar ventilation openings the formulation of pressure boundary conditions seems to be reasonable.

Most of the commercial and research CFD codes offer options to specify pressure boundary conditions. But the results of using it for natural ventilation are often unsatisfactory or at least very expensive concerning computational resources.

From the mathematical point of view finite elements are very flexible concerning boundary conditions. Therefore we investigated the implementation of pressure boundary conditions within a special finite element method. We do neither intend to present the final solution of the problem nor to definitely favour finite element calculations for ventilation problems. We simply would like to show some advantages of using finite element technology for calculation of natural ventilation.

MATHEMATICAL BACKGROUND

In order to explain what is meant with "flexibility" of a finite element formulation (or more exactly variational formulation) concerning boundary conditions we will look on the stationary Navier-Stokes equations for an incompressible fluid, dimensionless notation :

$$(u \circ \nabla)u - \nabla \circ \tau + \nabla p = f \quad \text{in } \Omega. \tag{1}$$

Here u denotes the velocity vector, p the pressure, f a body force, and τ stands for the stress tensor $\tau = v\left(\nabla u + (\nabla u)^T\right)$. A variational formulation of equation (1) is achieved by integration and multiplication with a test function v. For a mathematical detailed definition of functions and function spaces the reader is referred to Carey&Oden (1986). Using partial integration of the term with the stress tensor and the pressure gradient the following boundary integrals arise:

$$-\int_\Omega \nabla \circ \tau \circ v \ d\Omega = \int_\Omega \tau \mathbin{\overset{\circ}{\cdot}} (\nabla v)^T \ d\Omega - \int_\Gamma n \circ \tau \circ v \ d\Gamma, \tag{2}$$

$$\int_\Omega \nabla p \circ v \ d\Omega = -\int_\Omega p (\nabla \circ v) d\Omega + \int_\Gamma p\, n \circ v \ d\Gamma. \tag{3}$$

On one hand this is a so called weak formulation of the problem because it contains lower derivations than the original equation. On the other hand it is possible to use the boundary integrals to formulate boundary conditions in a natural way.

Using the pressure boundary condition according to Heywood et al.(1996) along the boundary Γ,

$$\sigma = p - n \circ \tau \circ n, \tag{4}$$

the momentum equation reads:

$$\int_\Omega (u \circ \nabla)u \circ v \ d\Omega + \int_\Omega \tau \mathbin{\overset{\circ}{\cdot}} (\nabla v)^T \ d\Omega - \int_\Omega p(\nabla \circ v) \ d\Omega = \\ \int_\Omega f \circ v \ d\Omega - \sigma \int_\Gamma n \circ v \ d\Gamma. \tag{5}$$

This means, the pressure boundary condition along the boundary Γ is included in the variational formulation.

IMPLEMENTATION IN A CFD CODE

The pressure boundary condition, equation (4), was implemented in the finite element code ParallelNS, a common research code of Göttingen University and Dresden University of Technology. This code was originally developed for testing a non-overlapping domain decomposition method, see Lube et al. (1998), Auge et al. (1998). Furthermore the code works with a stabilized finite element method (FEM) because standard methods are not suitable for calculating convection dominated flows. The method is called *Galerkin Least-Squares FEM* and includes stabilization to overcome the problems resulting from convection and velocity pressure coupling, Hughes et al. (1986), Hughes et al. (1989). The basic concept of the Least-Squares FEM can be used for a wide range of applications, Jiang (1998).

In co-operation with mathematicians from the Institute of Numerical and Applied Mathematics (University of Göttingen) the code was extended by a k-ε turbulence model especially for calculating flows within rooms and buildings. Basic equations were the time dependent Navier-Stokes equations for incompressible fluids. Effects of buoyancy were modelled based on the Boussinesq-Approximation. Wall functions for room air flows were implemented, see Neitzke (1998). Beside the boundary condition of the flow field for the temperature and the turbulence quantities appropriate boundary conditions at the openings have to be formulated. Thus a mix of boundary conditions of first and second kind is applied depending on the flow direction. That means, the incoming flow is specified by fixed values of temperature and turbulence quantities and for the outgoing flow the boundary condition is included in the variational formulation similar to equation (5).

EXAMPLES OF CALCULATION

We started our computations with simple test cases, see figure 1. A cube of 1 m x 1 m x 1 m with different openings and a temperature difference of 10 K was assumed. Thereby the structure of the flow at the openings and global balances were investigated. Horizontal openings are also possible to calculate but the highly unsteady flow is difficult to evaluate.

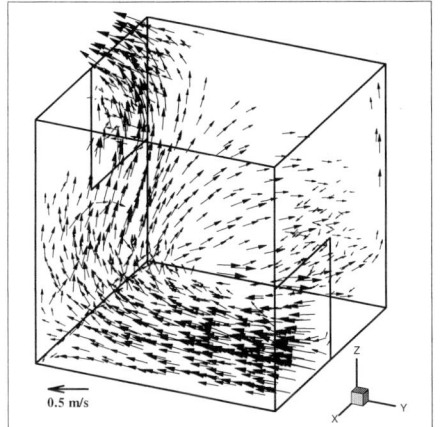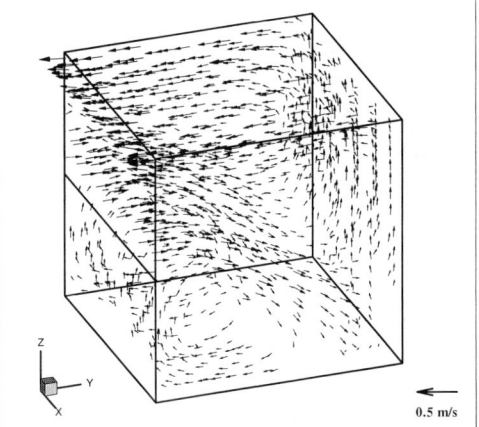

Figure 1: Velocity distributions in a cube with different openings

Our main efforts were dedicated to the computation of the flow through open windows at calm. In this case the pressure boundary condition as in (4) is equivalent to a stress free outlet, sometimes called "do nothing" condition.

Figure 2 shows a room of about 120 m^3 volume after a ventilation period of 20 seconds. The initial indoor temperature is 20 °C, the outdoor temperature is 10 °C. The walls are regarded as adiabatic. The illustrated cloud shows the temperature distribution field below 18 °C.

Figure 3 illustrates the above temperature distribution field 20 seconds later from another viewpoint. Reflecting and mixing of the cold air can be seen because the incoming air has reached the wall opposite to the window.

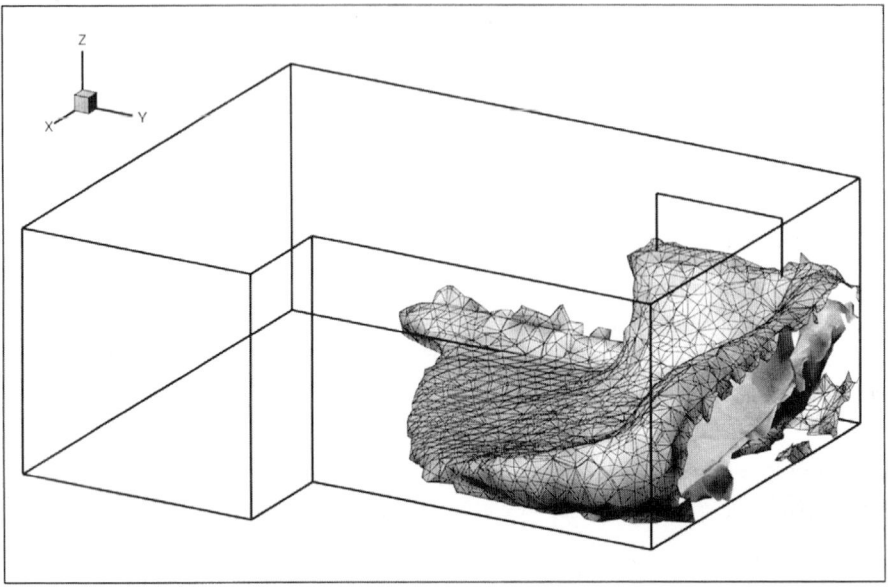

Figure 2: Region of temperature below 18 °C after a ventilation period of 20 seconds,
view from the wall opposite to the window

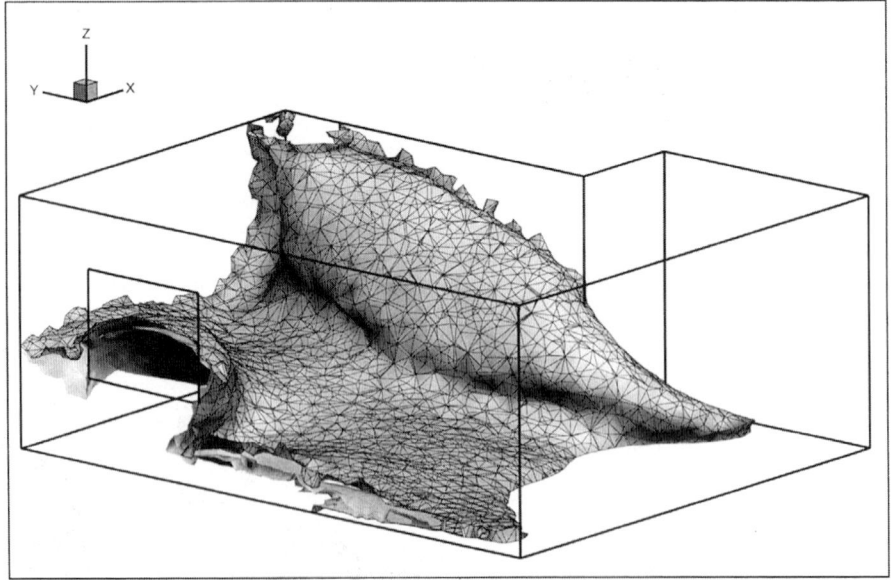

Figure 3: Region of temperature below 18 °C after a ventilation period of 40 seconds,
view from the wall with the window

Beside the permanent testing of balances we investigated the computational expense to carry out those calculations. For configurations shown in figures 1 to 3 we found a similar expense as working with specified velocity and temperature profiles at the openings.

OUTLOOK

To get some experimental data for comparison with the computational results a small test chamber (1m x 1m x 2 m) was built, based on the configuration used to develop special wall functions for natural convection, compare Neitzke (1998). In the next once experimental data are gained, this situation will be numerically investigated in detail.

In addition the code was extended by special procedures to calculate time dependent values of air exchange efficiency and the age of air. To get more realistic boundary conditions it is possible to interact with a thermal building simulation program by use of the procedures of PVM (Parallel Virtual Machine).

References

Auge, A., Kapurkin, A. Lube, G., and Otto, F.-C. (1998). A note on domain decomposition of singulary perturbed elliptic problems. *Proceedings of the Ninth International Conference on Domain Decomposition Methods*. John Wiley & Sons.

Carey, G.O. and Oden, J.T. (1986), *Finite Elements: Fluid Mechanics, Volume VI of the Texas Finite Element Series*, Prentice-Hall, Englewood Cliffs, New Jersey.

Heywood, J.G., Rannacher, R. and Turek, S. (1996). Artificial boundaries and flux and pressure conditions for the incompressible Navier-Stokes. *International Journal for Numerical Methods in Fluids* **22**, 325-352.

Hughes, T.J.R., Franca, L.P. and Balestra, M. (1986). A new finite element formulation for computational fluid dynamics: V. Circumventing the Babuska-Brezzi condition: A stable Petrov-Galerkin formulation of the Stokes problem accomodating equal-order interpolations. *Computer Methods in Applied Mechanics and Engineering* **59**, 85-99.

Hughes, T.J.R., Franca, L.P. and Hulbert, G.M. (1989). A new finite element formulation for computational fluid dynamics: VIII. The Galerkin/least-squares method for advective-diffusive equations. *Computer Methods in Applied Mechanics and Engineering* **73**, 173-189.

Jiang, B. (1998), *The Least-Squares Finite Element Method*, Springer, Berlin Heidelberg.

Lube, G., Otto, F.-C. and Müller, H. (1998). A non-overlapping domain decomposition method for parabolic initial-boundary value problems. *Applied Numerical Mathematics* **28**, 359-369.

Neitzke, K.-P. (1998). The Behaviour of the Flow in Rooms near Walls - Measurements and Computations. *Proceedings of the 6th International Conference on Air Distribution in Rooms*. KTH Stockholm, Sweden.

© 2000 Elsevier Science Ltd. All rights reserved.
Air Distribution in Rooms, (ROOMVENT 2000)
Editor: H.B. Awbi

A PARTICLE STREAK TRACKING SYSTEM (PST) TO MEASURE FLOW FIELDS IN VENTILATED ROOMS

D. Müller, U. Renz

Lehrstuhl für Wärmeübertragung und Klimatechnik
(Institute of Heat Transfer and Air Conditioning)
Aachen University of Technology (RWTH Aachen), Germany

ABSTRACT

The Particle Streak Tracking System (PST) is a fast method to measure two- and three-dimensional velocity fields in room air flows with measuring areas up to 5 m^2. The two-dimensional method works with a single pulsed white light sheet and one digital camera. For three-dimensional velocity measurements in planes a laser light sheet system using three separate laser sheets with two different wavelengths and two CCD-cameras is employed. To visualise the flow helium filled bubbles are used. A description of the set-up will be given and the data evaluation process will be explained. Measurements of two and three dimensional air flows will be presented.

KEYWORDS

Air flow pattern, full-scale experiments, measuring techniques, measuring error, particles, air flow distribution.

INTRODUCTION

Particle-Streak-Tracking method (PST) can be used in many technical applications to measure the velocity of tracer particles following the flow of interest. To get accurate measurements of the flow field the tracer particle have to be chosen with respect to the fluid, the desired velocity spectrum, the size of the measurement plane and the picture analysis technique. The PST method needs only one depicted particle in the light sheet for each velocity vector. Particles can be randomly distributed in the measurement area and thus, only low particle densities are necessary for this method making it suitable for room airflow measurements. Large particles can be used to decrease the necessary light intensity in the sheet resulting in less expensive and less dangerous light sources.

For airflow measurements in ventilated rooms small helium filled bubbles with a diameter of 1–2 mm are used as tracer particles. More than 500 bubbles per second are produced with a bubble generator. These particles follow the airflow almost perfectly and there diameters can be adjusted in order to get bubbles with a density comparable to air at different room temperatures. Depending on the air temperature and the humidity the tracer particles will evaporate within 2 to 5 min and do not cause pollution problems in field measurements. These tracer particles can be used to measure airflow velocities in planes of more than 5 m^2.

THE TWO DIMENSIONAL PST-SYSTEM

A Xenon lamp is the light source for the sheet. The light beam passes through a mechanical shutter and is focused into a set of three to six fibres with a length of 7 m. The light is emitted through cylindrical lenses forming the light sheet inside the enclosure. The location of the lenses can be adjusted to the geometrical requirements with respect to a uniform light intensity in the sheet. The thickness of the light sheet can be varied between 10 to 100 mm at distance of 1 m from the cylindrical lenses. A mechanical shutter system is implemented into the light source allowing a pulse duration Δt which may be adjusted to the flow situation of interest. The shutter system creates a pulse sequence consisting of one long pulse followed by a short break and then a short pulse to detect the flow direction. Due to the length of the fibres this light system maintains mobility. The light source can be placed outside the measurement area and no heat generation will disturb the flow.

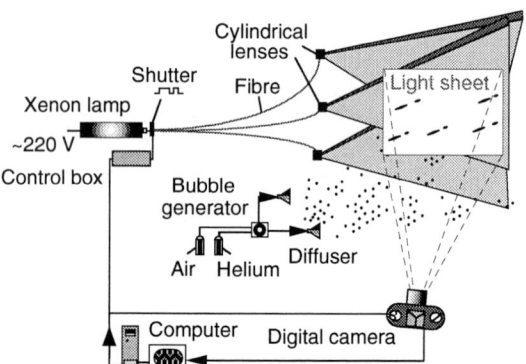

Figure 1: Two dimensional PST experimental set-up

A CCD-Camera is controlled along with the shutter system by the external computer system. The camera is connected to a frame grabber and is able to store up to 50 picture into the computer memory. This feature allows the measurement of instationary flow field. A detailed description of the system is given by Müller & Renz (1998). Further information about the two dimensional-PST system regarding the range of applications and the total cost is available at http://www.wuek.rwth-aachen.de/.

THE THREE DIMENSIONAL PST-SYSTEM

An Argon-Ion Laser is the light source for three sheets with two different colours. The laser beam passes a beam splitter and is divided into a blue (488 nm) and a green (514 nm) beam. The green laser light is pulsed either by a magnetic shutter system (min. pulse time $\Delta t = 5$ ms) or a chopper disc ($\Delta t = 10$ μs). The shutter system creates the same pulse sequence as for the two dimensional PST system.

The pulsed light passes an iris diaphragm, a lens system to increase the beam diameter three times and it is reflected by a mirror and a beam splitter to the cylindrical lens forming the middle pulsed light sheet. The green laser beam passes an iris diaphragm and a half wave plate to adjust the light polarisation for the beam displacer. Two parallel green laser beams leave the beam displacer. The distance of the beams is 2.7 mm and the light intensity difference of the beams can be controlled by the half wave plate. Because of the different beam diameters two parallel cylindrical lenses are used resulting in three comparable sized parallel light sheets. Two digital cameras with wavelength filters for each light sheet colour are used to record two different pictures from every tracer particle passing through the three light sheets during the camera exposure time, as shown in figure 2.

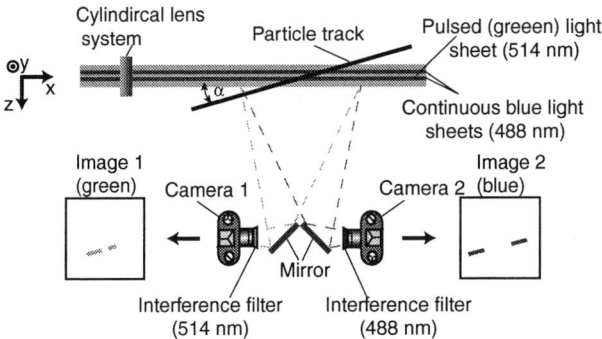

Figure 2: Three dimensional PST experimental set-up

One of these pictures (Image 1) shows the tracer lines from the pulsed light sheet. This picture will be used to determine the in plane velocity components. The other picture (Image 2) contains two tracks from each tracer with a velocity component perpendicular to the sheet passing the two continuous blue light sheets. To get the third velocity component perpendicular to the measurement plane the program analyses image 2 of camera 2. The angle α between the tracer line and the light sheet plane can be found by a simple geometrical relation using the known distance of the two light sheets and the displacement of the two track centres in image 2. First, all possible streak combinations in image 2 will be examined. The program looks for streak combinations in a search environment based on the light sheet parameters. Validation criteria and the related misalignments are applied to all possible streak combinations, see Müller & Renz (1998).

After a successful validation of the streaks in both images the program has to find counter parts of validated streak pairs in the corresponding images of camera 1 and camera 2. After applying a mirror operation to image 2 in order to fit the view of camera 1 the image has to be rotated and shifted using reference points in both images to get the same co-ordinate system for both images. To separate variables from image 1 and 2 a second index is used (C1, C2). All possible links of streak combinations from image 1 and 2 have to fulfil the following conditions.

$$\Delta l_p \leq \frac{\frac{1}{2}\delta_{ss} - \frac{3}{5}\Delta s_{C1}}{\tan\alpha} \; ; \quad \Delta l_p = \frac{(x_{m,C2} - x_{m,C1}) + (y_{m,C2} - y_{m,C1})\tan\beta_{C2}}{\cos\beta_{C1} + \sin\beta_{C1}\tan\beta_{C2}} \tag{1}$$

$$\Delta l_s < \Delta l_{3D,tol} \; ; \quad \Delta l_s = \frac{\Delta l_p \cos\beta_{C1} + (x_{m,C1} - x_{m,C2})}{\sin\beta_{C2}} \; ; \quad \Delta\beta = |\beta_{C2} - \beta_{C1}| < \beta_{3D,tol} \tag{2}; (3)$$

The geometric parameters Δl_s and Δl_p are shown in figure 3.

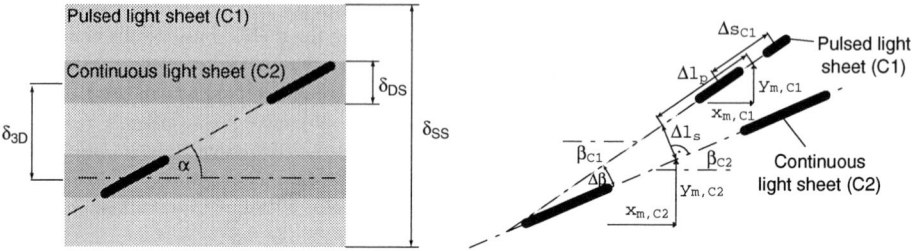

Figure 3: Top view three dimensional-PST light sheets

Still, more than one link of streak combinations from both images might appear. A check of the link quality is necessary.

$$Q_1 = \frac{\Delta l_s}{\Delta l_{3D,tol}} \; ; \quad Q_2 = \frac{|\beta_{i,C1} - \beta_{i,C2}|}{\beta_{3D,tol}} \tag{4}; (5)$$

After the angle α between the light sheet and a tracer line has been calculated, the velocity of the tracer particle perpendicular to the sheet can be determined. The direction of the particles passing the light sheet must be evaluated by an intensity analysis of the two continuous sheets.

TRACER PARTICLE BEHAVIOUR

Helium filled bubbles are used as tracer particles for the flow visualisation with diameters which can be adjusted in the range from 1 - 2 mm. To examine the behaviour of these bubbles in the flow of interest the average density of the bubbles is determined experimentally in a horizontal pipe flow. The horizontal position of more then 5000 bubbles pipe measured by taking digital pictures of each bubble in a light sheet at two different position of the pipe axis. The mean displacement of all bubbles is calculated. The known flow velocity inside the pipe allows to estimate an average falling velocity. Using a simple force balance on a sphere, the average density can be calculated.

The measurements shows that the density ratio $\sigma = \rho_P / \rho_F$ can be adjusted in the range of 0,8 to 1,2 a temperature of 20 °C by changing the air, helium and bubble liquid mass flow.
The dynamic behaviour of the bubbles will be examined with the equation of motion from Techem (1947) for a creeping flow around a sphere with corrections from Corrsin & Lumley (1956). Based on the analysis of Hinze (1959) Hjelmfelt & Mockros (1966) show several exact solutions for different tracer density ratios σ. Because of the their large diameter the assumption of a creeping flow is not valid for helium filled bubbles. Using measurements from Odar & Hamilton (1964) as correction factors c_A and c_H for higher particle Reynolds numbers Re, the equation of motion for a single spherical particle in a moving fluid read as follows.

$$m_p \frac{dv_P}{dt} = \underbrace{\frac{3}{4} \frac{A_F}{A_P d_P} m_P c_W |v_F - v_P|(v_F - v_P)}_{\text{Drag}} + \underbrace{m_F \frac{Dv_F}{Dt}}_{\text{Pressure}} + \underbrace{c_A m_F \left(\frac{Dv_F}{Dt} - \frac{dv_P}{dt}\right)}_{\text{Apparent mass}} \underbrace{-9 C_H \frac{m_P}{D} \frac{1}{\rho_P} \sqrt{\frac{\mu \rho_F}{\pi}} \int_0^t \frac{\frac{Dv_F}{dt'} - \frac{dv_P}{dt'}}{\sqrt{t-t'}} dt'}_{\text{Basset integral}} \qquad (6)$$

$$c_A = 1.05 - \frac{0.066}{A_C^2 + 0.12}, \quad c_H = 0.48 - \frac{0,52}{\left(A_C + 1\right)^3}, \quad A_C = \frac{|v_F - v_P|}{d_P \left|\frac{d|v_F - v_P|}{dt}\right|} \quad \forall \quad Re \leq 60$$

The terms on the right side of the equation correspond to the effects of drag for a viscous fluid, the pressure gradient of the undisturbed flow, apparent mass and the augmented viscous drag from the Basset history term. The drag coefficient depends on the Reynolds number and is calculated with the formula of White (1974). The influence of the Basset term can be not neglected for helium filled bubbles in air flows. The equation has to be solved numerically because of the non linear terms for the drag coefficient and the high Reynolds number corrections. A continuous changing of the flow direction is found in vortices. The motion of a particle in a rotational flow can be described approximately with a sinus curve for one local co-ordinate.

$$v_F(t) = A \omega \sin(\omega t) \qquad (7)$$

The particle acceleration is calculated with equation 6. During the numerical integration process the particle motion reaches a steady state after a few cycles. The program calculates the amplitude ratio $V = A_P / A_F$ and the phase shifting $\Delta\varphi = \Delta t \, \omega$ are calculated. These steady state values are shown in the frequency curves in figure 4. The variation of the frequency corresponds to a variation of the flow velocity.

The figure shows that particles with density ratio σ close to 1 will follow fluid velocity variation very well up to 100 Hz. If the density ratio σ is larger than 1.2 or smaller than 0.8 the particle motion will be damped or intensified when going to higher frequencies. In large scale fluid flows as in room air ventilation the PST system has to measure large scale turbulent motions with a low frequency. Helium filled bubbles have a reasonable frequency response for this kind of applications.

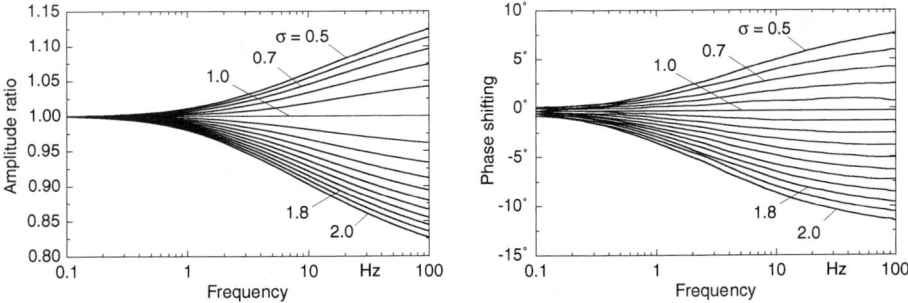

Figure 4: Amplitude ratio $V = A_P / A_F$ and the phase shifting $\Delta\varphi = \Delta t \, \omega$ of helium filled bubbles with varying density ratios $\sigma = \rho_P / \rho_F$

APPLICATIONS

As an example with the two dimensional PST system air flow velocities are measured in one plane of a 6 x 4 x 3 m³ test room with a swirl ventilation system. The two dimensional PST measurements are taken at three sub areas (constant x-co-ordinate) in the room with two heated dummies and a computer. During the post processing all sub planes are connect to one room plane and all velocity vectors are interpolated yielding to the flow map shown in figure 5. The PST data can be easily compared to the numerical predictions.

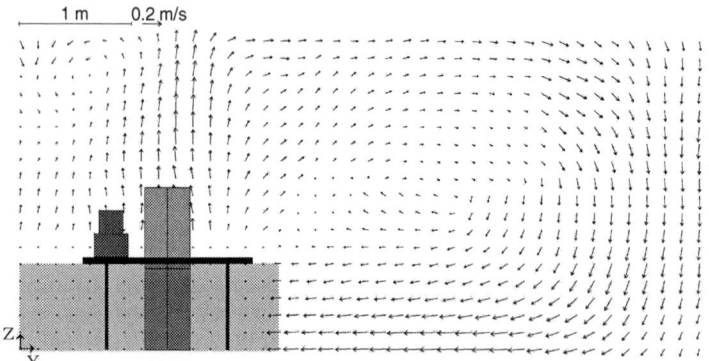

Figure 5: Measured vector flow map of a ventilated room

In order to test the accuracy of the three dimensional PST system the out of plane velocity component of a steady state heat exchanger intake flow is examined. The heat exchanger and the channel expansion are made of acrylglas except of the boiler tubes. The air flow velocities are measured in two vertical planes and one horizontal plane as shown in figure 6. The flow field is almost two dimensional and thus, averaging the measured out of plane velocity component in y-direction of the two vertical planes should show the same values as the related in plane measurements of the horizontal plane.

The plot in figure 6 shows good agreement of the average z-velocity of the vertical and the horizontal plane. The three dimensional PST system is able to measure correctly the out of plane velocity component.

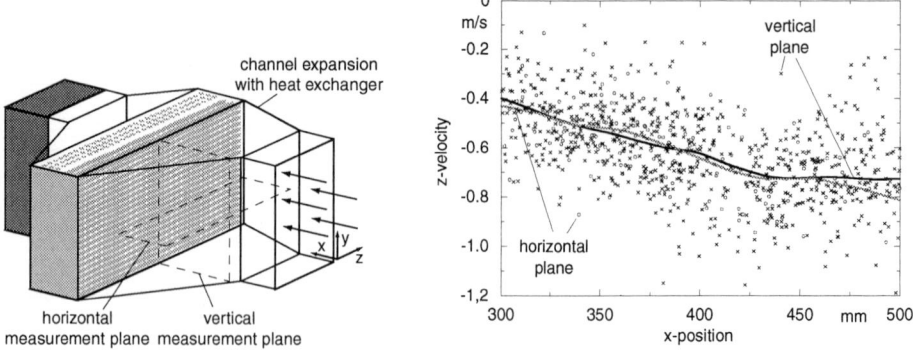

Figure 6: Test set-up and results for the three dimensional PST system

CONCLUSIONS

The 2D-PST system is able to measure room airflow velocity in large planes. The systems allows the recording of instationary flow field up to 50 frames. It is easy to handle and cause no safety or pollution problem in field measurements.

The developed 3D-PST measurement system is a fast and straight forward method to measure all three velocity components in a plane. Under stationary flow conditions it is possible to get velocity measurements of different planes within the measurement area and thus, it is possible to get an overview of the airflow pattern in a large field during a short measurement period. All measurements can be taken applying a global reference frame system for the reconstruction of the whole flow field.

The velocity field of a ventilated room and an heat exchanger intake flow has been measured with the PST system. For the 3D-PST a comparison of the velocities evaluated in the horizontal and vertical planes showed, that the system achieves an excellent accuracy of the velocity component perpendicular to the image plane.

ACKNOWLEDGEMENTS

This project was jointly funded by the German Federal Ministry for Research and Technology (BMBF 0329160A and 0329016B) and the Heinz Trox-Stiftung, Germany. The responsibility for the content of this paper lies solely with the authors. The support of the organisations which have co-operated in the project and the excellent support of Armin Knels, Bernhard Müller and Kurt Nährich are gratefully acknowledged.

REFERENCE

Abrahamson, S. D., Koga, D. J.; Eaton, J. K. (1988) An Experimental Investigation of the Flow Between Shrouded Co-rotating Disks, Report MD-50 of the Thermoscience Division of Mech. Eng. Dept., Stanford University, Stanford, Ca., USA

Corrsin, S., Lumley, J. (1956) Appl. Sci. Research, Vol. 6, 114.

Hinze, O.J. (1959) Turbulence, an Introduction to its Mechanism and Theory, McGraw-Hill Series in Mechanical Engineering, New York.

Hjelmfelt, A. T., Mockros, L.F. (1966) Motion of Discrete Particles in a Turbulent Fluid, Applied Scientific Research, Vol. 16, 149-161.

Müller, D; Renz, U. (1998) A Low Cost Particle Streak Tracking System (PST) and a New Approach to Three Dimensional Airflow Velocity Measurements, Proceedings ROOMVENT '98, Stockholm, Sweden.

Odar, F. Hamilton, W.S. (1964) Forces on a Sphere Accelerating in a Viscous Fluid, J. Fluid Mechanics, Vol. 18.

Rodieck, B. (1995) Best-fitting Ellipse Routines, IMAGE 1.58, Department of Ophthalmology, University of Washington, Seattle, USA.

Techen, C.M. (1947) Mean Values and Correlation Problems Connected with the Motion of Small Particles Suspended in a Turbulent Fluid, PhD-Thesis, Delft.

White, Frank M. (1974) Viscous Fluid Flow, McGraw-Hill Series in Mechanical Engineering, New York.

Air Distribution in Rooms, (ROOMVENT 2000)
Editor: H.B. Awbi

APPLICATIONS OF LASER VELOCIMETRY TECHNIQUES FOR AIR FLOW ANALYSIS IN POLLUTANT TRANSFER STUDIES

C. Prévost, N. Dupoux, J.C. Laborde

Institut de Protection et de Sûreté Nucléaire
Département de Prévention et d'Etude des Accidents
Service d'Etudes et de Recherches en Aérocontamination et en Confinement
CEA/Saclay – Bâtiment 383 – 91191 GIF-SUR-YVETTE CEDEX, France

ABSTRACT

The present paper concerns two different applications of the PIV (Particle Imaging Velocimetry) and LDV (Laser Doppler Velocimetry) optical techniques. These non-intrusive investigation methods are implemented in our research laboratory in order to visualise and understand the airflow structure and behaviour in various ventilated systems. The velocity mappings acquired during applications to air flow analyse under thermal operating conditions or around a dynamic confinement enclosure show the efficiency of these two complementary techniques to produce full of data.

KEYWORDS

Particle Imaging Velocimetry, Laser Doppler Velocimetry, airflow structure, pollutant transfer.

GENERAL CONSIDERATIONS

In normal or accidental conditions, airborne pollutant transfers are closely linked to the structure and the velocity of the air flowing inside rooms or containment enclosures. The PIV and LDV laser techniques applied to fluid flows analysis have undergone tremendous developments in the last years. They are used to give, with a good accuracy, a qualitative and quantitative insight into wide and complex flow structures in order to interpret and understand flows phenomena in various laboratory devices.

Particle Imaging Velocimetry

PIV is a quantitative flow visualisation technique which allows the mapping of instantaneous fluid field in space and time. The flow seeded by micron-sized particles is illuminated by a thin light sheet from a laser source. The processing techniques of the acquired cartographies are continuously improved. The variety of different computational Fourier transform and cross correlation techniques producing the velocity vectors, allows the operator to improve his experimental results.

Laser Doppler Velocimetry

This technique requires a tagging material to be able to measure the velocity of the fluid flow in any ventilated system. Particles moving with the fluid are illuminated with two-focused laser beams which form the measurement volume. The velocity of the particles is determined from the change in frequency (Doppler shift) of the light scattered by the particles. This technique is a very accurate but local method to follow in time velocity components, velocity averages, random medium squares, turbulence and others data.

FLOW ANALYSIS UNDER THERMAL CONDITIONS

The aim of the first application is to understand, and then to forecast, the behaviour of gaseous flows as in a ventilated room, especially in the case of nuclear surroundings, when a severe accident such as a fire occurs. In this case, the fire source gives off fumes and pollutant particles which can be released in the environment via the global ventilation network.

Experimental device

The device used is a scale model system ($1m^3$) representing a ventilated enclosure, as shown in figure 1. This reduced experimental model is designed from mathematical considerations to keep the dynamic similarity with a 100 m^3 full scale room. A geometrical similarity factor of 1/5 is also imposed for practical reasons. The location of the supply and exhaust openings can be modified according to the test conditions (locations near the ceiling or near the floor). The fire is simulated by a gas burner located at the floor, in the centre of the room. The heat release rate is constant during each experiment and equal to 1.2 or 3 kW. The transparent walls allow an optical access for the PIV technique inside the tested enclosure. A frequency-doubled Nd:YAG laser is used to illuminate micro-sized oil droplets (the seeded material for the tests). The laser-sheet energy is typically 200 mJ/pulse with a sheet thickness of 3 mm at the test section. The laser energy must be as risen as possible to provide enough light for the recorded image. The tested global area is effectively much larger than areas generally tested by this technique, and requires special optical lens with large divergence angle (40°). A 28 mm Nikon objective is used to record experimental frames under the charge coupled device camera with a 2 mm^2/pixel resolution. The flow seeding is one of the main difficulties encountered because the enclosure comprises several air flow distribution zones. Consequently, four simultaneous injection points are implemented in order to obtain a full seeding.

Results

Some main results corresponding to velocity mappings, respectively in the supply opening (plan A) and in the plume (plan B), are given in figure 2. Mappings in this areas are the trickiest because of the large velocity gradient present in the air jet and in the other parts of the tested map. Figure 3 shows the horizontal direction of the air velocity at the room inlet. Then, an incline becomes apparent due to the density effects in the ventilated room. Actually, a difference (approximately 60°C) between the supply air temperature and the gas temperature inside the room is observed. This phenomenon induces a full mixing of the air in the room. The mapping presented on the figure 4 is a 200 acquisitions mean. We note the plume incline, as well as its broadening, along the height of the room.

FLOW ANALYSIS AROUND A DYNAMIC CONFINEMENT ENCLOSURE

In this second application, the purpose of the study consists of going beyond the standardised tests generally implemented to control the dynamic confinement efficiency in a ventilated system, such as a fume cupboard for example. In many industrial sectors, the dynamic confinement is used in order to protect the operator from a toxic agent. In this work, the parameter studied is the back flow factor of various gas or particles tracers through a dynamic confinement barrier. We have underlined in this study the large influence of this factor on the gas flow structure around and inside the opening sash of the tested enclosure.

Experimental device

The test bench used for this application includes an enclosure equivalent to a laboratory fume cupboard (figure 5), equipped with a variable front opening sash of 1 m length and 0.6 m height, and with an air exhaust system. In front of this enclosure model, a manikin with a standard height simulates the presence of a working operator. A LDV one component system is used in order to acquire some global velocity mappings in the sash opening plan of the enclosure. It consists of an Helium-Neon laser which emits a 10 mW beam power. The optic probe is easy to handle because of its small dimensions and the use of optic fibers. The focal lens is typically 400 mm. The system allows the perpendicular and tangential velocities measurements in the same plan. During these LDV measurements, we have noticed some characteristic zones around the enclosure opening where the PIV investigations can provide complementary results. For this application, PIV instrument is a very compact system consisting of a frequency-doubled Nd:YAG laser with laser-sheet energy of 15 mJ/pulse. The laser-sheet is formed by means of a 15° divergence lens and a 60 mm Nikon objective is used to record captured images on the camera with a 0.002 mm^2/pixel resolution.

Results

The numerous LDV measurement points are represented in figure 5. They are located on the drawing grid represented at the opening place of the tested system. Several mappings are obtained with or without a manikin stationed in front of the fume cupboard. The main results presented here are 2D mappings of the perpendicular component U of the air velocity when standardised configurations are respected (opening height = 0.4 m and mean velocity at the opening = 0.5 m/s). The results show the U velocity components, particularly when the manikin is present (figure 7). The lower velocities which are associated with the stronger turbulence are observed near the enclosure edges. These phenomena increase with the presence of the manikin. In this case, the velocity outline takes a hollow shape especially in the middle of the opening area. On the other hand, velocity values detected on both sides of the manikin increase appreciably and decrease near the enclosure side edges. At the different zones notified Mh, Mb, D and N (figure 5), PIV measurements enable visualisation of the flow structure close to the enclosure edges and between the operator breathing ways and the top of the opening (figures 8 to 11). In most cases, we can appreciate the turbulent profile of the flow near the edges with the presence of many variable vortexes. The PIV mappings produced can be exploited to predict the probability of pollutants re-entering the test locations. In figure 12, the PIV method implemented at the Mb tested zone allows the visualisation and quantification of re-entering, seeding in terms of velocity, when the particles are injected inside the enclosure. This transfer is a random phenomenon which is quite difficult to characterise.

CONCLUSION

This paper illustrates a synthetic outline of the ability of two investigation techniques, PIV and LDV, to highlight air flows phenomena playing a great part on the airborne pollutant transfers. Under fire conditions, density effects involve an incline of the air supply and of the thermal plume. Around a fume cupboard, edge effects increased with the operator presence induce strong turbulence and transfer of the seeding particles from the inside of the fume cupboard to the outside.

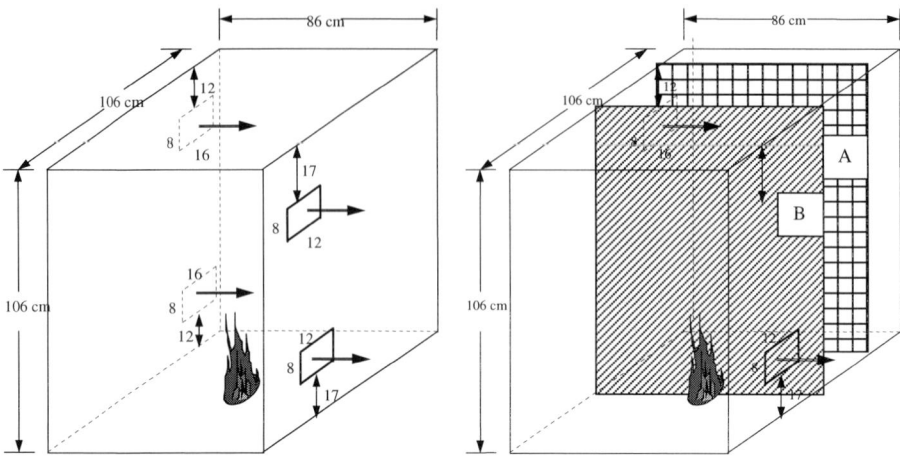

Figure 1: The ventilated scale model

Figure 2: Planes (A and B) tested in the model

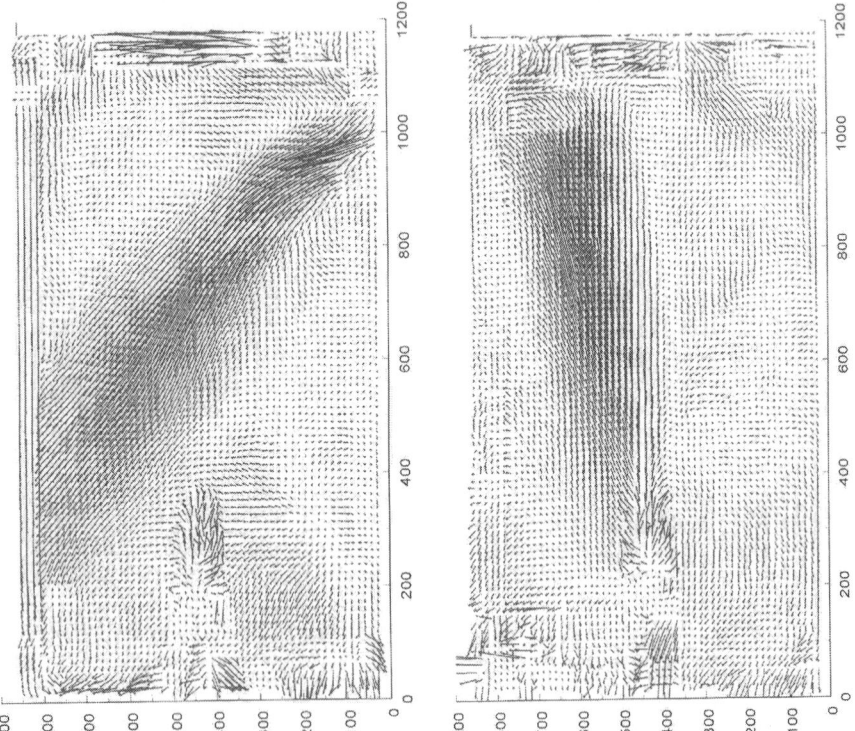

Figure 3: Air supply area (plan A)
(1060x860 mm²)

Figure 4: Thermal plume (plan B)
(1060x860 mm²)

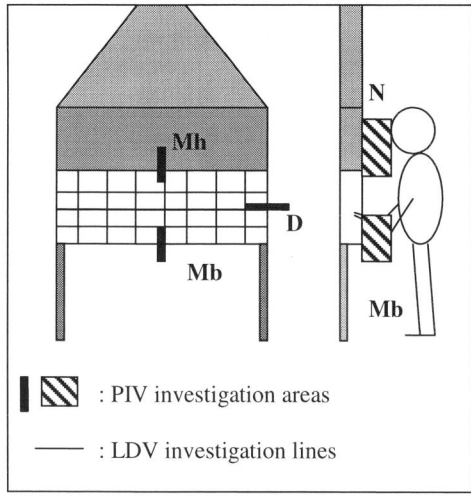

Figure 5: Front and side views of fume cupboard

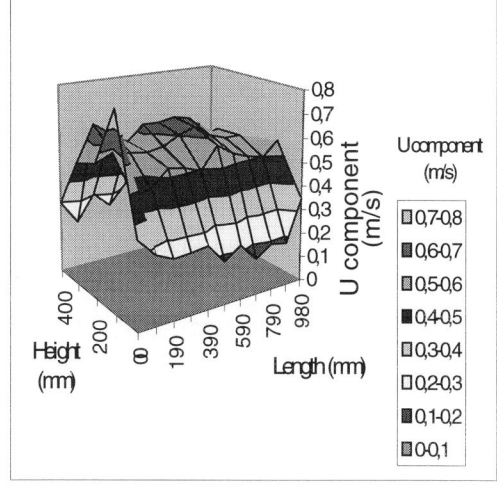

Figure 6: LDV investigation (without manikin)

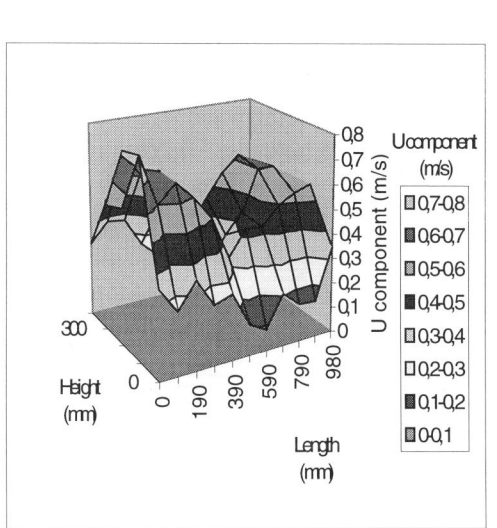

Figure 7: LDV investigation (with manikin)

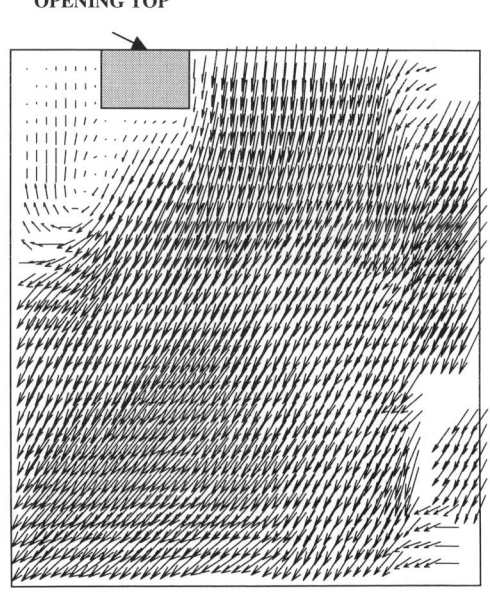

Figure 8: Mh PIV mapping (50x72 mm^2)

338

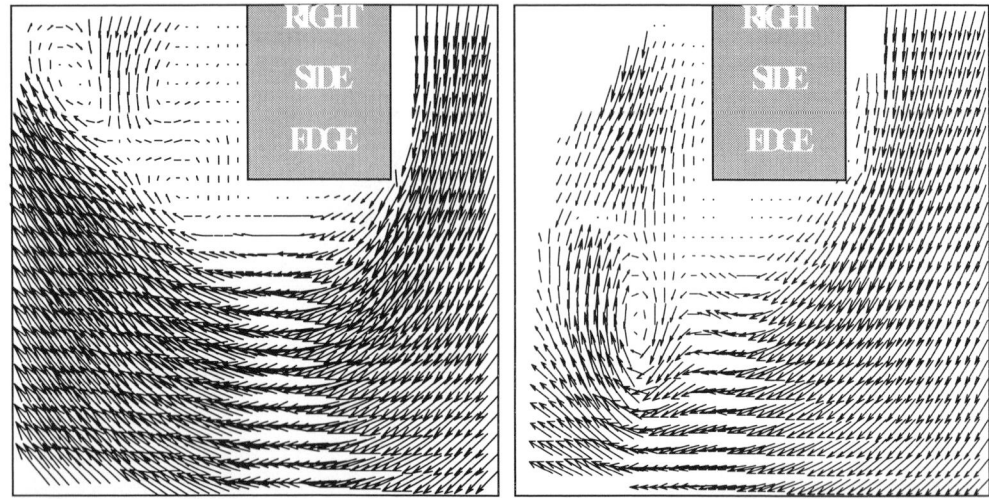

Figure 9: D PIV mapping (50x72 mm^2)

Figure 10: D PIV mapping (50x72 mm^2) acquired 100 ms after the mapping on figure 9

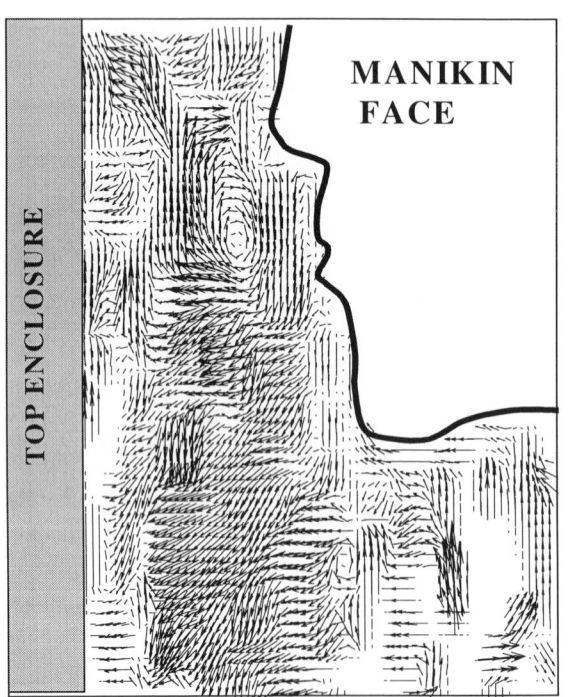

Figure 11: N PIV mapping
(160x200 mm^2)

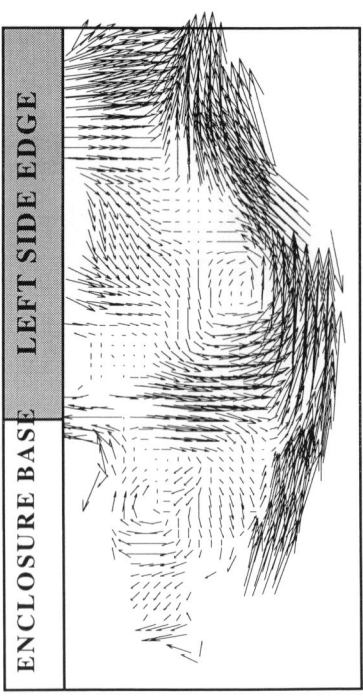

Figure 12: Mb PIV mapping
(50x72 mm^2)

Air Distribution in Rooms, (ROOMVENT 2000)
Editor: H.B. Awbi

VISUALIZATION AND MEASUREMENT OF AIR TEMPERATURE USING INFRARED THERMOGRAPHY

M. Cehlin[1,2], B. Moshfegh[1] and M. Sandberg[2]

[1] Division of Energy and Production Systems, Department of Technology,
University of Gävle, S-801 76 Gävle, Sweden
[2] Department of Built Environment, Royal Institute of Technology, S-801 02 Gävle, Sweden

ABSTRACT

The study of indoor climate in a large room requires the measurement with high resolution of many different parameters, such as velocity and temperature of the air, throughout the room. The simultaneous measuring of air temperature within a large area with present technology (point measuring technique) is very time-consuming and expensive; it also yields insufficient information. This paper reports research on a whole-field measuring technique to record and visualize air temperatures based on infrared thermography. The air temperatures are measured indirectly by use of a low thermal mass screen in conjunction with infrared thermography. Measuring screens that differ with respect to conductivity, structure, emissivity, etc, were studied in order to improve the performance of the method. The experimental result shows that this measurement method is very useful for visualization of the airflow pattern and temperature distribution in a room with displacement ventilation. However, measuring screens presented slightly higher temperatures than the air in the cold airflow stream. The source of error seems to originate from radiation from the surrounding objects and surfaces. A new dimensionless term, called Radiation number (Rad), was introduced in this paper as a measure of the radiation effect compared to convective effect on the measuring screen. The best infrared thermography measurement results were obtained with screen made of thin paper, which is also cheap and easy to apply. In order to optimize the performance of the method a homogeneous measuring screen should be used with high emissivity because of minimal background radiation and reflections from surrounding objects.

KEYWORDS

Thermography, whole field measurement, infrared camera, visualization, air temperature, airflow pattern

INTRODUCTION

People spend almost 90% of their time in different surroundings as houses, workplaces, and means of transportation. It is well known that the indoor climate has a strong influence on health and comfort. The thermal indoor climate is a complex function of a number of physical variables, such as air temperature, air velocity and pollutant concentration. It is therefore of great importance to be able to map indoor climate within large areas with simple and correct methods in order to optimize and arrange a pleasant and comfortable environment. But the simultaneous mapping of e.g. air temperatures within a large area is very time-consuming with present technology (point measuring techniques), and gives insufficient information as well. In today's design process the indoor climate is often only represented by a single temperature and velocity. Therefore, you can say that the indoor climate is invisible. This implies that it is more or less impossible to demonstrate the difference in performance between systems, and you cannot predict the consequences of a certain air distribution method. The design stage becomes in practice a trial and error process often resulting in malfunctioning systems. Consequently, there is a big need for modernization and development of new and better measurement methods in order to improve the indoor climate.

This paper reports research on a whole-field measuring technique to record and visualize air temperatures based on infrared thermography. The air temperatures are measured indirectly by the use of a low thermal mass screen in conjunction with infrared thermography. The measuring screen is mounted parallel to the airflow, acting as a target screen. This new method makes it possible to measure air temperatures quickly and easily at any cross section of a ventilated room, with very high spatial resolution. Another important advantage is that all temperature measurements are stored digitally and are presented as digital "pictures". A visualization of the temperature distribution in the room is obtained. This technique can also be used for visualization of airflow patterns, see for example Stetz (1993). A quite similar visualization technique of air temperatures has been developed by Kirkpatric (1995), which is based on liquid crystal sheets.

Infrared thermography has been developed for several decades and is now commonly used in industry and research activities such as building inspection, aircraft inspection, and electrical inspection. Actually, infrared thermography based on modern Focal Plane Array cameras is now currently used to measure the heat transfer coefficient with good accuracy, see Sargent (1998). But measurement of air temperature with infrared thermography is a quite new method. Some earlier studies are reported by Hassani (1994a), Hassani (1994b) and Stetz (1993). They measured air temperatures in regions with very high velocities (free jets) and the technique introduced by them can only be applied when the assumption of uniform background temperature is not violated. Sundberg (1993) has used thermography to make a rough estimate of the airflow pattern and the temperature distribution from an air supply diffuser.

This new measurement technique is nevertheless problem free. The need for some kind of a measuring screen assembled in the room can give rise to errors in measurement, such as:

- Measuring screen disturbs the airflow.
- Measuring screen does not obtain the same temperature as the local air temperature.
- The infrared camera does not accurately determine the screen surface temperature.

The objective of this work was to study how accurate air temperatures in a flow from a low velocity diffuser can be measured by infrared thermography. Low velocity diffusers in displacement ventilation supply ventilation air directly into the occupation zone, which makes the near zone of the diffuser very critical for the indoor performance. The work presented here is part of a research program with the aim of "Making the indoor climate visible at design stage". It includes whole-field measuring techniques for temperature, concentration (tomography), velocity (particle streak velocimetry) and Computational Fluid Dynamics.

EXPERIMENTAL PROCEDURE

The first purpose of the experiments described in this section was to identify possible effects of the measuring screen on the airflow field near the air supply diffuser. Thereafter, experiments were carried out to study possible differences between the local air temperature and the screen surface temperature and to compare the accuracy of air temperature measurements by infrared thermography relative to thermocouple measurements.

The experiments were carried out at the laboratory of Ventilation and Air Quality at the Royal Institute of Technology, Gävle, Sweden. The experiments were performed in a typical office room with displacement ventilation at steady-state conditions. This real-scale test-room was built inside a large room kept at same temperature as the test-room to minimize heat-flow through the test-room walls. For more information about the test room see reference Larsson (1999). All experiments reported were carried out with a 550mm high semicircular low-velocity diffuser placed centrally at the left wall down at the floor. The supply flow rate was varied between $10-30\cdot10^{-3}$ m^3/s and the supply air temperature was between ca 3-6 °C under room mean air temperature. The experimental region was 4.1x3.4x2.7m, which is like a typical office room.

Air temperatures in the room as well as surface temperatures on the walls and measuring screen were measured with copper-constantan thermocouples. The thermocouples on the wall and screen were covered with an adhesive tape almost with the same color and emissivity as the wall and the screen. Air temperatures were measured with the sensor uncovered. All thermocouples were individually calibrated and connected to a computer controlled data acquisition system to evaluate the values from the thermocouples. Measurements of air velocities in this paper have been carried out with a constant-temperature single-sensor wire probe type 55P81 from Dantec Measurement Technology. The probe is calibrated between 0.03-2.0 m/s with Dantec's calibration system. To measure air temperatures and air velocities the hot-wire probe and a thermocouple was placed on a computer controlled traverse, which made it possible to measure all point with the same probe and thermocouple and thus disturb the air flow minimally.

To identify the effect of the measuring screen on the air flow field and on the temperature distribution, air velocities and temperatures were recorded in the near-zone of the diffuser with and without a screen placed parallel with the airflow. A frame mounted perpendicular to the diffuser, parallel with the airflow, held the measuring screen in the air stream during the experiments. The measuring screen was about 600mm high and 1000mm long. Measurements were made at several points in all three dimensions.

Measuring screens that differ with respect to conductivity, structure, emissivity, etc, were employed and studied, in order to minimize temperature differences between the screen and air and hence improve the performance of the method, see table 1. Screen surface temperatures and the corresponding local air temperatures were measured at two horizontal distances from the diffuser (50mm and 300mm) for different indoor conditions.

TABLE 1

Properties of studied screens.

Material	Emissivity [1]	Conductivity (W/mK) [2]	Thickness (mm)	Surface type
Paper	0.92	0.1	0.2	solid
Rubber net	0.8	0.2	0.3	porous
Aluminium-foil	0.1	200	0.1	solid

[1] measured by infrared camera
[2] ref. Bejan (1993)

To identify the effect of the measuring screen on temperatures recorded by the infrared camera, different kinds of screens were employed and the temperatures were compared with thermocouples. The camera was an AGEMA 570 to measure the emitted infrared radiation from the screen. This type of camera has a FPA detector, with 320x240 pixels, sensitive to long-wave radiation (7.5-13 μm), which is suitable for detecting room temperature radiation (≈293 K). Infrared cameras measure and image the emitted infrared radiation from an object. The fact that radiation is a function of the object surface temperature makes it possible for the camera to calculate and display this temperature. However, the radiation measured by the camera does not only depend on the object temperature but is also a function of the emissivity. Radiation also originates from the surroundings and is reflected in the object. The radiation from the object and the reflected radiation will also be influenced by the absorption of the atmosphere. Therefore, to measure temperature accurately, it is necessary to compensate for the effects of a number of different radiation sources. This is done on-line automatically by the AGEMA 570, provided that emissivity, surrounding temperature and distance is supplied to the camera. These object parameters must be set correctly to achieve accurate measurements. The most important object parameter is emissivity. The hardest parameter to estimate is the surrounding temperature, because the temperature of surrounding surfaces is non-uniform. Therefore, some kind of a weighted mean surrounding temperature must be estimated and supplied to the camera. The measuring method introduced in this paper is in reality only valid at uniform surrounding temperature; as local heat sources or sinks can cause measurement errors. But if screen emissivity is high, the fraction radiation originating from the surroundings is small, as is detected by the camera. Under these circumstances local loads or sinks in the room cause negligible measurement errors.

EXPERIMENTAL RESULTS

Visualization of Flow Pattern and Temperature Distribution

Figure 1 and 2 show thermal images compared to smoke visualization, near the inlet diffuser at different Ar-number. As shown by the figures, this measurement method seems to be a very useful tool for the illustration and visualization of the relative temperature distribution and the airflow pattern in a room with displacement ventilation. However, as a measuring screen is interposed in the airflow, this screen can affect the actual airflow pattern and the temperature profile near the diffuser.

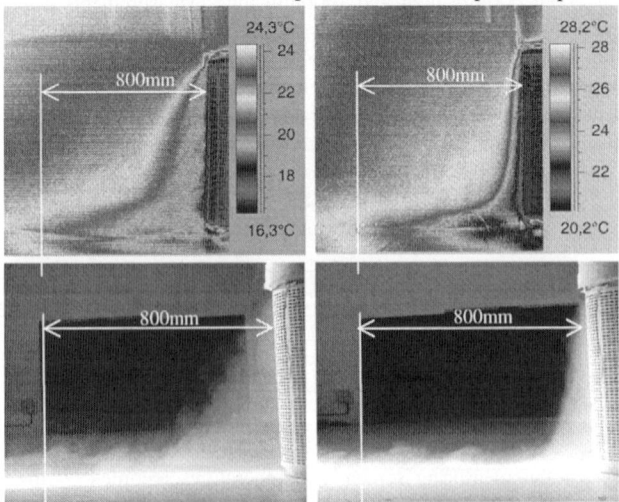

Figure 1. Thermal image compared to Smoke visualization. Ar = 1.25.

Figure 2. Thermal image compared to Smoke visualization. Ar = 3.5.

To identify the effect of the measuring screen on the air flow field and on the temperatures distribution, air velocities and temperatures were recorded in the near-zone of the diffuser with and without a screen placed parallel with the airflow. Several hot wire and thermocouple measurements indicate that a parallel-placed measuring screen doesn't disturb neither the airflow pattern nor the temperature distribution from a low-velocity diffuser in an enclosure. However, Hassani (1994a) found that in a free jet a porous screen created a thicker boundary layer than a smooth solid screen. This implies that there can be local perturbations in the actual airflow pattern as a result of the measuring screen at certain flow conditions. See ref. Linden (2000) for more information about comparison between infrared thermography and other visualization methods for airflow pattern from a low velocity diffuser.

Effect of Screen Material on the Temperature Measurements

In order to be able to optimize and improve the performance of the method, an energy balance was carried out for a measuring screen placed in a room with displacement ventilation. The screen gained energy by radiant heat transfer from warmer surrounding walls and objects. At steady state conditions, the radiant heat flow absorbed by the screen from surrounding objects was of the same magnitude as the energy leaving the screen by convective heat flow. The effect of the lateral heat conduction through the measuring screen is used to be negligible. The energy balance over the screen is shown in figure 3. The screen has radiant heat exchange with surrounding walls and objects. Assuming there were no heat sources or heat sinks in the room, energy was only transferred from the floor, roof and walls to the screen.

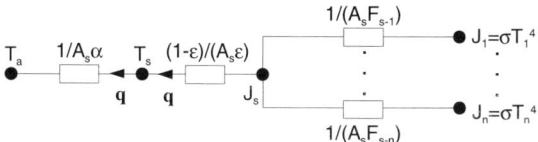

Figure 3. Energy balance over screen.

All the surrounding surfaces were well insulated and the areas were large compared to the screen. Under these circumstances the surface resistance of all walls is approaching zero, which makes them behave like blackbodies with an emissivity of 1.0. The radiant heat flux absorbed by the screen is defined by

$$\sum_{i=1}^{n} \frac{\sigma \cdot T_i^{\,4} - J_s}{\dfrac{1}{A_s \cdot F_{s-i}}} + \frac{\sigma \cdot T_s^{\,4} - J_s}{\dfrac{1-\varepsilon}{\varepsilon \cdot A_s}} = 0 \tag{1}$$

Energy gain is equal to energy loss, leading to a dimensionless relation between air temperature and screen temperature, by applying a number of new terms:

$$T_a^{\,*} = 1 - \mathrm{Rad}, \tag{2}$$

where

$$\mathrm{Rad} = \frac{\sigma \cdot \varepsilon \cdot T_s^{\,3}}{\alpha} \cdot \left(\sum F_{s-i} \cdot T_i^{\,*4} - 1 \right) \tag{3}$$

The new dimensionless term Rad, called Radiation number, is a measure of the radiation effect compared to the convective effect on the measuring screen. If Rad is close to zero, then the measuring screen will adopt a surface temperature near the local air temperature. As seen by Eq. 2 and 3, the surface temperature of a certain screen is highly depending on the screen emissivity. The screen emissivity should be as small as possible in order to minimize radiation heat exchange with surrounding objects. However a screen with low emissivity might by hard to use as a measuring screen due high reflection in the screen causing possible camera measurement errors. Another crucial parameter for the performance of the method is the convective heat transfer coefficient α. If α is large, convective heat transfer will be dominating and leading to a screen temperature close to the local air temperature. But in displacement ventilation air is supplied at low velocities leading to high Radiation numbers along the screen resulting in high thermal resistance between screen and local air. This can introduce surface temperatures not equal to the local air temperatures.

To correctly determine the local air temperatures from infrared thermography measurements, knowledge of α and all view factors (F_{s-i}) between screen and surrounding walls is essential. Research has been done to estimate α and the Nusselt number for special cases and well-defined airflow's, e.g a buoyant wall jet, see Kapoor and Julria(1989). But no research has been done to estimate α along a screen exposed to an airflow pattern from a low-velocity diffuser. The estimate of α and F_{s-i} is outside the scope of this work but will be examined in future research.

To compare the different screens, the temperature of the screens and corresponding local air temperature 10 mm from the screens was measured simultaneously by thermocouples. Figure 4 shows the temperature difference between the screen surface and local air, 50mm and 300mm from the diffuser, for two different supply air conditions. At 50mm the temperatures of the screens were all higher than the corresponding air temperature. At 300mm the screen temperature was again higher, but only near the floor. The experiments showed that all tested measuring screens exhibit slightly higher temperatures than the air in the cold airflow stream. The surface temperature of the paper and net screen was about 1°C and 2°C warmer then the air temperature in the cold region. The surface temperature of the aluminium foil was about 0.5°C warmer than the air temperature. Outside the cold airflow stream, the temperature of the air is almost at the same as the surroundings. In this region the screens adopted the same temperature as the air. The experiments indicate that radiation heat transfer is the dominating heat transfer mechanism and the lateral heat conduction seems to be negligible. This is in good agreement with Roots (1994), who reported no temperature difference between a painted aluminium foil ($\varepsilon=0.9$) and a paper screen ($\varepsilon=0.9$), thus aluminium has about 2 000 times higher

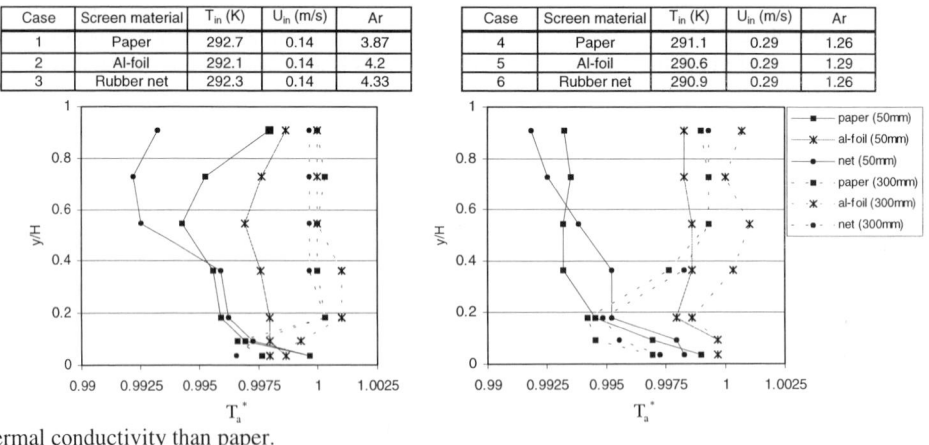

thermal conductivity than paper.

Figure 4. The dimensionless temperature T_a^* as a function of Ar number for different type of screens.

Comparison of Infrared Camera Measurements with Thermocouple

The comparison of screen surface temperatures measured by infrared thermography to those measured by thermocouples, were carried out simultaneously. Velocity profiles measured by hot wire were also presented. Figure 5 and 6 show the temperature profile and velocity profile, for two different supply air conditions, at a distance of 50mm and 300mm from the diffuser. All experiments reported very good agreements between infrared thermography measurements and thermocouple measurements of the paper screen surface temperature, as can be seen by figure 6 and 7. However, experiments showed that it is not suitable to use neither aluminium nor net as a measuring screen. Screens with low emissivity, as aluminium, were impossible to use as a target screen due high reflection in the screen leading to very high camera measurement errors. The figures also explore differences between hot wire and thermography measurement results presentation. Hot wire measurements yield mean velocities in a region with relatively low spatial resolution while an infrared camera measure air temperatures momentary with very high spatial resolution.

Screen material	T_{in} (K)	U_{in} (m/s)	Ar
Paper	290.0	0.29	0.72

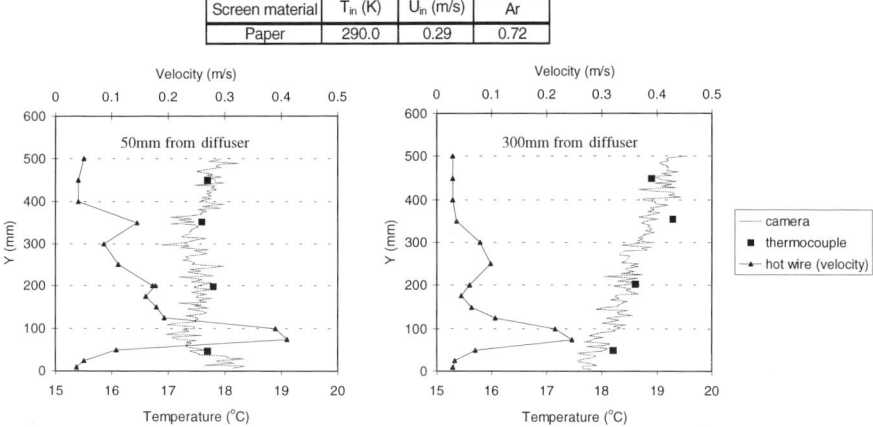

Figure 5. Infrared camera compared to thermocouple and velocity profile for Ar = 0.72. 50mm and 300mm from diffuser.

Screen material	T_{in} (K)	U_{in} (m/s)	Ar
Paper	290.0	0.20	2.10

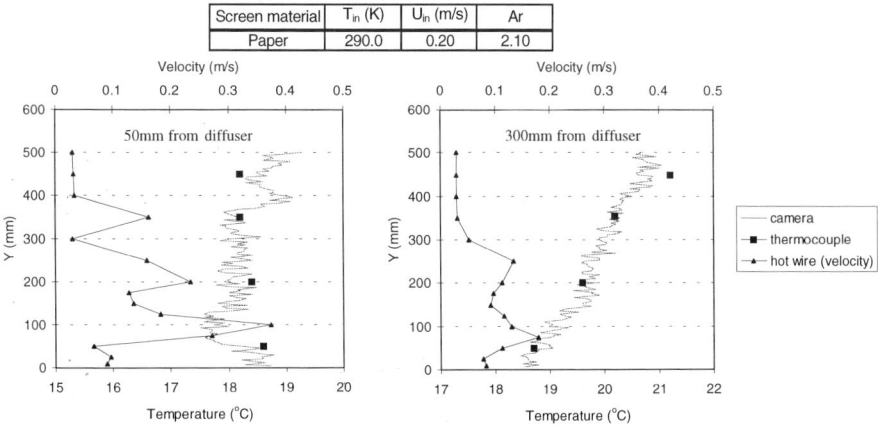

Figure 6. Infrared camera compared to thermocouple and velocity profile for Ar = 2.10. 50mm and 300mm from diffuser.

CONCLUSION

In this work, a whole field measuring method was introduced to record and visualize air temperatures. Experimental tests showed that a parallel placed measuring screen doesn't disturb neither the airflow pattern nor the temperature distribution from a low velocity diffuser in displacement ventilation. This implies that this measuring method is a very useful tool for visualization of the temperature distribution and the airflow pattern from a low velocity diffuser. However, measuring screens presented slightly higher temperatures than the air in the cold airflow stream. The source of error seems to originate from radiation from the surrounding objects and surfaces. A new dimensionless term, called Radiation number (Rad), was introduced in this paper as a measure of the radiation effect compared to convective effect on the measuring screen. The best infrared thermography measurement results were obtained with screen made of thin paper, which is also cheap and easy to apply. In order to optimize the performance of the method a homogeneous measuring screen should be used with high emissivity because of minimal background radiation and reflections from surrounding objects.

NOMENCLATURE

α = convective heat transfer coefficient (W/m^{2}K)
ε = emissivity (-)
σ = Stefan-Boltzman constant, 5.67 x 10^{-8} (W/m^{2}K^4)
A_1 = area of the measuring screen (m^2)
F_{s-i} = view factor between screen and surface i (-)
H = height of diffuser (mm)
J = radiosity (W/m^2)
q = heat flux absorbed by the measuring screen (W/m^2)
T_a = local air temperature (K), (°C)
T_i = temperture of surface i (K), (°C)
T_{in} = inlet temperature (K), (°C)
T_{room} = room temperature (K), (°C)
T_s = temperature of the screen (K), (°C)
U_{in} = inlet velocity (m/s)
y = distance from floor (mm)

$$Ar = \frac{g \cdot (T_{room} - T_{in}) \cdot H}{T_{room} \cdot U_{in}^2} \quad (-)$$

$$T_a^* = T_a / T_s \quad (-)$$

$$T_i^* = T_i / T_s \quad (-)$$

ACKNOWLEDGEMENT

The authors are thankful for the financial support from KK-Foundation (Stockholm, Sweden) and University of Gävle (Gävle, Sweden). The authors gratefully acknowledge the support received by personnel at the Department of Built Environment, Royal Institute of Technology, and especially to Hans Lundström for his helpful suggestions and comments.

REFERENCES

Bejan, A. (1993). *Heat Transfer*, Wiley & Sons, Inc., New York

Hassani, V. and Stetz, M. (1994a). Application of Infrared Thermography to Room Air Temperature Measurements. *Proceedings ASHRAE Transactions*, Part 2, 1238-1247.

Hassani, V. and Stez, M. (1994b). Effect of Local Loads of Negatively Buoyant Wall Jets in enclosed Spaces. *Proceedings ASHRAE Transactions*

Kirkpatric, A.T. (1995). Visualization of Diffuser Outlet Flow Using Liquid Crystal Sheets. *ASHRAE Journal*, August, 159-164.

Larsson, U., Moshfegh, B. and Sandberg, M. (1999). Thermal Analysis of Super Insulated Windows (Numerical and Experimental Investigations). *Energy and Buildings*, vol.29, 121-128.

Linden, E., Cehlin, M. and Sandberg, M. (2000) Temperature and Velocity Measurements of a Diffuser for displacement Ventilation with Whole-field Methods. *Roomvent 2000*.

Roots, P. (1997). Mätning av lufttemperatur inom ett stort område baserat på användning av värmekamera. *Working Paper No 44*, University College Gävle-Sandviken.

Sargent, S. R., Hedlund, C. R. And Ligrani, P. M. (1998) An Infrared Thermography Imaging System for Convective Heat Transfer Measurements in Complex Flows. *Measurement science & technology*, vol.9, 1974-1981.

Stetz, M. (1993). Characterizing cold air jets and diffuser performance via infrared thermography. M.S. thesis. Fort Collins: Colorado State University.

Sundberg, J. (1993). Use of thermography to register air temperatures in cross sections of rooms and to visualize the airflow from air-supply diffusers. *Thermosense XV*, vol. 1933, 61-66, SPIE.

Air Distribution in Rooms, (ROOMVENT 2000)
Editor: H.B. Awbi

VALIDATION TESTS FOR A PASSIVE
TRACER GAS TECHNIQUE

Viktor Dorer, Robert Gehrig, Matz Hill and Andreas Weber

EMPA, Swiss Federal Laboratories for Materials Testing and Research,
CH-8600 Duebendorf, Switzerland

ABSTRACT

In the frame of a Swiss research project, a passive tracer gas technique for the determination of multi-zone air flow and contaminant transport in buildings was tested, based on previous work in several other countries.

First, emission characteristics of the three different sources (PMCP, PMCH and o-PDCH) and the adsorption characteristics of the passive samplers (standard Perkin-Elmer AD400 adsorption tubes) were established. Then, tests were conducted in a $85m^3$ air flow chamber in order to determine the influence of the source and sampler position in the room on the concentration measurement accuracy. Test were made with constant and variable air flow and temperature conditions.

First results showed a non-negligible influence of sinks for these tracer gases in the air flow chamber. Additional tests then showed, that sink problems may also to be expected in typical office or residential building rooms.

The paper gives an outline of the research project, summarises the results of these first tests and discusses the influence of the findings on the applicability of the method.

KEYWORDS

Air flow and contaminant transport, passive tracer gas technique, characterisation, validation, sink effects, application

INTRODUCTION

For the determination of average air change rates a passive tracer method using perfluorinated hydrocarbons is widely used, known as the PFT-method (see Roulet &Vandaele (1991) and Roulet & Gehrig (1998) for an overview on the different methods developed).

Diffusive sources of the PFT with known emission rates are placed in the rooms, where an equilibrium concentration of the PFT is established after some time according to the air change rate. This concentrations are measured with passive adsorption tubes.

In the frame of a Swiss research project, it was planned to set up a passive tracer gas technique for the determination of multi-zone air flow and contaminant transport in buildings, based on previous work in several other countries, e. g. in the US, Dietz (1986) and in the Nordic countries, Sateri (1991), Bloemen (1989), Bergsoe (1989) and Stymne & Eliasson (1991).

METHODS

Characterisation of sources and samplers

PMCP, PMCH and o-PDCH were selected as tracer material (Table1). Sources were made using 2ml glass vials with 5mm opening and screwed cap. Permeation was through a silicon/teflon septum (Varian No. 03-949835-00). The vials contained approx. 0.5ml of one of the PFT materials.

TABLE 1: PFT MATERIAL USED IN THIS PROJECT

Name		Mol mass	Boiling point [°C]	Density [kg/l]
Perfluoromethylcyclopentan	PMCP	300	--	1.70
Perfluoromethylcyclohexan	PMCH	350	76	1.79
ortho-Perfluorodimethylcyclohexan	o-PDCH	400	101.5	1.85

HFE7100 (Methoxy-nonafluorobutan ($C_4F_9OCH_3$)) was tested as an alternative source, but no satisfactory chromatographic solution could be found. The ether group is the reason for the significant higher detection limits compared to the PFTs used.

For each source the emission characteristics were determined gravimetrically. All sources showed very constant emission rates in time. However the absolute permeation rates varied considerably from source to source, thus requiring an individual calibration of the emission rate of each source. Preliminary tests showed a temperature dependence of the emission rate of about 60% for 12K temperature increase.

Standard Perkin-Elmer AD400 adsorption tubes were used as samplers, with Carboxen569 (20/45 mesh, Supelco SA) as adsorbent medium, and analysed on a GC/ECD system Varian 3400. The adsorption characteristics of these passive samplers were established in a stainless steel chamber of 22.4 m^3 volume under a constant air change rate of 1 h^{-1}. The supply air was cleaned by active charcoal filter and conditioned to the desired temperature and relative humidity .

More details on the sources and samplers, on the analytical procedure as well as on the influence of temperature on the emission rates of the sources and on the adsorption rates of the samplers, are given in Hill et. al. (2000).

The relation between amount of PFT tracer adsorbed on the sampler and the PFT concentration in the room is characterised introducing the parameter 'virtual probe air flow' q_p [m^3/h], defined as

$$q_p = \frac{q \, ?m}{ER \, ?t_{exp}} \tag{1}$$

with q [m^3/h] supply air flow rate in the calibration test cell, m [µg] mass of tracer adsorbed on the sampler, ER [µg/h] emission rate of PFT source, t_{exp} [h] time of exposure.

From this, under stationary conditions, the PFT concentration in the room during the experiment can be determined by

$$c_m = \frac{m}{q_p \, ?t_{exp}} \tag{2}$$

Validation tests in EMPA's air flow chamber

Tests in the air flow chamber of EMPA (figure 1) were performed in one zone under constant as well as variable air flow and temperature conditions in order to simulate situations with mechanical ventilation as well as natural ventilation by window opening. All surfaces in this chamber can be conditioned individually. The zone was conditioned with one wall as an external wall including a window. The supply air inlet was under the window panel.

PFT source were positioned a) near the air inlet, b) on a leg of the table in the middle of the room, c) at an internal wall. Samplers were positioned on all walls, at the opposite table leg and in the exhaust.

Room air flow characteristics and air flow rates were determined online using a Bruel&Kjaer 7620 active tracer gas measurement equipment. N_2O was injected in the supply air and SF_6 at the location of one of the PFT sources. Concentrations were measured in the supply inlet, at the positions of the PFT samplers and in the exhaust. Such, the history of indoor air condition was fully recorded and analysed for later comparison with the PFT sampler results. In addition, a person dummy (80W) was placed besides the table (figure 2).

Figure 1: Air flow chamber at EMPA

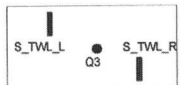

Figure 2: Positions of sources (Q..), passive samplers (S_..) and Bruel&Kjaer active tracer technique sampling points (BK_..)

Sink tests

In the air flow chamber

Preliminary results and checks of the PFT concentrations measurements showed discrepancies to the results of the Bruel&Kjaer tracer gas measurements. Crosschecks indicated a sink problem in the air flow chamber for all three PFT types, although not to the same degree. Therefore, the planned test were abandoned and sink test were performed under constant temperature and air flow rate condition. The PFT sources were placed in the test chamber and active adsorbent sampler probes (with a constant probing flow induced by a calibrated pumping system) were taken at regular time intervals of 12 to 24 h.

Sink tests in two office rooms

Since the air flow chamber with its two air conditioned transparent acrylic glass walls is not very representative for normal housing rooms, additional sink tests were performed in two office rooms at EMPA. The first room (LA125) was a fully operational office equipped by furniture, computer, books etc., and naturally ventilated through a small window gap. The second room (BA 245) was an empty office room with a supply fan installed to keep the outdoor air flow at a more constant rate (0.73 to 0.80 h^{-1}). Additional small fans were

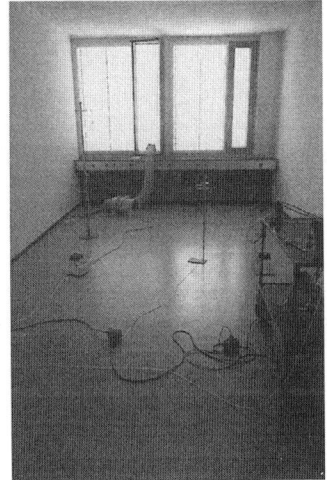

Figure 3: Test set up in the room BA 245

used in the room to assure complete mixing, see figure 3. The measurements were made during a time period of 200h.

RESULTS

At an outdoor air change rate of approx. 1 h^{-1}, PFT concentration in the air flow chamber were measured over a period of several days. As shown in figure 4, the measured concentrations were well below the expected theoretical steady state concentration c_0, as calculated from the PFT emission rate ER [µg/h] and the average supply air flow rate q [m³/h], assuming no sinks in the room:

$$c_0 = \frac{ER}{q} \qquad (3)$$

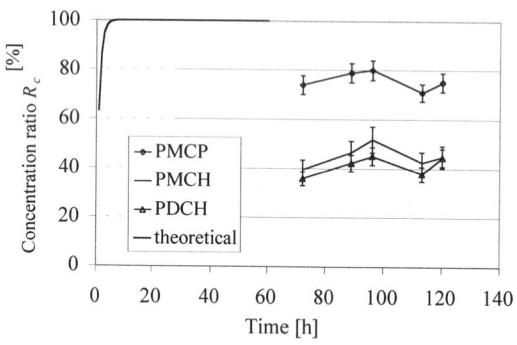

Figure 4: Concentration ratios R_c in function of time for the three PFT's used, as measured in the air flow test chamber, and the theoretical R_c.
Air change is constant at a rate of 1.0 h⁻¹.
Error bands are 95% confidence intervals.

Quite similar results could be observed in the two office rooms. Figure 5 gives results for the office room BA 245. Air change rate values are calculated in function of time from the concentrations of the different tracer gases. The values for N_2O represent the actual outdoor air change as measured continuously with the Bruel&Kjaer 7620 equipment by constant-emission method. The other values are determined from the results of the actively sampled adsorption probes. The starting values are too high because the concentrations have not reached steady state values yet.

Table 2 and figure 6 give values for the concentration ratio Rc as the ratio of the measured steady state concentration c_m to the theoretical expected steady state concentration without sink c_0, for the three PFT's and for the three measured spaces.

TABLE 2: RESULT SUMMARY OF THE SINK EXPERIMENTS

Room	Volume	Air change rate	Temperature	Concentration ratio R_c[%]		
	[m³]	[h⁻¹]*)	[°C]	PMCP	PMCH	PDCH
Calibration chamber	22.4	1.0	22	100 **)	100 **)	100 **)
Air flow chamber	85.4	1.0	20	80	52	45
Furnished office LA 125	77.9	0.12	22	84	74	62
Empty office BA 245	98.7	0.79	19	62	55	58

*)   Average air change rate determined by N_2O decay or constant emission technique
**) The method is calibrated in the stainless steel chamber, assuming no sinks in this chamber

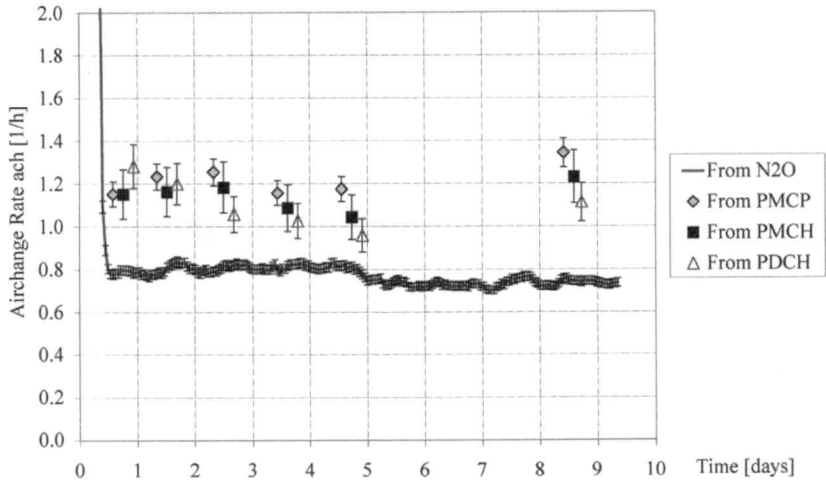

Figure 5: Air change rate values in function of time in the office room BA 245, as determined from the concentrations of the different passive tracer gases (with two adsorption tubes at one time, values separated in time in the figure for clarity). The values for N_2O represent the effective outdoor air change. Error bands are 95% confidence intervals.

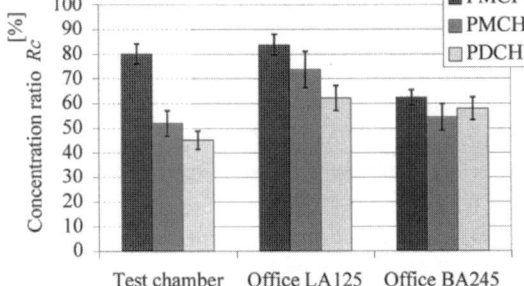

Figure 6: Concentration ratios R_c for the three PFT's used, as measured in test chamber and the two office rooms.
Error bands are 95% confidence intervals.

CONCLUSIONS

During validation tests for a passive tracer gas method in an air flow test chamber, sink effects for the PF tracers used have been observed and then further studied under steady conditions in two office rooms. Concentrations repeatedly measured over a period of several days are well below the expected steady state concentrations without sinks. Thus, when determining air flow rates by passive sampling with the three source gases used, the air flow rates will be overestimated by a factor in the range of 1.2 to 2.2. Additional tests, documented in Hill et al. (2000), have also shown significant temperature sensitivities of both tracer emission rates and sampler adsorption rates. This indicates that the tracer substances used in this project can hardly be further considered as useable for such applications.

While problems of errors associated with tracer gas measurements have been treated by a number of authors, also for passive methods, e.g. by D'Ottavio et al. (1987) and Sherman (1989), only few has been found in the literature on sink effects of passive tracers.

Therefore, further work has to focus on the sink characteristics and the underlying chemistry, on temperature effects, and on tests with other source gases. An extended evaluations of the calibrations and the experiences made with existing methods and tracers used may also be considered.

ACKNOWLEDGEMENTS

This work was conducted within the project ATEMAC (Application des traceurs passifs pour l'étude des mouvements d'air et de contaminants) and financially supported by the Swiss Federal Agency for Energy (Project No. 19'063).
We would like to thank H. Bloemen and H. Stymne for their the kind share of knowledge from their work on passive tracer methods and Claude-Alain Roulet, who leads this project, for his valuable contributions.

REFERENCES

Roulet C.A., Gehrig R. et al. (1998). *Project Atemac, Final Report Preliminary Phase*, EPFL LESO, Lausanne, Switzerland

Roulet C.A., Vandaele L. (1991). *Airflow Patterns within Buildings – Measurement Techniques*. AIVC technical note 34. AIVC. Coventry UK

Dietz R.N., Goodrich R.W, Cote E.A. and Wieser R.F. (1986). *Detailed description and performance of a passive perfluorocarbon tracer system for building ventilation and air exchange measurements.* STP 904, ASTM, Philadelphia

Sateri J (ed) . (1991). *The development of the PFT-method in the Nordic countries*. Swedish Council for Building Research, D9:1991

Bloemen H.J.T. et al. (1993). *Ventilation rate and airflow measurements using a modified PFT technique*. Proceedings 'Indoor Air '93', 1993, Vol. 5, pp 91-96

Bergsoe N.C. (1989). *Test and calibration procedures for the PFT laboratory equipment*. Seminar on ventilation measurements using the PFT method, held at the Helsinki University of Technology, May 1989

Stymne H., Eliasson A. (1991). *A new passive tracer gas technique for ventilation measurements*. 12th AIVC Conference, 1991, Ottawa, Canada

D'Ottavio T W, Senum G I, Dietz R N. (1988). *Error Analysis Techniques for Perfluorocarbon Tracer Derived Multizone Ventilation Rates*. Building and Environment, **23:3**, pp 187-194

Sherman M.H. (1987). *Analysis of errors associated with passive ventilation measurement techniques*. Lawrence Berkeley laboratory. Building and Environment **24:2**

Hill M., Gehrig R., Dorer V., Weber A., Hofer P. (2000). *Are Measurements of Air Change Rates with the PFT-Method Biassed by Sink and Temperature Effects*. Submitted to Healthy Building 2000 Conference

Air Distribution in Rooms, (ROOMVENT 2000)
Editor: H.B. Awbi

PRE-PROCESSOR FOR VENTILATION MEASUREMENT ANALYSIS

Heekwan Lee and Hazim B. Awbi

Indoor Environment & Energy Research Group
Department of Construction Management & Engineering
The University of Reading
United Kingdom

ABSTRACT

It is well known that the introduction of tracer gas techniques to ventilation studies has provided much useful information that used to be unattainable from conventional measuring techniques. Data acquisition systems (DASs) containing analog-to-digital (A/D) converters are usually used to perform the key role which is reading and saving signals to storage in digital format. In the measuring process, there are a number of components in the measuring equipment which may produce system-based noise fluctuations to the final result. These unwanted fluctuations may cause discrepancy in computations, especially when non-linear algorithms are involved.

In this study, a pre-processor is developed and used to separate the unwanted fluctuations (noise or interference) in raw measurements and to reduce the uncertainty in the measurement. Moving average, Notch filter, FIR (Finite Impulse Response) filters, and IIR (Infinite Impulse Response) filters are designed and applied to collect the desired information from the raw measurements. Tracer gas concentrations are measured during leakage and ventilation tests in a model test room. The signal analysis functions embedded in Matlab are used to carry out the digital signal processing (DSP) work.

KEYWORDS

Tracer Gas Measurement; Digital Signal Processing; Correlation Analysis; Signal Noise; Air Leakage

INTRODUCTION

In ventilation research, tracer gas techniques are very useful to quantify the physical phenomena occurring in room ventilation. The application of data acquisition system (DAS), equipped with an analog-to-digital (A/D) converter on a personal computer, reduces the difficulty in measuring and recording tracer gas concentrations and it provides more detailed information carried in the measured signal, in real-time domain.

Although this test facility produces much information, the actual measured data may contain some unwanted components in terms of system-based noise or interference from the tracer gas analyzer, the computer, and other components used in the tests. This system noise appears as fluctuations in the tracer gas measurements and could be propagated in the data analysis, especially in computation involving non-linear algorithms such as logarithmic or exponential functions.

In this study, several data pre-processing techniques have been performed to separate the potential error from the tracer gas measurements and to improve the ventilation analysis as a final goal. The moving average technique by Lee (1993) and a simple digital filtering technique by Lee and Awbi (1998) have already been studied and reported. A few more advanced digital filtering techniques are introduced in this paper. Finally, a tool to pre-process tracer gas measurement for further analysis is also presented.

DIGITAL SIGNAL PROCESSING

A *signal* is defined as any physical quantity that varies with time, space, or any other independent variable or variables. Mathematically, it is described as a function of one or more independent variables, as in Equation (1). In this study a signal $x(t)$ (real-valued or scalar-valued) is a function of the time variable t. The term real-valued means that for any fixed value of the time variable t, the value of the signal at time t is a real number.

$$x[n] = x(t)\,|_{t=nT} = x(nT) \tag{1}$$

where n is an integer number, t is a real number, T is a sampling time which is usually the reciprocal of sampling rate in Hz. By *sampling* (digitizing or windowing) process, an analog signal $x(t)$ is taken at discrete-time instants. A *digital signal* $x[t]$ is in turn generated and used for the digital signal processing. Analog-to-digital (A/D) converters conduct this sampling process with the help of computers.

A signal, in general, holds a characteristic called *duality* between the *time domain* and the *frequency domain*, which makes it possible to perform any operation in either domain. Usually one domain or the other is more convenient for a particular operation. Based on this characteristic, *Fast-Fourier Transformation* (FFT), which is one of useful mathematical tools in signal processing, decomposes a signal in time domain into a sum of sinusoidal components in frequency domain. *Inverse-FFT* (IFFT) also works for the reverse function from frequency domain to time domain, see Lee and Awbi (1998) for practical application of this.

Signal generation is usually associated with a *system* that corresponds to a stimulus or force. A system may also be defined as a physical device that performs an operation on a signal. For example, a *filter* used to reduce the noise and interference corrupting a desired information-bearing signal is also called a *system*. In this case the filter performs some operation(s) on the signal, which has the effect of reducing (filtering) noise and interference from desired information-bearing signal. When a signal is passed through a system, as in filtering, the output will be a processed signal. In this case the processing of the signal involves filtering the noise and interference from the desired signal, which is referred to as *signal processing*.

The major purpose of signal filtering is to remove or block unwanted components from a waveform. The mathematical foundation of filtering is *convolution*. A digital filter's output $y(n)$ is related to its input $x(n)$ by convolution of its impulse response $h(n)$:

$$y(n) = x(n) * h(n) \equiv \sum_{k=-\infty}^{\infty} x(k)h(n-k) \tag{2}$$

In the case of preparing waveforms for deconvolution or dressing up the results of deconvolution, the unwanted components are generally those of high-frequency noise. To reduce this noise contribution, a low-pass filter is used. There are a variety of filter functions that can be used, such as elliptical, Chebyshev, Butterworth, and so forth.

A more sophisticated window function called *Kaiser window* is generally used for the design of practical filters since it allows the designer the freedom to trade off the sharpness of the pass-to-stopband transitions with the magnitude of the ripples.

In signal processing, *correlation analysis*, which measures the correlation between observations at different distances apart, can be used to approve the results. Suppose that there exist two real signals $x(n)$ and $y(n)$ each of which has finite energy, the crosscorrelation of $x(n)$ and $y(n)$ is a sequence r_{xy}, which is defined mathematically as follows:

$$r_{xy}(l) = \sum_{n=-\infty}^{\infty} x(n) y(n-l) \qquad (3)$$

where the index l is the time shift (or lag) parameter, $0, \pm 1, \pm 2, \ldots$, and the subscripts xy on the crosscorrelation sequence $r_{xy}(l)$ indicates the sequences being correlated. Two signals correlated here could be input and output signals after the filtering process. The autocorrelation can be also achieved by correlating the same signal, xx or yy. The autocorrelation $r_{hh}(l)$ of the impulse response $h(n)$ exists if the system is stable. Furthermore, the stability insures the system does not change the type (energy or power) of the input signal.

TRACER GAS MEASUREMENTS IN VENTILATION TESTS

To obtain tracer gas measurements in time domain, model tests were conducted. Figure 1 shows the schematic of the test setup used. The model room, 1.6^m x 0.8^m x 0.7^m (H), has two ceiling-mounted openings to supply and exhaust ventilation air. An axial fan with speed controller was the ventilation source. Carbon dioxide (CO_2) was used as the tracer gas and measured by a gas analyzer which generates signals, 0~1 *DCV*, corresponding to the measured concentrations. The computer controlled the tracer gas generation through the A/D converter and the control box. Two mixing fans were used in the model for the leakage tests. The test procedure was coded and run using Matlab language, ver. 5.2.0.3084. The Matlab language was also used for the signal analysis work.

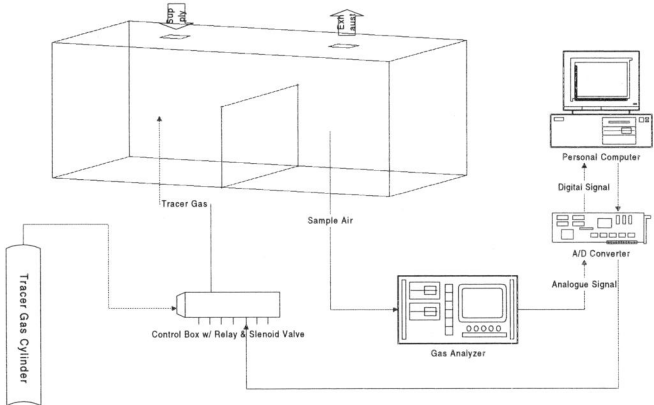

Figure 1: Schematic diagram of test setup for ventilation tests

The analogue signal from the tracer gas analyzer was continuously digitized by the DAS at a preset sampling rate 1 *Hz*. Different sampling rates were tested by Lee and Awbi (1998) and concluded that 1 *Hz* is fast enough to capture the variation in tracer gas concentrations occurring in room ventilation, although the sampling rate may go down further if necessary without influencing the final accuracy.

To improve the quality of the measured data, several pre-processing techniques were applied to the data. The moving average was the first technique used with several periods used for averaging. The 30-points average produced the best result and was used as the basis for comparison with other techniques. This technique was used for ventilation analysis by Lee (1993). It was, however, found that this technique cause the loss of desired information from the from the measured signals (Phillips and Braggs, 1994). The simple Notch filtering technique was introduced by Lee and Awbi (1998). The propagated fluctuation in ventilation calculation disappeared after applying this technique, although the pure Notch filter required some modification to achieve reliable crosscorrelation between the raw and the filtered signals.

In addition, a few more digital filters were applied in this study. The design of digital filter is classified into two categories; finite impulse response (FIR) and in-finite impulse response (IIR). The major difference between the two systems is the feed back sequence from the output to the input for subsequent iterations as shown for IIR below:

$$\text{FIR: } y(n) = \sum_{k=0}^{M-1} b_k x(n-k) \text{; IIR: } y(n) = -\sum_{k=0}^{N} a_k y(n-k) + \sum_{k=0}^{M-1} b_k x(n-k) \tag{4}$$

In FIR systems, the functions of Rectangular, Bartlett, Hamming, and Kaiser window were used to design the digital filters and the Kaiser window function was used for comparison. In IIR system, the functions of Butterwork, Chebyshev 1, Chebyshev 2, and Elliptic window are used to design the digital filters and the Chebyshev 1 window function was used for comparison.

The digitally filtered data using the several techniques mentioned above were then used for ventilation analysis, such as air change rate and mean age of air at the local sampling point under the outlet opening ($0.5H$). Considering continuity equity on a control volume in the test room gives Equation (5). The time-serial tracer gas measurement at a certain point in the model room can be used to obtain the local air change, see Lee (1993) for more detail, using the equation:

$$Q = -\frac{V}{t} \ln \frac{C(t) - C_{out}}{C(0) - C_{out}} \tag{5}$$

where Q is air change rate m^3/s, V is the room volume m^3, t is the elapsed time s, $C(t)$ is the measured tracer gas concentration ppm, C_{out} is the background tracer gas concentration, and $C(0)$ is the tracer gas concentration when the tracer decay begins in the test room. It implies that the air change rate can be estimated by conducting tracer gas measurements at a local point in the test room for local air change rate and at the outlet opening for room air change rate. The local mean age of air $\overline{\tau}_p$ was calculated by applying Equation (6) to the tracer pulse measurements in time domain, see Etheridge and Sandberg (1996).

$$\overline{\tau}_p = \frac{\int_0^\infty t C_p(t)dt}{\int_0^\infty C_p(t)dt} \tag{6}$$

RESULTS AND DISCUSSION

The model tests were conducted under two conditions, a leakage test and a forced ventilation test. Figure 2 shows a measured signal in the leakage test. For these tests, two mixing fans in the model test room were used to achieve a fully-mixed condition. Then tracer gas was injected into the model for a certain time period and the tracer decay was measured and recorded. The variation in the tracer gas concentration is easily observed from the measurement with small fluctuations, which could cause error in further analysis.

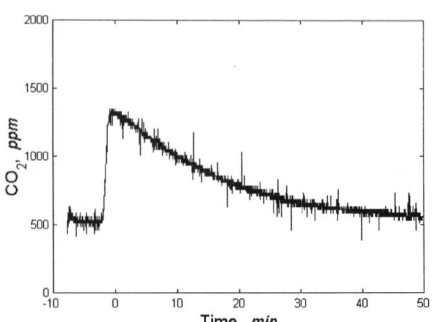

Figure 2: Raw tracer gas concentrations in the leakage test

Figure 3 shows the power spectrum of the measured tracer gas concentration using the FFT analysis. In the figure, the low frequency sources upto 50 Hz were collected by the Notch filter designed, while

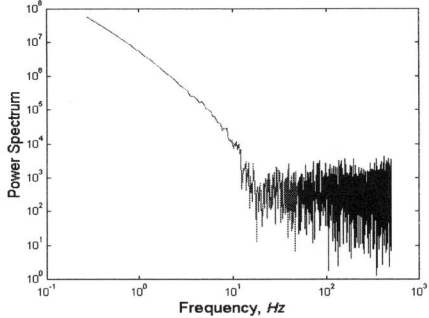

Figure 3: Power Spectrum of the raw measurement in the leakage test

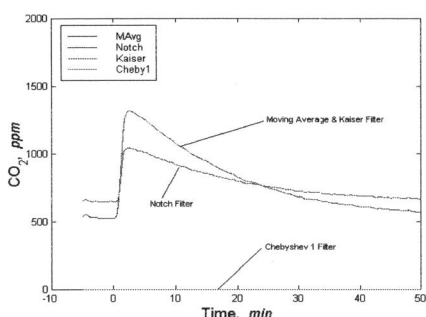

Figure 4: Digitally filtered measurements in the leakage test

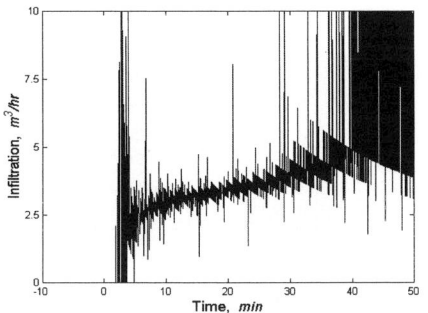

Figure 5: Air infiltration calculations using continuity for the raw measurement in the leakage test

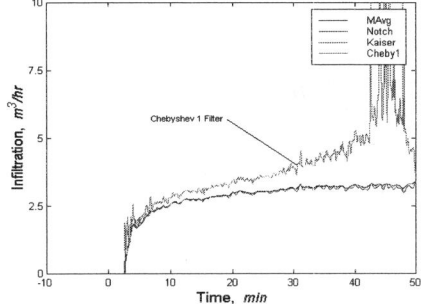

Figure 6: Air infiltration calculations for the digital-filtered measurement in the leakage test

frequency sources higher than 50 Hz were removed. The other digital filters were created in different manners and used to collect the desired power source in the measured signal.

Figure 4 shows the digitally filtered signal by the moving average, Notch filter, Kaiser filter (FIR), and Chebyshev 1 filter (IIR). The moving average and the Kaiser filter do not create significant change in the filtered data, but the Notch and the Chebyshev 1 filters do.

Figures 5 and 6 show the calculated air change rate, using Equation (5), due to infiltration in the test room. The ventilation system for these tests was running in pulling mode and formed negative pressure in the test room which caused air infiltration through invisible gaps. As the calculation for the air infiltration involves logarithms, the small fluctuations in the measured data are propagated and this causes difficulty in obtaining accurate value, as shown in Figure 5. The air infiltration calculations using filtered data in Figure 6 show improved results. Although the air infiltration was almost stable during the test, the calculations by the Chebyshev 1 filter show gradual increase with time.

Figure 7 shows a comparison between the calculated air infiltration rates. Frequency analysis was performed to find the air infiltration rate. The value having the highest frequency in the histogram analysis was taken to be the air infiltration rate. The tracer gas measurement was repeated for the ventilation test and the same analysis is carried out. Figures 8 and 9 show the results for the ventilation rate and the mean age of air, using Equation (6), respectively. The ventilation results show values close to that from the raw data except the moving average. In the mean age of air calculation the filtering processes gives almost identical results to that from the raw data except the *Notch* filter.

CONCLUSIONS

The digital signal processing techniques were applied to pre-process the tracer gas measurements in ventilation tests. In addition to the previous work referred to here, a few other techniques were used to provide a comparison. The major findings from this study were:

- Tracer gas measurements in ventilation tests are necessary to be pre-processed before further analysis to reduce uncertainty which may corrupt the desired information.
- Although the moving average technique used removes the random fluctuations from the measurements, it is not reliable enough to achieve stable ventilation calculation.
- The standard notch filter used in this study does not produce good results compared with other filters and may need some modifications before it is applied.
- Overall the FIR filter produces better results compared with other filters for ventilation rate and mean age of air calculations. Among those tested, the *Kaiser* filter was the best one for pre-processing the tracer gas measurements.
- Although the IIR filters help to reduce the random noise in the data, they cause considerable changes to the filtered data, which is undesired.

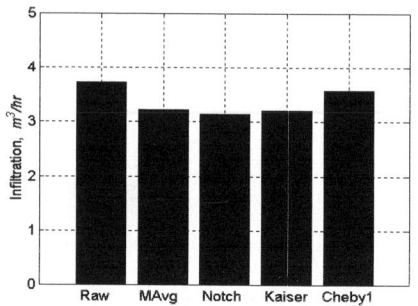

Figure 7: Air infiltration calculation in the leakage test

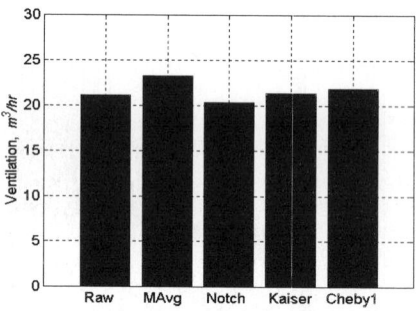

Figure 8: Ventilation calculation in the ventilation tests

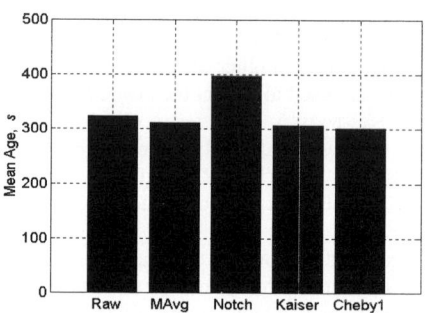

Figure 9: Mean age of air calculation in the ventilation test (Note: The time constant for the ventilation test is 72s.)

REFERENCES

Castleman K.R. (1996) *Digital Image Analysis*, Prentice Hall, New Jersey.

Chatfield C. (1996) *The Analysis of Time Series – an introduction*, Chapman & Hall, London.

Etheridge D. and Sandberg M. (1996) *Building Ventilation: theory and measurement*, John Wiley & Sons, Chichester, U.K.

Kamen E.W. and Heck B.S. (1997) *Fundamentals of Signals and Systems – using MATLAB*, Prentice Hall, New Jersey.

Lee H. (1993) *Study on the Influence of Ventilation on Indoor Air Pollutant Removal*, Master Thesis, the University of Seoul, Seoul.

Lee H. and Awbi H.B. (1998) Effect of data logging frequency on tracer gas measurement. In *Proceedings of the Roomvent '98*, KTH, Stockholm, Sweden, June 14-17, Vol. 2, pp. 477-482.

Phillips D. and Bragg G. (1994) The measurement of high frequency variations in concentratiOns of indoor air constituents. *ASHRAE Transactions*, Vol. 100, pp. 1225-1229.

Proakis J.G. and Manolakis D.G. (1996) *Digital Signal Processing*, Prentice Hall, New Jersey.

Ramirez R.W. (1985) *The FFT – fundamentals and concepts*, Prentice Hall, New Jersey.

Signal Processing Toolbox User's Guide (v4.2) (1998) Mathworks Inc., Massachusetts.

© 2000 Elsevier Science Ltd. All rights reserved.
Air Distribution in Rooms, (ROOMVENT 2000)
Editor: H.B. Awbi

EXPERIMENTAL STUDY ON INDOOR AIR QUALITY: INTEREST ON A LATERAL PLANE OF MEASUREMENT

S. Laporthe, J. Virgone, C. Teodosiu

Centre de Thermique de Lyon (CETHIL), UPRES A CNRS 5008, Equipe Thermique du Bâtiment, Institut National des Sciences Appliquées (INSA), Bât. 307, 20 avenue A. Einstein, 69621 VILLEURBANNE Cedex- Tel. : (33-4)72.43.87.77 - Fax : (33-4)72.43.85.22 - Email : stephanie.laporthe@insa-cethil-etb.insa-lyon.fr ; joseph.virgone@insa-cethil-etb.insa-lyon.fr

ABSTRACT

The airflow in a room is never two-dimensional even if we take care of setting a ventilation system and a tracer gas injection to have a symmetric system. So, the study of the vertical median plane could not give information for the entire room. For the ventilation efficiency calculation, it is necessary to take at the best the occupied zone into account. Otherwise, as regards a non-centre pollutant injection, it is interesting to study the dispersion of the pollutant in the room. We present in this paper the experiments carried out in the cell MINIBAT (CETHIL), and especially the measures in two vertical planes : the median plane and a lateral one. These results can be used as a reference for CFD codes. Air velocity, air temperature and tracer gas-concentration are presented in two cases indicative of the three-dimensional effect of the airflow. This phenomenon is particularly brought to the fore with the integration of a heating system in the cell, and a non-centre tracer gas injection. We conclude upon the difference between the two planes of the ventilation efficiency.

KEYWORDS

Experimental cell, heating system, tracer gas concentration, ventilation, ventilation efficiency.

INTRODUCTION

The aim of a ventilation system is to create a microclimate in the studied room. This microclimate must concern the indoor air in this room just as well as thermal conditions, and this in any place of the occupied zone. These notions should be considered as the two fundamentals conditions concerning comfort and occupants' well being. This work, along with the experimental measurements on which it was based, is carried out in the framework of the French programme of research undertaken by the "Groupe de Pilotage sur la Qualité des Ambiances" (GPQA), whose task is to initiate and coordinate work on air quality on a national scale; within this framework, our work is to study experimentally the efficiency of the ventilation system, taking into account different sources of disturbance (heating systems, people, etc.). A CFD code was used in a previous work (Brüel and Kjaer, 1991) in order to calculate the profiles of temperature and velocities in several planes of the test cell, and the differences observed in particular cases are significant. So, several modifications have been made on the apparatus in order to have measures over a lateral plane in complement to the median one. Two examples indicative of the differences between the two planes are presented in order to justify practically the necessity of a lateral plane.

THE EXPERIMENTAL SET UP

The experimental set up is the full scale test "MINIBAT" (CETHIL - INSA de Lyon in France), see Figure 1.This installation comprises two identical cells (cell 1 and cell 2) which measure 3.10 $\times$ 3.10 $\times$ 2.50 m^3 each. Cell 1 is separated by a glass wall from the climatic chamber, whose air-treatment system can produce temperatures of between -10 and 30°C. The thermal guard is maintained at a uniform temperature of 21°C, so as to represent adjacent spaces.

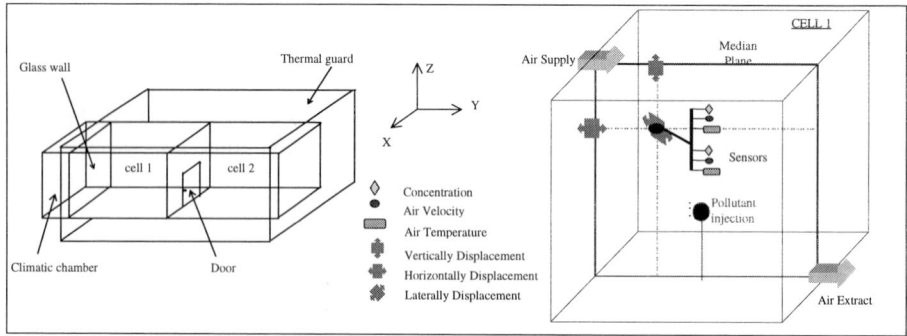

Figure 1: The experimental cell MINIBAT

Our study was carried out exclusively in cell 1 of MINIBAT. This cell's ventilation system has a fixed supply and a mobile extract. The user can set extract air flow between 1 and 5 ach (air changes per hour). The MINIBAT is equipped with a number of sensors for measuring surface and air temperatures, as well as air speeds and gas concentrations. Near the centre of each zone, relative air hygrometry and operative temperature are measured. Both the cells also have been equipped with a device, which can be used to move the sensors over different vertical planes. This device consists of three motors which actuate a metal arm on which is mounted an array of six sensors: two type-K thermocouples for air temperatures, two omnidirectional hot-wire probes for air speeds, and two measuring points for gas concentrations. A source pollutant injects tracer gas (SF$_6$) at the centre of cell 1, at a high of 1.20 m. We have choosen SF$_6$ (Laporthe, to be published) principally because of the sensibility of the measuring system Brüel and Kjaer (Brüel and Kjaer, 1991,Laporthe,1998).

Before beginning with the experiments, we have analysed a previous measurement (Castanet, 1998, Laporthe, to be published). This previous study concerning the definition of the mesh grid for the measurements has shown that it is not necessary to have a fine mesh grid (for example: 10 cm $\times$ 10 cm leading to 598 points of measurements) all over the plane and particularly on a lateral plane. So, the time of the experiments can be reduced in minimising the number of necessaries measuring points without affecting the accuracy of the field measurements (Laporthe, 1998). For the tests presented here we have 396 measuring points for the median plane and 168 measuring points for the lateral one.

DEFINITION OF VENTILATION EFFICIENCY

The ventilation efficiency index ε_C is defined by the ratio of pollutant concentration in the room to pollutant concentration at the exhaust. We thus obtain:

$$\varepsilon_C = \frac{C_e - C_s}{<C> - C_s} \tag{1}$$

where e and s refer to extract and supply, and $<C>$ is the average pollutant concentration in the room air. Ventilation efficiency expresses the ventilation system's capacity to evacuate pollutants. If we take it that the supply air is free of pollutants (i.e., $C_s = 0$), Eqn.1 becomes:

$$\varepsilon_C = \frac{C_e}{<C>} \tag{2}$$

EXPERIMENTAL STUDY FOR TWO SIGNIFICANT CASES

Results with a convector

The experiment is realised by adding a convector under the air supply (Figure 2). The climatic chamber is maintained at 10°C. The convector power is 310 W in order to compensate the losses of heat through the glass wall and the air renewal, and to maintain an air temperature in the room closed to 21 °C. The pollutant injection is in the centre of the cell at a high of 1.20 m.

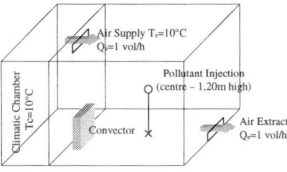

Figure 2

Figure 3 gives the repartition of concentrations on the median (X_1) and the lateral (X_2) planes.

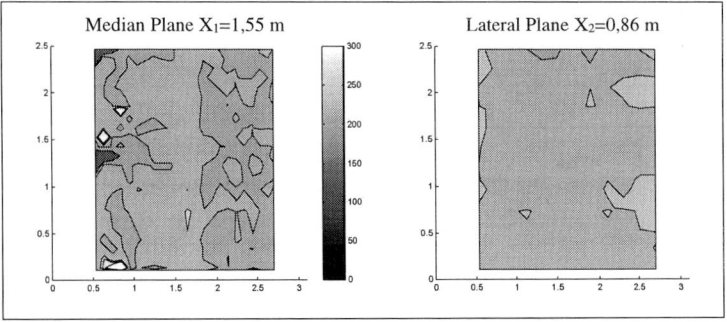

Figure 3 : Concentrations fields on the median and the lateral plane

Even if the mean ventilation efficiency for each plane is nearly the same : 1 for the median plane (X_1) and 1.03 for the lateral plane (X_2), Figure 4 shows a difference in the repartition of the ventilation efficiency in the occupied zone. A value closed to 1 means a perfect mixing in the cell. In reality, the convector's convective effects favour the mixing of the air in the cell. Indeed, concentration values are homogeneous on the lateral plane when the dispersion of these values are more important on the median plane (Figure 3). This phenomenon can be due to local turbulence effects activate by the convector, but it is explained, too by the fact that the median plane is the plane of the pollutant injection.

The observation of the air temperature and air velocity fields (Figure 5 and Figure 6) indicates significant differences between the two studied planes (X_1 and X_2), principally due to the convector's convective effect.

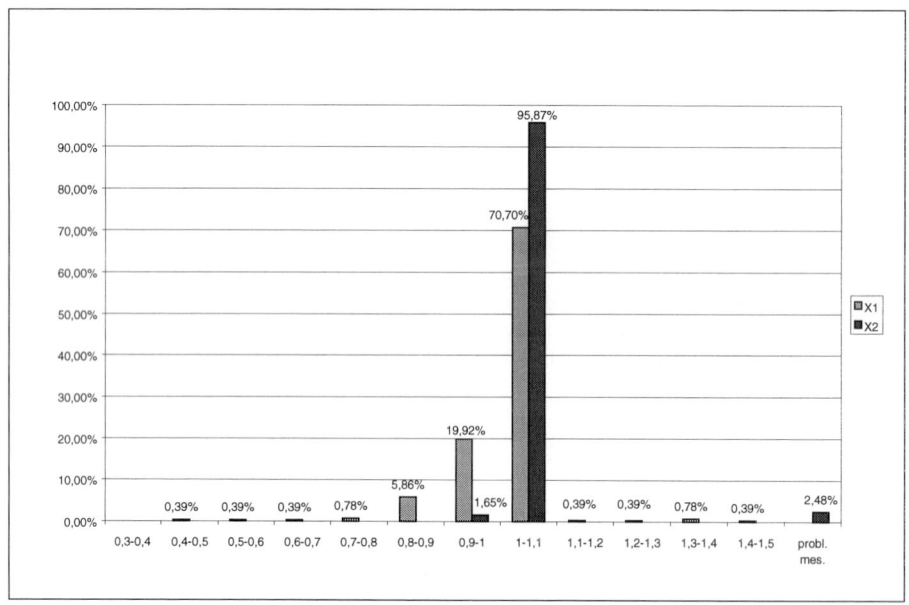

Figure 4 : Histograms of the ventilation efficiency

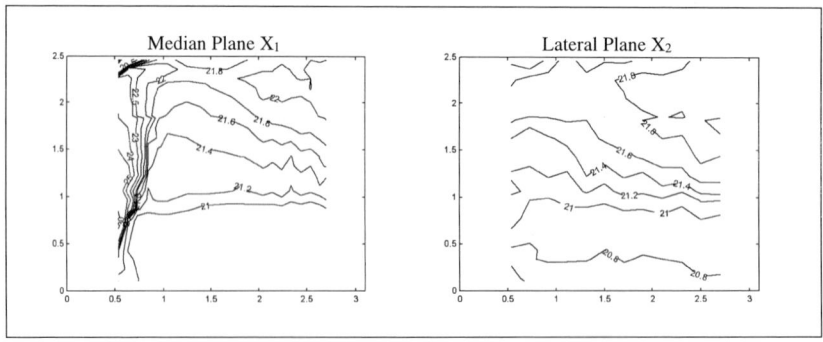

Figure 5 : Air temperature iso-curves

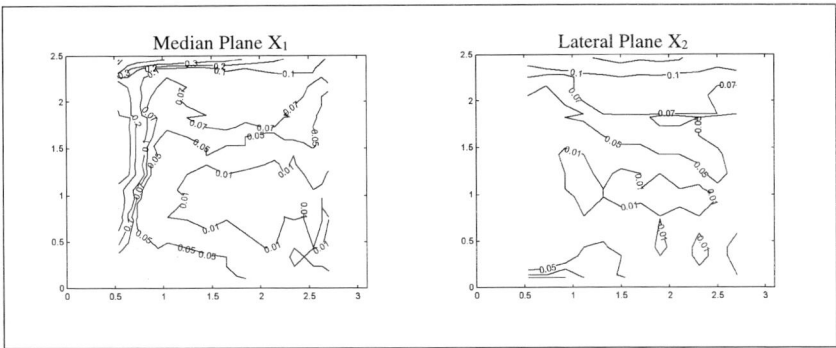

Figure 6 : Air velocity iso-curves

Results with a non-centre pollutant injection

It is the same case as previously, but with a non-centre pollutant injection. The pollutant is always injected at 1.20 m high, but in two different positions. First, the injection takes place at X=2.325 m and two planes are studied: the vertical median plane X_1=1.55 m and a vertical lateral plane X_2=0.86 m. The second injection position is at X=0.775 m, and we study the plane X_{2b}=0.86 m.

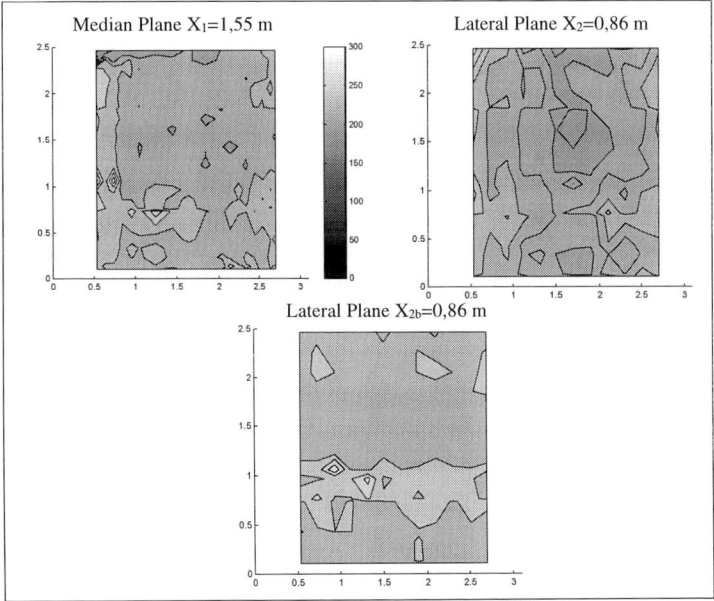

Figure 7 : Concentrations fields for a non-centred injection

Figure 8 shows that the repartition of the ventilation efficiency depends of the studied plane : more the pollutant injection is moved away from the studied plane, more the ventilation efficiency on this plane is good. Indeed, because of the ventilation, the pollutant is carried away to the extract, and the zones which are located far from the pollutant injection are less affected.

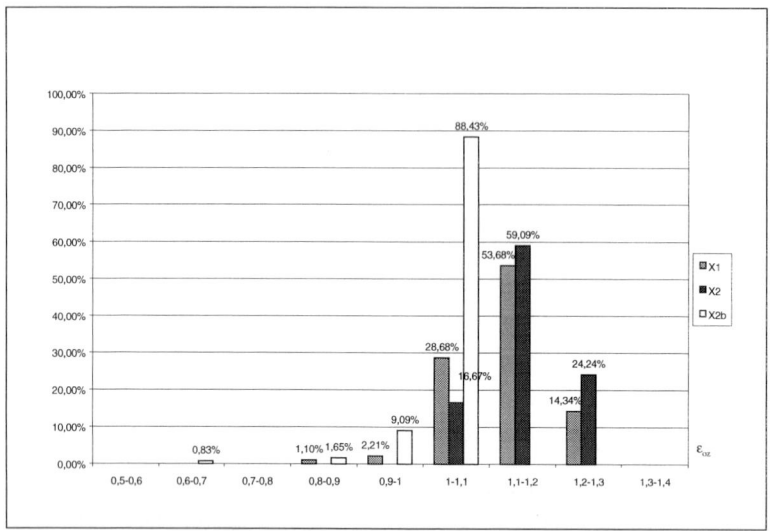

Figure 8 : Histograms of the ventilation efficiency

CONCLUSION

Temperature and velocity repartitions are quite different between the median plane and the lateral one because of the effect of the convector.

Concerning ventilation efficiency, the convector favours the mixing of the air inside the room, so the mean value of the ventilation efficiency is closed to 1, characterising a good pollutant extraction of a perfect mixing. This is true in the median plane and also in the lateral one in the case of an injection centred. But, in fact, we can observe a significant variation (0.7 to 1.1) in the median plane and a good homogeneity in the lateral one. With a non centred injection, we note more differences between the mean values of the three planes (from 0.9 to 1.3), the biggest value being obviously this one far from the injection.

We are working in the definition of an index of air quality bringing to the fore the differences of values of ventilation efficiency in a same room. If the air quality is good all over the room, then, the index will be good.

REFERENCES

Brüel and Kjaer (1991). Moniteur Multigaz Type 1302 Instructions.

Castanet, S. (1998). Contribution à l'Etude de la Ventilation et de la Qualité de l'Air Intérieur des Locaux. **Thèse de Doctorat de l'INSA de Lyon**, 289p.

Laporthe, S. (1998). Etude de l'efficacité de différents systèmes de ventilation. **Report of DEA, INSA de Lyon**, 125p.

Laporthe, S., Virgone, J., Castanet, S. (to be published). A comparative study of two tracer gases : SF₆ and N₂O. **Building and Environment**.

Air Distribution in Rooms, (ROOMVENT 2000)
Editor: H.B. Awbi

THE NATURAL VENTILATION OF A ROOM WITH AN AREAL SOURCE OF HEAT AND TWO OPENINGS

C. Gladstone and A. W. Woods

BP Institute, Bullard Laboratories, University of Cambridge, Madingley Road, Cambridge CB3 OEZ, U.K.

ABSTRACT

The ventilation and flow dynamics of a room with a heated floor and high and low level openings to the exterior are investigated using a mathematical model and analogue experiments. The heated floor generates an areal source of buoyancy. Openings allow displacement ventilation to operate. When combined these produce a steady state environment, where the heat provided by the floor equals the heat lost by displacement. We present new theory based on balancing heat fluxes and validate this with observations from small-scale laboratory experiments. At some time, a steady state temperature, T_{SS} is reached. We investigate the effects of varying heat fluxes and vertical positions of openings on T_{SS}. The new model suggests that T_{SS} is related to the temperature of the floor, T_F, and temperature of the exterior, T_{EXT}, as $(T_F - T_{SS})^{4/3} = (T_{SS} - T_{EXT})^{3/2}$. Theoretical predictions and experimental observations agree well.

KEYWORDS

Natural ventilation, displacement ventilation, buoyancy, areal heat source, modelling, experiments

INTRODUCTION

In this study we consider the natural ventilation of a room which is in contact with the exterior by only two openings while being heated over the whole lower surface of the room. This is an analogue model for both the heating of a room by the floor, and cooling of a room through the roof. Ventilation occurs naturally by convection and displacement, driven by pressure differences between air of different temperatures alone rather than mechanical systems. Much recent work has focused on the application of localised heat sources to natural ventilation (Linden, 1999). In many modern buildings however, an areal source of heating or cooling may be as common, particularly if there is a large portion of glass through which sunlight will act as an areal source of heat, while areal cooling will occur at night. We extend the work of Linden *et al.* (1990) who investigated the two-layer stratification which results from a point source of buoyancy combined with displacement ventilation. Observations on temperature changes and flow dynamics from laboratory experiments are presented. A mathematical model describing these is developed, and predictions from the theory are compared with the experimental data. A simple example showing application of the new model to modern buildings is given.

EXPERIMENTS

Method

As an experimental analogue of a confined room connected to a cold exterior by two openings and subject to uniform heating of the floor of the room, we have conducted experiments in a perspex tank (the 'room') with a square floor area of 311.5 cm² and height 28.7 cm (figure 1). This is filled with water to a depth of 27.5 cm and placed in a large reservoir of water filled to the same depth. The floor of the room comprises a steel plate containing tubes through which antifreeze is circulated. This areal heat source is adjusted by changing the temperature of the antifreeze. Type K thermocouples placed at different heights within the room and exterior monitor the water temperature by scanning every 10 seconds. One wall of the room contains three circular holes each of diameter 1.5 cm with their midpoints at 1.05, 13.50 and 25.95 cm above the floor. The floor is preheated to T_F and the room filled with fluid of the same temperature. To start an experiment bungs are removed from two openings. The two selected openings, temperature of the floor, T_F, and exterior, T_{EXT}, were varied from experiment to experiment. 28 experiments were performed and the steady state temperature of the fluid inside the room, T_{SS}, was monitored.

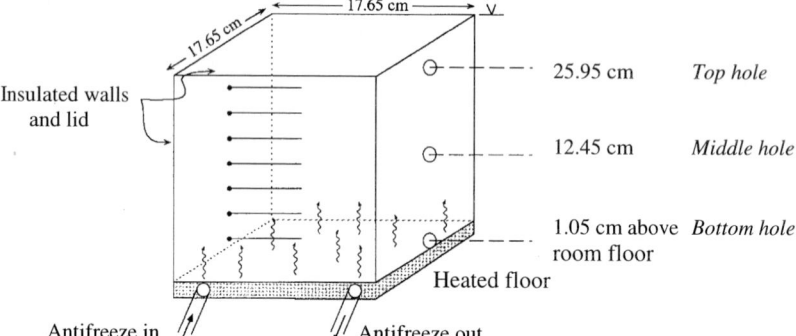

Figure 1: Analogue experiments are conducted in a perspex box with a floor heating plate providing an areal source of buoyancy. Thermocouples measure temperature changes.

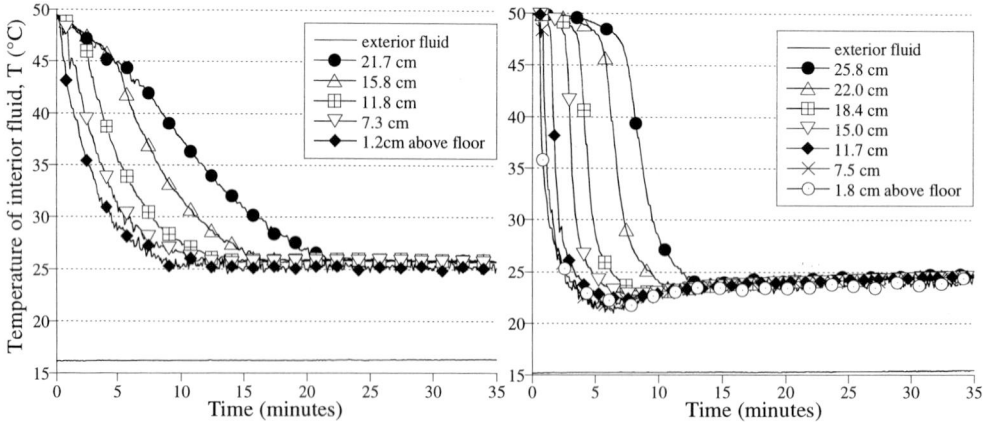

Figure 2a and 2b: In both experiments, T_F=50°C, T_{EXT}=16.5°C. The top and middle openings are employed in figure 2a while top and bottom openings are employed in figure 2b. T_{SS} is reached after 20 minutes in figure 2a, and 33 minutes in figure 2b.

Results

After removal of the bungs cold exterior fluid enters the room through the lower hole, displacing warm interior fluid upwards and out of the room through the upper hole. As the flow continues and the front between the warm and cold fluid ascends, the thermocouples in the room show a decrease in temperature sequentially from the bottom of the room to the top (figures 2a and 2b). However, as the lower layer deepens, convection driven by the heated floor becomes progressively more vigorous and the temperature in the room becomes more uniform with depth. A steady state temperature, T_{SS}, is attained once this front reaches the top of the room, typically after about 20 minutes. T_{SS} remains constant while T_{EXT} and T_F are held constant. The evolution of the temperature profile in the room also depends on the positions of the openings. Figure 2a shows temperature changes in a room ventilated through the top and middle openings, while in figure 2b the ventilation occurs through the top and bottom openings. The displacement ventilation, which dominates both experiments during early stages, is stronger when the distance between the openings is large. This process is so strong that the temperature of the room dips to a few degrees below T_{SS}, before convection from the floor becomes equally important, creating the steady state balance. The steady state temperature, T_{SS} depends on T_F and the positions of the two openings (figure 3). In all three configurations of openings, the higher the setting of T_F, the higher the value of T_{SS}.

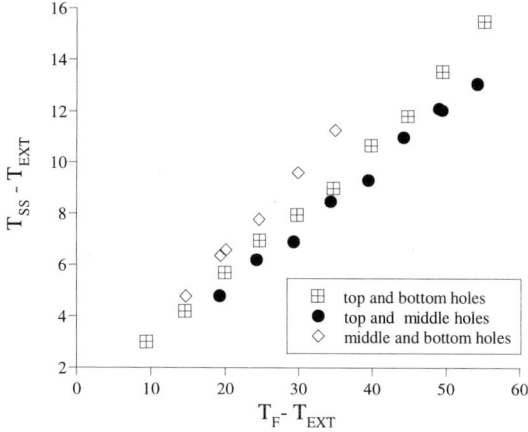

Figure 3: For a given configuration of openings, T_{SS} increases with T_F.

THEORETICAL MODEL

We now develop a leading order model of the heating and flow observed in the experiments. We assume that the fluid in the room is well-mixed by the convection produced at the heated floor. For openings whose vertical dimension is small compared to their separation, the volume flux through each of the openings during displacement ventilation, Q_V, is given from dimensional analysis by the relation (Linden *et al.* 1990)

$$Q_V = \left[\left(\frac{g(\rho_{EXT} - \rho_{INT})}{\rho_{INT}} \right) \frac{h}{2} \right]^{1/2} cA_{HOLE} \tag{1}$$

where g is the acceleration due to gravity, h the distance between the midpoints of the two openings, ρ_{EXT} and ρ_{INT} are the densities of the exterior and interior fluids respectively, A_{HOLE} is the average area of the openings and c is a constant describing the energy loss associated with flow through an opening.

The heat flux supplied by the hot floor may be estimated from previous work on the turbulent convection generated by horizontal heated plates. As a good approximation, if the heating plate has temperature T_F and the fluid in the room has temperature T, the heat flux supplied to the room, Q_H, is given by

$$Q_H = \lambda A_{ROOM} \left(\frac{\alpha g}{\kappa \upsilon} \right)^{1/3} \rho C_P \kappa (T_F - T)^{4/3} \tag{2}$$

where A_{ROOM} is the area of the room, ρ and C_P are the density and heat capacity of water respectively, λ is a dimensionless constant which characterises the heating from the plate, α, κ and ν are the expansion coefficient, thermal diffusivity and viscosity of water respectively (Denton & Wood 1979). Empirical measurements show that $\lambda = 2^{4/3} c_Q$ where c_Q is the heat transfer coefficient (Townsend 1959). The conservation of heat in the system at steady state gives a balance of the heating and cooling fluxes,

$$Q_H = \rho C_P (T_{SS} - T_{EXT}) Q_V \tag{3}$$

where T_{SS} is the steady state temperature to which the room evolves and T_{EXT} is the exterior temperature. It is convenient to scale Eqn. 3 to T_F and T_{EXT}, reducing it to the algebraic relation

$$(1 - \theta)^{4/3} = Z \theta^{3/2} \tag{4}$$

where $\qquad \theta = \dfrac{T_{SS} - T_{EXT}}{T_F - T_{EXT}} \qquad$ and $\qquad Z = \left(\dfrac{c}{\lambda} \right) \dfrac{A_{HOLE}}{A_{ROOM}} \left(\dfrac{g^{1/2} \alpha^{1/2} \upsilon}{\kappa^2} \right)^{1/3} \tag{5}$

Z is a dimensionless parameter incorporating the coefficients from the two transient processes. For small Z, corresponding to large openings and rapid displacement flow, the temperature in the room, T_{SS} is closer to that of the exterior fluid so $\theta \to 0$. In contrast, for large Z, corresponding to small openings and relatively rapid heating by convection, T_{SS} tends to the temperature of the heated floor so $\theta \to 1$.

EXPERIMENTAL CALIBRATION

In order to make quantitative comparisons between the model and the experiments we need to constrain the model parameters λ and c which quantify the efficiency of the heat transfer from the heated plate and the energy loss through the doorway associated with the displacement flow, respectively. We have therefore conducted two series of controlled calibration experiments.

Displacement Ventilation

In the first calibration experiment, the rate of ascent of a saline-fresh water interface inside the room, driven as a pure displacement flow, is examined so as to identify the value of the constant c. Extending the model of the displacement flow from Eqn. 1, the rate of ascent of the interface between light interior fluid and dense exterior fluid, z, is given by

$$A_{ROOM} \frac{dz}{dt} = c A_{HOLE} \left[g' \frac{(h - z)}{2} \right]^{1/2} \tag{6}$$

where $g' = g(\rho_{EXT} - \rho_{INT})/\rho_{INT}$. Eqn. 6 can be integrated, to give,

$$(h - z_0)^{1/2} - (h - z)^{1/2} = cBT \qquad \text{where} \qquad B = \left(\frac{A_{HOLE}}{A_{ROOM}} \right) \frac{g'^{1/2}}{2\sqrt{2}} \tag{7}$$

In these experiments, the room is filled with dyed fresh water and the exterior reservoir filled with 1%, 2% or 3% saline solution by mass. Bungs are removed from two of the three openings, either top and bottom, top and middle, or middle and bottom openings. The light interior fluid leaves the room through the upper hole while dense exterior fluid enters the room through the lower hole. To reduce the impact of mixing during initial stages on visualisation, the basal 2 cm of the room is filled with a layer of saline solution identical in concentration to that outside the room. The ascending interface is monitored using Digimage image analysis software. Twelve experiments were performed, using three configurations of openings and three salinities; three experiments were repeated. The interface rises with time although the rate of ascent decreases with time. The interface ascends more rapidly when there is a large density difference between the fluids (cf Linden et $al.$,1990). We scale our data on $(h-z_o)^{1/2}$ using Eqn. 7 (figure 4). The data for all experiments collapse to a straight line whose slope gives a value of $c=0.98\pm0.15$.

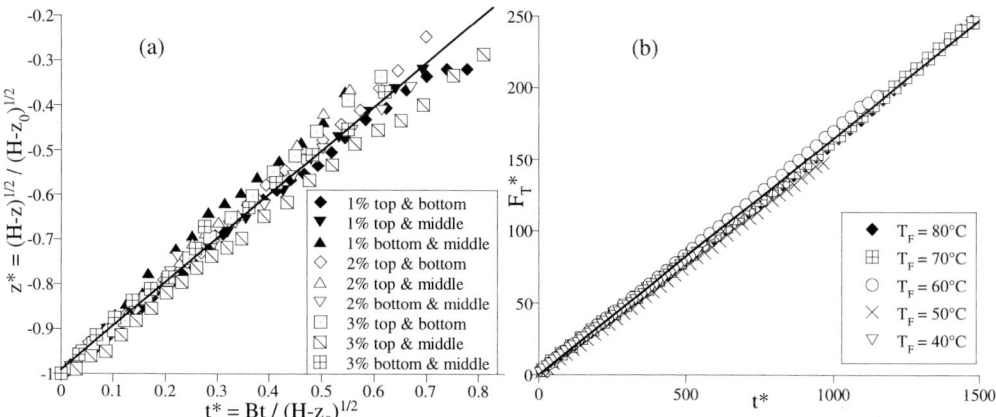

Figure 4: Scaled calibration experiments:
(a) displacement ventilation data collapse to a line with slope $c=0.98$ using Eqn. 7,
(b) transient heating data collapse to a line with slope $\lambda=0.166$ using Eqn. 8.

Transient Heating

The second calibration experiment involved heating the closed room by the floor plate and measuring the temperature of the room as a function of time, to identify the value of the coefficient λ from Eqn. 2 for our heating plate. Integrating Eqn. 2, and defining the resulting heating function as F_T gives,

$$F_T = \int_{T_{EXT}}^{T} \frac{1}{J_T (T_F - T)^{4/3}} dT = \lambda \int_{t_0}^{t} dt \qquad \text{where} \qquad J_T = \left(\frac{\alpha \kappa^2 g}{\upsilon} \right)^{1/3} \frac{1}{H} \qquad (8)$$

The room was filled with water of initial temperature $T_{EXT}=15°C$. Values for T_F of 40, 50, 60, 70 and 80°C were used. The water inside the room increases in temperature T rapidly at first and then more gradually as T approaches T_F. Temperature measurements at different vertical heights show that the room is well-mixed. Typically it takes five to eight hours for T to reach T_F.

Scaling Eqn. 8 using $J_{TF}(T_F-T_{EXT})^{4/3}$ provides expressions for the dimensionless heating function, F_T^*, with dimensionless time, t^*. When plotted until the time when $T=0.95T_F$ these data collapse to a line with slope λ where $\lambda=0.166\pm2\%$. Thus the heat transfer coefficient, c_Q is 0.066 comparing well with previous observations for single (Townsend 1959) and double plate (Denton & Wood 1979) turbulent convection experiments.

372

COMPARISON OF MODEL WITH EXPERIMENTS

The observations and theoretical predictions of T_{SS} agree well for our room when the top and bottom, or middle and bottom holes are opened (figure 5a). At high T_F, experimental measurements are slightly higher than values predicted by the theory. With the bottom hole open, cold exterior fluid entering the room passes directly over the heating plate allowing the floor to heat fluid of temperature T_{EXT} rather than the warmer mixed fluid. Consequently the transfer of heat is more effective, becoming pronounced at high T_F. When cold exterior fluid enters the room through the middle hole, experimental observations of T_{SS} are lower than theoretical predictions (figure 5b).

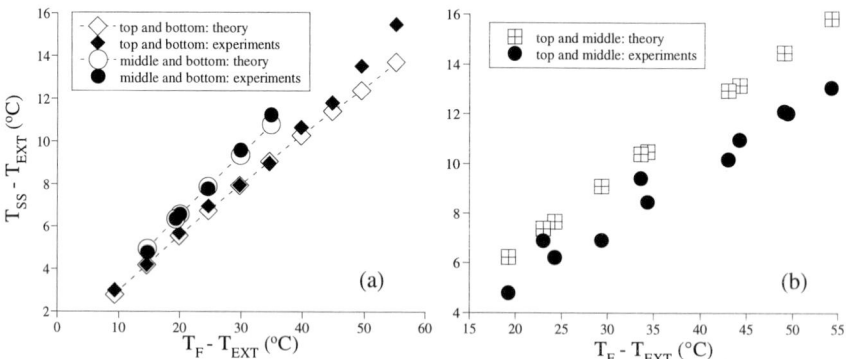

Figure 5: Theoretical predictions and experimental observations of T_{SS} using
(a) top and bottom, or middle and bottom openings, and (b) top and middle openings.

CONCLUSIONS

This experimental and theoretical study has shown that a fully-mixed flow regime becomes established in a warm room which is in contact with a cold exterior by two openings, and heated by convection from the floor. This contrasts with the two-layer stratification that results from localised sources of heating (Linden *et al.*, 1990). We have developed an analogue experimental system and show that the steady state temperature, T_{SS}, can be predicted well by a new model in terms of floor, T_F, and exterior, T_{EXT}, temperatures, based on balancing heating, Q_H, and displacement, Q_V, fluxes. The model may be applied to modern buildings with large surface areas of glass, leading to areal heating by day and cooling at night. For example, we consider a large atrium with a glass ceiling which at night leads to a temperature difference $\Delta T = 5°C$ between the interior and exterior. If $A_{HOLE}/A_{ROOM} \sim 0.01$, using the following typical values for air, $\alpha \sim 1/300$, $g \sim 10 \text{ ms}^{-2}$, $\upsilon \sim 10^{-5} \text{ m}^2\text{s}^{-1}$ and $\kappa \sim 10^{-6} \text{ m}^2\text{s}^{-1}$, we estimate $Z \sim 0.1$ and $\theta \sim 0.9$. Thus displacement ventilation is relatively rapid so T_{SS} tends to T_{EXT}, causing the atrium to be cold.

REFERENCES

Denton, R.A. & Wood, I.R (1979) Turbulent convection between two horizontal plates. *Int. J. Heat Mass Transfer.* **22**, 1339-1346.

Linden, P.F. (1999) The fluid mechanics of natural ventilation. *Ann. Rev. Fluid Mech.* **31**, 201-238.

Linden, P.F., Lane-Serff, G.F. & Smeed, D.A. (1990) Emptying filling boxes: the fluid mechanics of natural ventilation. *J. Fluid Mech.*, **212**, 309-335.

Townsend, A.A. (1959) Temperature fluctuations over a heated horizontal surface. *J. Fluid Mech.*, **5**, 209-241.

Air Distribution in Rooms, (ROOMVENT 2000)
Editor: H.B. Awbi

OPTIMIZATION OF THE DESIGN OF TWO TEST ROOMS BY MEANS OF CFD

B. Bjerg[1] K. Svidt[2] S. Morsing[3] G. Zhang[3]

[1]Dept. of Animal Science and Animal Health, The Royal Veterinary and Agricultural University,
Grønnegårdsvej 8, DK-1870 Frederiksberg C, Denmark
[2]Dept. of Building Technology and Structural Engineering, Aalborg University, Sohngaardsholmsvej
57, DK- 9000 Aalborg, Denmark
[3]Dept. of Agricultural Engineering, Danish Institute of Agricultural Sciences, Research Centre
Bygholm, DK-8700 Horsens, Denmark

ABSTRACT

This paper reports experience of using test rooms in development and validation of CFD-methods to predict airflow in animal houses. The work showed that it is extremely important to be aware of how the room geometry affects the airflow in the test room. Two test rooms were used. Both rooms were 8.5 m long and 3 m high. Test room one was 5 m wide and test room two was 10 m wide.

The purpose of test room one was to create a two-dimensional airflow as a reference case for later studies of the influence of simulated animals and pen partitions. Therefore, the room was equipped with two-dimensional inlet and exhaust conditions, i.e. slot inlet and exhaust. But the generated airflow was far from being two-dimensional. Succeeding CFD-calculations confirmed that the airflow was three-dimensional although the boundary conditions were two-dimensional. Other calculations showed that the aspect ratios of the room determined whether the airflow became two- or three-dimensional. Further studies based on CFD-calculations showed that guiding plates beneath the ceiling would be efficient to create the desired two-dimensional airflow in test room one.

Test room two was built to study the influence of inlet nozzles, and in this case CFD-calculations were used in the design process to determine a room dimension which would minimise a possible influence of the room geometry.

Overall the work showed that CFD is an efficient tool to avoid unintentional influence of room geometry in the design of test rooms for airflow studies.

KEYWORDS

Airflow, CFD, Test room, Three-dimensional airflow, Two-dimensional boundary condition, Symmetric flow, Symmetric boundary conditions. Room dimensions, Wall jet.

INTRODUCTION

This work is part of a project that aims to develop CFD-methods to predict airflow in animal houses. In the first part of the project the objective was to study how animals and equipment close to the floor affect the airflow in a livestock room. As reference case for this we wished to create a two-dimensional airflow near the floor in a empty space. Normally it is assumed that a symmetric room with two-dimensional boundary conditions will lead to a two-dimensional air flow pattern, and therefore we decided to build an orthogonal test room with an inlet slot beneath the ceiling in the full length of one wall - test room one.

Later in the project the objective was to develop boundary conditions for inlet devices used in livestock buildings. Many livestock buildings in Denmark and other countries are equipped with under pressure mechanical ventilation systems with automatically controlled prefabricated inlet openings. In this part of the project we needed a test room were it was possible to install one or a number of inlets. Taught by experience from the first part of the project we were very aware of undesired influence from the geometry of the room and, therefore, we made several simulations in different configurations before we decided to build a new test room - test room two.

METHODS

Test room one was 8.5 m long, 5 m wide and 3 m high. The inlet was located beneath the ceiling at one entire end wall, see figure 1. An outlet was equally distributed to the entire room width near the inlet wall. In this paper we discuss three experiments all carried out at 15 Pa pressure drop across the inlet.

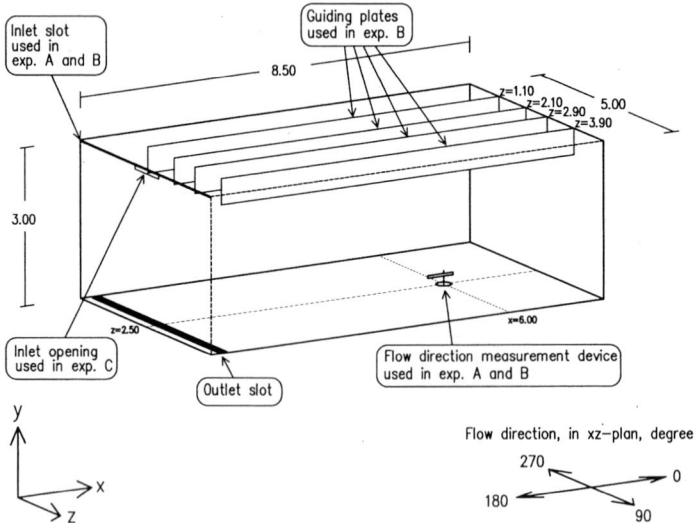

Figure 1. Test room one, with different configurations in experiment A, B and C.

In experiment A and B the inlet was 19 mm high and 5 m long (equal to the room width). A 0.10 m high and 1.00 m long symmetrically located opening was used in experiment C. To obtain two-dimensional flow in the occupied zone four guiding plates were mounted beneath the ceiling in experiment B. The plates were 7.65 m long and reached 0.5 m vertically down from the ceiling.

The overall airflow pattern was determined by smoke and the direction of the return air was measured with a flow direction sensor (Zhang *et al.* 2000) in experiment A and B. This sensor was located at $x=$ 6.5m; $y=0.15$ m; $z=2.5$ m. Air speed was measured with a multi-channel thermistor based omni-directional air velocity sensing system. (Zhang et al., 1996).

Test room two has equal height, length and exhaust configuration as test room one. The width is 10 m and there is one orthogonal inlet, 525 mm wide and 50 mm high. The inlet was orientated to point 20 degrees upward and the location of the inlet appears from figure 2. The same measurement system as in test room one was used.

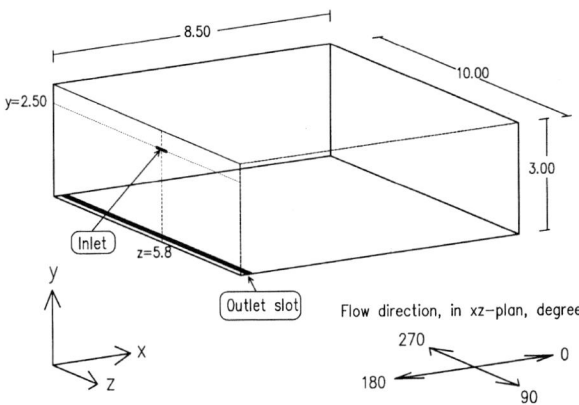

Figure 2. Test room two

The commercial numerical simulation codes CFX4 (AEA Technology Plc) and FLUENT5 (Fluent Incorporated) were used for simulation of airflow in the mentioned configurations of the test rooms and in a number of additional configurations. In the simulation the K-epsilon turbulence model was used with isothermal flow and wall functions at all surfaces.

RESULTS AND DISCUSSION

Three-dimensional airflow in a room with two-dimensional boundary conditions

Figure 3 shows calculated airflow beneath the ceiling and above the floor in the two-dimensional configuration used in test room one experiment A. Near the floor ($y=0.20$ m) it can be seen that the flow moves more or less diagonally from the back right to the front left corner of the room, and consequently, the flow is three-dimensional although the boundary conditions were two-dimensional. Beneath the ceiling ($y=2.8$ m) the flow turns to the right part of the back wall and a vertical vortex can be seen to the left near the back wall (top right corner of the figure).

In the experiment was found a corresponding airflow pattern by smoke visualisation, and the air flow direction at x, y, $z=6.50$, 0.20, 2.50 m was measured to 213 degrees. Depending on grid density the simulated airflow at the same point was between 218 and 223 degrees (Simulations were made with 2,040; 44,200 and 120,000 cells).

Beneath ceiling (y=2.8 m)   Above floor (y=0.2 m)

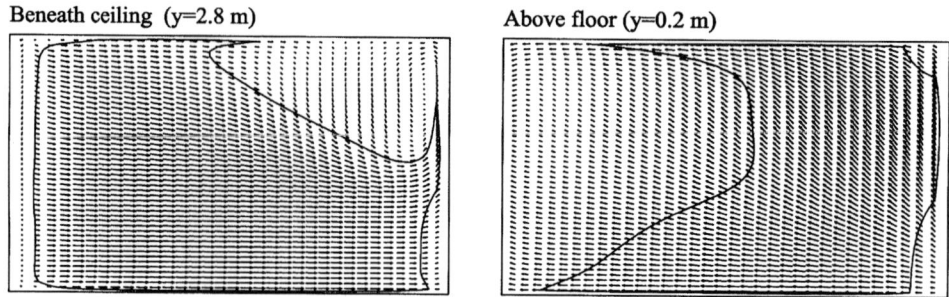

Figure 3. Calculated airflow beneath the ceiling and above the floor in the two-dimensional configuration used in test room one experiment A. The isolines show air velocity of 0.5 m/s.

Earlier investigations of wall jets in enclosures with two-dimensional boundary conditions have not reported three-dimensional effects on airflow analogous to that found in this investigation (see Bjerg *et al.* 1999). Differences in room dimensions might be an explanation for this and therefore, numerical simulations with different room dimensions were made. In this paper simulations with different ratios between room width and room height (W/H ratio) in 3 m high and 9 m long rooms is presented and discussed. Figure 4 shows the simulated air velocity 0.1 m beneath ceiling and 4.5 m from the inlet in rooms with W/H ratios of 1, 1.67, 3, 6 and 18. Two-dimensional inlet and exhaust conditions were used and the Reynolds number was 4,800.

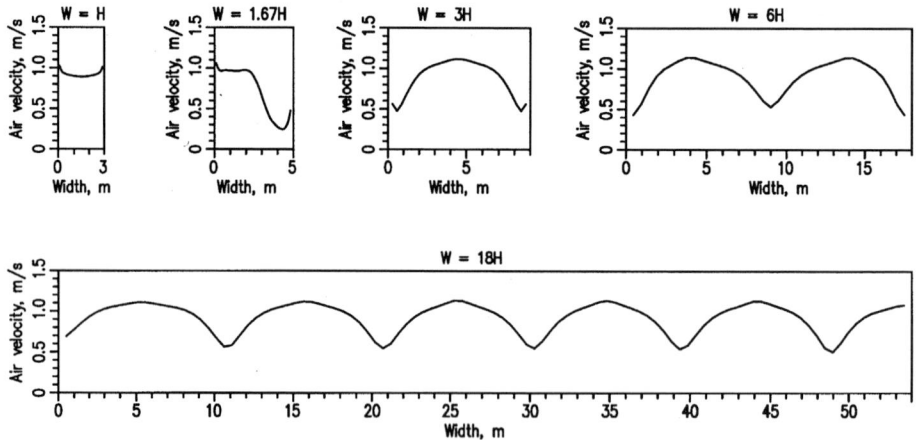

Figure 4. Simulated air velocities 0.1 m beneath the ceiling and 4.5 m from the inlet slot in a number of 3 m high (H) and 9 m long rooms with different room widths (W); the Reynolds number was 4,800

The diagrams show that the variation in air velocity was small for the room with W=H and the velocity profiles for rooms with W>3H describes periodic curves with a periodic length of about 10 m or 3H. To some extent the curve for W=1.67H looks like a part of a periodic curve. There has been no experimental confirmation of this phenomenon for a room width larger than 1.67H in this investigation, and none found elsewhere in the literature in connection with airflow in rooms.

Creation of two-dimensional air flow in the occupied zone.

The purpose of test room one was to obtain two-dimensional airflow above the floor and, therefore, CFD was used to analyse possible modifications of the room to avoid the three-dimensional effects. According to figure 4, one way would be to reduce the W/H ratio but this was not suitable in our case. Instead simulations were made with guiding plates beneath the ceiling as shown in figure 1 experiment B. The result of these simulations was a perfect two-dimensional airflow pattern, and therefore the guiding plates were implemented in the test room. Measurements confirmed the simulated effect off the guiding plates (Bjerg, *et al.* 2000).

Test room for studying airflow from an inlet used in livestock ventilation.

Later in the project the aim was to analyze how inlets as shown in figure 1 affect the airflow in livestock rooms. For this purpose we needed a room where it was possible to study three-dimensional flow from one or a number of openings without unintentional disturbances from the geometry of the room.

First trial was to equip test room one with a concentrated opening symmetrically located in the room as shown in figure 1, experiment C. The simulated airflow 0.2 m beneath the ceiling in this set-up is shown in figure 5. It can be seen that the jet deflects to one of the side walls and this was regarded as an unintentional disturbance from the geometry of the room. Smoke tests and measurement of air velocity in experiment C confirmed the simulated airflow pattern, and therefore it was necessary to design a new room.

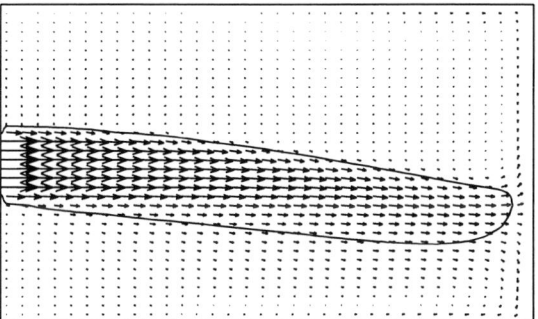

Figure 5. Calculated airflow 0.05 m beneath the ceiling in the configuration used in test room one experiment C. The isolines show air velocity of 0.5 m/s.

At that state we made simulation of a number of different geometries and this way we found two methods to reduce the unintentional disturbance of the jet. Either reduce the length or increase the width of the room.

Test room one was located next to an equal sized room, and therefore it was reasonable to evaluate the possibilities of joining the two rooms. A problem was that the room construction did it impossible to locate a inlet in the symmetric plane of this double size room and, therefore, was made simulations with one opening located 0.8 m from the symmetric plane. These simulations showed no significant disturbance from the room geometry and we decided to build test room two. Figure 6 compares simulated and measured air velocity in two distances from the inlet in test room two, and it can be seen that the measurement confirmed that the jet was symmetric around the center of the inlet.

378

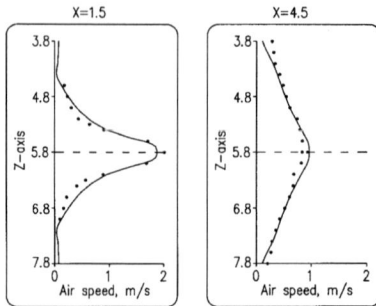

Figure 6. Comparison of measured (dots) and simulated (lines) maximum air velocity 1.5 and 4.5 m from the inlet wall in test room two. The dashed line shows the symmetric plane of the inlet.

Rooms with a number of inlets.

In real livestock rooms a number of inlets are used, and these inlets are often located with equal distance in one horizontal row in the wall. It is usually assumed that in some distance the jets from the inlets merge into a plane jet. In the on going project these conditions will be studied, and a result from a simulation with 4 opening is shown in figure 8. The assumed design of inlet and location above the floor was equal to the configuration used in figure 3. The distances between the center of the openings was 2.5 m and the outer openings was centered 1.25 m from the side walls.

Figure 7 shows that the jets from the 4 inlets did not merge to a plane jet in these simulations. Even in the distance of 7.5 m from the inlets there is a significant variation in the simulated air velocity in the shown profile. It is also distinct that the three shown profiles are symmetric around the mid plane of the room, and at long distance from the inlets (x=7.5) it seems like high air velocities are concentrated near the mid plane of the room. Concentration of high air velocities near the mid plane of the room was also found in the simulations in rooms with nearly the same room dimensions and two-dimensional inlet configuration (see figure 5, W=3H). This suggests that the periodic patterns found in simulation of wider rooms with two-dimensional inlet conditions also may appear in rooms with a number of inlets located in a row.

CONCLUSIONS

- Two-dimensional boundary conditions do not necessarily lead to two-dimensional airflow.
- The aspect ratio of the room has a decisive influence on the development of three-dimensional airflow patterns in rooms whit two-dimensional boundary conditions.
- Guiding plates beneath the ceiling are efficient to create two-dimensional air flow in a room with two-dimensional inlet and exhaust condition.
- In rooms with one inlet and symmetric boundary conditions, the aspect ratios of the room determine whether the flow will become symmetric or asymmetric.
- In a wide room with a single opening the jet maybe practically symmetric although the opening is located in some distance from the symmetric plane of the room.
- Periodic airflow patterns found in simulation of a wide room may also be found in rooms with a number of equally distributed inlets.
- CFD is an efficient tool to avoid unintentional influence of room geometry in design of test rooms for airflow studies.

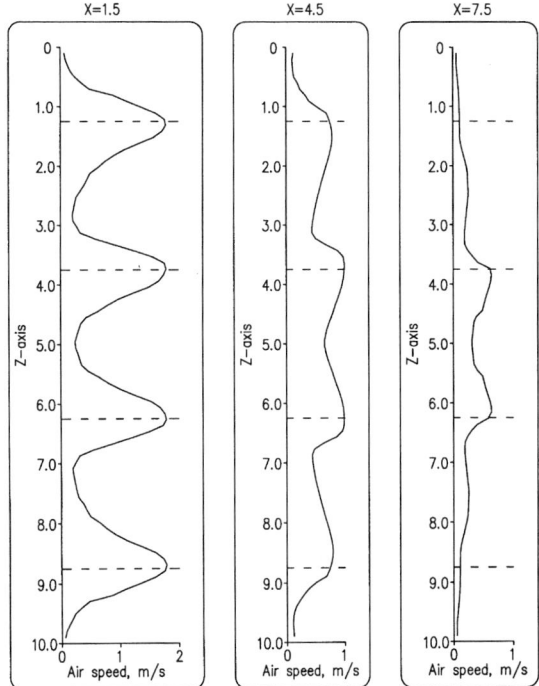

Figure 7. Simulated air velocity 0.05 m beneath the ceiling in test room two equipped with four inlets. The dashed lines show the symmetric planes of the inlets.

REFERENCES

Bjerg B; Morsing S; Svidt K; Zhang G. (1999) Three-dimensional Airflow in a Livestock Test Room with Two-dimensional Boundary Conditions . Journal of Agricultural Engineering Research, Vol. 74, No. 3, pp. 267-274

Bjerg B; Svidt K; Zhang G; Morsing S. (2000) The Effects of Pen Partitions and Thermal Pig Simulators on Airflow in a Livestock Test Room. Submitted for publication in Journal of Agricultural Engineering Research.

Zhang G; Strøm J S; Morsing S. (1996). Computerized multi-point temperature and velocity measurement system. Proceedings of ROOMVENT '96, Fifth International Conference on Air Distribution in Rooms, July 17-19, Yokohama, Japan.

Zhang, G.; Morsing, S.; Bjerg, B.; Svidt, K and Strøm J.S. (2000).The design of an air flow pattern test room for validation of CFD estimation. Submitted for publication in Journal of Agricultural Engineering Research.

2000 Elsevier Science Ltd. All rights reserved.
Air Distribution in Rooms, (ROOMVENT 2000)
Editor: H.B. Awbi

INTEGRATION OF HEATING MODE INTO VENTILATED COOLED BEAM

Risto Kosonen [1], Pekka Horttanainen [1] and Gordon Dunlop [2]

[1] Research & Development, Oy Halton Group Ltd,
Vantaa, 01510, Finland
[2] Ove Arup & Partners
London, W1P 6BQ, UK

ABSTRACT

Nowadays the ventilated cooled beam is one of the most popular air-conditioning system, e.g. in Scandinavia and Central Europe. With such beams, it is possible to create high-quality indoor climate conditions, including thermal comfort and a low noise level within reasonable life-cycle costs. The beam is suitable for spaces with a high cooling requirement, low humidity load and relatively small ventilation requirement. Typically, the beams are used in offices and conference rooms. Although the beam system is quite popular and well-known, there is still limited experience in using these beams for primary heating. The main objective of this paper is to present the results of a study to determine the heating power that can be achieved with the beams. The study comprised several tests carried out in laboratory conditions. In the laboratory measurements, the heating capacity, temperature gradient of the room, and the air velocities were analyzed when the parameters were the temperature of the inlet water and the flow rate of the supply air. The results obtained from the study, show that it is possible to create comfortable indoor conditions with the beams even during winter conditions. The measured air-flow velocities were lower than 0.2 m/s at all the measurement points. An inlet water temperature of 40 °C gives a reasonable temperature gradient in the room of under 3 °C. At the same time, the average air velocity in the occupant zone is under 0.1 m/s.

KEYWORDS

Ventilated cooled beams, heating, laboratory measurements, thermal comfort, air conditioning

BACKGROUND TO COMFORT AND HEATING

The heating and ventilation systems of a room are functions that contribute to comfortable, healthy and productive indoor climate conditions. However, increasingly there are higher expectations of indoor comfort conditions and the ability to control the room temperature, the air quality and limit drafts is now common. The basic guideline for thermal comfort is laid down in the ISO 7730 Standard (1990). The major part of this report concerns thermal comfort in terms of the temperature gradient in the room space and the room air velocities. ISO 7330 recommends the following:

- Air velocity (to avoid draughts):
< 0.15 m/s (winter)
< 0.25 m/s (summer)
- Vertical air temperature difference:
< 3°C from foot to head
when sitting, 0.1 m to 1.1 m, or standing, 0.1 m to 1.7 m.

Figure 1 shows some examples of the temperature gradients of common heating devices, (Heating and Ventilating Engineer 1985). The figure 1 illustrates that, with the traditional heating system, the maximum gradient is typically 3 – 5 °C. Of course, the existing conditions considerably affects a lot for the gradient, e.g. heat losses, positioning of the heating device, surface temperatures, and the sizing temperature levels of the heating device. All in all, the higher the heat losses and temperature level of emitters, the higher the temperature gradient in room space. Furthermore, the installation has a consequential effect on the gradient: it is more difficult to keep the gradient low when the installation height of the emitter is high.

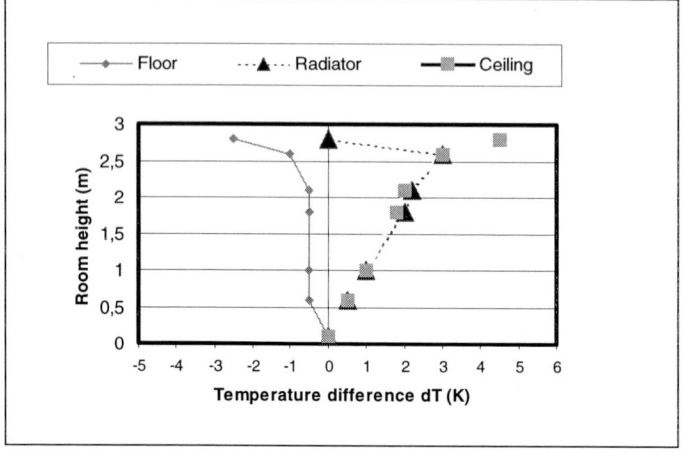

Figure 1: Temperature gradients of common heating systems

The required heating capacity is strongly dependent on the thermal properties of the building and the outdoor climate conditions. Figure 2 shows the required heating capacity as a function of the average thermal conductance and the outdoor temperature. In this case, the room dimensions are 2.8 m (room width), 4.5 m (room depth) and 3.0 m (height), giving a the floor area of 12.6 m^2 and a total external wall area of 8.4 m^2. In this case, the glazing ratio of the total outdoor wall area is fixed to 50 %. In the European case, there is a double-glazed window (U-value 3 W/m^{2}K) and the U-value of the wall is 0.45 W/m^{2}K. In the Scandinavian case, there is a triple-glazed window (U-value 2 W/m^{2}K) and the U-value of the wall is 0.28 W/m^{2}K. Figure 2 also shows an example which incurs high heat losses: this case is like a European one, except for having a single-glazed window. In addition, a case representing low heating demand is shown. Here, the heat losses are simply fixed 50 % lower than the Scandinavian case

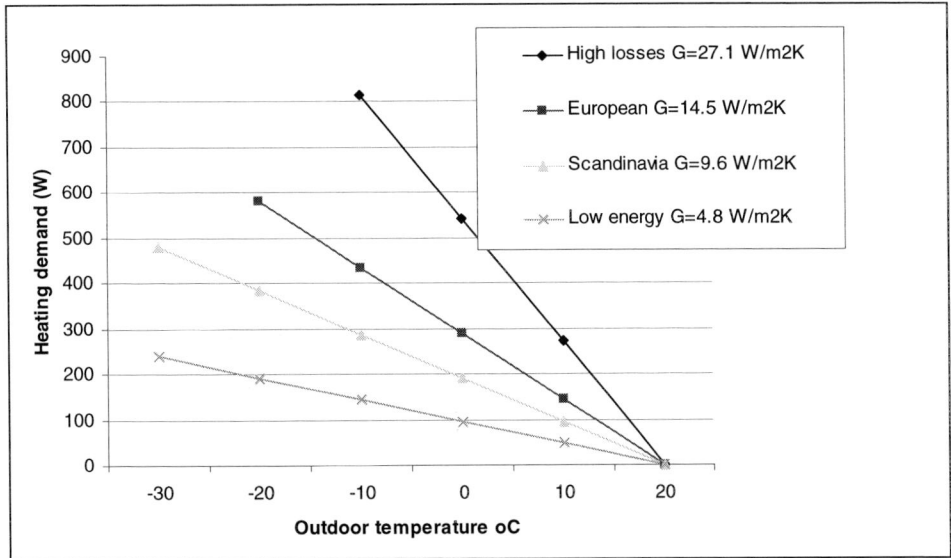

Figure 2. The required heating capacity as a function of thermal conductance and external temperature for a case room-module (room area 12.6 m² and external wall area 8.4 m²).

The main idea of Figure 2 is to show the trend-setting heating capacity for a typical room. Of course, during the design phase the required heating demand should be calculated case by case. Based on this calculation, the required heating capacity of 500 W per room-module is enough for the normal case.

HEATING WITH VENTILATED COOLED BEAMS

Nowadays the ventilated cooled beam is one of the most popular air-conditioning systems, e.g. in Scandinavia and Central Europe. With such beams, it is possible to create high-quality indoor climate conditions including thermal comfort and a low noise level within reasonable life-cycle costs. The beam is suitable for spaces with a high cooling requirement, low humidity load and a relatively small ventilation requirement. Typically, these beams are used in offices and conference rooms.

The ventilated cooled beams can be divided into passive and active types, according to their properties. In passive beams, the heat transfer mainly occurs by means of free convection and partly by radiation. With active beams, the supply air device is integrated in the beam and the heat transfer occurs by forced convection. In active beams, the air-flow supplied through the nozzles induces room air through the heat exchanger of the beam, Figure 3. The mixture of supply air and induced air is directed into the room through the longitudinal slot on one or both sides of the beam. In the open model of the beam, the induced air comes from the upper side of the beam. In the closed model, the induced air enters through the bottom plate under the beam (as shown in Figure 3).

Although the beam system is quite popular and well-known, there is still limited experience in using these beams as a primary heating device. Laboratory measurements were carried out to analyze the properties of the ventilated cooled beams for heating. This paper focuses on the results for the closed beam, which is designed to be installed with a suspended ceiling. In this beam, the heating is achieved by integrating separate cooling and heating water pipes in the same coil.

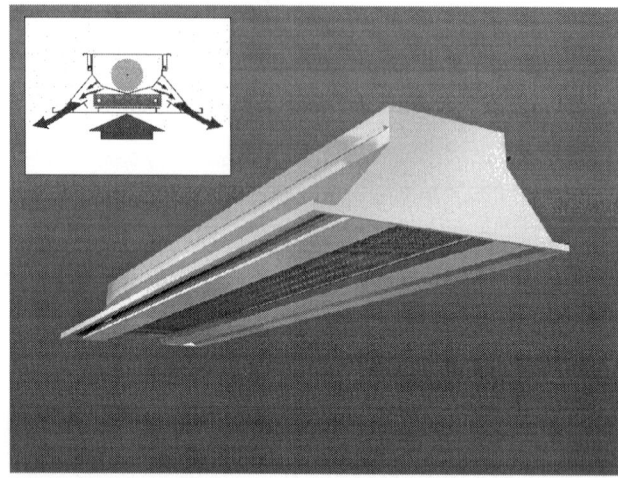

Figure 3: The studied ventilated cooled beam where heating and cooling modes are integrated.

In the laboratory measurements, the heating capacity, and the temperature gradient of the room and air velocities were analyzed when the parameters were the temperature of the inlet water and the flow rate of the supply air. The studied temperature levels of the inlet water were 40 °C, 60 °C and 70 °C, and the studied flow rates of the supply air were 13 l/s and 22 l/s.

The measurements were carried out in laboratory with dimensions of 4.5 m x 2.8 m x 2.7 m (height). The length of the studied beam was 1800 mm long (inside 1500 mm coil) and the width of the beam was 600 mm. The installation height of the beam was 2200 mm. In this measurement, the suspended ceiling was not used. In these measurements the air could freely stratify over the closed beam. In the normal case, when the suspended ceiling is used, the highest temperature is just under the beam. Table 1 gives the measurement results while Figures 4 and 5 show the temperature gradients.

TABLE 1
MEASURED DATA OF THE BEAM IN THE HEATING MODE.

Supply air rate q_v(l/s)	Supply Air temp. T_{supp} (°C)	Water output P_w(W)	Output per active length P_w(W/m)	Inlet water temp. T_{w1}(°C)	Outlet water temp. T_{w2}(°C)	Water temp. difference dT_w(°K)	Water flow rate $Q_{m,w}$ (kg/s)
22.3	18.7	561	373.7	38.6	34.2	4.4	0.0303
22.1	18.7	1154	769.1	60.7	51.4	9.3	0.0299
22.3	18.6	1519	1012.8	73.8	61.7	12.1	0.03
13.0	18.7	426	283.9	38.0	34.6	3.4	0.0301
12.7	18.7	824	549	57.3	50.8	6.5	0.0302
12.7	18.6	1233	822.3	73.2	63.1	10.1	0.0293

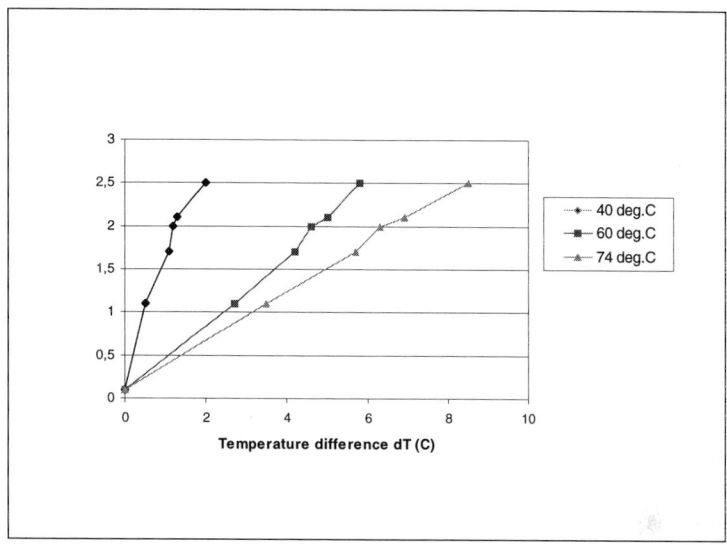

Figure 4: Room temperature gradient with three different inlet water temperature level. The supply air-flow rate is 22 l/s.

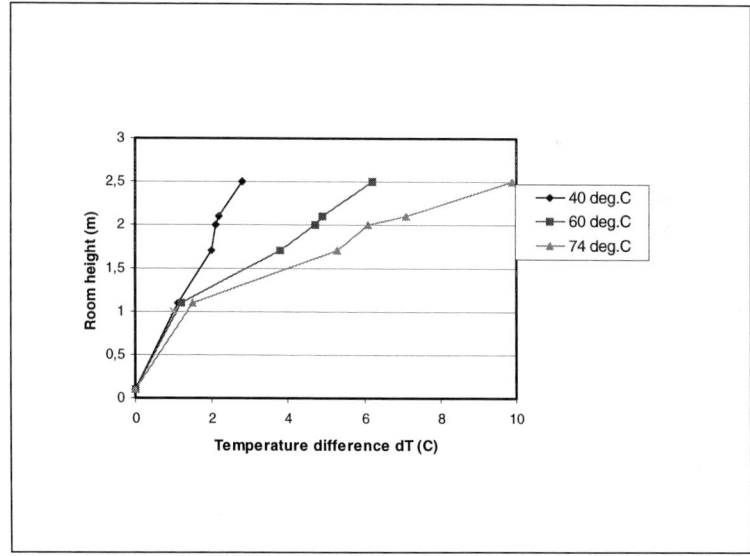

Figure 5: Room temperature gradient with three different inlet water temperature level. The supply air-flow rate is 13 l/s.

The measured data shows the effect of the inlet water temperature level on the temperature gradient. With relatively high temperatures, the gradient in the room rises: the temperature level of 74 °C leads to quite a high gradient 8 – 10 °C in the test room. Reducing the temperature level to 60 °C produces a 5 – 6 °C gradient in the room, which does not fulfil thermal comfort requirements. Using a 40 °C inlet water temperature gives reasonable gradients in the room of under 3 °C. At the same time, the average air velocity in the occupant zone is under 0.1 m/s.

Increasing the supply air-flow rate decreases the temperature gradient. In the measured case, using 22 l/s instead of 13 l/s reduced the gradient only about 1 °C. Therefore, the key factor is to control the inlet water temperature at a reasonably low level. Decreasing the inlet temperature is not a problem for fulfilment of the required heating capacity. With the temperature level 40 °C, it is possible to reach 500 W, which in most cases is enough.

DISCUSSION

Integration of the heating mode in the ventilated cooled beams is a challenging task. The main problem has been to control the temperature gradient at a reasonably low level, and so contribute to creating thermal comfort conditions in room spaces in an energy-efficient way. Increasing the gradient could be problematic for thermal comfort as well as increase energy consumption. On the other hand, meeting the heating capacity requirement is not a big problem in most cases.

The key factor is the temperature level of the inlet water. The lower inlet temperature leads to a lower gradient. Based on the measurements obtained, a temperature level of 40 °C or below, gives a reasonable gradient for the room space. For design practice, this also means the possibility to use low temperature level heating sources.

It should be noticed that these measured gradients show the worst case when there are no solar and internal heat gains. During working hours, there is normally enough or even too much of a heat load to keep indoor temperature within acceptable comfort levels. This means that this worst case exists in the beginning of the working day when the external temperature is also near sizing conditions. When the required heating capacity is lower the gradient is also smaller.

The meaning of the gradient should be analyzed case by case during the design process. If the period of high-heating peak load is short, the relatively high gradient could be acceptable and the inlet temperature could be higher. This analysis should be based on the life-cycle approach.

REFERENCES

International Standard ISO 7730, Moderate thermal environments- Determination of the PMV and PPD indices and specification of the conditions for thermal comfort, 1990.

Advances in Floor Heating in Europe. Heating and Ventilating Engineer number 674 1985. Pages 18-22.

Air Distribution in Rooms, (ROOMVENT 2000)
Editor: H.B. Awbi

WAVELET TRANSFORM OF TURBULENT VELOCITY FLUCTUATIONS

Matjaz Prek

University of Ljubljana, Faculty of Mechanical Engineering
Askerceva 6, 1000 Ljubljana, Slovenia

ABSTRACT

The central problem in fully developed turbulence is the energy-cascading process. This has resisted all attempts at a full mathematical formulation. The main reasons for this are related to the hierarchy of scales involved and non-linear character of the Navier-Stokes equations. The spatial intermittency of the dynamically active regions of flow presents an additional problem. A multifractal model of a fine-scale intermittency has been introduced to account for the more complex cascading process suggested by the experimental data on inertial range structure. The statistical models are insufficient, because they cannot give any information about the intermittent structures in real space. In the present work, the wavelet transform was used to analyze the velocity field obtained from the experimental data. The space - scale analysis showed evidence of the Richardson cascading process and its fractal character. Multifractals provide a rich framework in which to analyze the turbulent fluctuations of the dissipation field. Turbulence gives rise to random fractals and turbulence dissipation is also a process whose scaling properties are determined by random refinements. Scaling interpretation is determined using the self-similar properties of fractal measures at different length scales. The period-doubling transition to chaos is presented by the wavelet transform of golden-mean trajectory, calculated using the sine circle map. Both processes exhibits a similar fractal character, which is evident in wavelet transform map. The results indicate that the energy cascading process has remarkable similarities to the deterministic construction rules of the golden-mean sine circle map.

KEYWORDS

Turbulent flow, wavelet transform, sine circle map, intermittency, fractal, random.

INTRODUCTION

Fourier analysis is quite useful tool for determining the frequency components of a signal, but it cannot determine when those frequency components occured because it lacks time resolution. To analyze the local time and frequency content of a signal, more sophisticated techniques must be used. Several methods based upon multiscale decomposition have emerged in various fields as accurate and reliable techniques for analysing signals or for numerical simulations. Wavelet analysis is one method, which has been used succesfully in signal and harmonics analysis. Now it is commonly beleived that wavelet analysis was an important breakthrough, because it can be used to decompose arbitrary signals into localised contributions that can be labelled in terms of time and scale, thus overcoming the limitations of Fourier analysis.

The wavelet analysis is essentially used in two ways: as a technique to extract information about the process in time / scale domain, and as basis for representation or characterisation the process. In this paper, we investigate the application of wavelet analysis for experimental data. Since Argoul et.al. (1989) first time used the wavelet transform to analyze experimental turbulent data in the wind tunnel and provide the visual evidence of Richardson cascade (1922), wavelet transform has been widely used to reveal various turbulent or eddy structures, e.g. Farge (1992), Farge et al. (1996). It has become a well-known fact that the large-scale eddy motions exhibit a symmetric, periodic and apparent flapping motion, and that the evolution and interaction of large-scale structures play an important role in turbulent jet widening and momentum transfer. The local period of the turbulence and eddy motion with respect to space / time changes continuosly and the data, obtained with visualisation of motion, showed that the conventional correlation measurements had not reveal all features of turbulence. To develop a better understanding of vortex structure regarding both localization in wave number and physical space, the wavelet analysis is applied to velocity signals. In this work, the structure of the plane turbulent jet is investigated. The spatial structure of an eddy in time and physical scale is explored.

To study such multiscale processes, e.g. turbulent velocity fluctuations, the wavelet transform presents a suitable technique. Wavelets are able to decompose arbitrary signals into localised contributions that can be labelled in terms of time and scale and thus overcoming the limitations of Fourier analysis.

CASE STUDY

Each of the three primary spectral subranges, the energy-containing, the inertial and the dissipation, is characterised by its own length scale. In almost all relevant turbulent flows, the turbulent velocity fluctuations are a small fraction of the mean flow velocity. A stationary velocity probe in a turbulent flow registers a signal whose time dependency is primarily affected by spectral fluctuations of the turbulent velocity field, which is swept across the probe by the mean flow. If the turbulence level, expressed by the ratio of the root mean square value of the longitudinal velocity fluctuations to the mean velocity, is less than 0,3, then the use of Taylor's frozen turbulence is permitted, as stated, for example, by Herweijer (1996). In this case, the longitudinal structure functions can be computed from a time series of velocity readings, using the velocity differences.

In our experiment, the wavelet transform was performed on measured turbulent velocity fluctuations. To obtain the data, the velocity was measured in the centerline of a plane isothermal turbulent jet. A signal of total length 42 m (according to Taylor's frozen turbulence hypothesis), was analysed hierarchicaly. Figure 1 shows wavelet transform of measured signal and corresponds to the total span and to a range of scales from 3Λ to Λ/200 (Λ - internal length scale of turbulence).

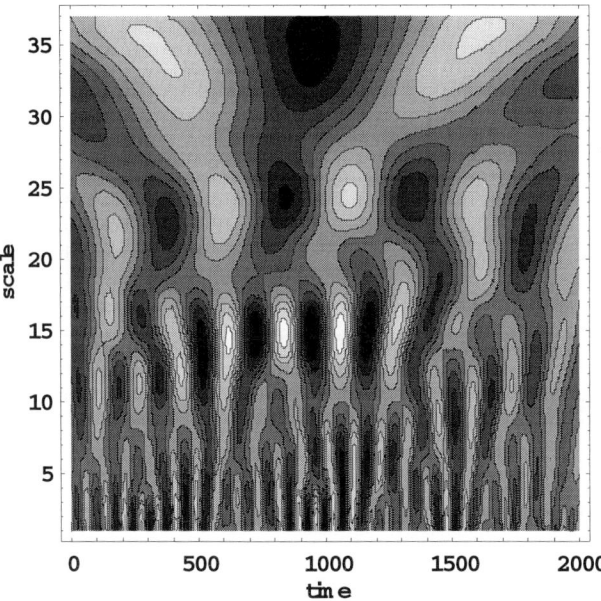

Figure 1: Wavelet transform of turbulent velocity fluctuations

The real part of wavelet coefficients are displayed, where the abscissa is time (or position, according to Taylor's frozen turbulence hypothesis) and ordinate is period (or scale). Light gray and white regions of the graph presents the positive wavelet coefficients, dark grey and black regions the negative values. From the figure are obtained the smooth velocity oscilations. The position of the maximum (white regions) gives the scale of the large eddies. From above, an eddy can be described as a function of time and scale (frequency). In the high-frequency region, the branching of eddies can be concluded from the positive values of wavelet transform coefficients. This reveal that a large eddy contain small eddies.

The displayed scale patterns are created by a process in which an eddy divides asymetrically (in intermittency) into two smaller eddies. This succesive forking process produce a fractal structure. The ratio between the scales of succesive generations takes different values, which is, according to Argoul et al. (1989), an indication of the multifractal nature of the process. Also, as emphasized by Arneodo, Grasseau & Holschneider (1988), the wavelet transform assist in the visualisation of the self-similar properties of fractal objects. In particular, it illustrates the complexity of the fractal under consideration, revealing the hierarchy that governs the relative positioning of the signal.

MODELING THE SELF-SIMILAR PROPERTIES OF FRACTAL MEASURES

In early studies, e.g. Arneodo, Grasseau & Holschneider (1989), most of the efforts were devoted to calculating the period-doubling Cantor set. A simple example is the standard uniform triadic Cantor set. It is constructed with dividing the initial unit interval [0,1] into two intervals, each of length $l = l_1 = l_2 = 1/3$; at the next stage, the same process is repeated on each of the two subintervals. Each of these subintervals receive the same probability and thus a single scaling index. Somewhat less simple is nonuniform Cantor set with a unique scaling parameter, but with two different probabilities. In the case of dissipative dynamical systems, which exhibits the cascade of period-doubling bifurcations, the process could be quite well descibed by one-dimensional map with a single quadratic extremum. Such a usefull tool for modelling presents e.g. logistic map. In this map the single parameter determines the infinite sequence of subharmonic bifurcations at each stage of which the period of the limit cycle is doubled. This period-doubling cascade accumulates at Feigenbaum number (F = 1,4011518...). Beyond this critical value the attractor becomes chaotic. This dynamical systems approach can be used to determine whether the onset of complexity in fluid flows is related to the onset of complexity in dissipative, non-linear dynamical systems in general. One aspect of this approach is the study of low-dimensional deterministic iterative maps. Two fundamental concepts have arisen out of these studies:
- low-dimensional maps with a low number of degrees of freedom can exhibit chaotic behaviour,
- stochastic or random sources are not required; the chaotic behaviour occurs in purely deterministic systems.

Chaos does not require many degrees of freedom and there are some features of the transition to chaos, controlled by the qualitative topological structure of an attractor in phase space, demonstrating the universal nature of the transition.

Among other one-dimensional maps, the circle maps has been studied in recent years as a tool for describing the transition from quasiperiodicity to chaos in dynamical systems. It implies the quantitative feature, that the transition to chaos depends only on map's qualitative nature. In this case, the transition was modelled with sine circle map, a one-dimensional nonlinear map with cubic inflection point:

$$\Theta_{n+1} = \Theta_n + \Omega - \frac{K}{2 \cdot \pi} \cdot \sin(2 \cdot \pi \cdot \Theta_n) \cdot \mathrm{mod}\,1 \tag{1}$$

It is a map of the circle on itself; the variable Θ_n presents the phase of oscilating system and is measured in discrete time intervals. The parameter K provides the strength of the nonlinearities and the parameter Ω sets the rate of rotation (frequency of the system in the absence of the nonlinear therm). For the $K = 0$ the Eqn. 1. becomes:

$$\Theta_{n+1} = \Theta_n + \Omega \tag{2}$$

which presents uniform rotation. As K increased, the intervals of frequency-locking increase and at $K = 1$ the quasiperiodic orbits constitute a Cantor set of measure zero. Also, at this value of parameter K, a critical line for transition to chaos exsists where the lock-on tongues overlap. This quasiperiodic route to chaos is generally studied following the path, ending at the golden mean point (the system moves from quasiperiodic state to a chaotic state). As the golden mean has the slowest possible convergence of all irrational numbers, it is implied that, on the K versus Ω plane, the golden mean point represents the state furthest away from the lock-on tongues approximating it.

The same wavelet transform, as used in the velocity signal analysis, was applied to sine circle map to reveal the hierarchy that governs the relative positioning of the singularities. For the analysis purposes, the velocity signal was replaced by golden mean trajectory, calculated with the sine circle map at the $K = 1$. This critical line is of interest since it marks the onset of chaos for quasiperiodic trajectories. Figure 2 shows the resulting wavelet transform of the map.

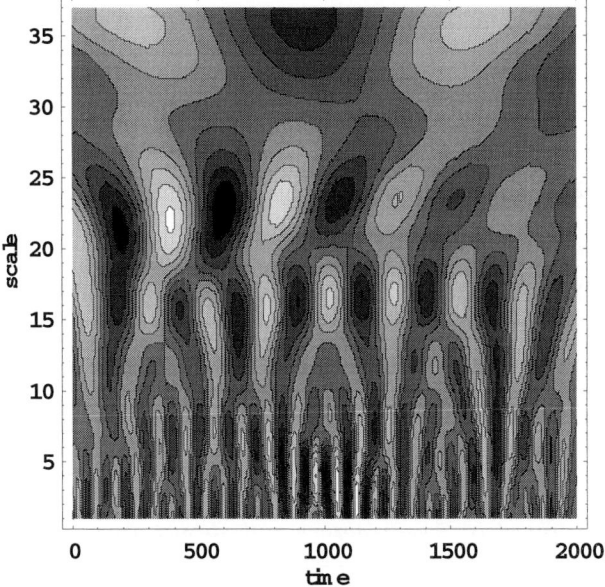

Figure 2: Wavelet transform of golden mean trajectory, calculated with sine circle map

The pitchfork branching are nonsymmetric (similar to those which would arise in the case of a non-uniform single scale Cantor set). Additionally, the ratio between the scales of succesive generations takes different values, which is an indication of the multifractal nature of process, quite similar as at turbulent velocity signal analysis.

CONCLUSIONS

Wavelet transform is presented as a mathematical microscope, which is well suited for studying the local scaling properties of fractal measures. This technique was applied to turbulent velocity signal, which reveals the Richardson's cascading process. The limitations of Fourier analysis were overcoming, decomposing signal into localized contributions in terms of time and scale. Also it indicated the multifractal nature of the process. The full complexity of the self-similar properties of fractal measures could be captured by visual inspection of wavelet transform. Fully developed turbulence gives rise to random fractals. Turbulence dissipation is apparently a process whose caling properties are determined by random refinements.

To model such a process (dissipative dynamical system that exhibits the cascade of period-doubling bifurcations), a one-dimensional map was used. The period-doubling transition to chaos is represented by a wavelet transform of golden mean trajectory, calculated using the sine circle map. The visual inspection of both wavelet transforms shows, that processes exhibit a similar fractal character. The results indicate, that the turbulent velocity fluctuations (as multiscale and multifractal process) has phenomenological similarities with the deterministic construction rules of the golden mean sine circle map.

REFERENCES

Argoul F., Arneodo A., Grasseau G., Gagne Y., Hopfinger E.J., Frisch U. (1989). Wavelet analysis of turbulence reveals the multifractal nature of the Richardson cascade. *Nature* **338**, 51-53.
Richardson L.F. (1922). *Weather prediction by numerical process*, Cambridge University Press, England.
Farge, M. (1992) „Wavelet transforms and their application to turbulence". *Annual Review of Fluid Mechanics*, **24**, 397-457.
Farge, M. Kavlahan, N., Perrier, V., Goirand, E. (1996) „Wavelets and turbulence".
Proceedings of the IEEE, **4**, 639-669.
Herweijer, J.A. (1996) *The small-scale structure of turbulence*. Thesis, Technical University Eindhoven.
Arneodo A., Grasseau G. and Holschneider M. (1988). On the wavelet transform of multifractals. *Physical Review Letters* **20**, 2281-2284.
Arneodo A., Grasseau G. and Holschneider M. (1989). Wavelet transform analysis of invariant measures of some dynamical systems. In: Combes, J.M., Grossmann, A., Tchamitchian, P. (eds), *Wavelets: time-frequency methods and phase space: proceedings of the international conference*, Springer-Verlag, Berlin, Germany

© 2000 Elsevier Science Ltd. All rights reserved.
Air Distribution in Rooms, (ROOMVENT 2000)
Editor: H.B. Awbi

THE INFLUENCE OF AIR INFILTRATION ON THE THERMAL DYNAMIC BEHAVIOUR OF BUILDINGS

D. Vytlačil

Faculty of Civil Engineering, Czech Technical University
Thákurova 7, Prague 166 29, Czech Republic

ABSTRACT

The thermal dynamic behaviour of buildings is solved by different methods; one of them is based on simulation by means of thermal node models. Computed results of the internal air temperature or the surface temperature are influenced by the used method, by the model for a solved problem situation, and by input values of model elements. The influence of the particular model element can be found by means of a sensitivity analysis. An example shows the calculation of the relative sensitivity for all parameters in a model describing the dependence of the interior air temperature changes on the external air temperature changes for one room of a building. Resultant values demonstrate the importance of conductivities related to the window, i.e. the thermal conductivity and the ventilation conductivity. The outcomes of this research reflect the necessity of using accurate air infiltration rate which influences the outputs obtained from a computer simulation much more than other parameters.

KEYWORDS

air infiltration, dynamic behaviour, sensitivity analysis, node model, computer simulation, heat transfer, lumped parameters

INTRODUCTION

The investigation of thermal dynamic behaviour of the building-external environment system leads to the application of thermal node models, Vytlačil (1993). These models help building designers to find optimal values of building parameters for different problem situations related to the summer or winter period. A well-designed building provides a low energy consumption and occupants' comfort.

Since thermal models started to be applied, there have been discussions about the accuracy of the method or used models. This paper is focused on the influence of input values on computed results which is usually an internal air temperature in rooms or a surface temperature of a building shell and other internal structures. Other models can find the optimal power of a heating system or an air-conditioning system, Vytlačil (1994).

THERMAL MODEL

The model describing the dependence of internal air temperature changes on external air temperature changes was developed. The one room model includes the most important building elements which have the influence on the thermal transfer: thermal conductivity of the building shell K_W, window conductivity K_O, ventilation conductivity K_A, thermal transfer conductivity between the shell and the interior K_I, thermal capacity of the shell C_W, thermal capacity of interior C_I, surrounding walls thermal capacity C_S, thermal transfer conductivity between interior and surrounding walls K_S. All elements are considered as lumped parameters. The relationship between thermal flows and temperatures is described in the matrix form in Eqn. 1. External temperature is T_1, internal surface shell temperature is T_2, interior temperature is T_3, internal wall surface temperature is T_4 and s is Laplace operator.

$$
\begin{bmatrix} Q_1 \\ Q_2 \\ Q_3 \\ Q_4 \end{bmatrix} =
\begin{bmatrix}
K_W + K_O + K_A & -K_W & -K_O - K_A & -K_S \\
-K_W & K_W + K_I + sC_W & -K_I & \\
-K_O - K_A & -K_I & K_A + K_O + K_S + sC_I & \\
K_S & & & K_S + sC_S
\end{bmatrix}
\cdot
\begin{bmatrix} T_1 \\ T_2 \\ T_3 \\ T_4 \end{bmatrix}
\tag{1}
$$

A solution of this matrix after Laplace transformation with algebraic complements is Eqn. 2.

$$
T_3 = \frac{\Delta_{13}}{\Delta_{11}} \cdot T_1 = \frac{N(s,x)}{D(s,x)}
\tag{2}
$$

$N(s,x)$ and $D(s,x)$ are polynoms with variable s and x.

SENSITIVITY ANALYSIS METHOD

The importance of model parameters is found by means of sensitivity analysis. Principles and a more detailed description of the method are given in Vytlačil & Moos (1993). For solving practical problems we usually use the relative sensitivity, see Eqn. 3, in which F is a function described in Eqn. 2, x is an investigated parameter and F_0 and x_0 are nominal values of the parameter and the function.

$$
Sr^F_x(s,x) = \frac{dF(s,x)}{dx} \cdot \frac{x_0}{F_0}
\tag{3}
$$

The relative sensitivity can be also determined by Eqn. 4.

$$
Sr^F_x(s,x) = x \cdot \left(\frac{N'}{N} - \frac{D'}{D} \right)
\tag{4}
$$

After substitution $s = j\omega_0$, where j is an imaginary unit, relative sensitivity can be calculated. The resultant value is a complex number, where the real part expresses the sensitivity of the temperature amplitude and the imaginary part expresses the sensitivity of the phase delay to the change of the parameter x, as written in Eqn. 5.

$$Sr^F_x(j\omega) = ReSr^F_x + j\ ImSr^F_x \tag{5}$$

CASE STUDY

In this section, relative sensitivities are calculated for all parameters in the thermal model. The building shell is made up of bricks. The wall thickness is 0,45 m. The input values of a model are given in Table 1.

TABLE 1

INPUT VALUES

K_W	K_O	K_A	K_I	K_S	C_W	C_I	C_S
		$(W \cdot K^{-1})$				$(Wh \cdot K^{-1})$	
14.7	4.2	5.6	84.4	322.0	1650	55	720

Results of simulation

Resultant values are shown in Figure 1. The relative sensitivity is the most important for the ventilation conductivity K_A that is derived from the air infiltration and depends on window and door properties. The important values are also the window conductivity K_O and the thermal capacity of surrounding walls C_S. Other parameters are less significant. The important elements for the phase delay are the thermal capacity and the thermal conductivity of the building shell C_W, K_W. A particular value means that if the parameter changes by 10% and the relative sensitivity is 0.2, the amplitude of a thermal curve of the internal temperature will increase by 2% and, similarly, from ImSr value we can calculate the change of a thermal curve on a time axis.

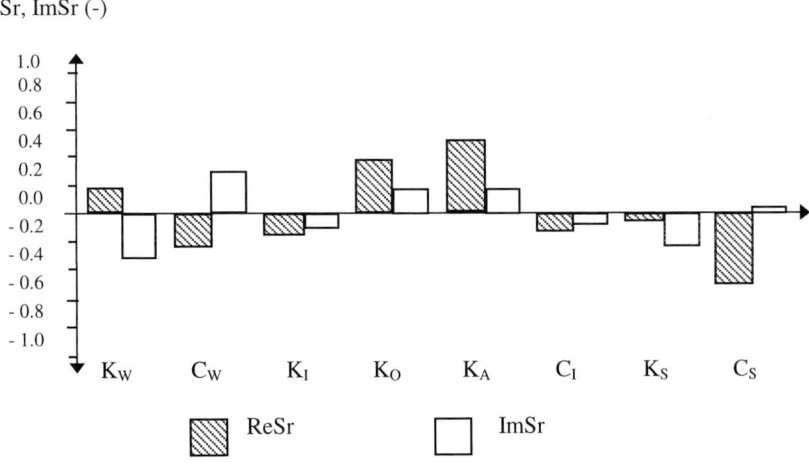

Figure 1: Relative sensitivity of model parameters

Influence of other parameters

The sensitivity is a variable depending on the nominal value of the investigated parameter and on other parameter values. Therefore the amplitude relative sensitivity of the ventilation conductivity was calculated for a range of shell thermal capacity values from 200 to 8000 Wh.K^{-1}. This dependence is presented in Figure 2. A graph shows high relative sensitivity values for medium-weight and heavy-weight building shells, which is the highest value among all sensitivity values. In light-weight structures the shell thermal conductivity is the most important parameter. The shell thermal capacity and the surrounding walls thermal capacity are also significant parameters in this model.

Calculated values of sensitivities demonstrate the importance of elements with an attribute of the conductivity that connect an external and internal environment, i.e. the thermal conductivity of windows and the ventilation conductivity.

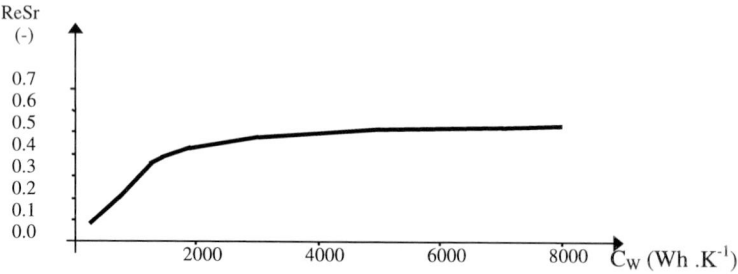

Figure 2: Relative sensitivity of the ventilation conductivity K_A for different values of the building shell thermal capacity C_W

CONCLUSION

This paper should help orientate research efforts in the computer simulation of the thermal dynamic behaviour of buildings. To obtain better outputs, it is necessary to focus the attention on the most sensitive elements. Finding the accurate input values may cause a difficult problem, especially in old buildings where the actual air infiltration rate is different from the building standard values. In the practical domain, it means the measuring of air infiltration in situ, or the investigating of the window condition. Only the correct input values guarantee a correct temperature-humidity building analysis. Massive walls of these structures also require the description by means of distributed parameters, Vytlačil (1997).

This is only one example of using the sensitivity analysis that has a wide range of applications in thermal models.

Vytlačil D. (1993). Model of Thermal Dynamic Processes in Buildings and HVAC Systems. Ph.D. thesis, Czech Technical University, Prague.
Vytlačil D. (1994). A Method for Assessing Optimal Heat Power. Proceedings of CTU Seminar, Czech Technical University, Prague, 233-234.
Vytlačil D. (1997). Using Distributed Parameters in Heavyweight Wall Buildings. Proceedings of Workshop 97, Czech Technical University, Prague, 555-556.
Vytlačil D. and Moos P. (1993). The Sensitivity Analysis of Building Thermal Network Elements. *Building and Environment* **28:1**, 63-68.

© 2000 Elsevier Science Ltd. All rights reserved.
Air Distribution in Rooms, (ROOMVENT 2000)
Editor: H.B. Awbi

NUMERICAL SIMULATION OF TRANSIENT EFFECTS OF WINDOW OPENINGS

G V Fracastoro, G Mutani, M Perino

Department of Energy Technology, Politecnico di Torino
Corso Duca degli Abruzzi, 24 – 10129 Torino, Italy

ABSTRACT

This work is centered on the transient analysis of natural ventilation provided by a single side opening when only indoor-outdoor temperature differences are present (no wind). Using both simplified "engineering" models and a CFD commercial code (2D), different cases have been examined by varying indoor-outdoor temperature difference, window size, and including or not a heating appliance in the room. The CFD results reproduce correctly the phenomenon, but the time-scale seems to be inconsistent with global mass and energy conservation principles, in spite of the fact that numerical convergence is always achieved. Re-scaling the time allowed the results of the simulation to substantially agree with zonal model results. Furthermore, they allowed to determine the values, varying with time, of ventilation efficiency when a window is opened. The air changes at steady-state are also expressed as a function of Grashof number.

KEYWORDS
Natural ventilation, windows opening, CFD, stack effect.

SYMBOLOGY
A = half window surface
A_p = wall surfaces
h_p = film coefficient
c = specific heat
H = window height, enthalpy
$\dot{Q}$ = heat flux
T = temperature
V = room volume

β = compressibility factor
ξ = accidental pressure loss coefficient
ρ = density
ν = kinematic viscosity
τ = time
Subscripts
o,i, = outdoor, indoor
p,v = constant pressure, volume

INTRODUCTION

Opening a door or window is a simple and common action to improve the IAQ in a room, yet its consequences when a certain indoor-outdoor temperature difference is present and even if no wind is present, are difficult to predict. Schaelin et al (1992) and Elsayed (1998) have already pointed out the difficulties which may arise when CFD techniques are used to simulate this phenomenon, which is actually complex, and strongly unsteady.

Difficulties start since the first stage of simulation, when the user has to choose the structure of the calculation domain. Confining the calculation domain to the heated space may, in fact, lead to surprisingly meaningless results under the physical point of view (e g, cold air entering the room through the upper part of the window and warm air exiting from below), even if the values of residuals would suggest a successful simulation.

Simplified models and equations (see for examples Etheridge and Sandberg, 1996, Andersen, 1996, ASHRAE Handbook of Fundamentals, 1997) are available for the calculation of air flow rates, but the user must provide one ore more "empirical" coefficients, whose value is not always known a-priori and may depend on the type of the opening as well as the temperature difference. Moreover, due to the unsteadiness of the phenomenon, the value of temperature difference to be used in these formulas has to be forecasted as an average between initial and final conditions.

The CFD results have been expressed as a function of Grashof number, $Gr = g \cdot \beta \cdot \Delta T \cdot H^3 / v^2$, where ΔT is the wall-outdoor temperature difference.

MODELS FEATURES

A naturally single-side ventilated enclosure 4.2 m (width) x 2.7 m (height) and 1 m (depth) has been taken in consideration. Only thermal effects have been taken into account. A total of 6 cases have been examined varying window height, temperature differences, and including or not a heating appliance in the room. Table 1 reports the main features of the simulations.

The initial air temperature has always been assumed equal to 20° C. The same value has been adopted for the wall temperatures, and considered constant.

TABLE 1

MAIN FEATURES AND BOUNDARY CONDITIONS OF SIMULATED CASES.

Case	Radiator	Window height [m]	$T_{i,o}-T_o$ [°C]	Grashof Number
1	No	1.5	20	$1.17 \cdot 10^{10}$
2	No	1.89	10	$1.13 \cdot 10^{10}$
3	No	1.5	10	$5.64 \cdot 10^{9}$
4	No	1.5	5	$2.77 \cdot 10^{9}$
5	No	1.5	2.5	$1.39 \cdot 10^{9}$
6	Yes	1.5	20	$1.32 \cdot 10^{10}$

CFD Models

A two-dimensional CFD transient analysis has been performed using a well known commercial software (FLUENT ®). The CFD model includes in the calculation domain a strip 2 m wide of outdoor environment, and is discretized using a non uniform grid made of 200 x 200 cells. The following assumptions were adopted: standard k-ε turbulence model, power-law interpolation scheme, standard log-law wall functions. A total number of about 100 time-steps varying from 0.5 s (in the first 20-30 s of simulation, to avoid numerical instability) up to 60 s were simulated, with 1000 iterations per time-step. The equivalent time span was about 600 s for all cases. The computational time was about 3 weeks on an HP Apollo 720 RISC WS (54 Mb RAM memory).

Engineering Models

a) A single-zone model has been developed, where the equation for mass flow rate ṁ, given by ASHRAE (1997), is coupled with the equation of conservation of energy for indoor air:

$$\dot{m} = \rho_o \cdot A \cdot C_d \cdot \sqrt{\frac{g \cdot H \cdot (T_i - T_o)}{T_i}} \tag{1}$$

$$h_p \cdot A_p \cdot (T_p - T_i) = \dot{m} \cdot c_p \cdot (T_i - T_o) + \rho_i \cdot V \cdot c_v \cdot \frac{\partial T_i}{\partial \tau} \tag{2}$$

where the discharge coefficient C_d is given by:

$$C_d = 0.40 + 0.0045 \cdot |T_i - T_o| \tag{3}$$

b) The two-zone model is described by the following equations

$$h_{p1} \cdot A_{p1} \cdot (T_{p1} - T_1) = \dot{m} \cdot c_p \cdot (T_1 - T_o) + \rho_1 \cdot V_1 \cdot c_v \cdot \frac{\partial T_1}{\partial \tau}$$

$$h_{p2} \cdot A_{p2} \cdot (T_{p2} - T_2) = \dot{m} \cdot c_p \cdot (T_2 - T_1) + \rho_2 \cdot V_2 \cdot c_v \cdot \frac{\partial T_2}{\partial \tau} \tag{4}$$

$$\dot{m} = A \cdot \sqrt{\frac{g \cdot H \cdot \left(\rho_o - \frac{\rho_1 + \rho_2}{2}\right)\rho_o \cdot \rho_2}{\xi_1 \cdot \rho_2 + \xi_2 \cdot \rho_o}} \tag{5}$$

The value of $\xi_1 = \xi_2$ has been determined assuming the initial air flow rate to be equal to the initial flow rate of the single-zone model. The air is supposed to flow only from the lower zone (1) to the upper zone (2), without recirculation.

These two models have been solved numerically, discretizing the ODE's and employing an explicit time integration. In order to comply with energy and mass balances a time step of 0.1 s has been used for the first 8-10 seconds, increasing it with time.

RESULTS

The results are strictly applicable only to a continuous "strip" type of window, but they may probably be extended to cases where the ratio of window height to width is not too large.

The profiles of ach's (determined considering 1 m depth) and air mean temperature versus time (not shown here for the sake of brevity) obtained by means of the single- and two-zone models are quite similar. Table 2 sums up the indoor air mean temperature and the ach's values at steady state conditions (and the corresponding time to reach them).

TABLE 2
STEADY STATE AIR CHANGES AND TEMPERATURES

Case	n [1/h]	T [°C]	n [1/h]	T [°C]	$\tau(\infty)$ [s]
	Single zone		Two zone		
1	35.4	4.7	37.0	3.7	≈150
2	34.8	12.5	37.1	11.9	≈150
3	27.1	13.0	29.0	12.3	≈220
4	20.8	16.8	22.6	16.4	≈220
5	15.9	18.6	17.5	18.4	≈200
6	36.2	5.2	-	-	≈200

The CFD results in terms of air flow patterns are consistent with conventional wisdom, as shown in Figure 1, where the profiles of mass flow rate across the window at different time steps are reported, as an example, for case 3.

On the opposite, the average room air temperature profiles versus time calculated by means of CFD are very different from those of the zonal models. Actually, the analysis of energy balance at each time step has revealed that CFD models predict correctly the enthalpy fluxes and convection

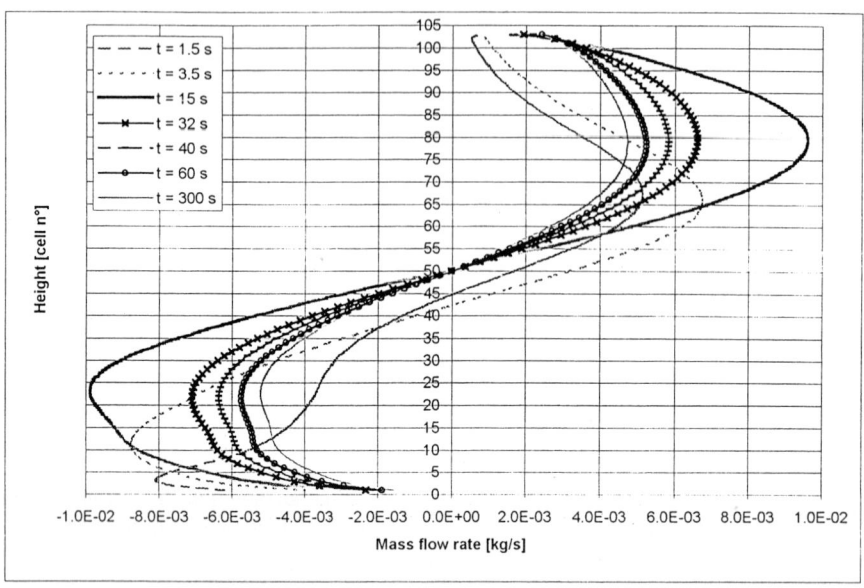

Figure 1 –Mass flow rates along the window height for different time steps– *Case 3*.

heat fluxes between walls and air, while the variation of temperature with time appears to be largely underestimated. The time step intervals adopted for the numerical solution procedure do not seem to fit with the physical time scale of the phenomenon. To overcome this problem the energy balance has been imposed at each time step, and the time-scale of the CFD results has been re-determined coherently from the following equation

$$\Delta\tau = \frac{\rho \cdot V \cdot c_v \cdot \Delta T}{\dot{Q} - \dot{H}} .$$
(6)

Figures 2 and 3 show the corrected CFD and the two-zone (2z) model results in terms of ach's and mean air temperature vs. time. The agreement is now fair for *cases 1, 3, 4 and 5*, while it is rather poor for *case 2*, probably because substantial re-circulation of warm exhaust air with cold air entering occurs, and lower indoor-outdoor temperature differences induce smaller air flows. Case 6 (same ΔT as case 1, but with radiator located adjacent to the wall opposite to the window) shows constantly lower ach's, but tending to the same value as case 1, and constantly higher room air temperatures, due to the higher equivalent surface temperature of the walls.

From the CFD results *ventilation efficiency* has been calculated for cases 4, 5 and 6. Its values start from one and tend to vary with time in a rather unpredictable way, decreasing for *case 4* increasing for *case 5* and maintaining perfect mixing conditions for *case 6*, as shown in table 3.

Table 3
Ventilation efficiency (-)

Time (s)	0	5	10	15	20	25
Case 4	1	0.98	0.96	0.91	0.74	0.61
Case 5	1	1.02	1.11	1.35	1.95	2.92
Case 6	1	-	-	0.98	1.02	1.01

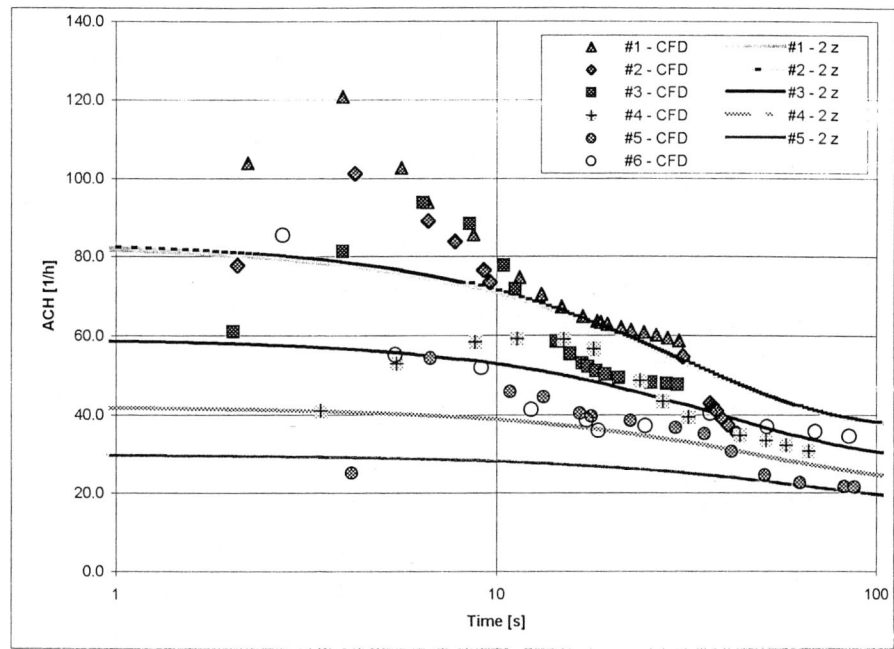

Figure 2 – Air changes per hour versus time (CFD re-scaled and 2-zone values)

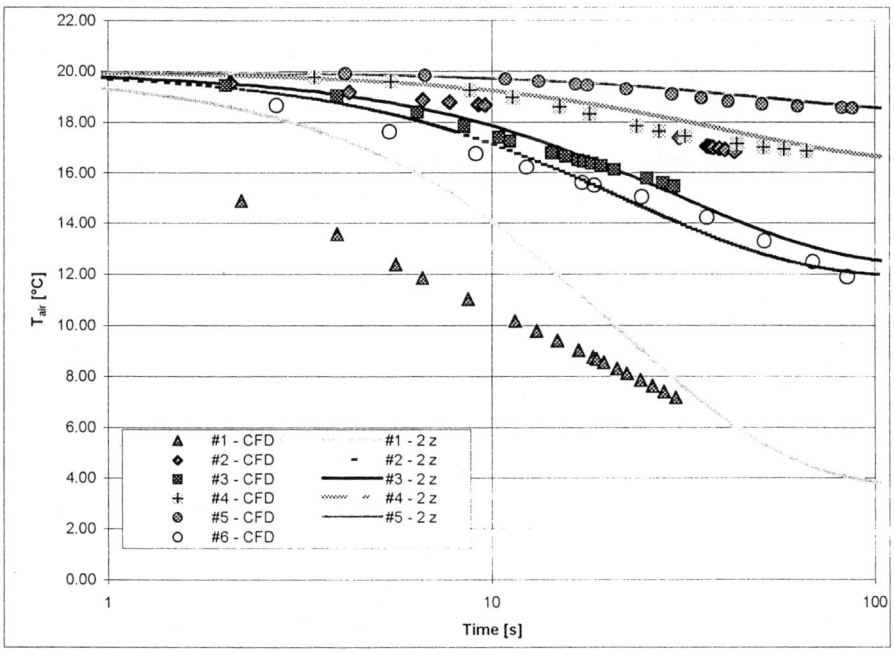

Figure 3 – Air temperature versus time (CFD re-scaled values and 2-zone)

402

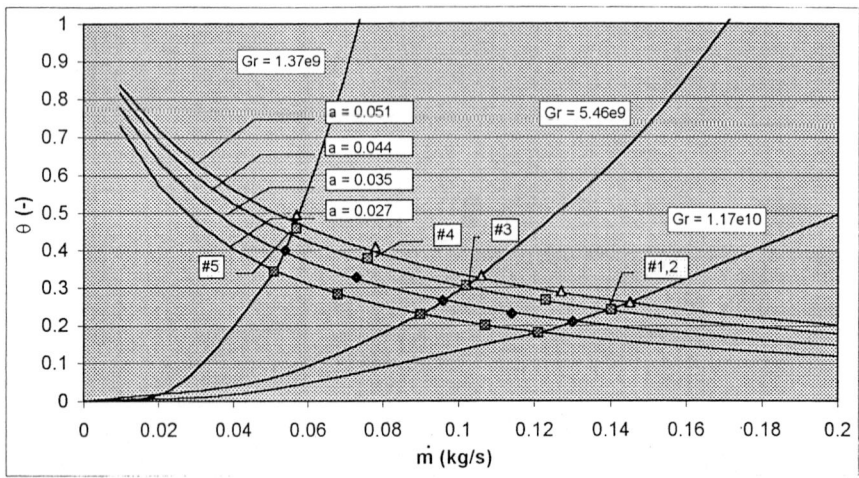

Figure 4 – Non-dimensional temperature as a function of mass flow rate and Gr at steady-state.

The relationship, derived from zonal models at steady state, between mass flow rate, non-dimensional temperature $\theta = (T_\infty - T_o)/(T_p - T_o)$, and Grashof number is shown in figure 4 ($a = hA_p/c_p$).

CONCLUSIONS

The aim of the paper was to verify the applicability of CFD transient 2D analysis to a simple, yet complex phenomenon such as the evolution of thermal and fluid dynamic fields following the opening of a window under the mere action of indoor - outdoor temperature difference.

Simplified "zonal" models have also been developed, to verify the time evolution of the phenomenon and the conditions at steady-state. A number of different configurations and boundary conditions have been simulated, including the presence of a radiator.

It has been shown that CFD should be used very carefully, with a suitable choice of the calculation domain. Furthermore, in order to comply with First Principle constraints, the time evolution had to be re-scaled. Once corrected, the CFD results fairly agree with engineering models. From CFD, the ventilation efficiency of airing by stack-effect has also been derived. Its evolution has shown controversial tendencies depending on the temperature difference.

References

- Andersen K.T., Design of natural ventilation by thermal buoyancy – theory, possibilities and limitations, 5th Int. Conf. ROOMVENT '96, July 1996, Yokohama.
- ASHRAE, Handbook of Fundamentals, Ch. 25, 1997.
- Elsayed M., Infiltration Load in Cold Rooms, HVAC&R Research, Vol. 4 No. 2, April 1998.
- Etheridge D., M. Sandberg, Building ventilation – Theory and Measurements, John Wiley & sons, Chichester, 1996, pp. 89-95.
- Schaelin, A., Van der Maas J., Moser, A., Simulation of air flow through large openings in buildings, ASHRAE Transactions, part 2, 1992.

Acknowledgments

This research activity has been developed in the frame of "*Indoor Environment Engineering*"– a National Project co-funded by the Italian Ministry of University and Research (MURST).

Air Distribution in Rooms, (ROOMVENT 2000)
Editor: H.B. Awbi

INTEGRATED MODELLING OF
FIRE, SMOKE AND OCCUPANTS EVACUATION OF BUILDINGS

D. Tang and P. Thompson

Integrated Environmental Solutions Limited, 141 St. James Road, Glasgow G4 0LT, UK

ABSTRACT

This paper describes the integrated modelling of building energy, fire and smoke distribution and evacuation occupants of buildings underlying the principles of the SAFE (Simulation Approach to Fire Engineering) project. Within SAFE, the time varying one-dimensional partial differential equations of continuity, state of gas, momentum, energy, concentration and solid conduction are solved simultaneously with the time dependent convective and radiative heat transfer coefficients. Parallel to this, it also calculates the movement and health conditions of the occupants in response to the development of fire and smoke. The solution can be applied to single or multi-storey buildings of any geometrical scale and shape. The theoretical basis is described and some of the simulation results are demonstrated.

KEYWORDS

Fire and Smoke, Evacuation, Modelling

INTRODUCTION

Each year fires in buildings in the UK result in more than 800 deaths and over 15,000 injuries. Fire causes major property loss - more than £1000 million direct loss/year in the UK - and can pollute the environment via contaminated fire-fighting water and the emission into the atmosphere of particulate and toxic fumes. In addition, fire can destroy our priceless heritage.

Also, smoke is the major cause of death and injury as it can overcome occupants, rendering them unconscious, long before they would be affected by the actual fire. Consequently as smoke spreads the time taken by occupants to evacuate the building is critical to save the occupants from injury, or worse. Therefore it calls for a greater understanding of a building's performance during a fire and the effectiveness of alternative fire engineering strategies. Other than physical modelling this can only be achieved by means of.

At present, the existing design methods are inadequate to accommodate new requirements and regulations in terms of fire and safety. Consequently, the existing simulation tools are either piecemeal each of which requires a different data structure, unable to handle the particular problems related to fire and occupant health or restricted by the scales of the models. Efforts for modelling various aspects of buildings have already been found in a number of software systems. For example, ESP-r [ESRU, 1997] and Apache-sim [IES, 1997] model detailed thermal building coupled with network based airflow based on conduction theory without modelling combustion. CFAST [Peacock, et. al. 1996] models detailed development of fire and smoke within building with simplified treatments for heat transfer of building structures. Jasmine, [Cox & Kumar, 1983] on the other hand is able to model combustion and conjugate airflow, gas radiation and conduction theory using CFD, but restricted by computing power. Simulex [Thompson, et. al. 1995] alone is able to model the movement of occupants towards the exits of the building in relation to fire.

In order to address the above issues an integrated software system has been developed as a joint research effort, namely the SAFE (Simulation Approach to Fire Engineering).

THE MATHEMATICAL MODELS OF SAFE

Within the development of SAFE, considerations had been focused on the provision an integrated simulation environment which encompasses the modelling concept underlying the best of the existing software systems. In addition to this the development of the Integrated Data Model (IDM) was considered as the prime need for data sharing among all the software modules.

To address the particular issues of combustion and species dispersion within in the building under fire, the following algorithms were incorporated:

- Time dependent combustion algorithm to take into account non-linear and transient behaviour of fire development and generation of combustion products;
- Ceiling jet theory to take into account the enhanced the convective heat transfer between the top layer of the hot air and the ceiling above the fire source.
- The extra mass flow entrained by the fire plume as a result of the dominant buoyancy effect of the combustion process;
- Transient species conservation and storage within each volume of the building;
- Grey gas radiative heat exchange between the mixture of combustion products and air in each zone.

System Equations

Generally speaking, SAFE models the combined effect of heat transfer, airflow, combustion of solid material and smoke distribution within building based upon the following assumptions:

- Two-dimensional airflow with buoyancy force coupling in z (height) direction. The density of air is modelled based on the law of ideal gas;
- One-dimensional thermal conduction, three-dimensional grey gas radiation;
- Lumped parameter representations of control volumes for all state variables, i.e. air temperature, velocity, concentration and pressure;

The momentum equation in which all variables vary with time and the x-dimension except the buoyancy force which acts in negative vertical direction:

$$\frac{\partial \rho u}{\partial t} + u \frac{\partial \rho u}{\partial x} = -\frac{\partial P}{\partial x} + \frac{\partial}{\partial x}\left(\mu \frac{\partial u}{\partial x}\right) - \rho g_z \theta \qquad (1)$$

The equation of continuity:

$$\frac{\partial \rho}{\partial t} + \frac{\partial \rho u}{\partial x} = 0 \qquad (2)$$

The equation of state of gas:

$$\frac{P}{\rho} = RT \qquad (3)$$

The energy equation of air:

$$\frac{\partial \theta_a}{\partial t} + u \frac{\partial \theta_a}{\partial x} = \frac{\partial}{\partial x}\left(\kappa \frac{\partial \theta_a}{\partial x} \right) + q_v \qquad (4)$$

The concentration diffusion equations of species:

$$\frac{\partial C_i}{\partial t} + u \frac{\partial C_i}{\partial x} = \frac{\partial}{\partial x}\left(D \frac{\partial C_i}{\partial x} \right) + S_i \qquad (i=1,2,...,n) \qquad (5)$$

The conduction equation:

$$\frac{\partial \theta_s}{\partial t} = \frac{\partial}{\partial x}\left(\alpha \frac{\partial \theta_s}{\partial x} \right) + q_s \qquad (6)$$

in which, ρ air density (kg/m^3), u air velocity (m/s), t time (s), x distance (m), P air pressure (Pa), μ dynamic viscosity (kg/ms), g_z gravitational acceleration (m/s^2), R gas constant ($J/molK$), T absolution temperature of gas (K), θ_a air temperature (K), κ diffusivity of air (m^2/s), q_v heat generation within air (W/m^3), θ_s temperature of wall (K), α diffusivity of wall (m^2/s), C_i concentration of gas (g/m^3), S_i source generation of gas (g/m³s), q_s heat generation within wall (W/m^3), i gas type.

Note algorithms for the modelling of combustion based convective and radiative heat transfer were embedded within the above equations in the forms of time varying coefficients.

Modelling of Building

Methods for solving the energy Eq. (4) in conjunction with transient conduction Eq. (6) were implemented in the thermal conduction module and solved simultaneously using the second-order time accurate implicit numerical scheme. The algorithms for the forced and natural convective heat transfer, short and long wave radiation heat transfer, shape and view factors, enhanced heat transfer coefficient between fire plume and surroundings, etc. were treated as source terms, time dependent coefficients and implemented as boundary conditions in the thermal conduction module.

Modelling of Network Airflow

In SAFE an unique network airflow representation and solution technique has been developed based on graph-matrix optimisation for the solution of the conjugate heat and mass transfer problem, i.e. Eq. (1). The new solver was able to transform the system network into weakly connected sub-systems to which only the much reduced scaled sub-systems were solved one at a time. The solution followed a network-dependent sequence using a predictor-corrector algorithm that calculated the first-order partial derivative, i.e. the Jacobian matrix, as first prediction. A straightforward substitution was then used to derive the solutions for the whole system. The technique was developed based upon the Prim's and Kruskal's algorithms for finding the minimal spanning tree from a system network. [Drozdek, et. al. 1995] Details of the derivation and solution procedures may be found elsewhere. [Tang, 1999]

In a room where fire and smoke occurs, the rising of mass of smoke plume was modelled by treating the plume as a point source, starting from the lower volume, rising through the middle volumes and finally arriving at the top volume of the room comprising the whole airflow network. The airflow in and out of the plume were calculated at each time step based on the distributions of the air temperature and pressure through the network.

Modelling of Contaminants

Contaminants in the SAFE were modelled as non-participating air-borne substances participating as mixture in the air. The generations of the contaminants were induced by the one-step combustion algorithms initiated by the threshold of temperature of ignition of the fire. For each volume comprising the airflow network, the transient storage effect of the concentration of each individual gas was modelled by solving the contaminant conservation equation, i.e. Eq. (5), simultaneously with the rest of the conservation equations of air and conduction for each volume of the space. The method for solving the species conservation, i.e. Eq. (5), was treated in analogue to the energy equation, Eq. (4).

Modelling of Combustion and Related Transfer Phenomena

In SAFE the transient process of combustion of fuel was modelled using the one-step reaction model based upon the availability oxygen and the parabolic fire growth rate obtained from experiments. [CIBSE, 1995] The combustion ended when all the available fuels were consumed. In the case where insufficient of oxygen occurred, only part of the fuel was consumed via complete combustion based on the quantity of oxygen that was available.

The fire and smoke plume was modelled indirectly using the 'virtual point source fire and smoke plume' model. [Peacock, et. al. 1996] The purpose of the calculation was to quantify the mass flow rate of the entrained air to be used by the momentum, energy and concentration conservation modules. In doing so the energy and momentum generated by the flow of the plume were conserved and mass, concentration and enthalpy were delivered to all the volumes comprising the airflow network. However in this one-dimensional approach the actual shape of the plume and its interaction with the surrounding surfaces was not taken into account.

Ceiling jet, which occurs during the early development of a fire and intensifies the gas-to-ceiling convective heat transfer is one of the most important issues in fire modelling. In SAFE the Cooper's [Peacock, et. al. 1996] ceiling jet model was used to calculate the enhanced convective heat transfer coefficient between the hot air and the ceiling of the room. The enhanced convective heat transfer coefficients were then used by the thermal module to establish the heat exchanges between the ceiling of the room and the adjacent air volume.

Modelling of Occupant Evacuation

In SAFE, the modelling of the movement of occupants in relation to the transient development of the fire and smoke and the available fire exits of the building was implemented in the evacuation module. The evacuation module had also been enhanced to take into account of the health-degradation condition of each individual person using the 'fractional effective dose' methods by which the accumulated toxic dose affected occupant health, behaviour, walking speed and decision-making ability. At present the evacuation module operated in arrears of the momentum, energy and contaminant simulations at each time step or at end of the simulation period. This was based upon the assumption that the movement of occupants had no effect on the dispersions of fire and smoke.

Fig. 1, 2 and 3 show some examples of output from the integrated simulation at various stages. The ground plan of the building is shown. The interim movements of the occupants from the upper levels towards the main exit at the ground level are shown in the stairs. The shaded areas show different concentration levels of contaminants varying with time. The movement of each individual person is shown and the health condition during the time of the evacuation may be queried at any stage.

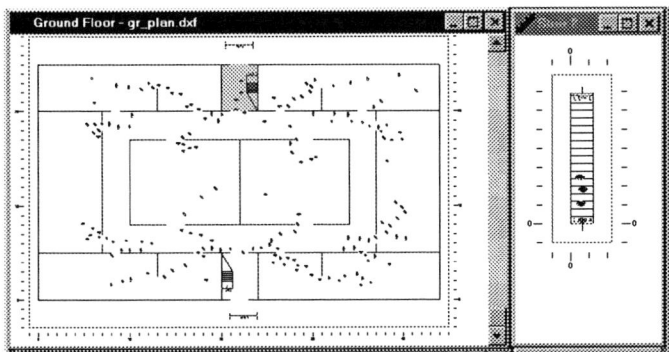

Figure 1. The early stage of an evacuation when fire starts at the main entrance.

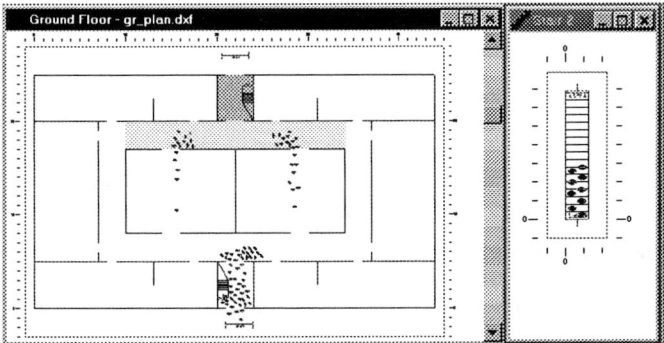

Figure 2. Occupants change route by avoiding the main entrance due to the concentration of smoke.

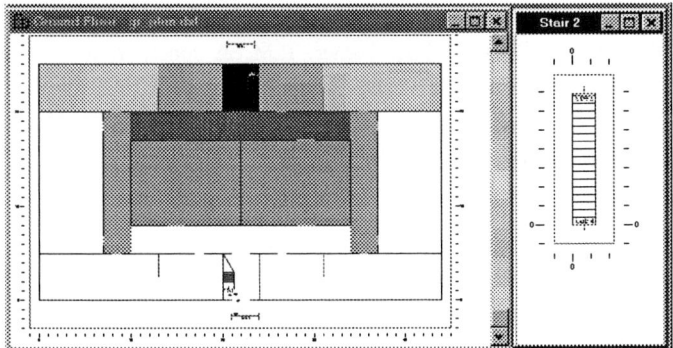

Figure 3. Towards the end of an evacuation when smoke becomes widely dispersed.

CONCLUSION AND FUTURE WORK

The work demonstrated an integrated approach for the modelling of building energy, airflow, combustion, smoke distributions and evacuation occupants of buildings. It did this by solving the time varying equations of thermal conduction, network airflow, air enthalpy and concentration dispersion, combustion of solid and occupant movement and health simultaneously and using a commonly shared data structure, the Integrated Data Model. New algorithms and solvers were implemented to improve the solution efficiency. Extensive user front-end had also been developed to enhance its user friendliness.

Unlike other zonal fire and smoke simulation systems, SAFE is able to model single or multi-storey buildings of any size and geometrical shapes with multiple fire sources which occur simultaneously or at different time located anywhere within the building. Further development and validation of SAFE is underway using published data from Steckler and FRS (*Jasmine*) to examine the performance of the network fire and smoke modules. Simulations of simple configurations have been compared with results obtained from CFAST.

ACKNOWLEDGEMENT

SAFE was a collaborative project developed by the IES Limited and the Fire Research Station funded by The Scottish Office under the SMART and SPUR Programmes.

REFERENCE

CIBSE. (1995). *Relationships for smoke control calculations*, CIBSE Technical Memoranda, TM19.

Drozdek, A. & Simon, D.L. (1995). *Data Structure in C*, PWS Publishing Co. pp.281

ESRU. (1997). *The ESP-r System for Building Energy Simulation, User Guide Ver. 9 Series.*

IES. (1997). *APACHE-sim User's Manual*. Integrated Environmental Solutions Limited.

Cox, G. & Kumar, S. (1983). Application of a numerical field model of smoke movement to the physical scaling of compartment fires, in *Proc. 3rd Int. Conf. on Numerical Methods. in Thermal Problems*, pp. 837-848.

Peacock, R.D. Forney, G.P. Reneke, P.A. Portier, R.M. & Jones, W.W. (1996). *CFAST, the Consolidated Model of Fire Growth and Smoke Transport*, NIST Technical Note 1299, NIST, U.S. Dept of Commerce.

Tanaka, T. (1978). "A Model on Fire Spread in Small Scale Buildings", *BRI Research Paper* No. 79, Building Research Institute, Ministry of Construction, Japan. pp.a-1

Tang, D. (1999). *Integrated Network Model for Fire and Smoke Dispersion with Interaction of Transient Thermal performance of Building*, Technical Report to the SPUR Award, Scottish Office.

Thompson, P.A. & Marchant, E.W. (1995). "A Computer Model for the Evaluation of Large Building Populations", *Fire Safety Journal*, V24, pp.149-166.

Air Distribution in Rooms, (ROOMVENT 2000)
Editor: H.B. Awbi

EFFECT OF EVAPORATIVE COOLING ON ATTIC RADIANT BARRIERS

Edwin P. Russo[1], Carsie A. Hall, III[1], Dionne DeBose[1], and Gerald A. Pitalo[2]

[1]Department of Mechanical Engineering, University of New Orleans
New Orleans, LA 70148, U.S.A.
[2]National Aeronautics and Space Administration
Washington, D.C., U.S.A.

ABSTRACT

A detailed transient computer analysis of an attic with radiant barrier system was performed. The computer model included radiation, forced and free convection, and conduction, as well as ventilation in the attic and evaporative spray cooling of the exterior roof. The solar flux and ambient temperatures were taken to be transient rather than constant. The purpose of the analysis was to determine the impact of varying numerous parameters. The following variations were studied: ventilation flow rates, ceiling R-values, roof shingle absorptivity/emissivity, barrier reflectivity, spray rates and humidity ratio. The evaporative cooling had a very pronounced impact on lowering the heat load into the conditioned space.

KEY WORDS

Evaporative Cooling, Spray Cooling, Radiant Barriers, Attic, Weatherization, Air Conditioning, Heat Gain, Cooling Load, Heat Shield, Attic Ventilation

INTRODUCTION

Radiant barriers are energy saving devices which reduce heat lost due to radiant heat transfer through an attic. In the late 1980's, tests conducted in the southern United States indicated that radiant barrier technology could effectively reduce air conditioning loads, and thereby conserve energy. These studies have been the subject of much controversy because claims made by many manufacturers were extreme (up to 100% shielding, Henke, 1989), with consumers paying high prices for ineffective devices. Other options such as increasing attic insulation and/or increasing attic ventilation should be taken into consideration before making a decision to install attic radiant barriers.

ANALYSIS

A detailed transient thermal computer analysis of an attic radiant barrier system was conducted. The electrical engineering code PSPICE was used for the study (see Figure 1 for the circuit computer model). In the model, an 8-hour sinusoidal solar irradiance (as given by equation 1) on a horizontal surface corresponding to latitude 30° (ASHRAE, 1985) was applied to the rooftop. It should be noted that an 8 hour (i.e. 16 hour harmonic) solar irradiance rather than a 12 hour (i.e. 24 hour harmonic) irradiance was used in the study since it was felt it would be more representative of daytime (working hours) cooling systems operations. This was considered typical of all latitudes in the southern United States. The above-mentioned solar irradiance can be represented on a per unit area basis, given by

$$Q_{solar} = 279 \sin\left[\frac{2\pi t}{16(3600)}\right] \qquad \text{[btu/hr/ft}^2] \qquad (1a)$$

$$Q_{solar} = 880 \sin\left[\frac{2\pi t}{16(3600)}\right] \qquad \text{[W/m}^2] \qquad (1b)$$

For a pitched roof the above equations should be multiplied by the cosine of the pitch angle. The outside ambient temperature also had an eight (8) hour sinusoidal variation beginning at 70°F (21°C) in the morning and increasing to 105°F (41°C) at noon thus representing a very hot day. Therefore

$$T_{outside} = 70 + 35 \sin\left[\frac{2\pi t}{16(3600)}\right] \qquad \text{[°F]} \qquad (2a)$$

$$T_{outside} = 21 + 10 \sin\left[\frac{2\pi t}{16(3600)}\right] \qquad \text{[°C]} \qquad (2b)$$

The heat flow due to a ventilation flow rate, F, was taken to be (Howell and Sauer, 1978)

$$Q_{ventilation} = F(60)(0.075)(0.245)(T_{outside} - T_{inside}) \qquad \text{[btu/hr/ft}^2] \qquad (3a)$$

$$Q_{ventilation} = 0.0037 \, F(0.245)(T_{outside} - T_{inside}) \qquad \text{[W/m}^2] \qquad (3b)$$

For radiation purposes, the temperature of the outside surroundings was taken to follow the above ambient temperature relationship. The ambient room temperature was fixed at 70°F (21°C) and the roofing material was taken to be 70 lb$_m$/ft^3 (1121 kg/m^3) asphalt with 15 pound (6.8 kg) felt paper. The combined R-value was 0.50 hr-ft^2-°F/btu (0.09 m^2-C/W) (ASHRAE, 1985). Roof sheathing was modeled as one half-inch (1.3 cm) Douglas fir plywood having an R-value of 0.62 hr-ft^2-°F/btu (0.11 m^2-°C/W) (ASHRAE, 1985). Rafters were taken to be 2 inch x 8 inch (5 x 20 cm) southern pine and joists were 2 inch x 6 inch (5 x 15 cm) southern pine. Ceiling sheetrock (drywall) was modeled as one half inch (1.3 cm) gypsum board having an R-value of 0.45 hr-ft^2-°F/btu (0.08 m^2-°C/W) (ASHRAE, 1985). The R-value (i.e., thickness) of ceiling fiber glass insulation (unfaced) was varied during the study. A radiant barrier was taken to be a thin sheet of reflective material with emissivities of both sides being a variable in the study, thus providing simulation of dust build-up. Natural/free convective heat transfer coefficients were all taken to be constant at .2 btu/hr/ft^2/°F (1.1 W/m^2/°C), for all surfaces within the attic and the conditioned space. The heat transfer coefficient on the outside surface of the shingles was taken to be 2.5 Btu/hr/ft^2/°F (14 W/m^2/°C). Evaporative cooling was included in the model by varying this outside heat transfer coefficient from 2.5 btu/hr/ft^2/°F (14 W/m^2/°C) to 1000 btu/hr/ft^2/°F (5678 W/m^2/°C).

Radiation within the attic between the barrier and the ceiling insulation or roof sheathing was predicted using the following:

$$Q_{radiation/attic} = \frac{\sigma\left(T_1^4 - T_2^4\right)}{1/\varepsilon_1 + 1/\varepsilon_2 - 1} \tag{4}$$

where σ is the Stefan-Boltzmann constant and ε is the emissivity. The subscripts 1 and 2 refer, respectively, to the rafter (and attic insulation) and radiant barrier surfaces. The areas of the ceiling and roof were taken to be approximately the same. The emissivity of the barrier was varied from 0.1 (reflective) to 0.9 (dusty). Radiation from roof shingles to the outside environment was modeled using

$$Q_{radiation/outside} = \varepsilon_s\sigma\left(T_s^4 - T_{air}^4\right) \tag{5}$$

where ε_s is the emissivity of the shingles, T_s is the surface temperature of the shingles, and T_{air} is the outside air temperature.

RESULTS

Ventilation flow rates of 1.0, 0.5, and 0.1 cfm per square foot of floor space (5.0, 2.5 and 0.5 liters/sec/m^2) were used in the model. For these rates, the ventilation did not have a significant impact on the heat flow into the conditioned space. The temperatures of various locations in the attic are shown in Figure 2. As previously mentioned, outside air temperature was assumed to follow a sinusoidal variation, peaking at 105°F (41°C) at solar noon. It can be seen that the maximum average temperature of the roof shingles without evaporative cooling was approximately 171°F (77°C). There was no noticeable difference in the temperature of the shingles for the barrier and non-barrier cases. With evaporative cooling the shingle temperature was approximately 55°F (31°C) lower. The average maximum air temperature in the attic without the barrier was 138°F (59°C). Evaporative cooling reduced the temperature by approximately 30°F (17°C). The temperature of the attic with the barrier was 11°F (6°C) less than without the barrier. Also shown in Figure 2 are the attic air temperatures with ventilation flow rates of 0.1cfm/ft^2 (0.5 liters/sec/m^2) with and without barriers and evaporative cooling, respectively. The ceiling insulation temperature (without barrier) reached a value of 98°F (37 °C) 3.5 hours after solar noon. With the barrier in place this temperature was reduced to 88°F (31°C). The ceiling surface (conditioned space side) temperature peaked four (4) hours after solar noon at values of 71.2°F (21.8°C) and 70.7°F (21.5°C) for the non-barrier and barrier situations, respectively.

SUMMARY AND CONCLUSIONS

A detailed transient computer analysis of an attic barrier was conducted. The model included radiation, convection, conduction and attic ventilation. Numerous parameters (e.g. barrier emissivity, ventilation flow rate, thermal resistance, etc.) were varied. Of primary interest in the study was the temperature increase of the roof material (shingles) due to a radiant barrier. A large temperature increase could reduce the life of the shingles. The model showed that the roofing material temperatures did not change significantly. Ventilation flow rates of 0.1 to 1 cfm/ft^2 (.5 to 5 liters/sec/m^2) were slightly effective in reducing the heat gain.

Table 1 lists the "Equivalent R-Values" for radiant barriers, evaporative cooling, and attic ventilation under various conditions. It should be noted that even though the evaporative cooling seems to be better, it is an active system, and requires addition energy to atomize the water particles. The attic venlation appears to be of negligible impact. This is in agreement with a study conducted b y the Texas Energy Conservation Center (1988), which indicated that it was more cost effective to operate the "cooling unit" in the conditioned space rather than run an attic ventilation system.

ACKNOWLEDGEMENTS

The authors acknowledge support given by the University Programs Office and the Propulsion Test Directorate at the National Aeronautics and Space Administration's John C. Stennis Space Center and the National Research Council's Research Associateship Program.

REFERENCES

ASHRAE (1985). *ASHRAE Handbook: Fundamentals*. American Society of Heating, Refrigeration and Air-Conditioning Engineers, Atlanta, Georgia.

Henke C. (1989). Barriers to Radiant Barriers. *Home Energy* v6:6, 22-26.

Howell R. and Sauer H. (1978). *Environmental Control Principles*. Supplement to *ASHRAE Handbook: Fundamentals*, 62-63.

Texas A&M University (1988). *Texas Energy and Mineral Resources* 14:4.

TABLE 1
EQUIVALENT RESISTANCE

ITEM /PROCESS	EQUIVALENT R-VALUE hr-ft^2-°F/btu (m^2-°C/W)
Radiant Barrier: $\varepsilon = .1$	8 (1.4)
$= .5$	2 (0.4)
$= .9$	1 (0.2)
Evaporative Cooling:	
$h = 10$ (57 W/m^2/ C)	8 (1.4)
$= 25$ (142 W/m^2/ C)	12 (2.1)
$= 100$ (570 W/m^2/ C)	15 (2.6)
$= 1000$ (5700 W/m^2/ C)	16 (2.8)
Attic Ventilation: cfm/ft$^2 = 0.1$ (0.5 liters/sec/m^2)	1 (0.2)
$= 0.5$ (2.5 liters/sec/m^2)	2 (0.4)
$= 1.0$ (5 liters/sec/m$^{2)}$	3 (0.5)

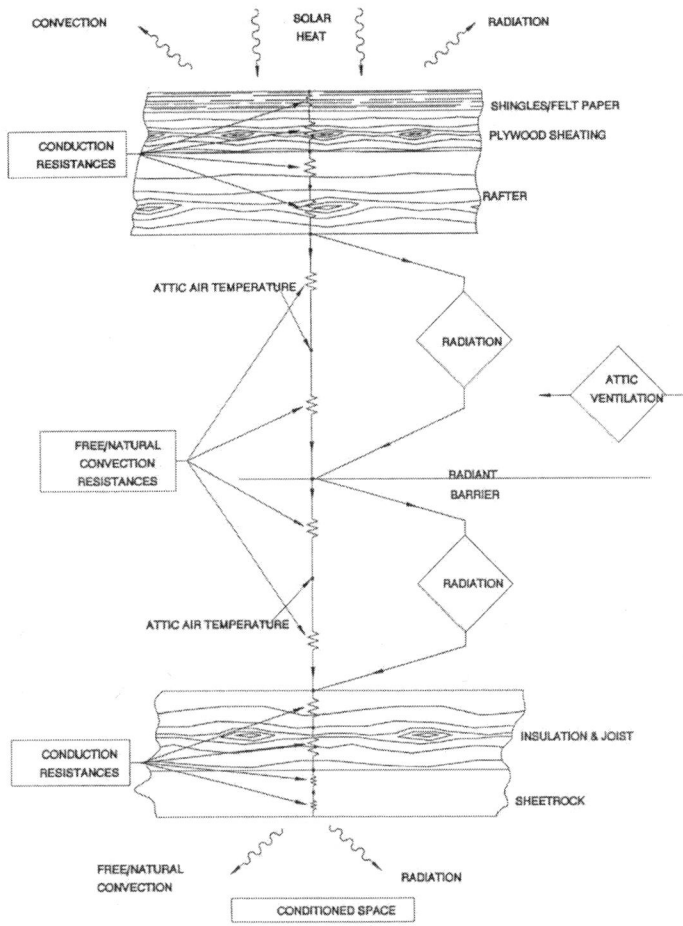

Figure 1: Computer model of radiant barrier system

414

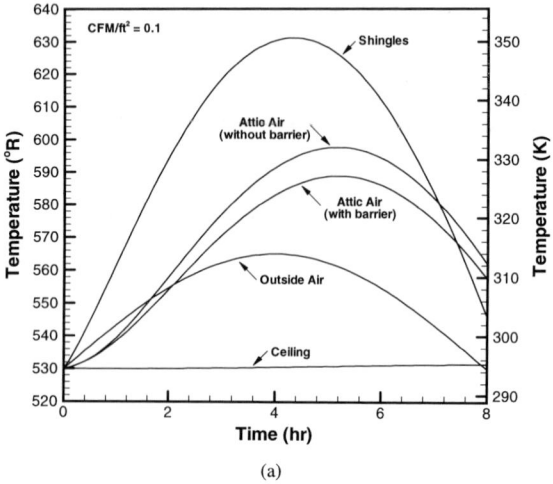

(a)

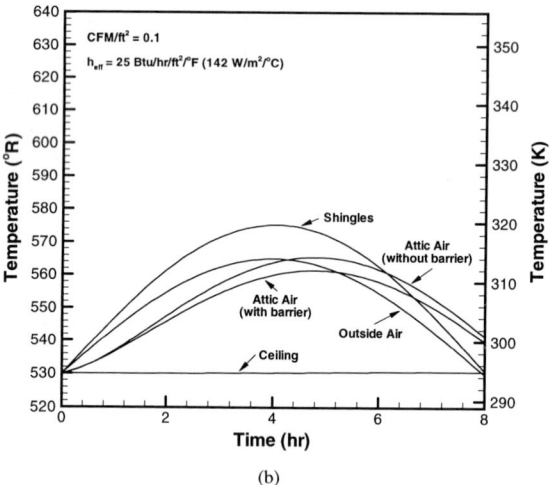

(b)

Figure 2: (a) Temperature variations without evaporative cooling and
(b) temperature variations with evaporative cooling

Air Distribution in Rooms, (ROOMVENT 2000)
Editor: H.B. Awbi

QUANTIFICATION OF UNCERTAINTY IN THERMAL BUILDING SIMULATION - PART 1: STOCHASTIC BUILDING MODEL

H. Brohus[1], F. Haghighat[2], C. Frier[1] and P. Heiselberg[1]

[1]Department of Building Technology and Structural Engineering
Aalborg University, DK-9000 Aalborg, Denmark (hb@civil.auc.dk)
[2]Department of Building, Civil and Environmental Engineering
Concordia University, Montreal, Quebec H3G 1M8, Canada (haghi@cbs-engr.concordia.ca)

ABSTRACT

In order to quantify uncertainty in thermal building simulation stochastic modelling is applied on a building model. An application of stochastic differential equations is presented in Part 1 comprising a general heat balance for an arbitrary number of loads and zones in a building to determine the thermal behaviour under random conditions. Randomness in the input as well as the model coefficients is considered. Two different approaches are presented namely equations for first and second order time-varying statistical moments and Monte Carlo Simulation. A simple test case is presented showing the mean value process and the standard deviation process (pursuing a confidence interval) for a naturally ventilated atrium during a winter week and during a summer week. Part 2 considers the generation of appropriate stochastic load input.

KEYWORDS

Thermal Building Simulation, Stochastic Differential Equations, Uncertainty, Stochastic, Deterministic, Natural Ventilation, Randomness, Monte Carlo Simulation, Statistical Moments.

INTRODUCTION

In the design of buildings, prediction of long term system performance with regard to energy consumption and thermal comfort is traditionally performed with thermal building simulation programmes based on deterministic building models and loads.

A deterministic approach implies that all input parameters and model coefficients are 100% certain with zero spread. In practice this is not the case, for instance inhabitant behaviour and internal loads may vary significantly and external loads as wind, external temperature and solar radiation are obviously stochastic in nature. One reason for ignoring randomness is the fact that mechanically ventilated heavy buildings are often highly "damped" and shielded toward external loads. This kind of buildings will also control the influence of the internal load effectively by means of the building energy management system and the HVAC system. However, lighter constructions that are naturally or hybrid ventilated are more sensitive to stochastic variations in the loads. The behaviour of the building depends thoroughly on natural driving forces characterised by external temperature, wind speed and direction, and the internal load of the building, etc. Due to the fact that this kind of buildings are increasing fast in number either as new built or retrofit stresses the importance of considering the effect of randomness in the design phase.

The present work is divided in two parts. Part 1 comprises the outline of a stochastic building simulation model and a simple test case. Part 2 considers the determination of stochastic loads.

STOCHASTIC BUILDING MODEL

A system of linear Stochastic Differential Equations (SDE) is formulated describing the temporal variation of the zone temperatures in a building model. The procedure of solving the SDE system can be regarded as an operator transforming the stochastic input quantities to stochastic output quantities (Haghighat et al., 1987). The input of the SDE system is modelled by a number of independent time varying stochastic processes representing internal and external loads applied on the building.

The stochastic building model comprises a general heat balance for an arbitrary number of zones, surfaces and construction parts. There are n nodes with unknown temperatures, θ_i, (output) and effective thermal capacities, C_i. There are m nodes with known temperatures, θ_j^b; denoted boundary nodes. There are k_i independent heat flux components, Φ_{ij}, applied on the i'th node with an unknown temperature. H_{ij} and H_{ij}^b are the specific heat losses related to the unknown temperatures and the boundary nodes, respectively. For node i at the time t the following heat balance apply

$$C_i \frac{d\theta_i(t)}{dt} = \sum_{\substack{j=1 \\ j \neq i}}^{n} H_{ij}(t)\big(\theta_j(t) - \theta_i(t)\big) + \sum_{j=1}^{m} H_{ij}^b(t)\big(\theta_j^b(t) - \theta_i(t)\big) + \sum_{j=1}^{k_i} \Phi_{ij}(t) \quad , i = 1, 2, \dots , n \qquad (1)$$

In the heat balance, Eqn. 1, all or some of the input parameters θ_j^b, H_{ij}, H_{ij}^b and Φ_{ij} can be regarded as time varying stochastic processes, the rest being deterministic. Each stochastic input component, denoted by $z(t)$, is modelled by a time varying mean value function, $\overline{Z}(t)$, superimposed by a fluctuating component comprising a time varying standard deviation function, $\sigma_Z(t)$, multiplied by a standard white noise process, $w(t)$, i.e.

$$z(t) = \overline{Z}(t) + \sigma_Z(t)w(t) \qquad (2)$$

The white noise process is usually regarded as an appropriate model for rapidly random fluctuating phenomena, when correlation in time becomes small rapidly (Arnold, 1974). The white noise process, $w(t)$, is defined as the time derivative of the so called Wiener process, $W(t)$, such that $dW(t) = w(t)dt$.

By rearranging Eqn. 1 and applying Eqn. 2 for each input process and by removing higher order terms of fluctuating components the following linear SDE system is obtained

$$d\theta = [\mathbf{X}\theta + \mathbf{x}]dt + \sum_{k=1}^{VAR} [\mathbf{Y}_k\theta + \mathbf{y}_k]dW_k \qquad (3)$$

where VAR is the number of stochastic processes. Assembly routines for $\mathbf{x}$, $\mathbf{X}$, $\mathbf{y}_k$ and $\mathbf{Y}_k$ can be found in Brohus et al., 1999. The SDE system, Eqn. 3, can be solved either for the statistical moments or for a number of response realisations of the output.

Statistical Moment Equations

A convenient description of the stochastic output parameters, i.e. the unknown temperatures, is their first and second order statistics given by the time varying expected values, $E[\theta_i]$, and the second order statistical moments, $E[\theta_i\theta_j]$. The corresponding standard deviations are easily derived from the first and second order moments by means of a statistical standard formula.

Arnold (1974) shows that the following deterministic differential equations for the first and second order statistical moments of the zone temperatures correspond to the SDE system Eqn. 3. Eqn. 4 and Eqn. 5 express the statistical moments and the initial conditions for the first and the second order moments, respectively. The first order linear differential equations can be solved by means of standard tools like the fourth order Runge-Kutta method.

$$\begin{cases} \dfrac{dE[\theta]}{dt} = \mathbf{X}E[\theta] + \mathbf{x} \\ E[\theta(t=0)] = E[\theta^0] \end{cases} \tag{4}$$

$$\begin{cases} \dfrac{dE[\theta\theta^T]}{dt} = \mathbf{X}E[\theta\theta^T] + E[\theta\theta^T]\mathbf{X}^T + \mathbf{x}E[\theta]^T + E[\theta]\mathbf{x}^T \\ + \displaystyle\sum_{k=1}^{VAR}\left[\mathbf{Y}_k E[\theta\theta^T]\mathbf{Y}_k^{~T} + \mathbf{Y}_k E[\theta]\mathbf{y}_k^{~T} + \mathbf{y}_k E[\theta]^T \mathbf{Y}_k^{~T} + \mathbf{y}_k\mathbf{y}_k^{~T}\right] \\ E[\theta(t=0)\theta(t=0)^T] = E[\theta^0\theta^{0T}] \end{cases} \tag{5}$$

Response Realisation

The SDE system Eqn. 3 has an infinite number of solutions. Every solution corresponds to one realisation of the solution process. The alternative approach of Monte Carlo Simulation uses realisations of the stochastic input processes generated according to their joint density functions. The corresponding output is then calculated from a deterministic model, which expresses the response of the building model. If the procedure is repeated a large number of times the resulting output data can be treated statistically. Eqn. 3 can be rewritten to obtain the following expression

$$d\theta = \mathbf{f}(\theta,t)dt + \mathbf{G}(\theta,t)d\mathbf{W} \tag{6}$$

The vector $\mathbf{f}(\theta,t) = \mathbf{X}\theta + \mathbf{x}$, the matrix $\mathbf{G}(\theta,t) = [\mathbf{Y}_1\theta + \mathbf{y}_1 \cdots \mathbf{Y}_{VAR}\theta + \mathbf{y}_{VAR}]$ and the $d\mathbf{W}$ vector is given as $d\mathbf{W}^T = [dW_1 \cdots dW_{VAR}]$. The components of the $d\mathbf{W}$ vector are generated as realisations of independent normally distributed variables with zero mean and variance dt. Eqn. 6 can be solved for instance by a stochastic version of the fourth order Runge-Kutta method, see Arnold (1974).

TEST CASE

A simple test case is chosen in order to demonstrate the SDE approach, see Figure 1. The test case comprises a naturally ventilated atrium surrounded by building parts exposed to internal and external loads. The thermal capacity of the atrium corresponds to a medium heavy building. The atrium is naturally ventilated with two equally sized openings fixed without control. The specific heat loss due to natural ventilation, H_{vent}, in this case is expressed by

$$H_{vent} = \rho c_p C_D A(g\Delta h)^{1/2} \left[\frac{(\theta_1 - \theta_{ext})}{\theta_1 + 273.15} \right]^{1/2} \quad (7)$$

where ρ is the density, c_p is the specific heat, C_D is the discharge coefficient, A is the opening area, g is the gravitational acceleration and Δh is the stack height. H_{vent} is found by assuming a constant value for the term $\rho c_p C_D A(g\Delta h)^{1/2} = 6 \cdot 10^4$ W/K during the simulations. If the $\left[(\theta_1 - \theta_{ext})/(\theta_1 + 273.15) \right]^{1/2}$ term is introduced directly in Eqn. 1 the system of differential equations becomes non-linear in θ. In order to avoid non-linearity, the values of θ_1 and θ_{ext} in Eqn. 7 are adapted throughout the simulations from the previous time step. Figure 3 shows the corresponding air change rate.

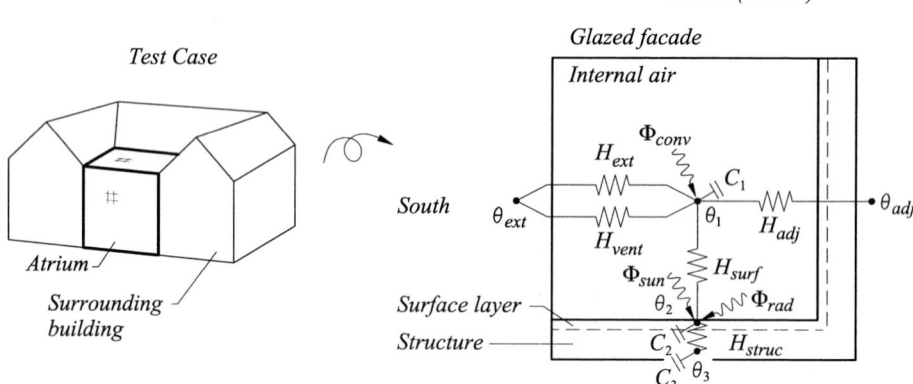

Figure 1: Test Case. An atrium surrounded by an adjacent zone and the external climate. Three unknown temperatures are to be determined, i.e. θ_1 - θ_3. The surface layer accounts for one fourth of the thermal capacity of the building and the "structure" accounts for the rest.

TABLE 1
TEST CASE PARAMETERS. θ_{ext} AND Φ_{sun} ARE DISCUSSED IN BROHUS ET AL., 2000

Parameter	Unit	Mean value	Standard deviation	Stochastic ?
θ_1, θ_2, θ_3	°C	Output	Output	Yes
θ_{ext}, Φ_{sun}	°C, W	Data from Danish DRY	Data from Danish DRY	Yes
θ_{adj}	°C	20	2	Yes
C_1, C_2, C_3	J/K	$6 \cdot 10^5$, $1 \cdot 10^7$, $3 \cdot 10^7$	0	No
H_{vent}	W/K	Calculated or 6000	0.3 times mean value	Yes
H_{ext}	W/K	800	0	No
H_{adj}	W/K	150	45	Yes
H_{surf}, H_{struc}	W/K	1400, 4800	0	No
Φ_{conv}, Φ_{rad}	W	Assumed data sets	0.3 times mean value	Yes

An assumed 24-hour load profile for the mean value of the internal sensible heat load is applied in the simulation. The sensible heat is divided into 50% convective heat, Φ_{conv}, and 50% radiative heat, Φ_{rad}. The convection heat flow is assumed to influence the internal air, i.e. θ_1, and the radiation heat flow is assumed to influence the surface layer, i.e. θ_2.

The calculations are performed during a warm week in summer and a cold week in winter, which can be thought of as a kind of design load periods. The simulations are started three days before the week in question in order to avoid unrealistic values due to the initial guess.

RESULTS AND DISCUSSION

Figure 2 shows the mean value process for the internal air during the winter week (left) and during the summer week (right). In addition the 95% confidence interval is shown corresponding to the mean value process ±1.96 times the standard deviation process. Due to a considerable air change rate, see Figure 3, the internal temperature is only slightly higher than the external air temperature, e.g. 3–6 °C.

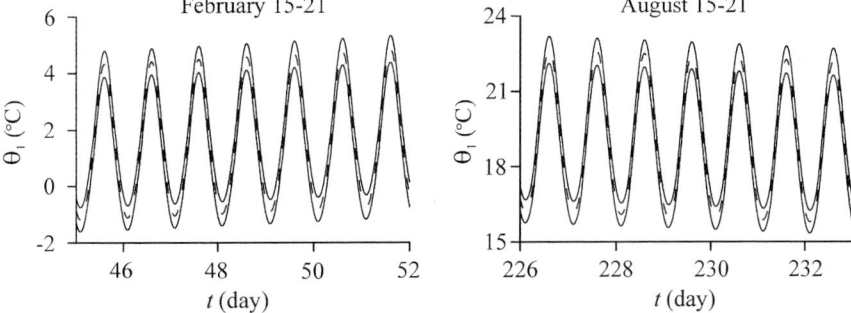

Figure 2: Internal air temperature in the atrium, θ_1, for a week during winter (left) and during summer (right). The figure shows the mean value and the 95% confidence interval (solid lines).

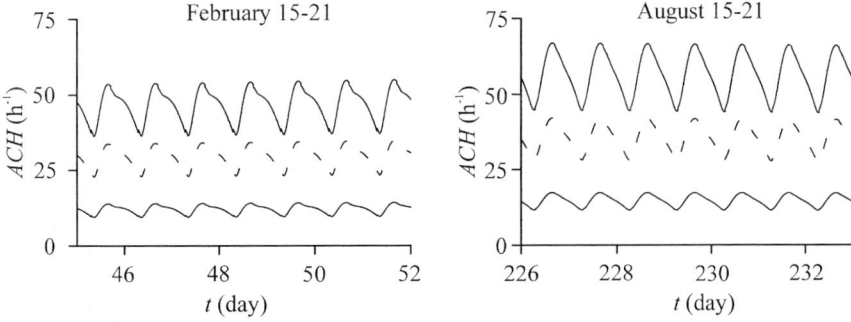

Figure 3: Mean air change rate and 95% confidence interval (solid lines) during a winter week (left) and a summer week (right). A fixed standard deviation of 0.3 times the mean value of H_{vent} is assumed.

In Figure 4 a realisation of the stochastic output process is presented together with the 95% confidence interval for the same temperature as shown in the right part of Figure 2. In this case a constant value of $H_{vent} = 6000$ W/K is chosen. The random fluctuations in this application are relatively small when compared with the mean values.

In this paper, the SDE approach is applied on a heat balance in order to examine the thermal behaviour of a relatively heavy building. However, if the approach is applied on a mass balance in order to determine for instance the contaminant concentration in a relatively light and hybrid or naturally ventilated building it is expected that randomness would play a more significant role.

420

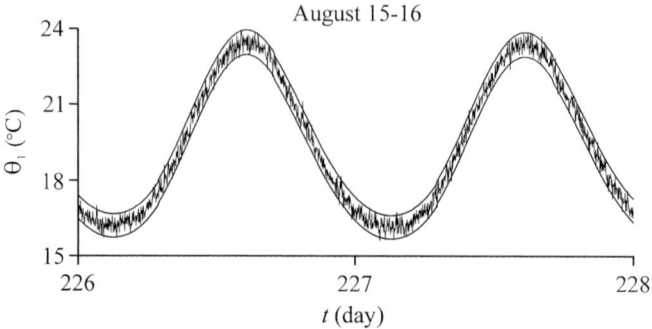

Figure 4: Stochastic realisation of internal air temperature in the atrium, θ_I. The figure also shows the corresponding 95% confidence intervals. In order to illustrate the stochastic fluctuations more clearly the plot is limited to show 48 hours of the summer week.

Future work may include a whole year case where it is possible to calculate the cumulative distribution and assess the energy consumption. Another further development is the inclusion of various control strategies in order to simulate realistic conditions both regarding thermal comfort and indoor air quality as well as energy consumption.

Probabilistic methods establish a new approach to the design of ventilation systems, which, apart from more realistic modelling, enable designers to include stochastic parameters like inhabitant behaviour, operation, and maintenance to predict the performance of the systems and the level of certainty for fulfilling design requirement under random conditions.

ACKNOWLEGDMENTS

This work is part of the Danish and Canadian contribution to the work of Annex 35 of the International Energy Agency "Hybrid Ventilation in New and Retrofitted Office Buildings". The Danish contribution was supported financially by the Danish Technical Research Council (STVF). Funding for the Canadian author participation was made possible through a contract provided by the CANMET - Energy Technology Centre of Natural Resources Canada. Their supports are highly appreciated.

REFERENCES

Arnold L. (1974). *Stochastic Differential Equations, Theory and Applications*, John Wiley & Sons.

Brohus H., Frier C. and Heiselberg P. (1999). Probabilistic Analysis Methods for Hybrid Ventilation - Preliminary Application of Stochastic Differential Equations, *Proceedings of HybVent Forum '99 - First International One-day Forum on Natural and Hybrid Ventilation*, ISBN 0-646-38043-5, pp. 171 - 180, 28 September, The University of Sydney, Darlington, NSW, Australia.

Brohus H., Haghighat F., Frier C. and Heiselberg P. (2000). Quantification of Uncertainty in Thermal Building Models. Part 2: Stochastic Loads. *Proceedings of ROOMVENT 2000,* Reading, UK.

Haghighat F., Chandrashekar M. and Unny T. E. (1987). Thermal Behaviour of Buildings under Random Conditions. *Applied Mathematical Modelling* **11:5**, 349-356.

Air Distribution in Rooms, (ROOMVENT 2000)
Editor: H.B. Awbi

QUANTIFICATION OF UNCERTAINTY IN THERMAL BUILDING SIMULATION - PART 2: STOCHASTIC LOADS

H. Brohus[1], F. Haghighat[2], C. Frier[1] and P. Heiselberg[1]

[1]Department of Building Technology and Structural Engineering
Aalborg University, DK-9000 Aalborg, Denmark (hb@civil.auc.dk)
[2]Department of Building, Civil and Environmental Engineering
Concordia University, Montreal, Quebec H3G 1M8, Canada (haghi@cbs-engr.concordia.ca)

ABSTRACT

In order to quantify uncertainty in thermal building simulation stochastic modelling is applied on a building model. Part 1 deals with the stochastic thermal building model and a test case. This paper deals with the determination of the stochastic input loads. The importance of obtaining a proper statistical description of the input quantities to a stochastic model is addressed and exemplified by stochastic models for the external air temperature and the solar heat gain.

Each of the external climate parameters is modelled as a stochastic process with time varying mean value function superimposed by a time varying standard deviation function. The statistics of the external air temperature is obtained by means of Fast Fourier Transform (FFT). A model of the solar heat gain is presented, considering the obvious fact that solar radiation is present only during daytime. The Danish Design Reference Year (DRY) is used as experimental data.

KEY WORDS

Thermal Building Simulation, Fast Fourier Transform, External Air Temperature, Solar Radiation, Solar Heat Gain, Design Reference Year, Experimental Data.

INTRODUCTION

In the design of buildings, prediction of long term system performance with regard to energy consumption and thermal comfort is traditionally performed with thermal building simulation programmes based on deterministic building models and loads.

A deterministic approach implies that all input parameters and model coefficients are 100% certain with zero spread. In practise, this is not the case, for instance, inhabitant behaviour and internal loads may vary significantly and external loads as wind, external air temperature and solar radiation are obviously stochastic in nature.

The purpose of this work is to quantify the uncertainty in thermal building simulation by means of a stochastic approach considering randomness in the loads as well as the model coefficients. In this connection, it is crucial to be able to quantify the uncertainty of the input parameters.

Part 1 comprises the outline of a stochastic building simulation model and a test case (Brohus et al., 2000). The stochastic thermal model presented in Part 1 requires that the loads be modelled by time varying stochastic processes. Hereby, the uncertainty in the load parameters is quantified in shape of time varying mean value and standard deviation functions. This paper considers the determination of those functions. As an example, statistics for the external air temperature and the solar heat gain are determined for instance by means of Fast Fourier Transform (Press et al., 1989).

Load data for the stochastic model is obtained from the Danish DRY (Jensen & Lund, 1995). DRY is an artificial data set, which is used in thermal building simulation to predict parameters like energy consumption and thermal response of buildings. In DRY a typical yearly weather climate is recorded in terms of hourly values of external climatic parameters, gathered from selected monthly data from a 15-year period. Although DRY is an artificial data set it will be used here to illustrate how measured weather climate parameters can be used for stochastic modelling.

EXTERNAL AIR TEMPERATURE

The mean value function, $\mu_{\theta ext}(t)$, of the external air temperature, $\theta_{ext}(t)$, is modelled by the sum of a constant and two cosine functions accounting for the systematic yearly and daily variations, respectively. Fast Fourier Transform (FFT) of the DRY data obtains the model coefficients

$$\mu_{\theta_{ext}}(t) = 7.76 + 8.93\cos(2\pi f_1 t + 2.74) + 2.45\cos(2\pi f_{365} t + 2.73) \text{ in } °C \tag{1}$$

where t is the time in seconds from the beginning of the year, $f_1 = 1/(365*24*3600)$ Hz is the frequency corresponding to the yearly variation and $f_{365} = 1/(24*3600)$ Hz is the frequency corresponding to the daily variation. When Eqn. 1 is subtracted from the original data the fluctuating part of $\theta_{ext}(t)$ is obtained. Based on visual inspection, the standard deviation function is assumed to be time independent and can be expressed as

$$\sigma_{\theta_{ext}}(t) = 3.42 \text{ in } °C \tag{2}$$

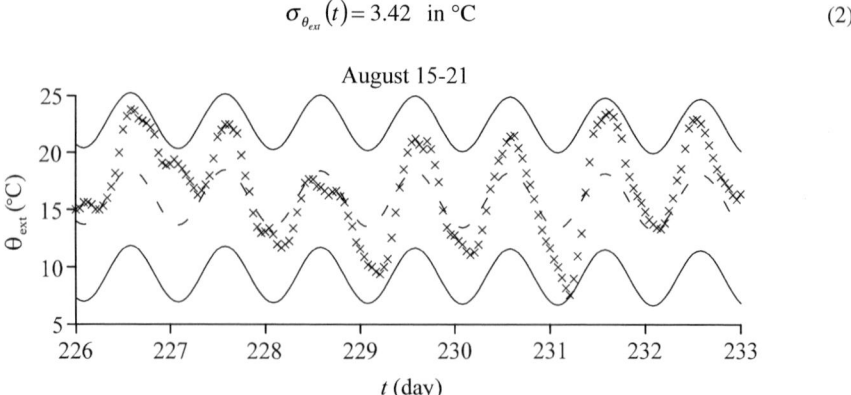

Figure 1: External air temperature data from DRY shown together with the modelled mean value function (dashed line) and the 95 % confidence interval (solid lines) during a week in August.

Figure 1 shows the DRY data for the external air temperature together with the mean value function and the 95 % confidence interval corresponding to the mean value function ± 1.96 times the standard deviation function.

SOLAR HEAT GAIN

The following model, partly adapted from Lund (1997), determines the solar heat gain, $\Phi_{sun}(t)$, as a function of reflection ratio of surroundings, window orientation, inclination, window area and effective reduction factor. The angles are defined in Figure 2.

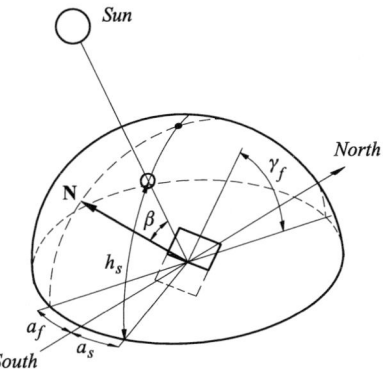

Figure 2: Definition of angles in the solar radiation model. β is the angle of incidence, h_s is the solar altitude, a_s is the solar azimuth, a_f is the surface azimuth, γ_f is the surface inclination, $\mathbf{N}$ is the normal vector to the surface (Lund, 1997).

A number of coefficients are defined and used in the calculation in order to facilitate the determination of the mean value function, $\mu_{\Phi sun}(t)$, and the standard deviation function, $\sigma_{\Phi sun}(t)$. Here, both direct, diffuse and reflection solar radiation are considered.

$$\cos(\beta(t)) = \cos(a_s(t) - a_f)\cos(h_s(t))\sin(\gamma_f) + \sin(h_s(t))\cos(\gamma_f) \tag{3}$$

$$c_1(t) = \begin{cases} \cos(\beta(t)) & \text{if} \quad \cos(\beta(t)) > 0 \\ 0 & \text{otherwise} \end{cases} \tag{4}$$

$$c_{aux}(t) = \begin{cases} 0.55 + 0.437\cos(\beta(t)) + 0.313(\cos(\beta(t)))^2 & \text{if} \quad \cos(\beta(t)) > -0.2 \\ 0.4751 & \text{otherwise} \end{cases} \tag{5}$$

$$c_2(t) = c_{aux}(t)(1 - \cos(\gamma_f)) + \cos(\gamma_f) \tag{6}$$

$$c_3(t) = 0.5\rho(1 - \cos(\gamma_f))\sin(h_s(t)) \tag{7}$$

$$c_4 = 0.5\rho(1 - \cos(\gamma_f)) \tag{8}$$

$$c_5 = A_f F_s \tag{9}$$

The coefficient c_5 considers the surface area, A_f, and the effective reduction factor, F_s, accounting for the frame-glass ratio and all kinds of solar shading.

In Eqn. 10 and Eqn. 11, statistical calculation rules for mean values and variances (and thus standard deviations) are applied on the deterministic formula. The standard deviation function, $\sigma_{\Phi sun}(t)$, is determined by assuming independence between $E_d(t)$ and $E_0(t)$

$$\mu_{\Phi_{sun}}(t) = c_5\left[(c_1(t)+c_3(t))\mu_{E_0}(t)+(c_2(t)+c_4)\mu_{E_d}(t)\right] \qquad \text{in W} \qquad (10)$$

$$\sigma_{\Phi_{sun}}(t) = c_5\sqrt{(c_1(t)+c_3(t))^2\sigma_{E_0}^2(t)+(c_2(t)+c_4)^2\sigma_{E_d}^2(t)} \qquad \text{in W} \qquad (11)$$

Both the diffuse solar radiation, $E_d(t)$, and the direct solar radiation, $E_0(t)$, exhibit the characteristics that both processes are zero during night and obtain a maximum at noon. The two processes are treated similarly, and only the results for the diffuse solar radiation will be shown here.

The mean value function, $\mu_{Ed}(t)$, is modelled by a parabola for each day as shown in Figure 3

$$\mu_{E_d}(t)=\begin{cases}\dfrac{4\mu_{E_{d,max}}(k)}{(t_{do}(k)-t_{up}(k))^2}\left[(t_{up}(k)+t_{do}(k))t-t_{up}(k)t_{do}(k)-t^2\right] & \text{for } t_{up}(k)\leq t\leq t_{do}(k) \\ 0 & \text{otherwise}\end{cases} \quad \text{in W/m}^2 \quad (12)$$

where $t_{up}(k)$ and $t_{do}(k)$ are the sunrise and sunset times, respectively, in seconds after the beginning of the year for day k, obtained from the DRY data by FFT analysis

$$t_{up}(k)=2.42\cdot10^4+8.89\cdot10^3\cos(2\pi f_1 86400(k-1)+0.12)+86400(k-1) \quad \text{in s} \quad (13)$$

$$t_{do}(k)=7.04\cdot10^4+8.95\cdot10^3\cos(2\pi f_1 86400(k-1)-2.88)+86400(k-1) \quad \text{in s} \quad (14)$$

The mean value function of the maximum diffuse solar radiation process, $\mu_{Ed,max}(k)$, is found by conducting a FFT-analysis on the data series consisting of the maximum values of $E_d(t)$ for each day to give

$$\mu_{E_{d,max}}(k)=206.26+139.74\cos(2\pi f_1 86400(k-1)-2.92) \quad \text{in W/m}^2 \quad (15)$$

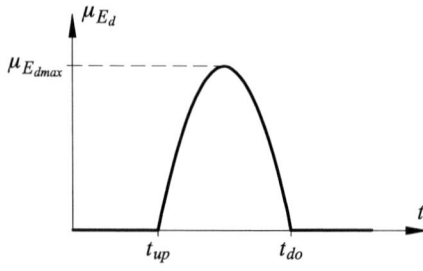

Figure 3: Outline of $\mu_{Ed}(t)$ parabola model shown for a single day

The standard deviation function of the diffuse solar radiation, $\sigma_{Ed}(t)$, is modelled as a scaled version of the mean value function as

$$\sigma_{E_d}(t) = \frac{\sigma_{E_{d,max}}(k)}{\mu_{E_{d,max}}(k)} \mu_{E_d}(t) \quad \text{in W/m}^2 \tag{16}$$

$\sigma_{Ed,max}(k)$ is the standard deviation of the maximum process of $E_d(t)$. This is obtained by sampling on a time window using the 11 points located closest to the time in question and FFT-analysis

$$\sigma_{E_{d,max}}(k) = 62.91 + 30.59\cos(2\pi f_1 86400(k-1) - 2.65) \quad \text{in W/m}^2 \tag{17}$$

Figure 4 shows the DRY data for the diffuse solar radiation together with the mean value function and the 95 % confidence interval.

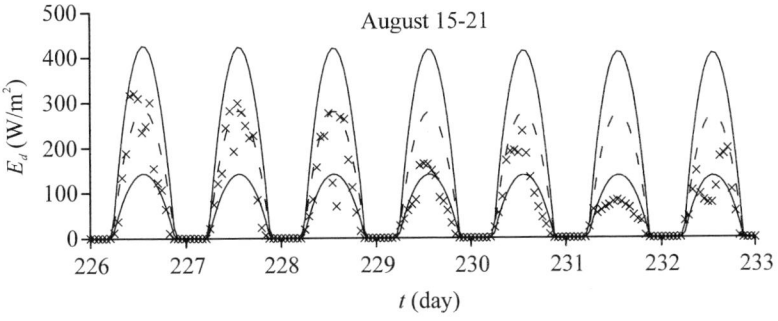

Figure 4: Diffuse solar radiation from DRY shown together with the modelled mean value function (dashed line) and the 95 % confidence interval (solid lines) for a week in August.

In Figure 5 an example is shown for one week in August for the solar gain corresponding to a vertical window, $(A_f = 1 \text{ m}^2, \gamma_f = 90\ °)$ at south $(a_f = 0\ °)$ with a reflection ratio $\rho = 0.25$ and a shading ratio $F_s = 0.5$. Data for the Danish solar altitude, $h_s(t)$, and solar azimuth, $a_s(t)$, are found in Lund, 1997.

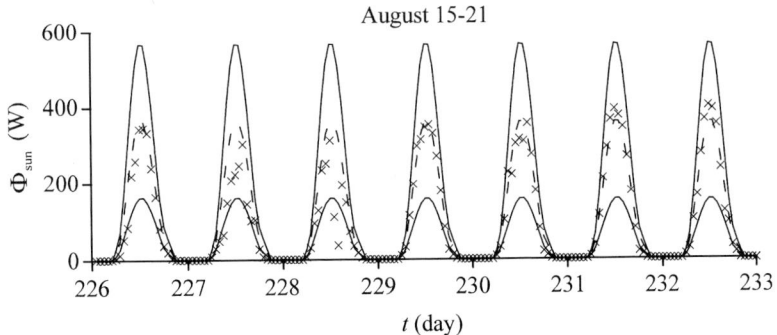

Figure 5: Solar heat gain for a 1 m^2 window facing south for a week in August. Results based directly on solar radiation data from DRY are shown together with the modelled mean value function (dashed line) and 95 % confidence interval (solid lines).

DISCUSSION

Stochastic models for the external air temperature and the solar heat gain have been developed on the basis of data from the Danish Design Reference Year. The results are applied directly in a stochastic thermal building simulation model by means of a Stochastic Differential Equation (SDE) approach as described in Part 1. Alternatively, the stochastic load models can be used for other kinds of probabilistic modelling.

The external air temperature and the solar heat gain are modelled as a time dependent mean value function superimposed by a time varying standard deviation function. Reasonable agreement between the DRY data and the stochastic models is obtained.

The fluctuating part of the processes are assumed independent both regarding mutual correlation and individual correlation in time, i.e. auto correlation, in order to be able to use the white noise assumption which is applied in the SDE approach presented in Part 1. In reality, the processes will to some extent be correlated both in time and mutually. Future modelling may include auto correlation functions, describing the time dependency of the parameters, and cross correlation functions, describing the mutual dependence.

ACKNOWLEGDMENTS

This work is part of the Danish and Canadian contribution to the work of Annex 35 of the International Energy Agency "Hybrid Ventilation in New and Retrofitted Office Buildings". The Danish contribution was supported financially by the Danish Technical Research Council (STVF). Funding for the Canadian author participation was made possible through a contract provided by the CANMET - Energy Technology Centre of Natural Resources Canada. Their supports are highly appreciated.

REFERENCES

Brohus H., Haghighat F., Frier C. and Heiselberg P. (2000). Quantification of Uncertainty in Thermal Building Models. Part 1: Stochastic Building Model. *Proceedings of ROOMVENT 2000*, Reading. UK.

Jensen J. M. and Lund H. (1995). *Design Reference Year, DRY - et nyt dansk referenceår*, Meddelelse Nr. 281, Laboratoriet for Varmeisolering, ISSN 1395-0266, Technical University of Denmark, Lyngby, Denmark (In Danish).

Press W. H., Flannery B. P., Teukolsky S. A. and Vetterling W. T. (1989). *Numerical Recipes, The Art of Scientific Computing*, ISBN 0 521 38330 7, Cambridge University Press.

Lund H. (1997). Udeklima, Chapter 2, In: Hansen H. E., Kjerulf-Jensen P., Stampe O. B., *Varme- og Klimateknik, Grundbog, 2. udgave*, ISBN 87-982652-8-8, Danvak ApS, Denmark (In Danish).

Air Distribution in Rooms, (ROOMVENT 2000)
Editor: H.B. Awbi

427

INTERACTIVE SIMULATION OF ROOM AIR TEMPERATURE AND ABSORPTION/EMISSION OF RADIATIVE HEAT WITH ROOM MOISTURE

Yasushi Kondo, Takeshi Ogasawara and Jun-ichi Fujimura
Department of Architecture, Musashi Institute of Technology
1-28-1 Tamazutumi Setagaya-ku, Tokyo, 158-8557 Japan

ABSTRACT

It is well known that water vapor and carbon dioxide in the air have several infrared bands where these gases absorb and emit radiative energy. Therefore the radiative heat absorbed by room moisture may not be neglected in the prediction of room air temperature. Glicksman and Chen et al. (1998) carried out numerical analysis and showed that the effect of radiative heat absorbed by room air moisture should be considered in the prediction of temperature distribution especially in large enclosed space. In this paper the analysis method of the radiant energy absorbed by room moisture is shown and some examples simulated by this method are presented. (1) The numerical analysis with a temperature of walls as a given boundary condition was carried out. When the simulated result with 50% relative humidity was compared with the case with absolutely dry air, the impact of radiative heat absorbed by room moisture on air temperature was confirmed. The effect of moisture on air temperature was remarkable especially when the temperature difference between walls was big and the volume of space was large. (2) The interactive simulation of convective and radiative heat transfer was carried out with the moisture's effect of moisture on air temperature simultaneously. The results showed that the impact of the radiative heat absorbed by room moisture appeared not only in air temperature but in the surface temperatures of walls.

KEYWORDS

Radiation, Absorption, Emission, Moisture, Room Air Temperature and Large Enclosure

RADIATIVE HEAT TRANSFER BETWEEN ROOM AIR AND WALLS

(1) Radiative Energy Absorbed with Water Vapor in Room Air

When an enclosed space is in vacuum condition and no air exists in it, all of radiant energy emitted from a certain wall reaches the other walls. However when the space is fulfilled with air including water vapor, a part of radiative energy emitted from a certain wall is absorbed with water vapor while the radiant energy goes through the space. Radiative energy emitted from a black body at 300[K] and the absorption ratios of radiative energy of the water vapor are shown in Fig.1. The definitions of wall temperature and the angles between walls in an enclosed space are illustrated in Fig. 2. The radiative energy emitted from the wall (i) in enclosed space can be described by equation [1] in Table 1. The radiative energy absorbed with the air in an enclosed space can be written in equation [2]. F_{ik}' appeared in equation [2] is defined as pseudo the shape factor including the absorption effect of radiant energy with water vapor as shown in equation [3].

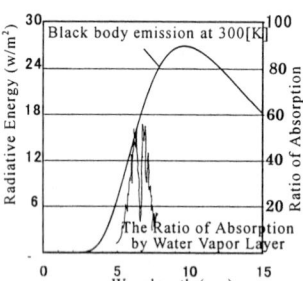

Figure 1 Radiative Energy from Black body at 300[K] and Absorption Rate of Water Vapor (see Ref.3)

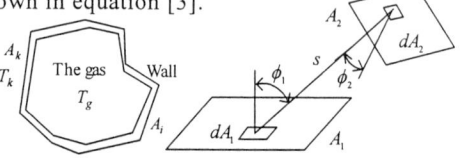

(a) Temperature of Enclosed Air with Wall
(b) Angle Relation among Two Walls

Figure 2 Definition of Temperature of Enclosed Air with Wall and Angle Relation Among Two Walls

<SYMBOLS>

σ : Stefan-Boltzmann constant [W/m²K⁴]

G_i :Radiosity of the wall (i) (self emission + reflective emission of irradiation) [W/m²] $G_i = \varepsilon_i \sigma T_i^4 + (1-\alpha_i)\sum_k G_k F_{ik}$

ε_i : Emissivity of the wall (i) [-]

α_i :Absorptivity of the wall (i) [-]

A_i : Area of the wall (i) [m²]

F_{ik} : Shape factor [-] s : Beam length of radiation [m]

F_{ik}' : Psuedo Shape factor including radiative energy absorption of the air [-]

$\alpha_G(s)$: Directional absorptivity of the air at the surrounding wall boundary [-]

$\varepsilon_G(s)$: Directional emissivity of the air at the surrounding Wall boundary [-]

ε_g :Local emissivity of the air at the surrounding wall boundary [-]

$\overline{\varepsilon}_{g,i}$: Mean emissivity of the air averaged over the wall (i) [-]

$\overline{\varepsilon}_g$: Representative value of the emissivity in a room [-]

Q_g : Net radiative heat transfer between walls and the air [W]

T_g : Temperature of the air[K]

T_i :Temperature of wall (i)[K]

L_{eff}:Equivalent beam length[m]

B_{ij} :Gebhart's absorption factor

Table 1 Basic Equations of Radiative Heat Emission and Absorption of Wall and Water vapor

(1)Radiative Energy Emitted from Walls

Radiative energy emitted from wall (i)	$G_i A_i \sum_{k=1}^{n} F_{ik}'$	[1]
Radiant energy absorbed with indoor	$\sum_{i=1}^{n}\left[\sum_{k=1}^{n} G_i A_i (F_{ik} - F_{ik}')\right]$	[2]
Pseudo shape factor including radiative energy absorption of indoor air	$F_{12}' = \frac{1}{A_1}\int_{A_1}\int_{A_2}[1-\alpha_G(s)]\cdot\frac{\cos\phi_1\cos\phi_2}{\pi s^2}dA_1 dA_2$ $= F_{12} - \frac{1}{A_1}\int_{A_1}\int_{A_2}\alpha_G(s)\cdot\frac{\cos\phi_1\cos\phi_2}{\pi s^2}dA_1 dA_2$	[3]
Shape factor	$F_{12} = \frac{1}{A_1}\int_{A_1}\int_{A_2}\frac{\cos\phi_1\cos\phi_2}{\pi s^2}dA_1 dA_2$	[4]
Summation law of shape factor	$\sum_{k=1}^{n} F_{ik} = 1$	[5]

(2) Radiative Energy Emitted from Water Vapor in Room Air

Radiative energy emitted from the air at a certain point on the surrounding boundary	$E = \varepsilon_g \sigma T_g^4$	[6]
Local emissivity of the air at the surrounding wall boundary	$\varepsilon_g = \int_A \varepsilon_G(s)\cdot\frac{\cos\phi_1\cos\phi_2}{\pi s^2}dA_2$	[7]
Radiative energy of finite area Ai on the surrounding wall boundary	$\int_{A_i} EdA_i = \sigma T_g^4 \int_{A_i}\varepsilon_g dA_i$	[8]
Mean emissivity of the air Averaged over finite area Ai	$\overline{\varepsilon}_{g,i} = \frac{1}{A_i}\int_{A_i}\varepsilon_g dA_i$	[9]
Total radiative energy emitted from the air	$\sum_{i=1}^{n}\sigma T_g^4 A_i \overline{\varepsilon}_{g,i}$	[10]

(2) Radiative Energy Emitted from Water Vapor

Radiative energy emitted from water vapor can be described in equation [6] (see Ref. 3) at a certain point on the boundary surface surrounding room air. Local emissivity of the gas on the boundary surface is shown in equation [7]. Radiative heat flux from the finite area on the boundary surface is shown in equation [8]. After averaging ε_g within the finite area A_i, the mean emissivity $\overline{\varepsilon}_{g,i}$ can be given by equation [9]. When the directional emissivity $\overline{\varepsilon}_g(s)$ can be supposed to be equal to the directional absorptivity $\overline{\alpha}_g(s)$, equation [11] is obtained from equations [3], [5], and [9] (see Ref. 3).

$$\sum_{k=1}^{n} F'_{ik} = \sum_{k=1}^{n} F_{ik} - \overline{\varepsilon}_{g,i} = 1 - \overline{\varepsilon}_{g,i} \qquad [11]$$

Radiant energy absorbed by water vapor can be estimated by equation [2'].

Radiant energy absorbed by water vapor $\quad = \sum_{i=1}^{n} G_i A_i \overline{\varepsilon}_{g,i}$ $\qquad$ [2']

By using equation [9], the total radiative energy emitted from room air can be derived as shown in equation [10].

(3) Radiative Heat Transfer between Walls and Room Air

The difference between equation [2'] and equation [10] is equal to the net radiant flux between walls and room air as shown in equation [12].

$$Q_g = \sum_{i=1}^{n} (G_i - \sigma T_g^{\,4}) A_i \overline{\varepsilon}_{g,i} \qquad [12]$$

(4) Mean Emissivity of Room Air

The directional emissivity of air can be regarded as the function of air temperature, wavelength of radiative energy, water vapor pressure and thickness of the air. The value of $\varepsilon_G(s)$ can be obtained from the chart arranged with two parameters, i.e. air temperature and the product of water vapor pressure and air thickness. Furthermore, in order to simplify the calculation process, the simplified method was proposed in which the mean emissivity is supposed to be represented as one value applicable to all walls in a room. In this method, mean radiant length in a room L_{eff} is required. The estimation of $\varepsilon_G(s)$ in some typical room shapes is tabulated in Ref. 3. The emissivity in a room $\overline{\varepsilon}_g$ can be evaluated as below.

$$\overline{\varepsilon}_g = \varepsilon_G(L_{eff}) \qquad [13]$$

Then the net radiant flux Q_g in equation [12] can be obtained with $\overline{\varepsilon}_g$ instead of $\overline{\varepsilon}_{g,i}$.

CFD SIMULATION WITH WALL SURFACE TEMPERATURE GIVEN AS BOUNDARY CONDITIONS

(1) Simulated Cases

Three cases were analyzed as shown in table 2. The simulation was conducted based on the

430

standard k-ε model with the boundary conditions in which the uniform temperature of wall surface in case 1 and case 2 was given as same as in Ref. 1. In case 1, a regular sized room without air-conditioning was simulated in winter condition. In case 2, a large room was investigated and compared with the results of case 1. In case 3, an atrium model in summer condition with air-conditioning was studied. In each case, room air was supposed to be kept in 50% relative humidity and the temperature distribution was simulated on the basis of the method shown above. These results were compared with the results of absolute dry room in which the effect of absorbed energy with water vapor was not considered. Table 3 shows the boundary conditions.

(2) Results and Discussions

In case 1, the simulated air temperature in a regular sized room with 50 % relative humidity was about 1 [^{o}C] higher than that with 0 % relative humidity as shown in Fig. 3. In case2, the large-scale room with 50 % was hotter than that with 0 % by about 2 [^{o}C]. It was the reason for these differences that a part of radiative heat was absorbed with water vapor in a room. The same results were also shown in Ref. 1. In case3 on a large-scale atrium, the effect of absorbed radiative energy with water vapor was appeared in air temperature by about 2 [^{o}C]. Therefore the impact of the radiative heat absorbed with water vapor was remarkable especially in the large-scale enclosed space.

Table 2 Simulation Models in Case 1, 2 and 3

Case		Room Models (L × W × H)	Relative Humidity RH	$\overline{\varepsilon}_g$	Qg (W/m^3)	Window Location (m^2)
1	a	3 × 6 × 3m	0%	0	0	Side Wall (9 m^2)
	b		50%	0.107	1.74	
2	a	9 × 18 × 9m	0%	0	0	Side Wall (81 m^2)
	b		50%	0.190	1.03	
3	a	15 × 15 × 20m	0%	0	0	All of Roof (225 m^2)
	b		50%	0.265	3.90	

Case 1,2: Natural convection
Case 3: Air-conditioned mechanically
Cool air at 15°C supplied 21600m^3/h   Velocity 4m/s
Supply opening FL+5m   Exhaust opening FL+0.5m

Grid	Case 1 15 × 30 × 20=9000
	Case 2 25 × 30 × 30=22500
	Case 3 15 × 32 × 40=19200

Table 3 Boundary Conditions in Case 1, 2 and 3
(Temperature[^{o}C]/convective heat transfer coefficient α_c [W/ m^2°C] [4])

	Window	Ceiling	Side Wall	Floor
1	5°C /2.65	30°C /0.49	20°C /1.0	20°C /1.0
2	5°C /2.65	30°C /0.49	20°C /1.0	20°C /1.0
3	50°C /4.7	50°C /4.7	Upper 5m 50°C /4.7 Lower 15m 30°C /4.7	20°C /4.7

The boundary condition for airflow is based on the generalized logarithmic

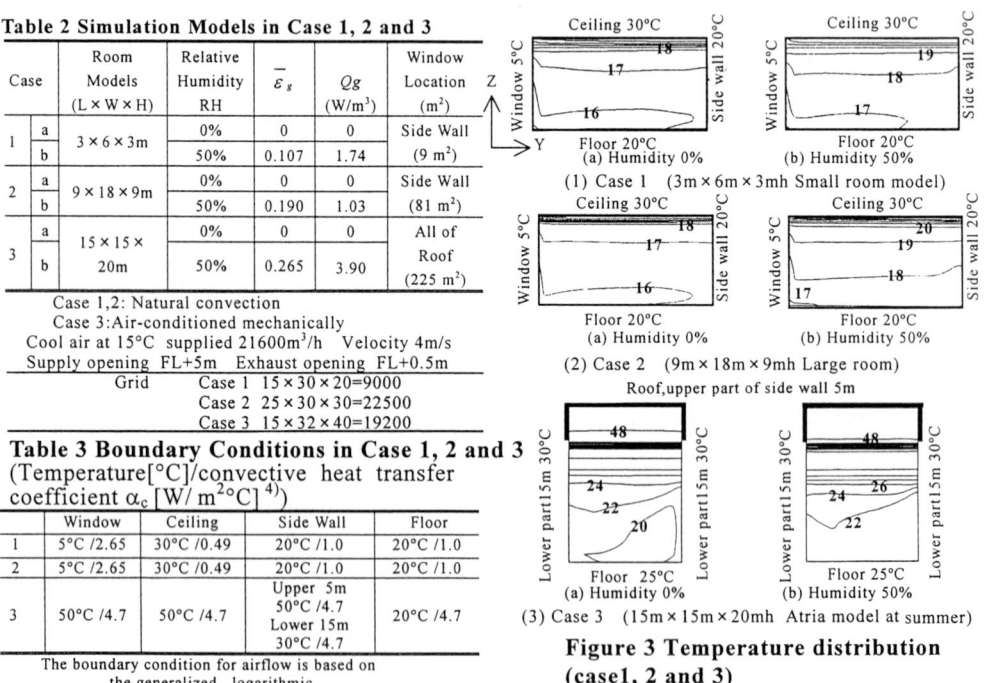

(1) Case 1   (3m × 6m × 3mh Small room model)

(2) Case 2   (9m × 18m × 9mh Large room)

(3) Case 3   (15m × 15m × 20mh Atria model at summer)

Figure 3 Temperature distribution (case1, 2 and 3)

COUPLED SIMULATION OF CONVECTION, CONDUCTION AND RADIATION INCLUDING ABSORBED RADIATIVE HEAT WITH WATER VAPOR

(1) Simulated Cases

Table 4 shows the outline of simulated cases in this section. The dimension of a room in case 4 and case 5 was the same as that in case 1 and case 2 respectively discussed in previous section.

(2) Simulation Method

In the previous section and also in Ref. 1, the uniform temperature of the wall surface was given as the boundary condition of the CFD simulation. On the other hand, the coupled simulation of heat conduction, convective and radiative heat transfer was also carried out in this section. Therefore the temperature of the wall surface was not necessary to given as the boundary condition in the simulation. In this section, the surface temperature of a window was treated as an unknown, and the surface temperature of the other walls was given as boundary condition. The total heat of absorbed radiative energy with room moisture was evaluated by equation [12] as shown in table 4 and it was treated as the internal heat generation of CFD simulation in entire volume of a room. Table 5 shows the boundary conditions. The temperature of a window made of glass was calculated by solving the heat balance of heat conduction, convective and radiative heat transfer. The outside air temperature, the convective heat transfer coefficient and thickness of glass were assumed as in table 5. The shape factor between walls was calculated by Monte Carlo method.

(3) Results and Discussions

In the simulated results of a regular sized room the case 4, the effect of the absorbed radiative heat with vapor on air temperature was not very large. Inside surface temperature of the window decreased about 1 [°C] because the radiative heat transfer between the window and the other walls was disturbed by water vapor. In this situation, the radiative heat from the ceiling, the floor and the side walls was received at the window, therefore the absorption of heat with water vapor decreased the window temperature.

Table 5 Boundary Conditions in Case 4 and 5

Window	(Velocity) Generalized logarithm law (Temperature) Temperature $\Theta^{(i)}{}_w$ of boundary mesh (i) was calculated from the following heat balance equation. $q^{(i)}{}_{CD} + q^{(i)}{}_{CV} + q^{(i)}{}_R = 0$ 1) Heat flux of conduction $q^{(i)}{}_{CD} = \lambda/\!\!/_S (\Theta^{(i)}{}_w - \Theta_{out})$ 2) Convective heat flux $q^{(i)}{}_{CV} = \alpha_c(\Theta^{(i)}{}_w - \Theta^{(i)}{}_l)$ 3) Radiactive heat flux $q^{(i)}{}_R = \sigma \sum \varepsilon_i B_{ij} A_i \{(1-\bar{\varepsilon_g})T_j{}^4 - T_i{}^4\}$ λ:Conductive heat transfer coefficient of a glass with 6mm thick. 0.696W/m·K α_c:Convective heat transfer coefficient outside 20W/m²·K inside 2.65W/ m²·K outdoor air temperature 2.6°C σ: Stefan-Boltzmann's constant 5.67×10^{-8}W/(m²·K⁴) $\bar{\varepsilon_g}$:Representative value of emissivity of the air in a room B_{ij} : Gebhart's absorption factor $\Theta^{(i)}{}_w$:Surface temperature of window cell (i) $\Theta^{(i)}{}_l$: Fluid temperature of first cell closed to window (i)
walls	(Velocity) Generalized logarithm law (Temperature) ceiling:30°C(uniform) α_c 0.49W/m²·°C other walls:20°C(uniform) α_c 1.0W/ m²·°C Emissivity 0.9 at all walls Shape factor was calculated by Montecalro method

Table 4 Simulation Models in case 4 and 5

Case		Room Models (L × W × H)	Relative Humidity RH	$\bar{\varepsilon}_g$	Qg (W/m³)	Window Location (m²)
4	a	3 × 6 × 3m	0%	0	0	Side Wall (9 m²)
	b		50%	0.107	1.74	
5	a	9 × 18 × 9m	0%	0	0	Side Wall (81 m²)
	b		50%	0.190	1.03	

Case 4, 5: Natural convection

Grid: Case 4, 5 20 × 42 × 30=25200

room with 50% RH was higher than the results with 0% RH by 0.5-1 [°C] and the inside surface temperature of window decreased about 3 [°C]. The impact of the radiative heat absorbed with water vapor was appeared in air temperature and wall temperature and it was remarkable in case of the large-scale enclosed space.

CONCLUSIONS

In this paper the method for estimation of the radiative energy absorbed with room air moisture was presented and the effect of absorbed heat with water vapor on air temperature was demonstrated with CFD simulation on some simplified room model. The results of this study were concluded as the following.

(1) From the simulated results it was confirmed that the effect of heat absorption by water vapor could not be neglected in some cases. This effect was remarkable especially in the case of a large enclosed space.

(3) The coupled simulation of heat conduction, convective and radiative heat transfer was also carried out in this paper. From the coupled simulation, the effect of the radiative energy absorbed with water vapor was appeared not only in air temperature but also in wall surface temperature.

APPENDIX

(1) Figure1 was made on the basis of data from Ref. 3 about water vapor layer with 220 [cm] thicknesses. However the data of Ref. 3 was limited from 5 to 8 [μm] wavelength.

(2) In this paper, when the radiative heat transfer between water vapor and wall surface was considered, the temperature of water vapor was assumed to be uniform. This is because that there is little variation in emissivity of water vapor under the room temperature distribution in a general condition.

(3) The absorption and emission of radiative energy with water vapor is occurred in some specific wavelength, however the emissivity of water vapor was assumed to be constant within all wavelength. Therefore the evaluation method using this assumption may lead some error in the results.

(4) The value of convective heat transfer coefficient was given the same as in Ref. 1 in case 1 and case 2.

(5) There exists two radioactive gases in air, i.e. water vapor and carbon dioxide. Emissivity of carbon dioxide is much less than that of water vapor in a general indoor condition. Therefore the effect of carbon dioxide was neglected in this paper.

REFERENCES

1) L. R. Glicksman and Q. Chen (1998): Interaction of Radiation Absorbed by Moisture in Air with Other Forms of Heat Transfer in an Enclosure, ROOMVENT98, pp. 111~118

2) M. Brandli, G. Schenler and B. Keller (1996): The Interaction of the Infrared Radiation from the Room Boundaries with the Humidity Content of the Enclosed Air, ROOMVENT96, pp. 145~152

3) Yosiro Katto "Den-netsugaku Gairon", in Japanese, YOKENDO 1964

4) H.C.Hottel and R.B.Egbert (1941) Trans. ASME. 63.297

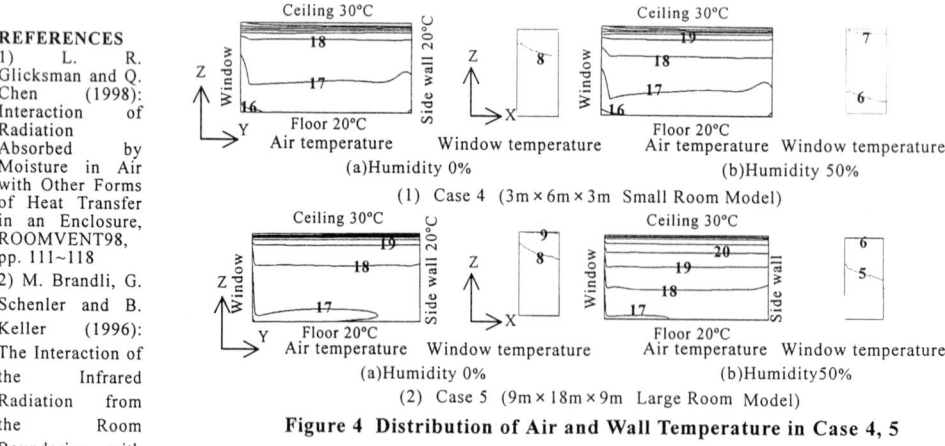

(1) Case 4 (3m × 6m × 3m Small Room Model)

(2) Case 5 (9m × 18m × 9m Large Room Model)

Figure 4 Distribution of Air and Wall Temperature in Case 4, 5

Air Distribution in Rooms, (ROOMVENT 2000)
Editor: H.B. Awbi

THE EFFECT OF OBJECT POSITIONS ON VENTILATION PERFORMANCE

Youngjun Cho and Hazim B. Awbi

Indoor Environment & Energy Research Group
Department of Construction Management & Engineering, The University of Reading, UK

ABSTRACT

The effect of the change in object positions (i.e. office furniture) on the air quality in a room was studied using zonal purging flow rates. In relation to the zonal purging flow rate in a room, the transfer probability from the inlet to a certain zone can provide information on the amount of fresh air from the inlet to the zone. In this study, the probability obtained from Markov chain theory was used to analyze the ventilation performance. Also, the mean ventilation effectiveness obtained from the transfer probability was compared with the traditional mean ventilation effectiveness for the room. The velocity fields in a two-dimensional parametric room (6m x 2.7m) were derived from CFD simulation under isothermal conditions. The velocity fields were used to calculate the interconnecting flow rate between zones, and to obtain the velocity data for the whole room and the occupied zone. Several different arrangements of objects (representing room furniture) were used to find the optimum location of the objects.

It was found that the mean ventilation effectiveness using the zonal purging flow rate concept is in agreement with the traditional mean ventilation effectiveness. Findings from this paper show that the object positions can play a key role in influencing ventilation performance, either in a positive or in a negative way. Depending on the location of object and airflow pattern, the presence of an object may produce two opposite effects either a hindering of the airflow in a room or an improvement in the air distribution. Also, an increase in the object lengths and a decrease in the free volume of the room do not always mean that the ventilation effectiveness decreases. The zone where the maximum velocity in the occupied zone occurs was the optimum location of an object in the room investigated.

KEYWORDS

Object positions, Mean ventilation effectiveness, Mixing ventilation, CFD, Zonal purging flow rate, Transfer probability from inlet, Free volume

INTRODUCTION

Even though a good ventilation performance may be predicted at the design stage, the furniture arrangement by the occupants can influence the environment in the room. However, owing to the arbitrary alteration of furniture (i.e. desks), the indoor airflow pattern can change and affect the ventilation condition predicted at the design stage.

Regarding the effect of an object in a room on ventilation performance, Sandberg and Claesson (1998) have found that the object is a factor to consider in the design of ventilation systems. Nielsen et al. (1998) investigated the changes in the air movement in a furnished room compared to an empty room, and also studied the change of velocity in the occupied zone due to the increase of furniture volumes. Although considerable research on the effect of obstructions has been done, no major study on the effect of changes in object locations has been carried out.

This paper focuses on the effect of changes in object location and volume on the ventilation performance. Using the zonal purging flow rate (U_p) as defined by Peng and Davidson (1997), the mean ventilation effectiveness, $<\varepsilon>_{pr}$, is obtained and compared with the traditional mean ventilation effectiveness, $<\varepsilon>_c$. The result obtained for $<\varepsilon>_c$ from the ratio of the exhaust concentration to the room mean concentration is very dependent upon the location of the emission source and the number of the emission sources. Thereby, there is some uncertainty in analyzing the ventilation performance when certain parameters inside the room (e.g. room furniture) are changed. In this paper, the mean ventilation effectiveness modified by the zonal purging flow rate (U_p) is introduced and used for investigating the influence of the change in object locations on the ventilation performance.

DEFINITIONS

Transfer Probability from Inlet (P_{sp}) and Zonal Mean Ventilation Effectiveness (ε_{cp})

The transfer probability, P_{sp}, can be obtained using Markov-chain model, which has been applied by Peng (1997). P_{sp} depends only on the air distribution and zone volume, and represents the fraction of fresh air from the inlet reaching a local zone. Zonal ventilation effectiveness, ε_{cp}, can be obtained from the ratio of the concentration in the exhaust, C_e, to the mean concentration in a certain zone, $<C_p>$. As a spatial average, the zonal mean concentration can be obtained from the ratio between the sum of a local contaminant concentration (C_{cp}) multiplied by a local volume (V_{cp}) in the zone and the zone volume (V_p).

$$\varepsilon_{cp} = \frac{C_e}{<C_p>} = \frac{C_e}{\sum C_{cp}V_{cp}/V_p} \tag{1}$$

Under condition of complete mixing from mass balance, the two indices, P_{sp} and ε_{cp}, can simply provide information on the amount of fresh air supplied to a zone and a contaminant removal from the zone.

Mean Ventilation Effectiveness, $<\varepsilon>_c$ and $<\varepsilon>_{pr}$

The mean ventilation effectiveness, $<\varepsilon>$, is defined as the ratio of the average purging flow rate, $<U>$ to the ventilation supply flow rate, (AIVC, TN34, 1991),

$$<\varepsilon> = \frac{<U>}{Q} \times 100 \tag{2}$$

Conventionally, the mean ventilation effectiveness, $<\varepsilon>_c$, has been defined as the ratio between the outlet concentration, C_e, and the room mean concentration, $<C_r>$ or the mean concentration in the occupant zone, $<C_{op}>$.

$$<\varepsilon>_c = \frac{C_e}{<C_r>} \quad \text{or} \quad <\varepsilon>_c = \frac{C_e}{<C_{op}>} \tag{3}$$

The zonal purging flow rate(or regional purging flow rate), U_p, calculated by Markov chain model, is defined as a measure of how much fresh air is supplied to a certain zone (Peng & Davidson1997). The zonal purging flow rate is obtained by multiplying the transfer probability from inlet to a certain zone (P_{sp}) with the supply flow rate (Q) : $U_p = P_{sp}Q$. If the average purging flow rate, $<U>$, is expressed as the spatial mean purging flow rate, the equation for $<\varepsilon>_{pr}$ can be expressed by the transfer probability (P_{sp}) from the inlet to a certain zone p in the room:

$$<U> = \frac{1}{V_r}\int_v U_p dv = \frac{\sum_{p=1}^{n} P_{sp}QV_p}{V_r} \quad , \text{thus,} \quad <\varepsilon>_{pr} = \frac{\sum_{p=1}^{n} P_{sp}V_p}{V_r} \times 100 \tag{4}$$

CFD SIMULATION

Under isothermal conditions, a two dimensional parametric room (6.0m x 2.7m) with mixing ventilation was simulated by the CFD code, VORTEX2D (Awbi,1996). Figure 1 shows a schematic of the room geometry. The supply device was centrally positioned at the ceiling, which had a baffle plate to produce two jets along the ceiling as shown in Figure 1. The supply velocity in each direction was 1.2m/s. Two outlets were placed on both end walls at 0.1m height above the floor. The room was divided into 10 zones with the equal volumes (1.8 m^2) except the two zone near the outlets (0.9 m^2). The size of the relevant object used was 1m x 0.8m. The location of object was restricted within the occupied zone from zone[2] to zone[6].

Several different arrangements of the object (representing room furniture) was investigated: (I) One object is placed (e.g.B2- the object in zone2); (II) Two objects are placed (e.g. B26 - the objects in zone2 and zone6); (III) Three objects are placed (e.g. B345 - the objects in zones 3, 4 and 5); (IV) Four objects are placed (e.g. B2345 - the objects in zones 2,3,4 and 5). Also, the room configuration is designed in a symmetric way and object arrangement that lies one upon another is avoided (e.g. case B5, it is the same as the case B2 and is thus excluded from the investigation). In each of the cases; B2, B23, B234 and B2345, the volume ratio between the object volume and the room volume was 4.9%, 9.9%, 14.8% and 19.8% respectively.

For evaluating the indices, i.e. $<\varepsilon>_c$ and ε_{cp}, the emission source was placed in the middle of the object length at a height of 1.2m above the floor. The number of emission sources used was the same as the number of objects, however, the amount of source emissions was the same as that for the empty room, i.e. 10ml/s. For the case of an empty room, the emission source is located in the centre of the room.

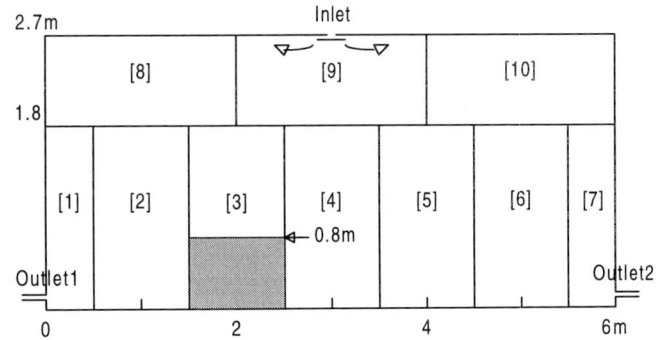

Figure 1: The room configurations and the imaginary zones when an object is placed at zone [3], i.e. B3

RESULTS AND DISCUSSION

Comparison between Mean Ventilation Effectiveness, $<\varepsilon>_{pr}$ and $<\varepsilon>_c$

Figures 2, 3 show different profiles of mean ventilation effectiveness, $<\varepsilon>_{pr}$ and $<\varepsilon>_c$, and the mean velocities in the whole room and in the occupied zone respectively. The trend of mean velocity at two cases is similar to the mean ventilation effectiveness.

As shown in Figure 2, the agreement between $<\varepsilon>_{pr}$ and $<\varepsilon>_c$ is generally good in spite of two distinctive definitions mentioned above. However, for the occupied zone in Figure 3, although the trend of $<\varepsilon>_{pr}$ and $<\varepsilon>_c$ is similar, the values are different. This discrepancy can be explained by the different methods used for calculating the two mean ventilation effectiveness.

$<\varepsilon>_{pr}$ is obtained from the probability distribution function (i.e. Markov Chain Theory) but $<\varepsilon>_c$ is calculated by the sampling distribution function. Therefore, in order to apply the transfer probability in $<\varepsilon>_{pr}$, the interconnecting flow rates between zones should be normalized with respect to the ventilation supply flow rate (Peng & Davidson 1998). Unlike $<\varepsilon>_{pr}$, the method of calculating $<\varepsilon>_c$ is based on the mean value for the

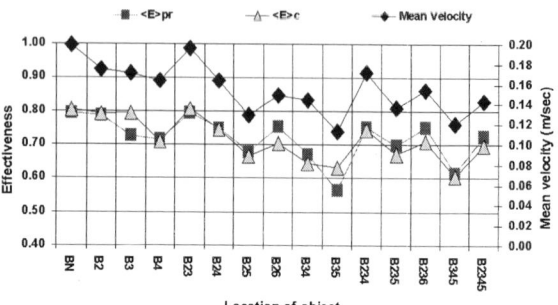

Figure 2: Variation of mean ventilation effectiveness and mean velocity in the room

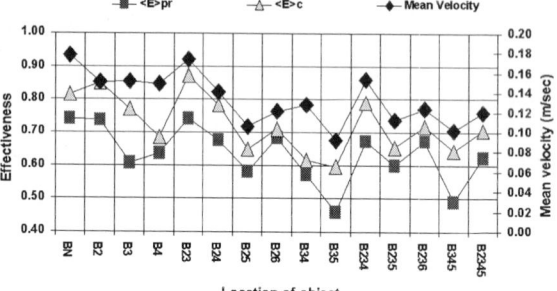

Figure 3: Variation of mean ventilation effectiveness and mean velocity in the occupied zone

occupied zone. In other words, the mean $<C_{op}>$ used for obtaining $<\varepsilon>_c$ is derived from the spatial average of the sum of the dimensionless concentrations in the occupied zone. As shown in Table 1 for the zonal ventilation effectiveness (ε_{cp}), the high and the low values indicate low and high concentration respectively in the zone. In the occupied zone, the difference between the zones with sources and zones without sources is substantial. The emission sources were located at a height of 40 cm above the object and the same number of sources as objects was used. The value obtained is thus dependent upon the location and the number of emission sources. The ventilation effectiveness $<\varepsilon>_c$ was found to be very sensitive to the object location in the occupied zone. In view of characterizing the air flow pattern and the contaminant dilution, the mean ventilation effectiveness $<\varepsilon>_{pr}$ and $<\varepsilon>_c$ can thus complement each other for examining ventilation performance.

Effect of Object Volume and Location on Ventilation Performance

Nielsen et al. (1998) found that the maximum velocity in the occupied zone (v_m) can be reduced by increasing the object volume in the room. As shown in Figure 4, all cases are classified into three regions according to the location of v_m. The v_m in the empty room (BN) occurs in the zone[2] and zone[6], and is higher than that in furnished room cases. For Region (b) where v_m occurs in zone[2], there is a good correlation between the maximum velocity and the increased object volumes. However, the regions (a) and (c) have no such a correlation.

The influence of increased object volume on the ventilation performance is also analyzed by the mean ventilation effectiveness, $<\varepsilon>_{pr}$, and the results can be seen in Figure 5 and Table 1. In order to investigate the effect of the reduction of free volume, four cases with free volumes of 95%, 90%, 85%, and 80% are defined on the basis of the empty room. The results show that $<\varepsilon>_{pr}$ is not proportional to the increase in object volume for all four cases. It is found that an increase in the length of an object and a decrease in the free volume of the room do not always mean a decrease in the ventilation effectiveness.

Table 1 shows the results of $<\varepsilon>_{pr}$ and v_m for the cases of separated and attached objects (i.e. B24 and B234; B25, B235 and B2345; B26 and B236; B35 and B345). There is a slight decrease in $<\varepsilon>_{pr}$ and v_m when the object volumes increase except the cases B35 and B345. Therefore, the object volume can be regarded as a minor parameter when studying its effect on the ventilation performance. Rather, the change in object location has more significant effect on ventilation performance.

As shown in Figure 5 and Table 1, the change in object location influences the airflow pattern and the ventilation effectiveness, $<\varepsilon>_{pr}$. When the object is located in zone[2] (B2), $<\varepsilon>_{pr}$ is similar to that of an empty room (BN) and it is the highest in Case [II]. In cases of B3 and B4, the results of $<\varepsilon>_{pr}$ for the whole room are different from those of $<\varepsilon>_{pr}$ for the occupied zone. In other words, more fresh air is supplied to the occupied zone in case B4 than in case B3. Also, the ventilation effectiveness of the upper zone (above the occupied zone) in case B3 is higher because the

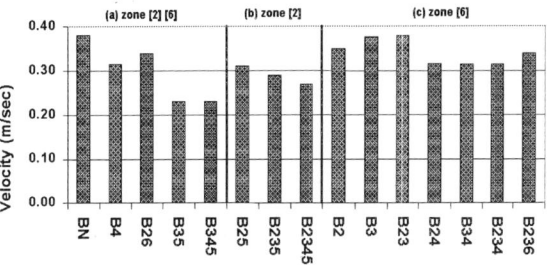

Figure 4: Maximum velocity in the occupied zone according to object location; (a) - v_m occurs at zone [2] and [6], (b) - v_m occurs at zone [2], (c) - v_m occurs at zone[6]

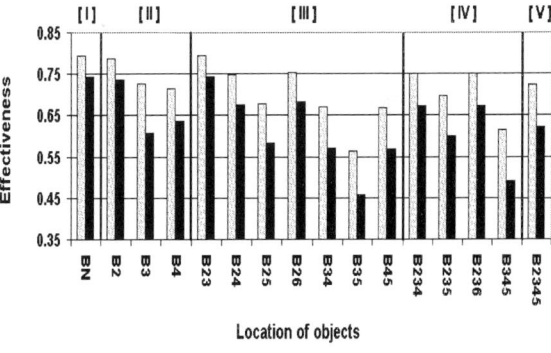

Figure 5: The mean ventilation effectiveness, $<\varepsilon>_{pr}$ for different cases; (I) Empty room, (II) One object, (III) Two objects, (IV) Three objects, (V) Four objects

Table 1: The velocity and the several indices of effectiveness in the occupied zone for different cases

Case	Free Volume	Object Location	Velocity (m/sec) Mean	Velocity (m/sec) Max.	Psp Low	Psp High	Ecp Low	Ecp High	<E>pr	<E>c
[I]	1	BN	0.18	0.38 [2] [6]	0.71	0.76	0.84	0.87	0.74	0.82
[II]	0.95	B2	0.15	0.35 [6]	0.57	0.76	0.32	1.23	0.74	0.85
		B3	0.15	0.38 [6]	0.44	0.75	0.22	1.27	0.61	0.77
		B4	0.15	0.31 [2] [6]	0.36	0.66	0.26	1.99	0.64	0.68
[III]	0.90	B23	0.17	0.38 [6]	0.66	0.74	0.38	1.60	0.74	0.87
		B24	0.14	0.32 [6]	0.62	0.67	0.44	1.90	0.68	0.78
		B25	0.11	0.31 [2]	0.46	0.66	0.39	1.27	0.58	0.65
		B26	0.12	0.34 [2] [6]	0.64	0.68	0.56	0.75	0.68	0.71
		B34	0.13	0.32 [6]	0.37	0.66	0.22	1.46	0.57	0.62
		B35	0.09	0.23 [2][6]	0.40	0.46	0.33	1.09	0.46	0.60
[IV]	0.85	B234	0.15	0.31 [6]	0.62	0.65	0.42	1.92	0.67	0.79
		B235	0.11	0.29 [2]	0.45	0.66	0.46	1.47	0.60	0.65
		B236	0.12	0.34 [6]	0.64	0.66	0.51	0.84	0.67	0.72
		B345	0.10	0.23 [2] [6]	0.37	0.66	0.33	1.10	0.49	0.64
[V]	0.80	B2345	0.12	0.27 [2]	0.48	0.66	0.24	0.82	0.62	0.70

object positioned in zone 3 deflects the main air stream upwards and away from the occupied zone.

When two objects are positioned in different zones (B23), $<\varepsilon>_{pr}$ is equivalent to that for an empty room (BN) and is highest in Case [III]. For the case B23, when the downward air jet attaches to the side walls, the objects in zones 2,3 help guiding the air to the occupied zone.

When the objects are in zone3 and zone5 (i.e.B35), $<\varepsilon>_{pr}$ is the lowest of all other cases, because the objects hinder the main air stream from reaching zones 3,4 and 5 and lead to a deflection of flow upwards in front of objects B3 and B5. The air movement in zones 3, 4 and 5 is stagnant. Through comparison between case B23 and case B35, it is found that, depending on the location of object, the object may play two roles: as a guidance or as a hindrance to air flow. Also, as can be seen in Case [IV], the objects in cases B234 and B345 have respectively guidance and hindrance effect to air flow.

For the case B2, B23 and B234, v_m and $<\varepsilon>_{pr}$ are similar to those for the empty room case (BN) or are respectively higher than the other cases with the same object volume. Therefore, the optimum object position for ventilation performance is thus strongly related to the position where the maximum velocity in the occupied zone occurs. When one puts furniture in an empty room, the region where v_m occurs (e.g. zones 2 and 6) should be considered as positioning the furniture. If the furniture is located in the zone where the velocity is small (e.g. zones 3 and 5), this may create stagnant regions in the room.

CONCLUSIONS

The main findings in this paper are:
1. There is a correlation between the mean ventilation effectiveness, $<\varepsilon>_{pr}$ and $<\varepsilon>_{c.}$
2. An increase in object lengths in a room and a corresponding decrease in free volume of the room do not guarantee that the ventilation effectiveness would decrease.
3. The zone, where the maximum velocity in the occupied zone occurs, can be considered as an optimum object position, i.e. can give good ventilation effectiveness.
4. Depending on the location of object and airflow pattern, the presence of an object may produce two opposite effects either a hindering of the airflow in a room or an improvement in the air distribution.

For future work, experimental measurements and 3D CFD simulations would be required to validate the results obtained from the two- dimensional simulation predicted in this paper.

REFERENCES

Awbi H.B. (1996) *Vortex User Manual - For Airflow, Heat Transfer And Concentration in Enclosures.* The University of Reading, UK

Nielsen J. R, Nielsen P.V. and Svidt K. (1998) The Influence of Furniture on Air Velocity in a Room - Isothermal Case, *Proceedings of 6th International Conference on Air Distribution in Rooms - Roomvent98,* **2,** pp. 281-286

Peng S.H. and Davidson L. (1997) Toward the Determination of Regional Purging Flow rate, *Building and Environment,* **32:6,** pp.513-525

Sandberg M. and Claesson L. (1998) The Effect of the Distribution of Air Blockage of Room Scale - An Experimental Study, *Proceedings of 6th International Conference on Air Distribution in Rooms - Roomvent98,* **2,** pp. 259-266

Air Infiltration and Ventilation Centre (1991) *Air Flow Patterns within Buildings* -Technical Note AIVC 34

Air Distribution in Rooms, (ROOMVENT 2000)
Editor: H.B. Awbi

INTELLIGENT COMFORT CONTROL SYSTEM

W. C. Lo and H. Y. Lai

Department of Electrical and Electronic Engineering, The University of
Hong Kong, Hong Kong SAR, China.

ABSTRACT

Most of the HVAC systems operating nowadays are not able to provide comfort condition for the occupants as they merely maintain a fixed air temperature, which is not adequate to realize the thermal comfort sensation of the occupants. Moreover, a thermostat or comfort sensor mounted on the wall cannot reflect the actual thermal condition in the occupied zone. The Intelligent Comfort Control System (ICCS) is developed to provide the desired comfort based on Predicted Mean Vote (PMV). The ICCS learns and generalizes the thermal comfort distribution of a room under different supply and environmental conditions from CFD simulation results. The ICCS has several input modules, which estimate the clothing insulation of the occupants and the mean-radiant temperature of the room. These modules enable the system to keep track with the environment and thus adjust the supplying parameters to avoid overcooling or overheating. It has been found that the ICCS is able to provide the desired comfort and has a better energy performance than conventional HVAC control strategies.

KEYWORDS

Thermal Comfort, Artificial Intelligence, Neural Network, CFD, HVAC Control, PMV

INTRODUCTION

Dissatisfactions with air-conditioned spaces are prominent and frequent. The reason is conventional HVAC systems are not able to provide a thermal comfort environment as they merely maintain a room fixed air temperature control, which is not adequate to account for the thermal comfort sensation of the occupants. Moreover, due to non-uniform thermal distribution of the conditioned space, a thermostat or comfort sensor mounted on the wall may not reflect the actual condition in the occupied zone.

The Intelligent Comfort Control System (ICCS) intends to provide the desired thermal comfort, based on the predicted mean vote (PMV) (Fanger, 1970), of the occupants. The ICCS has a neural network controller and four input modules which learns and predicts the thermal comfort characteristic of the environment from CFD simulations and keep track with the variation of the clothing and environment respectively.

CFD ROOM MODEL

Most of the HVAC design softwares available include building envelopes and room models for evaluation purpose. These models usually assume the room air is fully mixed and neglect the internal configuration of the room such as the location of supply, exhaust and the heat sources. However, these assumptions will be voided when thermal comfort of the room is being studied due to the non-uniform distribution of air temperature and velocity. Computational Fluid Dynamics (CFD) modeling offer a solution to the mentioned problem. It simulates the airflow pattern and thermal distributions of the room to a much detailed extend thus enabling us to estimate the PMV index at different locations. The fundamentals and techniques in CFD room modeling can be found in (Lo and Lai, 1999)

Fig. 1 shows the geometrical configuration of a 3-dimensional room model built by CFD2000 (Adaptive Research, 1996), with the wall named 'Front Wall' being exposed. The model simulates a typical small office environment with dimensions L×W×H = 5×4×3.05m. Ventilation of the room is provided by a ceiling mounted fan-coil unit and air returns through two exhausts at the back of the room. The nominal supply volume is 300cfm and the fan has three speed levels. It supplies air through a 0.8m × 0.15m grille which has a 70% effective area. The Momentum Method (Chen et. al., 1992) is applied to separately specify the mass flux and momentum of the supplying air jet as shown in Table 1. The room is symmetric about the vertical plane through the center of the ceiling mounted fan-coil unit, so only half of the room is simulated and the flow domain is divided into 25×12×17 cells. The humidity of the room air is assumed to be constant with a value of 50%RH.

FIGURE 1

CONFIGURATION OF THE ROOM MODEL

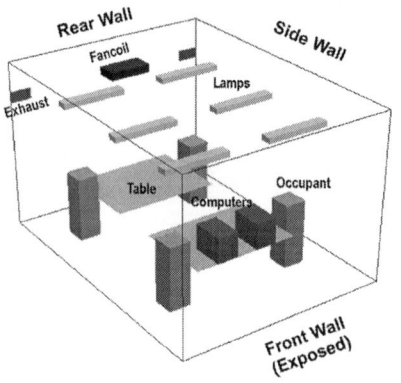

TABLE 1

SUPPLYING AIR PARAMETERS OF THE FAN-COIL

Speed	cfm	Mass Flux (kg/m^{2}s)	Velocity (m/s)	Air Change (ach)
High	300	1.416	1.686	8.36
Med	225	1.062	1.264	6.27
Low	150	0.708	0.843	4.18

TABLE 2

CONVECTIVE HEAT SOURCES IN CFD ROOM MODEL

Heat Source	Total heat	Convective part	Convective output
Occupant	120W	50%	60W
Computer	150W	70%	76.09W
Lamp	2x36W	70%	66.32W

The internal heat sources of the room are modeled by blocks with their corresponding convective heat outputs allocated evenly on each surface. The convective parts of the heat sources are estimated with reference to the volume to surface area ratio as shown in Table 2. The radiative parts of the heat sources will not affect the air temperature but it will affect the wall temperatures due to radiant heat exchange. This effect is handled and calculated by another software called HVACSIM+ externally. By choosing a heat flux boundary condition instead of a surface temperature, the heat transfer from the blocks will be independent of the grid layout in the CFD simulation.

In the model, the applied boundary conditions regarding heat transfer at the walls is the surface temperature. The temperatures of the walls are generated by HVACSIM+ based on the radiative heat flux from the internal sources and the outdoor weather condition.

The thermal comfort of the room can be evaluated by averaging the PMV of every cell in the CFD room model. However, if we have to compute the PMV of all the 5100 cells, it will be too computation expensive. Therefore, we adopted the practical procedure suggested by Fanger (1970). A total of 18 measuring points are chosen, distributed over the occupied zone. Since the room is intended for sedentary persons, the measurements are taken at 0.2, 0.5 and 1.0m above the floor level and averaged to give the PMV of the particular point. The room's PMV at particular supplying and environmental conditions is calculated as the average of the mean votes found from each location.

INTELLIGENT COMFORT CONTROL SYSTEM (ICCS)

Thermal distribution of a room is non-uniform and the distribution shows non-linear relationship to the supplying air parameters. Together with the non-linear characteristics of thermal comfort sensation with its contributing factors, it seems impossible for the traditional HVAC system to provide a comfort condition for the occupants. The proposed ICCS incorporates artificial intelligence into the HVAC control system which learns the non-linear and non-uniform characteristics of the room through detailed CFD investigation on thermal distribution.

The core of the ICCS is a neural network controller based on specialized learning. The inputs of the controller are: clothing insulation (I_{cl}), mean-radiant temperature of the occupied zone (MRT), supplying volume of the fan-coil (cfm) and the desired comfort setting of the occupants (PMV_d). The output of the controller is the required supply air temperature (T_s) which will provide the desired PMV.

Forward Modeling of CFD room

In order to train the neural controller, the calculated PMV of the room has to be compared with the desired value and the error is used to tune the weights and biases of the controller. However, each computation of the steady state results of the CFD model takes about 10 minutes, it will be impossible to use the CFD room model directly to train the neural controller. Therefore, in order to reduce the time needed for training the controller, a neural network model is built to emulate the CFD room.

The PMV index of the CFD room depends on four boundary parameters. Therefore a three layers feedforward network, known to be a universal function approximator (Funahashi, 1989), with 4 input nodes (I_{cl}, MRT, cfm, T_s) and a single linear neuron (PMV) in the output layer is used. The training samples are generated by taking all possible combinations of the following variables:

I_{cl} = [0.5, 0.7, 0.9, 1.1]
MRT = [21.9, 22.6, 22.9, 23.4, 23.9, 24.4, 24.9, 25.5, 25.9, 26.2]
cfm = [150, 225, 300]
T_s = [9, 13, 17, 21, 25]

An extra set of samples combined by T_s = 4 and 29 with the first 5 and last 5 values of the MRT respectively at the above I_{cl} and cfm. The I_{cl} range above represents the variation of office clothings from summer to winter. The values of MRT are calculated based on the wall temperature values generated by the $HVACSIM+$, which include several typical summer and winter weather conditions. The cfm values correspond to the supply volume of the fan-coil at 'Low', 'Med' and 'High' speed respectively. 180 CFD simulations were carried out and the steady state results of the room were used to calculate the room's average PMV with the 4 different clothing insulation values, which resulted in 720 training samples. The hidden layer has 7 tangent-sigmoid neurons which resulted in a mean-square-error of 0.0002 upon 200 training epochs. The neural room model was tested and it showed that the network is able to emulate the CFD model closely.

Specialized learning of ICCS Controller

Figure 2 shows the structure of the direct inverse control scheme used to train the ICCS controller. The neural network room model described previously is used to replace the CFD room and the output will be the PMV index of the room in steady state. The error between the desired PMV and the actual PMV of the room is used to train the neural network controller. The controller takes the comfort factors and the desired PMV index as inputs and outputs the supply air temperature set-point.

FIGURE 2
DIRECT INVERSE CONTROL SCHEME USED TO TRAIN THE ICCS CONTROLLER

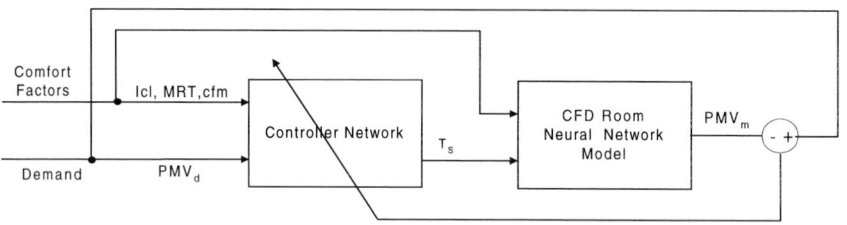

The neural controller is a 2-layer feedforward network, it has 10 hidden tangent-sigmoid neurons, which are determined by trial and error, and 1 output linear neuron. A combination network including both the controller network and room model network is built. The total network has two sets of inputs named as *Comfort Factors* = [I_{cl}, *MRT*, *cfm*] and *Demand* =[PMV_d], and four layers. The inputs *Comfort Factors* and *Demand* are feed into the first layer of the total network and the input set *Comfort Factors* is also connected to the third layer. The third and fourth layer, correspond to the room model network and their weights and biases are copied from the network trained previously. The corresponding learning parameters are set to zero and thus they are not adjusted during the training process.

FIGURE 3
REGRESSION ANALYSIS AND ERRORS OF THE ICCS TRAINING

(a)   (b)

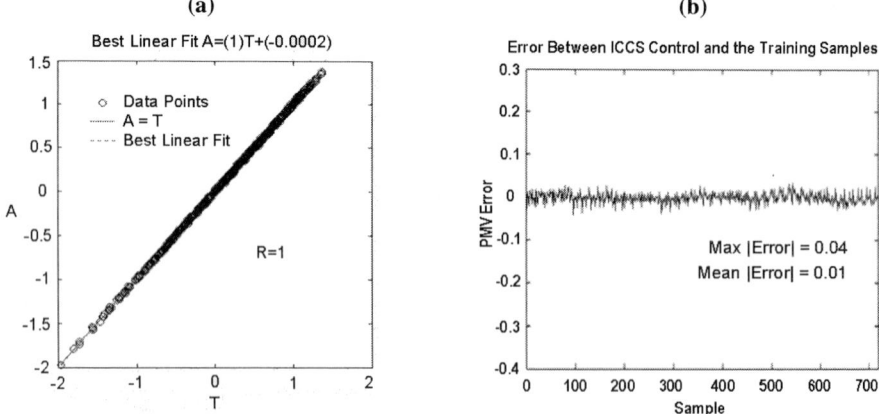

The network was trained using the Levenberg-Marquardt algorithm by 720 training samples. The mean-square error of the total network is 0.0001 after 200 epoch, the time needed is 66 seconds with a Pentium-II 300MHz computer. Figure 3(a) and (b) show the regression analysis and the errors of the training. The regression analysis shows the total network is a good model and has good correlation of the targets and outputs. The maximum and mean of the absolute error from the desired PMV is only 0.04 and 0.01 respectively. The weights and biases of the first and second layer of the total network are then copied to the controller network. The ICCS controller is thus trained and able to control the supply air temperature set-point which provide the desired comfort.

Input Modules of the ICCS

The ICCS has 4 input modules that generate the inputs to the neural controller:

(a) Clothing Insulation Prediction Module – predict the occupants' clothing of the day;
(b) MRT Approximation Module – estimate the mean-radiant temperature of the occupied zone from two wall temperature sensor
(c) Fan-coil Speed Switching Module – automatic supply-volume switching control
(d) Personal Comfort Preference Module – accept the occupants' personal preference setting.

Due to the limited space of this paper, they will not be discussed in detail, further descriptions of the modules can be found in Lai (1999).

ICCS PERFORMANCE

The complete Intelligent Comfort Control System is built by assembling the four input modules with the pre-trained ICCS neural network controller. The whole system is tested using the CFD room model with different weather conditions of the year generated by *HVACSIM+*.

The ICCS monitors the outdoor air temperatures and calculates the average temperature between 7-9 am, then it estimates the clothing insulation value of the occupants of the day. The average mean-radiant temperature of the zone is generated from the wall temperatures of the 'Front' and 'Rear' wall. The desired supply volumes and the comfort preferences of the occupants are accepted through the input modules. The neural controller then outputs the required supply air temperature set-point based on the input parameters. The Supply Volume Switching Module monitors the return air and outdoor air temperature and determines whether to switch into a higher or lower fan speed levels. If the supply volume has been changed, the supply air temperature set-point will be recalculated. The performance of ICCS was compared against conventional HVAC control strategies and the abstract of the results is shown in Table 3.

TABLE 3
COMPARISON BETWEEN FOUR DIFFERENT THERMALCOMFORT CONTROL STRATEGIES

			Set-points	PMV=0	PMV=0	23°C / 25°C	21°C / 24°C	Zone Temp when Control by Wall-mounted Thermostat
Day	Outdoor / °C	Time	Preferred cfm	ICCS	Wall–mounted Comfort Sensor	Return Duct Thermostat	Wall-mounted Thermostat	
Cool	13.3 (7am)	09:00	150	-0.01	-0.17	0.03	-0.46 (300cfm)	19.5 °C
	13.7 (8am)	14:00	150	0.00	-0.16	0.02	-0.41 (300cfm)	19.4 °C
	I_{cl} = 0.9 clo	19:00	150	-0.01	-0.17	0.04	-0.46 (300cfm)	19.5 °C
Warm	23.4 (7am)	09:00	225	0.01	0.02	0.27	0.12	23.8 °C
	23.8 (8am)	14:00	225	-0.01	0.07	0.32	0.13	23.6 °C
	I_{cl} = 0.7 clo	19:00	225	0.00	0.05	0.29	0.12	23.7 °C
Hot	28.4 (7am)	09:00	300	-0.01	-0.32	-0.01	-0.28	23.2 °C
	28.8 (8am)	14:00	300	-0.01	-0.31	0.00	-0.27	23.2 °C
	I_{cl} = 0.52 clo	19:00	300	-0.01	-0.31	0.00	-0.27	23.1 °C

(Data above are the PMV of the occupied zone)

The data inside the shaded area of Table 4 has been produced by a higher *cfm* than preferred, as the cooling capacity at the lower supply volume (150 *cfm*) is not capable to give the desired room temperature set-point (21°C), thus a higher supply volume (300 *cfm*) has been used to generate the results for comparison.

The measured *PMVs* of the simulations show that ICCS has a perfect control of the thermal environment. However, results from conventional HVAC control strategies reflect that there are difficulties to provide a comfort / desired working condition for the occupants even when the desired set-points are maintained. The *PMV* of the occupied zone deviate from the measured value at the comfort sensors, this can be accounted by the non-uniform mean-radiant temperature, air temperature and velocity distributions of the room. The results from the simulation controlled by thermostat show that thermal comfort condition can not be provided by a fixed air temperature control system. It is obvious because this kind of conventional control does not take into account of those environmental parameters which affect human thermal comfort sensation. Moreover, the zone air temperatures listed in Table 3 also conclude that the measured air temperature near at the wall cannot represent the occupied zone air temperature due to the non-uniform temperature distribution of the room.

Comparison on energy performance with conventional HVAC control systems has been conducted and the details can be found in (Lai, 1999). It has been concluded that the ICCS system has a higher energy efficiency while able to provide the desired thermal comfort of the environment accurately. However, it could not conclude exactly how much energy can be saved by operating the ICCS, as it depends on how the building administrator decide the set point of the room and it will be varied with different room configuration.

CONCLUSION

This paper briefly described the proposed Intelligent Comfort Control System which incorporates artificial intelligence to learn and generalize the thermal characteristics of the conditioned space and monitor the change of the environment and occupant's preference. The neural network controller was trained using CFD simulation results and it has demonstrated the power of learning the non-uniformity of the thermal distribution. There are four input modules of the ICCS which generate the inputs to the ICCS. These modules enable the ICCS to estimate the clothing of the occupants and the mean-radiant temperature of the occupied zone. They also accept personal thermal comfort preference of the occupants.. The ICCS was tested against traditional HVAC control systems, it has been shown that the ICCS has a very accurate control of the thermal comfort and is more energy efficient.

Detailed description of the energy performance of ICCS and further investigation on the variation of load pattern can be found in (Lai, 1999).

REFERENCE

Adaptive Research (1996). *Theoretical Background, CFD2000 Ver3.0 Manual*
Chen, Q., Moser, A. and Suter, P.A. (1992). Database for assessing indoor airflow, air quality and draught risk *International Energy Agency (IEA) Annex 20, Air Flow Patterns within Buildings* IEA.
Fanger, P.O. (1970). *Thermal Comfort: Analysis and Applications in Environmental Engineering.* McGraw-Hill Book Company.
Funahashi,K. (1989). On the approximatio realization of continuous mappings by neural networks. *Neural Networks.* Vol.2. p.183-192.
Lai, H. Y. (1999). MPhil Thesis. *Artificial Intelligence based Thermal Comfort Control with CFD Modeling.* The University of Hong Kong.
Lo, W.C. and Lai, H.Y. (1999). Thermal Comfort Control Study by CFD. *The 3rd International Symposium on Heating, Ventilation and air conditioning*, Vol 2. p.802-803

Air Distribution in Rooms, (ROOMVENT 2000)
Editor: H.B. Awbi

PROGRAM FOR CALCULATION OF THE THERMODYNAMIC PARAMETERS OF MOIST AIR

Yuriy Dobryansky

Department of technical sciences, University of Olsztyn
ul. Okrzei 1-A, 10-266 Olsztyn, Poland

ABSTRACT

The program calculates enthalpy, temperature or humidity of moist air if the atmospheric pressure and two of the three above-mentioned parameters are given. The humidity can be represented in any possible way: humidity ratio, relative humidity, wet bulb temperature, dew point temperature, partial pressure of water vapour. Output data also includes all kinds of the humidity parameters as well as pressure and humidity ratio of saturated air. The software is friendly for users: dialogue mode of the data input, diagnostics of incorrect data set, storage of the last data set till the next work session. The software can be used instead of *h-d* psychrometric chart and guarantees better accuracy and wider ranges of parameters.

KEYWORDS

Moist air, temperature, pressure, humidity parameters, software, dialogue.

INTRODUCTION

The well known *h-d* psychrometric charts, which usually are used for calculation of the thermodynamic parameters of moist air, are prepared only for certain values of the atmospheric pressure *B*. The interpolation of output data is necessary in this case. This and other circumstances, inherent to psychrometric chart, make its usage inconvenient. The software has been aimed to make the calculations easier and more accurate.

ALGORITHM

The calculations are based on the following equations and relations (Dobryansky Yu., 1991).
The enthalpy of the moist air:

$$h = 1005\,t + x\,(\,2501730 + 1758{,}5\,t\,),\ \text{J/(kg K)}; \tag{1}$$

The humidity ratio:

$$x = 0.622 \frac{p_{vapour}}{B - p_{vapour}} \text{ , kg H}_2\text{0/kg dry air} \tag{2}$$

The relative humidity:

$$\varphi = \frac{p_{vapour}}{p_s(t)} \tag{3}$$

The pressure of the saturated water vapour:

$$p_s(t) = P_0 \cdot \exp\frac{\alpha_s \cdot t}{\beta_s + t}, \text{ Pa,} \tag{4}$$

for $t \leq 0^\circ C$:	P_0=611,21Pa,	α_s=22,4893,	β_s=272,881
for $0 \leq t \leq 50^\circ C$:	P_0=611,21Pa,	α_s=17,5043,	β_s=241,2
for $50^\circ C \leq t$:	P_0=595,943Pa,	α_s=16,8150,	β_s=227,3977

The wet bulb temperature:

$$t_{w.t.} = f(h, \varphi=1), \tag{5}$$

The dew point:

$$p_s(t_{dew}) = p_{vapour} \tag{6}$$

Calculations of temperature t are accomplished by iteration method, if enthalpy h and relative humidity φ are given.

CHARACTERISTICS OF SOFTWARE

The variation range of the air parameters:
- temperature t: from -190 to +300 °C,
- humidity ratio x: from 0 to 1.0 kg/kg,
- enthalpy h: from $-2 \cdot 10^5$ to $3 \cdot 10^6$ J/kg,
- atmospheric pressure B: without restrictions.

Not all combinations of input data are allowable. Every input data is testing. Prompts of how the input must be corrected are given in case of incorrect data set.

The program calculates enthalpy h, temperature t or humidity of moist air if the atmospheric pressure B and two of the three above-mentioned parameters are given. The humidity can be represented in any possible way: humidity ratio x, relative humidity φ, wet bulb temperature $t_{w.t.}$, dew point t_{dew}, partial pressure of water vapour p_{vapour}. Output data also include every of the kinds of the humidity parameters as well as the pressure of the saturated water vapour $p_s(t)$ and the humidity ratio $x_s(t)$ of the saturated air.

It is possible to continue calculations with retaining of the previous input data and with changing of the value of any of them. It lets to calculate easily basic psychrometric processes: under the constant humidity ratio (x=const) for calculating of the dew point t_{dew}, under the constant enthalpy (h=const) for calculating of the wet bulb temperature $t_{w.t.}$, and under the constant temperature(t=const) for calculating of the parameters of the saturated air x_s and p_s.

An example of computer dialogue screen is shown on figure 1.

The results of the computer calculations are recorded in a file in the form of a table 1 shown below. Table 1 represents six examples of recording of one column from the computer screen as well as two examples of recording of both columns from the computer screen. The table has width of 80 symbols. It permits to print the table as a DOS test in ASCII codes.

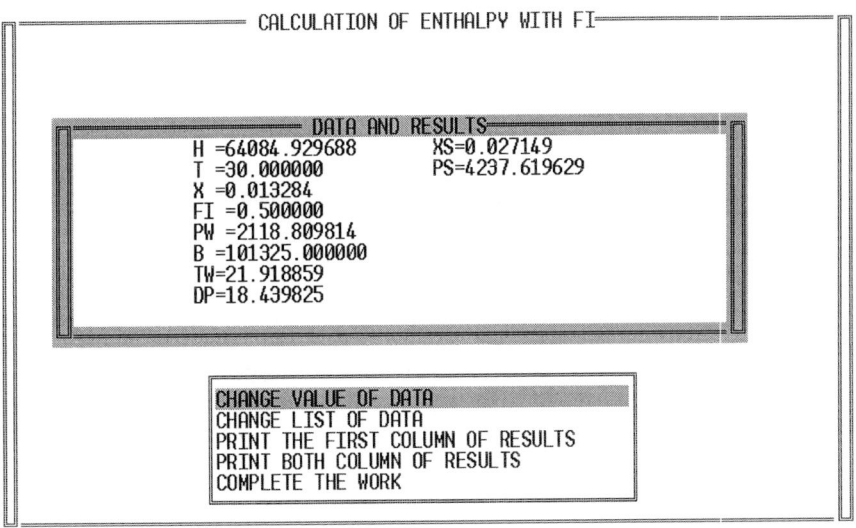

Figure 1. A view of one of the computer's dialogues pictures

TABLE 1
VIEW OF THE TABLE RECORDED IN THE RESULT FILE

```
CALCULATION OF PARAMETERS OF MOIST AIR
      T           H          FI         X          B         TW        DP         PW
{  25.000}    25125.{  .0000}  .000     {  101325.}   8.161  ********    .000
{  25.000}    35074.{  .2000} 3.908E-03{  101325.}  12.403     .476    633.
{  25.000}    45148.{  .4000} 7.865E-03{  101325.}  16.121   10.461  1.265E+03
{  25.000}    55350.{  .6000} 1.187E-02{  101325.}  19.412   16.694  1.898E+03
{  25.000}    65684.{  .8000} 1.593E-02{  101325.}  22.351   21.307  2.531E+03
{  25.000}    76150.{ 1.0000} 2.004E-02{  101325.}  25.000   25.000  3.163E+03
{  25.000}    50233.{  .5000} 9.863E-03{  101325.}  17.815   13.853  1.582E+03
                            XS= 2.004E-02                           PS= 3.163E+03
{  25.000}    76150.{ 1.0000} 2.004E-02{  101325.}  25.000   25.000  3.163E+03
                            XS= 2.004E-02                           PS= 3.163E+03
NOTE: Braces {...} mark input data.
```

The symbols used in the table 1 and in the equations (1)-(6) are:

T – t –temperature,

H – h – enthalpy,

FI – φ – relative humidity,

X – x – humidity ratio,

B – B atmospheric pressure,

TW – $t_{w.t}$ – wet bulb temperature,

DP – t_{dew} – dew point,

PW – p_{vapour} – partial pressure of water vapour,

XS – x_s – humidity ratio of the saturated air,

PS – p_s – pressure of the saturated water vapour.

The accuracy of program calculation in comparison with the data of psychrometric tables (*Ogrzewanie i klimatyzacja*, 1976) is near to round-off error, as it is obvious from table 2.

TABLE 2
ACCURACY OF THE PROGRAM CALCULATIONS

Tempera-ture t, °C	Relative humidity, φ	Partial pressure of water vapour, p_{vapour}:				Accuracy of humidity ratio	Accuracy of enthalpy
		program result		psychromet-ric table, mm Hg	accuracy		
		Pa	mm Hg				
0	0%	0	0	0	-	-	-
0	50%	306	2,2952	2,29	0,23%	0,11%	0,38%
0	100%	611	4,5829	4,58	0,06%	0,13%	0,24%
25	0%	0	0	0	-	-	0,02%
25	50%	1582	11,866	11,9	-0,29%	-0,07%	-0,84%
25	100%	3163	23,724	23,7	0,10%	-0,30%	-0,07%
50	0%	0	0	0	-	-	0,02%
50	50%	6172	46,294	46,3	-0,01%	0,12%	0,16%
50	100%	12340	92,558	92,5	0,06%	0,10%	0,12%

The software has a dialogue mode of the data input. It stores the last data set till the next work session.

The program includes two parts. The calculation part is written in FORTRAN computer language and the dialogue part is written in CI language.

The requirements to hardware: DOS operating system, the free place on hard drive - 100KB, coprocessor.

Software use: ventilation, air conditioning, drying, civil engineering as well as education.

REFERENCE

Dobryansky Yu. (1991). *Computer calculations of the heat and moist regime in the mine workings*, Naukowa dumka, Kiev, Ukraine.

Ogrzewanie i klimatyzacja, (1976), Arkady, Warszawa. (Translation of *Taschenbuch für Heizung und Klimatechnik, 57. Ausgabe.* (1972) by R. Oldenbourg, München).

Air Distribution in Rooms, (ROOMVENT 2000)
Editor: H.B. Awbi

INDOOR AND OUTDOOR AIRFLOW SIMULATION BY A ZERO EQUATION TURBULENCE MODEL

Bin Zhao, Xianting Li, Ying Li and Qisen Yan

Department of Thermal Engineering, Tsinghua University, Beijing, P.R. China

ABSTRACT

In order to simulate indoor air distribution and airflow around buildings quickly and accurately by CFD (Computational Fluid Dynamics) technique, a new zero-equation turbulence model and momentum method for inlet boundary condition are adopted. The new version of STACH-3, a three-dimensional CFD software is developed based on these. An example for outdoor airflow around an isolated building is given as well. For those high-density buildings with complex geometry, the TSM (Two Step Method) is proposed. Comparisons between the measured data and numerical results show that it is economic and engineering satisfied to simulate indoor and outdoor air distribution by the zero-equation turbulence model.

KEYWORDS

CFD, Turbulence Model, Inlet Boundary Condition, Momentum Method, TSM

INTRODUCTION

Nowadays it requires to simulate indoor and outdoor airflow more quickly and accurately in HVAC (Heating, Ventilating and Air Conditioning) industry. However, for most engineers who haven't large capacity computers, the $k - \varepsilon$ turbulence model for indoor and outdoor airflow simulation costs too much time, especially for those non-isothermal problems need LRN (Low-Reynolds Number) $k - \varepsilon$ turbulence model. Furthermore, the conventional method of describing supply opening for indoor airflow simulation can't introduce proper inlet boundary condition. Therefore, the new zero-equation turbulence model developed by Chen et al. (1998) is adopted to simulate indoor and outdoor airflow, with the momentum method (Chen et al., 1991) to describe the inlet boundary condition in the case of indoor airflow. As the model is developed by DNS (Directly Numerical Methods) data, it may get more accurate result. And, as consequence of not adding partial difference equation for Reynolds-Stress equations, it costs much less CPU time and computer capacity than those advanced models such as LRN $k - \varepsilon$ turbulence model.

THE ZERO-EQUATION TURBULENCE MODEL AND STACH-3

Indoor airflow is always mixed convection flow including thermal plumes, wall jets and flows with stratified temperature field. Nielsen pointed out that these flows require different turbulence models (Nielsen, 1998). To simulate natural convection and mix convection quickly and accurately, Chen et al. (1998) develop a new zero-equation turbulence model by DNS data. The model says:

$$\mu_t = 0.03874\rho Vl \tag{1}$$

where l is the distance to the nearest wall. With this, we can get the Reynolds Equations closed. All the control equations can be written in general format as following and details are listed in Table 1:

$$\frac{\partial}{\partial t}(\rho\varphi) + div(\rho \vec{u}\varphi - \Gamma_\varphi grad\varphi) = S_\varphi \tag{2}$$

TABLE 1
THE CONTROL EQUATIONS WITH ZERO-EQUATION TURBULENCE MODEL

φ	Γ_φ	S_φ
1	0	0
u	μ_{eff}	$-\dfrac{\partial p}{\partial x} + \dfrac{\partial}{\partial x}(\mu_{eff}\dfrac{\partial u}{\partial x}) + \dfrac{\partial}{\partial y}(\mu_{eff}\dfrac{\partial v}{\partial x}) + \dfrac{\partial}{\partial z}(\mu_{eff}\dfrac{\partial w}{\partial x}) + g_x(\rho - \rho_{ref})$
v	μ_{eff}	$-\dfrac{\partial p}{\partial y} + \dfrac{\partial}{\partial x}(\mu_{eff}\dfrac{\partial u}{\partial y}) + \dfrac{\partial}{\partial y}(\mu_{eff}\dfrac{\partial v}{\partial y}) + \dfrac{\partial}{\partial z}(\mu_{eff}\dfrac{\partial w}{\partial y}) + g_y(\rho - \rho_{ref})$
w	μ_{eff}	$-\dfrac{\partial p}{\partial z} + \dfrac{\partial}{\partial x}(\mu_{eff}\dfrac{\partial u}{\partial z}) + \dfrac{\partial}{\partial y}(\mu_{eff}\dfrac{\partial v}{\partial z}) + \dfrac{\partial}{\partial z}(\mu_{eff}\dfrac{\partial w}{\partial z}) + g_z(\rho - \rho_{ref})$
h	$\dfrac{\mu_{eff}}{\sigma_h}$	$S_h \qquad \sigma_h:1.0$
	$\mu_{eff} = \mu_l + \mu_t$	$\mu_t = 0.03874\rho Vl$

Based on these equations, the new version of STACH-3 is developed. The finite volume method and SIMPLE algorithm are adopted in the software. The boundary condition of walls is treated as source term of momentum equations and energy equation by wall function. (Zhao et al., 1999(b))

INDOOR AIRFLOW

Momentum Method for Inlet Boundary Condition Description

To introduce the correct momentum flow of inlets into the room, the momentum method (Chen et al., 1991) is used for indoor airflow. That is:

$$Jin = m\frac{L}{Ae} = m\cdot\frac{L}{A}\div\frac{Ae}{A} = m\cdot\frac{L}{A}\div\frac{Ae}{A} = m\cdot\frac{L}{A}\div f \tag{3}$$

where f is the ratio of effective area Ae to gross area A of the supply openings or diffusers, Jin is inlet momentum flow rate, m and L are inlet mass and volume flow rate respectively. Emvin et al. (1996) pointed out that the momentum method is not proper for refined grids. But to the new zero-equation model and engineering application, the grids are coarse enough to get less error.

Non-isothermal Ventilation

To compare with the measured data, we simulate the non-isothermal ventilation case in which the experiment was made by H. B. AWBI (AWBI, 1989) with STACH-3. The test room has a square floor of length 4.2m and height 2.8m. The air is supplied from a 24mm continuous slot diffuser in the ceiling spanning the width of the room and at a distance 1.2m from the wall. The room load was produced by electrically heated tapes laid over the floor area. So it's a 2-D case with uniform load distribution. Figure 1 shows the velocity and temperature distribution in the occupied zone of the room for 60 l/s/m flow rate. The numerical results agree very well with the measured data, where the maximal difference of velocity and temperature is only 0.08m/s and 0.8° C respectively. The mesh for the case is 28• 17• 3, which only costs 5 minutes to get convergence on a PIII 450 personal computer with 128M RAM. Therefore, it's cheap and quickly to simulate many cases in designing process for engineering application by STACH-3.

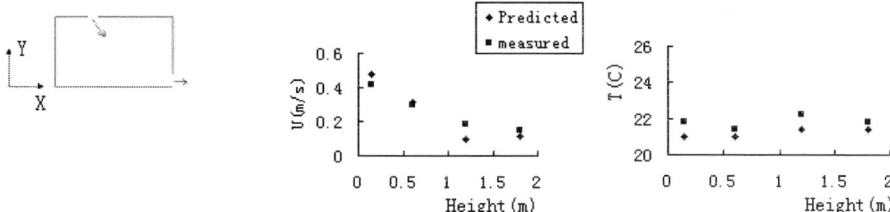

Fig.1 Mean velocity and temperature in the occupied zone

Displacement Ventilation

A displacement ventilation case is also simulated by STACH-3. The measured data is from the experiment of Yuan et al. (1998). There are 1 supply diffuser, 1 exhaust, 2 occupants, 2 computers, 2 tables, 2 boxes, and 6 lamps in the room. Figure 2 shows the comparison between the measured and calculated airflow pattern at Z=1.8m, while Figure 3 illustrates the calculated mean velocity and temperature VS. measured data at X=0.78m, Z=1.83m. They both show good agreement between the numerical and experimental results. The maximal difference of mean velocity and temperature is only 0.05m/s and 1.5° C respectively. For the mesh of 26• 26• 24, it costs about 12 hours to get convergence on the computer mentioned before. More comparisons at other place show that it is really engineering satisfied to simulate the problem by the new zero-equation turbulence model with momentum method to describe the inlet boundary condition.

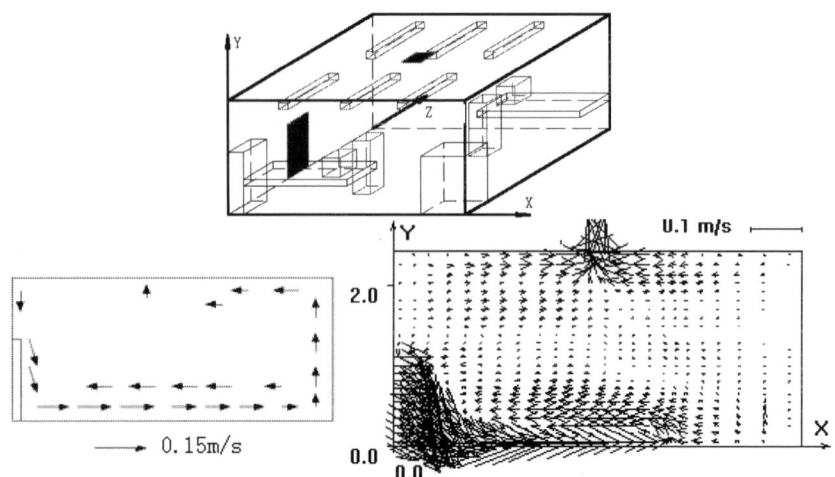

Fig.2 Airflow pattern at Z=1.8m (left: measured, right: calculated)

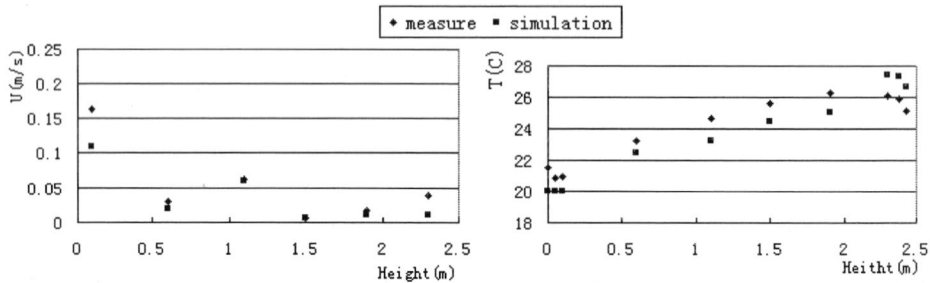

Fig. 3 Mean velocity and temperature distribution at X=0.78m, Z=1.78m

OUTDOOR AIRFLOW

Airflow around an Isolated Building

Airflow around buildings is also important for HVAC engineers. Here we use STACH-3 to simulate the airflow around an isolated building to validate the zero-equation turbulence model for outdoor airflow. The case is experimented by SHUZO et al. (1988) in wind tunnel. Figure 4, 5 shows the comparison between measured and simulated airflow pattern of the vertical and horizontal section at the middle of the building, respectively.

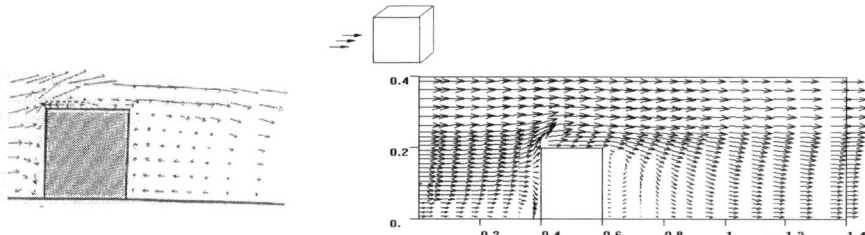

Fig.4 Airflow of the vertical section at the middle of the building (left: measured, right: simulated)

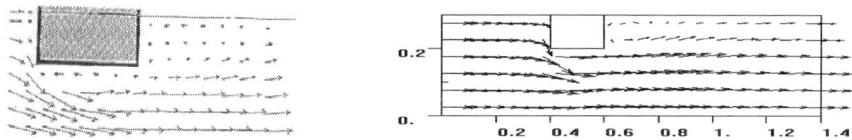

Fig.5 Airflow of the horizontal section at the middle of the building (l: measured, r: simulated)

The figures show that the zero-equation turbulence model can simulate the stagnation point at about 2/3 of the building's height and circumfluence behind the building rightly. On the same computer, it costs only 1 hours to get convergence for a mesh of 26• 28• 8.

TSM for High-Density Buildings with Complex Geometry

For those high-density building clusters, especially buildings with complicated geometry, a new method called TSM (Two Step Method, Zhao et al., 1999(a)) can be used. In TSM, the computational process is divided into two orders. In the first order, the chosen computational domain is large enough, maybe several times larger than the building area. Thus the boundary condition is easy to describe, as the buildings have no influence on it. Complicated-geometry buildings are simplified as regular blocks. In the second order, the computational domain is just the area we concerned. The boundary condition is provided by the results of the first order with interpolation method. Figure 6 shows the process. Detailed information about TSM can be found in reference (Zhao et al., 1999(b)).

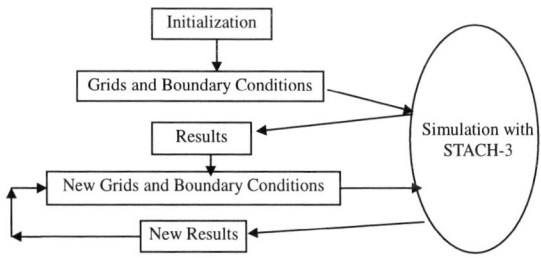

Fig.6 Flow sheet of TSM method

CONCLUSION

Non-isothermal and displacement ventilation simulations were made with STACH-3 which is based on the zero-equation turbulence model and momentum method of inlet boundary condition description. The results agree well with the measured data. The same thing happens to outdoor

454

airflow simulation. The conclusion that can be drawn is that it is effective and economic to simulate these problems by the zero-equation turbulence model. It takes rather short CPU time to get engineering satisfied results.

References

AWBI H.B. (1989). Application of Computational Fluid Dynamics in Room Ventilation, *Building and Environment*, **24:1**, 73-84

Chen Qingyan and Alfred Moser. (1991). Simulation of A Multiple-Nozzle Diffuser, *Proceedings of the 12th AIVC Conference on Air Movement and Ventilation Control within buildings*, **2**, 1-13

Chen Qingyan and Xu Weiran. (1998). A Zero-equation turbulence model for indoor air flow simulation, *Energy and Building* **28**, 137-144

Emvin Peter and Davidson Lars. (1996). A Numerical Comparison of Three Inlet Approximations of the Diffuser in Case E1 Annex20, *ROOMVENT'96*, 222

MURAKAMI, SHUZO and AKASHI MOCHIDA. (1988). 3-D Numerical Simulation of Airflow around a Cubic Model by Means of the k-ε Model. *Journal of Wind Engineering and Industrial Aerodynamics*, **30**, 283-303

Nielsen, P. V. (1998). The Selection of Turbulence Models for Prediction of Room Airflow, *ASHRAE Transactions*, **104:1,** 1119-1127

Yuan Xiaoxiong, Chen Qingyan, and Glicksman R.. (1998). Performance Evaluation and Development of Design Guidelines for Displacement Ventilation, *Final Report to ASHRAE TC 5.3*, 151-154

Zhao Bin, Li Xianting, and Li Ying (1999(b)). The manual of STACH-3 (new version), *Inner Report of Dept. Thermal Engineering*, Tsinghua University, P.R. China, 1-10

Zhao Bin, Li Ying, Li Xianting and Yan Qisen (1999(a)). Numerical Analysis of Wind Effect on High-Density Buildings with TSM, *Proceedings of Building Simulation'99*, **2**, 823-831

Air Distribution in Rooms, (ROOMVENT 2000)
Editor: H.B. Awbi

IMPROVEMENT OF CFD APPLICATION IN VENTILATED ENCLOSURES - A TEST CASE

C. Teodosiu[1], G. Rusaouen[1] and S. Laporthe[1]

[1]Centre de Thermique de Lyon (CETHIL), UPRES A CNRS 5008
Equipe Thermique du Bâtiment, Institut National des Sciences Appliquées
(INSA), Bât. 307, 20 avenue A. Einstein 69621 Villeurbanne Cedex – France

ABSTRACT

The aim of this study is to improve the utilization of CFD approach in the applications of air conditioning technology. More precisely, to establish principles and recommendations to follow in order to design air distribution systems in small enclosures at low room air changes per hour by means of CFD technique. By the use of a commercial code, Fluent, the accuracy and reliability of such a numerical simulation are elucidated in this work for a mixing ventilation system; the air supply terminal is a commercial diffuser which creates a complicated 3D - wall jet below the ceiling. We focus on the factors which have a major impact on the simulations: the description of the computational domain (particular emphasis is put on the supply airflow conditions), the turbulence model and the near wall treatment. Comparisons between predicted and measured values are given in terms of mean air velocity and temperature. The validation of the simulations is completed by an analysis related to analytical expressions for the velocity profile and centerline velocity decay of a three dimensional jet.

KEYWORDS

Numerical simulation, CFD, ventilation, turbulence models, near-wall flow, 3D –wall jet.

INTRODUCTION

At the beginning of the 21st century it is clear that the provision of a good air quality inside the ventilated enclosures as well as an acceptable level of energy consumption for this can not be achieved without modern computational techniques. In conjunction with the progress made in computer hardware, the CFD approach becomes more and more appropriate for the calculations concerning air distribution prediction in rooms, indoor air quality and thermal comfort. However, there is a need to ensure that these computational tools are correctly applied and the purpose of this study is to test the precision of such a numerical simulation by comparing the results obtained with those from experiments. In addition, this paper marked several guidelines that lead to a successful computation. The case taken into account is a mixing ventilation system for small enclosures at low room air changes per hour. We present a brief description of the experimental set-up in the next section.

PHYSICAL MODEL

The physical model is the test room 'MINIBAT' of the Thermal Centre Lyon – 'CETHIL'. Figure 1 illustrates the geometry of the model. The air supply terminal is represented by a commercial diffuser (a grille having an aspect ratio of 12.5) that was placed after a plenum – see figure 1.

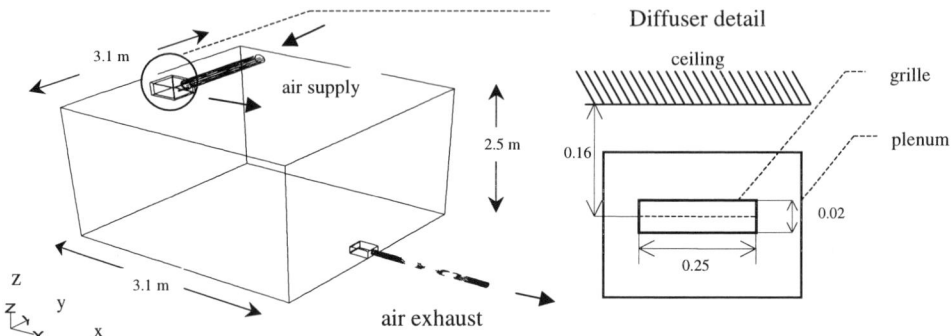

Figure 1: Test room MINIBAT and its ventilation system

Detailed descriptions of the test room and the diffuser are given in Castanet (1998). Measurements have been carried out in order to have a complete description of the boundary conditions (temperature and flow rate of the supply air, surface temperature on the inside walls). On the other hand, the experimentation allowed us to determine the velocity and temperature air distributions inside of the room in a vertical plane normal to the centre line of the air terminal devices. The experiments covered a range of Reynolds number, based on inlet dimensions, between 7000 and 14000 (namely 1-2 h^{-1} air changes rates) and that for non-isothermal conditions, the values of Archimede's number were in the range: $0.0032 - 0.015$.

NUMERICAL MODEL

The simulations were carried out by the use of the commercial CFD code Fluent – version 5.0 that is a general purpose, finite volume, Navier-Stokes solver – Fluent (1998).

In order to predict the airflow within the physical model already presented, we assume that the air is a Newtonian fluid, incompressible that has a constant viscosity and is affected by the gravity force. In the same time, the flow presents the following characteristics: 3D, turbulent, steady and non-isotherm. In our work, we focused on the main elements that have a major influence on the simulations. These key factors are discussed in the next sections.

Computational domain description and supply air flow conditions

This point is a critical one in the prediction of airflow in a ventilated room because it is well known that the flow in an enclosure is mainly governed by its properties at the supply inlet.

We considered several methods in order to reach a better description of the air diffuser and its inlet boundary conditions. First, we have taken into account the simplified methods presented in the studies within the work of the IEA (Annex 20 – 1993), Heikkinen (1991). Unfortunately, the results obtained, by applying these methods, in terms of air velocity in the jet zone were in disagreement with the experimentation. This surely means that these methods are related to a certain type of air terminal device. In fact, after a thorough study, we noticed the basic elements that must be taken into account for a successful simulation in our case. It is possible to replace the complicated diffuser with a simple opening but this opening must have the same aspect ratio as the real diffuser and must be located in the middle of the real diffuser. (The jet is a 3D-wall jet and every variation concerning the aspect ratio and the distance from the ceiling determines an altered behaviour of the jet). Moreover, in order to have a properly formed jet, we changed the real geometry (figure 2): we specified the inlet boundary conditions at a larger distance than the position of the virtual origin of the jet. Also, we imposed a thickness (0.02 m) for the air terminal device in order to have more computational cells in this region which represents now a flow passage.

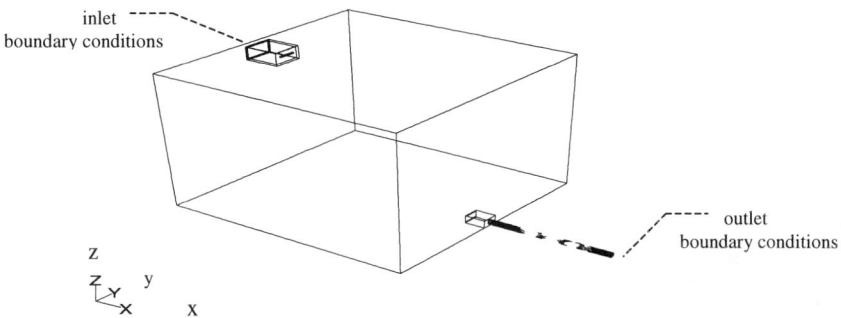

Figure 2: Numerical model geometry

Choice of turbulence model

The reliable representation of turbulence in airflow simulation modelling has proved to be an extremely difficult task and it is obvious that there is no single turbulence model which is able to cover all classes of problems. Besides, there are several parameters which could have an effect on the choice of a turbulence model such as the practice achieved for a specific domain of applications and the available computational resources.

There are four classes of turbulence models in Fluent: one-equation model (Spalart-Allmaras), two-equation models (k-ε), second order closure model (Reynolds stress model) and the large eddy simulation model (LES).

It is known that one-equation models are not always able to give accurate results when the flow changes rapidly from a wall-bounded to a free shear flow. On the other hand, the second order closure models have not found a very large application in simulations concerning ventilation problems since they don't allowed much better results than the k-ε model despite their complexity. If we add the more computer power required in the case of the stress models as well as for LES methods, we can understand why the two equation turbulence models still remain the preferred approaches for numerical simulation of ventilation flows even that the LES models become more and more of interest – Davidson & Nielsen (1996).

In our study, we used a variant of the well-known standard k-ε model – Launder & Spalding (1972), the so-called 'realizable' k-ε model which demand very little more computational effort than the standard k-ε model. This revised k-ε model includes new formulations for the turbulent viscosity and for the dissipation rate transport equation – for details see reference Fluent (1998). The comparisons carried out during our work between the standard k-ε model and its 'realizable' version have shown that this model leads to the best spreading rate predictions of the jet and this was a crucial point in our simulations knowing the particular behaviour of a 3D dimensional wall jet.

Near wall treatment

The correct specification of boundary conditions is a vital part of a numerical calculation of indoor airflow. There are always difficulties regarding the description of the near-wall region where the turbulent flows are affected by the damping influence of the solid walls. There are two techniques to deal with the near-wall region: the classical logarithmic wall functions and the two-layer approach.

The application of wall functions fails in our case as separation occurs – especially at the ceiling. Therefore we abandoned this approach despite its advantage (savings in computational effort) and we employed the two-layer near-wall model. This approach is based on the partition of the domain into a viscosity affected region and a fully turbulent region. The separation of the two regions is established by the turbulent Reynolds number (Re_y) – based on the normal distance from the wall at the cell. In the viscosity near-wall region (where $Re_y < 200$), a one-equation model is employed. Hence, in this case, the rate of dissipation of the turbulent kinetic energy (ε) is obtained algebraically by the use of length scales. The equations that allow to compute the turbulent viscosity and ε, as well as the formulas employed for the length scales can be found out in Fluent (1998).

Boundary conditions

- flow inlet boundaries:

velocity: fixed value across the opening located at the entree of the plenum (figure 2). The velocity component perpendicular to the opening was determined as the ratio of the measured airflow rate to the area of the opening.

temperature: uniform value using the experimental data

turbulence quantities: uniform specification, defining two parameters: the turbulence intensity – I, based on an empirical correlation for pipe flows, assuming a fully developed duct flow upstream, $I = 0.16(Re_{D_H})^{-1/8}$ (%) and the hydraulic diameter - D_H (m). The relationships used in order to compute the turbulent kinetic energy (k) and its rate of dissipation (ε) based on the I and D_H are presented in reference Fluent (1998). A brief parametric study was completed in order to see the influence of the inlet boundary conditions on the jet development. We noticed that the values specified for the turbulent parameters at the entree of the domain do not notably influence the flow pattern. At the contrary, the impact of the dynamic inlet boundary conditions on the flow within the enclosure is important. These two remarks are in accordance with the conclusions of other studies – Joubert (1996).

- wall boundary conditions: the thermal boundary conditions were imposed as fixed values of temperature at the wall internal surface using the measured values.

- outflow boundaries: a zero diffusion flux for all flow variables (except pressure). This condition presumes a fully developed flow and we located the outflow boundary away from the real exhaust device in order to fulfil this supposition – see figure 2.

Numerical scheme

The discretisation of the computational domain was realised by means of tetrahedral mesh elements and the main characteristics of our numerical computations were (the solver uses a co-located scheme, consequently the pressure and the velocity are both stored at cell centres):
- implicit-integration version of the solver
- diffusion terms are second-order central-differenced
- second-order upwind scheme for convective terms in order to reduce the numerical diffusion
- SIMPLE algorithm for the velocity-pressure coupling method

RESULTS

The experimental and numerical results were obtained for hot, cold as well as for isothermal jets. In our paper we developed only the comparisons for a heated jet (two room air changes per hour), the values of the Reynolds and Archimede's numbers being 13566 and 0.0032, respectively. It is worthwhile to note that the conclusions regarding this case are pertinent too for the rest of the configurations tested.

The comparisons between predicted and measured values are shown in figure 3 in terms of air mean velocity and temperature profiles in a median vertical plane for three sections located at different distances from the coordinate system presented in the figure 1.

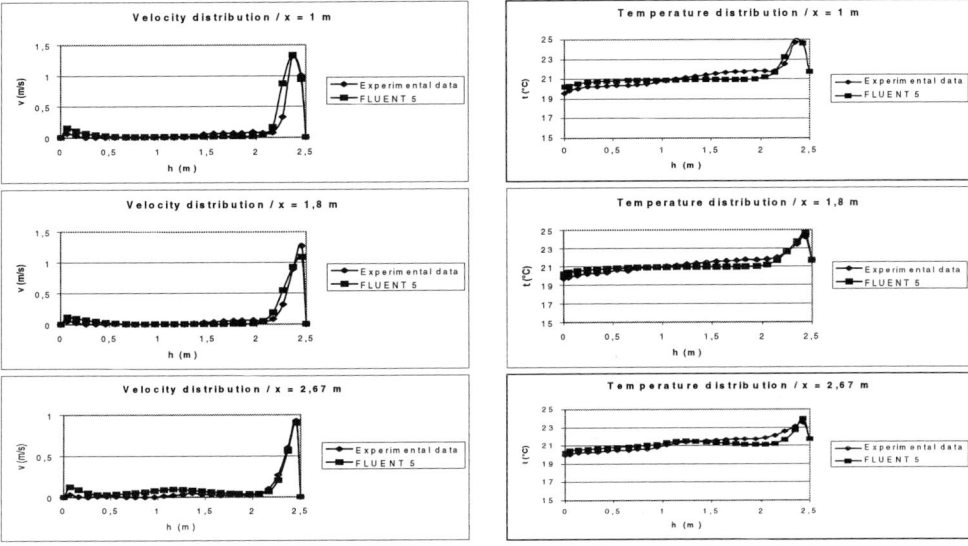

Figure 3: Vertical mean velocity and temperature profiles

They are related to the characteristic and axisymmetric decay region of the 3D jet. In fact, the first one (x = 1 m) is placed at the beginning of the characteristic region, the next one (at x = 1.8 m) is situated at the end of the same characteristic region and the last one (at x = 2.67 m) represents the axisymmetric zone of the jet. The length estimation of these regions was based on the study Trentacoste (1967). We completed our comparisons by a velocity profile jet obtained using empirical formulae. First, we determined the rise of the hot jet (the trajectory) applying the equation obtained by Frean and Billington presented in Awbi (1995). Afterwards, we employed the Sfeir's velocity decay equation – Sfeir (1976) - with a value of 6.25 for the throw constant, K_V. Finally, using the expression proposed by Sforza (1977), we obtained the velocity profile, taking into account the following relationship for the distance from the centre line of the jet where the velocity is equal to half the maximum velocity: 0.1 x – where x represents the distance from the opening. Figure 4 illustrates a comparison between numerical, experimental and empirical velocity profile of the jet at 1.4 meters from the air supply, in fact in the fully developed flow region of the three-dimensional jet. Based on the results shown in the figure 3, it can be demonstrated that the model agrees well to the experimental data, especially when compared with the thermal field (3,5 % maximal difference between simulated and measured values). Regarding the dynamic field, it can be concluded that the jet region is quite correctly predicted. Moreover, the penetration length is accurate performed and that means that the effect of the opposite wall is well taking into account. This fact provides an exact flow pattern in the room. However, the detailed comparison presented in the figure 4 reveals a larger spread of the numerical jet in comparison

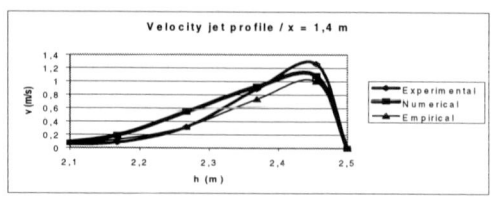

Figure 4: Velocity jet profile

with the experiment and empirical formulations. In our opinion, this is due to the turbulence model. On the other hand, in the region close to the ceiling where complex flows take place, the jet profile is properly numerical represented. This demonstrates that the simulations carried out in terms of near wall treatment and mesh generation in that particular region were correctly performed.

CONCLUSIONS

The test case analysed in this paper demonstrates that the representation of a complicated commercial diffuser and the prediction of a complex flow in small ventilated enclosure can be properly reached by the means of CFD simulation. There is a good agreement between numerical and experimental data. Therefore, the simulated solutions can be used to complete the measurements in works dealing with comfort studies in the occupied zone, ventilation efficiency, as well as prediction of pollutant diffusion in rooms. On the other hand, this study proves once more that the CFD approach is an interesting tool in the design of ventilation systems. The purpose of the guidelines presented here (the simplified description of the inlet conditions, the choice of an adequate turbulence model, the near wall treatment) is to improve the application of this technique in the design of air distribution systems for small ventilated rooms.

REFERENCES

Awbi H.B. (1995). *Ventilation of Buildings*, E & FN Spon, London, UK

Castanet S. (1998). *Contribution à l'étude de la ventilation et de la qualité de l'air interieur des locaux.* Ph.D. thesis. Insa de Lyon, France

Davidson L. and Nielsen P. V. (1996). *Large Eddy Simulations of the Flow in a Three-Dimensional Ventilated Room.* ROOMVENT '96, 161-168.

Fluent User's Guide, Version 5.0. (1998). Fluent Inc., Lebanon – NH, USA

Heikkinen J. (1991). *Modelling of a Supply Air Terminal for Room Air Flow Simulation.* 12th AIVC Conference, 213-230.

Joubert P.; Sandu A.; Béghein C.; Allard F. (1996). *Numerical Study of the Influence of Inlet Boundary Conditions on the Air Movement in a Ventilated Enclosure.* ROOMVENT '96, 235-242.

Launder B.E. and Spalding D.B. (1972). *Lectures in Mathematical Models of Turbulence*, Academic Press, London, England

Sfeir A.A. (1976). *The Velocity and Temperature Fields of Rectangular Jets.* Int. J. Heat Mass Transfer, Vol. 19, 1289-1297.

Sforza P. (1977). *Three-dimensional Free Jets and Wall jets: Applications to Heating and Ventilation.* Int. Seminar of the Int. Center of Heat and mass Transfer – Dubrovnik, 283-295.

Trentacoste N. and Sforza P. (1967). *Further Experimental Results for Three-dimensional free Jets.* AIAA Journal, Vol. 5, 885-891.

Air Distribution in Rooms, (ROOMVENT 2000)
Editor: H.B. Awbi

LARGE EDDY SIMULATION APPROACH FOR NON-ISOTHERMAL AIRFLOWS IN PARTITIONED ROOMS

P. Joubert [1], A. Sergent [1], P. Le Quéré [2] and F. Allard[1]

[1] LEPTAB, Univ. La Rochelle, 17026 La Rochelle, FRANCE
[2] LIMSI-CNRS, BP 133, 91403 Orsay cedex, FRANCE

ABSTRACT

We present in this paper a local subgrid viscosity model for Large Eddy Simulation approach of non-isothermal flows in cavities. This model is derived from the so-called "Mixed Scale Model" for subgrid viscosity, and allows to evaluate independently the subgrid thermal diffusivity and the subgrid viscosity at each grid point. We present here, as a first step, 2D results for turbulent flows in cavities in order to evaluate the performance of the proposed model when compared to a classical Reynolds analogy, based on a constant subgrid Prandtl number.

KEYWORDS
Large Eddy Simulation, time-dependent natural convection, partitioned room, subgrid scale model.

INTRODUCTION

Because conditions encountered in heated rooms can lead to non-stationary airflows, Large Eddy Simulation (LES) approach is a good alternative to Direct Numerical Simulation (DNS) approach which is too expensive when considering complex situations. In LES the dynamic of the flow is resolved as in DNS, but only for the large spatial scales of the motion. The effect of the filtered small scales is generally modelled by introducing a subgrid viscosity concept, based on the Boussinesq analogy with molecular viscosity.

LES is widely used in fluid mechanic problems, such as transitional or turbulent developed forced flows. Some applications can also be found for typical ventilated isothermal rooms (Bennetsen et al. 1996) (Davidson & Nielsen 1996), but at our knowledge, nothing have been done for pure natural convection which is an essential mechanism of motion in buildings. We present hereafter a comparison between results obtained with a LES approach for turbulent buoyant flows in cavities. The subgrid viscosity is evaluated with the "Mixed Scale Model" proposed by Sagaut et al. (1996). Two subgrid diffusivity models are used, a model based on a constant subgrid Prandtl number and a local diffusivity model we have adapted from the Mixed Scale Model. In this comparative study, which is a first step towards realistic situations, LES results have been obtained for the case of a 2D cavity differentially heated on its vertical sides and compared to DNS. Then, we present some results for the case of two rooms heated and cooled on their opposite sides and separated by a lintel, as a future application.

LARGE EDDY SIMULATION APPROACH

Governing equations

The flow in the cavity is governed by the momentum and energy equations under Boussinesq assumption. Introducing the components u_i of the velocity vector, one obtains the classical form of the filtered equations for LES, where overbared quantities stand for the resolved scales of the fields :

$$
\begin{cases}
\dfrac{\partial \overline{u_i}}{\partial x_i} = 0 \\[2ex]
\dfrac{\partial \overline{u_i}}{\partial t} + \dfrac{\partial \overline{u_i}\,\overline{u_j}}{\partial x_j} = -\dfrac{\partial \overline{P}^*}{\partial x_i} + \dfrac{\partial}{\partial x_j}\left[\left(\dfrac{Pr}{Ra^{1/2}} + \upsilon_{sgs} \right) \left(\dfrac{\partial \overline{u_i}}{\partial x_j} + \dfrac{\partial \overline{u_j}}{\partial x_i} \right) \right] + Pr\,\delta_{i2}\overline{\theta} \\[2ex]
\dfrac{\partial \overline{\theta}}{\partial t} + \dfrac{\partial \overline{\theta}\,\overline{u_j}}{\partial x_j} = \dfrac{\partial}{\partial x_j}\left[\left(\dfrac{1}{Ra^{1/2}} + \kappa_{sgs} \right) \dfrac{\partial \overline{\theta}}{\partial x_j} \right]
\end{cases}
\tag{1}
$$

Ra is the Rayleigh number defined on the height H of the cavity, Pr is the air Prandtl number ($Pr =$ 0.71). The reference quantities are H (lengths), $\dfrac{\kappa}{H}Ra^{1/2}$ (velocities) and $\dfrac{T-T_{mean}}{T_h - T_c}$ the characteristic difference of temperature (θ). κ is the thermal diffusivity of air. The dimensionless subgrid viscosity, ν_{sgs} is defined as $\tau^*_{ij} = -2\upsilon_{sgs}\overline{S}_{ij}$, with $\tau^*_{ij} = \tau_{ij} - \dfrac{\delta_{ij}}{3}\tau_{kk}$, $\tau_{ij} = \overline{u_i u_j} - \overline{u_i}\,\overline{u_j}$ and $\overline{S}_{ij} = \dfrac{1}{2}\left(\dfrac{\partial \overline{u_i}}{\partial x_j} + \dfrac{\partial \overline{u_j}}{\partial x_i} \right)$. The dimensionless thermal subgrid diffusivity, κ_{sm} is defined as $\pi_{i\theta} = -2\kappa_{sgs}\dfrac{\partial \overline{\theta}}{\partial x_j}$, where $\pi_{i\theta} = \overline{u_i \theta} - \overline{u_i}\,\overline{\theta}$.

Subgrid scale models

Subgrid viscosity model

The subgrid scale viscosity is evaluated in each point of the grid with the "Mixed Scale Model" (Sagaut et al., 1996). In this model, the subgrid scale viscosity is defined as :

$$
\upsilon_{sgs} = C|\overline{S}|^{\alpha}\, q'^{\frac{1-\alpha}{2}}\, \Delta^{1+\alpha} \quad \text{with} \quad 0 \le \alpha \le 1
\tag{2}
$$

$q' = \dfrac{1}{2}\left(\overline{u_i}\right)'\left(\overline{u_i}\right)'$ is the remaining kinetic energy at the cut-off ; $|\overline{S}| = \sqrt{2\overline{S}_{ij}\overline{S}_{ij}}$ and Δ is the size of the grid mesh, acting as the implicit filter. Following Bardina's similarity hypothesis (Bardina et al. 1980), the lowest fluctuating scales of the resolved fields $\left(\overline{u_i}\right)'$ can be extracted from the resolved scales by employing a test filter, ($\wedge$), with a characteristic cut-off lengthscale larger than the implicit filter of the mesh. Finally : $\left(\overline{u_i}\right)' = \left(\hat{\overline{u}}_i\right) - \left(\overline{u_i}\right)$. The main interest of this model is that the small scale dependency allows the model to vanish in fully resolved regions of the flow and at solid boundaries.

Subgrid thermal diffusivity

A classical way for evaluating κ_{sgs} is to use the Reynolds analogy : $\kappa_{sgs} = \dfrac{\nu_{sgs}}{Pr_{sgs}}$, where Pr_{sgs} stands for a subgrid Prandtl number. In this case, the subgrid diffusivity is driven by the dynamic of the flow, which is correct for forced boundary layers, but not for natural boundary layers for which the flow is governed by the heat flux at the wall, rather than the shear stress at the wall as for dynamic boundary layers. Therefore we propose hereafter a local subgrid diffusivity model based on the thermal characteristics of the flow, in order to evaluate independently the two subgrid quantities at each grid point.

The homogeneity of formulation relating the subgrid viscosity to the diffusivity is respected by mixing contributions based on the gradients of the resolved scales of temperature and velocity, and terms based on the subgrid heat flux Φ_{sgs}. The property that allows the subgrid quantities to vanish in fully resolved regions of the flow and at solid boundaries is thus preserved. The general form of this model is :

$$\kappa_{sgs} = C \cdot \frac{\Delta^{2\alpha+1}}{\Delta\theta} |\overline{T}|^{\alpha} |\Phi_{sgs}|^{1-\alpha} \quad \text{where } 0 \leq \alpha \leq 1 \tag{3}$$

$\Delta\theta$ being a characteristic temperature difference of the flow and C a constant. The norm of the subgrid

heat flux $|\Phi_{sgs}| = \sqrt{\frac{1}{2}\pi_{i\theta}\pi_{i\theta}}$ is evaluated using Bardina's similarity hypothesis, as $\pi_{i\theta} = (u_i)'(\theta)'$.

$|\overline{T}|$ represents the norm of the resolved heat flux dissipation, $|\overline{T}| = \sqrt{2\overline{T}_{ij}\overline{T}_{ij}}$ where

$\overline{T}_{ij} = \frac{1}{2}\left(\frac{\partial \overline{u}_i}{\partial x_j} \frac{\partial \overline{\theta}}{\partial x_j} + \frac{\partial \overline{u}_j}{\partial x_i} \frac{\partial \overline{\theta}}{\partial x_i} \right)$.

NUMERICAL PROCEDURE

The preceding set of instationary equations is treated with a finite volume description on staggered grids. The advance in time is obtained at each time step with a projection method, using a second order Euler scheme for time discretisation, an explicit Adams-Bashforth scheme for the non-linear terms, with a centred approximation of the velocities, and an implicit formulation for the viscous terms. The overall scheme, developed by P. Le Quéré, is formally second order accurate in space and time.

COMPARISON OF THE DIFFERENT L.E.S. MODELS

First, in order to test our LES model, we have considered the academic case of the "window problem", that is a square cavity filled with air and differentially heated on its vertical walls. The horizontal walls are supposed to be perfectly adiabatic. The air is heated on the left vertical wall at constant temperature T_h, and cooled on the right side at constant temperature T_c, producing a clockwise circulation in the cavity. This configuration has been widely used for comparison exercises of numerical procedures, with different turbulence models and DNS (Henkes & Hoogendoorn 1992, Xin & Le Quéré 1995, Liu & Wen 1999).

Table 1 presents the results obtained with DNS and LES using the Reynolds analogy or the local subgrid diffusivity model for a Rayleigh number of 5.10^{10}. We have used a regular grid for the NJ points in the vertical direction and a hyperbolic grid distribution for the NI points in the horizontal direction. The values reported for the mean Nusselt number along the hot wall, $\overline{Nu_h}$, the centreline stratification at mid-height, $S_{1/2}$, and the maximum vertical velocity at mid-height, $V_{max,1/2}$, are given in dimensionless units with respect to the reference quantities defined earlier. These statistical values have been obtained for a time interval longer than 200 units of dimensionless time, once the statistically steady state had been reached.

A first comparison of our DNS results (Sergent et al. 1999) has been made with the spectral DNS results of Xin and Le Quéré (Xin & Le Quéré 1995). This comparison has shown a good agreement, and we can consider the present DNS results as the reference results for the comparison with LES.

Looking at these results, one can see that the dynamic of the flow is quite well represented by LES, as the vertical velocity at mid height and its location from the wall are in good agreement with DNS (see Sergent et al., 1999, for more details). On the other hand, the thermal transfer at the wall is over-predicted when using LES with Reynolds analogy, but we can observe a very good agreement when using the local diffusivity model. In this case the Nusselt number profile is close to DNS and the general thermal pattern of the flow, presented on Figure1, is much more better than the one obtained with the Reynolds analogy, especially in the recirculation zones in the upper left corner and lower right corner of the cavity. This results in a real improvement in predicting the vertical stratification at the centre of the cavity, presented on Figure 3.

TABLE 1
RESULTS FOR DNS AND LES, $R_A = 5.10^{10}$

case #	1	2	3
Type	DNS	LES	LES
γ_{sgs} model	-	Reynolds analogy ($Pr_{sgs}=0.3$)	local subgrid model
NI x NJ	258 x 514	66 x 130	66 x 130
$S_{1/2}$	1.5	1.1	1.4
$\overline{Nu_h}$	152	191	158
$V_{max,z=1/2}$	0.237	0.234	0.235
$x(V_{max})$	0.0024	0.0028	0.0028

APPLICATION TO PARTITIONED ROOMS

We now consider the case of two square cavities separated by a lintel, in order to figure a partitioned room of aspect ratio Height/Width equal to 0.5 heated on its left vertical wall and cooled on its right vertical side. The height of the lintel is 0.3. The flow is investigated with LES, using the local diffusivity model, with a 130x130 grid points. The mean thermal pattern of the flow is presented on Figure 4. One can observe than the effect of the lintel is to create a blocked region in the upper part of the left cavity, with a stiff thermal stratification, while on the right hand side, the hot air goes around the lintel and then moves upwards, producing a recirculation zone behind the lintel and leading to a thermally homogeneous upper region (see Figure 5). We can notice that the influence of the lintel is limited to the upper part of the cavity, the vertical thermal stratification in the lower regions remaining the same as in the window problem.

CONCLUSION

We have presented a LES approach to investigate natural convective airflows in cavities figuring heated rooms. We have compared LES results for two subgrid diffusivity models. The first one is based on the Reynolds analogy with a constant subgrid Prandtl number, while the other is a local subgrid diffusivity model derived from the Mixed Scale Model for viscosity. This comparison has been performed for the case of a differentially heated square cavity at $Ra=5.10^{10}$. The LES results are in good agreement with DNS results for the general thermal and dynamic patterns of the flow, especially when the subgrid diffusivity is evaluated at each grid point. These results are encouraging, as we can expect that if our LES results are close to DNS for 2D simulations, they still will be in good agreement for a 3D complete description of turbulent buoyant flows.

Acknowledgements

This work is supported by ADEME, under AMETH project. Computations have been performed on C90 machines, at IDRIS, Orsay, France.

REFERENCES

Bardina J., Ferziger J.H., Reynolds W.C. (1980). Improved subgrid scale models for large eddy simulation. *AIAA Paper 80 – 1357*

Bennetsen J.C., Sorensen J.N., Sogaard H.T., Christiansen P.L. (1996). Numerical simulation of turbulent airflow in a livestock building. in *ROOMVENT-96 (Yokohama-Japan)* **2**, 169-176

Davidson L., Nielsen P.V. (1996). Large eddy simulations of the flow in a three-dimensional ventilated rooom. in *ROOMVENT-96 (Yokohama-Japan)* **2**, 161-168

Henkes R.A.W.M., Hoogendoorn C.J. (1995). Comparison exercise for computations of turbulent natural convection in enclosures, *Num. Heat Transfer* part B, **28**, 59-72

Liu F. , Wen J.X. (1999). Development and validation of an advanced turbulence model for buoyancy driven flows in enclosures. *Int. J. Heat Mass Transfer* **42,** 3967-3981

Sagaut P., Troff B., Lê T.H., Ta P.L. (1996). Large eddy simulation of turbulent flow past a backward facing step with a new mixed scales SGS model in *IMACS-COST Conference Computational Fluid Dynamics Three-Dimensional Complex Flows, (Lausanne-Switzerland), Sept 95*, 13-15

Sergent A., Joubert P., Le Quéré P., Allard F. (1999). Etude numérique d'écoulement de convection naturelle par une approche de simulation des grosses structures. *SFT99 (Arcachon-France)*, 173-178

Xin S., Le Quéré P. (1995). Direct numerical simulation of two dimensional chaotic natural convection in a differentially heated cavity of aspect ratio 4. *J. Fluid Mech.* **304**, 87-118

(1) DNS   (2) LES with local diffusivity model   (3) LES with Reynolds analogy

Figure 1 : Isotherms (-0.4 ; -0.3 ; -0.2 ; -0.1 ; 0 ; 0.1 ; 0.2 ; 0.3 ; 0.4) for DNS and LES

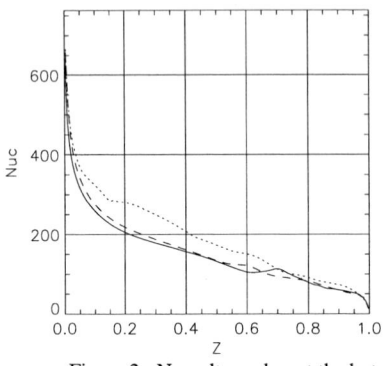

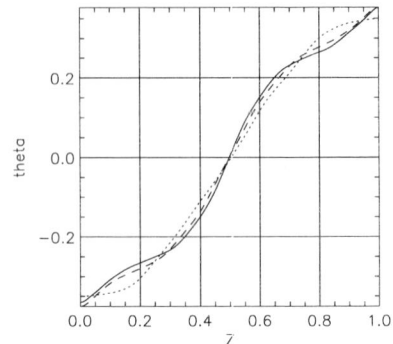

Figure 2 : Nusselt number at the hot wall
_____DNS ; ...... LES with Reynolds analogy ; _ _ _ _ LES with local diffusivity model

Figure 3 : Centerline thermal stratification

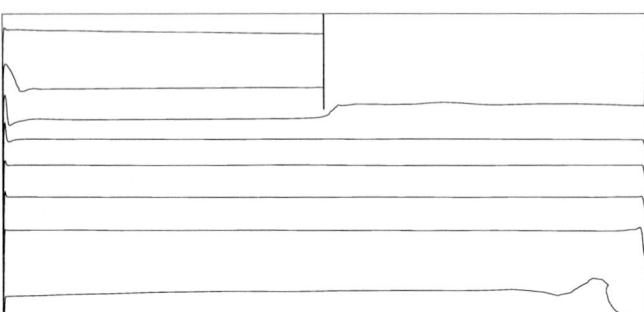

Figure 4: Cavity with lintel.
Isotherms
(-0.4 ; -0.3 ; -0.2 ; -0.1 ; 0 ; 0.1 ;
0.2 ; 0.3 ; 0.4)

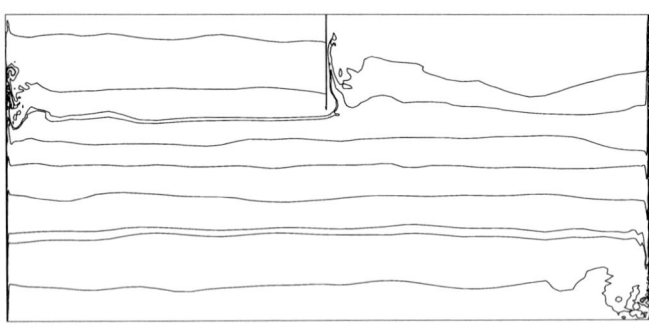

Figure 5 : Typical instantaneous
isotherms patterns for the cavity
with lintel.

2000 Elsevier Science Ltd. All rights reserved.
Air Distribution in Rooms, (ROOMVENT 2000)
Editor: H.B. Awbi

COUPLED SIMULATION OF CONVECTON, RADIATION, AND HVAC CONTROL FOR ATTAINING A GIVEN PMV VALUE

Taeyeon Kim[1] , Shinsuke Kato[1], and Shuzo Murakami[1]

[1] Institute of Industrial Science, University of Tokyo, 7-22-1 Roppongi Minato-ku Tokyo 106 Japan

ABSTRACT

A new CFD (Computational Fluid Dynamics) simulation for designing indoor climates is presented in this study. It is coupled with a radiative heat transfer simulation and HVAC (Heating, Ventilating, and Air-Conditioning) control system in a room. This new method can feed back the outputs of the CFD to the input conditions for controlling the HVAC system, and includes a human model to evaluate the thermal environment. It can be used to analyze the conditions of the HVAC system (e.g. temperature of supply air, surface temperature of radiation panel, etc.) and the heating/cooling loads of different HVAC systems under the condition of the same human thermal sensation (e.g. PMV, Operative Temperature, etc.)

To examine the performance of the new method, a thermal environment within a semi-enclosed space which opens into an atrium space is analyzed under steady-state conditions in the summer season. Using this method, the most energy efficient HVAC system can be chosen under the same PMV value. In this paper, two types of HVAC system are compared: one is a radiation-panel system and the other is an all-air cooling system. The radiation-panel cooling is found to be more energy-efficient for cooling the semi-enclosed space in this study.

KEYWORDS

Thermal Environment, HVAC Control, Coupled Simulation, PMV, CFD

INTRODUCTION

This paper presents a new CFD method coupled with a radiative heat transfer simulation and HVAC control in a room. It can provide precise information about the conditions of the HVAC system and heating/cooling loads which take into account the temperature distribution, etc. In the simulation, the conditions of the HVAC system (e.g., supply air temperature, supply air volume, etc.) are modified by changing the input B.C.s (boundary conditions) using a feedback system using HVAC control as shown in Figure 1. These modifications are determined based on simulations with modified input B.C.s to maintain the PMV value (Predicted Mean Vote, Fanger 1970) of a human model at the target value.

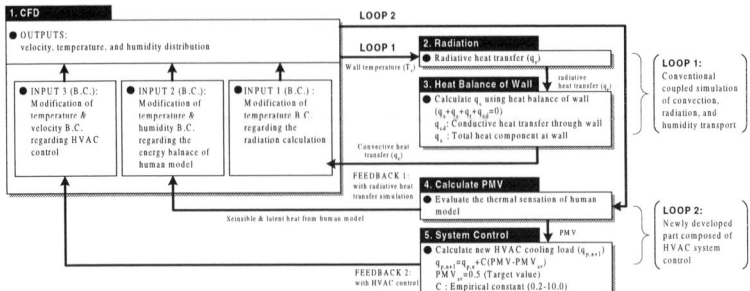

Figure 1: Procedure of Simulation including HVAC System Control

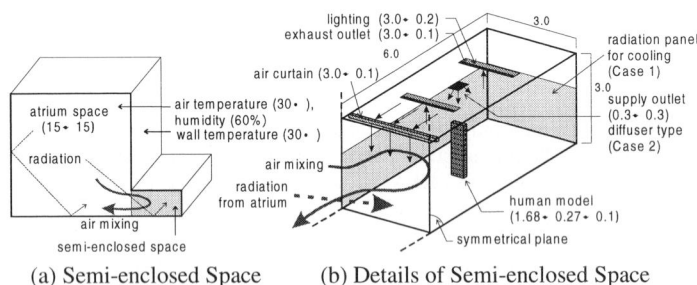

(a) Semi-enclosed Space   (b) Details of Semi-enclosed Space
Opening into Atrium

Figure 2: Semi-enclosed Space for Simulation (unit: m)
(half the space of the symmetrical room is illustrated)

To examine the performance of the new CFD method, the thermal environment in a semi-enclosed space (Figure 2), which opens into an atrium space, is analyzed. Using this CFD simulation method with a feedback system, the conditions of the HVAC system and cooling loads required for attaining the same thermal sensation are quantitatively estimated.

COUPLED SIMULATION OF CONVECTION, RADIATION, AND HVAC CONTROL

A feedback system, which returns the conditions of the HVAC control system, is added to the conventional coupled simulation of convection and radiation as shown in Figure 1. In the procedure of the simulation, the feedback system modifies the B.C.s of CFD to attain the given indoor thermal environment. In the usual HVAC system in a room, the air temperature at a specific point (e.g. exhaust outlets) is selected as the control target. However, it would be more rational for HVAC control systems to modify the HVAC outputs based on the thermal sensation of the occupants (S. Murakami et al. 1998). In this paper, the thermal sensation of an occupant is evaluated based on his PMV value. The HVAC outputs are thus modified to keep PMV at the target value (0.5 PMV).

CFD Simulation and feedback to boundary condition

Indoor air flow and temperature fields are calculated based on a 3D CFD simulation, using the standard k-ε model. In the radiation analysis, the view factor and the radiation heat transfer between the walls were calculated by the Monte Carlo method (T. Omori et al. 1990) and Gebhart's absorption factor method, respectively. Humidity distribution is solved based on the equation for humidity transport with the CFD method (S. Murakami et al. 1998).

TABLE 1
CASES ANALYZED AND CONDITIONS GIVEN

Case			Case 1: Radiation-panel system	Case 2 : All-air cooling system
Atrium	Wall and air temperature	[°C]	30.0	30.0
	Humidity	[%]	60.0	60.0
	Surface temperature of wall	[°C]	30.0	30.0
Semi-closed space	Cooling load imposed by human	[kW]	0.4	0.4
	Cooling load imposed by lighting	[kW]	0.4	0.4
	PMV	-	0.5	0.5
Radiation-panel for cooling	Area	[m²]	15.1	-
Supply outlet of all-air cooling system	Humidity	[%]		80.0
	Air flow rate	[m³/s]		907.0
		[h⁻¹]	-	17.0
Air-curtain	Air flow rate [ACH]	[h⁻¹]		60.0
	Outlet velocity	[m/s]		3.0
	Cooling load	[kW]		0.0

† Sensible and latent heats from human are assumed to be provided from one human model and the floor.
†† Values are given for half space of semi-enclosed space.

TABLE 2
CONDITIONS FOR CALCULATION

Pressure B.C.	Pressure	Static pressure 0 at atrium
	Inflow	$k_{in} = 3/2(U_{in} \times 0.05)^2$, $\varepsilon_{in} = C_\mu k_{in}^{3/2}/l_{in}$, l_{in} = width of the opening, $T_{in} = 30°C$, U_{in}: velocity of inflow [m/s], k_{in}: kinetic energy of inflow [m²/s²], ε_{in}: kinetic energy dissipation rate [m²/s³], l_{in}: specific length scale [m]
	Outflow	Free slip
Supply Outlet B.C.		$k_{in} = 3/2(U_{in} \times 0.05)^2$, $\varepsilon_{in} = C_\mu k_{in}^{3/2}/l_{in}$, l_{in} = width of the opening, Air temperature of supply outlet is modified in response to the HVAC control system during CFD
Wall, Radiation-panel, and Human Model	Velocity	Generalized log-law, free slip at symmetric plane
	Temperature	Convective heat transfer coefficient is fixed as: Radiation-panel: 5.5, adiabatic surface: 3.0, others: 4.0 [W/(m²°C)]
	Humidity	1. Human model : Emission rate of sensible/latent heat is changed based on energy balance of human. 2. Radiation-panel : AH (Absolute Humidity) is given corresponding to the saturated vapor pressure, when the surface temperature of radiation-panel is lower than dew point temperature of the air. In other cases, Gradient of AH = 0. Humidity transfer coefficient is calculated based on Lewis Relation. 3. Other walls : Gradient of AH = 0.
	Emissivity of radiation	Wall, human model: 0.9, Symmetrical plane: 0.0
Mesh System		CFD : 40 (x) × 20 (y) × 20 (z), Radiation : 22 (x) × 12 (y) × 5 (z)

Procedure of New Simulation Method

The procedure of the simulation is illustrated in Figure 1. In this paper, two types of HVAC system are compared: one is a radiation-panel cooling system and the other is an all-air cooling system. In the case of the radiation-panel cooling system, the heat flux at the surface of the radiation-panel is modified, while the temperature of the supply outlets is modified in the case of the all-air cooling system during the simulation. The cooling loads of the HVAC systems needed to achieve the target PMV of the human model (0.5 PMV) can be given through these simulations. The feedback system is stopped when this target point is achieved.

OPTIMAL DESIGN BASED ON CFD FEEDBACK MECHANISM

In this paper, only the B.C. of the heat flux of the radiation-panel and the air temperature of the supply outlet are modified in the simulation. However, with this method, other B.C.s (e.g., locations of supply outlets, number of supply outlets, etc.) can be modified so as to control the indoor climate at the target temperature. Through these simulations, the optimal design of the HVAC system can be achieved. It

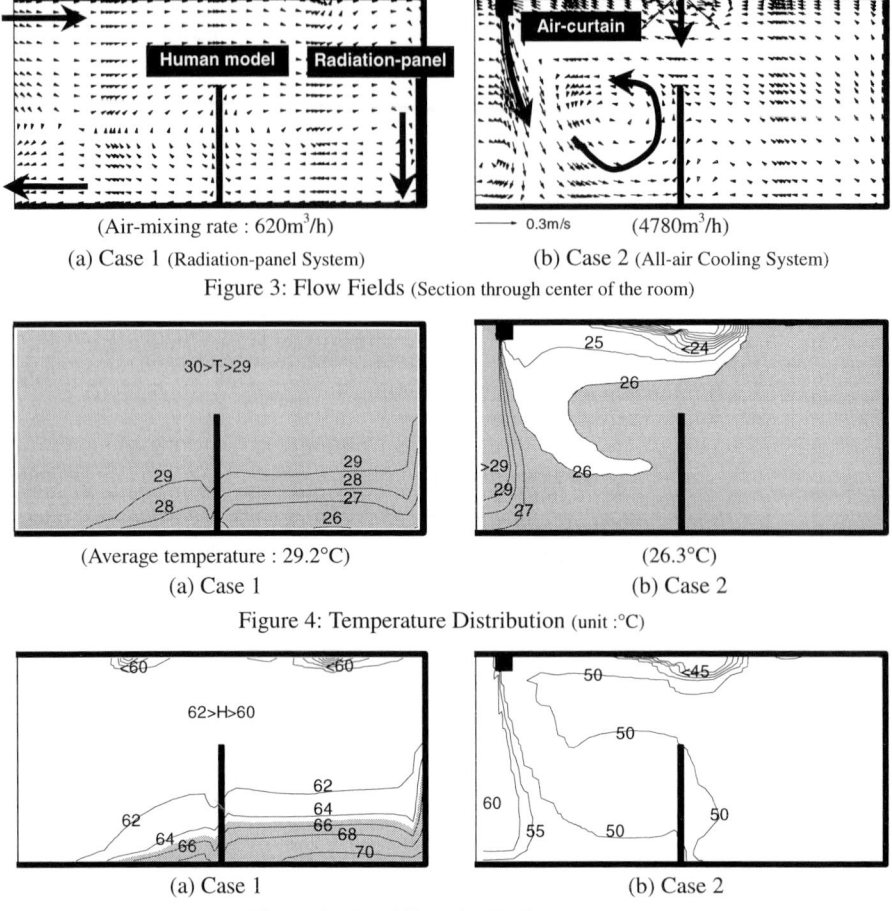

(Air-mixing rate : 620m³/h)

(a) Case 1 (Radiation-panel System)

0.3m/s (4780m³/h)

(b) Case 2 (All-air Cooling System)

Figure 3: Flow Fields (Section through center of the room)

(Average temperature : 29.2°C)

(a) Case 1

(26.3°C)

(b) Case 2

Figure 4: Temperature Distribution (unit :°C)

(a) Case 1

(b) Case 2

Figure 5 : Humidity Distribution (unit : %)

would be a very useful tool for environmental design. Although the description concerns cooling system control, the same control could be also applied to a heating system.

SEMI-ENCLOSED SPACE

To examine the performance of the new method, the thermal environment of a semi-enclosed space, which opens into an atrium space (15 m height), is analyzed under steady-state conditions (Figure 2). For simplicity, all the walls, ceiling, and floor are assumed to be adiabatic. In the CFD calculation, the thermal environment of the atrium is modeled simply; the air and wall temperature, and humidity of the atrium are fixed at 30°C and 60%, respectively. Radiation-panels operating below the dew-point temperature are installed in three walls (Figure 2 (b), Case 1). Consequently the panels can remove the humidity of the air in the room. In the case of the all-air cooling system, cold air is supplied through supply outlets in the ceiling (Figure 2 (b), diffuser type, Case 2). Half of the semi-enclosed space (3.0 m) is analyzed in consideration of the symmetrical configuration. The HVAC systems are controlled so as to keep PMV (i.e. thermal sensation) of the human model in the center of the room at the target value (0.5 PMV).

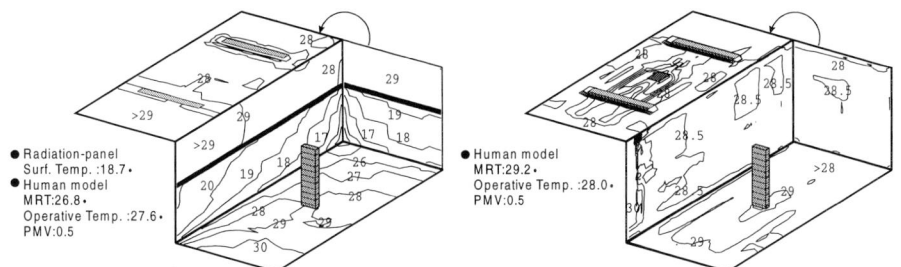

(a) Case 1 (Radiation-panel System)   (b) Case 2 (All-air Cooling System)

Figure 6: Wall Temperature Distribution and, OT & MRT of Human Model

TABLE 3
PREDICTED RESULTS OF COOLING LOADS

Case			Case 1 : Radiation-panel system	Case 2 All-air cooling system
Semi-enclosed space	Sensible cooling load due to air-mixing	[kW]	0.6	2.3
	Latent cooling load due to air-mixing	[kW]	0.3	5.6
	Sensible cooling load due to radiation	[kW]	0.1	0.0
	Air-mixing rate	[m³/h]	620.0	4780.0
	Average room air temperature	[°C]	29.2	26.3
Radiation-panel for cooling	Sensible cooling load	[kW]	1.4	-
	Latent cooling load	[kW]	0.4	
	Surface temperature	[°C]	18.7	
Supply outlet of all-air cooling system	Temperature	[°C]		16.1
	Sensible cooling load	[kW]	-	3.0
	Latent cooling load	[kW]		5.7

† Values are given for half of the semi-enclosed space.

CASES ANALYZED

Cases analyzed and the conditions of calculation are shown in Table 1 and Table 2, respectively. Two types of HVAC system are analyzed in this study. The radiation-panel cooling system is used in Case 1 and the all-air cooling system in Case 2. In Case 2, the air-curtain is installed at the entrance opening of the semi-enclosed space in order to minimize mixing of the air (Figure 2). The air-curtain does not contribute to removing the cooling load.

RESULTS

Flow Fields

In the radiation-panel system (Case 1, Figure 3 (a)), the air from the atrium enters at the upper part of the entrance opening and goes out near the floor. The cold air descends near the radiation-panel due to the negative-buoyancy effect. The air-mixing rate (620m³/h) is smaller than for the all-air cooling system (Case2, 4780m³/h). In Case 2 (Figure 3 (b)), the cold air from the supply outlet is mixed immediately with the room air by the effect of the air circulation produced by the air-curtain.

Temperature Fields

In Case 1 (Figure 4 (a)), temperature stratification is formed; the temperature difference between the upper and the bottom part is about 3-4°C. The average room air temperature (29.2°C) is higher than in Case 2. But the PMV value for the human model is the same for all cases (0.5 PMV) by the simulation process. In Case 2 (Figure 4 (b)), the temperature stratification disappears due to the air circulation.

Relative Humidity

In Case 1 (Figure 5 (a)), the relative humidity exceeds 60% over the whole area. Because of the cold air generated by the radiation-panel, a high RH is observed near the floor, and so a stratification of RH is produced. In Case 2 (Figure 5 (b)), a uniform RH distribution is observed. The RH around the human model is about 50%.

Distribution of Wall Surface Temperature and, Operative Temperature & MRT of Human Model

MRT (26.8°C) in Case 1 (Figure 6(a)) is much lower than in the all-air cooling case (Case 2, 29.2°C), despite having a much higher average room air temperature. Operative Temperature (OT) of human model is 27.6°C, which is lower than in Case 2 (28.0°C). Since the human models in both cases have the same PMV value, it is thought that the different distribution of RH causes the different OT. The temperatures of the ceiling and the floor in Case 1 are lower than those in Case 2, since a large amount of heat is removed by radiation.

Condition of the HVAC System and Cooling Loads in the Semi-enclosed Space

Cooling loads in the semi-enclose space are shown in Table 3. Since the all-air cooling system (Case 2) has a large air-mixing rate between the semi-enclosed space and the atrium, the sensible cooling load in Case 2 is larger than for the radiation-panel system (Case 1). The latent cooling loads in Case 2 are also larger than in Case 1. Since MRT in Case 2 is higher than in Case 1, the cooling load due to radiation exchange becomes lower. The temperatures of the radiation panel surface in Case 1 and the supplied air in Case 2 become 18.7°C and 16.1°C, respectively when the converged solution is obtained.

CONCLUSIONS

1. A new CFD technique, coupled with a radiative heat transfer simulation and HVAC control system, was proposed. With this new method, the conditions of the HVAC system and the required cooling loads for attaining the same thermal sensation for occupants (0.5 PMV) were quantitatively evaluated for different VAC systems.
2. In order to control the thermal environment of the semi-enclosed space, which opened into an atrium, the radiation-panel system was evaluated as very energy-efficient in this study.
3. The analyses showed that the CFD method coupled with the feedback system for HVAC control is a very useful tool for environmental design.

References

S. Murakami, S. Kato, T. Kim (1999). Indoor climate design based on feedback control of HVAC, Coupled simulation of convection, radiation, and HVAC control for attaining given operative temperature, Building Simulation '99.
Fanger, P. O. (1970). Thermal comfort, Danishi Technical Press.
T. Omori, H. Taniguchi, and K. Kudo (1990). Prediction method of radiant environment in a room and its application to floor heating, SHASE Transactions 42 : 9-18.
S. Murakami, S. Kato, and Jie Zeng, (1998). Combined Simulation of Airflow, Radiation and Moisture Transport for Heat Release from Human Body, ROOMVENT '98, Stockhom.

Acknowledgement

This study was supported by the Japan Science and Technology Corporation (JST), Japan.

2000 Elsevier Science Ltd. All rights reserved.
Air Distribution in Rooms, (ROOMVENT 2000)
Editor: H.B. Awbi

VOC DISTRIBUTION IN A ROOM BASED ON CFD SIMULATION COUPLED WITH EMISSION / SORPTION ANALYSIS

Shuzo Murakami[1], Shinsuke Kato[1], Yasushi Kondo[2], Kazuhide Ito[3], Akira Yamamoto[1]

[1] Institute of Industrial Science, University of Tokyo, 7-22-1 Roppongi Minato-ku Tokyo 106 Japan
[2] Musashi Institute of Technology, 1-28-1, Setagaya-ku Tokyo 158 Japan
[3] Tokyo Institute of Polytechnics, 1583 Iiyama Atugi Kanagawa 243 Japan

ABSTRACT

This paper presents physical models that are used for analyzing numerically the transportation of VOCs from building materials in a room. The models are based on fundamental physicochemical principles of their diffusion and adsorption / desorption (hereafter simply sorption) both in building materials and in room air. The performance of the proposed physical models is examined numerically in a test room with a technique supported by computational fluid dynamics (CFD). Two building materials are used in this study. One is a VOC emission material for which the emission rate is mainly controlled by the internal diffusion of the material. The other is an adsorptive material that has no VOCs source. The results of numerical prediction show that the physical models and their numerical simulations explain well the mechanism of the transportation of VOCs in a room.

KEYWORDS

CFD, VOCs, Diffusion, Adsorption, Desorption

INTRODUCTION

In this study, physical models of emission and sorption of VOCs are proposed. The models are based on fundamental physicochemical principles of diffusion and sorption of VOCs within both building materials and room air. To demonstrate the validity of the models, concentration distributions of VOCs in a room are numerically analyzed by a CFD technique. Here, the floor is covered with an emission material made of SBR (polypropylene styrene-butadiene rubber). The emission rate of VOCs from SBR is mainly controlled by internal diffusion in the material (Yang, X., et al, 1998). Coal-based activated carbon is spread over the sidewalls as an adsorbent material. The adsorbent material used here has no source of VOCs and affects room air concentration only through its sorption process. It is assumed that the composition ratio of VOCs does not change and a virtual VOC species (defined as simply VOCs) that represents the total property of VOCs emitted into the air is used in this study. The final goal of this study is to numerically predict the concentration of chemical pollutants in the air inhaled by the occupants of a room.

PHYSICAL MODELS OF VOC TRANSFER

Transportation in room air
VOCs emitted from building materials are transported by room air convection, diffused by molecular and turbulent diffusion, and taken out through an exhaust opening. The vapor phase concentration C [kg_{VOCs}/kg_{air}] of VOCs in room air can be described by the mass conservation equation shown in Eq. (1).

$$\rho_{air}\frac{\partial C}{\partial t} + \rho_{air}\frac{\partial(u_j C)}{\partial x_j} = \frac{\partial}{\partial x_j}\left(\left(\lambda_a + \frac{\rho_{air}v_t}{\sigma}\right)\frac{\partial C}{\partial x_j}\right) \tag{1}$$

Here, ρ_{air} [kg/m^3] is the air density, λ_a [kg/(m·s·kg/kg)] is the molecular mass conductivity, v_t [m^2/s] is the turbulence eddy viscosity, σ is the turbulent Schmitt number. Velocity u_j [m/s] and v_t are given by solving the flow field with the k-ε turbulence model of Low Reynolds number type (Murakami, 1996).

Diffusive and adsorptive transportation in materials

The transportation of VOCs in building materials (internal diffusion or permeation) is possible through the existence of fine pores within the building materials. Molecular and Knudsen diffusions in the vapor phase occur through the pores in the materials due to the concentration gradient. On the surface of the pores, vapor phase VOCs is adsorbed and desorbed as shown in Figure 1.

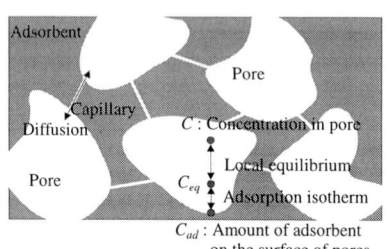

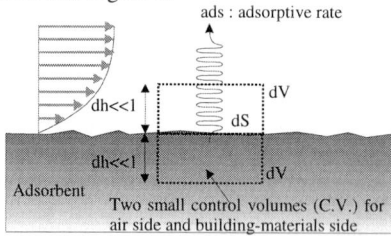

Figure 1 : Modeling of VOC diffusion and adsorption in the material

Figure 2 : Modeling of adsorption and desorption of VOCs on the surface of a sorptive material

It can be assumed that VOCs diffusion within the substance (the part with no pores in the material) of the material is so small that the diffusion here may be disregarded compared with that in the pores. The VOCs transportation through the pores and on the surface of them can be described by conservation Eqs. (2) and (3) respectively.

$$k\rho_{air}\frac{\partial C}{\partial t} = \frac{\partial}{\partial x_j}\left(\lambda_C\frac{\partial C}{\partial x_j}\right) - adv \tag{2}$$

$$\rho_{sol}\frac{\partial C_{ad}}{\partial t} = adv \tag{3}$$

Here, C_{ad} [kg/kg] is the solid (adsorbed) phase concentration on the surface of a pore. ρ_{sol} [kg/m^3] is the net density of adsorbent of the material, λ_C [kg/(ms kg/kg)] is the mass conductivity of VOCs in the air within the pores and k [m^3/m^3] is the porosity of the material. adv [kg/m^{3}s] is the mass transportation within the pores to the surface of the adsorbent.

Substituting Eq.(3) into Eq.(2), the diffusion and sorption equation in the material is obtained as Eq.(4).

$$k\rho_{air}\frac{\partial C}{\partial t} = \frac{\partial}{\partial x_j}\left(\lambda_C\frac{\partial C}{\partial x_j}\right) - \rho_{sol}\frac{\partial C_{ad}}{\partial t} \tag{4}$$

Generalized adsorption isotherm

In order to close Eq. (4), a so-called adsorption isotherm is introduced. For closed systems under steady conditions, the rate of adsorption becomes equal to the rate of desorption and thus an equilibrium state is achieved. Since this phenomenon of sorptive dynamics occurs much faster than with the molecular diffusion in gases, in a small CV (control volume) under isothermal conditions at constant pressure, an equilibrium relation between the concentration in the gas and that on the adsorbent surface is possible. The relation is expressed by the so-called general adsorption isotherm (Eq.(5)).

$$C_{ad} = f(C_{eq}, T) \tag{5}$$

Here, the function f is unique for a combination of adsorbed compound and adsorbent. T [K] is the absolute temperature. C_{eq} [kg/kg] is the vapor phase concentration in equilibrium with the solid phase concentration C_{ad} on the surface of the material. Since C in the pores becomes the same with C_{eq} (local equilibrirum), C_{eq} can be substituted by C in the small CV including the adsorbent surface.

Simple transportation model governed by effective diffusion

Eq. (4) of the diffusion and sorption equations in porous materials is re-expressed as Eq. (6) using the relation of the adsorption isotherm of Eq. (5).

$$\left(k\rho_{air} + \rho_{sol}\frac{\partial f}{\partial C}\right)\frac{\partial C}{\partial t} = \frac{\partial}{\partial x_j}\left(\lambda_C\frac{\partial C}{\partial x_j}\right) - \rho_{sol}\frac{\partial f}{\partial T}\frac{\partial T}{\partial t} \tag{6}$$

In Eq. (6), local equilibrium in a pore ($C_{eq}=C$) is assumed. When it is isothermal, $\partial T/\partial t$ can be omitted and Eq. (6) can be rewritten as simple diffusion equations in Eqs. (7) and (8).

$$\rho_{air}\frac{\partial C}{\partial t} = \frac{\partial}{\partial x_j}\left(\rho_{air}D_C\frac{\partial C}{\partial x_j}\right) \tag{7}$$

$$D_C = \lambda_C\bigg/\left(k\rho_{air} + \rho_{sol}\frac{\partial f}{\partial C}\right) \tag{8}$$

Here, D_C [m²/s] is the effective diffusion coefficient of VOCs in the material. D_C includes the adsorption isotherm f. Eq. (7) is closed when isotherm models such as Eqs. (13), (15) and (17) are introduced as described later. Vapor phase concentration C is related to the solid (adsorbed) phase concentration C_{ad} in the adsorbent with the adsorption isotherm in Eq. (5) through the assumption of $C = C_{eq}$. In this context, C represents the total VOC concentration in the material. In this study, C is used as the equivalent vapor phase concentration to express the concentration in the material, instead of C_{ad}.

The boundary condition at the air-material interface for analyzing the concentration in room air

Since the transportation of C in the material and in the room air are solved simultaneously, a boundary condition should be set at the air-material interface. The VOC emission rate at the air-material interface should be identical with the transportation rate by internal diffusion from the inside. This condition is expressed as a conservation law at the surface of the material, as shown in Eq. (9).

$$-\rho_{air}D_C\frac{\partial C}{\partial x}\bigg|_{B+} = -\lambda_a\frac{\partial C}{\partial x}\bigg|_{B-} \tag{9}$$

Here, $B+$ is the air-material surface in the material-side region, and $B-$ is the air material surface in the air-side region. Eq. (9) is used as the boundary condition with coupled simulation in the material and in the room air.

SIMPLE TRANSPORTATION MODEL ON SORPTIVE SURFACE

In general, the diffusion and sorption process in the material can be described using the effective diffusion coefficient D_C that includes implicitly the effect of sorption. If the material has no VOC source and it contributes to the room air concentration only with its sorptive process, we can simplify the modeling of the sorption process compared to the modeling stated above. For simple modeling, it is assumed that the adsorption occurs only on the surface (Figure 2). VOC transfer near building materials is also governed by Eqs. (2) and (3), as well as the phenomenon within the pores of the material. Here, two CVs (dV[m³]) are set at the interface for simply modeling the transportation on adsorbent material as shown in Figure 2. The thickness dh [m] normal to the surface element dS [m²] is very thin. This surface element dS is sandwiched by the CVs; i.e. the CV in the material and the CV in the air. Here, dh is assumed so small that the sorption process reaches equilibrium immediately in the CV. When volume integration is performed for Eqs.(2) and (3) in each CV neglecting the time differential term of the vapor phase concentration in air, following Eqs. are given.

$$0 = \lambda_a\frac{\partial C}{\partial x}\bigg|_{B-}dS - adv \cdot dV \qquad \text{(CV in the air)} \tag{10}$$

$$\rho_{sol}\frac{\partial C_{ad}}{\partial t}dV = adv \cdot dV \qquad \text{(CV in the material)} \tag{11}$$

With Eqs. (10) and (11), the VOC adsorption rate on the surface of the material is related to the molecular diffusion of VOCs close to the surface of the adsorbent as shown in Eq. (12).

$$-\lambda_a \frac{\partial C}{\partial x}\bigg|_{B-} = -adv \cdot \frac{dV}{dS} = -ads = -\left(\rho_{sol} \frac{\partial C_{ad}}{\partial t}\right)\frac{dV}{dS} = -\rho'_{sol} \frac{\partial C_{ad}}{\partial t} \tag{12}$$

Here, ads [kg/m^{2}s] is the sorption rate (positive/ negative of ads corresponds to the adsorption/ desorption rates respectively). $\rho'_{sol} (= \rho_{sol} \ dV/dS)$ [kg/m^2] is the plane density of a sorptive material.

Typical isotherm models

Henry model ; For the adsorption under low VOCs concentrations in the air, the Henry model is utilized.

$$C_{ad} = k_h \cdot C_{eq} = k_h \cdot C \tag{13}$$

Here, k_h [-] is the Henry's coefficient. The adsorption term in Eq. (12) is rewritten as Eq. (14).

$$\lambda_a \frac{\partial C}{\partial x}\bigg|_{B-} = ads = \rho'_{sol} \frac{\partial C_{ad}}{\partial t} = \rho'_{sol} \cdot k_h \frac{\partial C}{\partial t}\bigg|_{B-} \tag{14}$$

In the numerical simulation, the VOC concentration in the air-side CV adjacent to the material $C|_{B-}$ is assumed to be the same as the concentration in the equilibrium state; i.e, $C|_{B-} = C_{eq}$ .Eq. (14) is used as the boundary condition for the sorptive surface when solving Eq. (1).

Langmuir model ; When the VOC concentration in room air becomes high, the concentration of adsorbed VOCs, C_{ad} on the adsorbent saturates at a certain level of room air concentration. In this situation, the Henry model overestimates the amount of adsorbed VOCs. The Langmuir model is based on the model of monolayer adsorption that takes into account the concept of saturated concentration of C_{ad0}. The Langmuir model, which can be applied to higher VOC concentration fields, is more sophisticated than the Henry model.

$$C_{ad} = \frac{C_{ad0} \cdot k_l \cdot C}{(1 + k_l \cdot C)} \tag{15}$$

Here, k_l [1/(kg/kg)] is the Langmuir's coefficient, C_{ad0} [kg/kg] is the concentration of the saturated adsorption by monolayer adsorption. The adsorption term in Eq. (12) is rewritten as Eq. (16) in the same manner as Eq. (14). Eq.(16) is used as the boundary condition for the sorptive surface when solving Eq. (1).

$$\lambda_a \frac{\partial C}{\partial x}\bigg|_{B-} = ads = \rho'_{sol} \frac{\partial C_{ad}}{\partial t} = \frac{\rho'_{sol} \cdot C_{ad0} \cdot k_l}{(1 + k_l \cdot C|_{B-})^2} \frac{\partial C}{\partial t}\bigg|_{B-} \tag{16}$$

Polanyi DR model ; The Polanyi DR approach may be used to describe sorption equilibrium for whole classes of compounds on a particular adsorbent and the variation of the equilibrium with temperature. This relation is most often presented as a so-called characteristic curve of the form shown in Eq. (17).

$$C_{ad} = C_{ad0} \cdot exp\left(-k_p \left(\frac{T}{V_m}\right)^2 \cdot ln\left(\frac{C_{sat}}{C}\right)^2\right) \tag{17}$$

Here, k_p [(cm^3/mol$\cdot$K)^2] is the Polanyi's coefficient, C_{sat} [kg/kg] is the saturated concentration, V_M [cm^3/mol] is the molecular volume. The adsorption rate 'ads' at this time serves as Eq. (18).

$$\lambda_a \frac{\partial C}{\partial x}\bigg|_{B-} = ads = \frac{2 \cdot \rho'_{sol} \cdot k_p \cdot (T/V_m)^2 \cdot C_{ad0}}{C|_{B-}} \cdot ln\left(\frac{C_{sat}}{C|_{B-}}\right) \cdot exp\left(-k_p\left(\frac{T}{V_m}\right)^2 ln\left(\frac{C_{sat}}{C|_{B-}}\right)^2\right) \frac{\partial C}{\partial t}\bigg|_{B-} \tag{18}$$

ROOM MODEL AND BUILDING MATERIALS USED

The room model (2D) used for simulation is shown in Figure 3. As the VOC source, SBR plate was selected and used to cover the floor. The emission rate from the material is strongly related to both the initial concentration distribution C_0 and the effective diffusion coefficient D_C within the SBR. In this study, the initial VOC concentration distribution in SBR is assumed to be uniform, $C_0 = 0.16$ [kg/kg]. D_C is assumed to be 1.1×10^{-14} [m^2/s] (at 23°C), following Yang, X. et al. The thickness of SBR is $0.025 L_0$ (1.5×10^{-3}m). As the adsorptive surface, coal-based activated carbon was spread over the sidewalls. The capacity of adsorption depends on the surface area of the adsorbent. The activated carbon has a wide surface area of more than 10^6 [m^2] per 1 [kg]. In this study, the plane density ρ'_{sol} is set to be 1×10^{-6} [kg/m^2]; i.e. the surface area of the adsorbent is the same as that of the building materials. This value is quite small from the viewpoint of practical cases. However, one of the objects of this study is to examine the difference in various models of adsorption isotherm and such a difference will be clear in the condition where C_{ad} becomes close to its saturated concentration. Therefore the amount of the adsorbent material was decreased in this study. The parameters in the sorption isotherm are estimated by an experiment in which toluene is adsorbed on coal-based activated carbon (Yu, 1987). Here, we assumed that Henry's k_h [-], Langmuir's k_l [1/(kg/kg)] and Polanyi k_p [(cm^3/mol·K)^2] coefficients are 3.0×10^5, 1.5×10^6 and 1.3×10^{-3} respectively. Other estimated values are shown in Table 1.

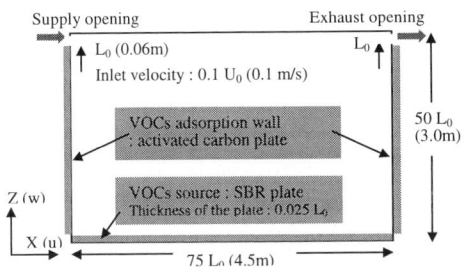

Figure 3 : Room model (2D)

TABLE 1
CASES ANALYZED

Case No.	Isotherm model	Applied Parameter
Case1	Carpet only	-
Case2	Henry model	k_h [-]= 3.0×10^5
Case3	Langmuir model	k_l [1/(kg/kg)]= 1.5×10^6 C_{ad0} [kg/kg]= 0.25
Case4	Polanyi model	k_p [(cm^3/mol·K)^2]= 1.3×10^{-3} C_{ad0} [kg/kg]= 0.43 C_{sat} [kg/kg]= 0.13 V_M [cm^3/mol]= 106

NUMERICAL METHODS AND BOUNDARY CONDITIONS

Table 2 shows the numerical conditions. Flow fields were analyzed with a low Reynolds number k-ε model (MKC model (Murakami, et al., 1996)) with an inflow velocity of 0.1 U_0 (= 0.1 [m/s]; air change rate = 1.6 [h^{-1}]). Using the results of flow field simulations, the behaviors of VOC transportation were analyzed. Table 1 shows the cases analyzed. Four cases were examined in total, under the different adsorption isotherm models.

TABLE 2
NUMERICAL CONDITIONS

Number of grid points (2D)		
Air region	:	68(x) × 64(z) (= 4.5m (x) × 3.0m(z))
Material region dominated by internal diffusion	:	68(x) × 41(z) (= 4.5m(x) × 1.5×10^{-3}m(z))
Width of the mesh adjacent to the surface	:	0.6 × 10^{-9} [mm]
Region of adsorptive surface for side walls	:	55(z) (= 2.94m(z))
Reynolds number	$U_0 L_0 / \nu$	= 4.2×10^3
Molecular diffusion coefficient of VOCs in air	λ_a / ρ_{air}	= 5.9×10^{-6} [m^2/s] (23°C)
Effective diffusion coefficient of VOCs in the material (Yang et. al.)	D_C	= 1.1×10^{-14} [m^2/s] (23°C)
Inflow velocity	0.1 U_0	= 0.1 [m/s]

RESULTS AND DISCUSSION

VOC concentration in room air

The history of the room-averaged concentration for 24 hours is shown in Figure 4. The concentration C is normalized by the initial concentration of the source material C_0. The room-averaged concentration after 24 hours decreases to 4.22×10^{-5} [g/m^3], and reaches steady state. The maximum room-averaged concentration in the early stages is evaluated as low in Case 2 and 3, which installed additional adsorption material compared with Case 1, and it turns out that the adsorbent materials have controlled the rise in the room-averaged VOC concentration. The maximum room average concentration is evaluated as low in Case 4 using the Polanyi model that evaluates the amount of adsorption greatly in a low concentration region.

Amount of adsorption on Adsorbent

In this analysis, since VOCs generated from the floor are mainly conveyed in the direction of the left wall by the room airflow, the amount of adsorption in the left wall side becomes large. The history of the average amount of adsorption (C_{ad}) by the left wall over 24 hours is shown in Figure 5. In Case 4 (Polanyi

478

model), the amount of adsorption is continuously estimated as large.

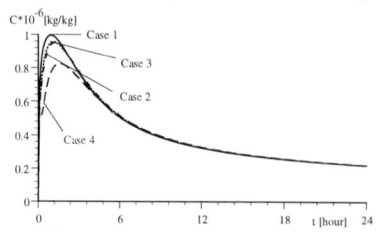

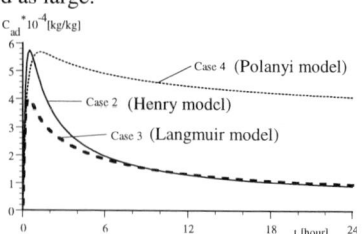

Figure 4 : Time history of room-averaged Concentration

Figure 5 : Time history of the amount of VOCs adsorbed by sorptive material

VOC concentration distribution in room

The concentration distributions of VOCs in Case 1 and 4 are shown in Figure 6. The VOCs concentration near the left wall is about 4 times higher than that near the right wall due to the influence of the airflow pattern. The VOC concentration distribution in the room is influenced by the position of the source and sink materials. The concentration is not uniform in the space, and rapidly changes near the floor of the VOC source. The VOC concentrations near the floor (4.67×10^{-4} [g/m^3]) are about 11 times higher than the average room concentration.

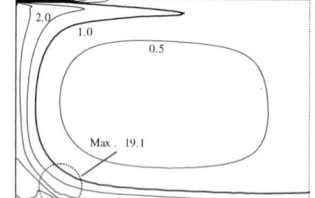

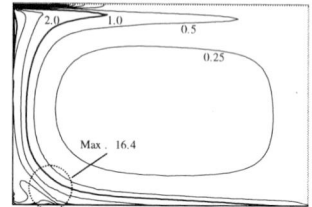

(a) Case 1 (SBR plate only, $*10^{-6}$ [kg/kg])   (b) Case 4 (Polanyi model, $*10^{-6}$ [kg/kg])

Figure 6 : Concentration distribution of VOCs (1 hour after the start of analysis)

CONCLUDING REMARKS

1. The room-averaged concentration decays gradually during the duration of the analysis. The adsorption effect on the room air concentration becomes clear when the room air concentration changes rapidly.
2. Since the concentration distribution in the room is not uniform, the amount of adsorption materials and their positions are important for the concentration.
3. The differences between 3 types of adsorption isotherm models were clear under the conditions of a small amount of adsorbent material.

References

Axley, J.W. (1995). New mass transport elements and compounds for the NIST IAQ model. NIST GCR pp. 95-676

Murakami, S. et al., (1996). New low Reynolds-number k-ε model including damping effect due to buoyancy in a stratified flow field. Int. J. Heat Mass Transfer, **39**, pp. 3483- 3496

Sparks, L.E. et al., (1996). Gas-phase mass transfer model for predicting volatile organic compound (VOC) emission rates from indoor pollutant sources., Indoor Air **6**, pp. 31-40

Yang, X. et al., (1998). Prediction of short-term and long-term volatile organic compound emissions from SBR bitumen-backed carpet at different temperatures. ASHRAE transaction

Yu, J.-W. (1987). Adsorption of trace organic contaminations in air. Stockholm: Royal Institute of Technology.

Acknowledgement

This study was supported by the Special Coordination Fund for Promoting Science and Technology of the Science and Technology Agency, Japan.

Air Distribution in Rooms, (ROOMVENT 2000)
Editor: H.B. Awbi

NUMERICAL MODELING OF THREE-DIMENSIONAL VENTILATION DUCT FLOW

N. Jovicic[1], D. Milovanovic[1], M. Babic[1], J.V. Soulis[2]

[1] Faculty of Mechanical Engineering, University of Kragujevac, Yugoslavia
[2] Democrition University of Thrace, Department of Civil Engineering, Xanthi, Greece

ABSTRACT

Presented in the paper is an efficient and accurate numerical method for simulation of ventilation duct flow. The mathematical method is based on the three-dimensional incompressible RANS equations with isotropic k-ω near-wall turbulence closures, written in generalized curvilinear coordinates in strong conservation form. The numerical method presented here is used to calculate the turbulent flow through a bend of rectangular ventilation duct featuring pressure induced secondary motions and rotation effects on turbulence. The problem has been the subject of an experimental study of Kim and Patel (1994). Numerical results are compared with experimental data.

KEYWORDS

Ventilation duct, three-dimensional flow, CFD, RANS, turbulence models

INTRODUCTION

Understanding of flow field details is crucial for designing of high efficiency ventilation systems. In order to achieve low energy loss, as well as low noise numerical simulation could be very attractive at an early stage in a project. From an engineer's point of view, the RANS equations in conjunction with two-equation turbulence models is an optimal choice for a reliable prediction of high Re turbulent flow through complex domains. In order to resolve phenomena in near-wall regions, numerical simulations require a very large number of grid points and extremely small grid spacing. Such meshes can dramatically deteriorate the convergence rate of numerical procedure. In order to make accurate, efficient and low-cost code for engineering purposes, it is necessary to apply the appropriate acceleration techniques. Among them, the multigrid method is the most efficient technique known today (Hirsch 1988). Objective of this paper is to present a numerical method for the analysis of 3D incompressible internal flows. The numerical procedure is based on the multistage Runge-Kutta scheme, proposed by Jameson et al., (1981), in conjunction with local time stepping, implicit residual smoothing and the multigrid method. In order to demonstrate the applicability of our numerical tool, it is applied to calculate a turbulent flow through a square sectioned 90 degree bend, for which Kim and Patel (1994) have provided detailed experimental data. Deng et al., (1997) and Sotiropoulos et al.,(1998) performed calculations on a curved duct turbulent flow calculations based on the numerical solution of the RANS equation. Deng et al., (1997), used two original near-wall Reynolds-stress transport models and very fine grid (480000 grid points). Sotiropoulos et al., (1998) performed computations on a slightly coarser-grid (350000 grid point) using two nonlinear variants of $k - \omega$ model based on the quadratic and cubic constitutive relations. We present numerical results obtained using a practical, coarser grid (202000 grid points), which corresponds to computing resources that are typical of those available to an industrial engineer, placing more grid points near the passage walls.

NUMERICAL METHOD

The three-dimensional incompressible RANS equations with isotropic two-equation, near-wall turbulence closures, are written in generalized curvilinear coordinates in strong conservation form (Soulis et al., 1998). The k-ω model (Wilcox, 1994) is chosen because it is the only available closure without restrictions on the distance from the wall to the near-wall grid point, which is very appropriate for complex geometrical configurations.

The artificial compressibility method is employed to couple the velocity and pressure field in the mean flow equations. The governing equations are discretized on a collocated mesh using the cell vertex finite volume method. The convective fluxes in RANS equation are discretized using second-order central-differencing in conjunction with a flux difference splitting upwind third-order accuracy scheme (Rogers&Kwak, 1988). Three-point central differencing is employed for the viscous fluxes and source terms in turbulence closure equations. The convective flux of turbulence closure equations is formed using Roe's upwind scheme (Hirsch, 1988) with monotone (TVD) interpolation at the cell face (Zijlema,1996).

The discrete mean flow and turbulence closure equations are advanced in time using a pointwise implicit (Zheng et al., 1997) four-stage Runge-Kutta scheme enhanced with local time stepping, variable coefficients implicit residual smoothing and multigrid acceleration. For most two-equation models, the source terms are usually dominant and become stiff near the solid wall and the efficiency of solution then totally depends on the treatment of those source terms. In our numerical model, the source terms in the k and ω equations are divided into two groups locally according to their sign and then only negative parts are treated implicitly. This so-called point-implicit technique is employed to improve the efficiency of the solution and to alleviate the stiffness of the governing equations in the near-wall region. In the present paper the local time step limit is computed with scaled spectral radii of the flux Jacobian matrices for the convective terms. A variable coefficients implicit residual smoothing is used to extend the stability limit and robustness of the basic scheme. The variable coefficients depend on the spectral radii of the flux Jacobian matrices as well as Courant numbers of the smoothed and unsmoothed scheme. A time-lagged or loosely-coupled approach is employed in solving the Navier-Stokes equations and two equation turbulence model in a time marching method, i.e., mean flow and turbulence closure equations are solved separately. A three level V-cycle multigrid algorithm with semi-coarsening in the transverse plane is applied only to the mean flow equations with one, two and three iterations performed on the first, second and third grid level, respectively. The turbulence closure equations are solved only on the finest mesh and eddy viscosity is injected to the coarser meshes and kept frozen during the multigrid process. Three single-grid iterations are performed on the turbulence closure equations per multigrid cycle (Lin et al., 1997).

APPLICATION

Kim and Patel (1994) provided detailed experimental data of the development of turbulent air flow with constant temperature through a 90 deg bend with rectangular cross-section. The duct geometry and the locations of the measured stations are shown in Fig. 1. The bend section was connected to straight inlet and outlet tangents, all of rectangular section 6H x 1H = 121.8 x 20.3 mm. Inlet bulk axial velocity corresponds to a fully developed turbulent flow in a rectangle section duct was $U_0 = 16$ m/s. Reynolds number based on the bulk axial velocity and hydraulic diameter was Re = $2.24 \cdot 10^5$.

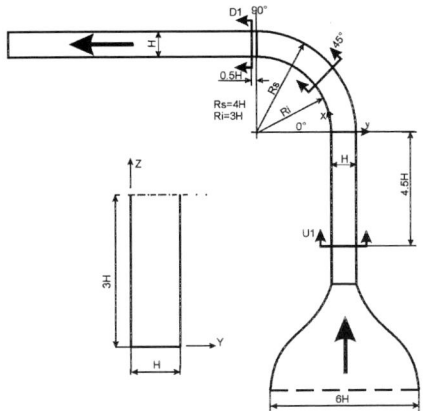

Fig. 1 Flow geometry and coordinate system used

The computational domain starts 4.5H upstream from the inlet of the bend (station U1) and extends up to 30H downstream from the exit of the bend. Due to flow field symmetry, only half of the domain is considered. Domain is discretized by 202581 grid points (81 x 41 x 61). For inflow conditions the measurement data at location U1 are used. Due to space limitations, only a sample of the computed results is compared with the experimental data at three station, using four Z=cons. lines (Fig. 2 - 4).

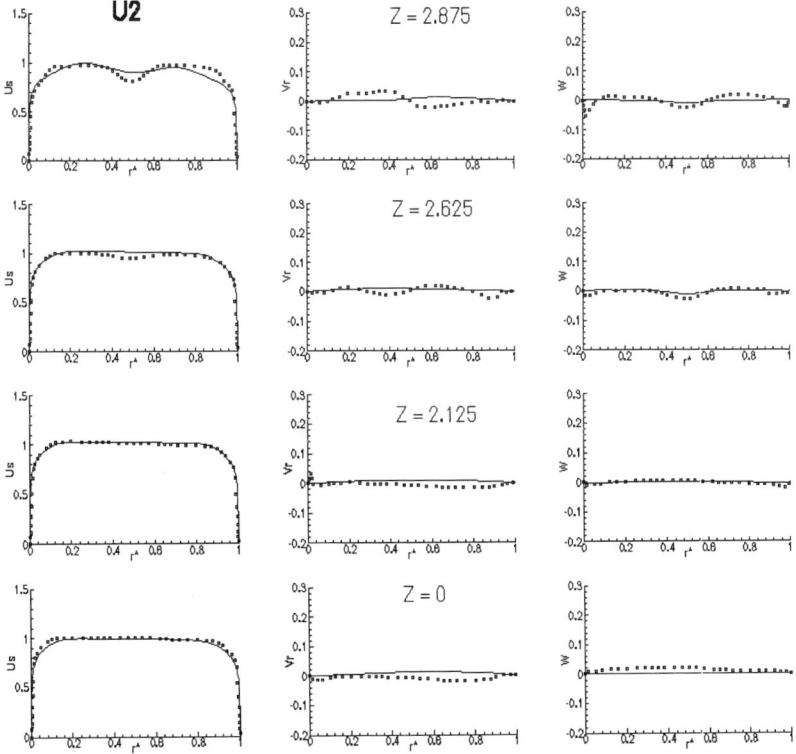

Fig. 2 Measured (Kim & Patel, 1994) and computed velocity profiles at U2 position.

At the station U2 (Fig. 2) the experimental results, as well as numerical results show distortion of streamwise velocity profile due to the inlet contraction-induced vortex pair. This is especially visible near the center of bottom wall.

The measured velocity profiles near the center of the bottom wall at 45 deg. (Fig. 3) are still distorted because the contraction-induced vortex pair continues to affect the flow within the bend. The favorable pressure gradient along the inner wall accelerates the flow and causes the maximum of the axial velocity profile to be shifted from the center toward the inner wall of the bend. The transverse pressure gradient, on the other hand, gives rise to significant secondary motion (sharp peak of the W profiles near the inner wall. This pressure-driven secondary motion acts to reduce the mean streamwise velocity near the inner wall by transporting there low-momentum fluid from the bottom wall boundary layer (U-velocity profile at Z = 2.875).

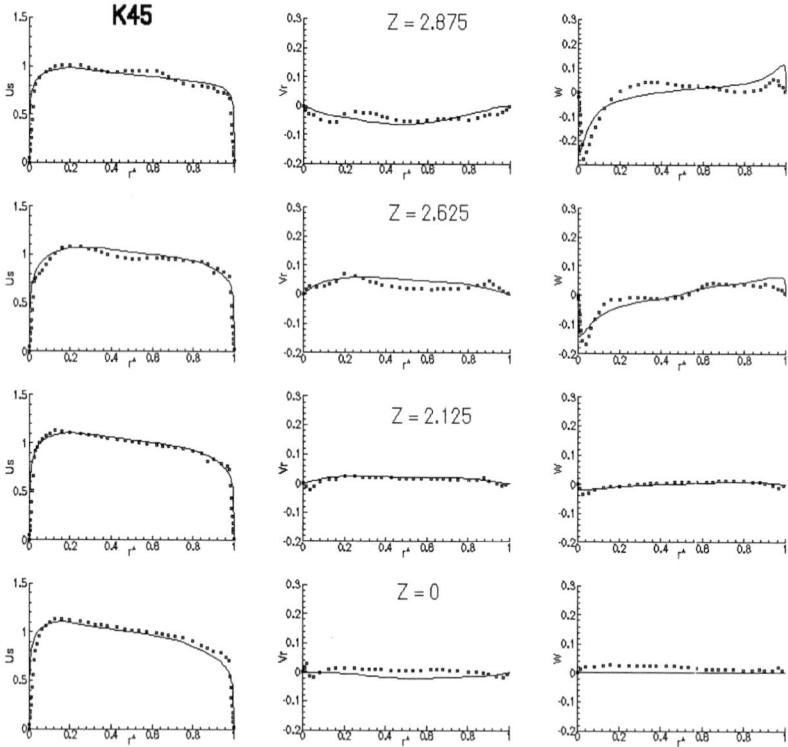

Fig. 3 Measured (Kim & Patel, 1994) and computed velocity profiles at 45 deg.

Unlike at 45 deg position (K45), the opposite pressure gradient at D1 position (Fig. 4) causes the flow acceleration near to the outer wall.

The pressure distribution along the channel at different r = cons. and z = cons. location is shown in Fig. 5. The pressure gradient induced by the curvature can be clearly seen. On the convex side, the boundary layer is subjected to a favorable pressure gradient starting upstream of the bend, and this is followed by an adverse gradient around the bend exit. The boundary layer on the concave side is subjected to pressure gradients of similar magnitude but opposite signs. Pressure distribution in the bend is visualized in Fig. 6.

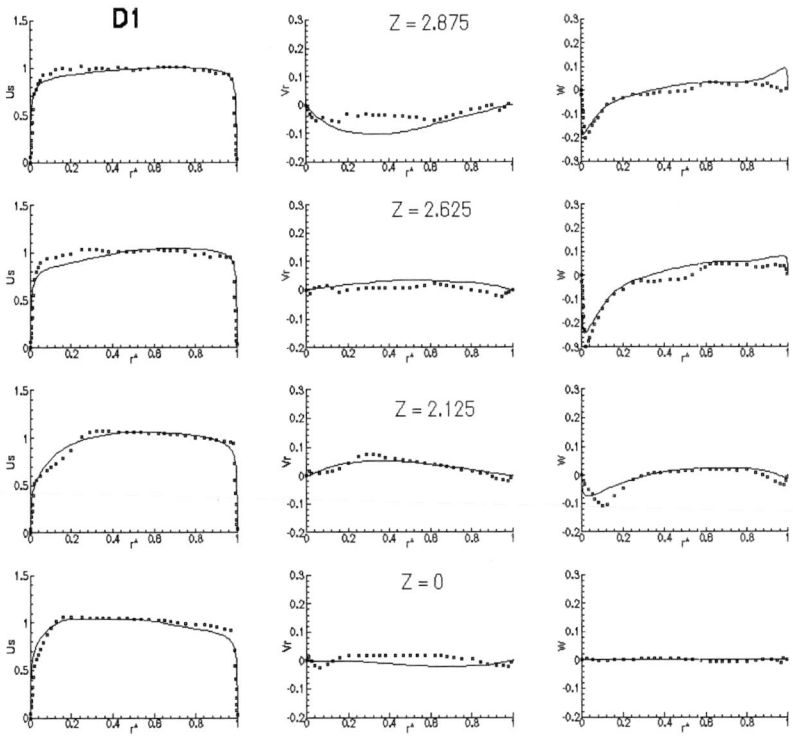

Fig. 4 Measured (Kim & Patel, 1994) and computed velocity profiles at D1 position.

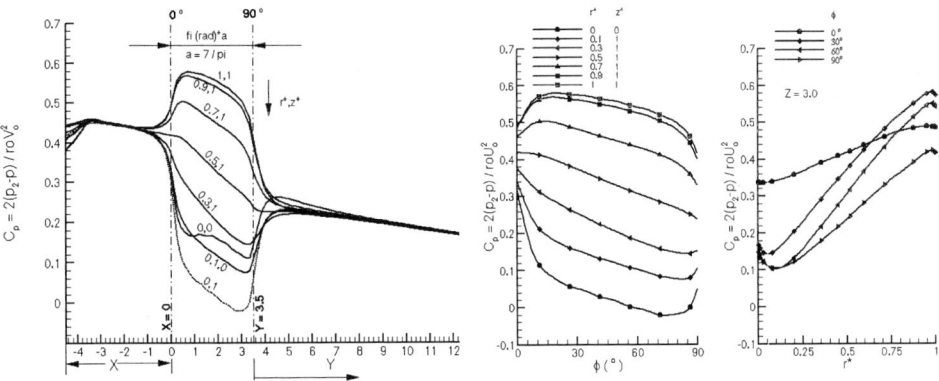

Fig. 5 Computed pressure coefficient distribution along the channel

CONCLUSION

The results obtained using numerical model based on RANS equation with isotropic two-equation near-wall turbulence model, show that global features of the flow field are correctly resolved but some significant differences between measured and numerical results still remain. The same is noticed in

484

Sotiropoulos et al., (1998) even in case a non-linear near wall turbulence closure is employed. A key factor in obtaining the correct primary flow is the secondary motion prediction with reasonable fidelity.

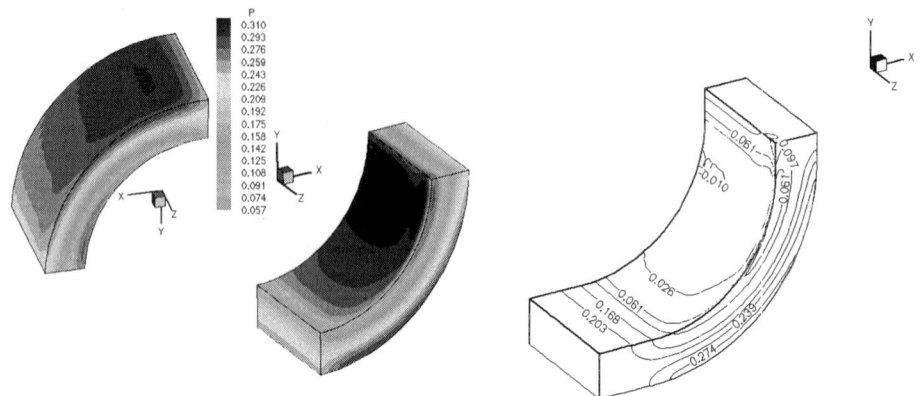

Fig. 6 Computed pressure distribution in the bend

Some differences between measured and computed values are, probably, due to failure of the isotropic two-equation turbulence model to resolve the turbulence structure in a bend that is significantly influenced by its curvature. The general conclusion is that the presented method needs some further improvement to resolve some subtle details that are not fully captured by the method.

REFERENCES

Deng G.B, Visonneau M. (1997), Assessment of Second-Moment Turbulence Closures for Three-Dimensional Vortical Flows, *ASME FEDSM97-3169*.

Hirsch C. (1988). *Numerical Computation of Internal and External Flows, Volume 2: Computational Methods for Inviscid and Viscous Flows*, John Wiley & Sons.

Kim W. J, Patel V. C. (1994), Origin and Decay of Longitudinal Vortices in Developing Flow in a Curved Rectangular Duct, *Journal of Fluid Eng.*, **116**, 45-52.

Lin F.B. and Sotiropoulos F. (1997). Strongly-Coupled Multigrid Method for 3-D Incompressible Flows Using Near-Wall Turbulence Closures, *Journal of Fluid Eng.*, **119**, 314-324.

Rogers, S E, Kwak D. (1988), An Upwind Differencing Scheme for the Steady-State Incompressible Navier Stokes Equations, *NASA* TM 101051.

Sotiropoulos F., Ventikos Y., (1998), Prediction of Flow 90-Degree Bend Using Linear and Nonlinear Two-Equation Turbulence Models, *AIAA Journal*, **36:7**, p. 7.

Wilcox D. (1994). Simulation of Transition with a two-equation turbulence model, *AIAA Journal*, **32:2**, p.285.

Zijlema M. (1996). Computational modeling of turbulent flow in general domains, *Ph.D Thesis*, Delft University of Technology, Netherlands.

Zheng, S, Liao, C, Liu, C, Sung, C H, Huang, T T. (1997). Multigrid Computation of Incompressible Flows Using Two-Equations Turbulence Models, *Journal of Fluids Eng.*, Vol. 119, pp. 893-899.6.

Jameson, A, Schmidt, W, Turkel, E. (1981). Numerical Solutions of the Euler Equations by Finite Volume Methods Using Runge-Kutta Time Stepping Schemes, *AIAA paper* No. 81 - 1259.

Soulis, J V, Jovicic, N, Milovanovic, D, et al (1998). Numerical Modeling of Incompressible Turbulent Flow in Turbomachinery, *Computational Fluid Dynamics '98* edited by Papailiou, K., et al., pp. 259-265, John Wiley&Sons.

Air Distribution in Rooms, (ROOMVENT 2000)
Editor: H.B. Awbi

LOCATING THE SEPARATION POINT OF NONISOTHERMAL WALL JETS USING COMPUTER VISION

C. Ghiaus[†], A. Meslem, F. Allard, M. Robitu[†]

LEPTAB - Laboratoire d'Etudes des Phénomènes de Transfert Appliqués au Bâtiment, Université de La Rochelle, France
[†]On leave from Technical University of Bucharest, Romania

ABSTRACT

Cold air jets introduced just below the ceiling attach to the horizontal wall and then separate from it. The separation point location, specific for a given type of nozzle, air jet speed and temperature difference between the supplied and ambient air, is an important design information. Light sheet flow visualization technique allows us to locate the separation point of nonisothermal wall jets through visual observation or computer vision. The instability of the separation point location requires readings sampled every second for several minutes, making the visual observation inefficient. As an alternative, a technique of computer vision can locate in a digital image the specific form of the detachment of the jet from the wall. A numerical method to locate features within an image is to compute the correlation between the image of the jet and a template that represents the characteristic form of the jet in the separation point. The correlation, performed using the Discrete Fourier Transform, gives a measure of the clarity of the unsteady separation pattern. The algorithm is faster and the location of the separation point is more accurate when the region of the image, on which the template is matched, is smaller. Consequently, the first step of the proposed method is to coarsely locate the separation point by determining the last frontier of the wall jet. Then, a gradient based algorithm finds the jet edges in the image. Finally, the correlation between the specific detachment pattern shape and the edge image is calculated. The separation point is located where the highest correlation is found. A simple method is used to scale the image and correct the lens distortion. The main drawback of the pattern matching in locating the separation point of wall jets is its sensitivity to jet clarity in the image.

KEY WORDS

Air conditioning, Nonisothermal jets, Flow visualization, Digital image processing

INTRODUCTION

In some air conditioning applications, cold air is supplied below the ceiling of the room. The jet detaches from the ceiling when its negative buoyancy is greater than the positive pressure difference caused by the proximity of the horizontal surface. The location of the separation point is important for

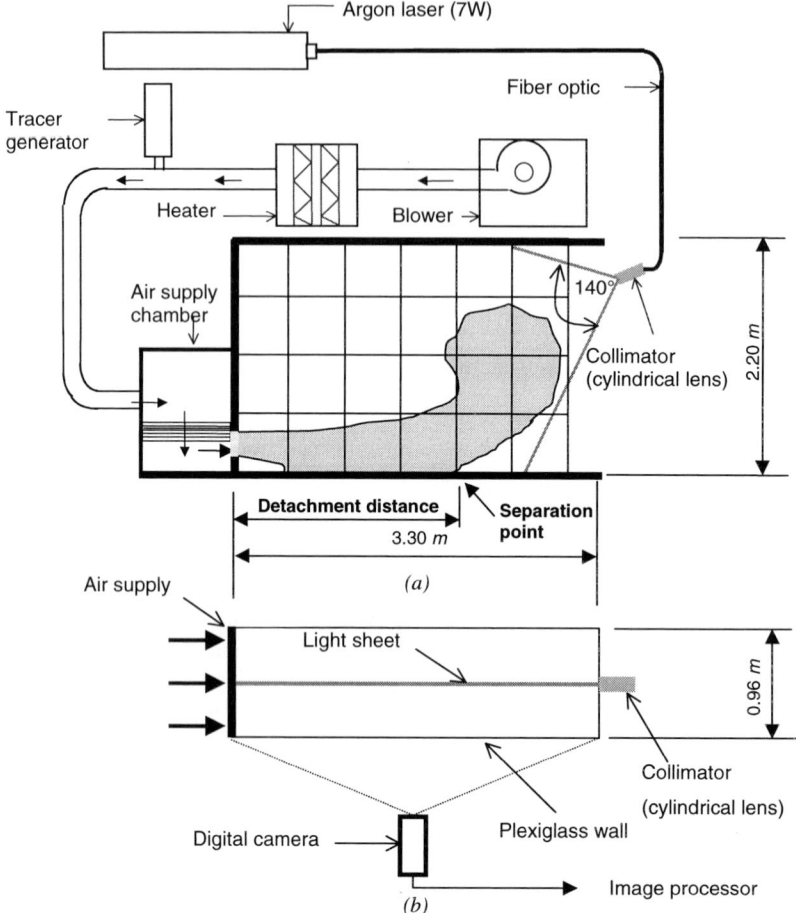

Fig. 1   **Test cell for airflow visualization in order to determine the separation point and to measure the detachment distance: (a) side view; (b) upper view.**

thermal comfort in the occupied zone of the room, as the air jet may separate from the ceiling before properly mixing, or it may have a too long trajectory.

From a physical point of view, a cold jet introduced below the ceiling is equivalent to a warm jet introduced above the floor; because the heating is easier to achieve technically than cooling, this equivalence is used to study the detachment distance of cold jets.

Several experimental procedures are used to measure the detachment distance of negatively buoyant wall jets. Actually, the separation point is located by measuring the velocity or pressure near or on the horizontal surface. To measure the velocity next to the surface, Myers *et al.* (1963) utilized a flattened, total-head tube with an opening of about 0.05 by 0.5 mm, Hassani *et al.* (1997) used boundary layer probes, and Meslem *et al.* (1999) a two dimensional hot wire anemometer. To measure the pressure on the surface, Sandberg *et al.* (1993) employed surface pressure tappings connected to a micrometer via a scanner-valve. Kim *et al.* (1996) exploit the change in color of liquid crystal surface to measure the heat flux. Hassani and Stetz (1994) located the jet separation point where the air velocity at the ceiling was small (0.25 m/s or less) and/or where the gradient of air temperature was sharp. They measured the velocity with hot wire anemometer and the temperature gradient by visual observation on images taken with an infrared camera.

Airflow visualization by introducing a stream in the room is a widely utilized technique [9 - 18]. Using this technique, Sandberg *et al.* (1993) visually located the detachment point of negatively buoyant wall jets. Laser tomography, or laser-light sheet, is particularly adequate for viewing a predetermined plane section while blurring out the images of other planes.

The principle of laser tomography is to widen a laser beam to obtain a light sheet of 2 mm to 3 mm in thickness; the light is diffused or reflected by particles admixed in the air flow that becomes visible. Both qualitative and quantitative measurements can be achieved with this technique, using visual observation or digital image processing of frames captured with a digital camera.

LOCATING IMAGE FEATURES

A digital image is a set of points in a plane, each with its own luminance and possible color. The image can be binary (with only two distinct values for luminance), gray-valued or colored. In flow visualization with laser light, only the gray images are considered; they are completely described by the luminance at each point. An image feature is located where a best match exists between a small pattern (template) and a set of patterns in a larger image. A way to quantitatively express the match is by performing the correlation.

Correlation, $g(x, y) = f(x, y) \circ h(x, y)$, is a mathematical operator on two functions, $f(x,y)$ and $g(x, y)$, defined as the sum of the products of the functions values for every relative position of the two functions:

$$g(x, y) = f(x, y) \circ h(x, y) = \sum_{a=-\infty}^{\infty} \sum_{b=-\infty}^{\infty} f(a,b) \cdot h(x+a, y+b) \qquad (1)$$

Correlation is closely related to convolution, $g(x, y) = f(x, y) * h(x, y)$, a mathematical operator on two functions, $f(x, y)$ and $g(x, y)$, defined as the sum of the products of the functions values, with one of the function mirrored in the origin:

$$g(x, y) = f(x, y) * h(x, y) = \sum_{a=-\infty}^{\infty} \sum_{b=-\infty}^{\infty} f(a,b) \cdot h(x-a, y-b) \qquad (2)$$

The "convolution rule" states that convolving f with h in space domain is equivalent to multiplying their corresponding Fourier transforms, F with H, in frequency domain:

$$f(x, y) * h(x, y) \Leftrightarrow F(u,v) \cdot H(u,v) \qquad (3)$$

This equivalence is important because a convolution product, which is a sum of products, may be performed using simple multiplication. Thus, in order to calculate the convolution product, we have to transform f and h in frequency domain, perform the product and transform the result back in space domain. Although the multiplication is faster then convolution product, this method needs a direct and an inverse Fourier transformation. However, in terms of number of arithmetic operations (considering the time of summation equal to the time of multiplication), the Fourier transform based convolution becomes attractive when the dimension of f and h are larger then 2^4; when the size is 2^{10}, the saving ratio is of about 200 (Strang and Nguyen, 1997).

Correlation can perform template matching, i.e. to find the position where a "best" resemblance exists between a small pattern (template) and an image (Theodorius and Koutroumbas, 1998; Fukunaga, 1990). In order to find the position in the image of features that resemble to the template, the correlation between the image and the template is performed through convolution between the image and the template mirrored in the origin. The maximum correlation is obtained for the objects that have the form of the template, whether or not they have other objects superimposed. The correlation value is lower where the resemblance with the template is lower.

LOCATING THE DETACHMENT POINT OF BUOYANT JETS BY TEMPLATE MATCHING

In the detachment point, the jet and the wall form an angle, which represents the feature to be located in the image of the jet boundary (Fig. 2). We observed that in the detachment point the jet rapidly shifts direction and the turbulence of the jet makes the separation point to be unstable and to drift with 15-20 cm, characteristics also mentioned by Hassani and Stetz (1994). The boundary of the jet in the separation region may be unclear and unstable. Template matching is identified to have advantages in dealing with such applications (Wu, 1995). The correlation gives a measure of the resemblance between a perfect form of the jet detachment, represented by the template (Fig. 2 (b)), and the image of the jet boundary (Fig. 2 (c)). Thus, the value of correlation indicates the clarity of the location of detachment point.

The jet boundary is found looking for places where the intensity changes rapidly. Canny's algorithm (Canny, 1986) looks for local maxima of the image intensity gradient. The method is suited for jet boundary because it may detect strong and weak edges. The weak edges are included in the output only if they are connected to the strong edges (Fig. 3 (c)).

Template matching may find the position in the image of patterns that resemble to the detachment of the jet from the wall. Other algorithms improve the reliability of separation point location. An

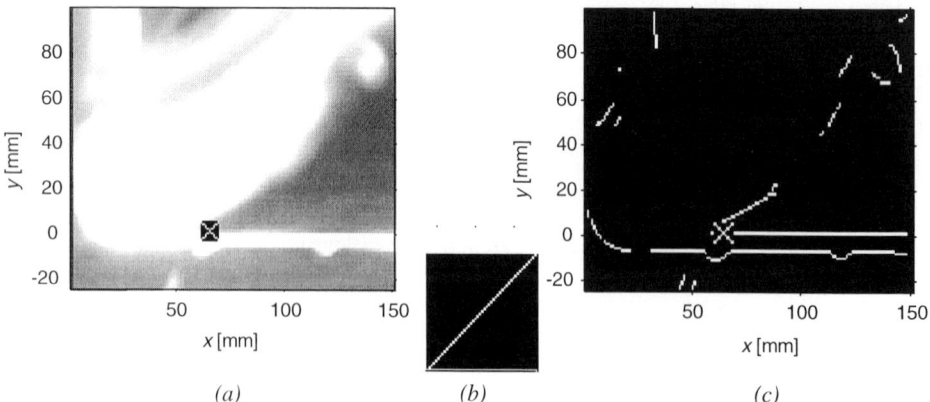

(a)   *(b)*   *(c)*

Fig. 2 Detachment point of a buoyant jet from the wall: (a) the image of the jet in the region of detachment; (b) the template of detachment of the jet from the wall; c) the image of the jet boundary which is matched with the template. The detachment point founded is marked with an "x" in a black square.

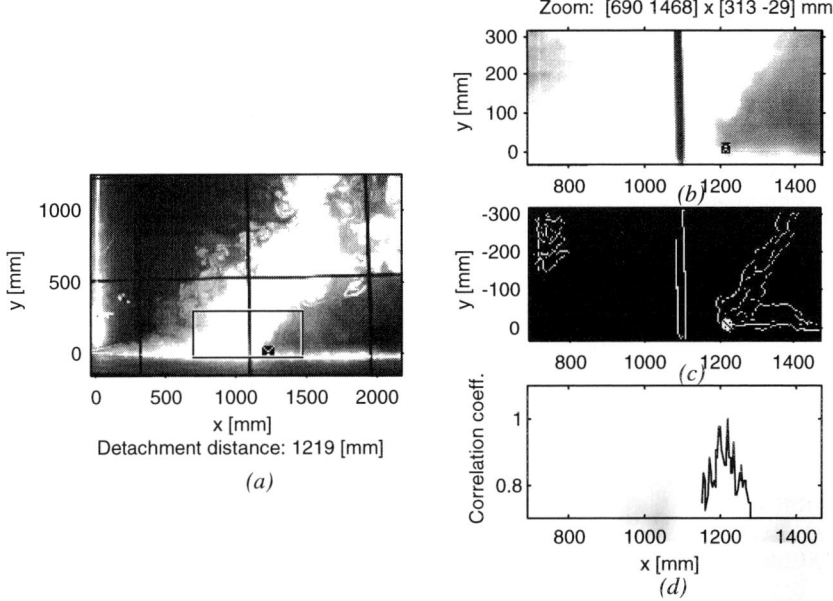

Fig. 3 Measuring the detachment distance of buoyant wall jets: (a) the image of the jet; (b) the detachment zone in which the template is matched; (c) the boundary of the jet; (d) correlation between the detachment template and the image of the boundary.

algorithm corrects lens distortion and scales the digital image. The scaling and the lens distortion correction are achieved by locating the position of cellophane tape markers stick on the horizontal wall. The tape reflects the laser sheet more than the black-painted wall; consequently, the marker positions can be easily located and used for scaling, without correcting for lens distortion, and thus avoiding the use time consuming non-linear transformation (Sawhney and Kumar, 1999). The location of the detachment point is speeded up by reducing the space for template matching to a zone around the detachment point coarsely located in the picture using a simple algorithm which detects the last frontier of the jet (Fig. 3 (a) and (b)). The complete algorithm decides, based on the intensity of the jet image, whether or not the image is appropriate for processing. The working algorithm finds the separation point in images considered difficult by a human expert. The correlation values give a measure of the range where the separation is produced (Fig. 3 (d)). The detachment point is chosen where the correlation is maximum (all the correlation coefficients are normalized by dividing with the maximum value of the correlation). Successive frames of a jet introduced in the same conditions may be processed statistically to locate the separation point of the jet from the wall.

CONCLUSIONS

The laser-light sheet technique provides a rapid means of observing nonisothermal jets distribution patterns in large spaces. The turbulent nature of the jet and the instability of the separation point are readily visualized. The jet separation point can be measured if the jet is observed for several minutes.

Template matching is an effective way to locate the separation point in digital images captured when laser-light sheet is used for flow visualization. In this method, a template of perfectly clear detachment pattern is matched through correlation with the image of the boundary of the jet, which are located where the maximum local gradient of the intensity of the image is found. The correlation is accomplished using multiplication in frequency domain instead of convolution product in space domain. The correlation coefficients give a measure of the clarity of jet detachment.

Scaling and correction of lens distortion of the distance is accomplished by visualizing markers with known location along the horizontal wall in the plane of the laser light and of the separation point.

The result of the detachment point location using pattern matching consists of a set of correlation coefficients. Their values may be used in statistical analysis of the detachment point drift.

ACKNOWLEDGEMENTS

The Poitou-Charentes Region, France and the European Commission supported this work through a research grant and a TEMPUS scholarship.

REFERENCES

Myers G. E., Schauer J.J., Eustis R.H. (1963). Plane Turbulent Wall Jet Flow Development and Friction Factor, *Journal of Basic Engineering, Transactions of ASME,* **85**, 47-54

Hassani A.V., Kirkpatrick A., Sforza P.M. (1997). Flow Separation Characteristics of Three-dimensional Negatively Buoyant Wall Jets, *HVAC&R Research,* **3 (2)**, 112-127

Meslem A., Beghein C., Inard C., Allard F. (1999). Lois de decroissance d'un jet turbulent tridimensionnel vertical de paroi impactant a forces de poussée defavorable, *Int. Comm. Heat Mass Transfer,* **26 (4)**, 487-498

Sandberg M., Wiren B., Claesson L. (1993). Attachment of a cold plane jet to the ceiling -- Length of recirculation region and separation distance, *Proc. Roomvent 93,* Aalhorg, Denmark, 487-499

Kim D.S., Yoon S.H., Lee D.H., Kim K.C. (1996). Flow and heat transfer measurements of a wall attaching offset jet, *Int. J. Heat Mass Transfer,* **39 (14)**, 2907-2913

Camuci C., Glezer B. (1997). Liquid crystal termography on the fluid solid interface of rotating systems, *Transactions of ASME,* **119**, 20-28

Hassani V., Stetz M. (1994). Effect of local loads on separation of negatively buoyant wall jets in enclosed spaces, *Proc. Roomvent 94,* Krakow, Poland, 345-362

Strang G., Nguyen T. (1997. *Wavelets and filter banks,* Welleslet-Cambridge Press, 263-271

Theodorius S., Koutroumbas K. (1998). *Pattern Recognition,* Academic Press

Fukunaga K. (1990). *Introduction to statistical pattern recognition,* Academic Press

Wu Q. X.(1995). A correlation-relaxation-labeling framework for computing optical flow -- Template matching from a new perspective, *IEEE Transactions on Pattern Analysis and Machine Intelligence,* **17 (9)**

Canny J. (1986). A Computational Approach to Edge Detection, *IEEE Transactions on Pattern Analysis and Machine Intelligence,* Vol. **PAMI-8 (6)** 679-698

Sawhney H.S., Kumar R. (1999). True multi-image alignment and its application to mosaic and lens distortion correction, *IEEE Transactions on Pattern Analysis and Machine Intelligence,* **21 (3)**

© 2000 Elsevier Science Ltd. All rights reserved.
Air Distribution in Rooms, (ROOMVENT 2000)
Editor: H.B. Awbi

TEMPERATURE AND VELOCITY MEASUREMENTS ON A DIFFUSER FOR DISPLACEMENT VENTILATION WITH WHOLE FIELD METHODS

Elisabet Linden[1]**, Mathias Cehlin**[1,2]**, Mats Sandberg**[1]

1. Division of Indoor Environment Ventilation, Department for Built Environment,
Royal Institute of Technology, Gävle, Sweden
2. Division of Energy and Production Systems, Department of Technology,
University of Gävle, Gävle, Sweden

ABSTRACT

In this study the instantaneous temperatures and velocities close to a diffuser for displacement ventilation have been recorded by using whole-field measuring techniques. The air temperatures were measured indirectly by the use of a low thermal mass screen in conjunction with infrared thermography. The measuring screen was mounted parallel to the airflow, acting as a target screen. By using the thermal images the size of the near zone was also calculated. To determine air movements a whole field method called particle streak velocimetry (PSV) was used. Images of tracks created by small, low-density particles, suspended in the air, were analysed using computerised image processing to obtain the velocities. The experiment took place in a climate chamber in which the wall and air temperatures were controlled. The diffuser was located in the centre of one of the walls. The tests were conducted for a supply flow of 15 l/s and a temperature difference between the inlet air and the room air of 4 $^{\circ}C$ and of 6 $^{\circ}C$. This paper deals with the results obtained from the two whole-field measurement methods. The results show that the two whole-field measurement methods can be good tools for visualising and measuring air velocities and temperatures in rooms. These techniques could be used in the work of improving the indoor climate.

KEYWORDS

Indoor climate, Whole field measurement methods, Air velocity, Air temperature, Digital infrared camera, Temperatures, Diffuser, Displacement ventilation, Infrared thermography, 2D Particle Streak Velocimetry (PSV), Digital pictures, Particle tracking.

INTRODUCTION

Low velocity diffusers, in displacement ventilation systems, supply air directly into the zone of occupation, making the near-zone of the diffuser very critical for comfort assessment. This type of air supply terminal may give rise to a supply Archimedes number, Ar(0), greater than unity. Therefore one can expect buoyancy to strongly influence the flow, leading to the formation of a gravity current with high air velocities close to the diffuser. It is therefore of great importance to determine the size of this zone, in order to be able to achieve acceptable indoor environment. It is rather difficult to investigate the near-zone of a low velocity diffuser in real situations.

Conventional methods for measuring both air velocities and temperatures are based on single point techniques using sensors. With these techniques the measurements are performed only at the spot where the sensor is placed. These conventional methods for measuring within the near-zone do not give adequate information. In this paper, two whole-field measurement methods were used to register accurate information from the whole near-zone instantaneously.

"Particle streak velocimetry" (PSV), or alternatively, particle tracking, is a whole-field method that is used for recording two- or three-dimensional air velocities. This is achieved with the use of images of tracks created by small, low-density particles suspended in the air. In this case the particles are injected into the supply duct. With the help of computerised image processing two-dimensional velocities can be obtained. This method is explained in more detail in Linden et al (1998).

New, high definition, digital infrared cameras have provided opportunities to develop a method for measuring temperatures within a large area. With this method the air temperature is measured indirectly by using infrared thermography and a measuring screen. The method is very useful for air temperature measurements as well as airflow pattern visualisation. For more detailed information about this measurement method see Cehlin et al (2000). The length of the near-zone could be determined with the use of the resulting thermal images.

The two above methods have been developed at the Centre for Built Environment in Gävle. The work presented here is part of a wider research program with the aim of "Making the indoor climate visible at the design stage". It includes whole-field measurement techniques for temperature (infrared thermography), concentration (absorption tomography), velocity (particle streak velocimetry) and computational fluid dynamics.

EXPERIMENTAL METHODS

Velocity and temperature measurements

The experiments took place in a climate chamber with displacement ventilation. The walls and the floor were painted black. The diffuser, with a height of 0.6 m and a free area of 0.03 m^2, was located at the centre of one of the walls. A number of thermocouples were installed in the room together with one in the diffuser. Tests were conducted for a supply airflow of 15 cm/s and temperature differences between the inlet air and the room air temperature at 4 °C and 6 °C. The instantaneous velocities and temperatures close to the diffuser were measured using the whole-field methods.

In this experiment some modifications have been made to the original procedures of the particle streak velocimetry method. These changes are explained below.

To be able to visualise and measure the air motion, particles are introduced into the inlet air. These particles must have a density low enough to follow the air and at the same time be big enough to be observed. In these measurements a new kind of particle has been used. The material consists of cellulose that emits fibres (particles) when it vibrates. A purpose made injector has been used to inject the particles. The injector was a loudspeaker where the cellulose material was placed on the membrane. Duo to the membrane vibrations the fibres were released into the air. This injector was placed inside the ventilation duct 3.5 m upstream of the diffuser.

A light sheet is used to observe the particle movements in the region of interest. The light sheet which has a thickness of about 4 cm to 5 cm is produced by a portable halogen lamp and a cylindrical lens.

Photographs were taken to register the particle movements close to the diffuser. In this work a standard SLR (single lens reflex) camera and black and white film have been used. An asymmetric "chopper" rotating at a known speed was placed in front of the camera. This made it possible to find the particles that have been inside the light sheet during the whole exposure and to identify their directions of movement.

While the camera-shutter is open, the air and therefore the particles, move a short distance. A few times while the camera-shutter is open the chopper will block the camera lens and no light falls on the film. On the photograph this results in streaks with three different lengths, shown in Figure 1. The length of these streaks depends on the chopper rotation speed and the particle velocities.

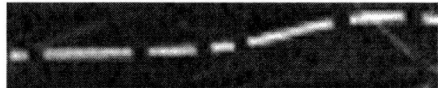

Figure 1: A streak made by a particle.

The photographs were digitised and then analysed in an image-processing program. The first step in analysing the images was to locate the streaks. Secondly, the co-ordinates of the end-points were measured. Only the streaks with the right proportions between the different sub-streaks, were chosen. The co-ordinates were measured in pixels in a two-dimensional system. Within the flow-field there was a system of three reference points with known co-ordinates. This co-ordinate system made it possible to convert the image co-ordinates to co-ordinates in the room.

By using the particle displacement ($\Delta x, \Delta y$) and the time-period Δt, the velocities were then be evaluated. With this method the velocities were registered at the points in the room where the individual particles were at the photographed moment.

The temperatures have been measured with the whole-field method explained in Cehlin et al (2000).

Near zone measurements

The horizontal distance, x_d, from the centre of a semicircular diffuser, with radius R and height H, to the point where the air drops down on the floor, is defined in a non-dimensional term, $x_d^* = x_d/H$. According to the theory we have

$$x_d^* \propto \sqrt{\left(\frac{R}{H}\right)^2 + 2\frac{R}{H}\frac{1}{Ar(0)^{1/2}}} \,. \tag{1}$$

See Etheridge and Sandberg (1996) page 384 for information about the derivation of Equation 1.

The purpose of the experiments described in this section was to study how accurate the dimensionless distance, x_d^*, from a low velocity diffuser can be measured by infrared thermography compared to the theory. The type of camera used was an AGEMA 570 with a resolution of 320x240 pixels, sensitive to long-wave radiation (7.5 µm to 13 µm). A paper screen was used as measuring screen throughout all of the experiments. Two light emitting diodes were placed out as position-markers. One was placed on the floor next to the diffuser-inlet and the other one was placed at a horizontal distance of 800mm from the diffuser. Experiments were carried out for 5 different supply air conditions with a semicircular diffuser of height 600mm and free area 0.0075m². x_d was redefined under the experiment because it was very hard to estimate the point where the air dropped down on the floor from the infrared camera measurements. Instead, x_d was defined for these experiments as the distance from the diffuser to the point where the airflow pattern seemed to change direction to a purely horizontal flow, as indicated in Figure 2.

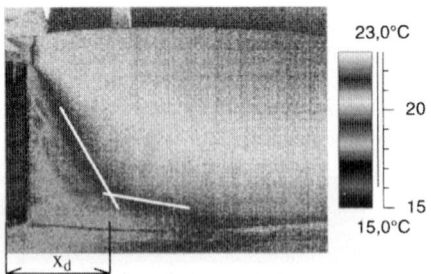

Figure 2. Thermal image, showing the distance, x_d, from the diffuser to the point where the airflow pattern seems to change direction to purely horizontal flow. $U_{in} = 0.13$ m/s, $T_{in}=17.0$ °C and $Ar(0) = 3.86$.

RESULTS

Velocity and temperature measurements

By using the PSV method a number of streaks were found and the position, speed and direction for the particles were calculated. The two-dimensional velocity field is presented as an airflow pattern diagram. The lengths of the arrows represent the speed of each particle. The results from the whole-field method using infrared thermography were images with different colours representing different temperatures of the air around the diffuser. Both whole-field methods were combined, as shown in Figures 3a and 3b.

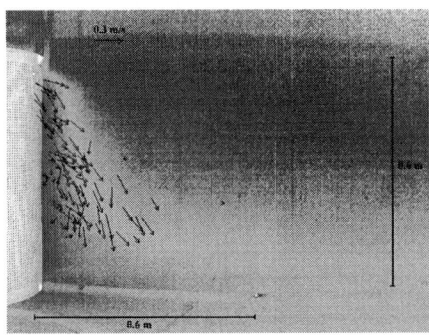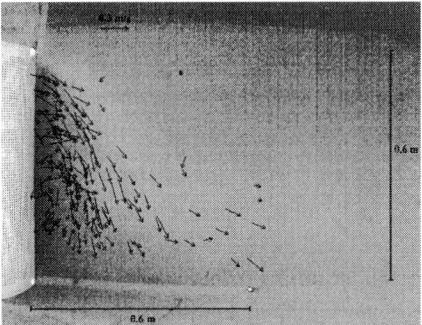

Figure 3: Temperatures and velocities around a diffuser.
3a (left): Supply flowrate =15 l/s, $\Delta T_{in} = 4\ °C$ and Ar(0) = 0.32.
3b (right): Supply flowrate =15 l/s, $\Delta T_{in} = 6\ °C$ and Ar(0) = 0.47.

Measurements of the near zone

As can be seen in Figure 4, the results show that the dimensionless horizontal distance, $x_d{}^*$, measured by infrared thermography conforms to Equation 1, which is derived from theory. However, $x_d{}^*$ was slightly less than predicted by the theory. One source of error is probably that the x_d defined in the experiment was not identical to the theoretical parameter. Another source of error is that the flow from the supply diffuser in the theory

Case	U_{in} (m/s)	T_{in} (°C)	Ar(0)
1	0.09	17.0	12.25
2	0.13	17.0	3.86
3	0.20	16.9	2.10
4	0.27	16.8	0.72
5	0.35	17.0	0.31

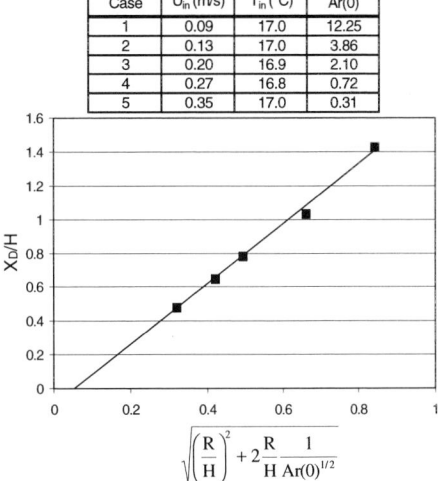

$$\sqrt{\left(\frac{R}{H}\right)^2 + 2\frac{R}{H}\frac{1}{Ar(0)^{1/2}}}$$

is assumed to be uniform, which is not the case in reality.

Figure 4. Recorded dimensionless distance for different supply air conditions.

DISCUSSION

"A photograph says more than thousand words" seems to be appropriate at least when looking at the results from the whole-field measurement techniques. Both the different colours representing temperatures and the velocity arrows clearly show the airflow in the near-zone of the diffuser. This kind of visualisation is of great assistance in understanding the indoor airflow. Furthermore, these techniques could be valuable for designing pleasant indoor environment systems.

NOTATIONS

g   Acceleration of gravity [m/s^2]
H   Height of supply device [m]
U_{in} Supply velocity [m/s]
T   Temperature [K]
ΔT_{in} Temperature difference between room air temperature and supply temperature T_{in}
Ar(0) Supply Archimedes number

$$Ar(0) = \frac{g \dfrac{\Delta T_{in}}{T} H}{U_{in}^2}$$

ACKNOWLEDGEMENT

The help from Mr. Hans Lundström is gratefully acknowledged.

REFERENCES

Piechocinski, J., Sandberg, M. and Linden, E. (1997). Accuracy of Stereo-Photogrammetry for determining three-dimensional velocities in rooms, KTH Byggd Miljö, TRITA-IMV, Technical Report 1997:1, ISSN 1402-5442

Linden, E. and Sandberg, M. (1997). Användning av färgbilder av partikelspår för att bestämma riktningen hos lufthastighets-fält., KTH Byggd Miljö, TRITA-IMV, Technical Report 1997:2, ISSN 1402-5442

Linden, E., Todde V. and Sandberg, M. (1998). Indoor Low Speed Air Jet Flow: 3-Dimensional Particle Streak Velocimetry, Proceedings Roomvent'98, July 14-17, 2, 569-576

Todde V., Linden, E. and Sandberg, M. (1998). Indoor Low Speed Air Jet Flow: Fibre Film Probe Measurements, Proceedings Roomvent'98, July 14-17, 2, 585-592.

Etheridge, D., and Sandberg, M. (1996). *Building Ventilation – Theory and Measurement,* JohnWiley & Sons, Chichester, UK

Cehlin, M., Moshfegh, B. and Sandberg, M. (2000). Visualisation and measuring of air temperatures based on infrared thermography, Proceedings of ROOMVENT 2000, Reading, UK.

Air Distribution in Rooms, (ROOMVENT 2000)
Editor: H.B. Awbi

MEASUREMENTS OF AIR CHANGE RATES IN INDOOR SPACES USING STEP-DOWN METHOD OF VIDEO IMAGE SIGNALS

OHBA, M. and IRIE, K.

Dept. of Architectural Engineering, Faculty of Engineering,
Tokyo Institute of Polytechnics, 1583 Iiyama, Atsugi, 243-0297 Japan

ABSTRACT

Video camera calibrations and field experiments have been performed to develop a new method of measuring air change rates using a video imaging technique. Calibration of a video camera used for broadcasting showed good correlation between image signals and luminous reflectance of achromatic color chips. This was achieved by appropriately adjusting the pedestal level of the video camera so that the image signals were made equal to zero for the black level of the picture. The 8mm video camera for home use caused non-linearity at low luminous reflectance due to narrow dynamic ranges. In the field tests, air change rates in the test house were measured from the decay curves of video image signals obtained by the step-down method assuming perfect mixing of tracer particles inside spaces. Smoke candles were used as tracers, and the mean diameter of the smoke particles varied from 0.77 µm to 1.06 µm. When adjusting camera f-stops for the experimental conditions, it was verified that the digital video method could measure air change rates with virtually the same precision as derived from the decay curves of smoke particle concentrations using an aerosol monitor.

KEYWORDS: Video image, Luminance signal, Air change rate, Step-down method, Test house

INTRODUCTION

The tracer gas decay procedure is widely used for measuring air change rates in buildings and other spaces (Japanese Industrial Standards A1406-1974). This procedure assumes perfect mixing of the tracer. Therefore, if uniform concentrations of tracer gas cannot be produced in large spaces such as atriums and factories, enough measuring points must be chosen both vertically and horizontally inside the spaces to obtain accurate average air change rates The air change rates occurring in natural ventilation change randomly, so at least five test runs must be conducted and averaged to comply with the Japanese Industrial Standards.

SF_6 gas rather than CO_2 has often been used as a tracer because of its nonexistence in the atmosphere. The measurement instrument can therefore accurately measure the tracer concentration up to ppb level, so SF_6 is a suitable tracer for measuring the air change rates for multi-zone or large indoor spaces. However, SF_6 contains F_2 gas, which causes global warming. So its usage will probably be prohibited in the near future. The current procedure requires a lot of time and labor to measure the average air change rates. Furthermore, a new tracer needs to be developed as a substitute for SF_6 and a new environmentally friendly procedure needs to be established that does not cause global warming.

MASS BALANCE MODEL

Figure 1 is a sketch of the mass balance in a room. Tracer gas is emitted at a rate of M from a floor source. Assuming instant and uniform mixing of tracer gas with indoor air, the following equation can be derived for the mass balance in steady state (Ohba and Irie 1998).

$$Mdt + C_oQdt - CQdt = Vdc \qquad (1)$$

Where
  V = indoor air volume, i.e., the entire air volume of a space or building (m^3)
  C = indoor concentration (m^3/m^3) C_o = outdoor concentration (m^3/m^3)
  Q = ventilation rate (m^3/s) t = time (s)

If C is equivalent to C_1 in the initial condition and tracer emission stops after mixing, the following equation is obtained:

$$C = C_o + (C_1 - C_o)\exp(-Qt/V) \qquad (2)$$

When the relationship between the concentration of $(C - C_o)$ and time t is plotted on semi logarithmic graph paper, the air change rate, i.e., Q/V, is derived from the slope. This ratio is considered to be an essential index indicating an acceptable level of indoor air quality.

LIGHT-SCATTERING BY PARTICLES

When a light beam irradiates a spherical smoke particle through an angle, the smoke particle scatters the light in proportion to the particle concentration of C according to the following equation as shown in Figure 2 using a Raleigh function of $R(\theta)$ if the absorption along the optical path is negligible:

$$i = i_o R(\theta) C / r^2 \qquad (3)$$

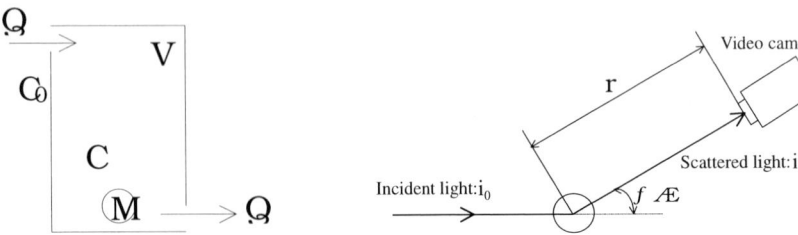

Figure 1   Mass balance of pollutant in a room   Figure 2   Light-scattering by particles

When the scattered light is collected by an optical system, the photosensitive device can convert the light signals to electrical signals. Using a photoelectric conversion system, Rosensweig et al.(1961) confirmed that the electrical signals were proportional to the seeded gas concentrations if a gas were seeded with the light-scattering smoke particles and the particles could mix perfectly with the gas.

VIDEO CAMERA CALIBRATION

Photoelectric Conversion of Video Camera

The brightness on the camera's image surface of the photosensitive device is related to illuminance, reflectance and camera f-stop by the following equation (Okazaki et al. 1998):

$$J = J_0 RT / \{4(m+1)^2 F^2\} \tag{4}$$

where
  J = illuminance on camera's image surface (lux)
  J_0 = illuminance of photographed objects (lux)
  R = reflectance
  T = camera transmittance
  m = camera magnification
  F = camera f-stop

The image illuminance on the photoemissive surface is converted to image signals through the photoelectric conversion system of the image sensor in the following form:

$$I = kJ^\gamma \tag{5}$$

where
  I = image signal or image intensity
  γ = gamma coefficient

The gamma coefficient is normally 0.5. The magnitude of the image signals or image intensity is proportional to the power of the color chip's reflectance, and also to the power of the color chip's illuminance.

Video Image Analysis System

Two types of video camera were used. One was a high-resolution charge-coupled-device for broadcasting with a 768 × 493 pixel imager, an S/N ratio of 60 dB, and an f-stop from 1.8 to 16. During the experiments, the automatic gain control was off so that consistent quantitative measurement could be made. A function of pedestal level could precisely adjust the black level in the picture. The other was an 8mm video camera for home use with an f-stop from 1.8 to 11. It did not have a pedestal level function. A standard video cassette recorder recorded the pictures on a Beta video tape at a recording speed of 30 Hz. The video image analysis system was composed of an image processor with 768 mono color frames of analogue-digital converters with 512 × 481 pixels and a computer with 32-bit precision. The image picture was digitized with 8-bit precision and normalized by 256, converting to image signals in the range from 0.0 to 1.0.

Results of Video Camera Calibration

The entire video system was calibrated as a single unit using 36 steps of achromatic color chips before the field tests were run. The chips, under non-reflecting glass on the desk, were illuminated by incandescent lamps and measured by the video camera above the desk. The image signals at a measuring point including the neighboring 8 pixels were calculated by the averaging process. Figure 3 shows the relation of the image signals to the luminous reflectance of the achromatic color chips. The illuminance on the chips was kept at 450 lux. The f-stop was set at 2.8 and the lens at 20 mm. When the chips were illuminated by brighter lamps for the home-use video camera, the dark current caused non-linearity at low reflectance. Thus, the 8mm video camera was found to indicate non-linearity in the range of image signals less than 0.065. However, for the broadcasting video camera indicated by M7 in Figure 3, the image signal for a luminous reflectance of zero was adjusted to zero by tuning the pedestal level. The linearity at the lower level was improved in comparison with the 8mm video camera. Figure 4 shows the relation of the image signals to the camera f-stops. For the M7 camera, the image signals were proportional to the inverse of the camera f-stop in the range of $F \leq 16$ with a power of 1.13. The gamma coefficient was 0.57, corresponding to a gamma coefficient that is normally 0.5. The performance of the 8mm camera in the range of $F \leq 3.4$ was lower than that of the M7 camera.

NEW PROCEDURE FOR MEASURING AIR CHANGE RATES USING VIDEO IMAGES

For enclosed spaces such as a test house, the smoke particles were stored inside spaces and mixed with the indoor air. The smoke particles diffused through the whole of the enclosure so that the light scattering might have worse effects on the decay curves of image signals. More than ninety runs for air change rates were conducted in the test house using the M7 camera. These field tests indicated that the performance of photoelectric conversion could be regulated well when the camera were adjusted to reduce the final image signals to less than 0.04 by tuning the f-stops and illuminances. The flow diagrams for the new procedure using video images are outlined in Figure 5 and Figure 6.

Adjustment of Camera and Illuminance

The photographed point is set to represent the ventilation in a room. The spotlights and video camera are arranged so that the smoke can be well observed and photographed. After adjusting the focal distance, white balance and black balance of the camera are set, smoke candles are lit and the emitted smoke is completely mixed with the indoor air using a mixing fan. When the smoke decreases and the photographed point is first visible to the camera, the image showing the maximum image signals of I_{max} is taken for various f-stops and illuminances. Next, the smoke is completely exhausted from the room by mechanical fans, and the final image of I_{end} is also taken for various f-stops and

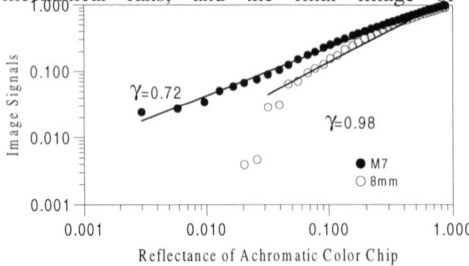 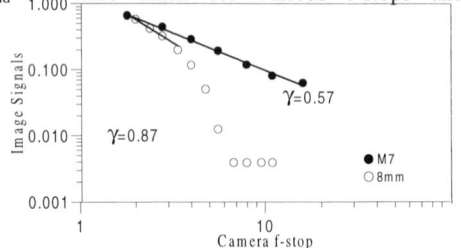

Figure 3   Relationship between luminous reflectance of Munsell achromatic color chips and image signals

Figure 4   Relationship between camera f-stop and image signals

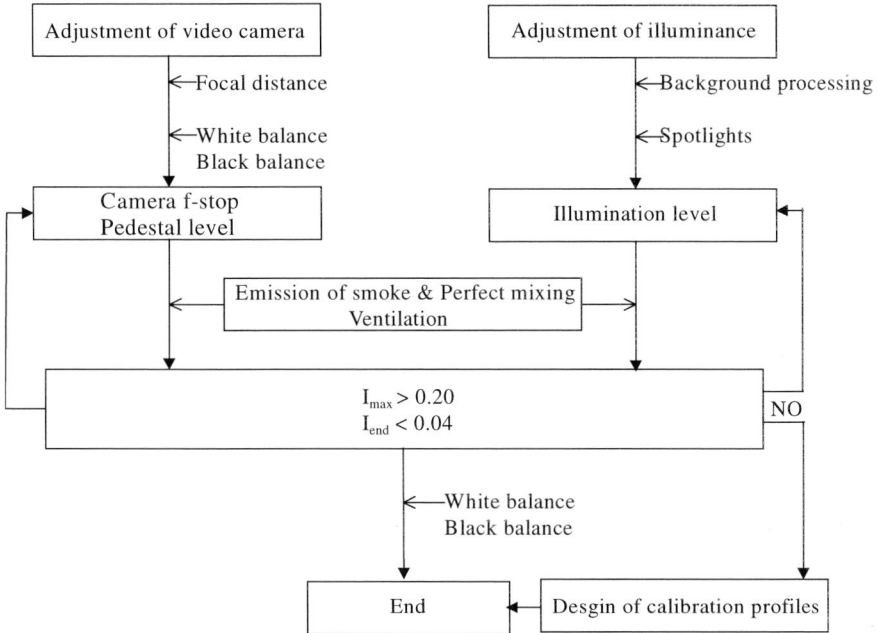

Figure 5   Procedure for adjustment of video camera and illuminance

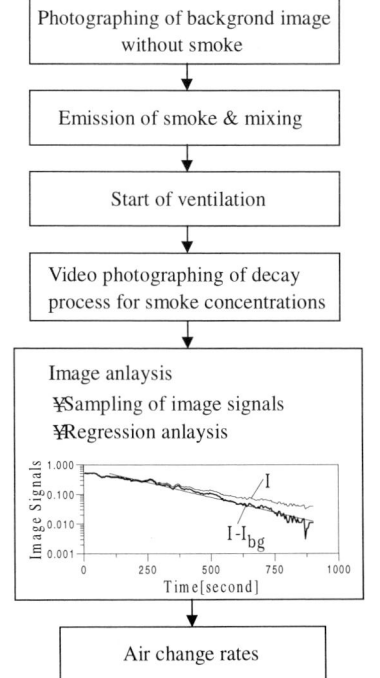

Figure 6   Procedure for measement of
air change rates

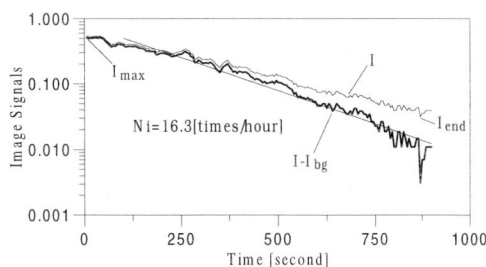

Figure 7   Decay distribution of image signals
for smoke candles as tracer

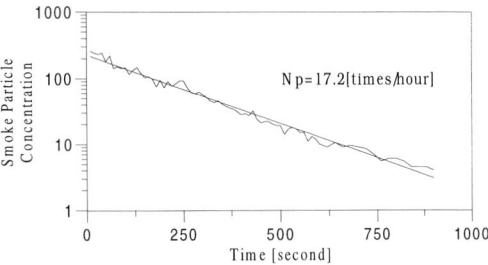

Figure 8   Decay distribution of smoke particle
concentrations for smoke candles
as tracer

illuminances. The illuminance and f-stop suitable for the photograhing are determined so that these image signals comply with the condition of $I_{max} > 0.2$ and $I_{end} < 0.04$. The camera and illuminance are adjusted until the condition can be sufficiently achieved. If the condition cannot be achieved, the numerical calibration profiles for the camera must be designed from the decay curves of the image signals and compared with those of particle concentrations by an aerosol monitor.

Procedure for Determining Air Change Rates

During smoke emission, the smoke is completely mixed with the indoor air using a mixing fan. After the mixing fan stops and the doors are opened, the smoke is exhausted by mechanical fans or natural ventilation. During recording, counter numbers are recorded on the tape to easily determine the start time and the end time. Figure 7 shows typical decay curves for the image signals. The background level indicated by I_{bg} in Figure 7 was subtracted from the original image signals. When calculating the regression curve by the least squares method, the start point was chosen as 90% of the maximum image signals, and the end point was set to the minimum image signal which could maintain linearity in the range. The air change rate was equal to 16.3 times per hour. Figure 8 shows the decay distribution of smoke particle concentrations. It was measured simultaneously with the video measurement.

Test House

The test house had a shed-roof with a volume of 208 m^3 and a door and three windows, as shown in Figure 9 (Ohba and Irie 1999). The windows were covered with black curtains to shield against solar radiation and to thereby keep the indoor illuminance constant during the field runs. Six sets of exhaust fans were attached to the opening of a window for mechanical ventilation. Spot-lights were projected from near the camera position into the photographed point. The mean diameter of the smoke particles varied from 0.77 μm to 1.06 μm during the measurement. The particle loss by the sedimentation occurred at the low ventilation, but no effect of buoyancy on the smoke movement due to the illumination was observed in the video pictures. The video images correspond to the horizontally-integrated concentration of smoke particles.

Measurement Accuracy of Air Change Rates

Figure10 and Figure 11 compare the air change rates obtained by the video image method with those obtained by the smoke particle method. To verify the measurement accuracy of the video method, an aerosol monitor measured smoke particle concentrations simultaneously with the video measurement. After perfect mixing of the candle smoke in the test house, the samples of air containing the smoke particles were drawn to an aerosol monitor through tubing by a small sample pump. The accumulated

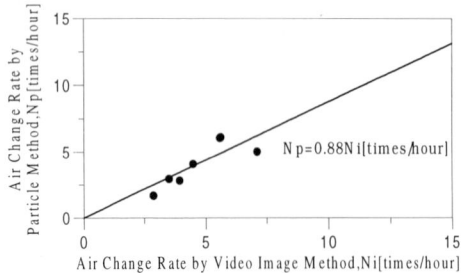

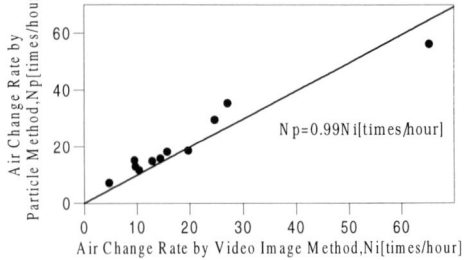

Figure 10 Relationship between air change rates by smoke particle method and by video image method using M7 camera

Figure 11 Relationship between air change rates by smoke particle method and by video image method using 8mm camera.

particle numbers relating to the particle concentrations were read at 10-second intervals. The image signals at the measuring point were calculated by the averaging process including the neighboring 8

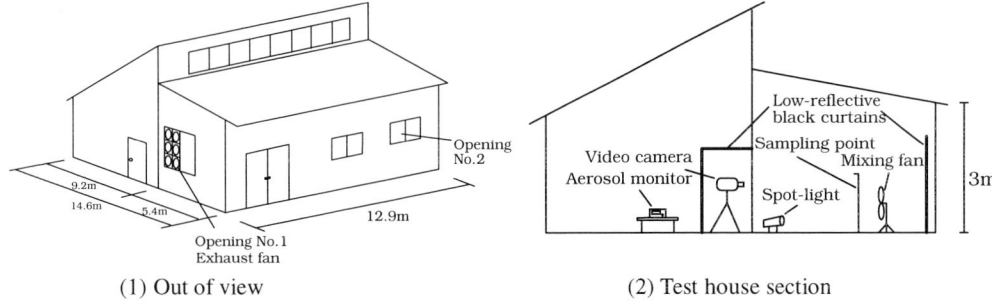

(1) Out of view　　　　　　　　　　　　　　　(2) Test house section

Figure 9 Test house

pixels. The sample interval was 5 seconds. The spatial resolution was 3 mm × 4 mm per pixel. The measuring point was the same as the sampling point by smoke particle measurements. The camera f-stop was set at 2.4 while the illuminance at the measuring point was kept at 600 lux. In Figure 10 for the M7 camera, the air change rates measured by the smoke particle method were in the range from 1.6 times an hour to 6.1 times an hour. The proposed method measured the air change rates with an error of less than 12%. Furthermore, as shown in Figure 11 for the 8mm camera, the video method measured the air change rates with an error of less than 1% while the air change rates was larger than 7.1 times an hour. These results show that the video method measured air change rates with virtually the same precision as the method using the decay curves of smoke particle concentrations if the cameras could be regulated well for the illuminance conditions.

CONCLUSIONS

The video image method can measure air change rates with virtually the same precision as the method using the decay curves of smoke particle concentrations, by adequately tuning the camera f-stops and adjusting the illuminance in a room. Compared to the precision of the tracer gas method, it is true that the proposed method has disadvantages, such as low image precision, because image pictures can be digitized with only 8-bit precision. However, even when considering this low resolution, the video image decay method is expected to be reliable for measuring air change rates. The major advantage of the video method is that it is possible to obtain detailed information about air change rates at one time in large spaces for evaluating ventilation performance. Another advantage is that the video method does not use SF_6 gas as tracers, which causes global warming. It will become an environmentally friendly procedure. Thus, in the near future, the video image method is expected to become useful for measuring air change rates.

Acknowledgements

The funding sources for this study were the Tokyo Institute of Polytechnics and the MONBUSHO Scientific Programs of the Ministry of Education.

References

JIS (1974). Method for measuring amount of room ventilation (Carbon dioxide method), Japanese Industrial Standards A1406-1974

Ohba M. and Irie K. (1998). A new technique for measuring air change rates in a cross-ventilation model using step down method of video image signals, Roomvent 98, 2, 453- 460

Ohba M. and Irie K. (1999). A new technique for measuring air change rates in a test house using video imaging, ASHRAE TRANSACTIONS, V. 105, Pt. 1, 103-109Okazaki H. and Tanigiti K. (1998). Imaging Process, Tokyo Kyoritu Press, Japan

Rosensweig R.E., Hottel H. C. and Williams G. C. (1961). Smoke-scattered light measurement of turbulent concentration fluctuations, Chemical Engineering Science, 15, 111-129

Air Distribution in Rooms, (ROOMVENT 2000)
Editor: H.B. Awbi

THE DETERMINATION OF AIR CHANGE RATE IN NATURALLY VENTILATED CATTLE BARNS

H.-J. Müller, B. Möller and M. Gläser

Institute of Agricultural Engineering
Max-Eyth-Allee 100, D-14469 Potsdam, Germany

ABSTRACT

The keeping of animals in livestock buildings requires the ventilation of these buildings. On the one hand good climatic conditions for the animals in the livestock building have to be provided, on the other hand the emissions have to be kept at a low level. The airflow through the livestock building plays an important role for both opposing requirements. The targeted control of the climate in the livestock building and for the minimization of emissions calls for knowledge about airflow and emission streams.

Especially for naturally ventilated buildings, the determination of air change rates leads to a few problems. For those cases, the Institute of Agricultural Engineering Bornim has developed a 40-sampling-points system which is using Krypton 85 as a tracer gas. All 40 sampling points are running simultaneously with a maximum sampling rate of one second. Air change rates of up to 1000 h^{-1} have been measured.

During the past 2 years more than 20 naturally ventilated cattle barns have been investigated. Apart from the measurement of volume flow, the concentration of gases and odours has been measured as well. In parallel to these inside investigations inside the barns, the climatic conditions outside have also been recorded. The results are extensive data on emissions. The measured emission flows of odours can be used - among other things - to calculate the minimum distance between livestock buildings and human living areas.

Keywords

Air change, livestock building, tracer gas method, emission flow

INTRODUCTION

The climate in livestock buildings shall be such that it meets the requirements of animals, that the animals can give their maximum performance, that the livestock building is protected against high humidity and that proper working conditions for human beings are provided. Therefore the ventilation of a livestock building is necessary. For some kinds of animals heating may be necessary for the winter period. For cattles, the ventilation of a livestock building is generally sufficient. Since cattle tolerate a pretty wide range of temperature, the requirements of ventilation and its control are low. Therefore most of the cattle barns in Germany are working only with natural ventilation. The control of volume flow is guaranteed by flaps which are situated in inlet and outlet openings. The position of those flaps can be changed either automatically or manually. During the summer season doors and windows are used for ventilation, too.

The necessary ventilation of livestock buildings leads to emissions. The targeted control of climatic conditions in livestock buildings and emissions requires knowledge about the ventilation of buildings. Therefore extensive measurements will be done at animal farms. These measurements determine the volume flow through the building and the concentration of gases and odours. With these data it is possible to evaluate the climatic conditions in a livestock building establish and the emission flows. The complicated air flow patterns of natural ventilation make it very difficult to get accurate measurements. According to wind conditions, temperature differences between inside and outside and the position of openings for ventilation, very different air flow patterns can be observed. Especially the measurement of volume flow through the livestock building is very difficult under such conditions. For the investigations the tracer gas method with Krypton 85 is applied. Experiences with that method have already been reported during the last ROOMVENT-Conference (Müller et al., 1998a). The following report continues to illustrate this research work and provides information about the measurements which have been conducted in various cattle barns.

METHODS

The measurements have been done in a wide variety of livestock buildings all over Germany and under different weather conditions. Altogether 31 cattle barns have been investigated. Types and sizes of buildings as well as ventilation systems vary a lot. Fig. 1 - 3 show examples of different livestock buildings for dairy cattle in which these measurements have been done.

Fig. 1: Livestock building with shaft ventilation (enclosed space: 5445 m³; number of animals: 202)

Fig. 2: Livestock building with eaves/ridge ventilation (enclosed space: 10694 m³; number of animals: 225)

Fig. 3: Front opened livestock building with spaceboard (enclosed space: 3411 m³; number of animals: 74)

The **airflow rate** through the livestock building is determined by means of the tracer gas method. Sulphur Hexafluoride (SF_6) or Krypton 85 have been used as a tracer gas. For the measurements of SF_6 concentration the multigas monitor is used. A multi-point sampler allows to switch among 12 measuring points (max.). The measuring cycle for one measuring point takes about 2 min and for 12 measuring points about 30 min. As far as time is concerned, this is a small resolution. Therefore SF_6 has to be dosed with a constant mass flow rate. The mass flow rate is adjusted by means of a rotameter and for the precise determination of the SF_6 mass flow, the mass reduction of the dosing bottle is determined dependent on time. The tracer gas is distributed as evenly as possible in the building through a hose system.

Especially in the summer, when air change rates are high and cattle barns ventilated naturally the small number of measuring locations and the small temporal resolution leads to higher measuring uncertainties. Therefore a special measuring system for Krypton 85 as tracer gas has been installed at ATB.

With this system it is possible to measure the concentration of Krypton 85 in the barn at 40 locations in parallel. The shortest measuring interval per location is 1 second. The tracer gas is distributed in the building in thinned solution and as fast as possible. This is done either with a distribution system (pumps and hoses) or manually, that means dosing equipment is moved evenly across the entire surface of the livestock building. Then the tracer gas is mixed with the air of the livestock building by waving big cardboard fans. After that, the decay of the concentration of Krypton 85 is measured. The exponent of the decay function is the local air change rate. In connection with the volume of the live-stock building, the volume flow can be determined. This method is described by Müller et al. (1998a). The **concentration of gases** (Ammonia - NH_3; Carbon dioxide - CO_2; Dinitrogen oxide - N_2O; Methane - CH_4; Sulphur Hexafluoride - SF_6; Water Vapour - H_2O) is measured with a Multi-Gas-Monitor inside the livestock building at a maximum of 12 locations (Müller et al., 1994).

For the determination of **odour concentrations** the Olfactometer TO 7 by Mannebeck is used. The air samples are taken in the livestock building by means of a special device. After that they are analysed at the laboratory by 4 test persons.

Results

In this article only a few results shall be demonstrated as examples. The examples are shown in Fig. 1 - 3. As the decay curves are measured at different measuring points in the building, we can show the varying distribution of air change above the entire surface of the barn (see Fig. 4 to 6). Fig. 4 - 6 show that fresh air distribution within livestock buildings is uneven. The distribution of air change depends on the design of the openings, wind direction and velocity.

These examples make clear that many measuring points are needed in order to get a relatively reliable statement about the air change. For every test the average value will be determined from the sum of all evaluated points.

In each livestock building several tests will be performed. From these tests, emission streams can be determined. The value of the emission mass flow which is shown in Fig. 4 - 6 is the product of the average value of the air change rate and the measured odour concentration. Parallel to these emission recordings, immissions outside of the livestock buildings are measured too. This makes it possible to see the emissions in direct connection to immissions.

Fig. 4 shows the distribution of air change in a livestock building with shaft ventilation. The high air change results from the opened doors and resultant longitudinal flow through the building.

In Fig. 5 the door in the east gable was opened. Although the air change is higher in the front opened livestock building (Fig. 5) than in the livestock building with eaves/ridge ventilation (Fig. 5), the emission flow is lower (Fig. 6) because of a significantly lower odour concentration.

The determined odour emission flow is an important input for the dispersion calculation. From this calculation (this is not the topic of this lecture) the necessary minimum distances between livestock building and human living areas is derived.

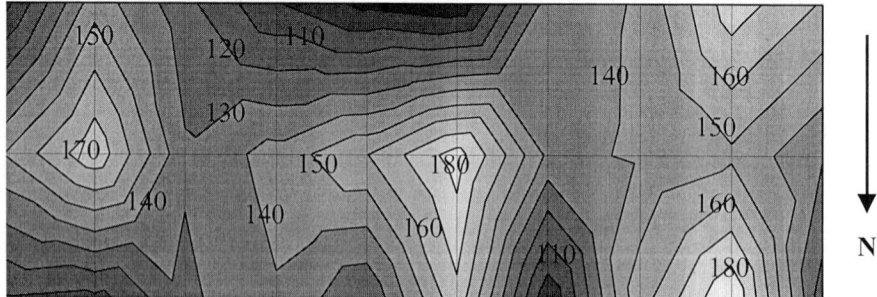

Fig. 4: Distribution of air change (measuring unit of the values: h^{-1}) across the surface area of the barn - example shaft ventilation
Average air change rate: 135 h^{-1}; maximum: 192 h^{-1}; minimum: 81 h^{-1}
Concentration of odours: 120 OU/m³; Emission flow: 119 OU/(sLU)
LU: livestock unit (500 kg); OU: odour unit

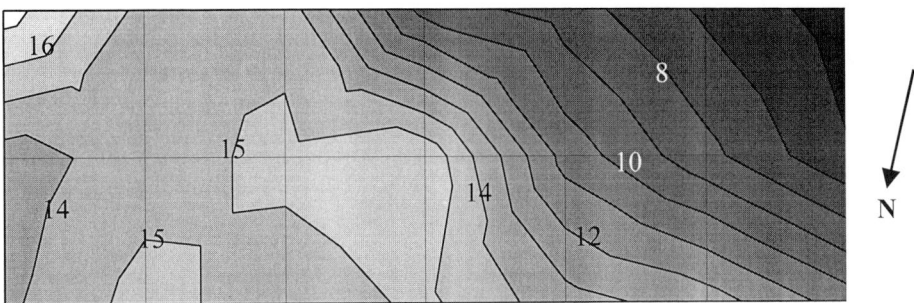

Fig. 5: Distribution of air change (measuring unit of the values: h^{-1}) across the surface area - example eaves/ ridge ventilation
Average air change rate: 12.2 h^{-1}; maximum: 17.5 h^{-1}; minimum: 5.5 h^{-1}
Concentration of odours: 157 OU/m^3; Emission flow: 22.8 OU/(sLU)

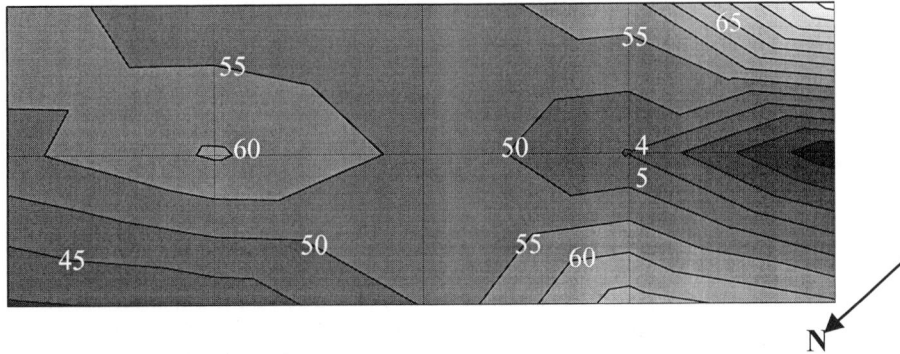

Fig. 6: Distribution of air change (measuring unit of the value: h^{-1}) across the surface area - in a front opened livestock building with spaceboard
Average air change rate: 54 h^{-1}; maximum: 93 h^{-1}; minimum: 26 h^{-1}
Concentration of odours: 18.8 OU/m^3; Emission flow: 20.6 OU/(sLU)

From publications (Hartung et al., 1994) it is well known that the emission mass flow in animal farming (livestock buildings) in general rises with rising volume flow. The representation of the relation between emission mass flow and volume (Fig. 7) shows, that this close relationship between the two parameters does not exist in the case of cattle barns.

With regard to the average odour emission and to the statistic mean variation our own measured values go along very well with the values obtained by Oldenburg (1989).

The connection between emission mass flow and volume flow is not as obvious as with turkey and duck farming (Müller et al., 1998b). Evidently, things like animal keeping, manure removal and feeding play a bigger role in cattle farming, when it comes to odour emission.

510

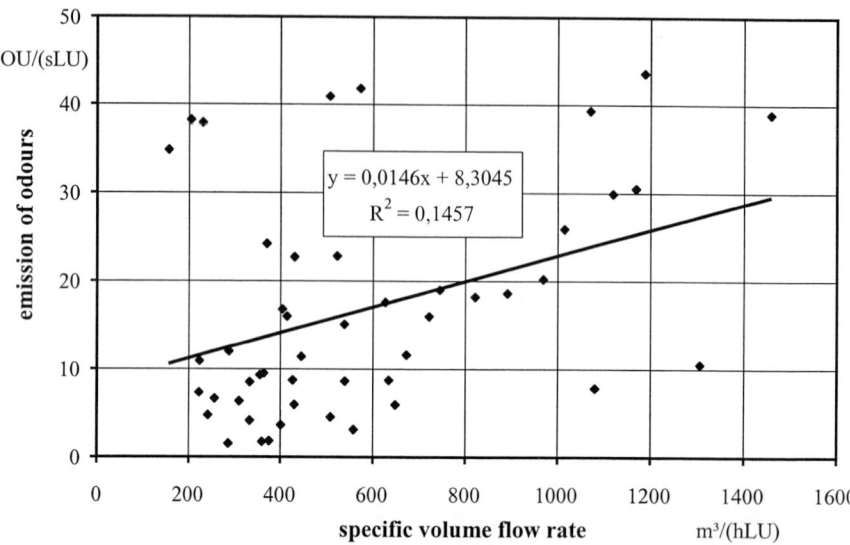

Fig. 7: Mass flow of emission depending on the volume flow rate

CONCLUSION

The emission of odours from livestock buildings can become a major environmental problem. Distances large enough between livestock buildings and human living areas will lead to sufficient rarefaction of odours in the atmosphere. For the determination of these distances, measurements have been carried out in 31 cattle barns. The distances are determined by means of a dispersion calculation. As an important input for the simulation model the emission flow is needed. With mainly naturally ventilated cattle barns, especially the determination of the volume flow is a problem. One suitable method to measure the flow volume is the tracer gas method using Krypton 85 as a tracer gas. The values of odour emission flows from cattle barns are to be found in a wide range. Nevertheless, the measured values are an important basis for simulation calculations in order to determine minimum distances between livestock buildings and human living areas.

REFERENCES

Hartung J. and Philipps V. R. (1994). Control of Gaseous Emissions from Livestock Buildings and Manure Stores. J. agric. Engng. Res. 57, 173-189

Müller H.-J. and Müller S. (1994). The Determination of Emission Streams from Livestock Buildings with different Tracer gases. ROOMVENT '94, Volume 2, 530-542

Müller H.-J. and Möller B. (1998a). Determination of Air Change Rates in an experimental Cattle Housing using Tracer Gas Methods. ROOMVENT '98, Volume 2, 511-516

Müller H.-J. and Möller B. (1998b). The determination of emission streams at different animal houses. AgEng '98, Paper no 98-E-059

Oldenburg J. (1989). Geruchs- und Ammoniakemissionen aus Tierhaltungen. KTBL-Schrift 333

AIR DISTRIBUTION

© 2000 Elsevier Science Ltd. All rights reserved.
Air Distribution in Rooms, (ROOMVENT 2000)
Editor: H.B. Awbi

INFLUENCE OF OUTLET CHARACTERISTICS ON FREE AXIAL AIR JETS

Tor G. Malmström, Zou Yue

Department of Building Science Engineering, Royal Institute of Technology (KTH), Sweden

ABSTRACT

The subject of this paper is grille influence on free, axisymmetric, isothermal air jets. Momentum flow rate is the most important factor for the jet flow. The most well defined outlet is a conical nozzle. Grilles compared to nozzles seem to have two effects on the momentum flow rate. The first is caused by changes in outlet flow rate and velocity. The second depends on a region of slightly decreased pressure just downstream of the grille causing an adverse force on the flow field resulting in loss of momentum flow rate. If the flow in or close to a nozzle is disturbed, induction of room air may increase. This typically is equivalent to the natural induction in the jet over a short distance. The influence can be described as a shift upstream in the location of the apparent source of the jet. The influence of grilles, vanes and other similar devices seems to be similar.

KEYWORDS

Air jet, Momentum, Entrainment, Grille, virtual origin, K_A-Value.

INTRODUCTION

Jets are widely used in air conditioning and ventilation applications for supplying air into room and spaces. For air supply jet behavior, the free axisymmetric isothermal jet is a reference case, which is a base for calculation also of other types of jets as wall jets (although the differences between the two jet types are big). The jets in HVAC applications differ from the "typical" jet studied in fluid mechanics literature by having rather big outlet dimensions and low outlet velocity.

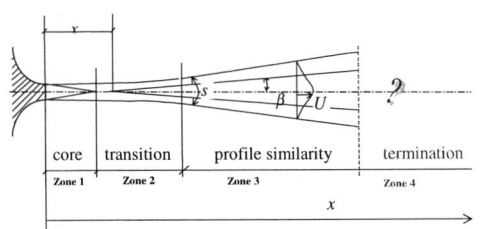

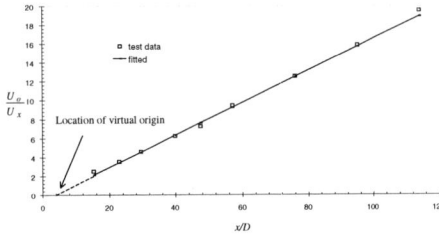

Figure 1: Symbols and denotations used in the model.

Figure 2: Inverse centerline velocities for a nozzle (D=1.5'') according to Malmström et al (1992).

In ASHRAE literature the development of a jet is divided into four zones, see figure 1. On a free axial jet, a grille influences the first two zones directly and the third zone indirectly. The direct influences typically are entrainment of room air at the grille, increased spread direct after the grille (if the grille has diverging vanes), or momentum losses. The indirect influences in the third zone mainly are due to the momentum losses close to the grille. Other types of jets may be directly influenced also in the third zone.

AXIAL JET MODELING

The most important part of jets in HVAC application is zone 3, which for an axial jet is the zone where the jet is fully developed. Here the centerline velocity decay characteristics can be described as:

$$\frac{U_x}{U_o} = K_A * \frac{\sqrt{A_o}}{(x - x_p)} \tag{1}$$

U_x: centerline velocity at x (m/s) $\qquad\qquad$ A_o: outlet flow area (m^2)

U_o: outlet velocity of air jet (m/s) $\qquad\qquad$ x: coordinate for distance from the outlet (m)

$\qquad\qquad\qquad\qquad\qquad\qquad\qquad\qquad\qquad$ x_p: coordinate for the virtual origin of

K_A: centerline velocity decay coefficient $\qquad\qquad\qquad\qquad$ the jet (m)

Figure 2 shows the quotient U_x / U_o plotted against the normalized distance from the outlet. For a jet from a nozzle, eq. (1) corresponds to the straight line in the graph, fitted to the measured values. The coefficient K_A (or rather $1/K_A$) is a measure of the slope of the line, and the distance x_p between the outlet and the "virtual origin" of the jet is defined by the intersection of the line and the x-axis. Figure 2 shows a case with typical values for high velocity ($U_o > 12$ m/s) jets. The part of the jet represented by the linear relationship in figure 2 corresponds to zone 3. Close to the outlet (zone 1 and 2) the velocity decay rate is slower, and at large distances faster (zone 4). As there are no well defined boundaries of zone 3, see Zou (2000), it is concluded (see figure 2) that decisions about which measured points should be included in the fit of the straight line have an influence on the evaluated value of K_A and a big influence on the evaluated value of x_p.

Earlier experiences show that similar jet velocities under different flow rates can be created by keeping the momentum of the jet constant, see Malmström et al. (1992). Therefore, a more general variation of *eq.(1)* is:

$$U_x = K_A * \frac{\sqrt{M_e/\rho_o}}{(x - x_p)} \tag{2}$$

M_e: momentum in the resulting jet (N) $\qquad\qquad$ ρ_o: air density at outlet (kg/m^3)

One reason that equation (1) are more common than equation (2) is that M_o is difficult to measure, but also U_o is difficult to evaluate in a representative manner. For a *diffuser* there is no single outlet velocity U_o but rather velocities within a big interval. When defining a characteristic U_o priority traditionally is given to the needs of field measurements of air flow rate V_o (m^3/s) through the diffuser, that is a velocity, which is easy to measure in a reproducible way, is chosen. The outlet flow area then is defined as:

$$A_o = \frac{V_o}{U_o} \tag{3}$$

V_o: air flow rate through the outlet (m^3/s)

The traditional way to choose U_O thus causes scatter in the evaluated values of the throw coefficient K_A and uncertainty about the real outlet momentum M_O. This can be expressed as:

$$M_O = b * \rho_O * V_O * U_O = b * \rho_O * A * U_O^2 \qquad (4)$$

b: outlet momentum coefficient

However, there is still another complication: the momentum losses due to pressure differences immediately downstream of the diffuser. These losses are related to mixing between supply air and room air (and to difficulties to entrain room air). Thus M_O, which is the momentum of the supply air, is not necessarily equal to the momentum in the resulting jet, M_e.

$$M_e = i * M_O \qquad (5)$$

i: momentum loss coefficient

For a very special type of "diffuser", perforated plates, "i" has the value between 0.3 and 0.7, i.e. by Malmström (1974) and Malmström and Hassani (1992). Results from these studies indicate that "i" is a function of Reynolds number for the flow through the perforated holes, of the number of holes, and of the percentage of perforation. More normal diffusers have smaller momentum losses, that is values of "i" closer to one, see e.g. Grimitlin (1970).

There are indications that the velocity decay coefficient K_A also is a function of outlet velocity, see Nottage's study from 1951 and ASHRAE Handbook (1997). This was studied in jets from nozzles by Malmström et al (1992) and Zou (2000). These studies indicated also that K_A was a function of U_O rather than Reynolds number (see figure 3).

The velocity decay coefficient evaluated from outlet and centerline velocity measurement in free jets for an air outlet is then influenced by at least three different mechanisms:
- how well the chosen outlet velocity U_O is a measure of the outlet momentum M_O (the factor "b")
- momentum losses immediately downstream of the outlet (the factor "i")
- low outlet velocity

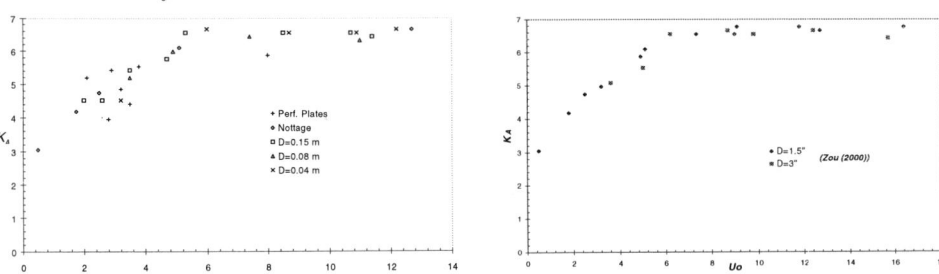

Figure 3: The factor K_A as a function outlet velocity U_O (m/s). Comparison between values from tests with nozzles [Nottage (1951), Malmström et al (1992) and Zou (2000)] and tests with perforated plates (Malmström 1974)

The value of the coefficient could also be influenced by the value of x_p, as K_A and x_p are evaluated together. This will be discussed later. The third mechanism is of course least understood. Anyhow, in figure 3 the values from tests with jets from nozzles reported by Malmström et al (1992) are plotted together with values evaluated from tests with perforated plates. These values are corrected for "i" and have $b=1$. Their scatter is big but can perhaps be interpreted as a scatter around the values for the nozzles. This matter needs further study.

THE VIRTUAL ORIGIN

In most test codes for air supply devices as well as in ASHRAE Handbook (1997) the influence of the virtual origin of the jet is ignored, that is jet characteristics as spread are considered proportional to x instead of $(x-x_p)$. This practice is well motivated as the distance x_p usually is small compared to the throw of the jet and thus x_p tends to unnecessarily complicate things. However, it is discussed here as it is related to basic jet physics.

The virtual origin is the evident location of the source of the zone 3 jet. The volume flow rate in the jet is proportional to the distance ($x-x_p$), see equation (1), and a change in the distance x_p will thus be reflected as a change in the jet volume flow rate at a distance x. An influence at the outlet, causing increased entrainment in a limited part of the jet, often can be described as an upstream shift of the virtual origin of the jet. As will be discussed below a free axial jet seems to have a tendency to attain "general" values of K_A and β, the spread of the jet. This is not generally true for jets and is probably associated with the natural tendency of the free axial jet to create "donut-formed" large eddies, governing the entrainment process. Thus an axial jet, which in some part has an artificially increased entrainment, will have a tendency to regain its natural development. Then the locally increased entrainment will not result in an increased spread of the jet (which would have caused a decreased value of K_A) but in an upstream shift of the virtual origin of the jet.

The tendency to a constant spread in axial jets was discussed by Becher (1949) and Nottage (1951). Nottage made experiments with jets from a 6 inch nozzle. For two of the outlet velocities (12.5 m/s and 1.8 m/s) he made tests with a "turbulence promoter" in the nozzles. This caused no change in the 12.5 m/s jet, but a rather dramatic change in the 1.8 m/s jet. The way this was reported by Nottage suggested a change in the K_A -value of the jet. However, figure 4 shows the centerline velocity measurement of Nottage for this case plotted as in figure 2. It is evident that the influence can be described as an upstream shift of the virtual origin, with 4-5 nozzle diameters, and that the slope of the centerline velocity decay at some distance seems to attain the same value as the slope of the undisturbed jet (see also Malmström (1974)). At large distances, velocities are very low in this test which probably explains why the tendency is less obvious there.

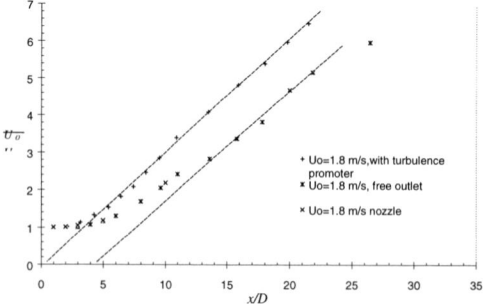

Figure 4: Comparison between the centerline velocities decreasing in jet from a nozzle, with and without "turbulence promoter". Data from Nottage (1951). The dotted lines illustrate the discussion about a shift in the location of the virtual origin.

An interesting question is : why was the 12.5 m/s jet not influenced when there was a dramatic influence in the 1.8 m/s (350 ft/min) jet? One explanation can be the "natural" frequency of large eddy formation downstream of the nozzle, which frequency varies with the outlet velocity. If Nottage's "turbulence promoter" initiated eddies within the nozzle close to this natural frequency in the case of the 1.8 m/s (350 ft/min) jet, this could explain the phenomenon. This natural frequency has been studied by various researchers including Crow & Champagne (1971). They found that "vortex puffs" were formed at an average Strouhals number of 0.3, based on S_r frequency f, outlet velocity and diameter (their tests were made with a 0.05 m nozzle and $Re = 10^5$).

$$S_r = \frac{f.D}{U_o} \hspace{3cm} (6)$$

Crow & Champagne also found that it was possible to influence the jet with an acoustic signal, and that an acoustic signal influenced the jet most on a frequency corresponding to the natural frequency. The effect in their test could be described as an upstream shift of the virtual origin of about 2 nozzle diameters.

Also pulsation in the outlet velocity in jets from round nozzles can create an effect which can be described as a shift in the location of the virtual origin, Remke (1973) and Binder & Favre-Marinet (1981).

GRILLE INFLUENCE ON THE VIRTUAL ORIGIN

Grilles have an influence on the momentum losses and the momentum transfer to the jets, see section 3, Grilles also have an influence in the location of the virtual origin, similar to the influences discussed in section 4.

The momentum losses caused by a grille are associated with entrainment of room air. Compared with a jet from a nozzle, a grille thus causes increased entrainment at the grille. The result may be taken into account as a shift in the location of the virtual origin.

A special case is grilles with diverging vanes. Becher (1966) and Grimitlin (1970) described their influence as a rapid spread close to the grille which gradually decreases downstream till the "normal" spread angle for free axial jets. This was demonstrated for a 90°-spreading outlet by Appelquist & Stengård (1972), see figure 5. In figure 5 result from a test with the same outlet mounted adjacent to a surface also is shown. The difference is obvious, illustrating the fact that a wall jet is very different from a free jet. For the wall jet spread increases downstream, compare Tuve (1953) and Sforza (1968).

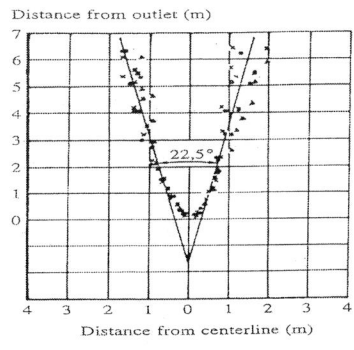

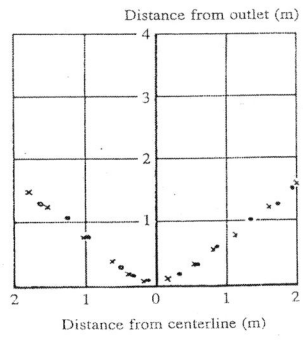

Figure 5: Jet spread from an outlet with diverging vanes for a free jet and for a wall jet (Appelquist & Stengård 1972).

DISCUSSION

The main characteristic of free axial air jets are jet momentum. A grille influences jet momentum by changing outlet velocities and by causing a low pressure zone downstream of the grille, which causes momentum losses. These losses depend on the velocity (or Reynolds number) of the individual jets from the different parts of the grille. There are also some indications that low outlet velocities can influence jet spread and the K_A –values in zone 3, at least for jets from nozzles.

The location of the virtual origin depends on initial effects on the jet as turbulence or vane setting. Grilles causing increased spread or entrainment in the initial parts of the jet seems to mainly influence the location of the virtual origin of the jet, in the equations describing jet behavior in zone 3.

The influence of outlet geometry has not been discussed in the paper. It has been investigated by e.g. Becher (1949), Tuve (1953), Trentacoste & Sforza (1966), and Sforza (1968). At big distances, jets from rectangular outlets adopt circular or elliptic form and then develop as round axial jets. Special phenomena may occur in the transition zone, see Zaman (1996).

REFERENCES

Appelquist, H.O. & Stengård, L. (1972) Luftstrålar från ett spridande galler. Examensarbete vid inst. för Uppvärmnings-och ventilationsteknik, KTH.

ASHRAE (1997) *ASHRAE Handbook Fundamentals.* Chap. **31**. pp 1-16, American Soc. Heating, Refrigerating, and Air Conditioning Eng., Atlanta.

Becher, P. (1949) *Om beregning af indblaesningsåbninger.* Gjellerup, Köbenhavn, Denmark.

Becher, P. (1966) Luftverteilung in gelüfteten Räumen. Heizung, *Lüftung- Haustechnik Nr* **7**.

Binder, G. & Favre-Marinet, M. (1981) Some Characteristics of Pulsating of Flapping Jets. *Conference of "Unsteady Turbulent Shear Flows",* Gothenburg, Sweden.

Crow, S.C. & Champagne, F.H. (1971) Orderly structure in jet turbulence. *J. Fluid Mech.* **48**, p 3.

Grimitlin, M. (1970) Zuluftverteilung in Räumen. *Luft-und Kältetechnik* **5**.

Malmström, T.-G. (1974) Om funktionen hos tilluftsgaller. *TM* **49**, Department for Heating and Ventilation, KTH, Stockholm, Sweden.

Malmström, T.-G., Christensen, B., Kirkpatrick, A.T. & Knappmiller, K. (1992) *Low velocity jets from round nozzles.* Bulletin **26**, Department for Building Services Engineering, KTH, Stockholm, Sweden.

Nottage, H.B. (1951) *Report on ventilation jets in room air distribution.* Case Inst. of Technology, Cleveland, Ohio.

Remke, K. (1973) Untersuchungen zum pulsierenden, Turbulenten Freistrahl. Schiftenreihe des Zentralinstituts für Mathematik und Mechanik bei der Akademie der Wissenschaften der DDR, *Heft 17.* Akademiverlag, Berlin.

Sforza, P.M. (1968) A quasi-axisymmetric approximation for turbulent, three-dimensional jets and wakes. *PIBAL-report no* **68-13**. Polytechnic Instutute of Brooklyn.

Trentacoste, N. & Sforza, P.M. (1966) An exipermental investigation of three-dimensional, free mixing in incompressible, turbulent, free jets. *PIBAL report no* **871**. Polytechnic Institute of Brooklyn.

Tuve, G.L. (1953) Air velocities in ventilating jets. *Heating, Piping and Air. Conditioning,* No **1**.

Zaman, K.B.M.Q. (1996) Axial switching and spreading of an asymmetric jet: the role of coherent structure dynamics. *J. Fluid Mech.* **316**, pp.1-27.

Zou, Y. Velocity decay in air jets for HVAC applications. *ASHRAE summer meeting 2000*

Air Distribution in Rooms, (ROOMVENT 2000)
Editor: H.B. Awbi

FLOW AND SEPARATION CHARACTERISTICS OF HORIZONTAL COLD AIR JETS INDUCED BY A REAL HVAC SYSTEM

E.Rutman[1] , C.Inard[2] , A.Bailly[1] and F.Allard[2]

[1] CIAT, Avenue Jean Falconnier, 01350 Culoz-France
[2] LEPTAB, Université de la Rochelle, Avenue Michel Crepeau, 17042 La Rochelle Cedex 1 France

ABSTRACT

Characteristics of a cold air jet induced by a real HVAC system in a climate chamber has been studied experimentally. General characteristics such as maximum velocity, temperature decay and the velocity and temperature half widths are analysed. At the end, we present results on the separation distance of the cold wall jets.

KEYWORDS

Air-conditioned offices, horizontal cold air jet, negative buoyant jet, experiment, HVAC systems

INTRODUCTION

To improve thermal comfort in buildings HVAC systems are often used. In summer the cold air is introduced into the room by the HVAC system, like the resulting air velocity and temperature distribution within the occupied zone is as uniform as possible to minimise local discomfort. Thus, in air conditioned-spaces, the cold air jet and plumes due to internal loads are usually the main driving flows. So, before discuss comfort conditions, we have to describe the driving flows and specially to provide the cold air jet laws. The purpose of this experimental work is to analyse the cold air jet induced by a real HVAC system.

EXPERIMENTAL APPARATUS AND CONDITIONS

The measurements were carried out in the CIAT's laboratory located in Culoz, France. Figure 1 presents the longitudinal section of our experimental test chamber. The volume is equal to 5.2x4.05x2.5 m^3 bounded on five sides by air volume regulated at a constant temperature level. The sixth side is submitted to the influence of a climatic housing, where we can simulate external air temperature variation. For these experiments, the value of the external temperature was set at + 30 °C. Furthermore, we simulate the presence of real internal heat loads, including computer, and human bodies. These thermal heat loads are balanced by a fan coil unit (1.2 m long and 0.6 m length) mounted on the ceiling. The dimensions of the fan coil unit supply grille are 1.19 x 0.091 m^2. In the jet, the air velocity and the relative intensity of velocity turbulence are measured with a Dantec laser Doppler anemometer unidirectional system. The temperature was measured with a Newport Omega, 0.051mm diameter, K-type thermocouple calibrate at the CIAT laboratory. For simultaneous measurements of velocity and temperature, we used a "Quatuor Industrie" two dimensional displacement device. We carried out 8 tests with various supply conditions. We have tested two mechanical configurations of the fan coil. With the first one, the air jet has one supply direction, for the second one, we have three directions for the airflow. Secondly, we modified the airflow rate, and we adapted the water conditions in order to get an operative temperature in the occupancy zone equal to 24.5°C ± 1.5°C. Table 1 gives the parameters used for the experiments. The temperature and velocity profiles across the discharge grille were measured and the temperature at the location in the front of the grille where the velocity is less than 0.02 m/s was assumed to be the reference temperature.

TABLE 1 : EXPERIMENTAL CONDITIONS

Test N°	Number of Supply direction	Q_0 m^3/h	U_0 m/s	T_0 °C	ΔT_0 °C	Ar_0 x10^3	Re_0
1	1	736	5.9	24.5	0	0	32737
2	1	732	5.5	14.1	9.1	0.91	30517
3	1	537	4.5	11.0	12.5	1.87	24969
4	1	366	3.0	12.5	12.1	4.05	16646
5	3	670	5.8	24.5	0	0	32182
6	3	722	5.6	11.8	11.8	1.14	31072
7	3	532	4.2	12.1	12.7	2.17	23304
8	3	366	2.9	11.5	13.5	4.85	16091

The Archimedes number, $Ar_0 = g\beta\Delta T_0 A_0^{0.5}/U_0^2$, and the Reynolds number $Re_0 = \rho U_0 A_0^{0.5}/\mu$ are used to characterise the supply conditions. We used the square root of the supply area (A_0) because the ratio effective length (l_0) / effective width (h_0) is equal to 13. This value is a characteristic of a three dimensional jet. Mean values of the exit temperature, T_0 and velocity, U_0 are obtained by measuring the temperature and velocity profiles over a 12 x 25 mesh (Δy = 47 mm and Δz= 5 mm). The integration of the velocity profiles over the grille area allows to calculate the air mass flow rate, Q_0 of the fan coil. The relative differences between these values and those obtained with a standard measurement of airflow, ISO (1997), are less than 7 %.

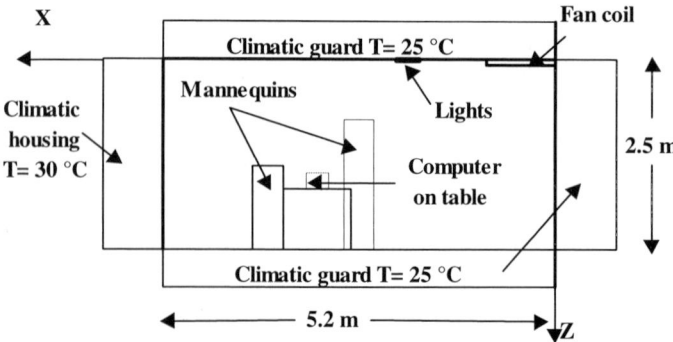

Figure 1 : Longitudinal section of the experimental cell

RESULTS AND DISCUSSION

The main results presented here consist in the maximum velocity and temperature decays in the horizontal wall jet. We also present the separation distance of the horizontal jet according to the inlet conditions.

Cold air jet characteristics

Figures 2 and 3 present the decay of the centerline velocity, U_m and temperature, ΔT_m. Usually, four distinct regions may be defined for a wall jet, as discussed by Hassani and al (1997) : the core zone, the characteristic zone, the transition zone and the fully established turbulent zone. The centerline velocity and temperature decays of the jet are given by:

$$U_m / U_0 = K_V (1 / (x+x_0))^{nv} \tag{1}$$
$$\Delta T_m / \Delta T_0 = K_T (1 / (x+x_0))^{nt} \tag{2}$$

The parameter x_0 is the virtual origin of the wall jet. Values of nv and nt depend of the zones discussed before. In accordance with figure 2 the centerline velocity and temperature decay for the one direction jet are given by :

$$U_m / U_0 \equiv (1 / (x+x_0))^{0.9}, \text{ the correlation coefficient is } 0.73 \tag{3}$$
$$U_m / U_0 \equiv (1 / (x+x_0))^{0.3}, \text{ the correlation coefficient is } 0.85 \tag{4}$$
$$\Delta T_m / \Delta T_0 \equiv (1 / (x+x_0))^{0.3}, \text{ the correlation coefficient is } 0.745 \tag{5}$$

Furthermore, in accordance with figure 3 the centerline velocity and temperature decay for the three direction jet are given by :

$$U_m / U_0 \equiv (1 / (x+x_0))^{0.9}, \text{ the correlation coefficient is } 0.75 \tag{6}$$
$$U_m / U_0 \equiv (1 / (x+x_0))^{0.3}, \text{ the correlation coefficient is } 0.95 \tag{7}$$
$$\Delta T_m / \Delta T_0 \equiv (1 / (x+x_0))^{0.9}, \text{ the correlation coefficient is } 0.85 \tag{8}$$

Thus with our real HVAC system, we can distinguish for the velocity decay two regions, a characteristic zone with nv = 0.3 and a transition zone with nv = 0.9. For temperature decay, the number of supply direction has a high influence on the nt value. Indeed, if the air jet has one direction we obtain a characteristic zone with nt = 0.3. On the other hand if the air jet has three directions we obtain a transition zone with nt = 0.9. The velocity profiles on Z axis and Y axis are shown on figure 4. We notice that the establishment of the jet is firstly achieved on the transverse axis Y. If we consider K_V and K_T, we got a K_T value higher than a K_V value. Thus the ratio K_T / K_V is equal to 1.13 when the air jet has one direction and 1.66 when the air jet has three directions. The usual value of this ratio is 0.84 as discussed by Meslem (1997). The values obtained with our experiments implies that the thermal diffusion is higher than the dynamic diffusion. We confirm this result with the figure 5 which shows the evolution of the velocity and temperature half widths, $Z_{0.5U}$ and $Z_{0.5\Delta T}$. For all tests, we obtained $Z_{0.5U} > Z_{0.5\Delta T}$. It is difficult to give an explanation to these results. Nevertheless, considering that the ratio between the grille length l_0 and the climatic chamber length is equal to 1/3, we may assume that the influence of the confined space is high on the development of the jet.

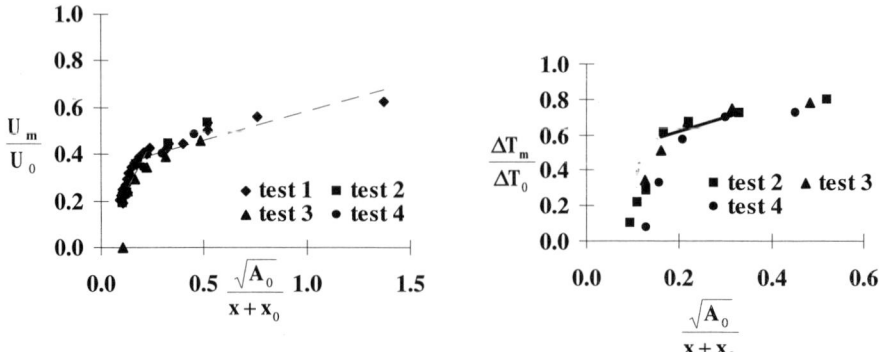

Figure 2 : Velocity and temperature decay for the one direction jet

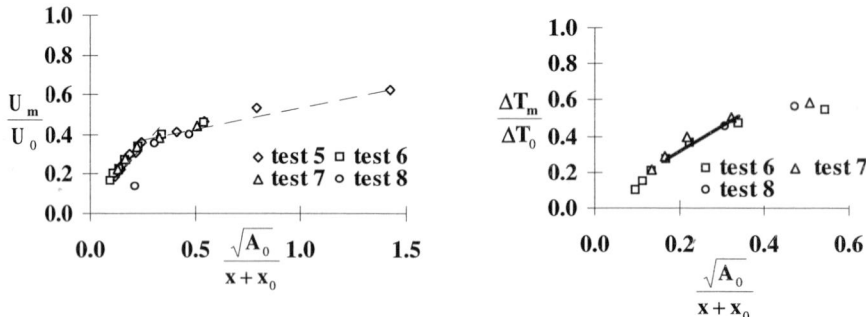

Figure 3 : Velocity and temperature decay for the three direction jet

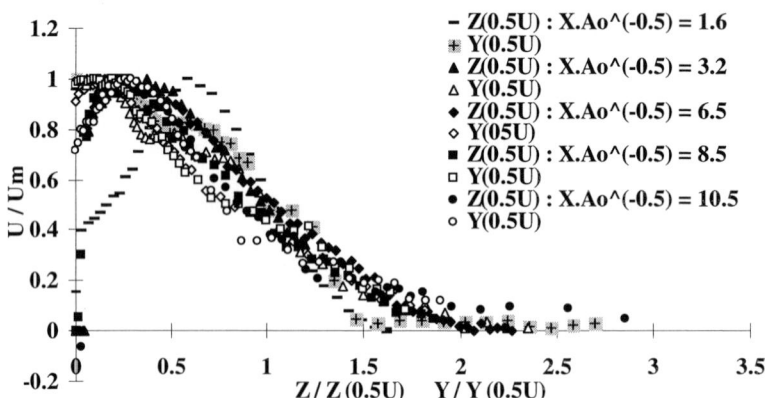

Figure 4 : Velocity profiles on axial and transverse axes

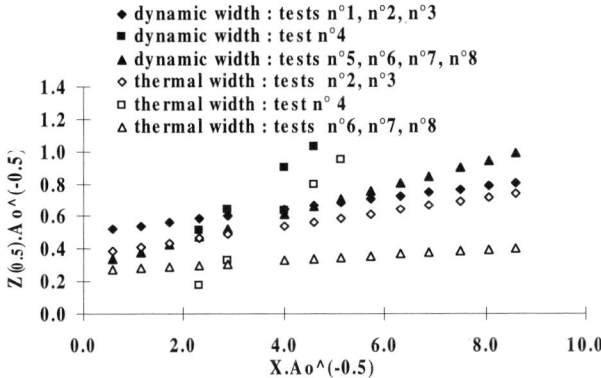

Figure 5 : Velocity, $Z_{0.5U}$ and temperature, $Z_{0.5\Delta T}$ half widths

Separation distance of the cold horizontal wall jet

The behaviour of the wall jets at separation is very complex as discussed by Hassani (1997). Figure 6 presents the separation distance X_S for all experiments. We can notice that the experimental values Xs are close to Grimitlyn and al (1993), Anderson and al (1991) and Rodhal (1977) values. The knowledge of the separation distance is important in order to find correlation between airflow and comfort in the climatic chamber. A relation between the detachment point and the initial Archimedes number is given by:

$$Xs / A_0^{0.5} = 0.23 \, Ar_0^{-0.6}, \text{ the correlation coefficient is 0.91.} \tag{9}$$

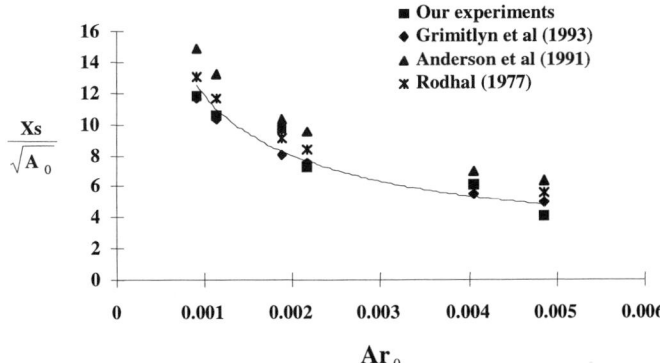

Figure 6 : Separation distance of the cold jet

CONCLUSION

Experimental studies of airflow characteristics in an air-conditioned chamber were carried out under various conditions. A first analysis made on the decay of the centerline velocity and temperature show the influence of the confined space. In discordance with others authors, we noticed a thermal diffusion more important than the dynamic diffusion. So, future studies on jets issued from real HVAC systems in confined spaces are needed. Furthermore, we proposed a relationship between the separation distance of the cold jet and the initial Archimedes number.

ACKNOWLEDGEMLENTS

This study was supported in part by the Research and Technology National Agency (ANRT), the Environment and Energy Management Agency (ADEME) and Electricity of France (EDF).

REFERENCES

ANDERSON R., HASSANI V, KIRKPATRICK A, KNAPPMILLER K and HITTLE D. (1991).Experimental and computational visualisation of cold air ceiling jet *ASHRAE Journal* **33:5**, 12-17.

GRIMITLYN M I and POZIN G.M. (1993). Fundamentals of optimising air distribution systems. *ASHRAE Trans*action **99:1**, 159-1365.

HASSANI A., KIRKPATRICK A. and SFORZA P. (1997). Flow and separation Characteristics of three-dimensional Negatively Buoyant Wall Jets. *HVAC&R RESEARCH* **3:2**, 112-127.

ISO (1997) Ventilateurs industriels-Essais aérauliques sur circuit normalisés. *ISO standard 5801:1997(F)*, ISO Geneva, Switzerland

MESLEM A. (1997). Contribution à l'étude du couplage thermique entre un jet et un local climatise. *These de Doctorat* , Institut National des Sciences Appliquees, Lyon, France

RODHAL E. (1977) The point of separation for cold jets flowing along the ceiling. *CLIMA 2000 Conference*, Belgrade, Yugoslavia, 219-228.

© 2000 Elsevier Science Ltd. All rights reserved.
Air Distribution in Rooms, (ROOMVENT 2000)
Editor: H.B. Awbi

A NUMERICAL ANALYSIS OF A GENERIC VORTEX DIFFUSER BY USING ZONAL HYBRID MESHES

S. C. Hu[1], Y. K. Chuah[1] and J. M. Barber[2]

[1] Dept of Air-Conditioning and Refrigeration National Taipei University of Technology
1, Section 3, Chung-Hsiao E. Rd, Taipei 106, TAIWAN
FAX: 886-2-27314919   Email: f10870@ntut.edu.tw
[2] School of Architecture and Building Engineering University of Liverpool
Liverpool PO Box 169, UK   Email: JMB1@Liverpool.ac.uk

ABSTRACT

In this study, a generic vortex diffuser is numerically analyzed by using zonal hybrid meshes. The regions around the disk and the guide vanes are meshed with tetrahedral cells. The inlet duct of the diffuser and the room space are meshed with hexahedral cells. The prismatic (or wedge) cells are used as the transitional cells to enable a change in cell type from hexahedral to tetrahedral. Such arrangement of the cell types has been found to be ideal for analysis of room air movement due to an air diffuser. For purposes of comparison, corresponding laboratory experiments were conducted. The measured vectors in the diffuser outlet vicinity were found to agree well with the results of the numerical analysis. The airflow patterns in the outlet region of the diffuser and the centerline velocity decay coefficient (K values) of the swirling jet were compared and evaluated. It is concluded that the zonal hybrid meshes are well suited for analysis of the complicated flow structure issuing from a vortex diffuser.

INTRODUCTION

Analysis of the room air movement including the influence of a diffuser is problematic because the computational grid in the room tends to be rectangular and therefore not well suited for a radial type of flow (if a radial ceiling diffuser is used). Another difficulty is that the size of a diffuser is usually much smaller than the room, thus resulting in very large mesh aspect ratio and/or mesh skewness in the location of the interface between the ceiling and the diffuser. Therefore "simple representations" of the diffuser were adopted in previous studies. However, the complicated airflow structures that created by complex diffusers, such as vortex diffusers, render the "simple representations" of the diffuser impractical.In view of the difficulty, the Annex 20 of the International Energy Agency (IEA) has evaluated and compared various techniques of the computational methods for predicting room air and contaminant flow. These include a modeling work for simulating room air motion from a wall-mounted diffuser with 84 small round nozzles (the diffuser was provided by HESCO Co. hence hereafter referred as HESCO-Type diffuser). There are at least four models that were used in describing the inlet boundary conditions at the supply opening of a room, as indicted by Nielsen (1992): the box method, the prescribed velocity method, the direct description method and the computer-generated supply method. The first two methods need either experiment data or more detailed information (Gosman et al. 1980, Nielsen 1990). The advantage of these methods is that the required computer storage and computational time are small. However, the data needed may often not be available. The direct description method is the most often used for modeling the influence of diffuser on air distribution in rooms. In using the direct description method, the flow information across the diffuser is simplified based on some physical hypothesis and geometrical simplification. For complex diffusers, such as vortex diffusers, the "simple representations" of the diffuser may become impractical. Previous studies using the direct description method include Skovgaard et al. (1990), Chen (1991), Jiang et al. (1992), Mizutani et al. (1994), Heikkinen and Piira (1994) and recently Hu et al. (1999). Heikkinen and Piira (1994) tested several simple representation methods to simulate the airflow in rooms using a multi-cone circular ceiling diffuser. Hu et al (1999) applied the momentum model for the modeling of cold air distribution systems using a nozzle type diffuser. The role of computers in research and development is changing. Just several years ago, it is commonly believed that a detailed analysis of airflow around a jet diffuser is difficult because the geometric configuration of a diffuser in practice is usually complex, and the flow is three-dimensional, turbulent, and with a considerably high Reynolds number. However, the rapid development of computer capacity and CFD techniques make it possible to perform the computation that takes into account of the geometry of diffuser. In such a case, detailed flow information around the diffuser that influence the flow pattern in a room can be included in the computation. The vortex diffusers, as shown in Figure 1, have a relatively high entrainment due to the swirl of the exit air, which creates a more thorough mixing of supply air and room air than with most other diffusers. In the diffuser, the swirl effect is caused by stationary twist guide vanes which increase the turbulence level by imparting a spiral twists or swirl to the supply air. Although currently on the market, the writer has been unable to obtain detailed performance data for them. The objectives of this paper are: to evaluate the capability of zonal hybrid meshes when applied in prediction of room air motion, and to obtain the detailed flow information of radial wall jet created by a vortex diffuser, to examine the effect of twist vanes on the turbulence level at the diffuser outlet region.

NUMERICAL METHODS

Airflow model

The flow is assumed to be at isothermal conditions. Most airflows in buildings are turbulent. The turbulent model used in this study is the standard k-ε turbulence model. With this eddy-viscosity turbulence model, the airflow transport can be described by the following time-averaged Navier-Stokes equations:

$$\text{div} (\rho \mathbf{V}\Phi - \Gamma_{\phi,\text{eff}} \, \text{grad}\Phi) = S_\phi \tag{1}$$

Where ρ is air density (kg/m^3), $\Gamma_{\phi,\text{eff}}$ is the effective diffusion coefficient (N.s/m^2), $\mathbf{V}$ is the air velocity vectors (m/s), S_ϕ is source term of the general fluid property, and ϕ can be any one of 1,u,v,w,k, and ε, where u,v,w is velocity components in three directions (m/s), k is turbulence kinetic energy (m^2/s^2), ε is dissipation rate of turbulence kinetic energy (m^2/s^3). When ϕ=1, the general equation changes into the continuity equation.

Space model and mesh generation

The studied space model includes the duct, diffuser and the room, as shown in Figure 2. The mesh generation was accomplished by using so called zonal hybrid meshes, where the regions around the disk (at outlet) and the guide vanes were meshed with tetrahedral cells, the duct (providing the inlet flow) and the room space were meshed with hexahedra cells, and the prismatic (or wedge) cells were used as transitional cells to enable a change in cell type from hexahedra cells to tetrahedral. Therefore, this arrangement enables one able to obtain the detailed flow information inside and outside the diffuser, without the need of simplified modeling the diffuser. Detailed mesh arrangement is depicted in Figure 3.

Boundary conditions

a. wall surfaces: duct walls, guide vanes, room ceiling and walls, vortex diffuser walls and disk.
b. velocity boundary condition: constant velocity boundary condition was applied at the duct inlet.
c. pressure boundary condition: open pressure condition was applied at the ground floor where the ground floor was connected to return duct.

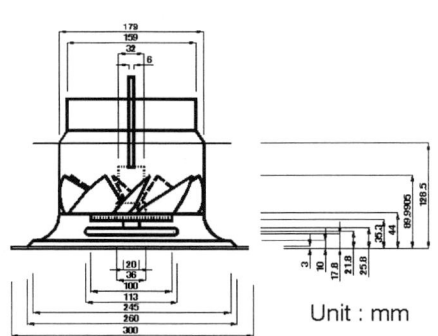

Unit : mm

Figure 1. The dimensions of the vortex diffuser.

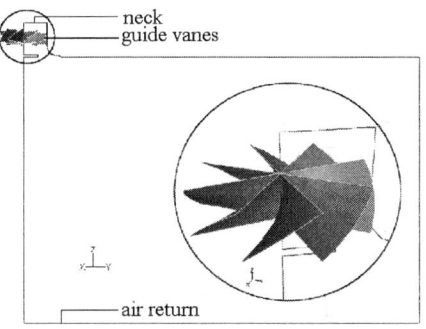

Figure 2. The environment chamber equipped with the vortex diffuser.

RESULTS AND DISCUSSION

Case setup
    There are four cases of study, as listed in Table 1.

Table 1. Case Setup

Case	Supply flow rate	With twist guide vanes ?
1	47.2 L/s (100cfm)	Yes
2	23.6 L/s (50cfm)	Yes
3	47.2 L/s (100cfm)	No
4	23.6 L/s (50cfm)	No

Experimental investigations were performed for the cases listed in Table 1. The supply flow rates used are less than the manufacturer's nominated supply flow rates. The reason for using less supply flow rate was to make the decay of the jet centerline velocity takes place in a shorter distance, so to fit the dimensions of the environmental chamber.

Data verification
    Airflow patterns in the outlet region of the diffuser and the centerline velocity decay coefficient (K value) of the swirling jet were compared and evaluated by experimental data. The flow patterns and turbulence characteristics in the diffuser outlet region and in the room were measured by using a three dimensional ultrasonic anemometer. In the region adjacent to the ceiling (0.05 under the ceiling), the velocity was measured by using a hot-wire anemometer. Figure 5 shows the comparison of predicted and measured velocity vectors for case 1 (flow rate = 47.2 L/s). The general congruence of the flow patterns in the diffuser outlet region is addressed.

526

Symmetric test

The symmetric test was performed at the diffuser outlet level (0.05 m under the diffuser). As indicated in Figure 4, a surprising octagonal shape uneven velocity profile (Vmax= 2.11m/s and Vmin=0.25 m/s) along the radial direction of the diffuser was observed. This means that the air distributions in a room produced by the vortex diffuser are uneven. This uneven velocity distribution is mainly due to the spin provided by the twist guide vanes. This observation was confirmed by corresponding experimental results. Therefore, in the following discussion, a vertical plane section with maximum velocity was selected.

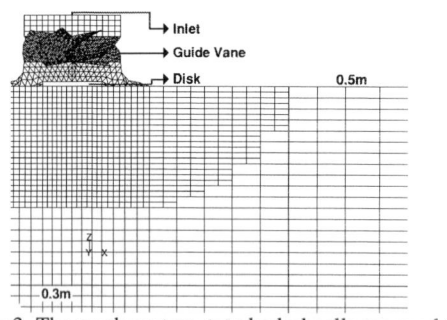

Figure 3. The mesh system, tetrahedral cells are used for the disk and the guide vanes regions, hexahedra cells are used for the connecting duct and the room space, the prismatic cells are used as transitional cells.

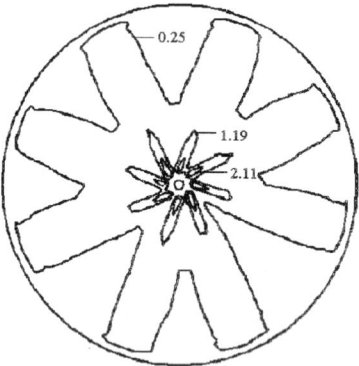

Figure 4. The predicted velocity contour at the level of 0.02 m under the ceiling.

Centerline velocity decay coefficient (K values)

The performance of the swirling jet created by a vortex diffuser can be characterized by a single constant, namely 'jet centerline velocity decay coefficient' (K value). The jet velocity decay coefficient is defined as:

$$K = \frac{V}{V_o} \frac{(x + x_p)}{\sqrt{A_o}}$$ (2)

When the measured data are rearranged in the following form, and by using linear regression techniques, then the K value and x_p value can be determined.

$$\frac{V_o}{V} = \frac{x}{K\sqrt{A_o}} + \frac{x_p}{K\sqrt{A_o}}$$ (3)

where V= centerline velocity at distance x from diffuser, (m/s), V_o= effective outlet velocity of the diffuser, (m/s), K= velocity decay coefficient, dimensionless, A_o= effective outlet area of the diffuser, (m), x = distance to centerline velocity Vx, (m), and x_p= the distance between the virtual origin of jet and the center of the diffuser, (m). It was very difficult to measure the effective outlet velocity of the vortex diffuser using the available instrument (an ultrasonic anemometer and hot-wire anemometer). For comparison purposes, K values were calculated based on the neck velocity/area and the predicted and measured results are listed in Table 2. Figure 6 shows the predicted and measured centerline velocity decay and the calculation details for case 1 and case 2.

Table 2. Comparison of the predicted and measured K values

	Predicted		Measured
Case 1 (Qs=100 [cfm])	Based on Vn	1.94	Based on Vn 2.5
	Based on Vo	1.62	
Case 2 (Qs=50 [cfm])	Based on Vn	1.70	Based on Vn 2.3
	Based on Vo	1.41	

It is noted that the measured K values based on the neck velocity (Vn) in the present study are in the range of 2.3 to 2.5, which are close to those of Shakerin and Miller (1995). The predicted K values based on the neck velocity are less than those of the measured ones. On the other hand, as the computational domain includes the diffuser, integrating the local effective velocity/area can easily derive the effective velocity/area in the diffuser's exit plane. The predicted K values based on the effective velocity/area are approximated to be about 1.62 and 1.41 for the cases 1 and 2, respectively. For the vortex diffuser studied, the predicted air entrainment ratio was calculated by the following equation:

$$\frac{Q_x}{Q_o} = 1.21 \frac{(x + x_p)}{KA_o^{1/2}} \tag{4}$$

Based on the K value predicted, the entrainment can be evaluated. Two computational runs (case 3 and case 4) were conducted for the cases without the twist guide vanes for comparison to that with the twist vanes. Figure 7 show the flow patterns of cases 3. Comparing Figure 7 and Figure 5 one finds that the effect of twist guide vanes on the flow patterns in the outlet region is very significant. A horizontal radial wall jet along the ceiling for the case 1 was observed while a "dumping" profile was obtained for case 3. The comparison between case 2 and case 4 exhibits the same trend. The "dumping" profile at the outlet vicinity of case 3 and case 4 makes it impossible to calculate the K value.

CONCLUSIONS

1. For the vortex diffuser studied, the K value based on the effective velocity is about 2.0 and the entrainment ratio is $\frac{Q_x}{Q_o} = 1.21 \frac{(x + x_p)}{KA_o^{1/2}}$.

2. The twist guide vanes affect the horizontal jet and turbulence characteristics at the diffuser outlet region very significantly.

3. The induction due to twist vanes is sufficiently high to have a significant effect on room air motion.

REFERENCES

ASHRAE Handbook - Fundamentals, 1997, Atlanta: American Society of Heating, Refrigerating and Air-Conditioning Engineers, Inc.

Gosman, A. D., P. V. Nielsen, A. Restivo, and J. H. Whitelaw. 1980. The follow properties of rooms with small ventilation openings, Journal

Heikkinen, J., and K. Pira, 1994. CFD computation of jets from circular ceiling diffuser. Proceedings of Roomvent '94, Krakow, Poland, Vol. 1, pp. 330-433.

Hu, S. C., J. M. Barber and Y. K. Chuah, 1999. A CFD study for cold air distribution systems. ASHRAE Trans. Part 1, CH-99-6-5.

528

Jiang, Z., Q. Chen, and A. Moster, 1992. Comparison of displacement and mixing diffusers. Indoor Air(2) 168: 179.

Mizutani, K. et al. 1995. Numerical simulation of airflow and ventilation efficiency in a cold air delivery room - study of cold air delivery systems (part 4). Proceedings of Technical Meeting. *SHASE-Japan*, Hiroshima, (1): 361-364,(in Japanese).

Nielsen, P. V. 1990, Working report of International Energy Agency Annex 20.

Nielsen, P. V. 1992. Description of supply opening in numerical models for room air distribution. ASHRAE Trans. Vol 98, Part 1, p 963-971.

Skovgaard, M, and P. V. Nelsen, 1991. Modeling complex inlet geometries in CFD-Applied to air flow in ventilated rooms. 12th AIVC Conference, Ottawa.

Shakerin, S., and Miller, P. L.,1995. Experimental study of vortex diffusers. NREL Report/TP-472-7331. Golden, Colorado: National Renewable Energy Laboratory.

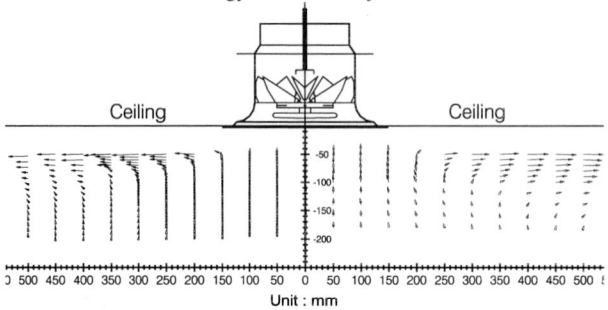

Figure 5. The predicted and measured (left & right side) flow patterns for case 1 (flow rate =47.2 L/s).

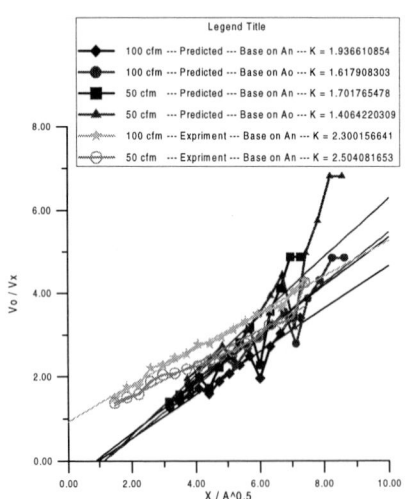

Figure 6. The predicted and measured centerline velocity decays for case 1 and case 2.

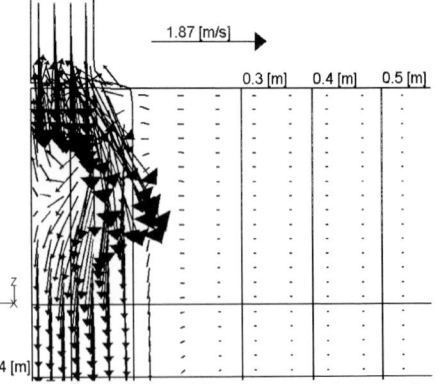

Figure 7 The predicted flow patterns for case 3 (flow rate = 47.2 L/s, without twist vanes).

© 2000 Elsevier Science Ltd. All rights reserved.
Air Distribution in Rooms, (ROOMVENT 2000)
Editor: H.B. Awbi

EXPERIMENTAL VALIDATION OF JET FORMULAE FOR AIR SUPPLY DIFFUSERS

J. Srebric, J. Liu and Q. Chen

Building Technology Program, Massachusetts Institute of Technology
Cambridge, MA 02139, USA

ABSTRACT

Indoor air quality, thermal comfort and contaminant distributions strongly depend on the flow characteristics from supply air diffusers. A correct prediction of the flow characteristics is most crucial for the prediction of the thermal comfort and the indoor air quality in an entire room. This investigation studied the performance of jet formulae in predicting the flow distributions from six commonly used diffusers, and measured room airflow in an environmental chamber with the diffusers. The measured data includes air velocity, temperature, and tracer gas concentration distributions, and the jet separation distances. The jet formulae can predict the air velocity and temperature profiles and the separation distance from the jets, if measured data is available to calibrate the formula coefficients, and if the jet can be classified as a free or wall jet. The study also found that the room air temperature distributions were uniform with the mixing diffusers, regardless of the type of mixing diffusers and the ventilation rates. However, the air velocity distributions can be different.

KEYWORDS

Diffusers, jet formulae, measurements, airflow pattern, environmental chamber.

INTRODUCTION

Air supply diffusers strongly influence the airflow patterns in an indoor space, and hence, they determine the distributions of indoor environmental parameters such as air velocities, temperatures, and contaminant concentrations. A correct prediction of the flow characteristics by the diffusers is most crucial for the design of the thermal comfort and the indoor air quality in a room. Predicting flow characteristics is a challenging task, since most of the diffusers used in buildings have a complicated geometry. Other important factors affecting the jet from a diffuser are room size, position, strength of heat sources, and obstacles in the room, such as furniture. Jet formulae are probably the simplest methods to calculate the air velocity, temperature, and species concentration distribution in a room. This paper reports the results of applying the jet formulae to study the indoor air distribution with six commonly used diffusers. These six diffusers are the slot (linear), grille, round ceiling, square ceiling, vortex, and displacement diffusers. The paper also discusses the characteristics of room air distributions with different ventilation rates and diffusers, according to the measured results from an environmental chamber.

JET FORMULAE

To design a room with an acceptable thermal comfort and indoor air quality level, one needs the information of air velocity, temperature, and species concentration distributions. Jet formulae have been developed in the past decades to calculate the velocity and temperature profiles along the jet. According to the flow conditions, jets can be classified as isothermal or non-isothermal, free or attached. If a diffuser is close to a wall, the jet will attach to the wall surface by the Coanda effect. On the other hand, free jets have unrestricted momentum diffusion and room air entrainment. The cross-sectional velocity distribution of a free jet can be represented by Gaussian profiles, while an attached jet has a sharp velocity gradient near the wall boundary layer. Jet velocity, temperature and species concentration profiles tend to be self-similar in the fully established jet region, which is much larger than the initial (core) and transitional jet regions. This profile similarity can be used to develop dimensionless formulae for the calculation of jet velocity, temperature, and species concentration profiles. For example, the following formulae are widely used to calculate the velocity profiles for free jets, attached jets (Verhoff 1963), and the separation distance for cold ceiling jets (Kirkpatrick & Elleson 1996), respectively:

$$\frac{u}{u_m} = \exp\left[-\frac{y^2}{2(cx)^2}\right], \quad c=0.082 \tag{1}$$

$$\frac{u}{u_m} = 1.48\eta^{\frac{1}{7}}\left[1 - erf(0.68\eta)\right], \quad \eta = \frac{y}{y_{0.5Um}} \tag{2}$$

$$x_s = a\,C_s\,K^{\frac{1}{2}}\left(\frac{\Delta T}{T}\right)^{-\frac{1}{2}} Q^{\frac{1}{4}}\,\Delta P^{\frac{3}{8}}, \quad C_s=1.2, \; a=48.04 \tag{3}$$

where u is the jet velocity at distance x from a diffuser, u_m is the maximum jet velocity at distance x, y is the coordinate normal to x, $y_{0.5Um}$ is the distance from a wall where jet velocity is half of the maximum value, x_s is the jet separation distance, K is the velocity decay coefficient, Q is the flow rate, ΔT is the room-jet temperature difference, T is the room temperature, and ΔP is the diffuser static pressure drop. More jet formulae, such as the jet centerline velocity, the temperature decay for radial jets, and the trajectory and drop of non-isothermal jets, can be found in Li et al. (1993).

TEST FACILITY

In order to validate whether velocity and temperature profiles from a commonly used diffuser can be determined by the jet formulae, this investigation conducted experimental measurements to obtain the corresponding data for the validation. The present study used a well insulated, full-scale environmental chamber (thermal resistance = 5.3 m²K/W) that is divided by a partition wall with a double glazed window into two parts - a test chamber and a climate chamber. The two chambers have separate HVAC systems. The control system allows a variable air supply rate ranging between 1 to 20 ach for the test chamber and 2 to 40 ach for the climate chamber. The climate chamber is used to simulate an outdoor summer condition at a constant temperature of 35°C. Figure 1 shows the test chamber, which is 5.16 m long, 3.65 m wide, and 2.43 m high. The chamber had two human simulators, two computers, two tables, two cabinets, and four fluorescent lamps. The chamber was equipped with six different diffusers. Only one of the diffusers is activated a time during the experiment.

The test facility has a flow visualization system for observing airflow patterns, a hot-sphere anemometer system for air velocity and temperature measurements, and a thermo-couple system for measuring wall and air temperatures. The measured velocities ranged from 0.05 to 5 m/s, and the repeatability was 0.01 m/s, or 2% of the readings. However, the hot-sphere probes had a great uncertainty for the measurements of air velocities lower than 0.1 m/s, due to self-induced natural convection. The smoke visualization was used to estimate low air velocities in some important room areas. The measuring errors for air temperature with the anemometers were 0.3 K, including the errors introduced by the data acquisition system.

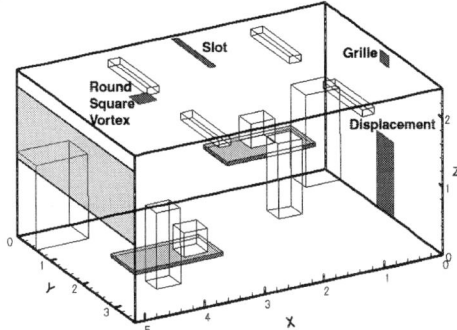

Figure1: The schematic of the test camber

The present study investigated room air distribution with different diffusers under summer cooling conditions. The diffusers supplied cold air to cool the heat generated by the occupants, computers, lighting, and window. Table 1 shows the supply air conditions and the cooling loads measured in the test chamber with different diffusers. For the grille diffuser, the study used three different ventilation rates. The internal heat sources were e same for all cases: each human simulator generated 75w, the heat from the computers was 108W and 173W, respectively, and each fluorescent lamp gave 34W. The total internal heat gain was 567W. Although the climate chamber was controlled at 35°C, the room airflow had an impact on the heat gain from the window and the partition wall. The heat gain difference from the window and the partition wall with different diffusers was very remarkable. In order to study indoor air quality, a tracer gas (SF_6) was released at the head level of the two human simulators to simulate contaminants.

TABLE 1

FLOW BOUNDARY CONDITIONS FOR DIFFERENT EXPERIMENTS

Diffusers	Air Exchange Rate (ach)	Supply Air Velocity (m/s)	Supply Air Temp. (°C)	Exhaust Air Temp. (°C)	Cooling Load (W)
	8.0	2.18	16.2	22.3	740
Grille	5.0	1.40	15.1	24.5	728
	3.0	0.86	10.6	25.3	673
Slot (Linear)	9.2	3.90	16.2	21.4	737
Vortex	5.1	7.50	16.4	24.8	667
Round Ceiling	5.0	3.58	16.1	24.8	670
Square Ceiling	4.9	5.20	14.5	24.1	725
Displacement	5.0	0.35	13.0	22.2	702

The experimental measurements of the air velocity, air temperature, and SF_6 concentration were conducted in the middle of the room, as well as in the vicinity of the air supply diffusers. Five poles with six anemometers were arranged in a matrix form in the middle of the room, and later grouped in front of the diffusers. Zero buoyancy smoke was used to visualize the airflow pattern. The smoke visualization showed that all six diffusers produced attached jets because of the air supply direction and the diffuser's close location to a wall.

RESULTS

The following section discusses how the jet formulae can be used to calculate the flow characteristics with the six commonly used diffusers. The jet from the displacement diffuser dropped immediately to the floor in front of the diffuser because of the low momentum and strong negative buoyancy. The jet then spread over the floor and reached the partition wall. Figure 2a shows a good agreement between the jet formula and measured data for a velocity profile 0.1 m from a displacement diffuser in the jet region, but the jet formulae can not predict velocity profile for the stratified flow above the jet region (> 0.2m). Jacobsen and Nielsen's (1992) study on displacement diffusers had the same conclusion.

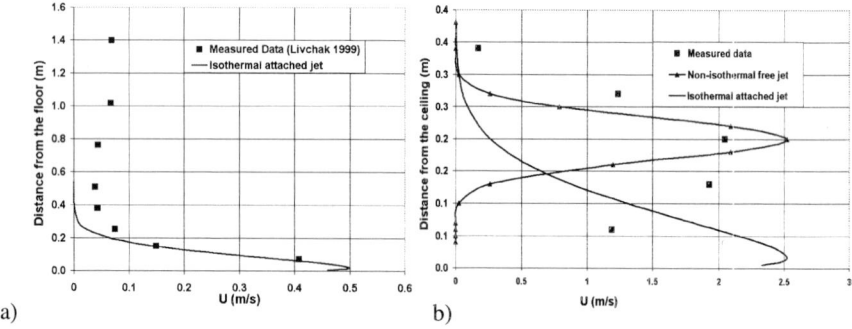

a)   b)

Figure 2: The jet formulae applied to a) displacement difuser b) grille diffuser

The grille diffuser generated a jet through deflection vanes that were adjusted in this case to be parallel to the air stream. The diffuser was installed 0.11m below the ceiling, but the jet from the diffuser can attach to the ceiling through the Coanda effect. Figure 2b represents the measured data at 0.4 m from the diffuser with an airflow rate of 5 ach. Similar profiles were measured for 0.6m and 0.8m from the diffuser. The results indicate that the jet is neither attached nor free. Therefore, both the attached and free jet formulae failed to predict the jet flow.

The slot diffuser had blades adjusted 45° down towards the upper right side of the room (see Figure 1). The initial jet flow was 45° downwards, but the jet attached to the ceiling and the north wall. Very similar to the results for the grille diffuser as shown in Figure 2b, neither the free jet formula nor the attached jet formula can correctly simulate the velocity profile along the jet.

The round, square, and vortex ceiling diffusers discharged the airflow horizontally in a radial direction, four directions, and a radial direction with swirl, respectively. The wall jet formula can predict the jet velocity profile if measured data is available to calibrate the formula coefficient. However, due to the geometric character of the diffuser, the formula cannot be applied to the region close to the vortex diffuser.

A cold jet would separate at a certain distance from the diffuser. In order to avoid a draft created by the jet, it is important to determine the separation distance. The dashed lines in Figure 3 show the separation distance in the test chamber with the grille diffuser for three different ventilation rates. Kirkpatrick and Elleson (1996) developed a formula to calculate the jet separation distance. The formula needs the diffuser static pressure drop as input. Unfortunately, many diffuser catalogues provide such pressure drop information only for a certain range of discrete airflow rates. Therefore, it is difficult to use the formula. We have applied the formula to grille diffusers for the flow rate with a known static pressure drop. The calculated separation distance for 8 ach was approximately 0.35m higher than that observed by smoke. This discrepancy probably comes from the velocity decay coefficient that could be only roughly estimated. Furthermore, factors including the room geometry, the heat sources can also have a significant impact on the jet development and separation. It is difficult to use the jet formulae to determine the velocity and temperature profiles and the separation.

In addition, jet formulae for velocity and temperature decay normally include coefficients that depend on the diffuser type. The coefficients are difficult to determine. Even if the jet formulae can be used, it is more reliable for determining the velocity and temperature decay within a small distance from the jet origin (the diffusers). Nevertheless, the jet formulae provide some important insight into the physical processes of diffuser jets, such as the jet diffusion and entrainment rates. Since the formulae are simple and easy to use, they deserve considerable attention in practical use.

Figure 3 shows dimensionless temperature profiles for the grille diffuser with 3, 5, and 8 ach at the middle vertical cross-section. The figure in the lower right corner illustrates the pole positions. The results indicate a very uniform temperature distribution in the lower part of the room, although the difference is significant in the upper part of the room. The difference is especially evident at the region close to the diffuser. In this case, a 3 ach ventilation rate can still create a reasonable amount of mixing in the occupied zone, although the jet separated from the ceiling very early. However, the velocity distribution in the occupied zone, especially near the floor, is rather different with different ventilation rates, because the separation distance and airflow pattern are different.

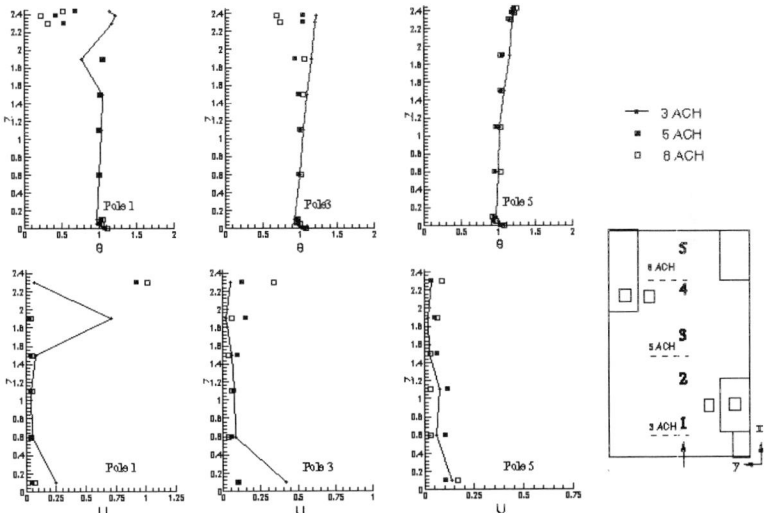

Figure 3: Dimensionless temperature and velodity profiles for the grille with different flow rates
$\theta=(T-T_{in})/(T_{out}-T_{in})$ and $U=V/V_{in}$

Figure 4 compares the dimensionless temperature profiles in the test chamber with square, round and vortex shaped ceiling diffusers with the same thermal and flow boundary conditions. Although the diffusers were very different, the temperature profiles were the same in the occupied zone. In fact, all three diffusers had attached jets. The smoke visualization showed that the airflow patterns differed only in the region close to the ceiling. The difference of the normalized velocities in Pole 5 is smaller than that in Pole 3. This indicates that the vortex diffuser has higher jet entrainment and faster velocity decay.

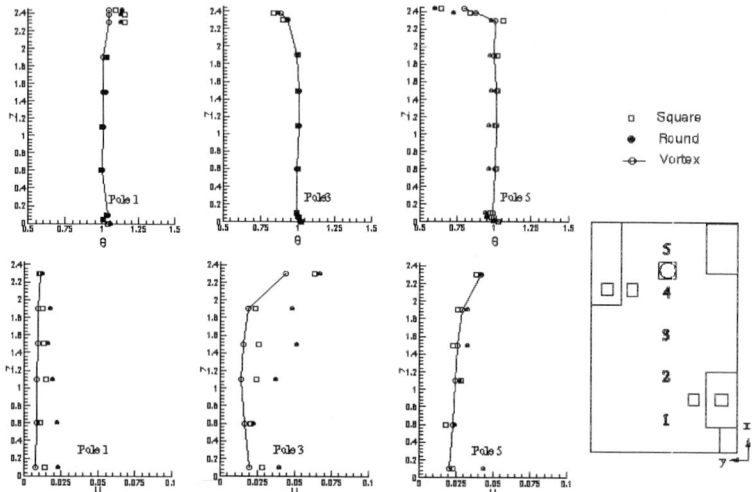

Figure 4: Dimensionless temperature, velocity and concentration profiles for the square, round and vortx ceiling diffusers $\theta=(T-T_{in})/(T_{out}-T_{in})$ and $U=V/V_{in}$

CONCLUSIONS

This study examined the applicability of the jet formulae for six commonly used diffusers. They are the slot (linear), grille, round ceiling, square ceiling, vortex and displacement diffusers. The jet formulae can predict the air velocity and temperature profiles, and the separation distance from the jets. However, they can work only if measured data is available to calibrate the formula coefficients, and if the jet can be classified as a free or wall jet. The jet formulae could not estimate the impact of room geometry and heat sources on the jet development.

The study also found that room air temperature distributions are uniform with the mixing diffusers, regardless of the type of mixing diffusers and the ventilation rates. However, the air velocity distributions can be different.

REFERENCES

Jacobsen T.V. and Nielsen P.V. (1992). Velocity and temperature distribution in flow from an inlet device in rooms with displacement ventilation. *ROOMVENT '92*.

Kirkpatrick A.T. and Elleson J.S. (1996). *Cold air distribution system design guidelines*. ASHRAE, Atlanta, Georgia, US.

Li Z.H., Zhang J.S., Zhivov A.M. and Christianson L.L. (1993). Characteristics of diffuser air jets and airflow in the occupied regions of mechanically ventilated rooms – a literature review. *ASHRAE Transactions* 99:**1**, 1119-1127.

Livchak A. (1999). Personal Communications.

Verhoff A. (1963). *The two-dimensional turbulent wall jet with and without an external free stream*. Report no.626, Princeton University 1963.

© 2000 Elsevier Science Ltd. All rights reserved.
Air Distribution in Rooms, (ROOMVENT 2000)
Editor: H.B. Awbi

EXPERIMENTAL AND NUMERICAL STUDIES OF AN ENCLOSED TURBULENT JET IMPINGING A POROUS LAYER

M. Prakash[1], J. Mahoney[2], Y. Li[2], O. F. Turan[1] and G. R. Thorpe[1]

[1] Faculty of Engineering and Science, Victoria University of Technology,
Melbourne, Victoria 8001, Australia
[2] Thermal and Fluids Engineering,
CSIRO Building, Construction and Engineering, Victoria 31d90, Australia

ABSTRACT

The interaction between turbulent flow and saturated porous media is investigated by studying an axisymmetrical turbulent jet impinging on porous foam in a cylindrical enclosure. Both flow visualisation and laser Doppler velocimetry techniques are used in the experimental study. A modified two-equation k-ε model, which considers the turbulence effect in the porous region, is used in the computational fluid dynamics study. Generally, good agreement is achieved between the numerical and experimental results. When both a modified turbulence model and a laminar porous model are used in the numerical simulations, the results obtained from the modified turbulence model give better agreement with the experimental results.

KEYWORDS

Computational fluid dynamics, porous media, turbulence, internal airflows, laser Doppler velocimetry.

INTRODUCTION

In computational fluid dynamics analyses of internal airflows, there are a number of applications which can be modelled as porous media, such as the seating area of large theatres, stored products in a cool or cold room, and furniture in an open-plan office. There are at least two important issues in modelling the porous flow systems with porous media: how the airflow in the porous region and the clear airflow region interact, and how to model the turbulence effect in the porous region if it exists. There have been a few attempts in the literature at modelling turbulence effects in porous media. Prescott and Incropera (1995) used a Darcy damping term in the turbulence kinetic energy equation to account for the damping of turbulence in the porous medium. Antohe and Lage (1997) derived a general two-equation (k-ε) macroscopic turbulence model for flow in a porous medium without any numerical or experimental results for validating their model. Chen *et al.* (1998) modified the model developed by Antohe and Lage (1997) and simulated a two-layer system consisting of a fluid and a porous medium, without any experimental validation of their modified turbulence model. A review of the literature on modelling turbulent flow in porous media by Antohe and Lage (1997) clearly illustrates the lack of experimental data to validate such models.

In this paper, some of the main features of an extensive experimental and numerical investigation are presented. A two-layer system is studied consisting of a turbulent fluid overlying a porous medium. Due to the opaque nature of the porous medium, flow visualisation and laser Doppler velocimetry (LDV) measurements were carried out only in the fluid layer. Since the flow in the fluid layer is turbulent, it is expected that this turbulence will persist in the porous medium for some height. Consequently, proper representation of the flow in the porous medium will lead to a better prediction of the flow in the fluid layer due to the interaction between the two layers.

EXPERIMENTAL SETUP

A schematic diagram of the experimental setup is presented in Figure 1 which shows the traversing mechanism and data acquisition system used for the LDV measurements. The same experimental rig was used for the flow visualisation experiments. An impinging jet (diameter = 0.019 m) enclosed in an acrylic cylinder (diameter = 0.39 m) was used to generate the turbulent flow. Details of the flow visualisation experiments are given in Prakash *et al.* (1999a). Measurements were carried out with a single-channel LDV system. A 1 W argon ion laser acted as the light source. The sampling rate of the instrument was adjusted to 10 MHz and a low pass filter of 5 MHz was applied. A total of approximately 10,000 samples were measured for each point, with the data rate varying between 100 and 1000 Hz. Repeatability experiments showed the experimental uncertainty to be within ±2% of the bulk velocity for mean velocities, and ±10% of the maximum turbulence intensity for the fluctuating velocities (Prakash *et al.* 1999b). For the experiments, water was used as the fluid, and porous foam was used as the porous medium. Foam grades G10, G30 and G60 corresponding to 10, 30 and 60 pores per inch were used for the experiments, with the lowest grade foam corresponding to the highest permeability value.

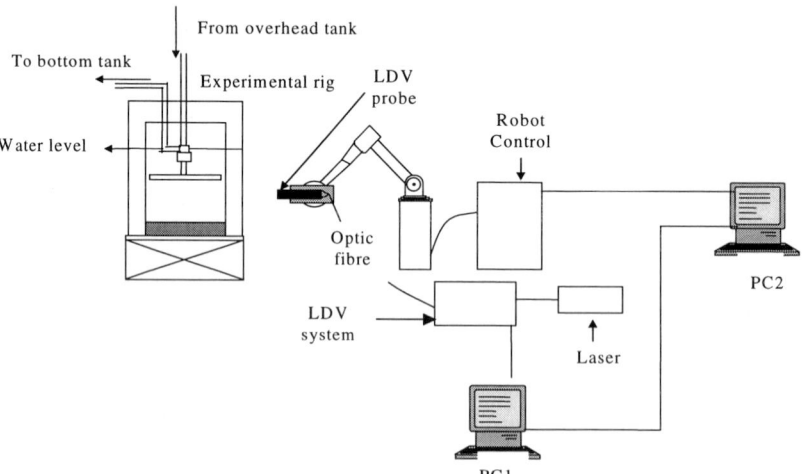

Figure1: Schematic diagram of experimental setup for LDV measurements

MODEL DEVELOPMENT AND NUMERICAL METHOD

In the simulations, the low Reynolds number k-ε model of Jones and Launder (1972) is used for the fluid layer. Laminar and turbulent flow models were used for the porous foam. For the porous medium with a turbulent flow model, it is assumed that turbulence is damped only due to the Darcy term as a first approximation. The effect of the Forchheimer term in the turbulence transport equations is currently under investigation. The flow is assumed to be axisymmetric for the computations. The conservation equations of mass, momentum, turbulence kinetic energy and its rate of dissipation can be summarised as:

$$\frac{\partial}{\partial x_i}(\rho u_i \varphi) = \frac{\partial}{\partial x_i}\left(\Gamma_{eff}\frac{\partial \varphi}{\partial x_i}\right) + S_\varphi \tag{1}$$

where φ = general dependent variable, u_i = velocities u_r and u_z in the r and z directions respectively, x_i = coordinate in the axisymmetric system, Γ_{eff} = effective diffusion coefficient for each dependent variable φ; and S_φ = source term for φ. The source terms have the usual form for each equation. Additional source terms in the momentum and turbulent transport equations are incorporated for the porous medium as in Table 1.

TABLE 1

SOURCE TERMS IN MOMENTUM AND TURBULENT TRANSPORT EQUATIONS DUE TO POROUS MEDIUM

Conservation equation	Variable	Darcy term	Forchheimer term
Momentum equations	u_r, u_z	$-\dfrac{\mu\phi u_i}{K}$	$-\rho\dfrac{c_F \phi u_i\left(u_j u_j\right)^{1/2}}{\sqrt{K}}$
k equation	k	$-2\phi\dfrac{\mu}{K}k$	Absent
ε equation	ε	$-2\phi\dfrac{2\mu}{K}\varepsilon$	Absent

In Table 1, ρ is the fluid density and μ is the fluid viscosity. ϕ is the porosity, K is the permeability and c_F is the Forchheimer coefficient for the porous medium. Details of the model can be found in Prakash et al. (1999b). The following boundary conditions were employed: $u_r = u_z = k = 0$ and the gradient of $\varepsilon = 0$ at the walls. A gradient equal to zero condition was applied at the symmetry plane for all variables, and only one half plane of the cylinder was solved. For the fluid layer, the conservation equations are used without the Darcy and Forchheimer terms in the momentum equations and without the Darcy modification in the turbulence transport equations with $\phi = 1$. For simulations with a laminar flow model for the foam, the interface between the fluid layer and the porous medium is treated like a solid wall for k and ε. For all computations with foam, a 200×200 non-uniform grid was found to yield grid-independent solutions. Details of the grid refinement study can be found in Prakash et al. (1999b). A hybrid, first-order upwind/second-order central difference scheme was used to discretise the convection terms, whereas the diffusion terms were discretised using the second-order central difference scheme.

RESULTS AND DISCUSSION

The numerical predictions and the LDV measurements are compared next for two heights of fluid region, $H = 0.15$ and 0.05 m, with a height of porous region $h_p = 0.05$ m for the G10 foam.

Vector Plots and Streamtraces

Figures 2 and 3 give a comparison of the vector plots and streamtraces for $H = 0.15$ and 0.05 m and h_p = 0.05 m. From the measurements in Figure 2(a), a two-cell pattern can be seen. However, the second cell closer to the wall is not well defined, indicating unsteady behaviour. The flow visualisation for the same case (Prakash *et al.* 1999a) showed a similar pattern. As shown in Figures 2(b) and 2(c), respectively, the simulations with the turbulent and laminar flow models for foam, show two cells. However, the present model is not able to capture the unsteady nature of the flow. From the streamtraces for the two simulations, one can see that the flow pattern with a turbulent flow model for the foam in Figure 2(b) is in closer agreement with the experimental pattern of Figure 2(a). Both the laminar and turbulent flow models predict that the dominant cell is closer to the jet than was either measured or visualised. Experimentally the two cell pattern was well established for the lower fluid heights of $H = 0.1$ and 0.05 m. Here, the two-cell pattern obtained with $H = 0.05$ m is illustrated in Figure 3(a). For the lowest fluid height, the simulation with the laminar flow model did not show the second cell close to the wall, as seen in Figure 3(c). The predicted results with the turbulent flow model, on the other hand, gave the size and position of the second cell for the lowest fluid height accurately, as seen in Figure 3(b).

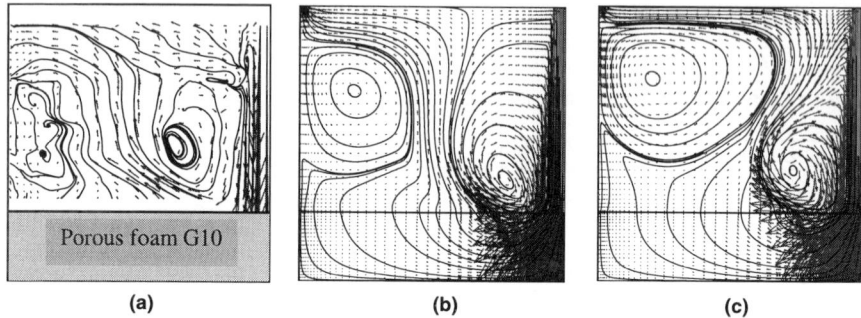

(a)   (b)   (c)

Figure 2: Streamtraces and vector plots for $H = 0.15$ m, $h_p = 0.05$ m, $U_b = 1.6$ m/s:
(a) LDV measurement; (b) numerical simulation, turbulent flow model for foam; and
(c) numerical simulation, laminar flow model for foam

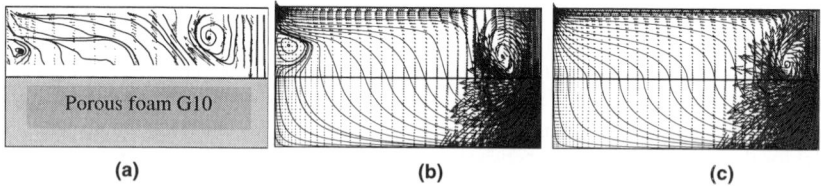

(a)   (b)   (c)

Figure 3: Streamtraces and vector plots for $H = 0.05$ m, $h_p = 0.05$ m, $U_b = 1.6$ m/s:
(a) LDV measurement; (b) numerical simulation, turbulent flow model for foam; and
(c) numerical simulation, laminar flow model for foam

Axial, Radial Velocity and Turbulence Kinetic Energy Profiles

Figures 4(a) and 4(b) show the mean axial velocity profiles for H=0.15 and 0.05 m, respectively, with $h_p = 0.05$ m for the G10 foam. In these figures, the abbreviations CFDL and CFDT refer to profiles obtained with a laminar and turbulent flow model for the foam. The results with the turbulent flow model for foam is in better agreement with the experimental points. The laminar flow model slightly overpredicts the strength of the dominant recirculation, as seen for the lower fluid height in Figure 4(b). Figures 5(a) and 5(b) show the mean radial velocity profiles for H = 0.15 and 0.05 m, respectively. With a laminar flow model, the peak in the radial velocity at positions such as y/H = 0.067 and 0.1 for H = 0.15 m in Figure 5(a) is predicted better than with the turbulent flow model. However, the turbulent flow model predicts the spread in the radial velocity better than the laminar flow model. The results for $0.7 \le y/H \le 0.9$ (Prakash *et al.* 1999b) are not shown here due to space limitations. With a decrease in the fluid height, the predictions of the turbulent flow model are closer to the experimental predictions. As the fluid height decreases, turbulence effects in the foam are expected to become more important for a lower fluid height. Figures 6(a) and 6(b) represent the turbulence kinetic energy profiles for the cases with foam for fluid heights of H = 0.15 and 0.05 m. The predicted turbulence levels, close to the nozzle exit are lower than the experimental values. This

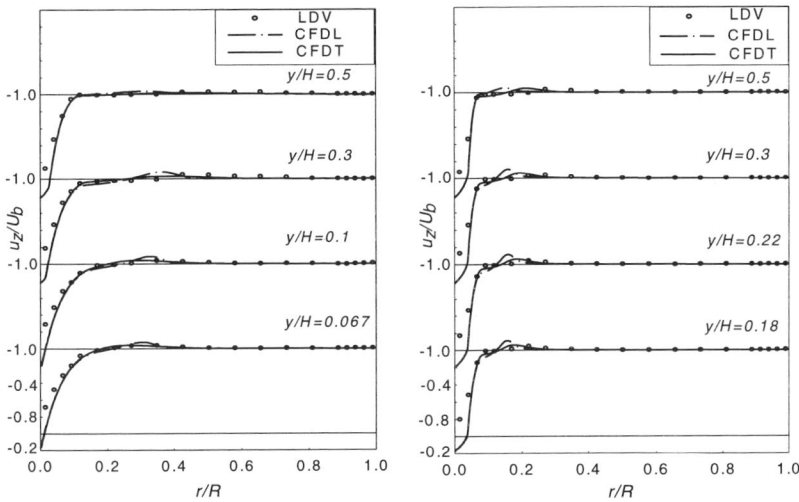

Figure 4: Axial velocity profiles: (a) H = 0.15 m; and (b) H = 0.05 m, h_p = 0.05 m, U_b = 1.6 m/s

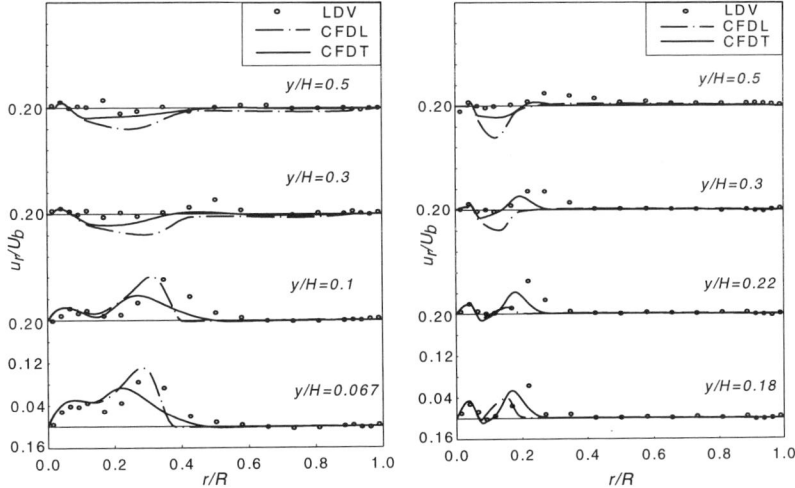

Figure 5: Radial velocity profile: (a) H = 0.15 m; and (b) H = 0.05 m. h_p = 0.05 m.

540

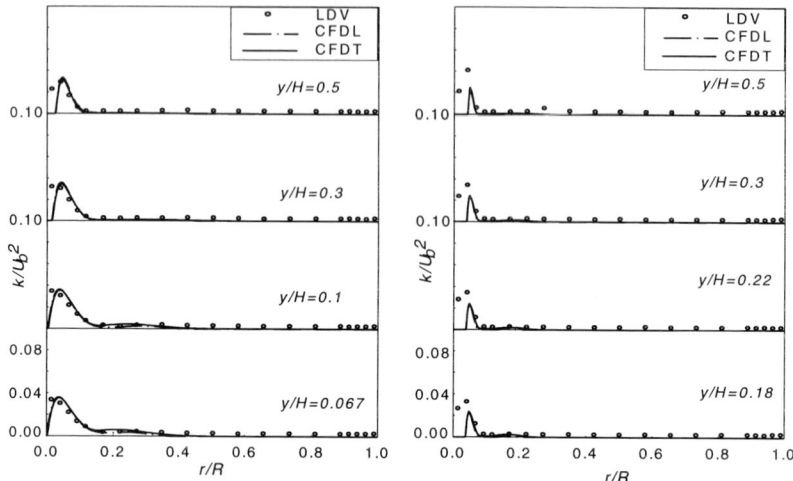

Figure 6: Turbulence kinetic energy profiles: (a) $H = 0.15$ m; and (b) $H = 0.05$ m. $h_p = 0.05$ m.

result does not affect the predictions in other regions of the flow. The hump in the profile without foam for the same fluid heights (see Prakash *et al.* 1999b) is reduced significantly for cases with foam. The reduction in the turbulence level is because the development of a wall jet is prevented due to the presence of the porous foam and its damping effect.

CONCLUSIONS

This paper presents the main features of an experimental and numerical investigation to study a system consisting of a turbulent fluid overlying a porous medium. LDV measurements are used to obtain the gross flow pattern for different fluid heights. Simulations with laminar and turbulent models for the foam indicate that the turbulent flow model leads to more accurate axial and radial velocity distributions. The laminar model is better for the higher fluid heights, especially close to the interface. Both models give acceptable turbulence kinetic energy profiles. The present turbulence model for the porous medium incorporates only Darcy damping.

REFERENCES

Antohe B. V. & Lage J. L. (1997). A General Two-equation Macroscopic Turbulence Model for Incompressible Flow in Porous Media. *Int. J. Heat and Mass Transfer* **40**, 3013–3024.

Chen L., Li Y. & Thorpe G. R. (1998). High-Rayleigh-number Natural Convection in an Enclosure Containing a Porous Layer. 11th Int. Heat Transfer Conf., Seoul.

Jones W. P. & Launder B. E. (1972). The Prediction of Laminarization with a Two-equation Model of Turbulence. *Int. J. Heat and Mass Transfer* **15**, 310–314.

Prakash M., Mahoney J., Li Y., Turan O. F. & Thorpe G. R. (1999a). Impinging Round Jet Studies in a Cylindrical Enclosure With and Without a Porous Layer: Part I – Flow Visualisations. *CSIRO Research Report*.

Prakash M., Mahoney J., Li Y., Turan O. F. & Thorpe G. R. (1999b). Impinging Round Jet Studies in a Cylindrical Enclosure With and Without a Porous Layer: Part II – LDV Measurements and CFD Simulations. *CSIRO Research Report*.

Prescott P. J. & Incropera F. P. (1995). The Effect of Turbulence on Solidification of a Binary Metal Alloy with Electromagnetic Stirring. *ASME J. Heat Transfer* **117**, 716–724.

© 2000 Elsevier Science Ltd. All rights reserved.
Air Distribution in Rooms, (ROOMVENT 2000)
Editor: H.B. Awbi

ROOM AIRFLOWS WITH LOW REYNOLDS NUMBER EFFECTS

C. Topp[1], P.V. Nielsen[1] and L. Davidson[2]

[1] Department of Building Technology and Structural Engineering, Aalborg
University, Sohngaardsholmsvej 57, DK-9000 Aalborg, Denmark
[2] Department of Thermo and Fluid Dynamics, Chalmers University of
Technology, S-412 96 Gothenburg, Sweden

ABSTRACT

The behaviour of room airflows under fully turbulent conditions is well known both in terms of experiments and numerical calculations by computational fluid dynamics (CFD). For room airflows where turbulence is not fully developed though, i.e. flows at low Reynolds numbers, the existing knowledge is limited.

It has been the objective to investigate the behaviour of a plane isothermal wall jet in a full-scale ventilated room at low Reynolds numbers, i.e. when the flow is not fully turbulent. The results are significantly different from known theory for fully turbulent flows. It was found that the jet constants are a strong function of the Reynolds number up to a level of $Re_h \approx 500$.

KEYWORDS

Room airflow, low Reynolds number effects, full-scale experiments, plane isothermal wall jet

INTRODUCTION

Supply flow rates for ventilation is often reduced due to more efficient ventilation or to reduce the energy consumption in mechanical ventilation. In natural ventilation where buoyancy and wind pressure are the driving forces there will be some periods during the year when the supply airflow rate is moderate.

Air for ventilation is often supplied as a jet above the occupied zone to achieve mixing with room air. When designing ventilation systems the jet and the entire room airflow is traditionally treated as turbulent flow although the airflow in some regions of a ventilated room is laminar or not fully turbulent.

The behaviour of turbulent jets is well known (Rajaratnam (1976) and Launder & Rodi (1983)) while little effort has been spend to investigate the behaviour of jets at low Reynolds numbers. It has been the objective of this work to investigate the behaviour of a plane isothermal wall jet in a full-scale ventilated room at low Reynolds numbers, i.e. when the flow is not fully turbulent.

As the wall jet enters an open space a boundary layer builds up along the wall. In the boundary layer the velocity changes from zero at wall to its maximum value, u_x, at some distance from the wall, see Figure 1.

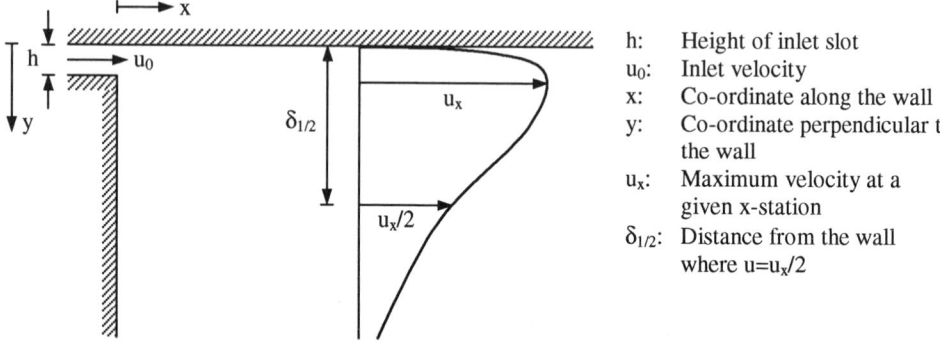

h: Height of inlet slot
u_0: Inlet velocity
x: Co-ordinate along the wall
y: Co-ordinate perpendicular to the wall
u_x: Maximum velocity at a given x-station
$\delta_{1/2}$: Distance from the wall where $u=u_x/2$

Figure 1: Outline of a typical velocity profile for a plane wall jet.

The velocity profile is generally represented by the maximum velocity, u_x, and the distance from the wall, $\delta_{1/2}$, at which the velocity has dropped to $u_x/2$.

For a plane fully turbulent wall jet the decay of the maximum velocity and the growth rate of the jet along the wall are given by

$$\frac{u_x}{u_0} = K_p \sqrt{\frac{h}{x + x_0}} \tag{1}$$

$$\delta_{1/2} = D_p (x + x_0) \tag{2}$$

where K_p is the velocity decay constant, D_p is the growth rate of the jet per unit distance from the inlet and x_0 is the virtual origin of the jet. K_p and D_p are both characteristic parameters for the diffuser.

At high Reynolds numbers (fully turbulent flow) K_p and D_p are constants and Eqn. 1 and 2 are valid. It has been chosen in the present work to apply Eqn. 1 and 2 to jets at low Reynolds numbers too, assuming D_p and K_p to be functions of the Reynolds number.

METHODS

A series of isothermal experiments were performed in a full-scale test room as shown in Figure 2 (left), corresponding to the room used in the IEA Annex 20 project but with full width slots (height h=0.02 m) as supply and exhaust openings. The supply opening was located at the top of the left end wall and thus generates a plane wall jet along the ceiling while the exhaust opening was located either at the top of the right end wall or at the foot of the left end wall, see Figure 2 (right).

The wall jet generated by the supply opening was considered plane due to the full width slot and velocity profiles were thus measured in the centreline of the jet. From smoke experiments similar recirculating airflow patterns were observed for the two ventilation set-ups indicating no influence of exhaust location.

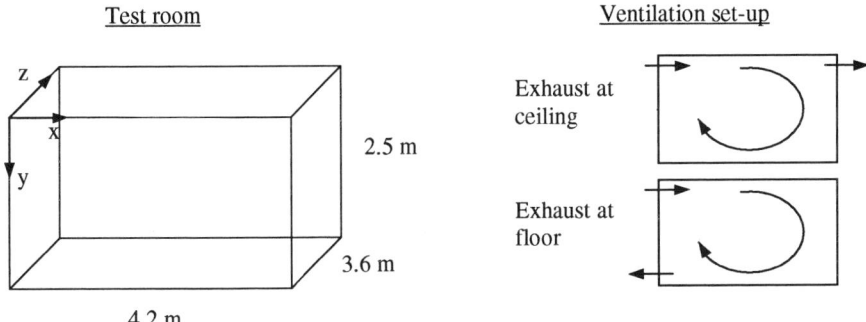

Figure 2: Outline of the full-scale test room (left) and the location of the full width supply and exhaust openings (right).

Velocity profiles were measured with hot-sphere probes in the centreline of the jet at different air change rates in the range from 0.4 h^{-1} to 4.0 h^{-1}, corresponding to supply Reynolds numbers, Re_h, ranging from 79 to 770.

RESULTS

In the following the Reynolds number refers to the diffuser and is based on the contracted slot height, h_0. The Reynolds number is thus given by

$$Re_h = \frac{u_0 h_0}{\nu} \qquad (3)$$

where ν is the kinematic viscosity.

Velocity Profiles

For a plane wall jet the velocity profiles at different x-stations and different Reynolds numbers are expected to express similarity when plotted in terms of u/u_x and $y/\delta_{1/2}$. The measured profiles are shown in Figure 3. Also included in the figure are the empirical relation by Verhoff (1963) for a fully turbulent wall jet and a laminar relation established from experiments by Quintana et al. (1997).

It is seen from the figure that the profiles express similarity to some extend as a high degree of similarity is found for the highest Reynolds numbers, i.e. $Re_h \geq 477$ while for $Re_h < 477$ there is no obvious similarity within the jet.

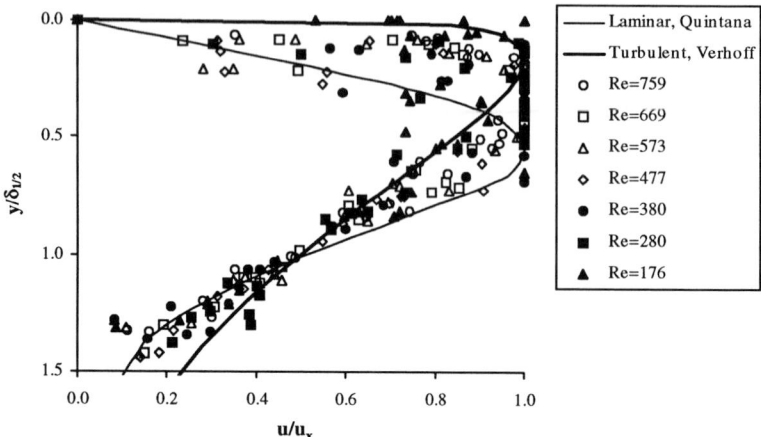

Figure 3: Velocity profiles in the wall jet plotted in non-dimensional co-ordinates.

From the figure it should further be noticed that the profiles neither fit the turbulent or the laminar relation but lies somewhere in between indicating that the jet is transitional.

Characteristic Parameters, D_p and K_p

The jet growth rate along the ceiling, D_p, has been established from a best fit of the experimental data to the relation of $\delta_{1/2}$ and x given by Eqn. 2. Figure 4 shows D_p versus Re_h for both ventilation set-ups. Results obtained by Nielsen, Filholm, Topp & Davidson (2000) from model experiments are included for comparison.

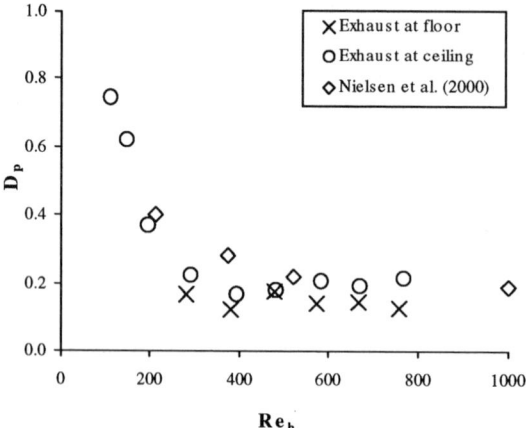

Figure 4: The growth rate of the jet per unit distance from the inlet, D_p, as a function of Reynolds number, Re_h. Results from model scale experiments by Nielsen, Filholm, Topp & Davidson (2000) are included for comparison.

It is seen that the jet growth rate is a strong function of Reynolds number. As Re_h increases D_p drops rapidly and seems to reach a constant level at $Re_h \approx 400$. It should be observed that there is no significant difference in D_p between the two ventilation set-ups. In addition, the results agree well with the model scale experiments.

The velocity decay constant, K_p, has been established from a best fit of the experimental data to the relation of the maximum velocity, u_x, and the horizontal distance, x, given by Eqn. 1 with $h=h_0$ (the contracted slot height). Figure 5 shows K_p versus Re_h for both ventilation set-ups. Results obtained by Nielsen, Filholm, Topp & Davidson (2000) from model experiments are included in the figure.

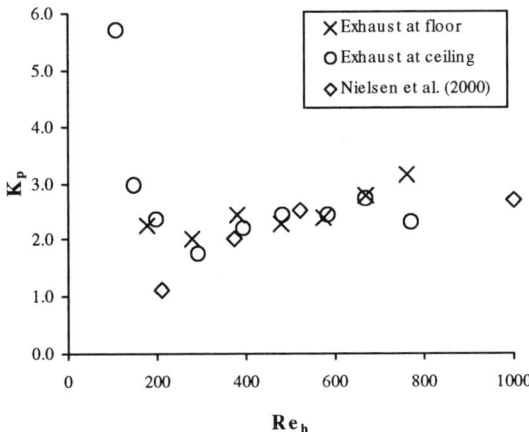

Figure 5: The velocity decay constant, K_p, as a function of Reynolds number, Re_h. Results from model scale experiments by Nielsen, Filholm, Topp & Davidson (2000) are included for comparison.

In similar to the jet growth rate the results show that K_p is a strong function of Re_h as K_p drops rapidly with increasing Re_h up to $Re_h \approx 400$ where K_p reaches a minimum. For $Re_h > 400$ K_p takes on a slight increase. Again, it should be observed that there is no substantial difference between the two ventilation set-ups. The results agree well with the model scale experiments except for at $Re_h \approx 200$.

DISCUSSION

A series of full-scale experiments were performed to investigate the behaviour of a plane isothermal wall jet at low Reynolds numbers, i.e. when the flow is not fully turbulent.

Designers of jets for ventilation purposes traditionally use turbulent relations that are independent of Reynolds number, Re_h. Care should be taken though, as the present work shows the jet to be a significant function of Re_h when the flow is laminar or transitional. The measured velocity profiles express similarity for $Re_h > 400$ while the characteristic jet parameters D_p and K_p are strong functions of Re_h up to a critical Reynolds number of approximately 500.

Two different locations of the exhaust opening were investigated and no substantial difference was observed indicating that the exhaust has no significant influence on the jet.

ACKNOWLEDGEMENT

This research was funded by the Danish Technical Research Council as a part of the International Centre for Indoor Environment and Energy at the Technical University of Denmark.

REFERENCES

Launder, B.E. and Rodi, W. (1983). The turbulent wall jet - measurements and modelling. *Annual Review of Fluid Mechanics* **15**, 429-459.

Nielsen, P.V., Filholm, C., Topp, C. and Davidson, L. (2000). Model experiments with low Reynolds number effects in a ventilated room (to appear). *Proceedings of ROOMVENT 2000*, Reading, UK.

Quintana, D.L., Amitay, M., Ortega, A. and Wygnanski, I.J. (1997). Heat transfer in the forced laminar wall jet. *Transactions of the ASME. Journal of Heat Transfer* **119:3**, 451-459.

Rajaratnam, N. (1976). *Turbulent jets*, Elsevier, Amsterdam, The Netherlands.

Verhoff, A. (1963). The two-dimensional turbulent wall jet with and without an external stream. Report No 626, Princeton University, Department of Aeronautical Engineering.

© 2000 Elsevier Science Ltd. All rights reserved.
Air Distribution in Rooms, (ROOMVENT 2000)
Editor: H.B. Awbi

SIMULATION OF DIFFUSERS IN SCALE MODEL EXPERIMENTS OF AIRFLOW DISTRIBUTION IN VENTILATED ROOMS

M. Hurnik, Z. Popiolek, S. Mierzwinski

Department of Heating, Ventilation and Dust Removal Technology,
Silesian University of Technology, Gliwice, Poland

ABSTRACT

Scale model experiments make it possible to analyse design concepts of ventilation, especially air distribution in large enclosures. The airflow structure similarity is fulfilled when experiment is carried out according to the principles of the approximate scale modelling. Special attention should also be paid to proper simulation of boundary and initial conditions. In a real ventilated object, the air is supplied with standard diffusers equipped with deflecting vanes. The question is how the supply opening should be constructed in the model in order to ensure the airflow similarity in the whole space modelled. The paper presents the results of experimental tests of supply jets generated by a standard diffuser and circular openings. An eight-channel omnidirectional thermoanemometer was used for the air mean velocity measurements. The jet characteristic parameters (origin distance, velocity distribution coefficient) were determined. Basing on the test results, a method for supply air jet reproduction in models is suggested. Satisfactory similarity of the air mean velocity field in the ventilated room and its models was acquired when real diffusers were simulated in the models by circular openings fitted with turbulizers and when the jet origin was properly positioned.

KEYWORDS

Scale model experiments, airflow similarity, jets, diffusers

AIRFLOW SIMILARITY IN SCALE MODELS OF VENTILATED ROOM

In many complex cases of room ventilation, scale modelling is an effective method for predicting air distribution in the real object. It can also well co-operate with numerical modelling (CFD) of turbulent ventilating flows. The airflow similarity in a real object and its model is fulfilled when experiment is carried out according to the principles of the approximate scale modelling. It is assumed that flows in the real object and in its scale model are fully turbulent and Re-number independent. Then, it is sufficient to maintain the same Ar and Pr numbers in the model (M) and in the prototype (P). Re and Gr- numbers should be over their threshold levels, Heiselberg et al. (1998):

$$Ar_M = Ar_P \quad Pr_M = Pr_P \quad \text{and} \quad Re_M \gg Re_l \quad \text{or} \quad (Gr\,Pr)_M \gg (Gr\,Pr)_l$$

Based on the equality of Ar and Pr numbers, the following relation between the scale factors of the representative velocity, length and temperature difference can be derived: $S_U=S_L^{0.5} S_{\Delta\Theta}^{0.5}$. Reynolds number in the model Re_M is equal to $S_L^{1.5} S_{\Delta\Theta}^{0.5} Re_P$. In practice, Re_M is about $10 \div 100$ times lower than Re_P. Therefore, the previous tests on the improvement in scale modelling, Hurnik et al. (1999) and Popiolek et al. (1998), were focused on determination of threshold Reynolds number, Re_l, in order to characterise the lower limit of the range of mean flow and turbulence spectrum similarity. The tests were carried out in three similar scale models of a sports hall. Satisfactory similarity of the mean velocity distributions in the whole area of the airflow pattern modelling i.e. both in the supply jets and in secondary flows was obtained at the threshold number Re_l about $4000 \div 2000$. The similarity of turbulence spectrum was observed for $Re > Re_{l2}$. Re_{l2} was identified as about $20000 \div 10000$. When doing the tests, high sensitivity of the mean velocity distribution to the way in which the boundary conditions were generated was found by Hurnik et al. (1999). When the boundary conditions were not reproduced properly, considerable distortions in the mean velocity field were found, e.g.: non-isothermal jet occurrence (the supply air warming in the fan), imprecise velocity reproduction in the supply openings (different types of supply system were used). However, if the boundary conditions were simulated properly, satisfactory similarity was observed. In the previous tests circular openings were used in all the models. In the real ventilated object, the air is supplied from standard diffusers fitted with deflecting vanes. The question is how the supply opening should be constructed in a model so that the airflow similarity will be ensured. In order to explain this problem, experimental tests of supply jets formed by various supply openings and mean velocity fields in the whole space modelled were carried out.

TESTS OF JETS FROM CIRCULAR OPENINGS AND STANDARD DIFFUSERS

The experiments included tests of jets from standard diffuser 80×180 mm and circular openings in scale 1:3 with equivalent diameter d=35 mm without and with turbulizers.

Description of the measurement stand

In the tests, basing on the mean velocity distribution measurements in the jet, its characteristic parameters (origin distance, velocity distribution coefficient) were determined. A scheme of the measurement stand is shown in Fig1a. Velocity distributions were measured in four cross-sections of the jet, at the beginning of the jet fully developed region. The distances between the measurement sections were assumed as multiplicity of the supply opening equivalent diameter, i.e.: 10d, 15d, 20d and 25d. The scheme of the grid used in the tests is shown in Fig.1b. An eight-channel omnidirectional thermoanemometer was used for the air velocity measurements. The averaging time was 5min. Movable systems for simultaneous measurement in eight points of the grid were constructed.

Analytical procedure for identification of the jet model characteristic parameters

The test results were approximated by a model of a free jet generated by a point source of momentum:

$$\overline{V}_x = \left(\frac{2 m \dot{I}}{\partial \tilde{n}}\right)^{0.5} \frac{1}{x+x_o} \cdot e^{-m\left(\frac{r}{x+x_o}\right)^2} \tag{1}$$

where: $\overline{V}_x$ - mean axial velocity component, • – momentum flux, ρ – density, x – distance from the opening, x_o –position of the jet origin (virtual point source of momentum), m - velocity distribution coefficient.

In order to identify the jet characteristic parameters: momentum flux • , velocity distribution coefficient m and origin position x_o, the computer optimisation was applied. At first the real actual position of the jet axis was identified.

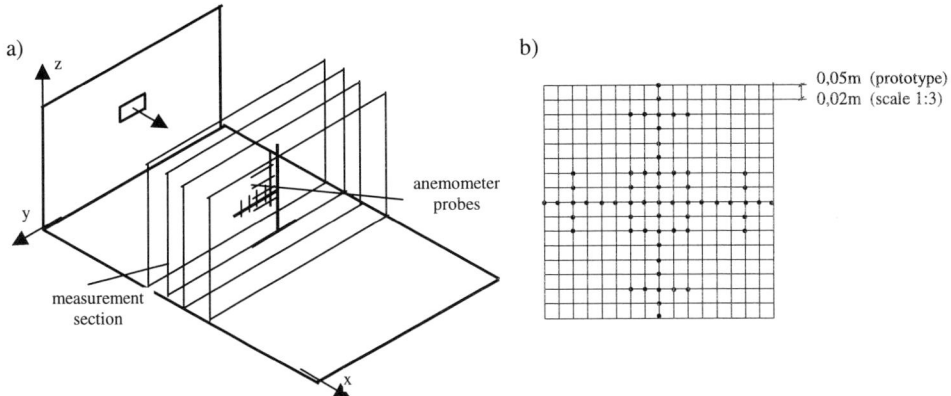

Fig.1.a) The measurement stand
b) The grid used in the tests of jets

Velocity distributions in two central axes of the measurement grid were approximated by Gaussian curves to find the co-ordinates of the jet axis y_a, z_a. Then real distances from the jet axis to the measurement points were calculated:

$$r_i = \sqrt{(y_i - y_a)^2 + (z_i - z_a)^2} \tag{2}$$

where: y_a, z_a – co-ordinates of the jet axis, evaluated separately at each cross-section. Measured velocity values were approximated by the model of jet from point source of momentum using least square method. The value of the approximation error was calculated as:

$$\Delta = \sum_{i=1}^{n} \delta_i^2 = \sum_{i=1}^{n} \left[\frac{\overline{V}_{x,i}}{\overline{V}_{x\,max,cal}} - \frac{\overline{V}_{x,cal}}{\overline{V}_{x\,max,cal}} \right]^2 = \sum_{i=1}^{n} \left[\frac{\overline{V}_{x,i}}{\left(\frac{2\,m\,\dot{I}}{\pi \rho} \right)^{0.5} \frac{1}{x + x_0}} - e^{-m\left(\frac{r}{x + x_0} \right)^2} \right]^2 \tag{3}$$

Then, the approximation error minimal value was sought by proper selection of $\dot{I}$, x_0 and m values. An example of the calculations is presented in Table1. An example of the normalised velocity distribution is shown in Fig.2. Velocity values lower then 10% of the axial velocity value were neglected in the approximation.

Next, the approximation error was minimised in another way: m and x_0 values were assumed and only $\dot{I}$ value was sought. The optimisation was carried out for all the combinations of the following m and x_0 values: $(m_{\Delta\,min} - 10) \le m \le (m_{\Delta\,min} + 10)$, $(x_{0,\Delta\,min} - 1.5 \cdot d) \le x_0 \le (x_{0,\Delta\,min} + 1.5 \cdot d)$, with the step equal 2 and 0.5d, respectively. Basing on those results, a map of approximation errors as a function of m and x_0 was generated by using a graphic computer programme. All the tested cases are shown as one map of approximation error fields limited by a line of equal error $(\Delta - \Delta_{min}) / \Delta_{min} = 1\%$, see Fig.3. The map gives information about m and x_0 values, which describe the jets with high accuracy. It represents sensitivity of approximation to velocity distribution coefficient m and position of the origin x_0.

Table 1.Example of measurement and calculation results

	r_i	$r_i/(x+x_0)$	$\overline{V}_{x,i}$ measured	$\dfrac{\overline{V}_{x,i}}{\overline{V}_{x\,max,cal}}$	$e^{-m\left(\frac{r}{x+x_0}\right)^2}$	δ_i^2	
-	m	-	m/s	-	-	-	$\Delta_{min} = \overset{233}{\underset{i=1}{\Sigma}}\, \delta_i^2 =$
1	0.006	0.01793	2.319	0.98324	0.97952	1.4E-05	
2	0.014	0.04241	2.159	0.9154	0.8907	0.00061	$= 0.9494$
3	0.020	0.05985	1.917	0.81287	0.79413	0.00035	
4	0.022	0.06531	1.760	0.74624	0.75997	0.00019	
5	0.026	0.07777	1.364	0.57847	0.67757	0.00982	$I\big|_{\Delta min} =$
6	0.050	0.14913	1.144	0.48510	0.23901	0.06056	$= 0.018\ kg\cdot m\cdot s^{-2}$
....	....	....	....	....	....	....	
232	0.132	0.18713	0.122	0.11017	0.10418	3.6E-05	$m\big|_{\Delta min} = 64.35$
233	0.137	0.19476	0.116	0.10434	0.10504	4.9E-07	
234	0.148	0.20933	0.098	0.08827	0.09372	3E-05	$x_0\big|_{\Delta min} = -0.042\ m$
235	0.152	0.21485	0.096	0.08639	0.05021	0.00131	
....	....	....	....	....	....	....	

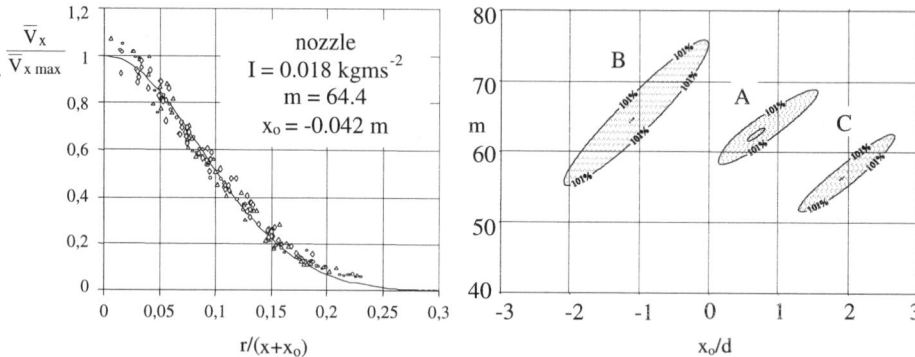

Fig.2. Normalised velocity distribution   Fig.3. A map of approximation errors

Test results

Jets from the following openings were tested: standard diffuser 80×180 mm, see Fig.4a, V_0= 6 m/s, case A, circular opening in scale 1:3 d=35 mm, see Fig.4b, V_0= 3.6m/s, case B.

The test results are shown in Fig.3 as the areas A and B. The ranges of the origin distance x_0 and the velocity distribution coefficient m are different for the standard diffuser and circular opening. However, when assuming the mean value of the velocity distribution coefficient as m=60, the difference in the jet origin position is about 2d, i.e.: x_0=+0,5d for the standard diffuser and x_0=-1,5d for the circular opening. It suggests that similarity of jets may be acquired when the position of the jet origin in room is the same. In order to simulate the jet from the standard diffuser, circular openings with various turbulizers were tested. Basing on the velocity distributions at the distance 20d from the outlet plane, the velocity profiles widths were determined, see Table 2. The toothed ring turbulizer, placed inside the cylindrical extension of the nozzle at the distance of 2d from the outlet plane, generated the turbulent jet at the widest spreading angle.

Table 2. Velocity profiles widths in jets generated by a nozzle fitted with various turbulizers

Type of turbulizer	Relative width of the velocity profile R/d
nozzle without turbulizer	1.83
grid 5x5 mm in the outlet plane of the nozzle	1.80
toothed ring placed in the outlet plane of the nozzle	1.63
toothed ring placed inside the cylindrical extension of the nozzle at the distance of 2d from the outlet plane	2.14

The nozzle was used in further tests. The result of the tests is shown in Fig.3 as area C. For m=60 the origin position is $x_o=2.5d$.

The test results show that it is possible to generate jets in which the origin position is the same as in the case of jets generated by diffusers when nozzles with turbulizers are placed at the right position in reference to the wall. The nozzles should be put at the distance of 2d before the wall (Fig.4c).

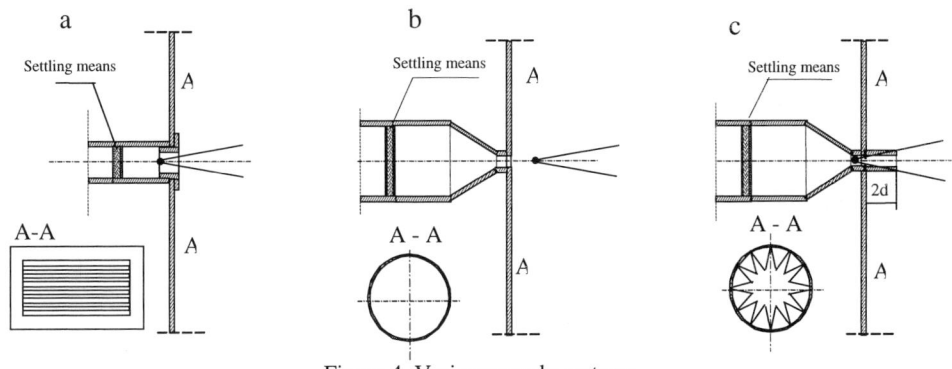

Figure 4. Various supply systems

THE EFFECT OF SUPPLY OPPENING FITTING ON THE MEAN VELOCITY FIELD IN SCALE MODELS

In order to determine the effect of the supply opening fittings on the mean velocity field in the whole region of the flow modelled, air velocity measurement was carried out in the cross-sections of the models according to the method described by Hurnik (1999). The measurement series data are as follows:

1. Prototype, room dimensions - 9×3.3×5.4 m (length, height, width), with three standard diffusers (Fig.4a), V_o=6 m/s
2. model 1:3, supply from circular, nozzle openings (Fig.4b), V_o=3.6 m/s
3. model 1:3, supply from circular opening with turbulizer (Fig.4c), V_o=3.6 m/s

The airflow was tested in isothermal conditions to avoid difficulties in the simulation of thermal boundary conditions and air velocity measurement. Although practical situations are all non-isothermal such the simplification was possible since the aim of the tests was not to provide information on the velocity field for a particular existing ventilated room but only to find out how the supply opening should be constructed in the model. The test results are shown as normalised mean velocities isolines maps - V/V_o (Fig.5). For the cases compared, quantitative correlation of normalised velocities was determined. Regression and correlation coefficients were calculated (Fig.6). Fields of large velocity gradients close to the supply openings were neglected in the analysis.

552

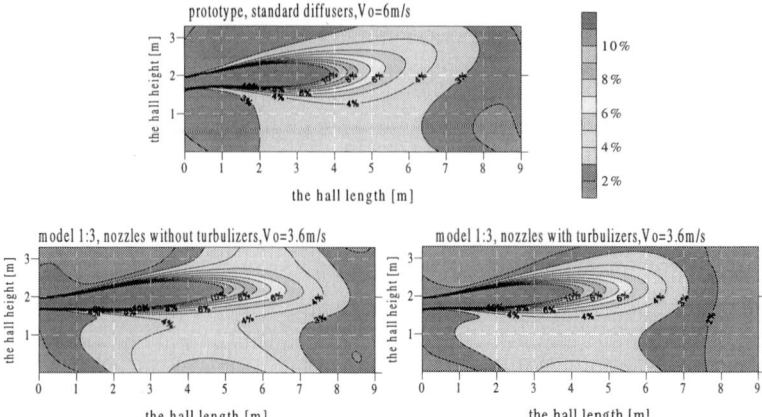

Figure 5. Comparison of results of three measurement series

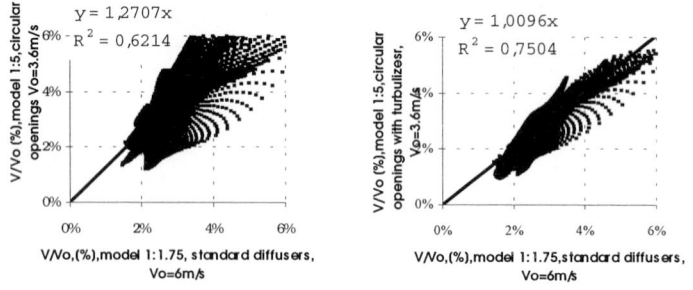

Figure 6. Convergence diagrams of the measurement series compared

CONCLUSION

Satisfactory similarity of the air mean velocity field in the ventilated room and its models was acquired when real diffusers were simulated in the models by circular openings fitted with turbulizers and when the jet origin position was adjusted to be the same in the prototype and its model.

ACKNOWLEDGEMENTS

The research was carried out within the Research Project No 8 T10B056 17, supported by the Polish State Committee for Scientific Research (KBN).

REFERENCES

Heiselberg P., Murakami S., Roulet C., - editors (1998). Ventilation of Large Spaces in Buildings – Analysis and Prediction Techniques, chapter 2.5 Scale model Experiments, IEA Annex 26, Aalborg University, Denmark

Hurnik M., Mierzwinski S., Popiolek Z. (1999). Problem of air flow pattern reproduction in scale models of ventilated rooms. *Proc. Indoor Air '99*, Edinburgh

Popiolek Z., Mierzwinski S., Hurnik M., Wojciechowski J. (1998). Air flow characteristic in scale models of room ventilation. Proc. Roomvent '98, Stockholm

2000 Elsevier Science Ltd. All rights reserved.
Air Distribution in Rooms, (ROOMVENT 2000)
Editor: H.B. Awbi

DEVELOPMENT OF A DUCTLESS AIR SUPPLY SYSTEM USING LOW TEMPERATURE AIR

Y. Hashimoto[1], Y. Tsubota[3], Y. Nakano[4], W. Urabe[4],
T. Oka[5], M. Kaneko[5], Y. Hazama[2], T. Imai[1]

[1] Technical Research Institute, TONETS Corporation,
Ichikawa, Chiba, 272-0142, JAPAN
[2] Technical Headquarters, TONETS Corporation,
Tokyo, 104-8324, JAPAN
[3] Power Engineering R&D Centre, Tokyo Electric Power Company
Yokohama, Kanagawa, 230-0002, JAPAN
[4] Customer Systems Department, Central Research Institute of Electric Power Industry
Komae, Tokyo, 201-8511, JAPAN
[5] Construction Department, Faculty of Engineering, Utsunomiya University
Utsunomiya, Tochigi, 321-0904, JAPAN

ABSTRACT

This paper proposes a new ductless air supply system with a ceiling plenum chamber using low temperature air as a secondary HVAC system for an ice thermal storage system. The proposed air supply system mixes low temperature air with return air from a room using a mixing fan unit (MFU), pressurizes a plenum chamber with the mixed air and supplies the air to the occupied room from diffusers on the ceiling. The purpose of this study is to develop a new HVAC system to utilise low temperature air, to prevent cold draught and dew condensation, to keep thermal comfort in the room and to save fan energy consumption. Room temperature is fed back to controllers that regulate fan motors of an air handling unit and an MFU with inverters. Low temperature air from an air-handling unit is controlled at 10°C by regulating flow rate of chilled glycol. A 7m and 10m and 3m height room is provided to evaluate the proposed system. 24-hour time series data are obtained by data loggers. Vertical temperature distribution is within setpoint $\pm$ 1K and horizontal temperature is controlled between 25.7 and 26.2°C at 1.5m height above floor. Time-averaged PMV at the centre of the room is +0.24 and PPD is 6.2%. 24-hour time-averaged temperature is 26.2°C at the sensor for control and the root mean square error is 0.18K. Temperature difference between mixed and room air is averaged to 9K. It prevents cold draught due to low temperature air supplied directly to the room from ceiling-mounted air diffusers. Air velocity at the centre of the room is averaged to 0.22m/s and turbulence intensity is 8.6%. The results meet ISO7730-1994. This system is also expected to reduce the fan energy consumption by 70%. In addition, use of night electric power for ice storage systems saves energy cost of a chiller. The ductless air supply system using low temperature air is appreciated satisfactory for thermal comfort and energy saving.

KEYWORDS

Ductless, Ice Storage Systems, Air Supply, Large Temperature Difference, Energy Saving, Variable-Air-Volume (VAV), Ceiling Plenum Chamber, Low Temperature Air, PMV, PPD

INTRODUCTION

Ice storage systems are widely applied to shift on-peak cooling load to off-peak nighttime. Low temperature water and air is available in ice storage systems. In a decade, energy saving strategy has been studied to use large temperature difference water and air by using low temperature water and air. It should be noted that problems in an application of low temperature air supply are dew condensation and thermal discomfort. To solve the problems, the authors proposed a new air supply system using a ceiling plenum chamber and mixing fan units (MFUs). MFUs mix low temperature supply air from an air-handling unit with return air from the occupied room, and pressurise the ceiling plenum chamber with the mixed air. The mixed air is supplied to a room from diffusers mounted on the ceiling. The mixed air temperature is almost between 16 and 18 °C. The final temperature differences between supply air and room air are from 8 to 10K. The air supply system uses the insulated plenum chamber instead of branch ductworks in order to lead air from MFUs to the ceiling-mounted diffusers. The proposed system reduces first cost and ductwork spaces above ceiling.

It may also be helpful to renewal works.

To control room temperature, supply air volume varies according to the thermal load. This control strategy is expected to save fan power energy consumption using variable-air-volume (VAV) air-handling unit and MFUs.

This paper will show that the proposed air supply system satisfies thermal comfort conditions in a occupied room according to ISO standard 7730 and helps fan energy saving.

EXPERIMENTAL SETUP AND TEST CONDITIONS

Tests are conducted in a full-scale room simulator. The test room has dimensions of 7m by 10m by 3m high and is furnished with simulated perimeter load heaters. 6 slot diffusers and 6 circular diffusers are mounted on the ceiling. Plenum chamber upper the ceiling is fully insulated with fibreglass boards. See Figure 1 for the room tested.

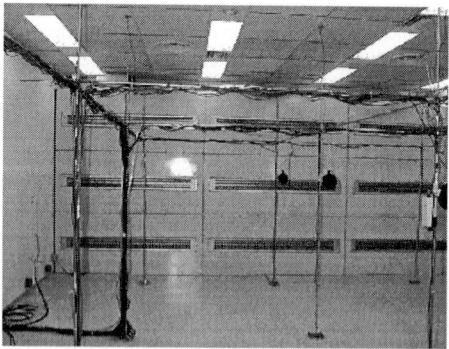

Figure 1: The room tested for the experiments

To give equivalent cooling load through the envelope wall and windows, electric heaters are installed on the wall. The heating capacity of the heaters can be adjusted continuously (maximum 5kW).

Air temperature and velocity are measured using T-type thermocouples and thermal anemometers. Mean radiant temperature is measured by T-type thermocouples inserted in globe thermometers. Power consumption of a brine chiller, air-handling unit fan, electric heaters on the wall and lighting is calculated by pulse counters. Data are scanned by data loggers and transported to a PC through RS-232C serial port every minute.

PID controllers vary fan speed of the air-handling unit and MFUs using pulse-width-modulated (PWM) inverters. They achieve variable-air-volume control for the room air temperature. Maximum total air volume of the slot and circular diffusers mounted on the ceiling is 2630m^3/h. It is 12.5 air changes per hour. To maintain the requirement for fresh air, the minimum air volume is set as 30Hz of inverter frequency for the air-handling unit and MFUs.

A brine chiller produces 2°C glycol as simulated low temperature water from an ice storage system. Glycol is supplied to the cooling coil of the air-handling unit. Heating operation is also available to use an electric heater installed in a brine tank. To supply low temperature air of 10°C from the air-handling unit, flow rate of glycol through the cooling coil is controlled by two-way valves. A variable-air-volume (VAV) air-handling unit and MFUs control room temperature with inverters as shown in Figure 2.

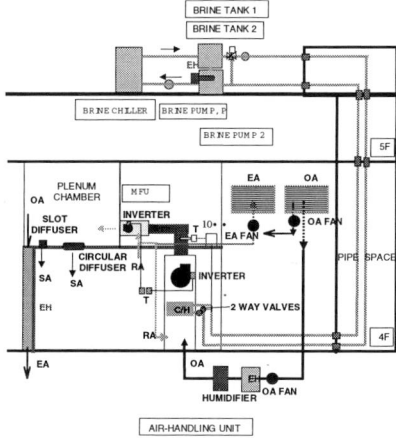

Figure 2: Ductless Air Supply System

Air temperature is measured vertically at the surface of the floor, 0.1m, 0.5m, 1.0m, 1.5m, 2.0m and 2.5m above floor, at the lower and upper surface of the ceiling and at 0.5m above the ceiling (3.5m above the floor) horizontally in 12 points of the room tested. Figure 3(a) shows the horizontal location of measurement points. Radiant temperature is measured at points of No.2, 5, 8 and 11 at 1.5m above floor.

EXPERIMENTAL RESULTS

The main goal of the experiments is to demonstrate the thermal comfort and energy saving which this new air supply system can achieve. The following results are obtained by experimental data of cooling mode. The heaters give 2.5kW cooling load as the perimeter load. Cooling load of overhead lighting is totally 1.6kW. Setpoints of 25 and 26• are given to the room air temperature. For all cases, tests are conducted over 24-hour periods continuously.

Horizontal Air Temperature Profiles

Horizontal temperature profile in the room tested is described in Figure 3(b). Setpoint of the room temperature is 26°C. Horizontal room temperature difference at 1.5m above floor is kept within 0.5K.

556

in a steady-state. Space-averaged room temperature of 12 points at 1.5m above floor is 26.0°C. Such a tendency is also observed at a different setpoint. It may be evaluated favourable for thermal comfort.

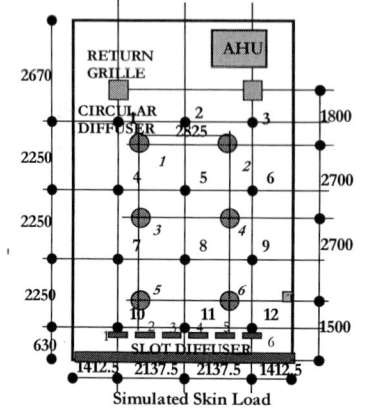

(a) Measurement Points and Diffusers Location

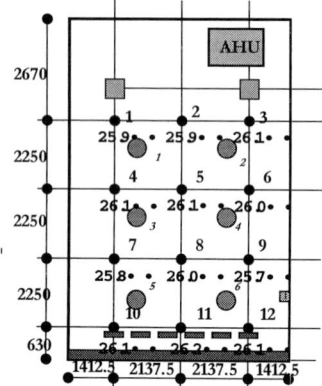

(b) Horizontal Temperature Profile (1.5m above Floor)

Figure 3: Horizontal Temperature Profiles

Vertical Air Temperature Profiles

Vertical temperature profiles in the room tested are presented in Figure 4. Temperature difference at 0.1m and 1.1m above floor is less than 2K at each setpoint temperature. According to ISO 7730, the vertical air temperature difference between 0.1m and 1.1m above floor shall be less than 3K. The experimental results for 12 measuring points meet the recommendation of ISO7730.

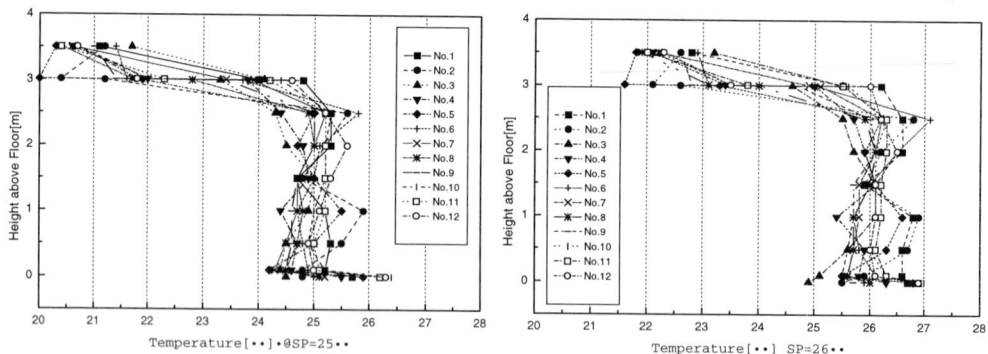

Figure 4: Vertical Temperature Profiles

Stability of the Room Temperature

Time series data of air temperature and humidity over 24-hour are presented together in Figure 5. Fluctuations of the room air temperature are within 0.5• from the setpoint of the room air temperature at the centre of the room tested. The standard deviation of the room air temperature is 0.18K. The room temperature is controlled considerably at a constant level. Thus, controllability of the room temperature is evaluated to be excellent.

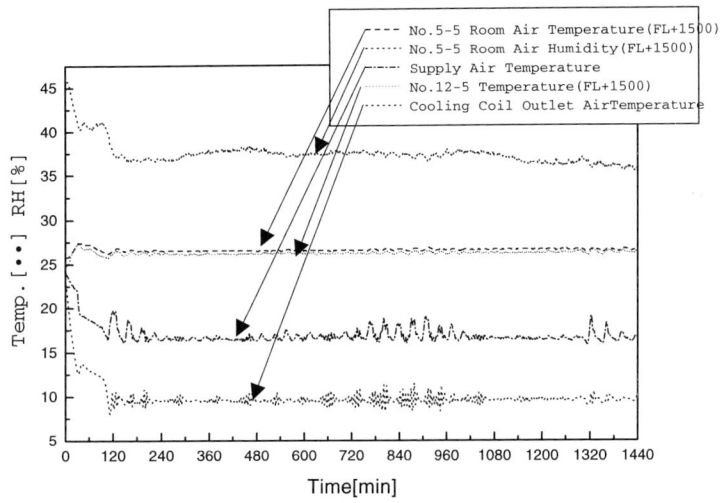

Figure 5: Time Series Data of Temperature and Humidity

Air Velocity

Mean air velocity in the centre of the room tested is 0.22m/s and the standard deviation is 0.018m/s over 24-hour period. It may be estimated rather constant in spite of fluctuations of the supply air velocity caused by the fan of air-handling unit. Turbulence intensity (Tu) is defined as a percentage of the standard deviation divided by mean air velocity. In this case Tu is 8.6%. Allowable air velocity at Tu of 10% is 0.32m/s at 26• according to ISO 7730. It meets the recommendation of ISO 7730.

Energy Consumption

The energy consumption considered here is based on the operation of the air-handling unit fan and MFUs. A VAV air-handling unit and MFUs are expected to reduce fan power. Fan power is assumed to be proportional to squares of fan frequency by using inverters. Fan power consumption for the air-handling unit is 5.8kWh over 24-hour period. For the MFUs, power consumption is not measured here, but it is estimated approximately the same value as the air-handling unit. The nominal fan power output of the air-handling unit is 1.5kW and that of the MFUs is 1.5kW. The conventional constant volume air-handling unit might consume 1.5kW over 24 hours and it might lead to 36kWh. The energy consumption of 5.8kWh plus 5.8kWh is nearly 30% of that, compared to
the conventional air supply system.

Time-averaged cooling load is calculated by measuring the flow rate and temperature difference of chilled water. Mean cooling load over 24-hour period is estimated 50% of the cooling capacity of the air-handling unit. It is assumed to be equal to a seasonal cooling load factor of a commercial building. Thus, the test condition is evaluated to be equivalent to a real building.

DISCUSSIONS

Thermal Comfort

Thermal comfort in the room tested is assessed according to ISO7730 and ANSI/ASHRAE55. PMV is calculated according to ISO7730. Room air temperature, relative humidity, mean radiant temperature and air velocity are measured. Activity level and clothing of occupants are given as 1.2Met and 0.6clo in summer conditions.

The room air temperature, relative humidity, mean radiant temperature and air velocity averaged over 1-hour period in a steady-state is 26.1•, 35.6%,26.0• and 0.21m/s relatively in the centre of the room tested, when the setpoint is 26•. PMV is +0.24 and PPD is 6.2%, while ISO7730 recommends – 0.5<PMV<+0.5 and PPD<10%.

Economical Evaluation

A variable-air-volume air supply system can reduce the fan energy consumption using the proposed ductless system. In addition, the proposed system can work as a secondary HVAC system for an ice storage cooling system, which employs night electric power at lower price. In Japan, night power for thermal storage systems is charged at about 1/5 compared to daytime.

CONCLUSIONS

A new ductless air supply system using large temperature difference air is proposed to optimise secondary HVAC systems for ice storage systems. Two goals has been accomplished during this study:

1. Thermal comfort in the room tested is appreciated satisfactory according to ISO 7730-1994 in summer conditions.
2. Fan power consumption can be reduced by using a VAV air-handling unit and MFUs by 70%. The proposed system can save energy cost for fan power.

The proposed air supply system may be expanded to a wider range of buildings.

References

ISO Standard 7730 (1994). Moderate thermal environments - Determination of the PMV and PPD indices and specification of the conditions for thermal comfort.

ANSI/ASHRAE 55 (1992). Thermal Environmental Conditions for Human Occupancy, ASHRAE, USA.

Knebel D. E. and John D. A. (1993). Cold Air Distribution, Application, and Field Evaluation of a Nozzle-Type Diffuser. *ASHRAE Transactions* **99:Pt.1,** 1337-1348.

Elleson J. S. (1993). Energy Use of Fan-Powered Mixing Boxes with Cold Air Distribution. *ASHRAE Transactions* **99:Pt.1,** 1349-1358.

Chow W.K., Wong L.T., Chan K.T. and Yiu J.M.K. (1994). Experimental Studies on the Airflow Characteristics of Air-conditioned Spaces. *ASHRAE Transactions* **100:Pt.1,** 256-263.

Air Distribution in Rooms, (ROOMVENT 2000)
Editor: H.B. Awbi

AIR CHILLED SUSPENDED CEILING IN COMBINATION WITH SUPPLY OF VENTILATION AIR

H. M. Mathisen

Department of Refrigeration and Air Conditioning at SINTEF Energy
Research, 7465 Trondheim, Norway

ABSTRACT

A suspended ceiling covering a part of the ceiling area, integrated with a ventilation system has been tested in a full-scale laboratory model. The air cools the suspended ceiling before it is supplied to the room. The air flows into the room through a 10 millimetres high slot across the ceiling. The aim of the study was to find how much excess heat this system was able to remove from a typical modern office module of 2.4 times 4.8 meters. Three different shapes of the ceiling were tested. The measurements show that for air velocities of up to 0.15 m/s in the occupied space, heat loads up to 80 to 90 W/m^2 could be removed without draft risk for the best of the tested designs. This is comparable to what was obtained with a water chilled ceiling with one ceiling mounted diffusers in the same test room. The shape and the size of the ceiling make less of an impact on the result.

KEYWORDS

Ventilation, Air supply, Suspended ceiling, Thermal comfort, Measurements, Full-scale tests.

INTRODUCTION

In offices with south facing façades and extensive use of technical equipment, the internal heat load per square meter is high. A water-chilled suspended ceiling in combination with ventilationis often used to cool the room in buildings with a high comfort level. This is a relatively costly installation, which also includes a risk of water leakage. Therefore, air-cooled suspended ceilings could have advantages. On the other side, water based systems may give lower costs for production and transport of cooling energy. According to Makulla (1997) water based systems use less energy for removing heat loads of more than 50 W/m^2. For Norwegian conditions, this number might be higher due to lower outdoor temperatures. If low air temperature is used the number could also be increased. As an alternative to water based systems and air systems using more air, ABB Environment in Norway has developed a new ceiling system. This system utilises low temperature ventilation air to cool a part of the ceiling, before it blows the air into the room.

The aim of this paper is to investigate if this new ceiling system is able to remove relatively high amounts of excess heat from the room without draft risk for the occupants. It was not the intention to study details of the air flow pattern, but the installation, geometry, materials and heat flows should be identical to a real office room.

DESCRIPTION OF THE CEILING SYSTEM

The ventilation air supply device consists of a curved suspended ceiling covering a part of the ceiling. It is made in different shapes. See Figure 1. The suspended ceiling consist of 0.7 mm thick aluminium sheets with 18% perforation. On the inside, an acoustic sheet is glued to the aluminium. Ideally, the acoustic sheet should be airtight. The suspended ceiling joins close to the inner wall and two sidewalls, while it at the fourth side makes a 10 mm high slot between the ceiling and the sheet. The ventilation air is supplied above the ceiling and flows through the slot into the room. Since cold dry air surrounds the duct condensation will be avoided without insulating the duct.

For summer conditions cold air is supplied. The cold air cools the aluminium sheet which then cools the room through radiation and convection. By doing this the ventilation air temperature will also increase before it flows into the room. This means that colder air could be used than if it was blown directly into the room, because the drop of the cold jet is reduced and condensation is avoided. Three different shapes of the ceiling have been tested, as shown in Figure 1.

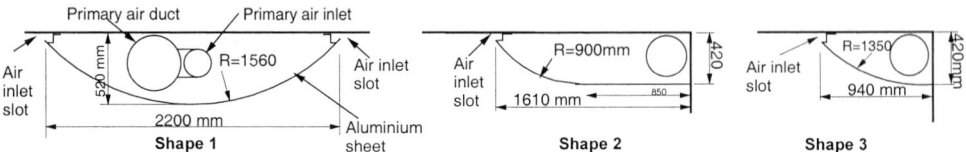

Figure 1. Three different shapes of the ceiling system. Shapes two and three have primary air inlets equal to shape one.

MEASUREMENT EQUIPMENT

The tests were performed at the Refrigeration and Air Conditioning laboratories of SINTEF Energy Research/Norwegian University of Science and Technology. Air temperatures and velocities were measured in the occupied part of the room. The tests were carried out for summer conditions with different heat loads and airflow rates. The construction was the same as for a real office building and the room was furnitured as an office room. Figure 2 shows the test facilities.

The exterior wall has the same construction as for a typical modern Norwegian building. The interior walls were made of plaster wallboards with 3cm insulation. The neighbouring room was kept at the same temperature as the test room to avoid heat transfer through the walls. The primary air temperature was adjusted to maintain equivalence between the supplied heat and the heat removed by the air.

Figure 3 shows the position of the columns with the measuring points. In all points, a thermocouple type T was used to measure the temperature. In 10 of the positions TSI anemometers of type TSI 8475-300 and TSI 8465-300 were used. For the lowest velocities reported in this paper, the accuracy is therefore limited.

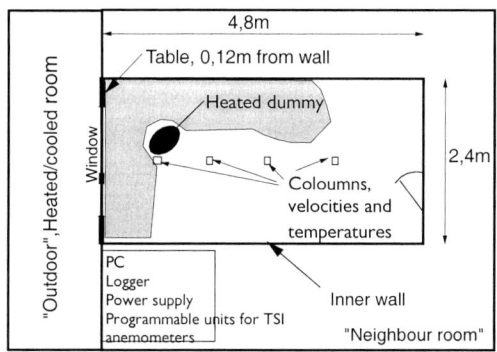

Figure 2. Test room. The ceiling height is 2.9 metre.

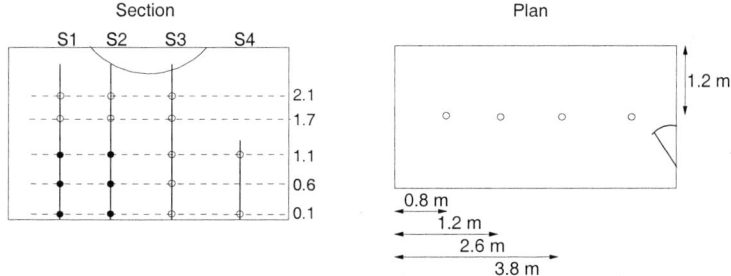

Figure 3. Location of columns with measuring points

TABLE 1 shows the tests that have been carried out with their airflow rates, heat loads, and under-temperature of primary air. The same heat loads applies to all shapes of the ceiling system. The letters in the table corresponds to the heat sources described below the table.

TABLE 1. *COMBINATIONS OF VENTILATION AIRFLOW RATES, HEAT LOAD AND DIFFERENCE BETWEEN SUPPLY AIR TEMPERATURE, AND EXHAUST AIR TEMPERATURE.*

<table>
<tr><td colspan="2"></td><td colspan="3">ΔT (supply air temperature minus exhaust air temperature)</td></tr>
<tr><td colspan="2"></td><td>13 °C</td><td>9 °C</td><td>5 °C</td></tr>
<tr><td rowspan="3">Airflow rate</td><td>7 m³/hm²</td><td>30.4 W/m² g, d, e, c, b</td><td></td><td>11.7 W/m² b, c</td></tr>
<tr><td>13 m³/hm²</td><td></td><td>39.2 W/m² g, d, e, c, b, f</td><td></td></tr>
<tr><td>20 m³/hm²</td><td>87 W/m²
g, d, e, c, b, f, a</td><td></td><td>33.5 W/m² g, f(40W), b,
c, d, e</td></tr>
</table>

Heat sources:
a) Solar radiation, 560W (Simulated by electric heated sheets underneath the carpet)
b) Fluorescent tubes, suspended lamp, 116 W
c) Low energy bulb, suspended lamp, 18W
d) Table lamp, 18W
e) Extra heat source, 100 W
f) Dummy person, 100W or 40W
g) PC with screen, 92 W

RESULTS

Figure 4, 5,Figure 5 and 6Figure 7 show some of the measured velocities for shape one, two, and three respectively. The graphs show the velocity in six different measurement positions. These positions were located at columns S1 and S2 at levels 0.1 m, 0.6 m and 1.1m above the floor. See Figure 3 and the points in black. The graphs on the left show the mean of the velocities in the six points, while the graphs on the right show the velocity in the point that has the highest velocity.

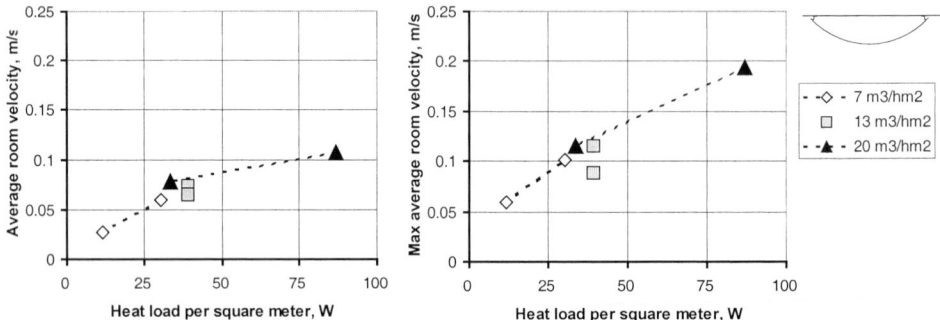

Figure 4. Air velocities for different airflow rates and heat loads for shape one. Each column contains average or maximum for six different measurement positions in the occupied space.

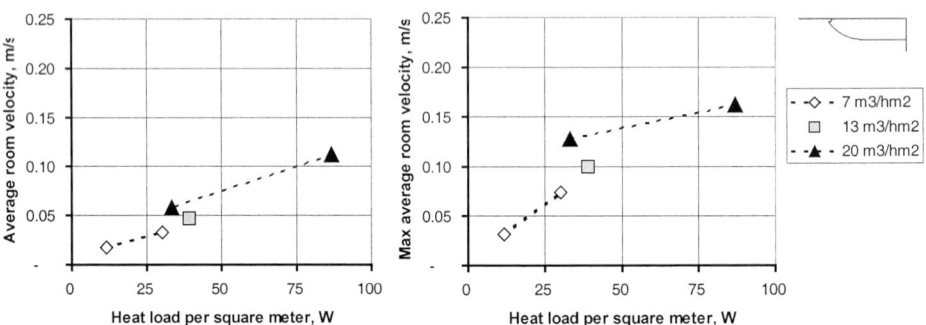

Figure 5. Air velocities for different airflow rates and heat loads for shape two. Each column contains average or maximum for six different measurement positions in the occupied space.

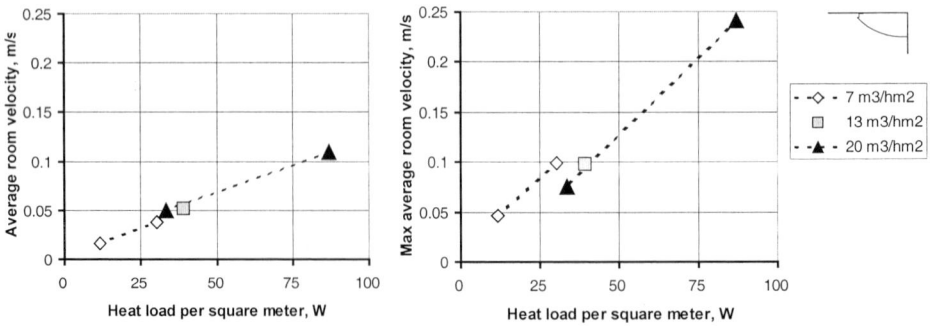

Figure 6. Air velocities for different airflow rates and heat loads for shape three. Each column contains average or maximum for six different measurement positions in the occupied space.

A much used draft criterion is that the air velocity should be lower than 0.15 m/s for summer conditions. From the graphs in Figure 8, one can see that the maximum heat load that could be removed from the test room without draft risk was about 80 to 90 W/m². Any of the three shapes could remove up to 75 W/m² if one allows a velocity of 0.20 m/s.

How does the suspended ceiling contribute? Figure 9, 8, and 9 show the temperature increase due to the suspended ceiling. As one would expect, the results show that the lower the airflow rate is the greater the temperature increase is. Comparing the gradients in the three figures confirm that the smaller the ceiling area the lower the temperature increase. However, also the smallest module gives a considerable temperature increase.

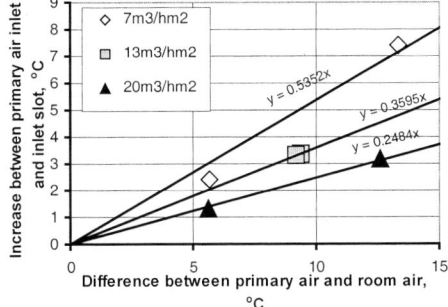

Figure 7. Temperature increases for shape one.

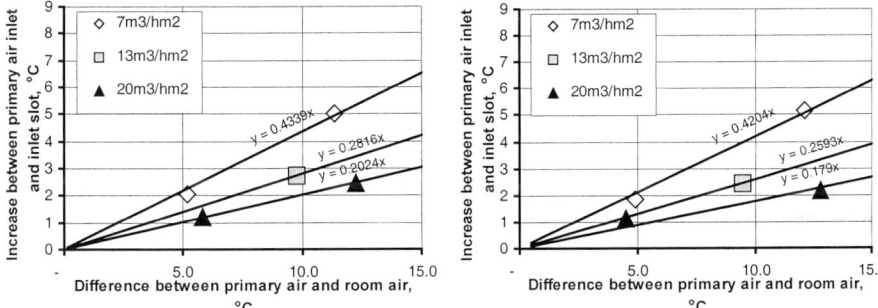

Figure 8. Temperature increases for shape two.   *Figure 9. Temperature increase for shape three.*

A calculation of the heat transfer coefficient based on the ventilation airflow rates and measured temperature differences indicates that especially for the largest airflow rates there must be leakage of air. This occurs through the acoustic sheet and the perforation together with the joints between the sheets.

Comparison with other tests

Figure 12 shows a comparison of the ceiling system with tests of other solution for climatization, done in the same test room with the same inventory and heat sources. In these tests a water chilled suspended ceiling covering the room area was used. The air supply was through a ceiling mounted diffuser; five different brands from four manufacturers were used. 15 m³/hm² supply airflow was used for all these tests. With an airflow rate of 20 m³/hm², the air cooled ceiling system gives about equivalent or lower velocities.

564

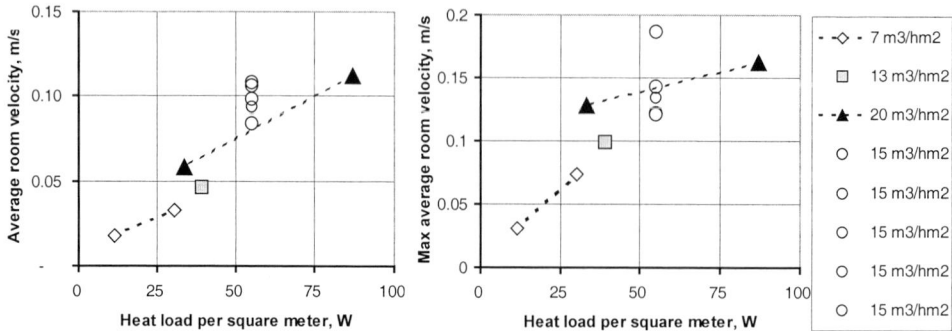

Figure 10. Comparison with other tests.

DISCUSSION AND CONCLUSIONS

The air velocity is very low in most of the tests. That means that the accuracy is limited, but the relative velocity ratio between the tests should give reliable information. For the more critical velocities around 0.15 m/s, the accuracy is reliable.

From the graphs showing the velocity for different airflow rates and heat loads (Figures 4 to 6) it is evident that the heat load has a greater impact on the air velocity than the airflow rate. I.e. the impulse due to thermal forces is greater than the impulse from the inlet jet. The velocity is as low as it could theoretically be. (We can not obtain lower velocity than the thermal movements give).

The air cooled ceiling's thermal comfort and cooling capacity is equal to water chilled ceilings in combination with a ceiling mounted diffuser for single module office rooms. If two ceiling mounted diffusers were used, the cooling capacity would probably improved considerably.

One can conclude that the shape and the size of the ceiling have lessof an impact on the efficiency since the result for the three modules does not differ much. For the smallest ceiling the interior velocity is higher and the temperature lower, this increased the heat transfer.

As a general conclusion, the air cooled ceiling seems to be a well working system in rooms with high heat loads

ACKNOWLEDGEMENT

This paper is based on projects carried out by SINTEF for ABB Environment from 1997 to 1999. The author would like to thank ABB Environment for their support and their permission to use the results from the projects in this paper.

REFERENCES

Mathisen, H.M. (1997). Test of supply unit for module office (In Norwegian: Prøving av tilluftsenheter for modulkontor). *SINTEF report no STF84 F97213*, Trondheim, Norway

Makulla,D. (1997). Energetischer Vergleich einer LuftKühldecke mit einer Wasserkühldecke und Quellüftung. *Ki Luft- und Kältetechnik* **7**, 299-304.

Air Distribution in Rooms, (ROOMVENT 2000)
Editor: H.B. Awbi

NUMERICAL STUDY OF A NEW VENTILATION TOWER SYSTEM FOR FRESH AIR SUPPLY IN AN AIR-CONDITIONED ROOM

K. HIWATASHI[1], S. AKABAYASHI[2], Y. MORIKAWA[1] and J. SAKAGUCHI[3]

[1]Technology Research Institute, Taisei Corporation,
344-1, Nase-cho, Totsuka-ku, Yokohama-city, 245-0051, JAPAN
[2]Graduate School of Science and Technology, Niigata University
8050, 2nocho, Igarashi, Niigata-city, 950-2181, JAPAN
[3]Department of Human Life and Environmental Science, Niigata Women's College
471 Ebigase, Niigata-city, 950-8680, JAPAN

ABSTRACT

In an air-conditioned office building, the ventilation air is normally mixed with the return air from the room in the air-handling unit. Therefore, the value of the air exchange efficiency defined by age of air is usually about 1.0, which is close to the perfect mixing case. If the fresh air and air-conditioning air are supplied separately, it is possible to increase the value of the air exchange efficiency at the breathing zone if the former is supplied directly to the breathing zone.

In this paper, the results of the CFD investigations for the ventilation tower system are described. In this system, fresh air is directly supplied to the breathing zone without mixing it with the return air from the room. This is achieved by supplying fresh air through a vertical duct rising from the floor and distributing fresh air to breathing zone height. With this system, the indoor air quality at the breathing zone will be improved and the requirement of the fresh air will be reduced in comparison with usual air-distribution systems.

The CFD simulation for isothermal conditions was carried out to compare the value of the air exchange efficiency between a conventional air-conditioning system and the ventilation tower system.

The results from CFD simulation show that the age of air at the breathing point is about 37 minutes for the conventional air-conditioning system and about 20 minutes for the ventilation tower system when the nominal time constant for the room was about 31 minutes. This result shows that the IAQ at the breathing zone provided by the ventilation tower system is better than the conventional systems.

KEYWORDS

Fresh Air, Ventilation and Air-Conditioning System, Ventilation Tower, IAQ, Smoking, Saving Energy, Age of Air, Scale for Ventilation Efficiency 3 (SVE3), Computational Fluid Dynamics

566

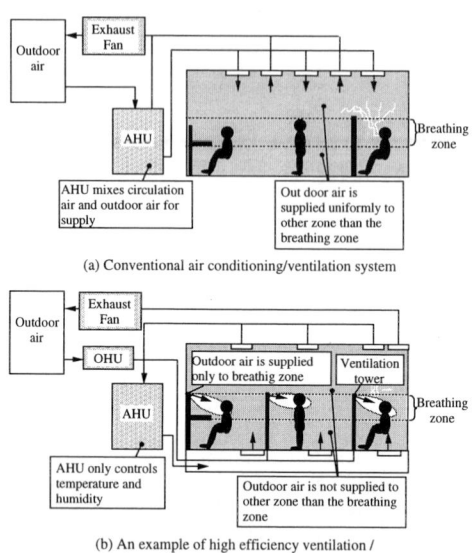

(a) Conventional air conditioning/ventilation system

(b) An example of high efficiency ventilation /
air conditioning system

Figure 1 Concept of high efficiency ventilation/air conditioning systems
taken as the object of this study

INTRODUCTION

Because of the concept of ventilation efficiency indices established recently, both the distribution efficiency of fresh air at each room point and the local air exchange efficiency at each room position can be obtained by the experiment and the numerical simulation.(e.g. Kuwahara et al 1999) These items were considered as difficult to obtain in the past. In the revised edition of the HASS102 Ventilation Standard and Description issued by The Heating, Air-Conditioning and Sanitary Engineers of Japan, the ventilation design employing the ventilation efficiency is proposed.

The indices of ventilation efficiency include the age of air as one of the items. The age of air is defined with a time required for supplied fresh air to reach any desired point such as the human breathing zone. Using this index, therefore, the supply efficiency of fresh air can be clarified.

In this study, high efficiency ventilation/air conditioning systems are investigated by considering the age of fresh air. Figure 1 shows the concept of the system. With the system shown in Figure 1 (b), the fresh air is independently supplied from the ventilation tower, that can be considered to reduce the required fresh air volume in comparison with the conventional system leading to a success in saving energy. Further this may provide a new solution in relation with smoking problems. Faulkner et al (1999) have studied a method for supplying outdoor air to a human body directly by mounting air supply outlets on a desk. However this system may require laborious work in changing the office layout. Therefore in this paper, the ventilation tower independent from a desk was studied for convenience in changing the office layout.

As the first step of the present study, this paper conducts a numerical simulation on the different air conditioning/ventilation systems shown in Table 1 under isothermal condition for comparison and investigation of the results obtained. The simulation results were evaluated by adapting the Scale for Ventilation Efficiency 3 (SVE3)($=\overline{\tau_p}$: local mean age of air / τ_n: nominal time constant) proposed by Murakami and Kato (1986).

OUTLINE AND PROCEDURE OF ANALYSIS

The object of the simulation employs a part of the laboratory where the experiment will be conducted. The analysis model is shown in Figure 2 and the boundary and calculation conditions are in Table 2. For the simulation, the airflow field was applied with steady state simulation beforehand. Then tracer gas is generated from the fresh air supply outlet by the step-up method, and the age of the room air is calculated by unsteady state calculation with a time step of 60 seconds until the mean room concentration reaches the concentration of the gas generated.

TABLE 1
AIR CONDITIONING / VENTILATION SYSTEM FOR COMPARISON

	Air conditioning system	Ventilation system
Case 1	Floor supply / ceiling return	Supplied after mix into circulation air
Case 2	Floor supply / ceiling return	Supplied from ventilation tower directly into room
Case 3	Ceiling supply / ceiling return	Supplied from ventilation tower directly into room

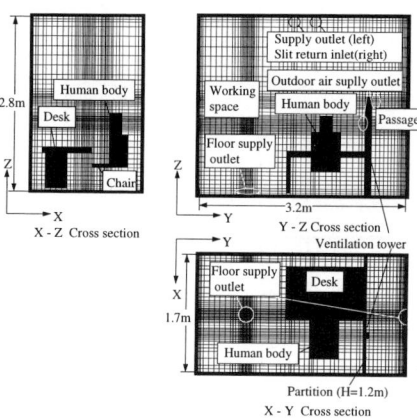

Figure 2 The analysis model

TABLE 2
CALCULATION AND BOUNDARY CONDITION

CFD grid points	47(X) x 70(Y) x 55(Z) Supposing a part of interior space, 1 occupant
Turbulent flow model	Standard k-ε model
Difference scheme	Advection term QUICK
Simulation method	Finite volume method based on SIMPLE method
Flowing boundary	$kin=(Uin/100)^2$ ε in=$C\mu kin^{3/2}$/Lin, Lin = Supply outlet opening width $C\mu$ =0.09
Boundary condition	Free slip supposing continuous space for X-face wall, Y-face wall at passage side, Standard log-law for others
Nominal ventilation time	Fresh air volume: 30m³/h Fresh air volume + Circulation air volume: 150m³/h Air exchange rate: 2.0 times/h Circulation rate: 9.8 times/h Nominal ventilation time: τ_n = 30.5 min.
Floor supply outlet	Mesh splitting 10 x 10 per one spot. Reference 5) Supply air volume 80m³/h.set (Case 1) ,100m³/h.set (Case 2)
Ventilation tower (for Fresh air)	Mesh splitting per one supply air outlet: 3 x 3 Two supply air outlets per 1 tower (For working space and passage) Each 15m3/h supposing a case when occupied and moving hours are equal Supply outlet dimension: 0.058m x 0.058m Set to Uin = 1.24m/s to suppress wind velocity supposed breathing zone less than 0.5m/s according to Act for Maintenance of Sanitation in Buildings of Japan
Ceiling supply outlet	0.025m wide slit outlet, U in = 0.78m/s (Case 3)
Ceiling return inlet	0.025m wide slit inlet, U out = 0.98m/s (Case 1-3)

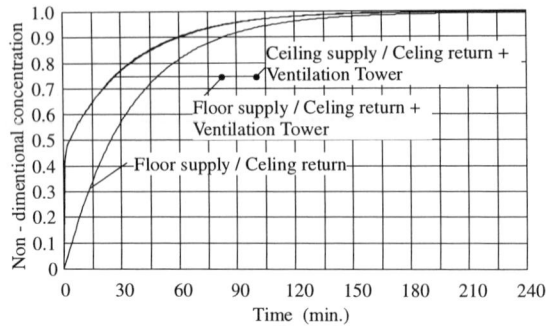

Figure3 Time-based change of concentration rise at breathing point supposed
in each air conditioning/ ventilation system

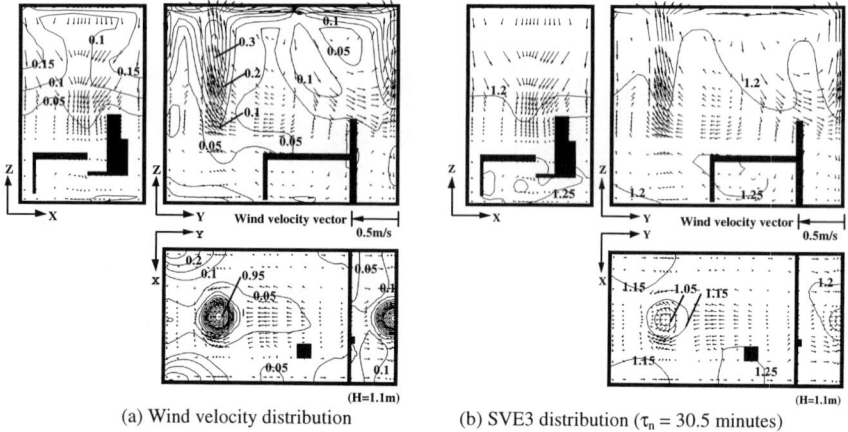

(a) Wind velocity distribution   (b) SVE3 distribution (τ_n = 30.5 minutes)

Figure 4 Wind velocity and SVE3 distribution when fresh air is mixed into circulation air
(Floor supply/ceiling return)

SIMULATION RESULTS

Figure 3 shows the time-based change of an increase in the non-dimensional concentration of the tracer gas at the supposed breathing point in each air conditioning/ventilation system. The supposed breathing point represents the location of FL + 1.1m at the position where the occupant is sitting in a chair. In the two systems where the ventilation tower was installed to supply fresh air directly, the concentration exceeds 0.4 at 1 minute after the tracer gas was generated. Although the increasing speed of the concentration in the case of the floor supply was higher, no significant difference was found between the two systems both showing quick concentration increase. After about 145 minutes elapsed, it reaches a concentration of 99% under an early steady state. In the case when fresh air is mixed into the circulation air on the other hand, about 205 minutes were required to reach a concentration of 99%, 1.4 times of the former value approximately.

Figures 4~6 show the wind velocity distribution and SVE3 distribution in the supposed breathing point of each air conditioning/ventilation system. The SVE3 in the supposed breathing point counts for 1.2 approximately when fresh air is mixed with circulation air, while it counts for 0.65 and 0.7

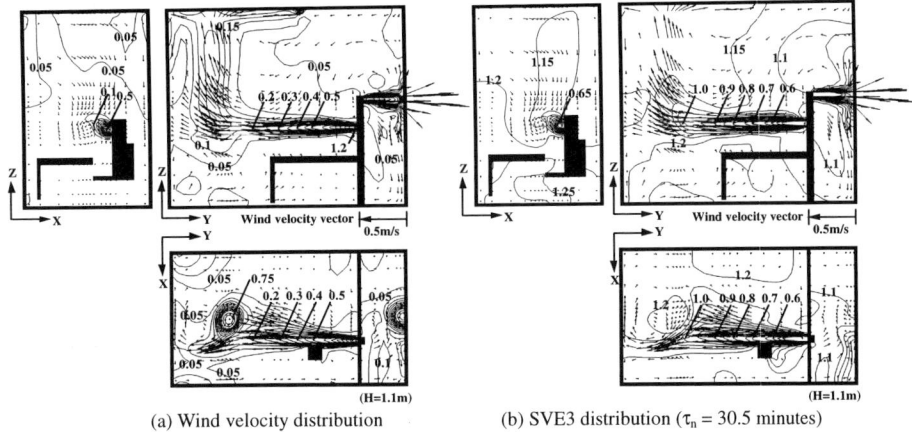

(a) Wind velocity distribution   (b) SVE3 distribution (τ_n = 30.5 minutes)

Figure5 Wind velocity and SVE3 distribution when fresh air is directly supplied
(Floor supply/ceiling return + Ventilation tower)

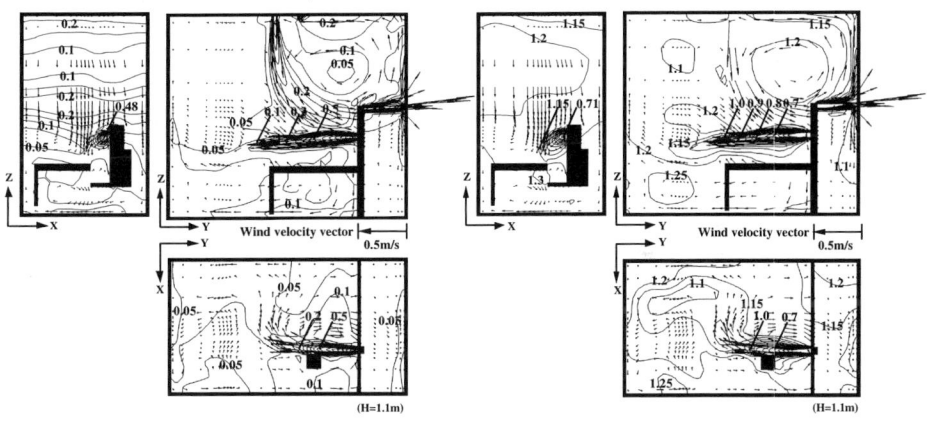

(a) Wind velocity distribution   (b) SVE3 distribution (τ_n = 30.5 minutes)

Figure6 Wind velocity and SVE3 distribution when fresh air is supplied directly
(Ceiling supply/ceiling return + Ventilation tower)

approximately in the floor supply and ceiling supply respectively when using the ventilation tower. From this, it can be judged that the direct supply of fresh air to the breathing point will be effective in improving the IAQ.

Table 3 shows the mean SVE3 by height and the room mean SVE3. The room mean SVE3 shows comparatively high values counting for 1.1~1.2 in any air conditioning/ventilation system. Since the position of the supply outlet and that of the return inlet used in this simulation locate nearby, it likely causes short-circuiting which may lead to such result.

When using the ventilation tower, it would likely be resulted in a longer age of air for most regions although a very short age of air is presented locally. Here, the plane distribution of SEV3 at FL + 1.1m of each air conditioning/ ventilation system is compared. While the region exceeding 1.25 is observed in the case of the ceiling supply/ceiling return + ventilation tower system, it is also observed in the case when fresh air is mixed into the circulation air. The mean SVE3 at FL + 1.0~1.8m counts for 1.1~1.2 without any significant difference in each system, while the mean SVE3 shows a lower value in the case when using the ventilation tower.

Although the influence caused by the layout of the supply outlet and return inlet should be taken into account when using the ventilation tower, it can be considered that the age of air in the major region will not be longer significantly as the region shown with a shorter value locally is limited.

TABLE 3
MEAN SVE3 BY HIGHT AND ROOM MEAN SVE3
IN EACH AIRCONDITIONING / VENTILATION SYSTEM

	Air conditioning system	Ventilation system	FL 0 - 1.0m	FL+1.0 - 1.8m	FL+ 1.8 - 2.8m	Entire room
Case 1	Floor supply / ceiling return	Supplied after mix into circulation air	1.20	1.18	1.16	1.18
Case 2	Floor supply / ceiling return	Supplied from ventilation tower directly into room	1.17	1.13	1.14	1.15
Case 3	Ceiling supply / ceiling return	Supplied from ventilation tower directly into room	1.18	1.16	1.17	1.17

CONCLUSION

For high efficiency ventilation/air conditioning systems that take the age of fresh air into account, a numerical simulation was applied for investigation to a method that supplies fresh air directly to the breathing zone. In the simulation, different ventilation/air conditioning systems were compared under isothermal condition. As a result, it was proved that the IAQ can be upgraded by directly supplying fresh air into the breathing zone. Also proved was that energy can possibly be saved while keeping the IAQ in the breathing zone almost at the conventional level.

The authors wish to conduct simulations in more detail by taking room thermal load into account in future. In addition, the results of the numerical simulation will be verified by real scale experiments.

REFERENCES

Faulkner, D., Fisk, J.D., Sullivan, D. P. and Wyon, D. P. (1999) Ventilation Efficiencies of Task/ambient Conditioning Systems with Desk-mounted Air Supplies. *Indoor Air 99* 2, 356-361

Kuwahara, R., Akabayashi S., Mizutani K., and Sato H., (1999) Relationship between Ventilation Efficiency and Inlet-outlet System of Office Rooms - Air Conditioning Systems and Ventilation Efficiency in Buildings Part1. *J. Archit. Plann. Environ. Eng.* **517**, 29-36

Murakami, S. and Kato S. (1986) New Scale for Ventilation Efficiency and Calculation Method by Means of 3-dimensional Numerical Simulation for Turbulent Flow – Study on Evaluation of Ventilation Efficiency in Room. *Transactions of the Society of Heating Air-Conditioning and Sanitary Engineers of Japan* **32(Oct., 1986)**, 91-102

Subcommittee on Ventilation and Standards of The Society of Heating Air-Conditioning and Sanitary Engineers of Japan (1995) *Symposium on HASS102-1995 Ventilation Standards* The Society of Heating Air-Conditioning and Sanitary Engineers of Japan, Tokyo, JAPAN

Air Distribution in Rooms, (ROOMVENT 2000)
Editor: H.B. Awbi

THE IMPACT OF THE "COANDA EFFECT" ON THE INDOOR AIR MOTION GENERATED BY A WINDOW-TYPE AIR-CONDITIONER

Y. C. Shih*, H. Chiang and R. J. Shyu

Energy & Resources Laboratories, Industrial Technology Research Institute
Chutung, Hsinchu, Taiwan 310, R.O.C.
*E-mail: f860824@itri.org.tw

ABSTRACT

The effect of the installed position of a window-type air-conditioner on the indoor air motion was investigated by both experimental and CFD methods in this paper. The temperature distribution of a room conditioned by a window-type air-conditioner was measured in an environmental chamber at the Indoor Environment Research Laboratory, Industrial Technology Research Institute (ITRI). Also, a CFD program, PHOENICS, was adopted to predict the indoor air distribution. In practice, a window-type air-conditioner was usually installed close to a room corner. The results of this study showed that strong short circulation of airflow occurred when the window-type air-conditioner had its air return opening close to the adjoining wall. But, when the air-conditioner was shifted to the opposite corner of the room with its supply opening close to the other adjoining wall, the "Coanda Effect" might cause the supply air stream to deflect and attach to the bounded wall. By inspecting the results, it was found that the "Coanda Effect" had important influence on the overall performance of the indoor air distribution.

KEYWORDS

Coanda Effect, Indoor Air Motion, Window-type Air-conditioner, CFD Simulation, Experimental Measurement, HVAC System

INTRODUCTION

Window-type air-conditioners are widely used in Taiwan for the space cooling of residential buildings. This type of building has relatively small spaces and is more sensitive to the building envelope loads. Although the increase of space cooling loads can be balanced by increasing the supply air volume or reducing the supply air temperature, this change might risk room occupant's thermal comfort because of the increasing draught rate and uneven air distribution.

The performance of room air distribution can be affected by the parameters, such as supply air speed, air temperature, the inclined angle of airflow, the shapes and the sizes of air supply and return openings as well as the installed position. Among these factors, the installation of the air-conditioner has been long neglected by architectures. As a part of effort to create pleasantly living environments, the purpose of this paper is to evaluate the effect of the installed position of the window-type air-conditioner on the room air distribution.

In the past, many researches have been done for the indoor air movement. The main points of those studies focused on analyzing the indoor air distribution, turbulence characteristics, thermal comforts and indoor air quality, etc. (ASHRAE Handbook 1997, Murakami et al. 1992). Two methods, including experimental measurement and CFD simulation, were commonly used in those studies. Experimental measurement customarily took a lot of time and expense; thus, CFD simulation became the most economic and efficient tool to study the indoor air motion. In order to make the simulation results reliable and accurate, lots of researches (Chen 1997, Henkes and Hoogendoorn 1995, Li and Baldacchino 1995, Murakami et al. 1994, Neilson 1998, Xu et al. 1998) have been performed on inspecting (1) the influence of different turbulence models and various numerical algorithms on the simulation results; (2) the verification of simulation and experimental results. Unfortunately, there was no unique turbulence model or numerical scheme valid for solving all indoor air motions until now. Hence, choosing an appropriate turbulence model and a suitable numerical scheme is the first priority for simulating the indoor air motion.

In practice, a window-type air-conditioner is usually installed near one of the interior corners of the conditioned room. In such cases, the adjoining wall might restrict the development of the jet flow discharged from the supply opening of an air-conditioner. In order to investigate the effect of wall on the indoor air distribution, a window-type air conditioner with air supply opening on the right and air return opening on the left was installed in a model room as the test sample. It was placed on two different places close to the associated corners in the room to test the influence of the installed position. The results of experimental measurement were also verified by CFD simulation in this paper.

EXPERIMENTAL FACILITY

In this study, the temperature distribution of a room ventilated by a window-type air-conditioner was measured inside the indoor unit of a large-scale climate simulation chamber at ITRI. The indoor unit has two rooms (A & B) with the same size of 4.6m(length) x 3.7m(width) x 2.7m(height), as shown in Figure 1(a), to be used as the test rooms of the window-type air-conditioner. It is surrounded by the outdoor unit equipped with temperature and humidity controls. Both test rooms are symmetrical to the adjacent wall and have the same boundary conditions.

A window-type air-conditioner displayed in Figure 1(b) was placed at the indicated corners of both rooms consecutively for doing the test. The reason for placing the same air-conditioner at different rooms was to measure the temperature distributions of each room under the same boundary conditions. The temperature distributions for each room were measured by the 480 points of T-type thermocouple sensors, which formed a 3-D matrix uniformly spanning over the entire room. The steady-state temperature distributions were recorded by a discrete data-logger network system.

CFD METHOD

In this paper, the CFD software-PHOENICS (1991) was utilized to simulate the indoor air distribution. The numerical model of PHOENICS is based upon the "Finite Volume Method". The governing equations used in PHOENICS obey the principle of conservation and can be expressed as the following general form:

$$\frac{\partial}{\partial t}(\rho\varphi) + \nabla \bullet (\rho\vec{V}\varphi - \Gamma_\varphi \nabla\varphi) = S_\varphi \qquad (1)$$

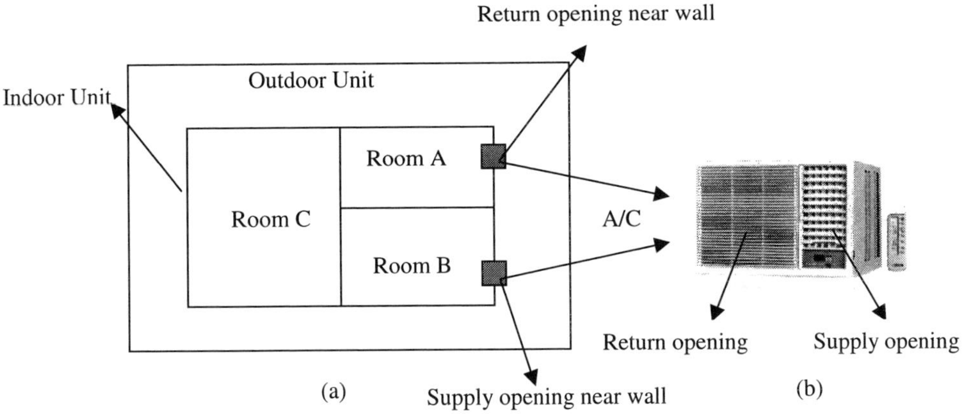

Figure 1: Schematic of the Indoor Environment Research Laboratory and the installed positions of a window-type air-conditioner (A/C)

The equations of continuity, momentum and energy can be derived from Eqn. 1. The standard k-ε turbulence model was adopted in this study to predict the turbulence behavior. To account for the buoyancy effect, the indoor air was assumed to follow the Boussinesq's approximation.

Since Eqn. 1 follows the principle of conservation, it is easy to discretize it into the finite difference equation from the finite volume assumption. The general discretized equation can be written as

$$a_p\Phi_p = \sum a_n\Phi_n + b \qquad (2)$$

By coupling the staggered grid system with the iteration method, the pressure, velocity and temperature fields can be solved from Eqn. 2. During the iteration procedure, the SIMPLEST algorithm was employed to solve the pressure-velocity coupling equations.

RESULTS AND DISCUSSIONS

On both experimental and CFD studies, the window-type air-conditioner was operated at the cooling mode with low air supply speed. The inclined angle of supplied air was set horizontally. The temperature of the outdoor unit was controlled at 35 °C. The air-conditioner was installed on the mounted walls at 1.8 m above floor and 0.3 m away from the adjoining walls in both test rooms, as depicted in Figure 1, respectively. The results of both experimental and CFD studies included the indoor air distributions of the same air-conditioner operated at rooms A and B. When the air-conditioner was placed at room A, its return opening was near the adjoining wall. If it was shifted to room B, it became that its supply opening was near the neighboring wall. The results of both cases are discussed as follows:

Experimental Measurement

The experimental results, as shown in Figure 2, were divided into Cases A and B which showed the room air temperature distributions of the air-conditioner operated at rooms A and B, respectively. Figures 2 (a) & (b) displayed the temperature contours of both cases at the central cross-section of supply opening. From the results, it was found that the average temperature of Case A at this cross-section was higher than that of Case B. The results of Figures 2 (c), (d), (e) and (f) depicted the temperature contours of both cases at different locations above the floor. They also disclosed that the overall temperature distributions of Case B were lower than those of Case A. To compare Figure 2 (c) with Figure 2(d), it can be seen that the cold air jet of Case A exiting from the air-conditioner did not spread widely and was restricted to a limited region, while that of Case B deflected toward the adjoining wall and moved along it. Those interesting phenomena resulted in different temperature distributions; therefore, the installed position of a window-type air-conditioner has important influence on the air distribution performance.

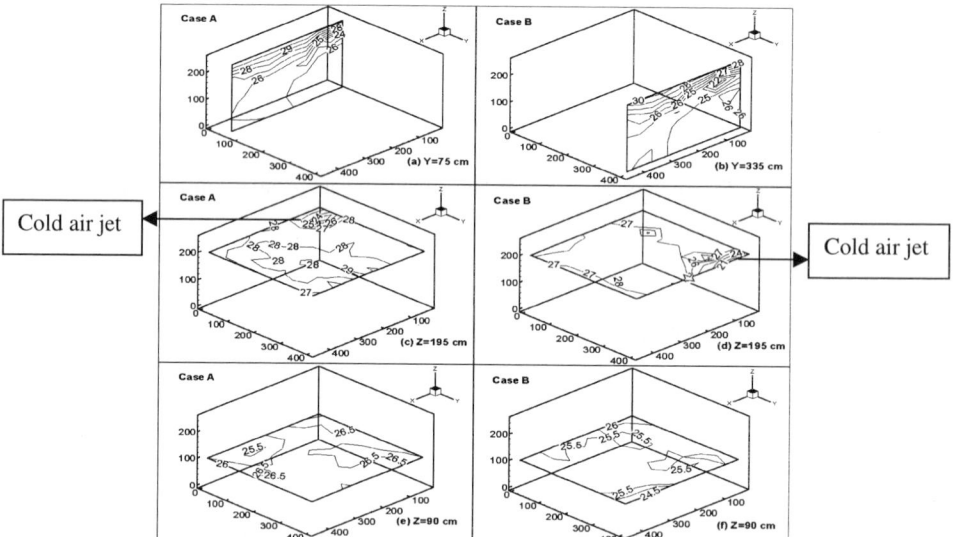

Figure 2: The comparison of the indoor air temperature distributions between Cases A and B by experimental measurement

CFD Simulation

In the CFD simulation, the inlet parameters of supply opening, including air velocity, air temperature, and turbulence intensity were given by the experimental data. The thermal boundary conditions were calculated by

$$q'' = U(T_0 - T) \tag{3}$$

In Eqn. 3, the overall heat transfer coefficient U and the surrounding temperature T_0 were estimated by the experimental measurement. 70000(40x50x35) grids were adopted in this study to simulate the indoor air motion.

Figure 3 showed the temperature distributions of the indoor air for Cases A & B. By comparison Figure 2 with Figure 3, it was found that similar results existed between experimental and simulation results. Figures 3 (a) and (b) revealed that the average temperature of Case B at the central cross-section of supply opening was lower than that of Case A. Also, the cold air jet of Case B diffused more distant. From the results of Figures 3 (c) and (d), it can be seen that the supply jet of Case A had the inclination to run a short circulation back to the return opening and that of case B deflected toward the adjoining wall obviously. Although there still existed some discrepancies between experimental and simulation results, which could result from the inaccuracy of experimental measurement, incomplete turbulence model and complex boundary conditions, the overall physical phenomena agreed quite well.

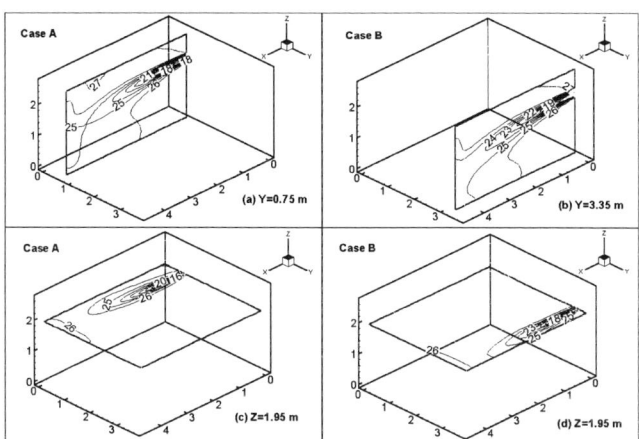

Figure 3: The comparison of the indoor air temperature distributions between Cases A and B by CFD simulation

The results of Figures 4 and 5 were the velocity fields and the pressure contours of both cases at the planes of 1.95 m above the floor, respectively. By inspecting Figures 4(a) and 5(a), it was known that there was a low-pressure region near the return opening; therefore, the supply jet of Case A was easier to divert to the direction of the return opening. The velocity field of Figure 4(b) showed that the development of supply jet for Case B was restricted to the neighboring wall; hence, there was a recirculation region near the corner, which resulted in a low-pressure region, as depicted in Figure 5(b). Apparently, the supply jet inclined to deflect toward the wall. This is the so-called "Coanda Effect" which usually happens to the ceiling jet owing to the pressure difference. From the results of this study, it was found that the "Coanda Effect" also occurred in the indoor air motion of the window-type air-conditioner.

Recirculation region

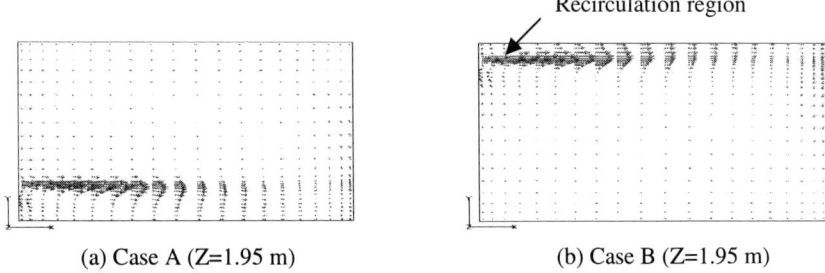

(a) Case A (Z=1.95 m)　　　　　　　　　　　(b) Case B (Z=1.95 m)

Figure 4: The comparison of the velocity fields between Cases A and B by CFD simulation

576

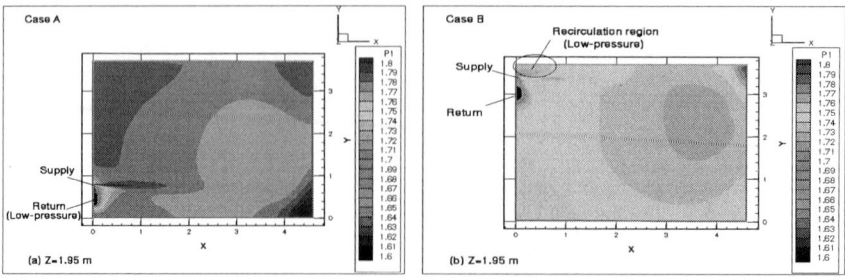

Figure 5: The comparison of the pressure contours between Cases A and B by
CFD simulation

CONCLUSIONS

Both experimental and CFD methods were used to study the indoor air motion generated by a window-type air-conditioner in this paper. From the experimental results, it was found that the average air temperature was lower when the supply opening of the air-conditioner was near the adjoining wall while that was higher if the return opening was near the corner wall. This extraordinary phenomenon was explained by detailed CFD simulation results. From the results, it was known that the "Coanda Effect" tended to move the air stream toward the adjoining wall and avoided the occurrence of short circulation if the supply opening was near the neighboring wall. When the return opening was near the corner wall, the phenomenon of short circulation was easier to occur. This study revealed that the "Coanda Effect" had great impact on the indoor air distribution performance of a window-type air-conditioner. In the future, large speed and different inclined angles of supply air will be performed to investigate more.

REFERENCES

ASHRAE Handbook (1997). *Fundamentals*, American Society of Heating, Refrigeration and Air-Conditioning Engineers, Inc.
Chen Q. (1997). Computational Fluid Dynamics for HVAC: Success and Failures. *ASHRAE Trans.* **103:1**, 178-187.
Henkes R.A.W.M. and Hoogendoorn C.J. (1995). Comparison Exercise for Computations of Turbulent Natural Convection in Enclosures. *Numerical Heat Transfer* **28:B**, 59-78.
Li Y. and Baldacchino L. (1995). Implementation of Some Higher-Order Convection Schemes on Non-Uniform Grids. *International Journal for Numerical Methods in Fluids* **21**, 1201-1220.
Murakami S., Kaizuka M., Yoshino H., and Kato S. (1992). *Room Air Convection and Ventilation Effectiveness.* American Society of Heating, Refrigeration and Air-Conditioning Engineers, Inc.
Murakami S., Kato S., and Ooka R. (1994). Comparison of Numerical Predictions of Horizontal Nonisothermal Jet in a Room with Three Turbulence Models- $k - \varepsilon$ EVM, ASM, and DSM. *ASHRAE Trans.* **100:2**, 697-704.
Neilson P.V. (1998). The Selection of Turbulence Models for Prediction of Room Airflow. *ASHRAE Trans.* **104:1B** 1119-1127.
PHOENICS (1991). *The PHOENICS Reference Manual CHAM TR/200.* CHAM Limited, UK.
Xu W., Chen Q., and Nieuwstadt F.T.M. (1998). A New Turbulence Model for Near-Wall Natural Convection. *International Journal of Heat and Mass Transfer* **41**, 3161-3176.

Air Distribution in Rooms, (ROOMVENT 2000)
Editor: H.B. Awbi

CFD-MODELLING OF ACTIVE DISPLACEMENT AIR DISTRIBUTION

Hannu Koskela

Finnish Institute of Occupational Health
Hameenkatu 10, FIN 20500 Turku, Finland

ABSTRACT

Active displacement air distribution is a relatively new low-impulse method, which is mainly used in large spaces such as industrial halls. The supply air is distributed with nozzle ducts, which are placed above the occupied zone. The air velocity in the nozzles is high, but it is quickly reduced due to circulation between the jets. The supply air mixes with the room air in the near zone of the duct. The flow pattern in the near zone is complex and can not be modelled in detail in practical applications. A simplified model for the nozzle duct inlet device is therefore needed for CFD-modelling of room air flow patterns. The purpose of this study was to develop and test such a model. The flow field near the supply device was first investigated with laboratory measurements in order to determine the correct boundary conditions. The developed model is based on momentum sources describing the induction of secondary air caused by the jets. The model was validated by simulating a laboratory test case and comparing the simulation results with the measurements. The comparison showed good agreement between the simulated and measured flow patterns, but some differences were found in velocity levels.

KEYWORDS

CFD, Modelling, Air distribution, Air terminal units, Air velocity, Full-scale experiments

INTRODUCTION

Active Displacement

Active displacement air distribution is a relatively new low-impulse method, which is mainly used in large spaces such as industrial halls. The supply air is distributed with nozzle ducts, which are placed above the occupied zone (Figure 1). Air velocity in the nozzles is high, but it is quickly reduced due to circulation between the jets. The supply air mixes with the room air in the near zone of the duct. The air flow pattern and the momentum supplied by the nozzle duct depend on the distribution of the nozzles on the duct. In this study, a CFD-model was developed for a device with nozzle sector 240° upwards. This type of nozzle duct device is typically used for cooling applications. It creates a flow pattern that is a combination of mixing and displacement. The unit takes secondary air from below and the introduced momentum is high enough to prevent the direct downfall of the mixed subtempered air.

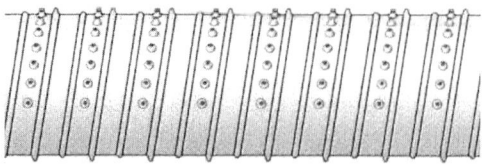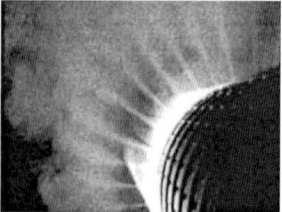

Figure 1: Activent nozzle duct device and visualisation of the supply air flow pattern.

CFD-modelling of Inlet Devices

The geometry of an inlet device is usually complex and, therefore, it is not possible to use an exact geometrical model in practical CFD-simulations. Instead, a simplified model, which describes the performance of the device, is needed. The model has to introduce the correct mass and momentum flow into the room. The flow pattern and induction of secondary air have to be in agreement with the behaviour of the real device. Simplified models for inlet devices have been studied e.g. in the IEA research programme Annex 20 (Moser 1991).

The effective area of an inlet device is usually smaller than the total area, which increases the momentum flow. Therefore, either extra momentum has to be added to the inlet air flow or the size of the opening has to be reduced in the model (Chen & Moser 1991, Heikkinen 1991). Other possibilities for modelling the inlet device include the 'prescribed velocity method', where the velocity profile is fixed in a selected area, or the 'box method', where the boundary conditions are given on the surfaces of an imaginary box around the inlet device (Nielsen 1988).

METHODS

The Activent nozzle duct modelled in this study had a diameter of 0.315 m and a nozzle sector of 240° upwards. The nozzles were 5 mm long and had a 5 mm diameter. The distance between the nozzles was 22 mm and the distance between the spiral nozzle rows was 74 mm.

The laboratory test arrangement is shown in Figure 2 (Koskela & al. 1998). The test room was thermally insulated from the surrounding hall. Air was supplied through two 3 m long nozzle ducts located at a height of 3 m symmetrically in the room. Three convective heaters with dimensions of 1.2 m x 0.4 m x 0.1 m were placed in the central plane, two with a 600 W heating power at a distance of 0.5 m from the walls and one with a 1200 W power in the middle of the room.

The air velocity was measured with two 3-dimensional Kaijo-Denki WA-390 ultrasonic anemometers, which have an accuracy of ±0.02 m/s. The ultrasonic anemometers measure the velocity vector with three pairs of ultrasonic transducers, each pair having a distance of 50 mm between the transducers. The air temperature was measured with Craftemp thermistors. The measurements were carried out by traversing the sensors in the central plane of the test room with an automatic traversing system (Figure 2). The traversing was done in half of the central plane with a 0.1 m spacing between measurement points. The averaging time in each point was 3 minutes and the sampling frequency was 1 Hz. The number of measurement points was circa 2000 and the number of samples in each point 180 for each variable. The supply air flow rate and the temperature were monitored and kept constant during the traversing.

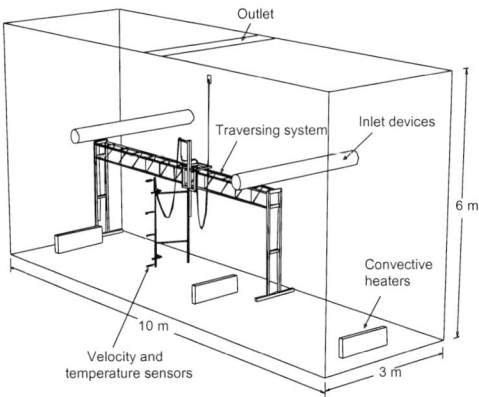

Figure 2: Laboratory test room.

The CFD-simulations were carried out with CFX 4.2 by using QUICK-discretisation, k-ε turbulence model and Boussinesq approximation for buoyancy. A computational grid consisting of 59 000 cells in 12 blocks was set up for one quarter of the test room with symmetry boundary conditions in two horizontal directions. The grid dependence was tested with double grid density. The heat sources were modelled as thin surfaces with a constant heat flux. The room surfaces were defined adiabatic with logarithmic wall functions.

RESULTS

Near Zone Measurements

The flow field in the near zone of the nozzle duct was studied by carrying out velocity measurements with ultrasonic anemometers. A typical operating condition with a temperature difference of –5°C between the inlet air and the room air and a flow rate of 0.058 m³/s per metre of the duct was selected as a basic test case for modelling. The measured near zone flow field is shown in Figure 3. The nozzle area in the Activent unit is 1.1 % of the duct surface. Therefore, the momentum flow rate in the jets is increased by a factor of 90 compared to the situation where the inlet area is 100% of the surface. The velocity of the jets is, however, substantially reduced within a distance of 0.2 m and a strong circulation is formed between the nozzle rows. Secondary air flows to the jet system from below and the flow continues between the jet rows along the surface of the duct.

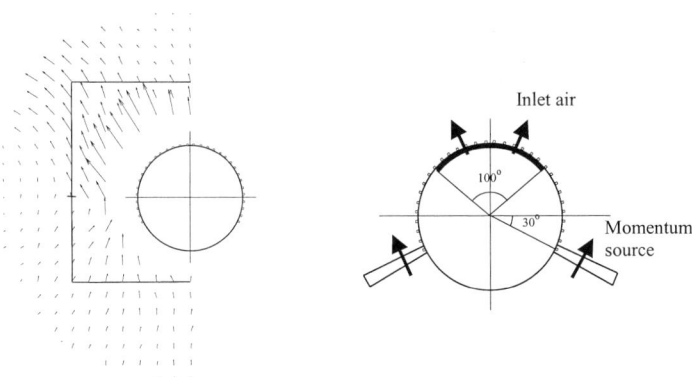

Figure 3: Measured near zone flow field and the momentum source model.

Momentum Source Model

Several principles for simplified models were tested for giving the boundary conditions for the nozzle duct. However, problems emerged with each of the principles. Adding extra momentum by using either a momentum source, 'porous media' or by reducing the inlet surface area did not yield satisfactory results. The problem with the 'box' and the 'prescribed velocity' models was that the flow pattern in the near field of the device changes with different inlet air temperatures and flow rates.

The developed momentum source model for the nozzle duct is based on the idea that the suction of secondary air caused by the jet system is the governing phenomenon for the resulting momentum flow of the device. This was modelled by adding momentum sources on the sides of the duct in the position of last nozzle, where the secondary air enters the jet system (Figure 3). The direction of the momentum source was set tangential to the duct surface. Because the momentum of the jets is strongly reduced by the circulation, the inlet air is blown with low momentum from the upper surface of the duct.

The strength of the momentum source was determined from the velocity measurements. The momentum flow rate of the secondary air can be calculated from

$$M = qv\rho = \frac{q^2 \rho}{A} \tag{1},$$

where q is the volume flow rate and v is the velocity of the secondary air, A is the cross-section area of the flow and ρ is the density. The velocity is assumed to be constant in the cross-section.

The flow rate of the secondary air was determined by integrating the normal velocities along the line shown in Figure 3. The flow rate was calculated from both the inward flow in the lower part and the outward flow in the upper part. The difference of these flows should be the inlet flow rate 58 l/s·m. The results are shown in Table 1.

TABLE 1
MEASURED SECUNDARY FLOW RATE

	Flow in	Flow out
Left	38 l/s	108 l/s
Top		132 l/s
Bottom	113 l/s	
Secondary	151 l/s	182 l/s
Inlet	58 l/s	58 l/s
Total	209 l/s	240 l/s
Secondary / Total	72.2 %	75.8 %

The two measured secondary flow rate values differ by 20%, which is probably due to the coarse measurement grid. From the results, the secondary air flow rate can be estimated to be approximately 75% of the total flow rate or three times the inlet flow rate. If the width of the secondary flow is estimated to be 0.1 m, the momentum flow of secondary air from Equation 1 is 0.18 N per meter of the nozzle duct. The original momentum flow in the jets is correspondingly 0.54 N. The momentum flow of the secondary air is thus approximately 1/3 of the original momentum flow of the jets.

Simulation and Measurement Results

The momentum sources were defined as two 'user 3d patches' in CFX 4.2, which were located as shown in Figure 3. The cross-section area of the sources was 0.00284 m^2, width 0.142 m and length 3 m. The vertical and horizontal components of the momentum source were set as 30 N/m^3 and 15 N/m^3, which gives the total source strength of 0.19 N per meter. The inlet boundary condition was set for normal velocity as 0.21 m/s, turbulence intensity as 10% and dissipation length scale as 1 m.

The simulated and measured flow fields near the nozzle duct in the centreline of the test room are shown in Figure 4. Eddies were formed on both sides of the duct and the flow pattern remained stable during the 63-hour measurement period. The position and form of the eddies were correctly predicted in the CFD-simulation.

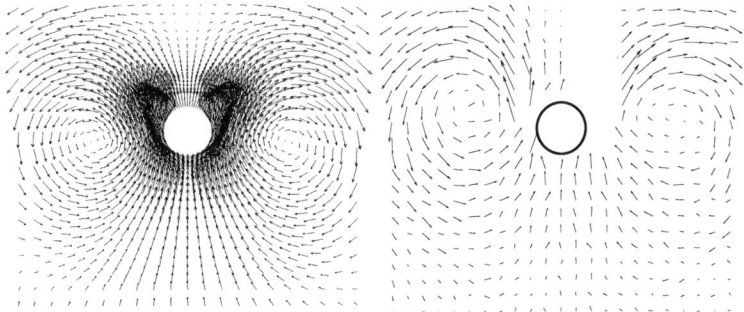

Figure 4: Modelled and measured velocity fields near the inlet device.

The predicted and measured distributions of air velocity are shown in Figure 5. The velocities in the occupied zone are low both in simulations and measurements, but the predicted values are higher than the measured ones. This can be seen in Figure 6, where the distribution of velocity is presented in two horizontal cross-sections at the levels of 1.1 m and 1.8 m. The predicted maximum velocities in the plumes are double compared to the measured ones. The distribution of air temperature in the simulation results was close to the measured distribution.

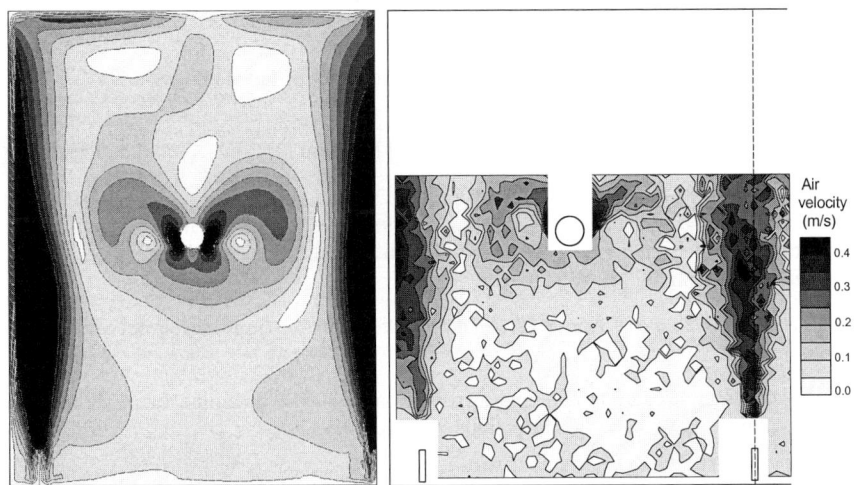

Figure 5: Modelled and measured distributions of air velocity in the central plane of the test room.

582

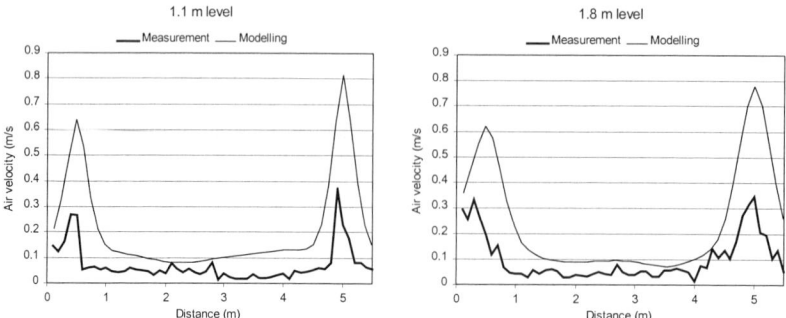

Figure 6: Velocity profiles in two horizontal lines in the central plane of the test room.

DISCUSSION

A simplified CFD-model was developed for the nozzle duct inlet device. The supply air is distributed through small nozzles and a complex flow pattern is created in the near zone of the device. The model developed in this study consists of momentum sources describing the secondary air flow induced by the nozzle duct. It is based on the idea that the suction of secondary air caused by the jet system is the governing phenomenon for the resulting momentum flow of the device.

The shape of the flow pattern near the inlet device and in the room was correctly predicted by the model. However, differences were found in the absolute velocity levels. Especially the velocities in plumes were higher than the measured ones. On the other hand, the plume velocities given by the empirical plume equations and measurement results of undisturbed plumes (Welling 2000) were found to be close to the predicted velocities. This discrepancy can be due to the disturbance caused by the inlet air on the plumes. The interaction between the inlet air and the plumes may be a time dependent phenomenon and therefore cannot be correctly predicted by a steady state simulation.

The developed model is being validated in other operating conditions in laboratory. In addition, the model will be tested in real buildings for studying the applicability for design purposes.

REFERENCES

Chen Q., Moser A. (1991). Simulation of a multiple-nozzle diffuser. Proceedings of the 12th AIVC conference, Ottawa. Vol.2, 1-13.

Heikkinen J. (1991). Modelling of the supply air terminal for room air flow simulation. Proceedings of the 12th AIVC conference, Ottawa. Vol. 3, 213-229.

Koskela H., Hautalampi T., Sandberg E., Lindgren M. (1998). Test room and measurement system for active displacement air distribution. Proceedings of Roomvent 98, KTH Building Services Engineering, Stockholm, 225-232.

Moser A. (1991). The message of Annex 20: Air flow patterns within buildings. Proceedings of the 12th AIVC conference, Ottawa. Vol. 1, 1-26.

Nielsen P.V. (1988). Representation of boundary conditions at supply openings. Internal report for IEA Annex 20, University of Aalborg, ISNN 0902-7513 R8902.

Welling I. (2000). Experimental study of natural-convection plumes from a heated horizontal square plate and a vertical cylinder. Experimental Heat Transfer, Vol 13:1, 7-20.

Air Distribution in Rooms, (ROOMVENT 2000)
Editor: H.B. Awbi

PREDICTION OF TWO-PHASE FLOW OF AIR AND WATER IN A CLOSED-WET COOLING TOWER FOR COMFORT COOLING OF BUILDINGS

S. B. Riffat and G. Gan

Institute of Building Technology, School of the Built Environment, University of Nottingham
University Park, Nottingham NG7 2RD, UK

ABSTRACT

Chilled ceilings are used for comfort cooling of industrial and office buildings. Evaporative cooling in closed-wet cooling towers is an environment-friendly method for providing cold water for the chilled ceilings without the need for refrigerative cooling. This paper presents a CFD technique for predicting two-phase flow of air and water in closed-wet cooling towers. The technique involves a continuous phase model for air flow and a dispersed phase model for the droplet trajectories. The continuous phase model for air flow consists of the conservation equations for mass, momentum, enthalpy and concentration as well as turbulence. The trajectory of a water droplet with evaporation is computed using the Lagrangian method. The impact of the dispersed phase on the continuous phase flow is accounted for by coupling the momentum, heat and mass transfer between the two phases. The predicted thermal performance of a prototype closed-wet cooling tower is compared with experimental measurement.

KEYWORDS

Cooling tower, Chilled ceiling, Comfort cooling, CFD, Two-phase flow, Air flow, Droplet trajectory, Heat transfer

INTRODUCTION

Chilled ceilings are used for comfort cooling of industrial and office buildings. In a chilled ceiling system, cooling is accomplished mainly by radiation heat transfer to provide thermal comfort. Radiative heat transfer allows chilled ceilings to remove a considerable amount of heat at a relatively small temperature difference between room air and the ceiling. The system can therefore operate with supply water temperatures as high as 18°C to 20°C. Evaporative cooling in closed-wet cooling towers is an environment-friendly method for providing cold water for the chilled ceilings. A closed-wet cooling tower employs a heat exchanger as a heat transfer medium and the fluid inside the tubes of the heat exchanger is cooled by means of evaporation and convection. Fig. 1 is a schematic diagram of a closed-wet cooling tower. It is a indirect-contact counter-current flow cooling device where water through a distribution system is spread onto the tube surfaces of the heat exchanger while air flows upwards over the wetted tube surfaces and through the falling water droplets. Evaporation of part of the water into the unsaturated air results in cooling both the tube surfaces and the water droplets simultaneously.

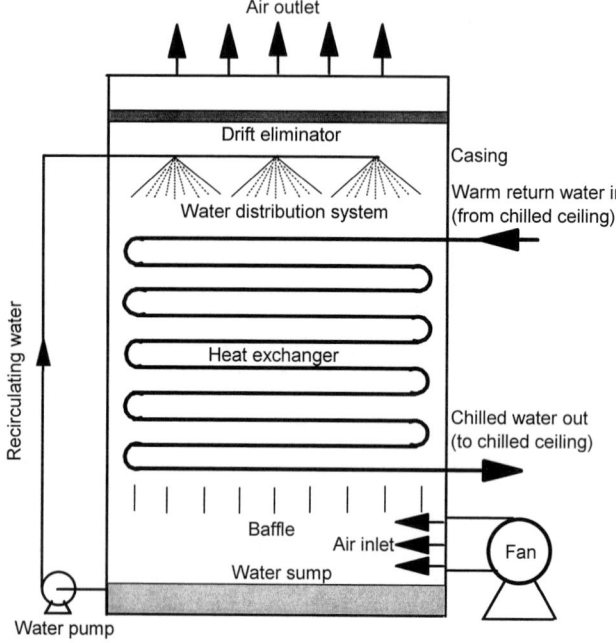

Fig. 1: Schematic diagram of the closed-wet cooling tower

The thermal performance of a closed-wet cooling tower depends on its design and operating conditions. Traditionally, the thermal performance is evaluated by experimental measurement or simplified one-dimensional solution of heat, mass and momentum transfer processes in the cooling tower. The experimental measurement of these processes would be not only costly and time consuming but also difficult whereas a simple one-dimensional solution may not be accurate enough for optimising the design and operation of a closed-wet cooling tower. Besides, the results from such methods cannot be extended to all geometries or air flow patterns. The objective of the present work is to use computational fluid dynamics (CFD) as an alternative means to optimise a CFC-free energy-efficient cooling system that combines a closed-wet cooling tower with chilled ceilings.

METHODOLOGY

The numerical technique involves simulation of the two-phase flow of gas and water droplets in the cooling tower. A commercial CFD software package (FLUENT 1993) is used for this purpose. The CFD package employs a continuous phase model for air flow and a dispersed phase model for water droplets. The impact of the dispersed phase on the continuous phase flow is accounted for by coupling the momentum, heat and mass transfer between the two phases (Gan and Riffat 1999).

Continuous Phase Model

The continuous phase model consists of the conservation equations for mass, momentum, enthalpy and concentration as well as turbulence of air. For steady-state incompressible air flow, the model can be represented by the following general equation

$$\nabla \bullet (\rho \overline{V} \phi - \Gamma_\phi \nabla \phi) = S_\phi + S_\phi^p \tag{1}$$

where ϕ is the flow variable, $\overline{V}$ is the mean air velocity (m/s), ρ is the air density (kg/m³), Γ_ϕ is the diffusion coefficient (Ns/m²), S_ϕ is the source for the continuous phase and S_ϕ^p is the source due to the interaction between air and water droplets.

Dispersed Phase Model

The dispersed phase consists of spherical water droplets dispersed in the continuous phase. The flow of dispersed droplets is characterised by the trajectories.

The trajectory of a water droplet is computed using the Lagrangian method. This involves integrating the following equation for the balance of forces on the droplet including inertia, drag, buoyancy, apparent mass and pressure gradient:

$$\rho_p \frac{dV_p}{dt} = \frac{3}{4} \frac{\rho C_D |V - V_p|}{d_p} (V - V_p) + g(\rho_p - \rho) + \frac{1}{2}\rho \frac{d}{dt}(V - V_p) + \frac{\partial P}{\partial r_p} \tag{2}$$

where V is the instantaneous air velocity (m/s), V_p is the instantaneous velocity of the droplet (m/s), C_D is the drag coefficient, d_p is the droplet diameter (m), ρ_p is the droplet density (kg/m³), P is the static pressure (Pa) and r_p is the droplet trajectory (m).

Assumptions

The cooling tower for simulation is a prototype 1.2 m long, 0.61 m wide and 1.55 m high. The heat exchanger is comprised of 228 staggered tubes of 10 mm in outside diameter. Fig. 2 shows the tube arrangement for half of the heat exchanger. It is 24 rows deep and has longitudinal and transverse pitches of 0.02 m and 0.06 m, respectively. The performance of the cooling tower has been experimentally tested for different operating conditions (Riffat, et al. 1999).

The following assumptions are used for the prediction of the thermal performance of the cooling tower:

Tower configuration: The tower is quite long compared with the width, so only a two-dimensional flow is considered. Fluid flow through the tower is symmetrical. The tower casing is adiabatic.

Heat exchanger: For a real heat exchanger, the temperature of the fluid inside tubes varies along the coils. For the two-dimensional flow in one cross-section of the tower, the tubes representing the heat exchanger become separate entities as shown in Fig. 2. Each tube is modelled as a smooth circular cylinder with heat generation, the rate of which is calculated from the experimental data of cooling load. The heat transfer from the surface to air involves evaporation and convection. However, the software does not allow evaporative cooling of tubes to be directly taken into account. Therefore, only the sensible heat transfer is used in terms of a surface heat flux. The estimated sensible heat transfer is about 24% of the total heat transfer for high mass flow ratios. The heat flux is assumed to be the same for each row of tubes.

Spray water: The spray water in one cross-section is composed of 100 trajectories of droplets and injected into the tower through a nozzle at the centre line above the heat exchanger. The injection velocities of the water droplets from the nozzle vary such that the spray water would cover the entire width of the heat exchanger. The mean diameter of the water droplets is estimated from the terminal velocity of the droplets at a mean velocity of air flowing over the heat exchanger.

On the basis of the above assumptions, a nonuniform computational grid of 110 x 342 is used for the simulation of two-dimensional flow in the tower, with dense grid cells distributed over the tubes of the heat exchanger. Fig. 3 shows the mesh around some tubes of the heat exchanger.

RESULTS AND DISCUSSION

The thermal performance of the cooling tower is assessed according to the temperature distribution of the tubes representing the temperature of chilled water in the heat exchanger. The CFD technique for predicting the tower performance is demonstrated for the following test conditions:

The flow rate of supply air (Q_a) was 0.48 m^3/s. The measured dry bulb temperature of supply air (T_a) was 13.1°C. The relative humidity of supply air (RH) was 85%. Spray water at a temperature (T_{sw}) of 16.2°C was injected onto the heat exchanger at a rate (m_{sw}) of 1.38 kg/s. The calculated mass flow ratio of spray water to supply air (m_r) was rather high (= 2.34). The measured temperatures of chilled water at the inlet (T_{wi}) and outlet (T_{wo}) of the heat exchanger were 18.53°C and 17.02°C, respectively. The temperature drop of chilled water ($\Delta T = T_{wi} - T_{wo}$) was then 1.51 K. The flow rate of chilled water in the heat exchanger for the cooling tower was at the design rate of 0.8 kg/s.

Figure 4 shows the predicted air flow pattern in the cooling tower. The velocity vectors are plotted for every one in four grid cells on both directions. The supply air is forced into the tower and then flows upwards through the heat exchanger. The upward air flow is reasonably uniform with the help of baffles. Above the heat exchanger, the injection of spray water results in air near the tower centre under the spray water nozzle flowing downwards while the exit air being forced towards the tower casing at a much higher upward velocity than the average value. Fig. 5 shows the predicted trajectories of water droplets in the cooling tower. It is seen that water flows downwards by gravity across the whole width of the tower, thus wetting the heat exchanger for effective evaporative cooling. The downward flow of a water droplet is not straight but rather randomly zig-zagged due to the presence of tubes, increasing the effectiveness of surface wetting.

The predicted mean air temperature in the tower is 15.4°C. The air temperature between tubes varies from 15.4°C to 16.4°C. The air throughout the heat exchanger is completely saturated due to high levels of the mass flow ratio and supply air humidity. The predicted mean tube temperature is 18.8°C. The predicted temperatures for the top and bottom rows of tubes are 19.1°C and 17.5°C, respectively, which are higher than the measured chilled-water temperatures for the inlet and outlet of the heat exchanger. This may be attributed to over-estimation of the rate of sensible heat transfer from tube surfaces. The predicted temperature difference between the top and bottom tube rows (1.6 K) is close to the measured temperature drop of chilled water (1.51 K).

Further simulations are performed to compare with experimental measurement. Table 1 shows a comparison between the predicted and measured tower performance in terms of inlet and outlet water temperatures and their difference at the design flow rate. It is seen that in general the predicted thermal performance is in good agreement with the measurement.

TABLE 1
PREDICTED AND MEASURED THERMAL PERFORMANCE OF THE COOLING
TOWER AT THE DESIGN WATER FLOW RATE

Case	Supply air			Spray water			Chilled water		
	Q_a (m^3/s)	T_a (°C)	RH (%)	m_{sw} (kg/s)	T_{sw} (°C)	m_r	T_{wi} (°C)	T_{wo} (°C)	ΔT (K)
Test 1	**0.48**	**13.1**	**85**	**1.38**	**16.2**	**2.34**	**18.53**	**17.02**	**1.51**
Pred. 1	0.48	13.1	85	1.38	16.2	2.34	19.1	17.5	1.6
Test 2	**1.08**	**13.0**	**87**	**1.38**	**13.6**	**1.04**	**15.86**	**14.35**	**1.51**
Pred. 2	1.08	13.0	87	1.38	13.6	1.04	16.3	14.9	1.4
Test 3	**0.48**	**15.7**	**51**	**1.38**	**15.0**	**2.36**	**17.24**	**15.76**	**1.48**
Pred. 3	0.48	15.7	51	1.38	15.0	2.36	18.1	16.8	1.3
Test 4	**0.48**	**20.7**	**45**	**1.38**	**18.1**	**2.41**	**20.38**	**18.90**	**1.48**
Pred. 4	0.48	20.7	45	1.38	18.1	2.41	20.6	19.2	1.4

CONCLUSIONS

CFD has been used to predict the thermal performance of a closed-wet cooling tower. The predicted thermal performance is in good agreement with experimental measurement at the design water flow rate. Therefore, the CFD technique can be used to optimise the design and operation of cooling towers.

However, for a given cooling load, because the temperature drop of chilled water through the heat exchanger varies with the water flow rate, the thermal performance would be under-predicted when the water flow rate is lower than the design rate. To predict the thermal performance accurately, a CFD software should be capable of modelling the heat transfer between and within tubes of the heat exchanger and the evaporative cooling of tube surfaces as well as air.

ACKNOWLEDGEMENTS

This work is in part supported by the EU JOULE III contract JOR3-CT97-0195.

REFERENCES

FLUENT. (1993). *User's Guide*, Fluent Inc., USA.

Gan G. and Riffat S.B. (1999). Numerical simulation of closed wet cooling towers for chilled ceiling systems, *Applied Thermal Engineering*, **19:12**, 1279-1296.

Riffat S.B., Oliveira A., Facao J., Gan G. and Doherty D. (1999). Thermal performance of a closed-wet cooling tower for chilled ceilings - Measurement and CFD simulation, *International Journal of Energy Research* (in press).

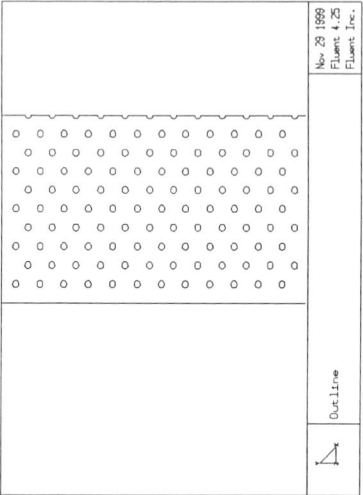

Fig. 2 Tube arrangement of the heat
exchanger

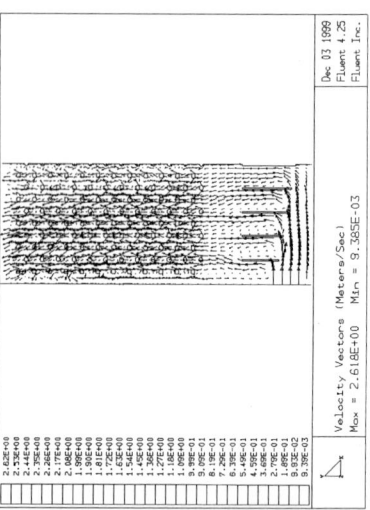

Fig. 4 Predicted air flow pattern in the
cooling tower

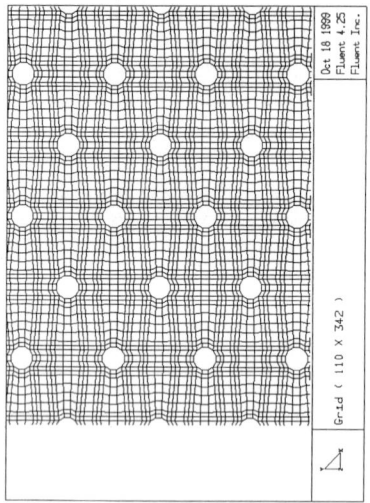

Fig. 3 Mesh pattern around some tubes of
the heat exchanger

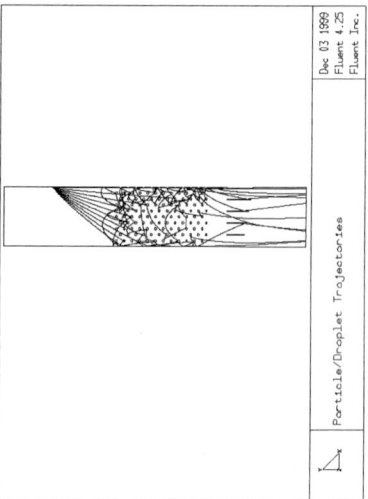

Fig. 5 Predicted trajectories of water
droplets in the cooling tower

Air Distribution in Rooms, (ROOMVENT 2000)
Editor: H.B. Awbi

RADIANT COOLING SYSTEM
USING LARGE EFFECTIVE TEMPERATURE DIFFERENCE
-EXPERIMENTAL SYSTEM AND TEST RESULTS-

Y. Nakano[1], W. Urabe[1], Y. Tsubota[2], T. Oka[3], M. Kaneko[3], Y. Hazama[4], Y. Hashimoto[5], and T. Imai[5]

[1]Komae Research Laboratory, Central Research Institute of Electric Power Industry (CRIEPI),
201-8511, Japan
[2]Power Engineering R&D Center, Tokyo Electric Power Company (TEPCO), 230-8510, Japan
[3]Department of Architecture, Utsunomiya University, 321-0912, Japan
[4]Technical Headquarter, TONETS Corporation, 104-8342, Japan
[5]Technical Research Institute, TONETS Corporation, 272-0142, Japan

ABSTRACT

The authors are investigating radiant cooling system integrated with ice storage system to cope with load leveling, energy saving, and thermal comfort. We are expecting not only the advantages of load shifting by ice storage and energy saving by using large effective temperature difference through ice storage, but also better thermal comfort by the combination of the two. The paper describes an experimental system to develop a better secondary air-conditioning system based on our concept. It consists of a room of 7m (W) × 10m (D) × 3m (H), a ductless ceiling plenum chamber along with radiant ceiling panel, and an air-handling unit to supply air of low temperature and low humidity on the assumption of using ice storage system. The air primarily supplied into the ceiling plenum chamber cools the metallic radiant panels installed on the ceiling. Then it is introduced into the room from the end of the ceiling on the perimeter side. The continuous measurement is carried out and the results show that the system produces a good thermal environment as we expected.

KEYWORDS

radiant cooling system, panel air system, large effective temperature difference, ductless air supply system, ice storage system, load leveling, energy saving, thermal comfort

INTRODUCTION

Load leveling and energy saving in HVAC systems, especially of commercial buildings, are remarkably important issues for Japan's electric power companies. On the other hand importance of thermal comfort in buildings is being well recognized these days. The authors are investigating radiant cooling system integrated with ice storage system to cope with these three requirements. We are expecting not only the advantages of load shifting by ice storage and energy saving by using large effective temperature difference through ice storage, but also better thermal comfort by the combination of the two. We have already made a comprehensive evaluation of the system from the perspectives of load leveling, energy saving, and economy, Nakano, Miyanaga, and Oka (2000).

The proposed system is illustrated in Figure 1. In a rooftop ice storage tank, the cooling medium is stored in the form of ice, which is produced using a heat pump during the night in summer. For daytime cooling, chilled water is supplied from the ice storage tank to the air conditioners on each floor. The air conditioners then use the chilled water to generate air of low temperature and low humidity, which is blown into the space between the ceiling slabs and metallic radiant panels installed on the ceiling. The air cools the panels by convection. The cooled panels remove heat by radiation from sources such as the occupants, floor and wall faces in the room. The cool air in the plenum above the suspended ceiling is introduced into the room from the end of the ceiling on the window side. It is used for heat removal in the perimeter zone and for ventilation.

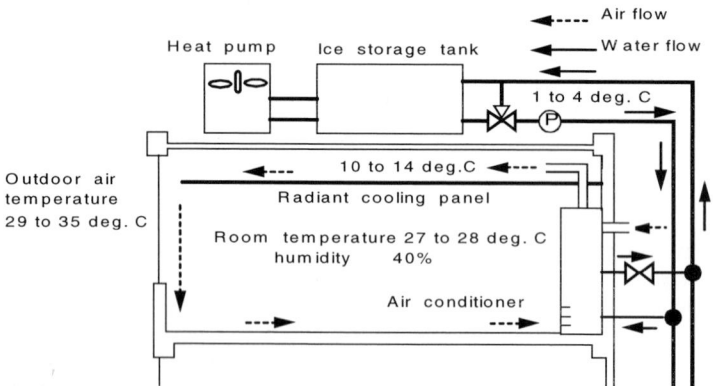

Figure 1: Concept of radiant cooling system integrated with ice storage system

In this system the temperature of the chilled water supplied to the air conditioners can be maintained at 1~4°C, which is lower than that in a conventional system (6~7°C). The temperature of the air at the outlet of the air conditioners is 10~14°C, which is also lower than that in a conventional system (16~17°C). Consequently, we can expect a reduction in the power required for the pumps and the fans. At the same time, the improved dehumidification capacity makes it possible to supply less humid air to a room. It is expected that the radiant cooling effect and the less humid air will bring about a similar thermal environment to that obtained using a conventional cooling system, even with a higher room temperature. We assume that the room temperature of 27~28°C and the relative humidity of 40% accompanied with the cooled radiant panels will produce a similar thermal environment to that of 26°C and 50% relative humidity, Matsuki, Nakano, Miyanaga, Yokoo, and Oka (1999).

EXPERIMENTAL SYSTEM

The outline of the experimental system based on the concept described above, which is to simulate only the secondary part of the system, is shown in Figure 2 and TABLE 1. It consists of a room of 7m (W) × 10m (D) ×3m (H), a ductless ceiling plenum chamber along with ceiling radiant panel, and an air-handling unit (AHU). The AHU supplies air of low temperature (10°C) and low humidity (6.8g/kg) on the supposition of using ice storage system. The metallic radiant panels are made of aluminum on which special painting is made to enhance emissivity. We have totally about 200 measuring points to acquire spatial temperature distribution of the room and the chamber, air velocity at the air outlets and in the room, the brine temperature and flow, the electricity consumption of the chiller and the AHU, and so on. The target of room temperature, relative humidity, and radiant panel surface temperature is 27°C, 40%, and 24°C respectively. They are decided based on the results of the subjective evaluation carried out by Matsuki et al. (1999), previously referred to.

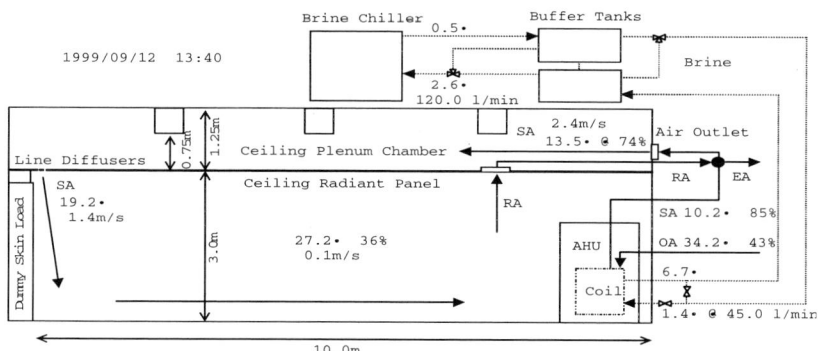

Figure 2: Experimental system

TABLE 1
OUTLINE OF EXPERIMENTAL SYSTEM

Room	7.0m(W) × 10.0m(D) × 3.0m(H)
Ceiling plenum chamber	7.0m(W) × 10.4m(D) × 1.25m(H) (0.75m(H) at beams)
Ceiling radiant panel	Aluminum coated with high emissivity paint Emissivity 0.93 (wavelength 8~13μm)
Line diffuser	(600mm(W) × 15mm(D) × 2deep) × 6
Chiller	Cooling capacity 7.5kW
AHU	Air volume 580~1,050m³/h (VAV) Air change rate 2.8~5.0 changes per hour
Thermal load	Total (average) 4.6kW (66W/m²) Lighting 1.6kW ((40W × 2deep) × 20) AHU 0.4kW (average) Dummy skin load 0 ~ 5.2kW (variable) (Electrical heaters)
Temperature settings	Supply brine temperature for AHU coil 2.0°C Supply air temperature from AHU 10.0°C

TEST RESULTS

Temporal Variation

Figures 3 and 4 illustrate typical 24-hour temporal variations in temperature and relative humidity at important measuring points. It shows that the room temperature and the room relative humidity are well and stably controlled. The average room temperature measured is 27.2°C with variance of 0.1°C for the setting of 27°C. That of the room relative humidity is 37% with variance of 1% for the target of 40%.

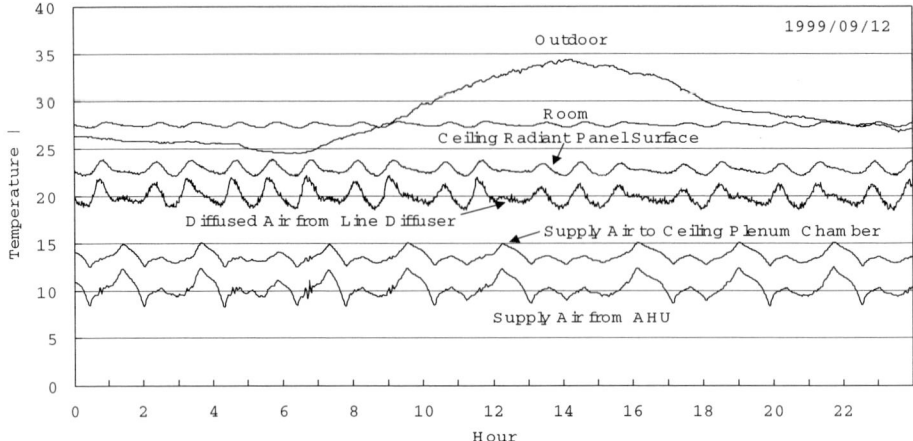

Figure 3: Temporal variation in temperature

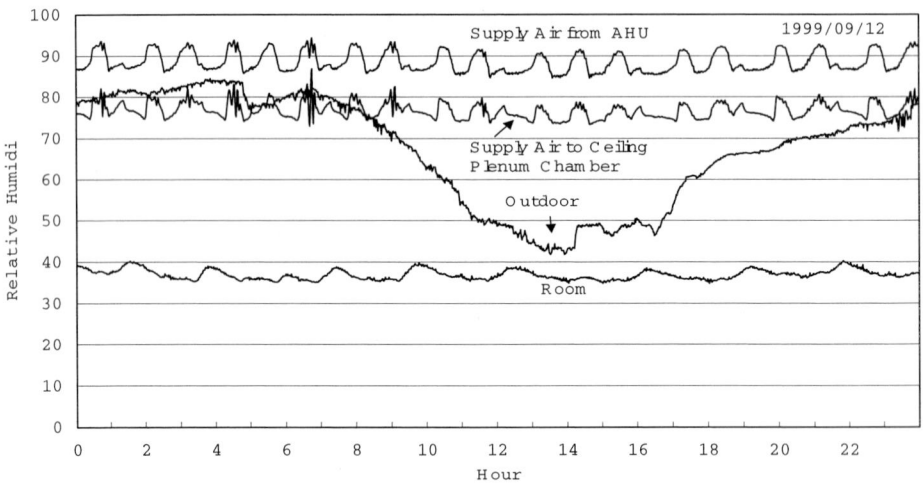

Figure 4: Temporal variation in relative humidity

Horizontal Temperature Distribution

Figures 5 and 6 show horizontal temperature distributions of the room and the ceiling plenum chamber respectively. They are for the moment when the outdoor temperature reaches the peak of the day.

Figure 5 includes the measured temperatures at the ceiling radiant panel surface ("PS" in the figure), 1.5m above the floor ("1.5H"), and the floor surface ("FS"). The average temperature at "PS" is 24.0°C with variance of 1.6°C for the target of 24°C. Those at "1.5H" level and "FS" are 27.2°C with variance of 0.1°C and 26.7°C with 0.5°C respectively. The temperature difference between "PS" and

"FS" is 2.7°C. The horizontal temperature difference between the maximum and the minimum at "PS" is 4.8°C. Those at "1.5H" level and "FS" are 0.9°C and 2.4°C respectively. The reason why the temperatures at the measuring points of No. 10, No. 11, and No. 12 are slightly higher than other points is that they are closer to the dummy skin load (electric heaters). The panel surface temperature at No. 2, which is the closest measuring point to the air outlet, is 2°C to 5°C lower than other points.

Figure 6 contains the data at the back of the radiant panel ("BP" in the figure) and 0.5m high from the panel ("0.5HB"). The average temperature of the former is 22.3°C with variance of 1.4°C. That of the latter is 18.6°C with 0.4°C.

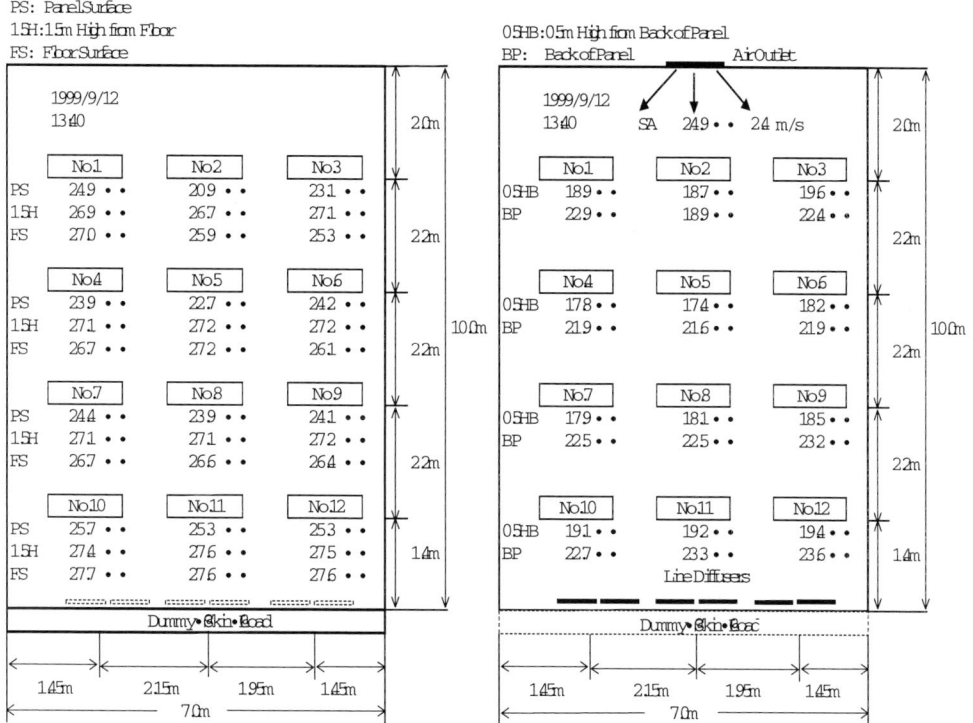

Figure 5: Horizontal temperature distribution in room

Figure 6: Horizontal temperature distribution in ceiling plenum chamber

Vertical Temperature Distribution

Figure 7 illustrates the vertical temperature distributions on the center cross section at the moment when the outdoor temperature reaches the peak of the day. The vertical temperature difference between 0.1m and 2.5m above the floor is below 1.5°C at every measuring point.

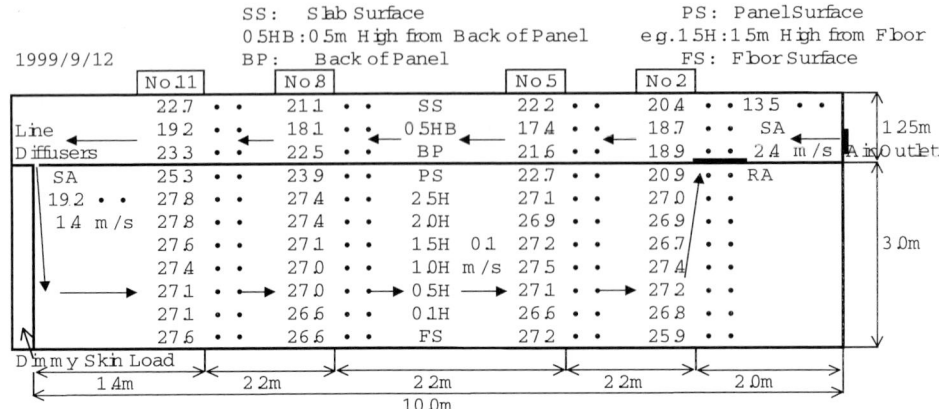

Figure 7: Vertical temperature distribution at center cross section

Air Flow

The supply air from the AHU is diffused to the ceiling plenum chamber from the air outlet at the velocity of below 3 m/s, as shown in figure 7. Then it is introduced to the room through six line diffusers at the other end of the ceiling. The air velocity at the line diffusers is below 1.5 m/s. That in the interior is below 0.2m/s.

CONCLUSION

We set up the experimental system which simulate the secondary part of the radiant cooling system integrated with ice storage based on our concept. The measurement results show that the system produces a thermal environment we expected. The data also demonstrate that the obtained thermal environment is good from the viewpoints of asymmetric thermal radiation, vertical air temperature difference, and draft, referring to ASHRAE Handbook Fundamentals (1993).

References

ASHRAE (1993)., *1993 ASHRAE Handbook Fundamentals SI Edition*. ASHRAE, USA
Matsuki N., Nakano Y., Miyanaga T., Yokoo N., and Oka T. (1999). Performance of Radiant Cooling System Integrated with Ice Storage. *Energy and Buildings* **30:2,** 177-183.
Nakano Y., Miyanaga T., and Oka T. (2000). Comprehensive Evaluation of Radiant Cooling System Integrated with Ice Storage System. *2000 ASHRAE Winter Meeting*.

Air Distribution in Rooms, (ROOMVENT 2000)
Editor: H.B. Awbi

INFLUENCE OF VENTILATION SYSTEM ON THE PERFORMANCE OF COOLING CEILINGS: APPLICATION TO CHILLED BEAMS

J. Lebrun[1], A.Ternoveanu[1] and Q.Wei[2]

[1]Laboratory of Thermodynamics, University of Liège, Belgium
Campus du Sart-Tilman, Bat. B49, Parking P33, 4000-Liège, Belgium
[2]Department of Thermal Engineering, Tsinghua University, 100084, Beijing, China

ABSTRACT

Cooling ceiling systems are controlling only the sensible heat balance of the rooms; they are always combined with a ventilation system foreseen to control indoor humidity and to cover air renewal requirements. Between the types of cooling ceiling in use, the passive chilled beams seem to be the most sensitive to ventilation air influence. In most of the cases, the ventilation outlets are located in the ceiling void, and consequently this generates a penalty on the beam cooling power. The work presented aims at estimating this influence, through results issued from experimental studies. The values obtained show that in usual cases, the beam cooling power can decrease down to 20-25% from its nominal value. Moreover, by using simple modeling assumptions, one should be able to make a preliminary evaluation of this effect. The results of this study are regarded as helpful for design of cooling systems.

KEYWORDS

Free convection, chilled beams, ventilation air, temperature difference, modeling, system performance.

INTRODUCTION

The passive chilled beam system is one of the most common cooling ceiling equipment in use. It is made of a set of horizontal finned water coils installed in the ceiling void of the room and connected in parallel. They are using mainly natural convection heat transfer; the radiation heat transfer is negligible, due to the location of the beam, above a false ceiling (figure 1). In comparison with radiant cooling ceilings, the difference in heat transfer is compensated thanks to the enhancement of external heat transfer area which can be up to 5 times bigger [ASHRAE 1992]. A very easy air circulation must be also maintained between the room and the ceiling void through the false ceiling.

Chilled beams are not reversible for heating regime like the radiant panel ceilings, because of the thermal gradient which nullifies the natural convection effect.

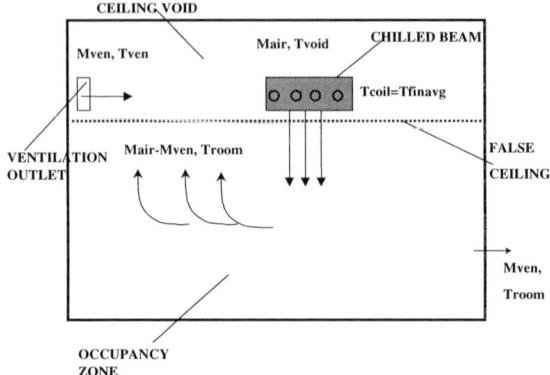

Figure 1. Schematic of a passive chilled beam system.

HEAT TRANSFER AND PERFORMANCE OF CHILLED BEAMS

The chilled beam system can be considered as a heat exchanger (the finned coil) supplied on one side by water and on other side by air aspirated in the ceiling void. In order to define an effective cooling capacity, the air temperature is not taken as reference, but replaced by the space average of globe temperatures measured in the occupancy zone. This temperature is the supply temperature of the beam on air side. A convenient approximation consists in considering a contact factor of 100%, i.e. in assuming that the exhaust air temperature is the same as the average temperature of the finned surface. This temperature can be determined in function of the fin effectiveness. Though, the so-defined cooling capacity is mostly affected by the temperature difference between water and room, by the fin efficiency and by the ventilation mode.

It is proposed here to neglect the water flow rate influence.

Consequently, for a given beam and room configuration (geometry and loads) working in given (nominal) temperature conditions, the performance will still depend on the ventilation mode.

The global heat transfer coefficient of the beam can be defined as :

$$AU = \frac{\dot{Q}_{cool}}{\Delta t_{LOG}} \qquad (1)$$

where $\dot{Q}_{cool}$ is the cooling capacity of the system

The logarithmic mean temperature difference is defined as :

$$\Delta t_{LOG} = \frac{\left(t_{room} - t_{wex}\right) - \left(t_{finavg} - t_{wsu}\right)}{\ln\left(\frac{\left(t_{room} - t_{wex}\right)}{\left(t_{finavg} - t_{wsu}\right)}\right)} \qquad (2)$$

Where :

t_{wsu} and t_{wex} - are the water supply and exhaust temperatures.

t_{finavg} - is the average temperature of the fins.

For any combination of the tested conditions the cooling capacity can be calculated using the heat transfer effectiveness :

$$\dot{Q}_{cool} = \varepsilon \dot{M}_{air} c_{pa} \left(t_{room} - t_{wsu}\right) \qquad (3)$$

where :
$$\varepsilon = \frac{\left(t_{room} - t_{finavg}\right)}{\left(t_{room} - t_{wsu}\right)} \qquad (4)$$

The fin effectiveness can be defined as :

$$\Phi = \frac{t_{room} - t_{finavg}}{t_{room} - t_{fino}} \qquad (5)$$

with t_{fino} is the temperature at fins root (pipe contact temperature) and t_{finavg} is the average temperature of fin surface. According to heat transfer theory [Kreith F.1976] it is depending on fin geometry, material (conductivity), and on global heat transfer coefficient between fin and room air .

$$\Phi = \frac{tanh\left(L_{fin}\sqrt{\frac{2h_{air}}{k_{fin}\, b_{fin}}}\right)}{L_{fin}\sqrt{\frac{2h_{air}}{k_{fin}\, b_{fin}}}} \qquad (6)$$

L_{fin} , b_{fin} are fin length and width and h_{air} is the heat transfer coefficient between fin and air. The global heat transfer coefficient of the beam can be defined by associating, three thermal resistances in series:

$$AU = \frac{1}{R_w + R_m + R_a} \qquad (6a)$$

These resistances correspond to the convection heat transfers on water and airsides and to metal conduction respectively (the last one includes the effect of fin effectiveness).

The resistance on the airside is the most important term in the definition of the heat transfer, as demonstrated in previous work [M.Bravo et al. 1995]. Consequently, the following empirical equation can be used to describe the relationship between cooling power and air-water mean temperature difference:

$$\dot{Q}_{cool} = K\Delta t_{LOG}^{1+n} \qquad (7)$$

The two constants K and n of this equation are related to coil geometry and to heat transfer mode respectively.

For horizontal finned tubes the free convection heat transfer coefficient is defined as [Giblin R. 1974]:

$$h = h_{conv} = K_2\Delta t_{LOG}^{n_2} \qquad (8)$$

In pure free convection mode, the exponent "n" might float somewhere between 0.25 and 0.5. Except for metal and water side resistances, he main factor which can "damp" the exponent is the effect of a lower temperature in the ceiling void than in the room. This can be due to the ventilation air blown in the void. The results obtained for cooling capacity confirm these assumptions as "n" value found are in the range 0.15 to 0.3 [A.Ternoveanu et al 1999].

EFFECTIVE INFLUENCE OF THE VENTILATION AIR FLOW

As chilled beam systems are mounted above perforate plate ceilings, the ventilation supply openings are often located in the ceiling void. One must take into account that when using cooling ceilings, the ventilation contribution is limited to cover only air renewal requirements, which corresponds to a small flow rate. However, this small flow rate can modify the temperature in the proximity of the beam and modify its performance, by cooling the ceiling void.

Experimental results on chilled beams exhibit a decrease of the cooling power related to the mentioned cooling effect.

For a chilled beam the cooling power is given by eqn. (7) above. We assume the ventilation flow rate is essentially acting on the effective temperature difference between the two fluids, and not on the shape of the heat transfer law, as the air flow is too small to generate induction effect on the beam.

In that case, the real effective cooling capacity is :

$$\dot{Q}^*_{cool} = K \, \Delta t^*_{LOG}{}^{1+n} \tag{9}$$

The mean temperature difference is defined this time by using the ceiling void temperature :

$$\Delta t^*_{LOG} = \frac{\left(t_{void} - t_{wex}\right) - \left(t_{finavg} - t_{wsu}\right)}{\ln\left(\frac{\left(t_{void} - t_{wex}\right)}{\left(t_{finavg} - t_{wsu}\right)}\right)} \tag{10}$$

In order to determine the average ceiling void temperature one should make the following assumptions :
-   Perfect mixing between ventilation air and room air aspirated in the void ;
-   Negligible heat losses between ceiling void and adjacent rooms.

Thus the air flow through the chilled beam is given by eqns. (3) and (4) :

$$\dot{M}_{air} = \frac{\dot{Q}^*_{cool}}{c_{pa}\left(t_{void} - t_{finavg}\right)} \tag{11}$$

According to the above assumptions, the temperature of the void can be determined as (Figure 1) :

$$t_{void} = \frac{\dot{M}_{ven}\, t_{ven} + \left(\dot{M}_{air} - \dot{M}_{ven}\right)t_{room}}{\dot{M}_{air}} \tag{12}$$

The uncertainty of this evaluation is acceptable, as it appears from figure 2 where are plotted measured cooling capacities on water side against calculated cooling capacities from eqn. (9).

The results above are gathered from tests performed on three different chilled beams systems.

In order to extrapolate the ventilation influence for all types of chilled beams one should define a set of non-dimensional terms:

- The ratio between the cooling capacity of the ventilation air and ceiling system :

$$\omega = \frac{\dot{Q}_{ven}}{\dot{Q}^*_{cool}} \tag{13}$$

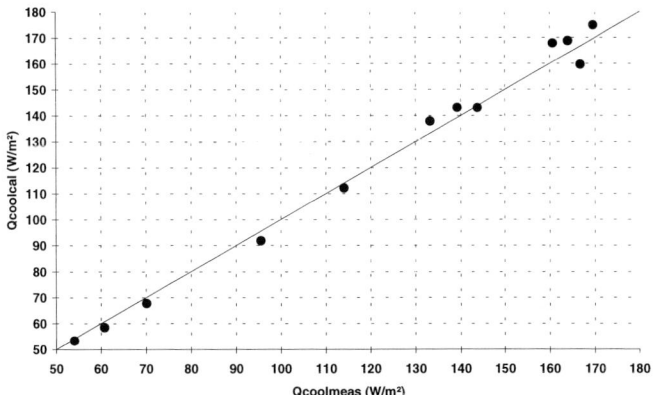

Figure 2. Capacities determined using ceiling void temperature against measured values.

- The ratios between air-water mean temperature difference, capacity, AU and effectiveness, with and without ventilation:

$$\delta = \frac{\Delta t_{LOG}^{*}}{\Delta t_{LOG}} \quad ; \quad \zeta = \frac{\dot{Q}_{cool}^{*}}{\dot{Q}_{cool}} \quad ; \quad \alpha = \frac{AU^{*}}{AU} \quad ; \quad \beta = \frac{\varepsilon^{*}}{\varepsilon} \qquad (14\text{-}17)$$

The diagrams of Figure 3 show how the factors of temperature difference (δ) and of cooling capacity (ζ) may vary against the ventilation cooling factor ω. It appears that the penalty on the cooling capacity is almost the same as the ventilation cooling capacity itself. This means that it is useless to blow the air at a temperatures lower than room average temperature. The ventilation flow rate is only depending on the number of occupants.

In the case of a small office room, the penalty may reach 40 % of the nominal capacity.

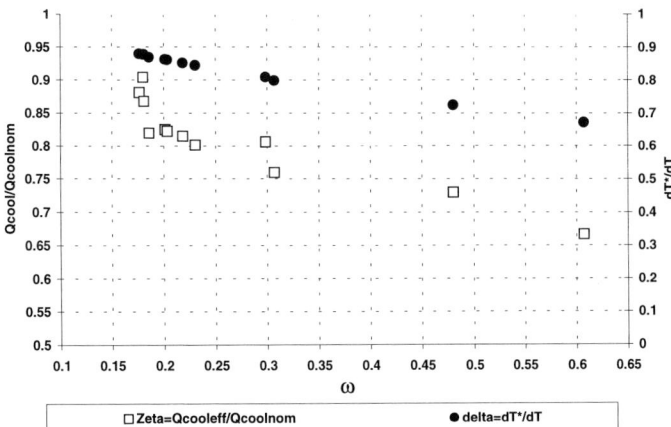

Figure 3. Mean temperature difference and cooling capacity ratios against ventilation energy flow ratio.

600

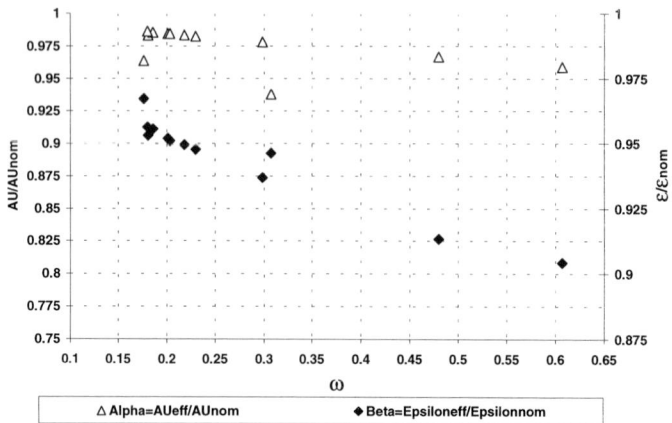

Figure 4. AU and effectiveness ratios against ventilation energy flow ratio.

The figure 4 shows the slope of AU (α) and effectiveness (β) ratios against ω. The decrease is much smaller than for cooling power as the global heat transfer coefficient is less affected by the temperature difference variation. The dispersion of certain values for α is due to the exponent "n" in the equation (7) and (9) which varies slightly for the different types of chilled beam system studied.

CONCLUSIONS

Introducing ventilation air in the ceiling void provides a penalty on the chilled beams performance, as the system is working at lower temperature difference. Consequently, for the same room and water supply average temperatures the chilled beam will provide less cooling power. The penalty is proportional to the ventilation energy flow rate, and may reach up to 40% of the total cooling capacity of the beam.

This influence of ventilation air should be taken into account during the design phase when choosing the optimal compromise between the cooling capacity needed and ventilation parameters as: outlets location, air flow rate and temperature.

Two alternative solutions should be considered:

1) To provide the air directly to the occupancy zone, but with some risk of draught;

2) To use adiabatic air drying and supply the air directly inside the ceiling void without any preliminary cooling (this introduces a supplement of sensible load in the ceiling void, which is easily compensated by the chilled beam.

REFERENCES

ASHRAE (1992). Systems and Equipment Handbook.
Kreith F. (1976). Principles of Heat Transfer. McGraw-Hill.
M. Bravo, J.Lebrun, A.Ternoveanu (1995). Synthesis Study on Cooling Ceiling Systems for Office Rooms. 3rd Tsinghua HVAC Conference, Beijing, China.
Giblin R. (1974). Transmission de la chaleur par convection naturelle. Collection A.N.R.T. de la SFT.
A.Ternoveanu, Q.Wei (1999). Preliminary analysis on a research project for cooling ceilings – Synthesis of Available Information. University of Liège, Faculty of Applied Sciences -Report for "CLOSE"

© 2000 Elsevier Science Ltd. All rights reserved.
Air Distribution in Rooms, (ROOMVENT 2000)
Editor: H.B. Awbi

MODELLING AND INVESTIGATION OF THE INTERACTIONS BETWEEN THE ENVIRONMENT AND OPEN REFRIGERATED DISPLAY CABINETS IN RETAIL FOOD STORES

W. Xiang and S. A. Tassou

Department of Mechanical Engineering
Brunel University
Uxbridge, Middlesex, UB8 3PH, UK

ABSTRACT

The cold air spillage from vertical open multi-deck display cabinets into the aisles of retail food stores, causes discomfort to the shoppers and staff and leads to energy wastage. Over the years, a number of approaches have been tried to overcome this problem with varying degrees of success. This paper reports on results of investigations into the 'cold aisle effect' using Computational Fluid Dynamics (CFD). The results show that under-floor heating, which is a popular method of heating large modern retail food stores, is not effective in reducing the 'cold aisle effect'. The paper proposes an alternative method of aisle air management and heating which can lead to a considerable reduction or elimination of the 'cold aisle effect' and a reduction in the overall energy consumption of the store.

KEYWORDS

CFD modelling, retail food stores, display cabinets, HVAC systems.

INTRODUCTION

The open refrigerated display cabinet is a very popular merchandising fixture in retail food stores because it allows the customer unrestricted access to the displayed products. A standard design feature of open display cabinets is the air curtain, which is designed to provide a barrier between the chilled air in the cabinet and ambient store air. When a refrigerated display cabinet operates in a store environment it exchanges heat and moisture with the store air. Although the air curtain may reduce the heat and moisture exchange between the air in the cabinet and the ambient air in the sore, it cannot completely eliminate these exchanges. Warm air from the store is entrained into the cold air of the air curtain and, as a result, cold air spills out of the cabinet into the aisles of the retail store, producing the 'cold aisle effect'.

The low temperature air in the aisles between rows of vertical open refrigerated cabinets reduces the thermal comfort of the shoppers and staff retail food store chains are keen to develop solutions to the problem without increasing significantly the overall energy consumption of the store. Over the years a number of approaches have been adopted and reported in the literature with varying degree of success. One of these methods is to extract air through the base of the cabinets through refrigerant pipe trenches or ducts or use fans behind the cabinets to draw the air from the floor and discharge it at high level through the back of the cabinet (ASHRAE, 1999). Another approach which aims to reduce the impact of the cold aisle effect on the shoppers is to increase the temperature of the air in the aisles at low level through the use of under-floor heating.

The design of refrigerated display fixtures has largely been based on experience and trial and error. With the advent of Computational Fluid Dynamics (CFD) and the reduction in the cost of owning and operating CFD software packages over the last few years, however, a number of researchers and commercial organisations have applied the technique to the design optimisation of refrigerated display cabinets. W S Atkins (1992) built a steady state 2-D model of an open refrigerated display cabinet in order to investigate the effect of the air curtain on the temperature and energy performance of the cabinet. Stribling *et. al.* (1996, 1997) presented a 2-D steady-state model of a vertical dairy display cabinet that could be used for the design optimisation of such equipment. Researchers and building services designers have also used CFD to study air flow patterns in retail food stores and the effect of air flow control devices on cabinet performance and the `cold aisle effect' Krueger (1996). A difficulty with the above analyses is that those which dealt with optimisation of cabinet design did not consider in detail the effect of the cabinet and cold air spillage on the environment whereas those that considered the environment in retail food stores did not consider the effect of the environment on cabinet performance and product temperature.

Xiang *et. al.* (1998) developed a 3-D dynamic model of display cabinets to investigate the effect of time varying conditions on the performance of the cabinet and product temperature. The product in the cabinet was modelled as a conducting wall and the model was validated against monitored data obtained from controlled tests in an environmental chamber. This model has now been extended to cover the interactions between the HVAC system and store environment with the refrigerated cabinets and has been used to investigate the effect of different local environment control methods on the cabinet performance and `cold aisle effect'.

MODELLING VERTICAL DISPLAY CABINETS IN RETAIL FOOD STORES

Display cabinets in retail food stores interact on a continuous basis with the store environment. To model the cabinet in the store, ideally the whole store together with display cabinets and the product in the cabinet should be modelled. If the product in the display cabinet is not considered in detail, then fewer grids would be needed and it would be possible to model the whole store with modern CFD software. However, if the product is to be modelled in detail, the grid for the display cabinet should be much finer and a much bigger grid size would be required to model the whole store. This is still not practical using modern workstations and simplifications are required to satisfy the requirements of both the cabinet and store environment modelling. In the analysis presented in this paper, in order to simplify the problem and reduce the total number of grids required, it has been assumed that the display cabinets will exchange heat and moisture only with their local environment and that the store air outside the calculation domain will be quite still. The interfaces between the display cabinet zone and other areas of the store can then be modelled as pressure boundaries which allow the effects of air movement in the zone caused by the operation if the display cabinets and the HVAC system in the zone to be taken into consideration.

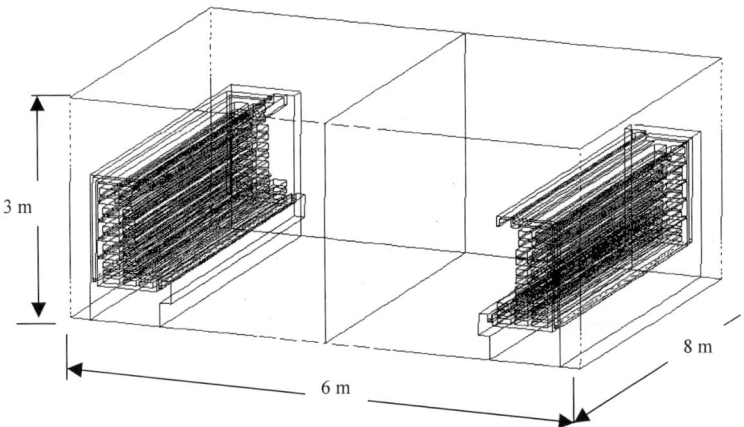

Figure 1: Two rows of display cabinets in the retail food store to be modelled

In a large food retail store the display cabinets will normally be arranged in rows as shown in Figure 1. Since the space to be modelled is symmetrical in two directions, only a quarter of the system needs to be modelled. The grid size adopted in the analysis was 118000 and at the interface between the display cabinet and the store environment the grids were assumed to have the same temperature and moisture content as the store environment.

THE COLD AISLE EFFECT IN RETAIL FOOD STORES

To illustrate the problem of the 'cold aisle effect' in retail food stores, the display cabinet system shown in Figure 1 was modelled assuming no heating in the aisle area. In the simulations, the bulk air temperature in the vicinity of the cabinets as well as the rest of the store was assumed to be 23 °C. The initial air curtain temperature was assumed to be – 4 °C and air curtain flow rate 0.1 m^3/s per metre length of display fixture. Figure 2 shows the temperature map obtained.

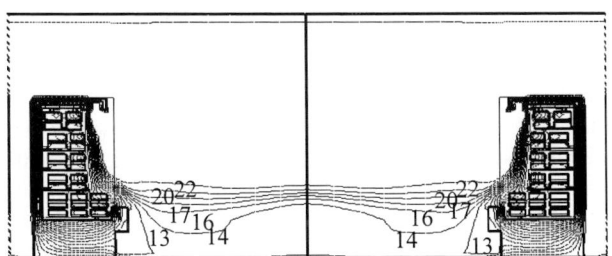

Figure 2: 2-D temperature map without space heating in the aisle

It can be seen that close to the floor the temperature is quite low, around 13 °C rising to 22 approximately 1.2 m above floor level, giving a vertical temperature difference of 9 °C which will cause cold feet discomfort, Olesen *et. al.* (1979).

INFLUENCE OF UNDERFLOOR HEATING ON THE COLD AISLE EFFECT

As mentioned earlier, a large number of modern retail food stores employ underfloor heating. Underfloor heating increases the floor surface temperature which should lead to improved comfort and reduction of the cold aisle effect. Where the total store floor area is heated by underfloor heating the practice is to design for a higher heat flux in the chilled food area to account for cold air spillage and a lower heat flux in the 'dry' goods area. Results from modelling the effect of underfloor heating in the aisle between the cabinets are shown in Figure 3. These are based on a floor heat flux of 150 W/m^2.

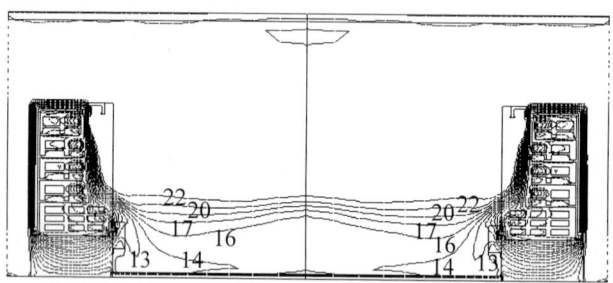

Figure 3: Temperature distribution when the aisle is heated by an underfloor heating system

It can be seen that when the aisle is heated by the underfloor heating system the temperature of the air at floor level in the centre of the aisle increases by one or two degrees and the overall vertical temperature difference decreases slightly. Close to the cabinets, however, the air temperature remains almost unchanged at around 13 °C. Customers and staff will still experience the cold aisle effect when they approach the cabinets to view or remove products from the shelves. An important finding as far as the products in the cabinets are concerned is that the heating system will cause a temperature rise on the surface of the products facing the aisle of around 2.2 °C and a rise at the core of the product of approximately 0.8 °C. This is due to heat transfer by radiation between the warm floor and the chilled products in the cabinet. Another influence of the floor heating system is to increase slightly the total cooling load of the cabinet. Increasing the heat flux on the floor would reduce cold discomfort but would have a detrimental effect on the chilled products in the cabinets.

AIR SUCTION AT THE BASE OF THE CABINETS

The use of air suction at the base of the cabinet to reduce the cold aisle effect is being recommended by ASHRAE (1999). In the UK, although the technique has been applied in a number of stores (Taylor ,1994) it has not gained universal acceptance due to the unavailabilty of design guidance and well documented design examples. In this paper CFD modelling has been used to investigate the effect of air suction at the base of the cabinet. The results for a suction volume flow rate of 0.05 m^3/s per m length of cabinet are shown in Figure 4.

It can be seen that with air suction at the base of the cabinet the temperature distribution in the aisle is improved considerably. The vertical temperature difference is reduced down to 5 °C which although still exceeds the recommended value of 3 °C (Olesen et. al., 1979), it is much better than for the floor heating system considered above. It has also been determined that base air extraction has no effect on the product temperature and cooling load of the cabinet.

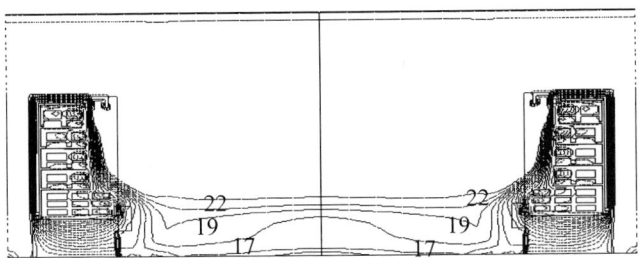

Figure 4: Temperature distribution with air suction at the base of the display cabinets

DISPLAY CABINETS WITH AIR SUCTION AT THE BASE AND SUPPLY AT HIGH LEVEL IN THE AISLES

A problem with the solution described above is how to utilise the cold air withdrawn from the aisles through the base of the cabinets. In the summer, when the dry goods area may require cooling the cold air can be ducted and distributed to the dry goods area. This is the approach adopted at the new Saindsbury's millenium store in Greenwich, Bunn (1999). In winter, the easiest option would be to use mixing fans to mix the cold air with the store air at the top of the cabinets. This would reduce the space temperature in the chilled food area and heating would need to be provided to maintain the design temperature. Requirements for the heating system would be to maintain the required bulk space temperature in the chilled food area with minimum impact on the air curtains of the display cabinets and product temperature and minimum vertical temperature difference in the aisles.

A number of alternative solutions have been investigated using CFD in terms of fulfilling the above criteria. The results indicated that the most promising solution would be the supply of air at low velocity and at the same temperature as the bulk space temperature from linear slot diffusers mounted on the ceiling in the aisles. In such a strategy, the air withdrawn from the base of the cabinets would be heated to the bulk space temperature by electric or hot water heaters and supplied at high level and at low velocity to the aisle from two linear diffusers running along the aisle. The resulting velocity vectors and temperature distribution for a supply flow rate of 0.05 m^3/s per metre length are shown in

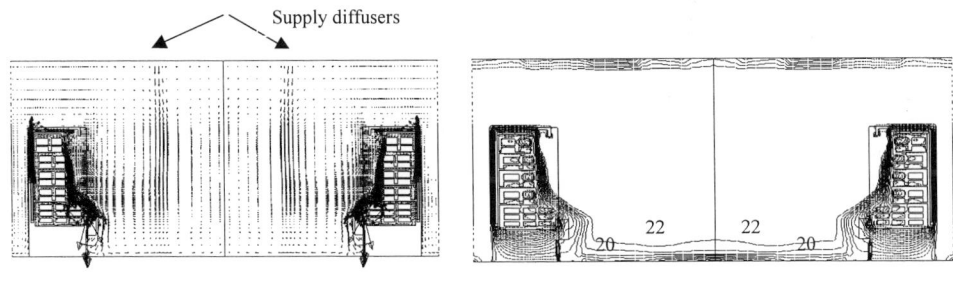

a) Velocity vectors   b) Temperature distribution

Figure 5: Velocity and temperature distribution when bottom suction and top supply is used

From Figure 5(a) it can be seen that most of the cold air overspill is drawn out of the aisle and the air supplied from the diffusers pushes the cooler air in the aisle close to the floor, making the temperature in the aisle much more uniform and reducing the vertical temperature difference to less than 3 °C, Figure 5(b).

CONCLUSIONS

Table 1 compares the performance of the alternative solutions for air management in the aisles. The findings can be summarised as follows:

TABLE 1
COMPARISON OF PRODUCT TEMPERATURES AND COOLING LOADS

System	Max. Product Surface temperature (°C)	Average product temperature (°C)	Sensible cooling load (kW/m)	Vertical temperature difference °C
No heating	9.19	5	1.004	9
Floor heating	11.4	7	1.045	8
Bottom suction	9.37	5	1.001	5
Bottom suction and top supply	8.8	4.7	0.96	2

- It can be seen that if the aisles are not heated, the air temperature in the aisles will be too low and uncomfortable for the customers and staff even though the air in the rest of the store is maintained at 23 °C.
- Underfloor heating which is commonly practiced in superstores in the UK is not very effective in reducing the cold aisle effect
- Suction of cold overspill air from the base of the cabinet can reduce the cold aisle effect but cannot completely eliminate it.
- Reheating and supplying the withdrawn air to the aisles from high level at low velocity and at a temperature approximately equal to the bulk air temperature in the store can improve thermal comfort and reduce energy consumption.

REFERENCES

ASHRAE (1999). *Handbook of HVAC Applications*, Section 2.4, *American Society of Heating Refrigerating and Air Conditioning Engineers*, Atlanta, USA

Bunn, R. (1999). Going to ground, *Building Services Journal*, October 1999, 20-24.

Krueger (1996), An Investigation of air outlet throw and spread in supermarket cold aisles, *Application note,* Form GA0001, Krueger USA.

Olesen, B. W., Scholer, M and Fanger, P. O. (1979). Vertical air temperature differences and comfort. *Indoor climate*, edited by Fanger and Valbjorn, Danish Building Research Institute, Copenhagen, 561-579.

Stribling, D., Tassou, S. A. and Marriott, D. (1996). Optimisation of the Design of refrigerated Display Cases Using Computational Fluid Dynamics. *Proceedings of the Institute of Refrigeration*, **92**, 7.1-7.7.

Stribling, D., Tassou S. A. and Marriott D. (1997). A two–dimensional CFD model of a refrigerated display case, *ASHRAE Transactions,* V.103, Part 1, pp.88-94.

Taylor, G. (1994). An evaluation of methods of controlling the environment adjacent to multi-deck refrigeration display cabinets, *Proceedings CIBSE National Conference*, 2-4 October 1994, Brighton, UK, Vol. II, 216-226.

W. S. Atkins Engineering Sciences (1992). Chilled Multi-deck Cabinet Airflow Study, *Report for Safeway Stores PLC*, UK, 31 p.

Xiang, W., Tassou, S. A. (1998). A dynamic Model for Vertical Multideck Refrigerated Display Cabinet. To be published in *Proceedings of International Conference on Advances in the Refrigeration Systems, Food Technologies and Cold Chain*, 23-26 September 1998, Sofia, Bulgaria

© 2000 Elsevier Science Ltd. All rights reserved.
Air Distribution in Rooms, (ROOMVENT 2000)
Editor: H.B. Awbi

FLOW PATTERN IN VENTILATED ROOMS WITH LARGE DEPTH AND WIDTH

Zou Yue[1], Peter V. Nielsen[2]

[1]Department of Building Science Engineering, Royal Institute of Technology (KTH), Sweden
[2]Department of Building Technology and Structural Engineering, Aalborg University, Denmark

ABSTRACT

In many buildings, for instance tunnels, underground, parking areas and industrial halls, the *L/H* is so large that the flow pattern induced by a two dimensional supply air jet along the ceiling can be completely different from that in rooms of normal sizes. Earlier model experiments indicate that, in this case, the supply jet will have a limited penetration length (l_{re}) because the entrainment generates a backward flow in the lower part of the ventilated space which at a given distance will disperse or deflect the jet.

In this study both model experiments and Computational Fluid Dynamics (CFD) are employed to study the isothermal flow pattern in a ventilated room with different *L/H* and inlet velocities. The maximum size of the model is 1.4* 0.72*0.0714*m* and the measurement is made with a Laser Doppler anemometer.

The CFD simulation is carried out by Flovent code with a *k-ε* model. Although some discrepancies occur it is clear that the simulation correctly represents the general features of the measurements. Both measurements and CFD simulations show the tendency that l_{re}/H may be independent of *W/H* when W/H >5. The velocity decay of the wall jet and the maximum velocity in occupied area are also studied in this paper.

KEYWORD

Deep ventilated room, penetration length, CFD, model, wall jet, maximum velocity

INTRODUCTION

In mixing ventilation, wall jets are extensively used for supply of ventilation or conditioned air to rooms and spaces. Before the air enters the occupied area of the room velocities and temperature differences must have decreased to an acceptable level. Considering that the test of supply devices is always done in rooms different from the room where the supply device finally is installed it is of practical interest to study the characteristics of flow in rooms of different sizes.

For rooms with small characteristic dimensions (*W/H* < 2, *L/H* < 3), where *W*, *L* and *H* are width, length and height of the ventilated space, the wall jet undergoes a number of deflections at the corners it meets during its course from the supply to the floor. When the jet is approaching an opposing side (room corner), an adverse pressure gradient is built up and the jet "restarts" again, see Sandberg (1998). When the ratio *L/H* is larger than a certain value, entrainment implies that the air must be led back along the floor and this will disperse the jet. The distance from the wall with the inlet to the stagnation point where the flow diverges is called penetration length l_{re}, see figure 1. The penetration length is a significant parameter for proper room air distribution design. At a distance from the supply opening which is larger than l_{re} the velocity is very low since the supply air is distributed over the whole cross area, while the velocities are very high at distances smaller than l_{re} because large volume of air is set into motion by the entrainment below the wall jet. For a normal room the ventilated section should always be smaller than the calculated penetration length so a rotary airflow pattern can be established. However, some applications, such as mining tunnels, require a penetration length as long as possible. Therefore, there is a need for a procedure which allows the penetration length and the velocity distribution to be determined as a function of the room geometry and the inlet condition. The results presented here demonstrate that this can be achieved by the CFD simulation.

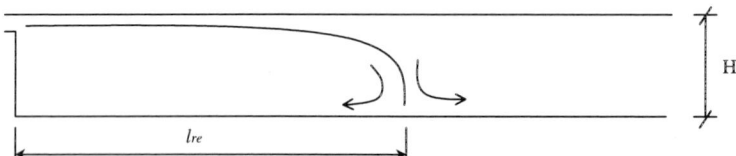

Figure 1: Penetration length of long room.

Experimental investigations of the penetration length can be found in e. g. Urbach (1971), Katz (1974), Förthmann (1934), Nielsen (1976) and Nielsen et al. (1987). With the aid of smoke visualization Urbach (1971) makes tests in a model with $W/H = 1$ and found value of l_{re}/H about 3 for h/H between 0.02 and 0.1 and Reynolds number between 3500 and 12000. Katz's (1974) experiments were also carried out with small W/H but in an open water channel. His tests showed that the penetration length is somewhat dependent on the location of the end wall and l_{re}/H was found between 3 and 4.5. In Förthmann (1934) the velocity profiles in a deep model with h/H = 0.17 and W/H = 3.6 were measured. Based on these data l_{re}/H was calculated as 5.3. Nielsen et al. (1987) investigated air distribution in rooms with ceiling-mounted obstacles and found that penetration length can be strongly influenced in some cases. Detailed tests of penetration length with different model geometric parameter can be found in Nielsen (1976). Some of them will later be discussed.

EXPERIMENT SETUP AND SIMULATION

The measurements were performed in a model as shown in figure 2 which also indicates the coordinate system used. This model was made from perspex and a full-width slot was placed just below the ceiling. Supply air was led into the model by a ventilator located downstream of the exhaust air terminal device to make sure the tests were held under isothermal conditions. The supply outlet was also preceded by a smooth curved area contraction to produce a uniform velocity profile and reduce the turbulence level. To determine the influence of width and length on the velocity characteristics two different model sizes were applied in the tests. Geometric parameters of these two models are summarized in table 1. For all tests the supply air velocity is 15.1 m/s, corresponding to a Reynolds number about 4000. The measurements of mean velocities and the corresponding turbulence intensities were carried out by a Laser Doppler anemometer. The minimum measurement time for each point is 30 seconds.

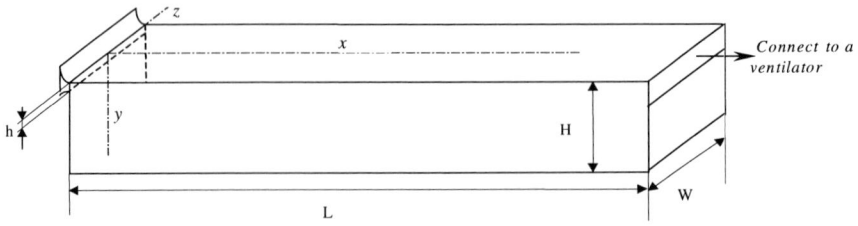

Figure 2: Layout of model

TABLE 1
MODEL DIMENSIONS

	H	h	h/H	L/H	W/H
Model 1	7.14 cm	4mm	0.056	20	10
Model 2	7.14cm	4mm	0.056	15	5

The CFD simulations were carried out under steady state conditions. The flow field was divided into a collection of small rectangular cells and the flow at each cell is determined by numerically solving the governing conservation equations of fluid dynamics. The grid density was 80*50*80 grids along the length (x), height (y), width (z), respectively. The grids have smaller spacing at locations where more flow detail are needed (the supply and the jet flow area). Turbulence is modelled by a standard k-• model. For all cases, in the absence of turbulence data, the turbulence intensity at inlet is assumed to be 4% and the turbulence kinetic energy k and its dissipation rate ε at the inlet are determined by

$$k = \frac{3}{2}(IU_o)^2 \tag{1}$$

$$\varepsilon = \frac{0.09k^{3/2}}{h} \tag{2}$$

where

I = Relative turbulence intensity (-)
U_O = Outlet velocity evaluated from the estimated mass flow (m/s)

RESULTS AND DISCUSSION

Flow pattern and comparison

Figure 3 presents profiles of the longitudinal velocity component at three x and z positions measured in model 2. As it can be seen, the flow pattern is two-dimensional at $x/H=2.8$. However, the flow is not always symmetric around the mid plane (within measurement precision) at other positions. This fact indicates that care must be taken when making tests which only represent a part of a room.

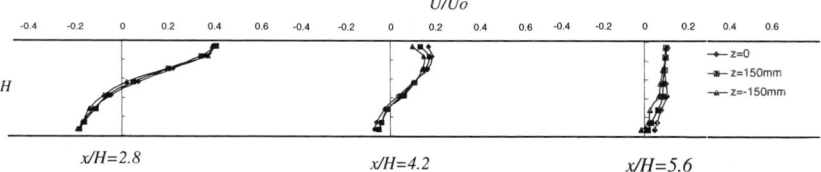

Figure 3: Measured mean velocity profile at different x and z positions

In a model with similar characteristic size ($W/H=4.7$) to model two, Nielsen (1976) observed that an instantaneous flow occurred in the z direction for flow between $x/H =4.2$ and 5. In our test, such evidence is not found. However, the RMS (*Root Mean Square*) value of the longitudinal velocity component at $y=4mm$ of model 2 had a peak value in this area, see figure 4.

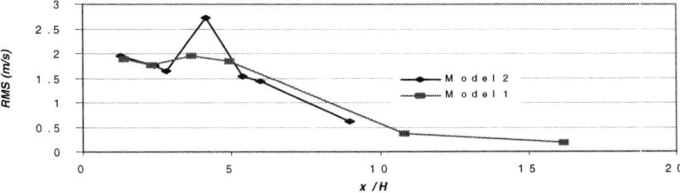

Figure 4: RMS value distribution along then centreline at 4mm from the top for model 1 and 2

Three-dimensional CFD simulations show that the flow patterns in both models are close to be two-dimensional. No evidence of non-symmetry around the mid plane can be found. In figure 5 comparisons of both velocity profile and velocity distribution at ceiling level and floor level along the centre line in the model 1 are presented. Although some discrepancies occur it is clear that the simulation correctly represents the general features of the measurements.

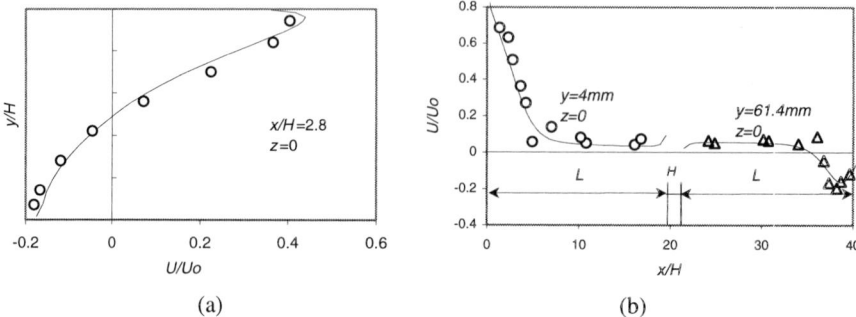

(a)   (b)

Figure 5: Comparisons in velocity profile (a) and velocity distribution (b) along the centreline in the model 1

Penetration length

Dimensional analysis indicates that the flow in a ventilated room can be described by the geometrical parameters, the Reynolds number Re and the boundary conditions, see Nielsen (1976). Therefore the penetration length can be expressed as:

$$\frac{l_{re}}{H} = f(\frac{W}{H}, \frac{L}{H}, \frac{h}{H}, Re) \qquad (3)$$

where three variables on the right hand side represent the geometry normalized by the height of the room. The Reynolds number is given by

$$Re = \frac{U_o h}{v} \qquad (4)$$

where v is the kinematic viscosity

For a fully developed turbulent flow the normalized velocity distribution in ventilated rooms is independent of the Reynolds number, see Nielsen (1976) and Sandberg et al (1998). Figure 6 shows that the normalized return flow in the model expressed by the ratio U/U_o in model 1 at $y = 6.64$cm and $x = 20$cm is almost independent of the Reynolds number for $Re \geq 4000$.

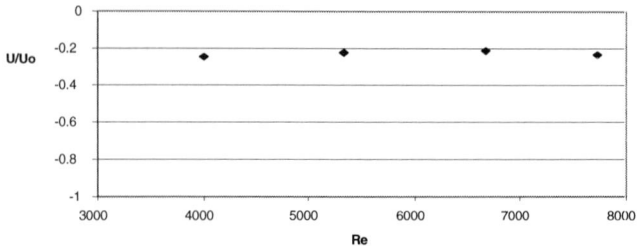

Figure 6: Normalized velocities in model 1 at 5mm from bottom and 20cm from supply outlet for different outlet Reynolds numbers

According to measurement data at 0.5 cm from bottom the penetration lengths in both model 1 and model 2 are estimated to be between 4.5 H and 5 H while CFD calculations give 5.9 H and 5.6 H for model 1 and 2, respectively. In figure 7 available penetration length data is presented for different W/H and a best-fitted line for penetration length for all measurements with $h/H = 0.056$ is also given. This line shows a tendency that l_{re}/H may be independent of W/H when W/H >5.

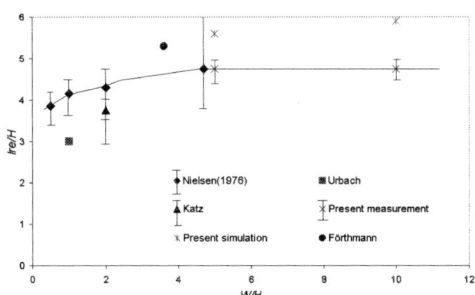

Figure 7: Penetration length versus W/H

Velocity Distribution

In figure 8 the normalized maximum measurement velocities U_m/U_o of model 2 are plotted log-log as a function of the normalized distance from the inlet x/h. The velocity decay has a slope of –0,5 in the first part. Obviously this part belongs to the zone of fully established turbulent flow where the maximum velocities can be expressed as:

$$\frac{U_m}{U_o} = K\sqrt{\frac{h}{x + x_o}}$$ (5)

where

K = Velocity decay coefficient for wall jet
x_o = Distance to virtual origin

In this test, K and x_o are found to be 3 and 25mm which is in agreement with other experiments, see Malmstrom (1974).

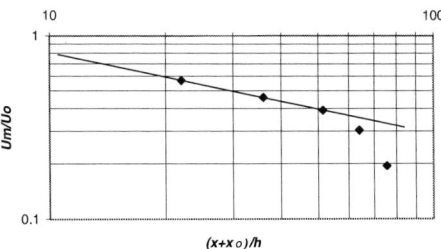

Figure 8: Maximum velocity decay for wall jet

The maximum velocity in the occupied zone U_{rm} is another important parameter in the design of an air distribution system. In a "shorter" room with mixing ventilation this maximum velocity is located close to the floor at $2/3L$ from the inlet, see Zou (1999). Table 2 shows that data of U_{rm} and the locations of

U_{rm} from the inlet x_{rm} are related to the penetration length. It is interesting to note that the measured data of x_{rm}/l_{re} are also close to 2/3.

TABLE 2
MAXIMUM VELOCITY IN OCCUPIED AREA

		U_{rm}/U_o	x_{rm}	l_{re}	x_{rm}/l_{re}
Model 1	Measurement	-0.2	2.8H	4.5H-5H	0.56-0.63
	simulation	-0.17	2.4H	5.9H	0.41
Model 2	Measurement	-0.19	2.8H	4.5H-5H	0.56-0.63
	simulation	-0.17	2.4H	5.6H	0.43

Conclusion

In this study both model experiments and Computational Fluid Dynamics (CFD) are carried out to study the isothermal flow pattern in the ventilated room with different L/H and supply air velocities. Although some discrepancies occur it is clear that the simulation correctly represents the general features of the measurements. Both measurement and CFD simulation show a tendency that l_{re}/H may be independent of W/H when W/H >5. The maximum velocity decay of the wall jet and the maximum velocity in the occupied area are also studied in this paper.

Acknowledgement

The measurement part of this work is made in cooperation with one student group in Aalborg University, Denmark. The first author would also like to thank Aalborg University that sponsored this project.

REFERENCES

Förthmann E. (1934). Über turbulente Strahlausbreitung. Ing. Archiv 5.

Katz P. (1974). Wurfweite, Eindringtiefe und Lauflänge von Zuluftstrahlen im klimatisierten Raum. *HLH* **No. 3**.

Malmström T.-G. (1974). Om Funktionen hos Tilluftsgaller. *TM* **49**, Department for Heating and Ventilation, KTH, Stockholm, Sweden.

Nielsen P.V.(1976). *Flow in Air Conditioned Rooms*, PhD thesis, DTH, Copenhagen

Nielsen P.V., EvensenL., Grabau P. and Thulesen-Dahl J. H. (1987). Air Distribution in Rooms with Ceiling-Mounted Obstacles and Three-Dimensional Isothermal Flow. *ROOVENT 87*, Stockholm

Nielsen P.V. (1995). *Lecture Note on Mixing Ventilation*. Dept. of building technology and structural engineering, Aalborg University, Demark

Sandberg M. (1998). Some Confinement Effects of Jets in Ventilated Rooms. *ASHRAE Transaction*, **SF-98-28-2**, pp 1748-1754

Sforza P.M. (1977). Three-Dimensional Free Jet and Wall Jet: Application to Heating and Ventilation. *ICHMT Congress*, "Heat transfer in buildings", Dubrovnik

Urbach, D.(1971). *Modelluntersuchungen zur Strahllüftung*, Diss. RWTH Aachen

Zou, Y. (1999). A CFD Study for Airflow Distribution at Floor Level in a Slot-Outlet and Slot-Inlet Ventilation Room. *The 3rd International Symposium on HVAC*, Shenzhen, China

Air Distribution in Rooms, (ROOMVENT 2000)
Editor: H.B. Awbi

INVESTIGATING DISPERSION OF AN AIRBORNE CONTAMINANT IN A LARGE SPACE: EXPERIMENTS AND CFD PREDICTIONS

A. J. Gadgil, E. U. Finlayson, M. L. Fischer, P. N. Price, T. L. Thatcher,
M. Craig, K. H. Hong, J. Housman, C. A. Schwalbe, D. J. Wilson,
E. Wood, and R. G. Sextro

Indoor Environment Department, Mailstop 90-3058, Lawrence Berkeley
National Laboratory, Berkeley, CA 94720 USA

ABSTRACT

It is of scientific interest to investigate and understand the dispersion of an airborne pollutant following a short-duration release in a large indoor space. We have applied a novel and powerful experimental technique to full scale chamber experiments, conducted small scale table top experiments, and performed computational fluid dynamic (CFD) simulations. We briefly characterize the full scale test facility, and describe our novel experimental technique for mapping two dimensional gas concentrations non-intrusively and rapidly in large indoor spaces. We describe the table top experiment, and compare the experimental results with predictions from a commercial CFD code.

KEYWORDS

Room Airflow, Pollutant Dispersion, Contaminant Control, Computed Tomography, Remote Optical Sensing, Full-scale Experiments, Small-scale Experiments, CFD, Large Indoor Spaces

INTRODUCTION

It is of scientific interest to investigate and understand the near term dispersion of an airborne pollutant following a short-duration release in a large indoor space. Such understanding would have bearing on measures to reduce health risks in the event of an accidental release or spill of a pollutant in an occupied large space (Rasouli and Williams 1997), and also on design of effective ventilation systems to improve energy efficiency and occupant comfort in large spaces. Owing to the paucity of published experimental data and validated computational models for this dispersion regime, we have undertaken our investigations with full and small scale experiments as well as computational fluid dynamic (CFD) models.

In the past 18 months, we have (1) designed and commissioned a full-scale test facility for an experimental study of dispersion in a large indoor space from a point source, (2) developed and commissioned a powerful new experimental technique for rapid non-intrusive mapping of airborne tracer gas concentrations in the full-scale test facility, (3) conducted selected small scale experiments on dispersion in a table top model of the full scale facility, and (4) selected a suitable commercial CFD code, conducted limited CFD simulations of the small scale experiments, and compared the CFD predictions with experimental results.

In the section below, we briefly characterize the full-scale test facility; in the later section we describe our novel experimental technique for non-intrusive rapid mapping of two dimensional gas concentrations in large indoor spaces. We then describe the table-top experiments and in the section after that we compare the CFD predictions with small-scale experimental results. More details of the research summarized in this paper are available in Gadgil et al. (2000). In the last section, we summarize our long-term goals and future plans.

FULL SCALE TEST FACILITY

The full scale chamber at LBNL has dimensions 7m x 9m x 11 m high. Its inside surface is lined with galvanized steel sheets with seams sealed. A schematic of the chamber, with the HVAC air supply and exit registers, is shown in Figure 1. An instrumented HVAC system can supply air at temperatures ranging from approximately 10 to 30 C, within 0.5 C of specified value, at approximately 1.5 to 6 air changes per hour (ACH), and its configuration can be varied from full recirculation to all outside air. The temperatures of the chamber walls, roof, and floor are measured at 2 positions on each surface and logged.

We use 4% CH_4 in N_2 as the tracer gas. This is the highest concentration of CH_4 considered explosion-safe for release into open air. A mass flow controller regulates the tracer release rate with a maximum flow rate of 100 lpm. The concentration being supplied is monitored continuously. Presently, the tracer gas is released uniformly over a 1m x 1m area source from near the room floor.

COMPUTED TOMOGRAPHIC EXPERIMENTAL TECHNIQUE

Most of the past work on measuring the dispersion of indoor air has been conducted using conventional pump-and-tube sampling. Since 1993, we and others have developed a non-intrusive optical method combining open-path (OP) remote sensing (typically with Fourier transform infrared spectroscopy) and computed tomography (CT) (Yost et al. 1994; Drescher et al. 1997; Todd and Bhattacharya 1997; Piper et al. 1999; Price 1999). New technologies have now allowed us to make substantial improvements over these past efforts.

The data in the mapping plane are collected by measuring path-integrated gas concentrations, determined by the attenuation of infrared radiation as a function of frequency in a narrow band containing an absorption line of the tracer gas of interest (here, methane). The optical gas concentration measurement system is called LasIR, and was manufactured for us by Unisearch Inc., London Ontario, Canada.

The LasIR employs a 10 mW tunable diode laser (TDL) operating near a vibrational rotational line of CH_4. Methane is detected using two-tone frequency modulation (TTFM), whose fundamentals have been reviewed by Brassington (1995). The resulting detection limit for the LasIR instrument for CH_4 is about 2 ppm-m in a one second integration, with a dynamic range of about 10,000. LasIR provides two optical channel outputs, each of which is multiplexed onto one of 30 fiber optic lines that lead to the open-path source lenses.

We use two types of open optical paths. Thirty long optical paths (from 2-30 m), each path comprising separated source and detectors pairs positioned at the edges of the space, are termed "remote" sensors. Thirty short optical paths (1 m), each comprising a single folded optical path using a source and detector pair and a return mirror, all located on a mounting rod, are termed "point" sensors. The optics are mounted so that all paths are horizontal and positioned within about 10 cm of a horizontal plane 2 m above the chamber floor. The 30 remote optical sensors are mounted to the chamber walls, while the 30 point sensors are suspended from cables, (positioned so as to avoid blocking any of the optical paths of remote sensors). One full measurement sweep through all 60 channels takes about 6.5 seconds. Data from the long paths are used to produce computed tomographic (CT) reconstructions. Data from the short paths are currently used to validate the maps produced with CT. Our experimental plan calls for eliminating the point sensors once validation is complete.

CT has found common application in medical imaging, specifically in Computer Aided Tomography and Positron Emission Tomography (Natterer 1986). Most standard methods of computed tomography (CT) are based on picture elements (pixels). Unfortunately the computational methods that work very well for medical imaging do not work well for our problem (see Drescher et al. 1996, and Todd and Ramachandran 1994).

For these reasons, the Smooth Basis Function Method (SBFM) developed by our group (Drescher et al. 1996) was chosen for performing the reconstructions. This method assumes that the methane concentration in the measurement plane can be written as a superposition of smooth basis functions. The parameters of these basis functions are chosen so that the path integrals for the reconstruction match the measured path integrals as well as possible. In our work, we use two-dimensional Gaussian profiles. A major advantage of this approach over any current pixel-based approach is that the resulting reconstructions are smooth, and have a number of local maxima less than or equal to the number of Gaussian basis functions. In the actual implementation of the CT method, we have made a number of mathematical and algorithmic improvements that must be omitted here for brevity. Please see Gadgil et al. (2000) for more details.

Before July 2000, we will place on our website (http://eetd.lbl.gov/ied/APT/APT.html) a video of the time-evolution of the concentrations obtained by CT reconstructions, and their comparison with point-sampled measurements.

TABLE TOP SMALL-SCALE EXPERIMENTS

An approximately 30:1 scale model of the full-scale chamber is built inside a 20 gallon water tank which is divided with a partition. One side of the partition houses the scale model, while the other side provides a container from which the overflow water can be removed. The upper surface of the tank is a free surface with no lid or ceiling. For the experiments summarized here, only mechanical ventilation effects are investigated. Thus, the experimental plan was restricted to a neutrally buoyant plume with isothermal flows and surfaces.

During an experiment, dye-free water flows into the model through the water inlets. Experiments so far operate in a single pass configuration (no recirculation) with a water flowrate of 28 lpm, which is approximately one tank volume per minute. A pollutant release is simulated by introducing a uranine (sodium fluorescein) dye solution at a concentration of 10 mg/l and a flowrate of 1 cc/s into the tank, through a 5 mm diameter foam ball positioned just above the tank floor. The water and dye exit the model through the overflow weir.

A sheet of intense blue-green light, 1 cm thick and centered 6 cm from the floor of the water model, is formed by spreading the beam from a 5 watt, argon ion laser. The laser light causes the uranine dye to fluoresce, with the fluorescence at any location directly proportional to the product of the local dye concentration and the local intensity of laser light. A black and white linear video camera records the fluorescence from (and therefore the dye concentration in) the plane. A selective light filter prevents laser light from reaching the camera, so that only the fluorescence is measured. The images from the camera are digitized as either high resolution still images at 1 frame per second or lower resolution video at 5 frames per second. The digitized images are subsequently processed, calibrated, error corrected, and smoothed with appropriate software on a PC.

Ideally, a scale model would exactly match the Reynolds numbers for the larger space being emulated. However, in practice this is not always easily achievable. We operate the model at 28 lpm (20% of the flow needed to match Reynolds numbers). This leads to an acceptably small loss of resolution with regard to the smallest eddies in the flow. Based on this flowrate, 1 second in the water model is equivalent to 15 seconds in the full scale room.

For brevity, we limit this report to fully developed flow without any interior obstructions except the dye source, and a small tube connected to the peristaltic pump. Transient release experiments, and those with internal obstructions in the space, are described in Gadgil et al (2000).

For all measurements, the flow field is established prior to beginning dye injection, which is initiated 5 minutes prior to beginning image collection. Then, one image is collected every 3 seconds for 50 minutes (1000 images). Comparisons between the results from replicate sets of images show that such 1000 images are sufficient to characterize the concentration and fluctuations accurately.

Even after the concentration distribution becomes fully established, the stochastic nature of the flow results in large changes in the instantaneous concentration profile in the measurement plane over time. However, the time-averaged concentration profile is stable. Meaningful time-average concentration profiles can be obtained by averaging a large enough number of independent images. This can be beautifully illustrated by comparing results of averaging larger and larger numbers of images from independent image sets, however we omit this demonstration here for brevity.

COMPUTATIONAL FLUID DYNAMICS SIMULATIONS

We selected a commercial CFD code modeling of the pollutant dispersion. The process for code selection is described in Gadgil et al (1999). We carried out the 900,000 node simulation in about 2 days on an Origin 2000 machine with two processors. 3000 iterations were required to obtain a steady converged velocity field. Predictions of dye concentrations in the measurement plane are compared to the experimental observations (average of 1000 images) from the table top experiment in Figure 2. Qualitatively the agreement is excellent. A 21 X 21 uniform rectangular grid is overlaid on the two figures. The spatial average of the concentration in corresponding grid pairs is plotted as a scatterplot. A straight line fit to these data, through the origin, has a slope of 1.1 and R-square value of 0.66.

FUTURE PLANS

In the future, we intend to extend our explorations to heavier-than-air pollutant releases. We will extend the CFD predictions to non-isothermal flows. We will explore extending CT capabilities to three dimensions. We will also consider conducting CT experiments and CFD simulations of tracer gas dispersion in real large indoor spaces, such as an auditorium. Our long term research objectives are to (1) develop and experimentally verify a CFD model, and use it to make predictions of point-source pollutant dispersion in realistic large indoor spaces, with non-isothermal flows and negatively buoyant tracers, and (2) develop and extend a novel experimental measurement method for possible use in field situations, and for mapping indoor airborne concentrations in three dimensions.

Illustrations

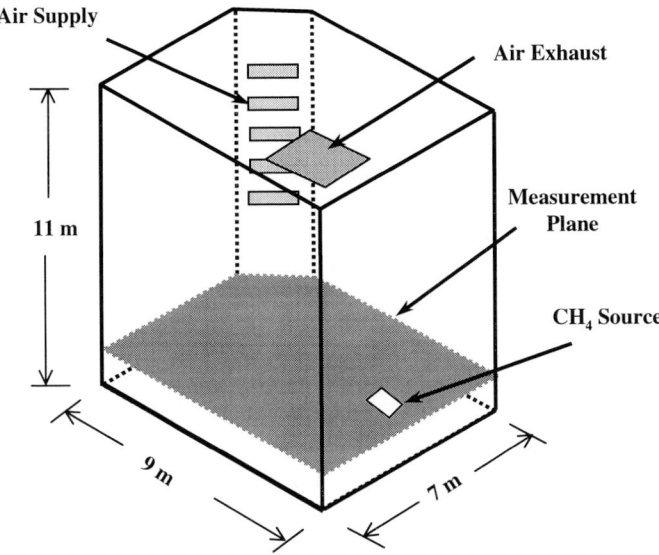

Figure 1: Schematic of experimental chamber showing air supply and return registers, tracer gas release source, and open-path gas measurement plane.

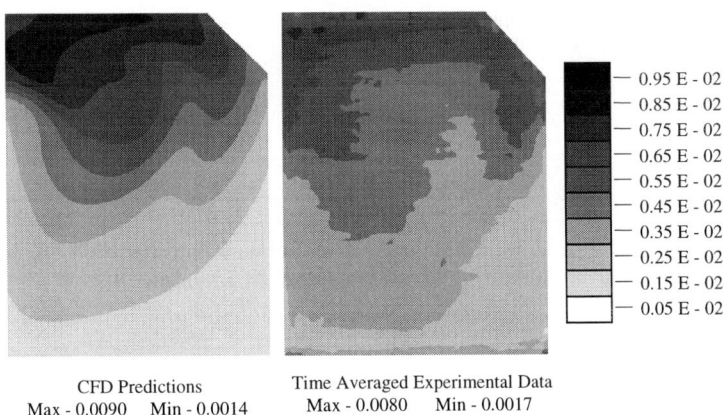

Figure 2: Comparison of predicted dye concentrations with CFD (left), with the experimentally observed dye concentrations (right) under steady state dye release. The experimental data is the mean of 1000 independent images. Dye concentration scale is dimensionless (liter of dye/liter of solution).

618

References

Brassington D.J. (1995). Tunable Diode Laser Absorption Spectroscopy for the Measurement of Atmospheric Species. *Spectroscopy in Environmental Science*. R. J. H. Clark, and R. E. Hester, John Wiley & Sons: 85-148.

Drescher A.C., Gadgil A.J., Price P.N., and Nazaroff W.W. (1996). "Novel Approach for Tomographic Reconstruction of Gas Concentration Distributions in Air: Use of Smooth Basis Functions and Simulated Annealing." *Atmospheric Environment*, **30** (6): 929-940.

Drescher A.C., Park D.Y., Yost M.G., Gadgil A.J., Levine S.P., and Nazaroff W.W. (1997). "Stationary and time-dependent indoor tracer-gas concentration profiles measured by OP-FTIR remote sensing and SBFM-computed tomography." *Atmospheric Environment* **31**(5): 727-740.

Gadgil A., Finlayson E., Hong K-H, and Sextro R. (1999), "Commercial CFD Software Capabilities for Modeling Burst Release of a Pollutant in Large Indoor Spaces," Proceedings Indoor Air 99, 8th International Conference on Indoor Air Quality and Climate, August 8-13, Edinburgh, UK. Published by IAIAS. LBNL-42716.

Gadgil A.J., Finlayson E.U., Fischer M.L., Price P.N., Thatcher T.L., Craig M., Hong K.H., Housman J., Schwalbe C.A., Wilson D., Wood J.E., and Sextro R.G. (2000) "Status Report January 2000: Investigating Dispersion of an Airborne Contaminant in a Large Space" Lawrence Berkeley National Laboratory Report LBNL-44791

Natterer, F. (1986) *The mathematics of Computed Tomography*. John Wiley and Sons, New York.

Piper, A.R., et al. (1999). "A field study using open-path FTIR spectroscopy to measure and map air emissions from volume sources." *Field Analytical Chemistry and Technology* **3**(2): 69-79.

Price P.N. (1999). "Pollutant tomography using integrated concentration data from non-intersecting optical paths." *Atmospheric Environment* **33**(2): 275-280.

Rasouli, F. and Williams T.A. (1997). "Application of dispersion modeling to indoor gas release scenarios - Response." *Journal of the Air & Waste Management Association* **47**(7): 738-738.

Todd L.A. and Bhattacharya R. (1997). "Tomographic reconstruction of air pollutants: evaluation of measurement geometries." *Applied Optics* **36**(30): 7678-7688.

Todd L.A. and Ramachandran N. (1994). "Evaluation of algorithms for tomographic reconstruction of chemical concentrations in indoor air." *American Industrial Hygiene Association Journal* **55**:403-417.

Yost M.G., Gadgil A.J., Drescher A.C., Zhou Y., Simonds M.A., Levine S.P., Nazaroff W.W., and Saisan P.A. (1994). "Imaging Indoor Tracer-Gas Concentrations With Computed Tomography - Experimental Results With a Remote Sensing FTIR System." *American Industrial Hygiene Association Journal* **55**(5): 395-402.

Air Distribution in Rooms, (ROOMVENT 2000)
Editor: H.B. Awbi

EXPERIMENTAL INVESTIGATION OF THE VELOCITY FIELD AND AIR FLOW PATTERN GENERATED BY COOLING CEILING BEAMS

Jan Fredriksson*,†, Mats Sandberg† and B. Moshfegh*

* Division of Energy and Production Systems
University of Gävle
S-801 76 Gävle, Sweden

† Department of Built Environment
Royal Institute of Technology
S-801 02 Gävle, Sweden

ABSTRACT

In the modern office environment there are numerous heat generating equipment. In addition there are loads from solar radiation and heat produced by people. Therefore, the loads will often exceed the load the ventilation system can cope with. To meet this demand on extra cooling capacity the commercial market provides cooling ceiling panels and cooling beams. A literature review shows that until now the majority of the research has been focused on the cooling performance and only a minor part on the thermal comfort and air quality.

A cooling beam is a source of natural convection, creating a transport of cold air into the occupied zone. Furthermore, natural convection flows are vulnerable to disturbances.

Experiments have been conducted in a mock up of an office room. A large variety of heating loads have been used. Qualitative information has been obtained by the use of visualisation, registered by a digital video camera. The time history of velocity and temperature has been registered. The change in position (oscillations) with time of the air-flow generated by the chilled beam has been documented by the use of small and fast thermo-couples and thermistor anemometers.

The results show that the air-flow from the chilled beam exhibits strong oscillations both sideways and along the chilled beam. Furthermore, air-flow generated by heat sources in the room may reverse the flow generated by the chilled beam.

KEYWORDS

Cooling ceiling beams, Measurements, Natural Convection, Plume flow.

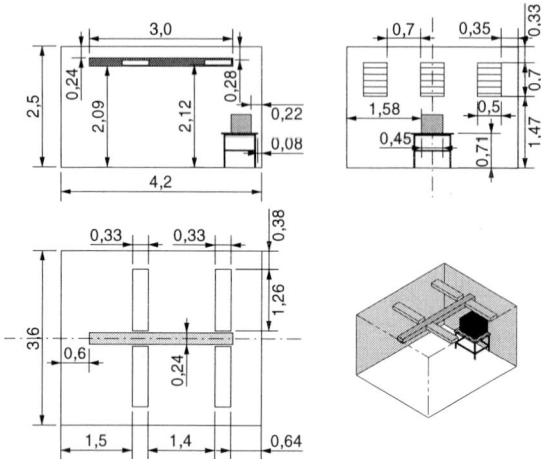

Figure 1. Room layout. All measures in meter.

EXPERIMENTAL SET UP, ROOM LAYOUT

A test room (L x W x H, 4.2 m x 3.6 m x 2.5 m) was set-up to provide a realistic office environment, see figure 1.

To mimic a common office environment the room was furnished with a PC-model with variable heat generation and a "mannequin" with approximately the same area and heat load as a human. In the test room a chilled beam (L_B x W_B x H_B, 3.0 m x 0.24 m x 0.17 m) was installed 0.24 m below the ceiling. The inlet water to the chilled beam was held at a nearly constant temperature of 14.7 °C and the water flow-rate was kept constant at $3,6 \times 10^5$ m^3/s. Electric heating foil was used to simulate solar irradiated windows and the lighting was provided by four fluorescent tube fittings. For the quantitative measurements the following three heat-sources were used; fluorescent tube fittings (347 W), PC-model (443 W) and electric heating-foil (345 W).

The heat-loads were combined into three cases as shown in table 1 together with the corresponding cooling power of the chilled beam. The differences between heat-load and cooling power is due to heat transmission through the room walls.

Case	Heat-sources	Heat-load [W]	Beam Cooling Power [W]
A	Fluorescent tube fittings	347	301
B	Fluorescent tube fittings, PC-model	790	609
C	Fluorescent tube fittings, PC-model, heating foil (window)	1135	777

Table 1. Heat balance for the three test cases.

The room surface- and air-temperature were measured with a large number of thermo-couples. The cooling power of the chilled beam was assessed by measuring the temperature difference between inlet and outlet together with the flow-rate.

FLOW PATTERN MEASUREMENTS

The qualitative properties of the global characteristics of the flow pattern were obtained by visualisation by smoke together with laser light or white light illumination. This was documented with the use of a digital camera or a video recorder.

To quantitatively examine the velocities and fluctuations a 26-channel thermistor anemometer system were used. The anemometer system, CTA88, is specially designed for the typically low velocities in room flow, see Lundström et. al. (1990). The data were acquired with a sampling frequency of 2 Hz. The numbers of samples for all cases were 10240 giving a total measuring time of 5120 seconds. The probes were calibrated in downward flow and fitted with a least-square relation according to the well-known Siddal-Davies correlation. The probes were all mounted in a row attached on a tube and the probe spacing was selected in relation to the actual plume width. Traversing of the probes were made with a computer-controlled traversing system. Measurements were carried in downstream direction to a distance where the plume was beginning to dissolve and no further reliable measurements could be made.

THEORY

The air-flow generated by the chilled beam is treated as a negatively buoyant plume. The following reasoning can be found in Sandberg and Etheridge (1996).

The driving force for the air-flow generated by the chilled beam is caused by the buoyancy forces occurring due to temperature differences between the chilled beam and the surrounding air. The theoretical velocity for a two-dimensional line source can be expressed as a function of the specific convective buoyancy flux at the source, $B(0)$. After an initial developing phase, assuming a normal distribution of velocity and temperature distribution, the following expression is obtained,

$$u(c) = 2.45B(0)^{1/3} \qquad (1)$$

where
$$B(0) = \frac{g\beta q}{\rho c_p} \qquad (2)$$

and
$u(c)$ = centre-velocity of the plume [m/s]
g = acceleration by gravity [m/s^2]
β = volume coefficient of expansion [K^{-1}]
q = convective heat transfer rate per meter [W/m]
ρ = density [kg/m^3]
c_p = specific heat at constant pressure [J/kg·K]

In the analysis of plumes in the atmosphere has Scorer (1978) presented a relation for the acceleration of a buoyancy-driven plume with non-negligible area source. This relation indicates that the width diminution within the nonentrainment zone is varying with distance from the source, i.e.:

$$W \propto z^{-1/2} \qquad (3)$$

where
W = width of the plume [m]
z = distance from source [m]

RESULTS

GLOBAL CHARACTERISTICS

The results indicate that the flow through the chilled beam is rather unstable. Figure 2 shows a sequence from the video recording. Smoke has been introduced into the beam. Close to the beam there

Figure 2. Laser light visualisation from Case B, 25 seconds between each frame.

is a slow drift across the width of the beam. At the beginning the flow is laminar (white continuos sheet) and exhibits a meandering motion until the point where there is a transition to turbulent motion with associated vortex motions and entrainment of ambient air. The dark areas within the plume are interpreted as consisting of entrained ambient air. The point of transition oscillates occurs within an interval equal to 1 W_B -2 W_B.

These oscillations become stronger when the cooling power is reduced and less pronounced at higher heat-loads. The chart in figure 3 shows the standard deviation of the change of position. It can be seen that the standard deviation decreases from case A to case C. Similar oscillation phenomenon can also be observed along the

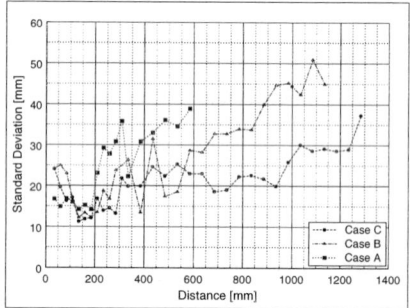

Figure 3. Standard deviation of change in position below the beam.

beam. The air-flow is not uniformly generated along the beam. The air flow-rate is sometimes higher through one end of the chilled beam than through the other.

The interaction with other heat-sources is strong. Heat-sources located near the boundary of the plume or within the plume will influence the plume strongly. It is e.g. very easy to turn the flow from downwards to upwards by placing a heat-source below the beam. Furthermore, the plume gets attracted towards warmer surfaces and the air-flow will thereafter turn upward along the surfaces.

LOCAL CHARACTERISTICS

If the maximum velocity is interpreted as the centre of the downward plume, the results from the velocity measurements verifies the observation from above. There are fluctuations in the direction of the plume centre-line, however, the fluctuations are of rather slow nature. Examples from the cross-wise velocity-measurements are shown in figure 4. There are also shown the fluctuations in the mean velocity. It can be noted that the velocity fluctuations, compared to the other examples, are of largest magnitude at a distance of 235 mm from the beam.

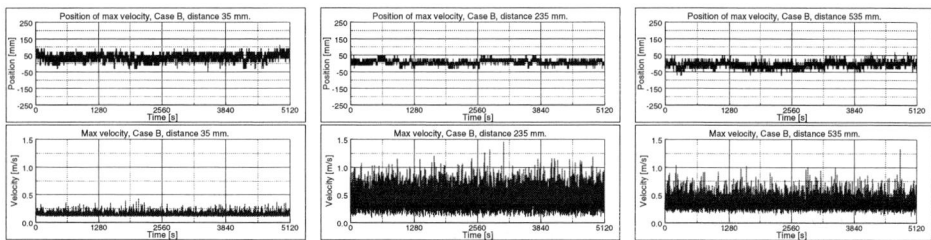

Figure 4. Change in position and maximum velocity fluctuations, Case B at distances of 35 mm (left), 235 mm (middle) and 535 mm (right) from the chilled beam.

The width of the plume is defined as the distance from the position of the local mean maximum velocity to the point, where the velocity has decreased to half the value of the local mean maximum velocity. It can bee seen in figure 5 that there is an acceleration phase very close to the chilled beam. The width decreases and the plume becomes thinner in the near region of the beam. After a certain distance the width starts to grow linearly. In the same region the velocity at first increases to obtain a maximum value and thereafter it exhibits a slow decay.

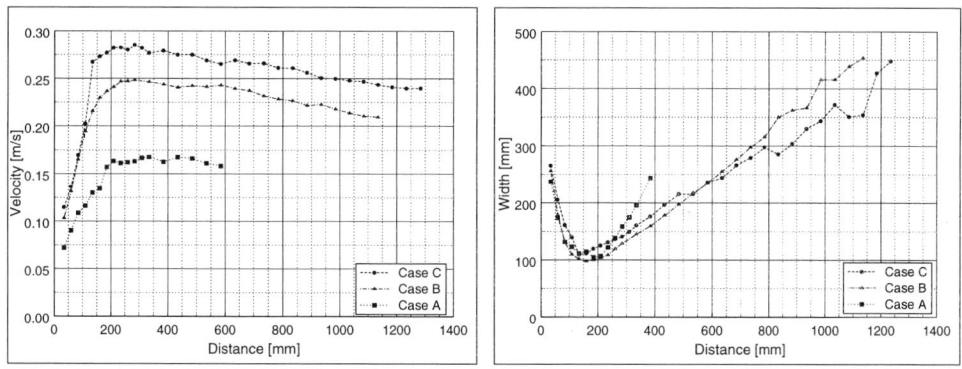

Figure 5. Mean maximum local velocity (left) and plume width (right).

As discussed in previous paragraph, the plume is attracted by nearby surfaces with higher temperature than the plume. This phenomenon is observed in all three test cases when the plume gets attracted to the walls of the room. In figure 6 the profiles of the plumes are shown. In case B and C the plume is deviated to the left-hand side but, in case A, the plume bends to the opposite side. This is the result of different boundary condition at the walls, in case B and C the left wall is warmer than the right and in case A it is the opposite.

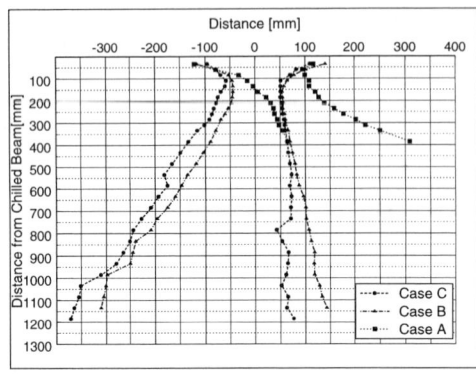

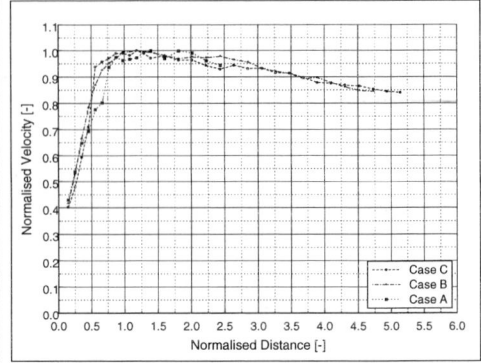

Figure 6. Profiles of the plumes. *Figure 7. Normalised local mean velocity.*

DISCUSSION AND CONCLUSION

Figure 7 shows the normalised local maximum mean velocity as a function of the normalised distance from the chilled beam. The normalised local maximum mean velocity is the ratio of local maximum mean velocity and the maximum velocity at the corresponding test case, the normalised distance is the ratio of the distance from the beam and the beam with, W_b.

The graphs from the different test cases coincide quite well. This indicates that neither the acceleration phase nor downstream decelerations, with a possible exception for Case A, are particularly dependent of the cooling load. The dependency is in the magnitude of the velocity. As noted in formula (1), the theoretical centre-velocity should become constant and proportional to the convective specific buoyancy per meter raised to one-third. The heat transfer in this case is in fact both convection and radiation.

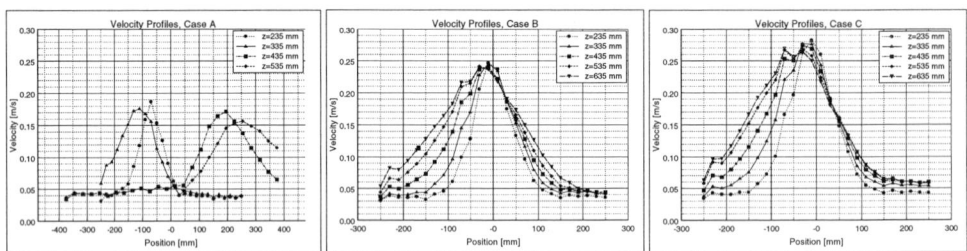

Figure 8. Velocity-profiles at different location from the chilled beam.

In figure 8 are the velocity-profiles from the three case shown at different distances from the beam. It is notable that the measurements indicate velocities next to the beam. Therefore an additional reasonable reduction factor is the underlying circulation of air that the beam generates in the room. This theory is summarised in table 2 with the assumptions; the convection heat transfer is reduced with 50% due to radiation and the background velocity is deducted with estimations from figure 8. The slow decay in velocity can probably be attributed to that the plume is not an ideal two-dimensional plume because it is generated by a cooling beam of finite length.

Case	Convective Specific Buoyancy Flux, $B(0)$ [m^3/s^3]	Theoretical Velocity [m/s]	Background velocity reduction [m/s]	Resulting velocity [m/s]	Measured Velocity [m/s]
A	0.00136	0.271	0.035	0.235	0.16
B	0.00275	0.343	0.040	0.303	0.24
C	0.00351	0.372	0.050	0.322	0.28

Table 2. Assessment of velocities assuming 50% convective heat transfer.

Interesting observation can be made by comparing the actual distribution of the location of the local maximum velocity and the Gaussian normal distribution. This is shown in figure 11 for three examples from case B, at different locations from the beam. It can be seen that the actual distribution can be interpreted as normal distributed. The discrepancies are caused by insufficient spatial resolution.

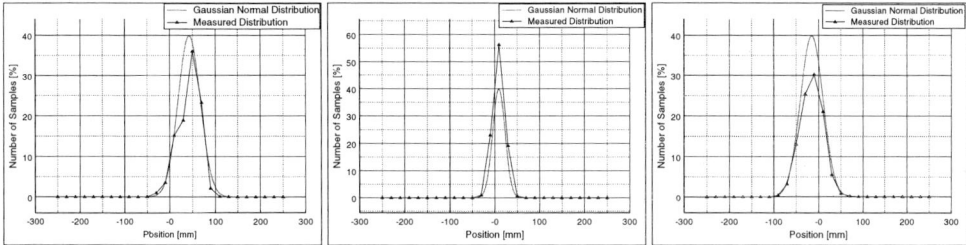

Figure 9. Gaussian normal distribution of change in position at different location from the chilled beam.

To conclude this paper it can be said that the air-flow generated by a chilled beam involves strong (two-dimensional fluctuations) of a rather slow nature. The fluctuations are dependent on the cooling power produced by the chilled beam, as the stability of the plume will increase with increasing cooling power. The spreading of the plume is sensitive to the presence of other heat-sources. The plume is attracted towards surfaces with higher temperature and a heat-source placed underneath the beam can reverse the plume flow generated by the plume.

ACKNOWLEDGEMENT

The help from Mr. Claes Blomqvist, Mr. Hans Lundström and Mr. Ragnvald Pelltari in various shapes of the project is gratefully acknowledged.
This project is sponsored by KK Foundation, Stockholm Sweden, and University of Gävle, Gävle Sweden.

REFERENCES

Hans Lundström, Claes Blomqvist, Perolof Jonsson and Ingemar Pettersson. (1990)
A Microprocessor-Based Anemometer for Low Air Velocities,
Proceedings of Roomvent 1990, Session B1:27

Etheridge, David W. and Sandberg, Mats. (1996)
Building Ventilation: Theory and measurement. Wiley, Chichester

Scorer, Richard Segar. (1978). *Environmental Aerodynamics*
Halsted, New York

© 2000 Elsevier Science Ltd. All rights reserved.
Air Distribution in Rooms, (ROOMVENT 2000)
Editor: H.B. Awbi

APPLICATION OF PHYSICAL MODELLING TECHNIQUES TO OPTIMISE A CLOSE TEMPERATURE CONTROL ROOM

W. B. Booth and A. Topaltziki

MicroClimate & Refrigeration Centre, BSRIA, Bracknell, RG12 7AH,
UK

ABSTRACT

A full size physical and architectural mock-up of a room in which temperature was required to be controlled to very tight tolerances (20±0.1°C) was constructed in BSRIA's laboratory. The design requirement for the close temperature control room and its achievement is believed to be unique. BSRIA's brief was to provide an independent assessment of the achieved performance as well as assisting in optimising the performance of the room supply system. Surveys over a horizontal grid were carried out using a vertical array of anemometers. Analysis of the measurements enabled the performance of the room supply system to be optimised. The application of physical modelling and new 3D data visualisation techniques are described. The two main development phases of the model are summarised. The first phase involved changes to the number of supply diffusers, extract locations and air change rates. The second phase concentrated on optimising the performance of the room including counter-measures to the influence of heat loads (simulated equipment). The benefit to multi-disciplinary design teams of the application of the data visualisation techniques is stressed.

KEYWORDS

Airflow, models, rooms, mechanical ventilation, laboratory testing, climate chambers, temperature, anemometers, measuring, visualising, air change rate, diffusers.

TEST PROGRAMME

Overview

A full size physical and architectural mock-up of a room in which temperature was required to be controlled to very tight tolerances (20±0.1°C) was constructed in a laboratory at BSRIA on behalf of a prestigious client. It is believed the design requirements were uniquely demanding. As an illustration of the scale of the challenge, 1 kW of sensible heat uniformly distributed within the room with a mechanical supply flow rate of 40 air changes per hour will produce a temperature rise of 0.8°C, assuming ideal and complete mixing. BSRIA's brief was to provide an independent assessment of the achieved performance as well as assisting in optimising the performance of the room supply system.

The physical mock-up was 5.8 m by 4.3 m by 4 m high internally. It was constructed of the actual building materials to be used and was housed in an insulated test chamber 10 m by 6 m by 6 m high. A volume within the room was defined, within which the specified conditions had to be produced. The volume was 0.5 m from the walls and extended from 0.5 m to 2 m above the floor, ie 24 m^3. A schematic of the final supply and extract ductwork is shown in Figure 1.

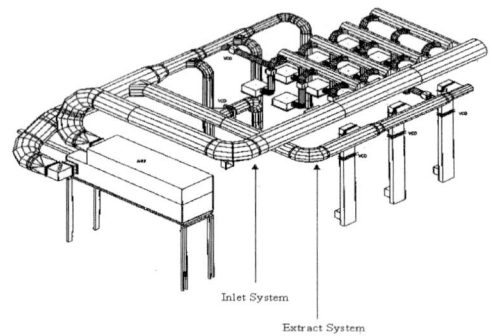

Inlet System

Extract System

Figure 1 : Schematic Of Supply And Extract Ductwork

Test Method

A room survey was carried out using an array of anemometers mounted at fixed heights on a pole supported in a wheeled base (see Figure 2). A regular grid was marked out and each point surveyed in turn. The engineer entered the room to move the pole, returned to the outer chamber, and after a six minute settling period, initiated the next reading. The primary performance criterion was the close control of temperature. There was a requirement for mean air velocity within the control volume to be below 0.25 m.s^{-1}. Although the pole-push surveys enabled this to be demonstrated, it was the diagnostic interpretation of the airflow patterns which proved to be a highly valuable feature of the test programme.

Because the integral temperature sensor fitted to the anemometers recorded data with a resolution of 0.1°C, the temperature measurement capability was upgraded. A platinum resistance temperature (PRT) sensor was mounted parallel to each anemometer, as shown in Figure 2. The exposed elements were to 1/10th DIN standard and manufactured in batches to optimise the matched response. To facilitate accurate calibration, the full set of sensors were mounted on a support frame in a circular array. A reference probe was mounted in the centre of the array. The whole assembly was positioned inside a Dewar flask (itself within the close temperature control room) and readings were taken over a twenty-four hour period. Satisfactory agreement was achieved, with a maximum variation of 0.06°C between all sensors.

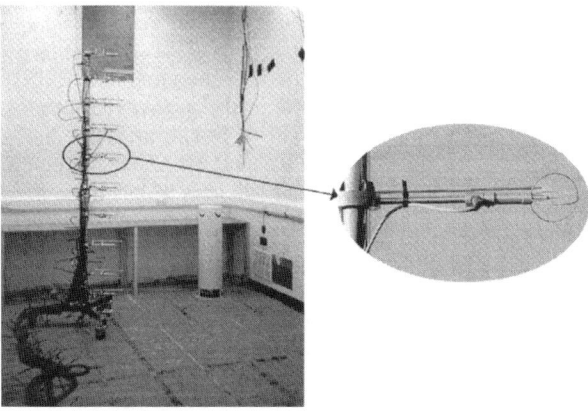

Figure 2 : Instrumentation As Used

Data Interpretation And Visualisation

Analysis of the temperature and velocity measurements enabled the performance of the room air supply system to be optimised. Data visualisation techniques were employed to enable the client and their multi-disciplinary design team to grasp the strengths and potential problems for a given configuration. These techniques are applicable to any physical measurements taken over a grid. Rosén and Andersson's (1988) work provides an example of using computer graphics to visualise physical data. Hanibuchi & Hokoi (1996), Kherrouf, et al. (1996) and Li et al. (1994) have all used 2-D shaded contours to interpret 3-D physical data. The BSRIA project team developed software to plot velocity and temperature contours in real time during a survey. Where appropriate, calibration factors were applied to the measured data, thereby, eliminating the need for post-processing and allowing unambiguous interpretation of the data plots. The plots were a three-dimensional representation of the data. The user interface allowed contour values and colours to be selected and the data to be viewed from any horizontal or vertical angle. The software incorporated default values so that a viewable image was guaranteed. An important feature was the ability to produce a sequence of images which could be viewed using an Internet browser.

A key performance indicator was the variation in locally measured temperature with respect to the temperature recorded at the centre of the room alongside the sole control sensor. Routines were set up to process the data by height and to display the percentage of visited points falling within the ±0.1°C pass band. The software allowed a wire frame representation of the room and key features. This, in conjunction with the contour plots of data collected at several hundred measurement points, enabled the identification of problem areas such as high velocity regions or adverse temperature profiles.

DEVELOPMENT PHASES

First Phase

The primary purpose of the physical model was to assess the performance improvement obtained by increasing the complexity of the mechanical ventilation system. The cost implications of the more complex options had to be assessed by the design team. The generic design had to be applied with modifications to over twenty close temperature control rooms of different layouts and size.

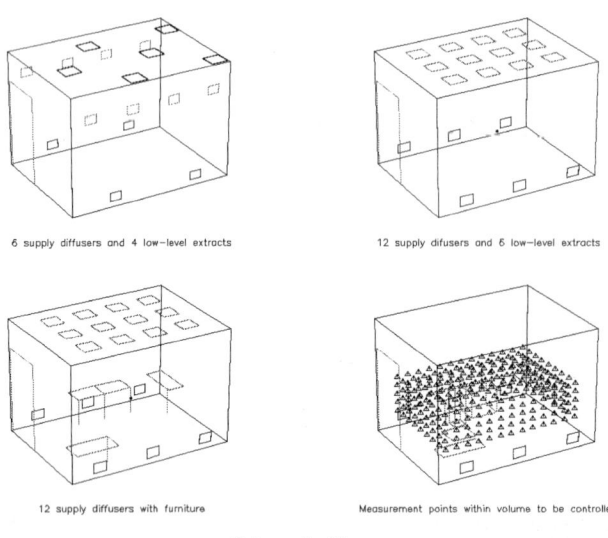

6 supply diffusers and 4 low-level extracts

12 supply difusers and 6 low-level extracts

12 supply diffusers with furniture

Measurement points within volume to be controlled

Schematic Views

Figure 3 : Schematic Of Room Layouts

The first development phase involved quantifying the performance improvement achieved by increasing the number of supply diffusers from six to twelve. The number of low level extracts was increased. Figure 3 provides a schematic view of some of the room configurations investigated. A threefold increase in the percentage of visited points falling within the pass band was achieved with representative height loads and furniture installed. Flow visualisation using smoke was carried out. Considerable use was made of the velocity and temperature contour plots during the first development phase. Two different colour schemes were used in the contour plots. Three-colour contour plots (e.g. white below pass band, green within pass band and red above pass band) were effective in identifying pass/fail areas within the room. Graduated colours (e.g. forty shades at 0.01°C intervals) were useful in assessing whether adjustments or remedial measures were likely to be productive. Examples of the three-colour contour plot are given in Figure 4 and Figure 5 where three monochromatic shading patterns have been used. The examples have been selected to demonstrate how changes in performance are readily identifiable.

Second Phase - Optimising Performance

Once the generic design concept of the close temperature control room had been established, the second phase in the physical model concentrated on optimising the potential overall performance. Prototype changes were made to the supply diffusers by the manufacturer. In due course, a pre-production batch of supply diffusers were installed which were subsequently assessed. A considerable period of time was spent on investigating the zones of influence of various heat loads (typically 2 kW) and the effectiveness of counter-measures such as extract canopies. The handling of heat loads was of paramount importance. Part of the management philosophy was to minimise the overall effect of local variations. The significance of effective commissioning techniques was quantified.

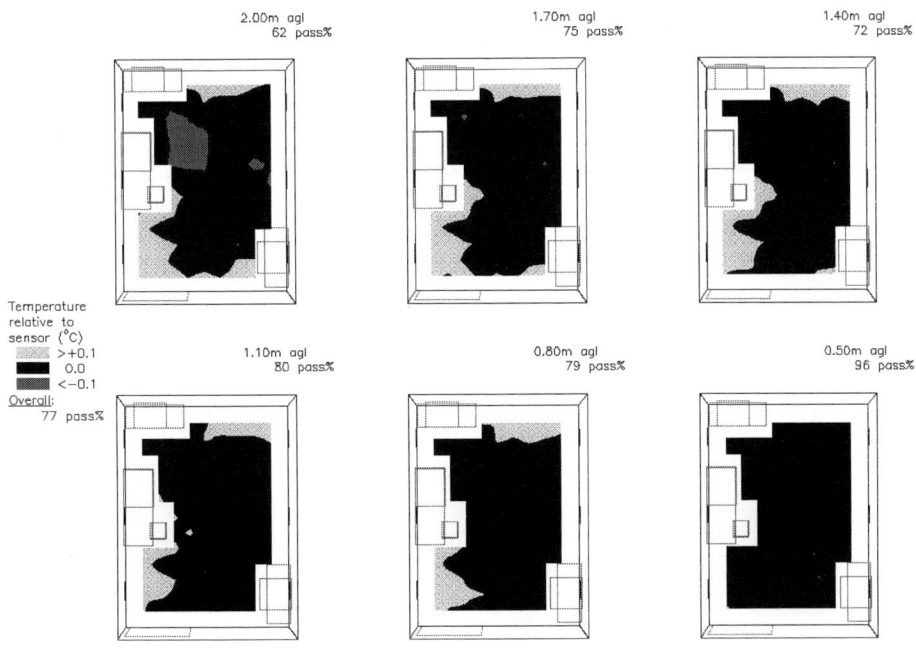

2.00m agl
62 pass%

1.70m agl
75 pass%

1.40m agl
72 pass%

Temperature
relative to
sensor (°C)
>+0.1
0.0
<−0.1
Overall:
77 pass%

1.10m agl
80 pass%

0.80m agl
79 pass%

0.50m agl
96 pass%

2 shielded heat loads and 1 occupant

Figure 4 : Three-Colour Pass Band Contour Plots - Condition A

CONCLUSIONS

Surveys using a vertical array of anemometers and specially calibrated PRT temperature sensors enabled assessments to be made throughout the room. Data was plotted in real time using temperature and velocity contour plots within a wire frame representation of the room. The format of the data plots was matched to either clearly indicate pass/fail grid points or to assist in diagnostic interpretation. BSRIA has found the data visualisation techniques described above to be of considerable value in explaining the findings of physical models. Draught risk, effective draught temperature and comfort indices (eg PMV and PPD) can all be plotted if required. The enhanced presentation of physical data matches techniques normally associated with mathematical modelling. The importance of this parity is the trend towards to the use of VR techniques, see e.g. Aoki (1996), to aid the interpretation of results by non-specialists. Clearer understanding of the findings of physical modelling techniques is a major benefit for multi-disciplinary design teams of architects, building services consultants, quantity surveyors and manufacturers.

632

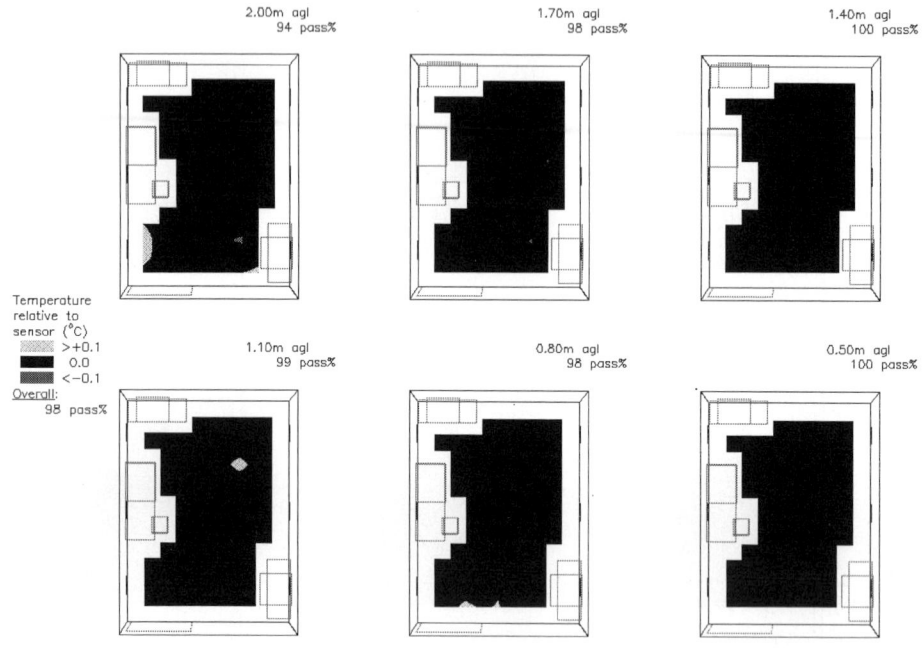

Furniture in situ (no heat loads)

Figure 5 : Three-Colour Pass Band Contour Plots - Condition B

References

Aoki M. (1996). Cohabitation in Cyberspace to Experience Indoor Air Conditions on Internet. *5th International Conference on Air Distribution in Rooms* 2, 209-214.

Hanibuchi H. and Hokoi S (1996). An Analysis of Velocity and Temperature Fields in a Room Heated by a Room Air-Conditioner. *5th International Conference on Air Distribution in Rooms* 1, 445-452.

Kherrouf S., Ribéron J. and Millet J-R (1996). Prediction of Air Distribution in an Air-Conditioned Room: An Experimental and Numerical Approach. *5th International Conference on Air Distribution in Rooms* 3, 111-118.

Li Z., Christianson L.L., Kulp R.N. and Sparks L.E (1994). Outdoor Air Delivery Rates and Air Contaminant Movement in Offices. *Roomvent '94 Air Distribution in Rooms Fourth International Conference.*2, 30-45.

Rosén G. and Andersson I.M. (1988). Visualisation of Air Pollution with Computer Graphics. *Ventilation '88 Proceedings of the Second International Symposium on Ventilation for Contaminant Control* 155-159.

Air Distribution in Rooms, (ROOMVENT 2000)
Editor: H.B. Awbi

THE INFLUENCE OF POSITIONING A PORTABLE AIR CLEANER ON ITS EFFECTIVENESS

H. Chiang*, D.Y. Liu, Y.C. Shih and J.S.Juang

Energy & Resources Laboratories, Industrial Technology Research Institute
Chutung Hsinchu, Taiwan 310, R.O.C.
*E-mail: hchiang@itri.org.tw

ABSTRACT

A portable air cleaner was positioned in several locations in a real-scale test chamber to study the corresponding effectiveness in dust removal. A multiple channel particles counter was employed to measure the particle concentrations. The decay rates were then used to calculate the CADR values according to the ANSI/AHAM standard. Our results reveal an up to ten-percent maximum discrepancy of CADR values among these tests. It occurs especially when the air change rates of air cleaner are small. CFD simulation results of flow fields and concentration distributions yield a reasonable explanation for the difference in air cleaning effectiveness as well as suggest the importance of air cleaner positioning for optimum performance.

KEYWORDS

Air cleaner, Dust removal, AHAM, CADR, CFD, Positioning, Cleaning effectiveness

INTRODUCTION

The Association of Home Appliance Manufacturers has developed an American National Standards Institute (ANSI)-approved standard for portable air cleaners (AHAM 1986). Under this standard, room air cleaner effectiveness is rated by a clean air delivery rate (CADR) for each of three pollutants in indoor air: tobacco smoke, dust and pollen. The AHAM standard is very practical because it takes into the consideration of both the efficiencies of the filtration device itself and the amount of air handled by the device.

Although the efficiencies of filtration devices dominate the overall air cleaning performance, it is always suspicious that the effectiveness of air cleaner might decrease if clean air can not fully mix with room air before reentering the device. The standard does not rigorously state how to arrange air cleaner and pollutants measuring equipment in the test chamber. Offermann *et al.*, (1985) also said that the equilibrium decay rates measured at any location in the room were equal to the decay rate in average concentration and can be used to characterize the performance of an air cleaner.

Researches on the performance testing of air cleaner in an environment chamber (Niu et al. 1998), or in the field with larger space (Nelson et al. 1999), have increased rapidly in this few years. It is mainly because air cleaning is considered as a feasible alternative to reduce outdoor air exchange in maintaining a good indoor air quality. However, the influence of the ways using air cleaners in a room was not considered in most papers.

In this paper, an air cleaner was placed in three locations in the test room and the decay rates of particle concentration were measured at two points, *i.e.* at the center and near the wall. Both experimental and numerical methods were used. The experiment was proceeded according to the ANSI/AHAM standard for testing air cleaners' CADR values. Although air cleaners are rated by the dust, smoke and pollen removal efficiencies in the standard, this paper considers only the dust removal test. It is because the standard tested dust can generate the particle sizes from 0.5 to 3.0μm, covering the most interesting ranges, and does not experience any significant aggregation or decomposition of particles during the testing.

Computer simulation was not widely applied for air cleaner study. The difficulty in simulation arises from the fact that the motion of particles is the result of the action of gravity and the resistance of the air to particle motion. Busnaina and Abuzeid 1991 studied submicron particle deposition in a clean room by solving the mean flow field using a two equation κ-ε turbulence model, and then calculating the particle trajectories by solving the Lagrangian equation of motion. While many effective simulation models are still under developing, this paper intends to use a more simplified model, the Stokes-Einstein diffusion model, in the corresponding advection-diffusion equation. The simulation results give a reasonable explanation for the change in air cleaning effectiveness due to different air cleaner positions in the room.

EXPERIMENTAL METHOD

The AHAM/ANSI air cleaner performance testing uses an environment chamber located at ITRI with the size of 3.2m x 3.0m x 2.8m (1110 ft^3). The particle monitor, aerosol generator, tested dust and environment controls were complied with the AHAM standard. The test facility is illustrated in Figure 1 but the test procedures should refer to the standard or Offermann *et al.* (1985) and Niu *et al.* (1998) for more detail.

In order to investigate the influence of airflow patterns created by the air cleaner. Three locations were chosen to put the air cleaner and two points were selected to collect the particle concentrations. The layouts of air cleaner, particle counter and aerosol generator in the room are shown in Figure 2. These components were all placed on the corresponding tables with the height of 0.9 m above the floor. The tested air cleaner uses a HEPA filter and its volume is 40cm x 30cm x 12cm. The cleaned air is supplied through the top opening (15cm x 8cm) at a 12 degree inclined angle with respect to the vertical line and the room air is drawn in from the front opening (36cm x 24cm).

In proceeding the test, the recording of particle concentration is initiated, after the average particle concentration in the room attains a specified range (*eq.* 200~400 particles/cc for dust removal test). The measured decay rate can then be fitted by the first-order decay model as,

$$C_t = C_i e^{-kt}$$

where C_i, C_t = concentration at initial and time t;

   k = decay constant (time^{-1});

   t = time (minutes).

The above calculations are used for finding out natural decay (when no cleaner is operating) constant k_n and cleaner removal performance decay constant k_a. The natural decay effect is excluded, and the equation for calculating the CADR is:

$$CADR = V(k_a - k_n)$$

where V = volume of the test chamber, ft^3 ;

   k_a = measured decay rate, min^{-1};

   k_n = natural decay rate, min^{-1}.

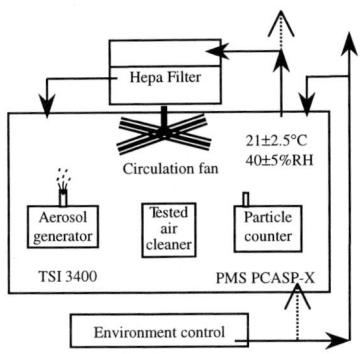

Figure 1: AHAM/ANSI air cleaner
CADR experimental Facility

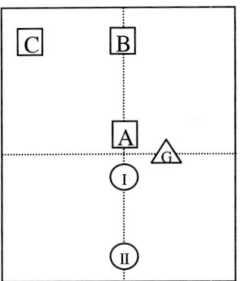

A,B,C-Air cleaner   I,II-Particle counter
G-Aerosol generator

Figure 2: Component layouts in
the test room

NUMERICAL METHOD

The airflow patterns and particle concentration distributions are solved by using a finite volume based software-FLOVENT. The governing equations including the continuity, momentum, turbulence (κ,ε) and concentration equations can be represented as the following general form,

$$\frac{\partial}{\partial t}(\rho\varphi) + \nabla \bullet (\rho\vec{V}\varphi - \Gamma\nabla\varphi) = S$$

where φ = dependent variables, $\vec{V}$, κ,ε and C;

   Γ = diffusion coefficients, μ and D;

   S = source term;

   ρ = air density.

The diffusion coefficient of an aerosol particle can be expressed in term of particle properties by the Stokes-Einstein derivation. It is assumed that the diffusion force on the particles, which causes their net motion down the concentration gradient, is equated to the stokes drag force. The diffusion coefficients and some typical properties of particles calculated by Hinds 1998 are shown in Table 1. Since the settling velocities are very small, the simulation does not include the effect of gravity.

TABLE 1

PROPERTIES OF STANDARD-DENSITY SPHERES AT 20°C

Particle Diameter (μm)	Diffusion Coefficient (m²/sec)	Terminal Settling Velocity (m/s)
0.00037 (air)	2.0×10^{-5}	—
0.1	6.9×10^{-10}	8.8×10^{-7}
1	2.7×10^{-11}	3.5×10^{-5}
10	2.4×10^{-12}	0.0031

When doing the simulation, the initial particle concentration for the whole room is set to the value measured by the experiment. The particle concentration specified at the air cleaner outlet is equal to zero because the HEPA filter is supposed to '100%' clean out the particles from the air moving through it. The surface concentration is zero and a linear diffusion model for particle surface adhesion is applied.

RESULTS AND DISCUSSION

The influence of positioning an air cleaner on the effectiveness of air cleaning rates in a room was expressed in terms of the CADR values measured by the experiment. The measured CADR values are shown in Table 2 for the test conditions of three air cleaner positions (A, B, C), two probe locations (I, II) and two air change rates (3.16, 6.31 ach/hr). The coefficients of determination for these tests were greater than 0.98 indicating that the possibility of equipment or operator error was diminished. Most cases were repeatedly tested for several times and the discrepancies were less than 1.5% for ach/hr=3.16 and 1.0% for ach/hr=6.31 respectively.

TABLE 2 CADR VALUES MEASURED

Probe located at the center (I)				Probe located near the wall (II)			
Cleaner position	Air change rate (ach/hr)	Measured CADR values (cfm)	Discrepancy w.r.t. the Case A (%)	Cleaner position	Air change rate (ach/hr)	Measured CADR values (cfm)	Discrepancy w.r.t. the Case A (%)
A	3.16	53.6	—	A	3.16	56.2	—
B	3.16	51.1	-4.7	B	3.16	54.8	-2.5
C	3.16	49.8	-7.1	C	3.16	50.8	-9.6
A	6.31	107.9	—	A	6.31	109.1	—
B	6.31	105.2	-2.5	B	6.31	108.4	-0.6
C	6.31	103.8	-3.8	C	6.31	107.5	-1.5

The results indicate that the largest CADR values present at the cleaner position A (*i.e.* room center). In that position, the air cleaner tends to eject a high-speed air stream upward, and withdraws room air at slow speed from most directions. The air movement is considered to be able to clean out the whole space in the shortest time to achieve the best performance. When air cleaner is shifted to the envelope of room, the room air motion loses its symmetry and results in poor air movement for some portions of the room. The degrading of efficiency might be as large as 9.6%, especially for those cases with small air circulation rates. It is believed that not enough momentum and short circulation of airflow are the causes for deteriorating the performance.

A reasonable explanation for the experimental results can be obtained by observing the flow fields and particle concentration distributions from the numerical simulation results. The transient variations of particle concentrations in the room are shown in Figure 3 and Figure 4 for air cleaner positions A and B. The air change rate was 3.16 ach/hr and the diffusion coefficient for 1 μm particles was selected. The results clearly show that the particle concentration decays faster when the air cleaner is placed at the position A instead of B. In fact, when the air cleaner is placed close to the wall, the flow field has shown that a lot of clean air finishes its circulation in a rather short route. Hence some clean air is not fully mixed with the room air.

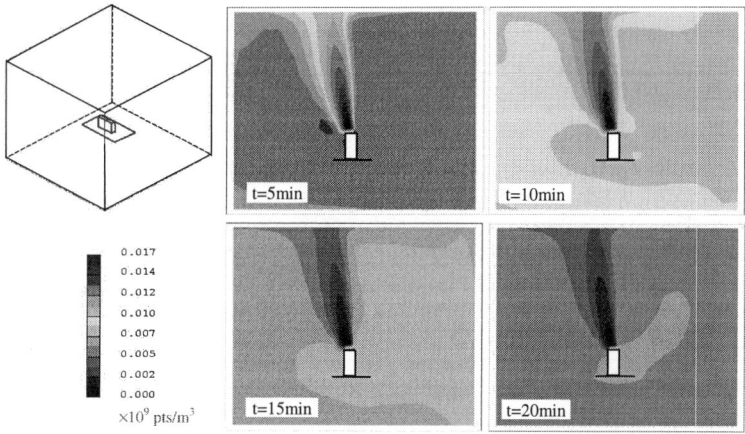

Figure 3: Decay of particle concentration for air cleaner at the center (d=1μm, ach/hr=3.16)

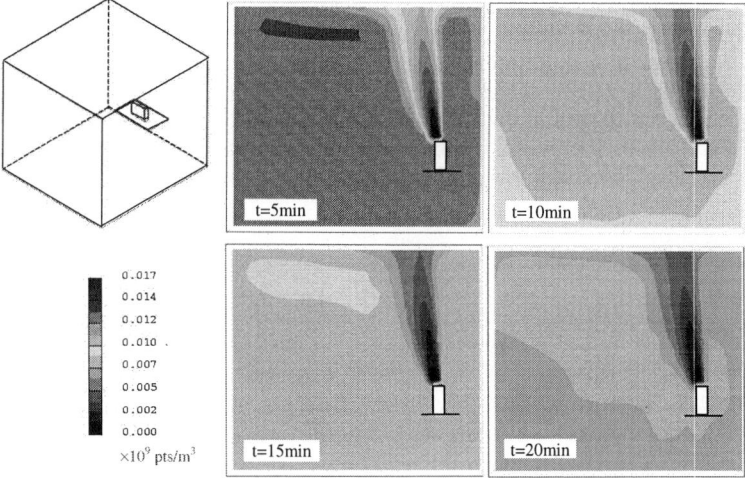

Figure 4: Decay of particle concentration for air cleaner near a wall (d=1μm, ach/hr=3.16)

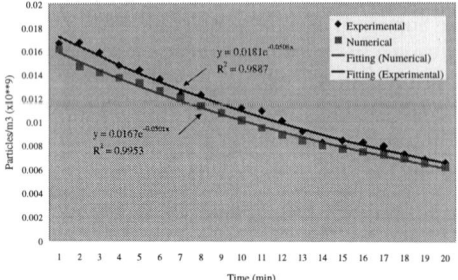
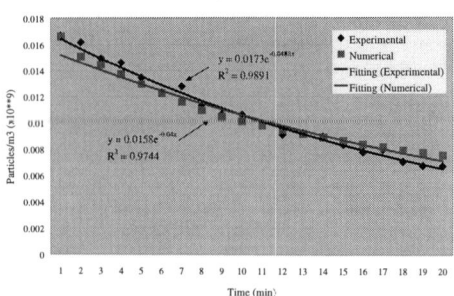

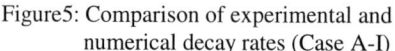

Figure5: Comparison of experimental and numerical decay rates (Case A-I)

Figure6: Comparison of experimental and numerical decay rates (Case B-I)

Theoretically, the air cleaner CADR values can also be obtained numerically. The decay rates of particle concentration are collected from the simulation results at the same probe positions. Figures 5 and 6 show the comparison of particle decay rates obtained from the numerical and experimental data. Depending on the cases, the discrepancy for air cleaner positioned near the wall (B) is larger than that positioned at the center (A). The larger difference might come from the turbulence model, particle diffusion and surface adhesion models used in the simulation as well as the negligence of gravity force.

CONCLUSIONS

Experimental measurements of air cleaner CADR values have shown that the positions of air cleaner have definite impact on the decay rates of particle concentration in a room, especially at low air change rates. In the paper, numerical studies attempt to yield a reasonable explanation for the variation of performance with respect to the air cleaner locations. Observing from the field distributions, it is believed that if the clean air can more evenly mix with room air and avoid short circulation, better performance can thus be achieved. Numerically prediction of air cleaner performance is a promising method for improving the design. However, more research is required in future to fine-tune the simulation models to further reduce the errors.

REFERENCES

AHAM (1986). Method of Measuring Performance of Portable Household Electric Cord-Connected Room Air Cleaners. Standard AC-1-1986: Association of Home Appliance Manufacturers, Chicago.

Offermann F.J. et al. (1985). Control of Respirable Particles in Indoor Air with Portable Air Cleaner. Atomspheric Environmentm **19:11**, 1761-1771.

Niu J. et al. (1998). Using Large Environmental Chamber Technique for Gaseous Containment Removal Equipment Test. ASHRAE Trans., 1998, **104:2**, Paper number TO-98-23-3, 1289-1296

Nelson P.R. et al. (1999). In Situ Measurement of Air Cleaner Ventilation Effectiveness. Indoor Air 99, Edinburg, England, vol. 4, 81-86.

Busnaina A.A. and Abuzeid S. (1991). Submicron Particle Deposition on a Surface in Laminar and Turbulent Flow. Proceedings, Institute of Environment Sciences. 130-139

Hinds W.C. (1998) Aerosol Technology: Properties, Behavior, and Measurement of Airborne Particles, 2nd. John Wiley & Sons, Inc., New York.

Air Distribution in Rooms, (ROOMVENT 2000)
Editor: H.B. Awbi

DIFFUSION OF AIR WITH CONTROLLED VELOCITY PROFILE

Gregory Gottschalk

Hesco (Schweiz) AG
CH- 8630 Rüti / ZH, SWITZERLAND

ABSTRACT

This paper presents a concept to produce air jets for ventilation purposes. The approach named Profile Controlled Diffusion (Procondif [®]) treats an air outlet as a precise generator of the desired velocity field in its close proximity. Velocity is controlled in the sense of a vector value i.e. both features: direction and speed are adjusted independently.
The motivation for this development finds its origin in various studies and publications from different authors. The technical aspects of realisation the task are shortly discussed.
A ceiling outlet using the new principle is used as an illustration of the general method.

KEY WORDS

Outlet, Velocity distribution, Velocity gradients, Velocity profile, Positive buoyancy, Honeycomb

INTRODUCTION

Knowledge of the human thermal sensation and care of quality of the inhaled air are guidelines for new trends in ventilation technology.

The common feature of these modern techniques, especially in the comfort ventilation, is the introduction of supply air with low velocity through large high permeable surfaces. Specific elements in the design of outlets homogenise the outflow or on the contrary privilege certain directions. The main progress has been done in recent years in the field of displacement ventilation. The outlets reveal good performance for air quality and comfort.

Nevertheless the presence of relatively large objects on the floor or the walls may cause area problem for the user or aesthetic refuse from an interior architect. The ceiling solutions for mixing ventilation have preserved their attractiveness. A downward low velocity supply of air in form of a falling wall film is also used nowadays.

On the other hand there are applications where a high speed jet is of great utility: stability at even large distances is required , e.g. long range nozzles and air curtains.

From the engineering point of view the question arises what are the technical possibilities for optimal constitution of jets for various applications.

In this paper the velocity distribution in the outlet opening is of principal interest.

JETS IN VENTILATION

The classical studies in ventilation field were mainly focused on turbulent jets. A zone of fully established turbulent flow and terminal vanishing in room air were the main points of interest .
The improved generalisation formulas are still developed and conditions at the outlet play an important role for the downstream previsions, e.g. by Schädlich (1993) and Sefker (1989).
Nevertheless the particular features of jets in the initial, laminar zone, even if less applicable in HVAC were also observed. Hanel & Richter (1979) reported own experiments and presented data from Vulis, see Figure 1.

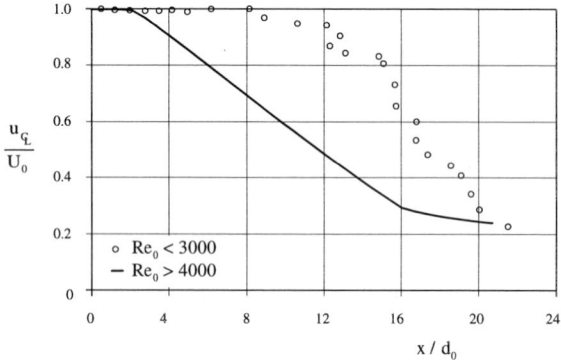

Figure 1: Decay of centreline velocity. Data from Vulis, by Hanel & Richter (1979)

Long laminar phase and more rapid centreline velocity decrease by smaller Reynolds numbers is commentated that the jet was issuing from a long tube: $L = 100 \cdot d_0$. The rest of experiments at the same range of Reynolds number but with a different jet conditioning indicates instable character: the length of the initial laminar zone varied strongly in unpredictable manner.

The relative turbulence intensity along the centreline is shown to increase more steeply in the core region than at the zone of fully established flow. Measurements presented by Sandberg (1996) across the jet from a commercial nozzle reveal coincidence of high velocity gradients and locally increased standard deviation,

Schädlich (1993) identified the ratio: "specific momentum flux / length of opening perimeter" and geometry of the edge (sharp or rounded) as influencing the core region and velocity decay.

A recent, precise and well documented study from Todde (1998) on the low velocity jet where the precautions were made (short steepless reduction) to get top hat velocity profile shows no significant elongation of the potential core.

These are the indications that the velocity distribution at the outlet influence propagation of a jet.

PROCONDIF METHOD

Models of smoothing out of velocity discontinuity are analysed by Schlichting (1987) and Batchelor (1992). There are numerous details depending on admitted hypothesis but in first approximation adjusting of the apparent viscosity ν allows to describe the phenomenon by Eqn. 1.

$$u(y,t) = U \cdot erf\left\{\frac{y}{(4 \cdot \nu \cdot t)^{\frac{1}{2}}}\right\} \tag{1}$$

The presented solutions indicate strong decrease in time t of the initial step like discontinuity 2U towards more stable low gradient profiles. The resulting rapid evolution in time of the velocity gradients at the interface between the moving and not moving fluid is shown in Figure 2.

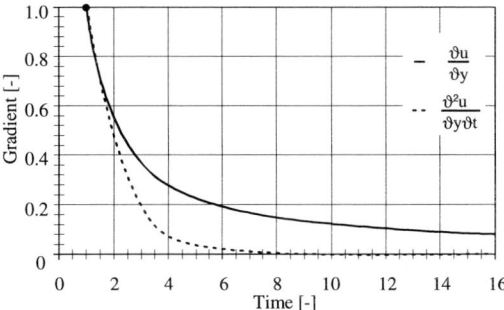

Figure 2: Relative gradients by smoothing out of velocity discontinuity

The proposed method is aiming to attenuate this process. The smoothed velocity distribution already at the outlet is more similar to the natural profile at a further distance or otherwise speaking at a later time moment. As a consequence of this, a jet is undergoing less essential remodelling: interfering with ambient air is dampened and entrainment at short distances reduced.

The applied technical means consist of putting together small jets issuing from adjacent narrow ductings. The well known geometry of honeycomb (e.g. from rotary wheels) is used for the purpose. The velocity distribution in each of the individual jet disappears as the common face velocity is formed at the outlet, see Figure 3.

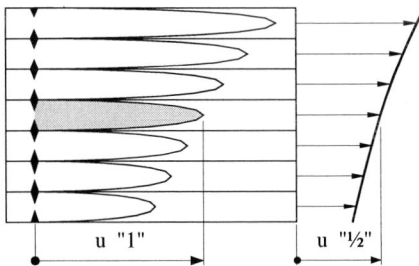

Figure 3: Profiling of velocity in the honeycomb wafer

The quiescent character of singular jets is carried over to the resultant large jet. Moreover, the adjusted throttlings yield distinct pressure drops in the individual ductings: the smoothed face velocity is shaped at the outlet. The momentum significant width comprises several cells of honeycomb leading to larger characteristic length of the outflow.

CEILING OUTLET

The aim was to develop the ceiling outlet for quiescent, vertical diffusing of colder air. The perform-ance should be possibly similar to displacement ventilation.
Otherwise like isothermal plug flow from a parabolic textile diffuser described by Holmberg (1994), the downward transport of primary air occurs here in a limited space leaving most of the area free for ascending convective currents.
The appropriate honeycomb wafer had been made for the purpose, see Figure 4.

Figure 4: Detail of the honeycomb structure

A few millimetres large divergent ductings (1), Figure 5, cover a full circle and point about 30 degrees downward from the ceiling.

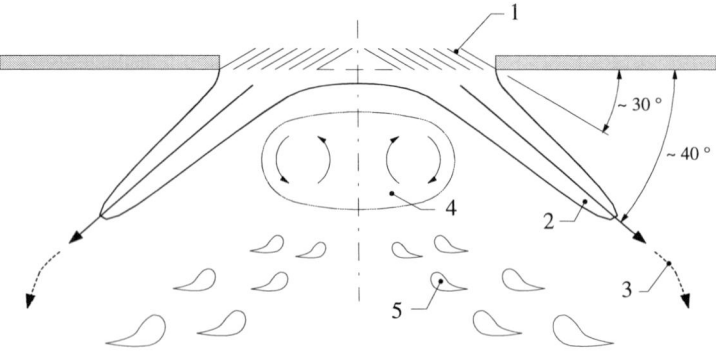

Figure 5: Ceiling outlet with smoothly shaped face velocity (sketch)

The intentionally shaped velocity profile (2) has a well pronounced carrying part for efficient expanding of the outflow. The highest velocity (about 2 m/s, local Reynolds number about 1500) is undergoing gradual decrease towards the periphery and the centre (ratio 5:1). Supply air a few degrees colder than the room air is creating a diffusion pattern resembling a big church bell (3). A toroidal vortex (4) is generated under the outlet. It recirculates locally the flow and looses downward the mass (5) filling the bell volume.

Dispersing of powder tracer outside the main outflow or inside the bell showed no smudging at the outlet. It indicates the repulsive flowfield in proximity of the diffuser.

The overall character of the flow shows influence of both driving elements: an initial momentum and a positive buoyancy, see Figure 6. The downward flow depends on the outlet volume rate V and temperature difference ΔT between primary and room air. The resulting shapes of the velocity profile at further distances from the outlet are to be associated with respective graphical symbols: "a" for positive buoyancy governed, "c" for momentum governed and "b&d" for equilibrated, balanced cases.

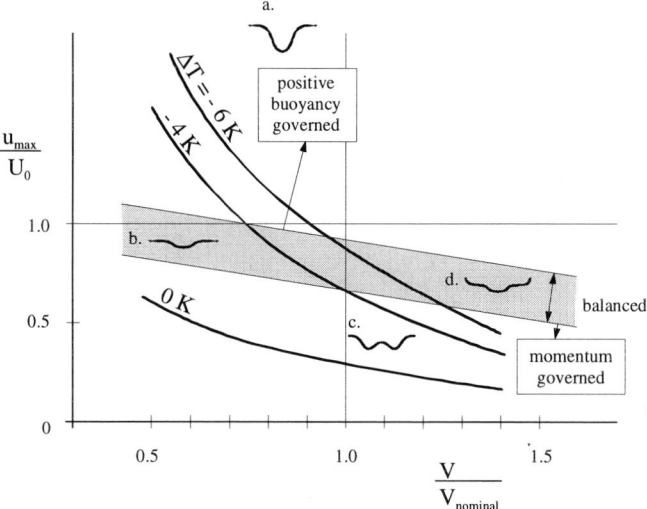

Figure 6: Downward flow at a further distance from the outlet
(a, b, c, d : shapes of velocity profiles)

The maximum velocity of downward movement u_{max} is proportional to to the average velocity at the outlet U_0 but its acceleration along the axis was not observed in the investigated ΔT range. The constant downward movement is an important feature of the applied here particular face velocity distribution.

DISCUSSION

The downward diffusion of colder air calls for balancing the positive buoyancy with friction (case of a wall film) or with flow spreading (stand alone outlet). Otherwise the discomfort area would be too large. A sharp flow extension involves intrinsic intense entrainment. In the presented example precautions were taken to promote gentle expanding and the internal recirculation of the flow.

The specific momentum flux from the examined outlet is about 1/3 of comparable mixing situation with ceiling jets. That explains tranquil character of air movement in the main part of the room. The observations with a smoke tracer showed that ascending convective plums could be better preserved than in a mixing ventilation experiment when they were irrevocably transported along the ceiling and recirculated into the occupational zone.

The standard solution to extract air with ceiling inlets leads often to a short-circuit flow. The down-ward diffusing introduces firstly all the mass of primary air into the occupational zone, elongates the path of the inflow and eliminates the drawback of a short-circuit.

The particular divergent, plane honeycomb wafer was developed. It revealed to be a silent and a low resistance diffusing device : 4 dB(A) less and 30 % less of pressure loss compared to a standard mix-ing outlet.

Generally, the proposed method increases the precision of shaping the outflow to get the desirable air flow pattern in the room. Total porosity makes possible repulsive outflow eliminating contamination of the outlet with dust suspended in room air.

Data on propagation of jets with initial specific velocity profile are rare. The reported experiments concern mostly small sizes and laminar parabolic distribution. The characteristics and limits of diffu-sion with intentionally shaped velocity profile are estimated to be worth further investigations.

REFERENCES

Batchelor G.K. (1992). *An Introduction to Fluid Dynamics*, University Press, Cambridge, Great Britain

Etheridge D., Sandberg M. (1996). *Building Ventilation: Theory and Measurement*, John Wiley & Sons, England

Hanel B., Richter E. (1979). Das Verhalten von Freistrahlen in verschiedenen Reynolds – Zahlbereichen. *Luft und Klimatechnik* 1979:1, 12-17.

Holmberg S., Parocki A., Tang Y-Q. (1998). Diffusers for horizontal air flow. ROOMVENT '94, 4th International Conference, Krakow, Poland, June 15-17, vol. 2, pp. 229-241.

Sefker Th. (1989). *Verallgemeinerte Darstellung des Verhaltens isothermer Freistrahlen*, Uni.–GHS-Essen, Germany

Schädlich S. (1993). *Der Einfluss Verschiedener Luftdurchlassgeometrien auf das Freistrahlverhalten*, DKV, Stuttgart, Germany

Schlichting H. (1987). *Boundary Layer Theory (Inc. Reissued)*, McGraw-Hill,Inc., New York

Todde V., Linden E., Sandberg M. (1998). Indoor Low Speed Air Jet Flow: Fiber Film Probe Meas-urements. ROOMVENT '98, 6th International Conference, Stockholm, Sweden, June 14-17, vol. 2, pp. 585-592.

Air Distribution in Rooms, (ROOMVENT 2000)
Editor: H.B. Awbi

REINFORCED EXTRACTION SYSTEM: EVALUATION OF OPTIMIZED OPERATING PARAMETERS USING CFD

D. Gubler and A. Moser

Air&Climate Group, Institute of Building Technology, ETH Zürich,
ETH Zentrum LOW, CH-8092 Zürich, Switzerland

ABSTRACT

This paper describes the evaluation of a set of optimized operating parameters for a jet enhanced local extract system. The field of useful operating parameters for a specific exhaust design is investigated using CFD. The calculations are validated by measurements done in a similar setup to the CFD model. The goal of the study is to validate a 3D-model of an enhanced extract system and to evaluate limitations for the operating parameters. Results of the study are: calculation of the 3-dimensional flow pattern of a jet enhanced extract system is possible, the results correspond well to measured data. The evaluation of optimized operating parameters with CFD was successful. It is not possible to define one best operating point but a range of operating parameters with good performance of the system and rules for limitations of the set of operating parameters.

KEYWORDS

CFD, Local Ventilation, Industrial Application, Exhaust Range, Capture Distance

INTRODUCTION

Illness of the respiratory tract may arise from poor air quality at the working place. The most effective way to protect workers' health is capture of contaminants near their source. Recommendations for the evaluation of optimized operating parameters for a jet enhanced extract system are reported in this paper. In 1965, the basic concept of a jet enhanced extract system was presented for the first time. It has extended suction range, and captures airborne contaminants at their origin without obstructing the work area. The basic idea is to use a secondary air jet to entrain and thereby attract a large volume of ambient air to the suction opening of the extract system. Shrouded in this general streamline pattern, the contaminated air is guided to the extract. Different experimental and theoretical studies, Pedersen & Nielsen (1991), Saunders & Fletcher (1993), have demonstrated the performance of the system. Nevertheless, it has not been applied to industrial ventilation so far. The objective of the ongoing project is to find limitations for the range of operating parameters for this kind of enhanced extract system. The study should provide knowledge about parameter sensitivity of the device. Computational

Fluid Dynamics (CFD) are used as a design tool along with measurements on prototypes. The reliability of CFD-calculations is verified and the numerical method validated by comparison with experimental data.

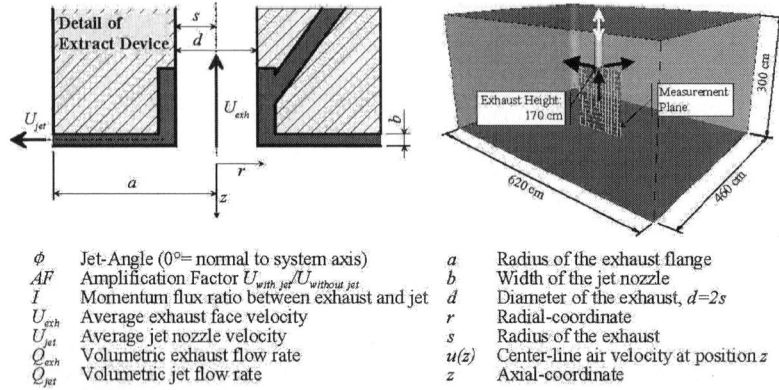

ϕ	Jet-Angle (0°= normal to system axis)	a	Radius of the exhaust flange
AF	Amplification Factor $U_{with\ jet}/U_{without\ jet}$	b	Width of the jet nozzle
I	Momentum flux ratio between exhaust and jet	d	Diameter of the exhaust, $d=2s$
U_{exh}	Average exhaust face velocity	r	Radial-coordinate
U_{jet}	Average jet nozzle velocity	s	Radius of the exhaust
Q_{exh}	Volumetric exhaust flow rate	$u(z)$	Center-line air velocity at position z
Q_{jet}	Volumetric jet flow rate	z	Axial-coordinate

Figure 1: Principal schematic and nomenclature

CFD-MODEL

The 3-dimensional CFD simulation is carried out using a commercial code with multi-grid capability. The exhaust device is mounted vertically in the center of a room with the exhaust face 1.7m above the floor. An analog setup was used for measurements of the velocity field in front of the extract device. Turbulence is modeled by the standard k-ε model for all calculations within this paper. The design of the exhaust used in this study is the same as it was used by Pedersen & Nielsen (1991). The setup shown in figure 1 is defined by four main boundary conditions:

1. The exhaust opening with radius s=51.5mm and a mean exhaust face velocity $U_{exh}=Q_{exh}/A_{exh}$
2. The supply jet nozzle is a circular slot nozzle with radius a=112.5mm, height b=6mm and an average exit velocity of $U_{jet}=Q_{jet}/A_{jet}$
3. To equalize differences between the mass flow of extracted and supplied air, a third opening is specified, directed towards the roof of the chamber. This boundary is defined as a zero-pressure opening where air exchange to the outside of the chamber is possible
4. All other boundaries are rough walls (using logarithmic wall functions)

The influence of the walls on the flow pattern of the extract is significant. Due to the limited room expansion (6.2x4.3x3m), increased air velocities occur for larger distances to the exhaust face (compared to calculations using infinite space expansion). For this study, only the volumetric flow rate and thus the air velocities are varied. All geometric parameters, especially the jet nozzle height b are constant. The influence of specific geometric parameters on the behavior of the system is also very important and will be investigated in further studies.

VALIDATION

The verification of the (numerical) accuracy of CFD Results with measured data and other mathematical methods is important. Thus, the model is first verified using suction alone (I=0.0) by comparing a potential flow model for circular flanged exhausts, Drkal (1970), with the CFD calculation. Then it is compared to measured data of a similar experimental setup, Weber & Sprecher (1999), and measurements by Pedersen & Nielsen (1991) and finally the sensitivity of the model is tested for different influences like turbulent boundary conditions and grid resolution.

Suction alone

Figure 2 shows a comparison between the potential flow model for the circular flanged exhaust by Drkal and the CFD calculation. The potential flow model is given by the equation:

$$\frac{U(z)}{U_0} = 1 - \frac{z/d}{\sqrt{0.25 + (z/d)^2}} \tag{1}$$

The extracted flow rate for this comparison is $Q_{exh}=465 m^3/h$ ($U_{exh}=15.5 m/s$). Good agreement between the potential flow model and the CFD results for the conventional exhaust was achieved for distances less than 3 diameters. For larger distances, the potential flow model does not correspond very well to the CFD result because it does not consider the influence of limited room space and so the axis velocities are underestimated by the potential theory.

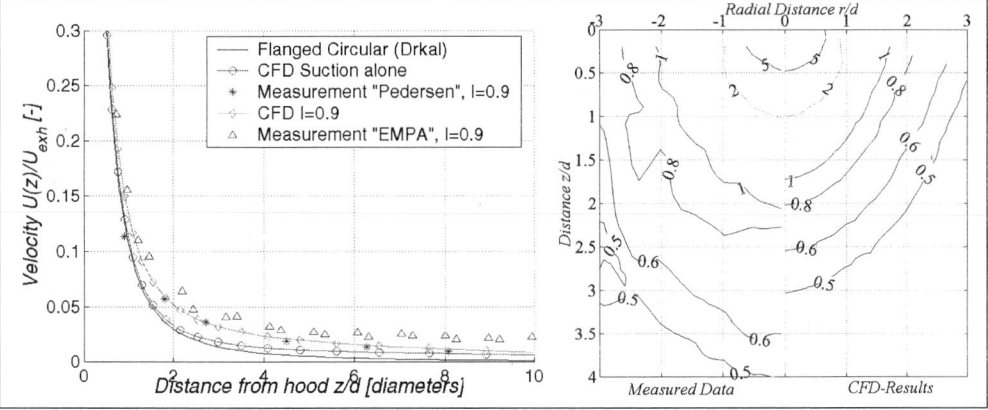

Figure 2: Centerline velocities for suction alone (I=0.0) and I=0.9; 2-D velocity field (units [m/s])

Enhanced Suction

The axis velocities for a momentum ratio of I=0.9 are illustrated in figure 2, left part. The right part of figure 2 shows the measured velocity distribution in a half plane compared to the corresponding CFD calculations. Both show a good agreement between measurements and calculation but with increasing distance to the exhaust face, the air velocities are underestimated by the CFD calculation. Geometric asymmetries in the experiment (installations and measurement equipment, which are not modeled in CFD) may induce cross drafts in the room which raise the general velocity level in the experiment. However flow visualization with smoke showed airflow normal to the exhaust axis, which were not observed in CFD. The agreement of CFD results with the measurements of Pedersen & Nielsen (1991) (axis velocity only), is good.

PARAMETER ANALYSIS

It is the objective of this work to evaluate sets of operating parameters (OP) with good performance for the present exhaust design. Thus the field of possible operating parameters is surveyed and the performance of the system across this field is investigated.

Field of Operating Parameters

The operating point of a jet enhanced extract system is defined by its geometry, the extracted and the supplied flow rates OP=f(Geometry, Q_{exh}, Q_{jet}). Pedersen used the ratio of the momentum flux $I=Q_{jet}U_{jet}/(Q_{exh}U_{exh})$ to describe the performance of the enhanced extract system. In figure 3 the straight lines of constant values of I are illustrated as well as iso-velocity lines and iso-amplification lines respectively. The values of I can be seen in figure 4.

Axis Velocities

The axis velocity in a certain distance to the exhaust face is a good indicator for the suction range of an extract system. In figure 3, the velocities are drawn over the field of operating parameters for a distance of 40cm (approximately 4d) to the exhaust face. Increasing the extract flow rate or the momentum ratio I produces higher axis velocities at constant extract flow rates.

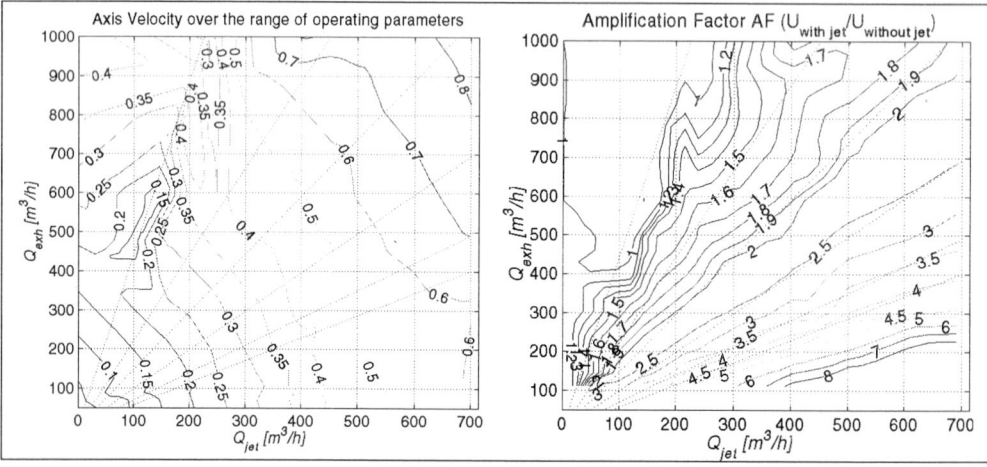

Figure 3: Iso-velocity and iso-amplification lines for the field of operating points OP for a distance of 40cm (approx. 4d) to the exhaust face. Straight lines are iso-I-lines, values see figure 4.

Pedersen & Nielsen (1991) observed that the supplied jet is directly extracted by the exhaust and thus the performance of the system is reduced if the value of I is below I=0.1. The CFD calculations do reproduce this effect very well. For applications, a certain level of air velocities is needed to avoid sensitivity to disturbances (e.g. 0.5m/s for welding), depending on the behavior of the contaminant.

Amplification Factor

An even better indicator for the performance of an enhanced extract system is the amplification factor $AF=U_{with\_jet}(z)/U_{without\_jet}(z)$. The ratio AF expresses directly the performance of the enhanced extract system in comparison to a conventional exhaust with the same volume of extracted air. The iso-amplification lines do follow the iso-I-lines over a wide range of operating parameters. The area of short circuit and thus a bad performance is indicated by an amplification of AF=1 or less. Although AF increases with increasing I, the range of useful operating parameters is limited towards large momentum ratios. The opening angle of the capture zone decreases with increasing I. Thus it is necessary to find an operating point with enhanced suction range but still a sufficiently expansion of the capture zone.

Limitations for operating parameters

The analysis of the velocity- and amplification distribution showed some obvious limitations for operating parameters of enhanced extract systems. The limitations are illustrated in figure 4 and discussed below:

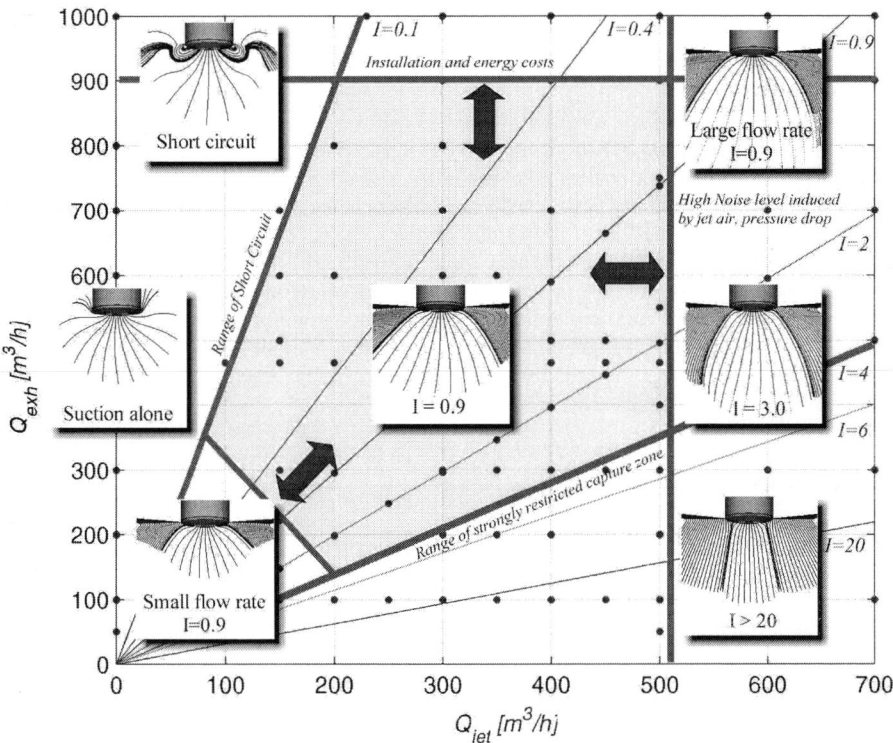

Figure 4: Limitations of operating parameters for enhanced extract system

1. Short circuit: The operating points where short circuit and thus a reduced performance occur are well defined by a line following the value of I=0.1. All operating points above I=0.1, do increase the capture range of the system.

2. Insufficient capture velocity: The second limitation for useful operating parameters is the minimal desired capture velocity. It highly depends on the application, the environment where the extract system is used and the exhaust geometry. It must be adapted to specific needs.

3. *Strongly restricted capture zone:* If the momentum ratio I is chosen too large, the expansion of the capture zone is restricted strongly. This is illustrated in figure 4 for a value of I>20. Considering turbulent mixing effects along the dividing streamline, the capture efficiency of such a system will be very bad. The upper limit for I must be defined from case to case but should not exceed I=4 (In practice maybe smaller).

4. *Costs for installations and energy:* The cost for installations and energy as well as industrial standards limit the range of extracted air volume to the upper side. Conventional ventilation systems usually do not exceed extracted flow rates of Q_{exh}=1200 m^3/h. This limit also must be adapted for work bench applications.

5. *Noise level induced by the air jet, pressure drop:* Finally, the field of operating parameters is limited by the noise level induced by the air jet and the pressure drop for producing it. Both noise level and pressure drops depend on the design of the jet nozzle and the average exit velocity. Improvement of the nozzle design might reduce the noise level significantly.

CONCLUSIONS

Calculations of the 3-dimensional flow pattern of a jet enhanced extract system using CFD are possible and agree fairly well with measurements. The resolution of small geometric details with large gradients of the air velocity need a carefully developed grid for the calculation. The amount of CPU time used for a fully 3-dimensional calculation is big due to the extreme differences in length scales. The influence of the laboratory walls in a confined space is significant. It is not possible to identify one optimized operating point for the specific exhaust, but a range of useful operating parameters has been defined. We have also shown rules for limits of useful operating points. It is important to specify the optimized OP according to the application of the extract system. Further investigations on the amount of entrained air by the jet and the sensitivity to disturbances are needed to bring the jet enhanced extract system to industrial applicability.

REFERENCES

Drkal F. (1970). Theoretische Bestimmung der Strömungsverhältnisse bei runden Saugöffnungen mit Flansch. *Heizung Lüftung / Klimatechnik Haustechnik* **21**,271-273.

Hunt G.R. and Ingham D.B. (1993). A Three-dimensional Axisymmetric Model of the Air Flow Pattern Created by a Local Exhaust Ventilation System Reinforced by a Turbulent Radial Jet Flow. *Aerosol Science* **24:1**, 15-16.

Pedersen L. G. and Nielsen P. V. (1991). Exhaust System Reinforced by Jet Flow, *Ventilation'91. Int. Symposium on Ventilation for Contaminant Control, Cincinnati.*

Saunders C.J. and Fletcher B. (1993). Jet Enhanced Local Exhaust Ventilation, *Ann. Occup. Hyg.* **37:1.**

"TASCflow User documentation", Advanced Scientific Computing Ltd., October 1995.

Weber R. and Sprecher P. (1999). Local Exhaust Ventilation with REEXS, *Ventilation'2000. Int. Symposium on Ventilation for Contaminant Control Helsinki.*

© 2000 Elsevier Science Ltd. All rights reserved.
Air Distribution in Rooms, (ROOMVENT 2000)
Editor: H.B. Awbi

EFFECT OF SASH HEIGHT AND OPERATOR
ON AIRFLOW IN A FUME HOOD

Allan T. Kirkpatrick and Robert Reither

Department of Mechanical Engineering, Colorado State University,
Fort Collins, CO 80523 USA

ABSTRACT

A three dimensional computational fluid dynamics (CFD) analysis has been used to predict airflow patterns in laboratory fume hoods. The simulation includes bypass fume hood primary operational features including the top and bottom bypasses, front airfoils, and rear slotted baffles. The study included the effects on the fume hood airflow of sash height changes, an operator positioned outside the fume hood, and equipment within the main fume hood chamber. It was shown for conditions of a fully open sash height, a person in front of the fume hood, and an object inside the fume hood, the fume hood experiences a loss of containment of the flow.

KEYWORDS

Fume hoods, airflow modeling, containment, sash effect

INTRODUCTION

The laboratory fume hood is a safety device used in chemical, research, and teaching laboratories. It provides a location at which to work on toxic substances with reduced risk to users. The primary function of the hood is to prevent contaminants from entering the laboratory. The fume hood basically provides an exhausted enclosure, operating at a negative pressure relative to the room, which is intended to vent contaminants away from the user and the laboratory space. There are several types of fume hoods including constant volume or bypass, restricted bypass, and auxiliary air fume hoods.

The subject of this paper is the effect of the fume hood sash position and the presence of the operator in front of the hood on the airflow patterns in a bypass fume hood. The model in the study simulates a bypass type bench fume hood with a vertical sash, as shown schematically in Figure 1. Major features of the bypass type fume hood are a top mounted exhaust duct, top bypass supply, bottom bypass supply, variable height vertical sash, back baffles, and front airfoils. Air enters the main fume hood chamber through one of three locations: sash opening, top bypass, and the bottom bypass formed by the bottom airfoil. The purpose of the top bypass is to maintain a constant volume of air entering the fume hood, regardless of the sash height. The back baffles are positioned such that air is exhausted directly from the work surface as well as the top of the main fume hood chamber. The front airfoils reduce the amount of turbulence and eddy motion entering the fume hood. In addition, the bottom airfoil is extended beyond the vertical plain of the sash to provide a direct flow (also called floor sweep) across the bottom of the fume hood main chamber.

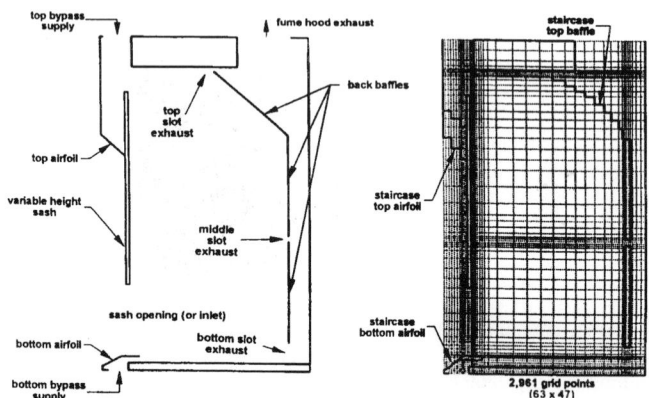

Figure 1: Bypass-type fume hood features

At present fume hoods are designed empirically and tested experimentally. For example, Chang (1994) measured the airflow around an operator standing in front of a fume hood. At present only a small number of fume hood studies have been numerical. Durst and Pereira (1991) performed a numerical analysis of an empty two-dimensional fume hood model, which included a back baffle, to study containment capabilities of fume hoods at selected sash openings. Inclusion of the airfoils, bypasses, and angled slotted baffles in this study furthers the previous numerical studies performed on fume hoods.

COMPUTATIONAL MODEL

A turbulent finite difference model was used for the simulation (Kurabuchi, T., J.B. Fang, and R.A. Grot. 1989). The computational model uses a standard two-equation κ–ε model with buoyancy terms included. The κ–ε model was selected because it is relatively stable and computationally efficient compared with the more complicated Reynolds stress models and is applicable to a wide range of turbulent flows (Neilsen, Peter V., 1998). There are seven non-dimensional equations used by the model: two vector partial differential equations (continuity, momentum), two scalar differential equations (turbulence kinetic energy, dissipation rate of turbulence energy), and one algebraic relationship for turbulent viscosity. There are also seven required empirical constants determined from correlations of experimental data (Launder, B.E., and Spalding, D.B. 1974).

To solve the system of non-linear partial differential equations, the code uses an explicit time marching algorithm, pressure relaxation method, and a central or upwind differencing scheme (Kurabuchi et al. 1989). A marker and cell method is used for the staggered grid. Velocity components are defined at the center of their normal cell faces and the scalar variables (pressure, temperature, turbulence kinetic energy, and dissipation rate of turbulence energy) are defined at the center of the cell. The wall boundary condition for the momentum boundary layer used by the code is based upon the power law assumption of the velocity profile. A detailed description of these boundary conditions can be found in Knapmiller and Kirkpatrick (1994). A contaminant dispersion study was also performed, using a turbulent contamination diffusion model (Kurabuchi et al. 1989). The converged flow field output file from the flow field calculation is used as input for computing the contaminant dispersion, since the velocity field is independent of the concentration field.

3D COMPUTATIONAL GEOMETRY

The airflow in a fume hood is three dimensional due to the finite width of the main chamber, the airfoils, and the presence of a person and object in front of and inside the fume hood, respectively. The three dimensional computational grid used to model the room and the laboratory fume hood is shown in Figure 2. The model has a room containing the fume hood resting on a cabinet, a person standing on the floor in front of the fume hood, and an object inside the fume hood. In addition, the model contains a variable height sash, top, bottom, and side airfoils, top, middle, and bottom baffles, and a rectangular exhaust. The room is approximately 109.5 inches (2.78 m) high, a half width of the fume hood wide (36", 0.91 m), and three fume hood interior depths long (83.5", 2.12 m). Two sash heights were studied, a nominal working height of 18" (46 cm) and a fully open position at 30.25" (77 cm). The three slot exhaust heights were fixed for all computations at: 0.75" (1.9 cm) top, 1.0" (2.5 cm) middle, and 3.375" (8.6 cm) bottom. Three fume hood depths were selected for the room length because beyond 2.5 fume hood depths the room airflow streamlines are unchanged.

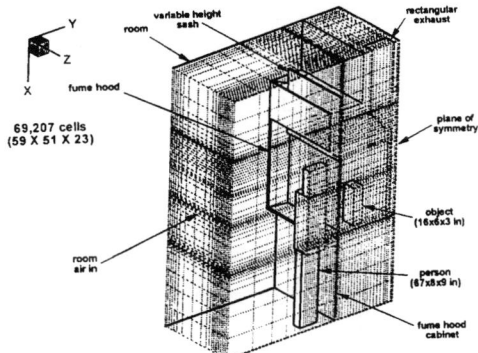

Figure 2: 3D room and fume hood computational grid

As seen in Figure 2 the room, fume hood, person, and object are split down the center in the computational model to take advantage of symmetry. Boundary conditions to start the simulation are as follows: air flows into the room, on the x-z plane at the front of the room, as a uniform flow. This room inlet air is given an initial y-velocity, turbulence energy, and a turbulence dissipation rate. The non-dimensional turbulence energy of the room inlet air is 0.005 and the non-dimensional turbulence dissipation rate is 0.00125; both values are typical of room air motion computations (Kurabuchi et al. 1989). The non-dimensional turbulence energy corresponds to a turbulent intensity of 5.8 percent. The room exhaust is through the rectangular exhaust at the top of fume hood, where only the pressure is specified and the velocity is computed by the model. Power law wall boundary conditions are specified on the ceiling and floor of the room as well as on the back surface of the fume hood. All other surfaces, including those of the fume hood, the person, and the object, have a zero normal velocity boundary condition with no shear. The top of the object in the fume hood releases contaminates, with a non-dimensional boundary value of 1.0.

Grid development of a fume hood model proved to be a very challenging problem. The actual fume hood contains three very thin airfoils, all at different angles, as well as a very thin top baffle at yet another angle. All these angled surfaces are modeled as staircases with an infinitely thin surface between adjacent cells, which have zero normal velocity. There are also three different sized flow entrances each in having their own direction and four exhausts of different sizes in two different directions. The grid contains 69,207 non-uniform cells (59 – x direction, 51 – y direction, 23 – z direction) optimized by size, location, and number of cells to include all of the above features, as well as the person, object, and variable height sash in the same grid. A grid sensitivity study was performed on a two-dimensional model of the room and fume hood. To validate the computations at each sash height, a smoke test and mass balance were performed on an empty fume hood. Further information about the validation experiments is given in Kirkpatrick and Reither (1998). The nominal computation time was 22 hours on a 200 MHz Pentium II desktop PC.

RESULTS

The results for the case of the working sash height set at 18" (46 cm) are shown in Figure 3. Figure 3 and subsequent similar figures show a vertical plane cut at the plane of symmetry of the model. As can be seen in the Figure, the flow enters the room from the left and enters the fume hood at any of the three entrances; sash opening, top bypass, or bottom bypass. The flow then exits the main fume hood chamber through the top, middle, or bottom slot exhausts. As shown in Figure 3 a vortex appears downstream of the sash, and the operator's body generates a small clockwise vortex downstream of the person at approximately waist level. Details of this small vortex are shown in Section A of Figure 3. The vortex is generated by the flow over the top of the person and down their body and the flow coming up between the person's legs. As shown in Figure 4, the computation also predicts that containment is maintained by the fume hood, i.e., there is no reverse flow out of the fume hood entrance, when the sash height is maintained at the nominal 18 (46 cm) working height.

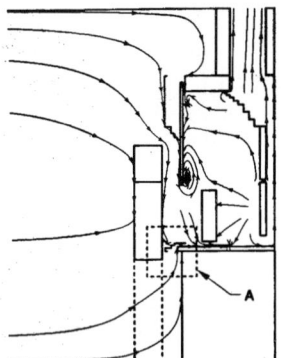
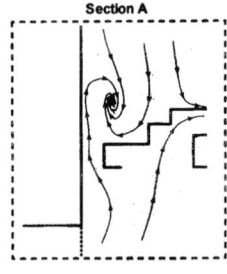

Section A

Figure 3: Vertical slice of airflow pattern with sash at nominal 18" (46 cm) sash opening

The airflow for the fully open sash is shown in Figure 5. Similar flow structures (vortices behind the sash and person and recirculation behind the object) are predicted as in the 18" (46 cm) sash height case. Figure 4 shows flow moving from the front of the object inside the fume hood to the small vortex downstream of the person outside of the main fume hood chamber.

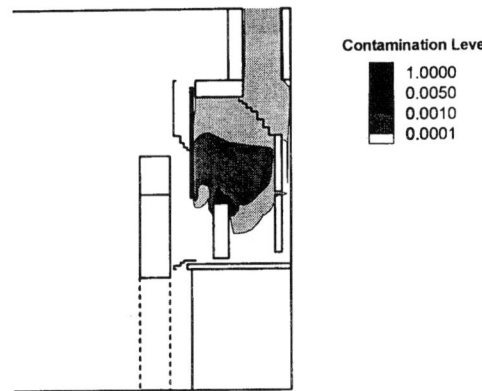

Contamination Level

1.0000
0.0050
0.0010
0.0001

Figure 4: Vertical slice of contamination profiles with sash at nominal 18" (46 cm) sash opening

The flow behind (downstream of) the object in the 30.25" (0.77 m) is predicted to flow over the top of the object and down, with just a slight sweep away from the back baffle. With the sash all the way up, more flow is allowed to pass over the top of the object and less flow is forced around the sides, as seen before in the 18" (0.46 m) sash height case. Section A of Figure 5 shows a small clockwise vortex behind (downstream of) the person of the same size as before, but located a little higher above the waist. Vortices are also seen in a horizontal slice for this configuration. A strong vortex now appears just downstream of the person and a weak flow recirculation exists just downstream of the object. For the fully open sash height, flow is no longer forced down the body of the person by the sash, consequently the flow around the sides of the person generates a vortex. The lowest pressures in the flow field are just downstream of the operator.

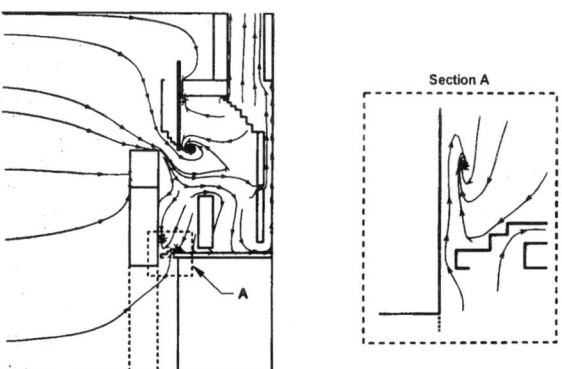

Section A

Figure 5: Vertical slice of airflow pattern for fully open sash at 30" (77cm)

Figure 6 displays the results of the contamination computation for the fully open sash. The most significant result for the fully open sash case is the prediction of loss of containment in the main fume hood chamber. This is a consequence of the low pressure vortex structure in front of the operator. As seen in Figure 6, the contaminant follows the airflow patterns of Figure 5. A small concentration of contaminant moves from the top of the object, down the front face of the object, and out of the fume hood main chamber, and into the laboratory space. This represents a failure mode of the fume hood, so that operation of the fume hood with the sash at a fully open height should be avoided.

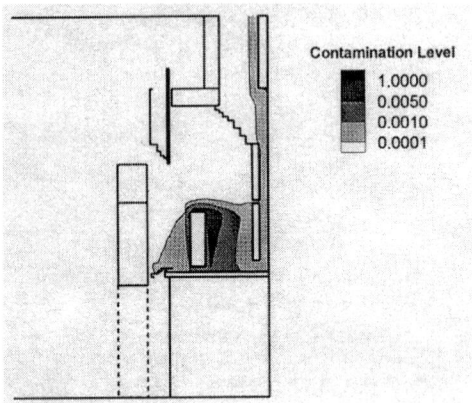

Figure 6: Vertical slice of contamination profiles for fully open sash at 30" (77cm)

SUMMARY AND CONCLUSIONS

The numerical study included the effects on the fume hood airflow of sash height changes, an operator positioned outside the fume hood, and equipment within the main fume hood chamber. With the fume hood sash at a nominal working height, there was no loss of contamination from the hood. However, with a fully open sash in a by-pass fume hood , the presence of a person in front of the fume hood will cause a loss of containment of the flow in the fume hood.

REFERENCES

Chang, S. (1994). Air Velocity Profiles Around a Person Standing in Front of Exhaust Hoods. *ASHRAE Transactions*, **100:2**, 439-447.

Durst, F. and Pereira, J. (1991). Experimental and Numerical Investigations of the Performance of Fume Cupboards. *Building and Environment*. **26:2**, 153-164.

Kirkpatrick A. and Reither R. (1998), Numerical Simulation of Laboratory Fume Hood Airflow Performance, *ASHRAE Transactions*, **104:2**.

Knapmiller, K. and Kirkpatrick A. (1994). Computational Determination of the Behavior of a Cold Air Ceiling Jet in a Room With a Plume. *ASHRAE Transactions*. **100:1**, 677-684.

Kurabuchi, T., Fang J., and Grot R. (1989). A Numerical Method for Calculation of Indoor Airflows Using a Turbulence Model. NISTIR 89-4211. Gaithersburg, MD: National Institute of Standards and Technology.

Launder, B. and Spalding, D. (1974). The Numerical Computation of Turbulent Flows. *Computer Methods in Applied Mechanics and Engineering*. **3**, 269-289.

Neilsen, P. (1998). The Selection of Turbulence Models for the Prediction of Room Airflow. *ASHRAE Transactions*. **104:1**.

Air Distribution in Rooms, (ROOMVENT 2000)
Editor: H.B. Awbi

THERMAL AND DYNAMIC STUDY OF A NEW CONCEPT OF HVAC SYSTEM USING AN ANNULAR AIR JET

F. Penot [1] and Ph. Meyer [1]

[1] Laboratoire d'Études Thermiques, E.N.S.M.A., UMR CNRS 6608,
Avenue Clément Ader, BP 40109, FUTUROSCOPE CHASSENEUIL CEDEX, FRANCE

ABSTRACT

This paper deals with an experimental study of a new concept of air conditioning using an annular air jet. A model has been realised and instrumented in order to check the behaviour of such a device in different temperature conditions and thermal orders.
A description of the device and the instrumentation are done. Visualisations, thermocouple measurements as well as 2D velocity fields obtained with the use of a LDV system are presented. The visualisations show the circulatory effect created by the annular jet. Temperature fields exhibit also a large isothermal region at a medium temperature between the hottest and the coldest temperatures (jet or ambient). This area corresponds to a protected area in which comfort conditions can be reached. Velocity measurements show the amplitude of the vortex located between the annular exhaust and the central air return. This device can be very useful to achieve a small microclimate in large volumes without any material confinement and a small consumption of energy.

KEYWORDS

Air conditioning , annular jet, flow visualisations, temperature fields, LDV, mixed convection.

INTRODUCTION

A new concept and a model of air conditioning system is presented. The major property of this device is to confine a volume, the protected area, in which thermophysical properties of air can be controlled without the use of any material wall. The confinement effect is created by an annular jet inclined from the horizontal direction. Heat exchangers induce the necessary temperature differences. The experimental device and the measurements associated are described. Flow visualisations, local temperature fields obtained with a very thin thermocouple and velocity measurements made with a 2D LDV system are compared in the case of warm jets flowing down in a fresh ambience and also for cold jets in a warm air at rest. The discussion deals with the global behaviour of the air flow in these different configurations. Secondary phenomena due to aiding mixed convection regimes or opposing mixed convection flows are also presented. This kind of air conditioning system constitutes a new approach which is not described in classical literature (ASHRAE (1995), Haines R. W. and Wilson C. L. (1994), COSTIC (1995)). Most of the systems use the development of jets to create conditions of comfort and a complete description of jets behaviour is given in Rajaratnam (1976). The influence of density variations on the jet is similar to the phenomena described in Rodi W. (1982), but we do not have any information concerning annular jets in the mixed convection regime.

THE ANNULAR JET SYSTEM

Description of the device

The principle of the annular jet system is to provide a free circulatory air flow coming out of a peripheral circular nozzle and drawn up to the centre where the air return is located. As shown in Figure 1, the jet looks like a torus located just under the device. Due to the entrainment of ambient air by the annular jet and to the conservation of mass between the air return and the nozzle, a vertical air flow is created below the primary stream. This secondary air flow completes the protected area which can be defined as the zone influenced by the dynamical and thermal conditions imposed by the annular jet system. In all cases studied the downstream air flow behaviour is influenced by the temperature of the air discharge. In fact, the jet works in the mixed convection regime, so the protected area can be modified by buoyancy forces acting locally. For some combinations of temperature and air velocity at the discharge, the circulatory effect can be destroyed and the protected area disappears.

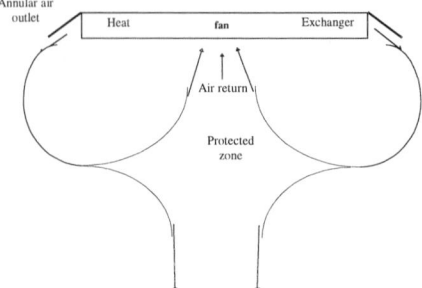

Figure 1: Schematic view of the annular air jet.

The device producing the annular jet is constituted of two elements: an air flow circuit and a heat exchanger as shown in Figure 2. A centrifugal fan draws up the air from the protected area and flings it through the copper pins of the heat exchanger with a large tangential component of velocity. So each particle of air has a helical trajectory that increases the distance passed inside the heat exchanger and consequently, improves the heat flux transmitted to the air flow. The electrical power applied to the fan is adjusted in order to get to a velocity of approximately 1.5 m/s at the annular outlet of the device. The corresponding flow rate varies between 0.02 and 0.04 m³/s.

Heat can be introduced or removed from the air flow through a heat exchanger connected to Peltier cells. The heat exchanger is made of a thin circular flat plate of copper (2 mm thick), on which, on one side, copper wood screws have been glued with a special thermal glue. Approximately 2 000 screws of 40 mm long and 3 mm in diameter are necessary to cover a wreath of 480 mm in outer diameter and 230 mm in inner diameter. The average heat transfer coefficient calculated with our experimental results reaches values very close to 400 W/m²/°C, which is a very good performance with the air as a working fluid and with this low velocity range.

The air flow is canalised between the copper plate and a plexiglas transparent sheet (2 mm thick) flattened against the pins. All around the circular outlet, a deflector made of plexiglas is mounted to produce the required inclination angle. These transparent walls give the opportunity to visualise the air flow and also to do local velocity measurements. In this model, 12 small square Peltier cells were used. Their characteristics, given by the manufacturer (Melcior), are the following: 62 mm long, 4,6 mm thick, maximum intensity and voltage of respectively 14 A and 15.4 V, reference CP2-127-06L. They were wired in series, and located all around a circle of 250 mm in diameter and tightened with nylon screws between the copper disc and a square aluminium plate of 5 mm thick and 420 mm by 420 mm long.

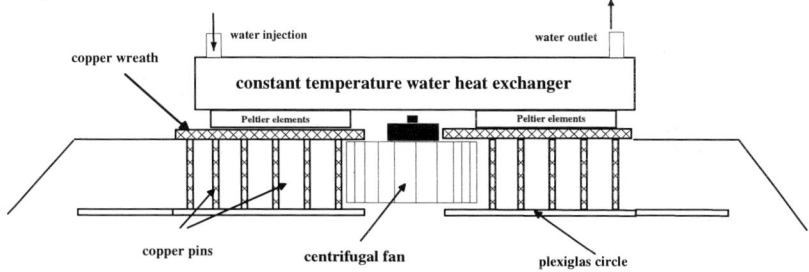

Figure 2: Vertical section of the device.

Thermal insulation is increased everywhere it is possible in order to reduce thermal shorts. Eventually the heat produced or removed on the other side by the Peltier cells is transported, via the aluminium plate, to another heat exchanger at constant temperature, in which water coming from a tank regulated in temperature circulates.

Thermal and dynamic instrumentations

This model was equipped with several devices to perform measurements of air characteristics, such as velocity and local temperature. Numerous thermocouples were also introduced to control the temperatures and the thermal behaviour of the device. As far as our interest is concerned, 12 thermocouples were distributed all around the nozzle (wired up in parallel to measure a bulk temperature) and 10 were located in the horizontal section of the air return. By the same way, 5 thermocouples were distributed in the test room to measure an average ambient temperature. The walls of the test room are maintained at a constant temperature by an exterior air stream regulated in temperature.

Temperature measurements were done with a very thin thermocouple (0.12 mm in diameter). A LDV system was used for velocity measurements. The light back-scattered by small sending particles (bubbles of edible oil, of 1 µm of diameter) is amplified and analysed by a BSA system (DANTEC) which provides the Doppler frequency of the signal and then the instantaneous and local component of the velocity at right angels to the plane of fringes. Both laser lens and the thin thermocouple are gripped on a displacement device controlled by a PC. This system enables us to scan a region of approximately 0.8 m high and 0.8 m wide. These quantitative measurements were complemented by flow visualisations to check the behaviour of the system in different configurations. White smoke is introduced into the air flow by a specific device and illuminated by a laser sheet coming from a semicylindrical lens. It is possible to obtain vertical or horizontal sections of the torus air stream. The amplitude of the protected area can be determined very easily and the influence of such parameters (blowing temperature, ambient temperature, inclination or amplitude of the velocity in the nozzle…) can be shown quickly, but in a qualitative way.

RESULTS

Flow visualisation

A vertical section of the jet is presented in Figure 3, with an air flow characterised by a difference of temperature of 5 °C between the nozzle and the ambient, and a flow rate of 0.037 m³/s.

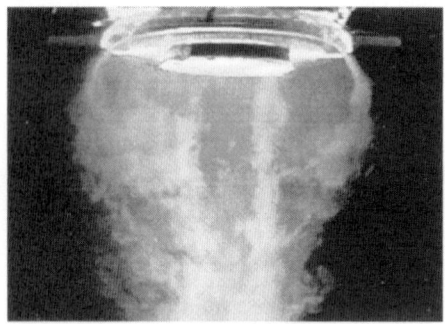

Figure 3 Flow visualisation of the annular jet and its surrounding

Figure 4 Lateral view of the annular jet with an obstacle.

Figure 5 Front view of the annular jet with an obstacle.

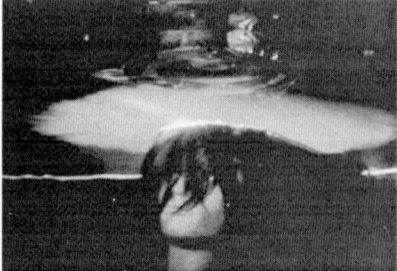

Figure 6 Bottom view of the annular jet with an obstacle.

One can see the area covered by the smoke injected at the inlet of the fan, which corresponds to the volume affected by the jet. The peripheral torus can be recognised and the laser sheet illuminates also the residual flow falling down under the jet. When an obstacle is introduced into the air stream, it is obvious that no any important perturbation of the main flow can be seen. For example, in Figure 4 and 5, two illuminated sections are shown (respectively a vertical section and a horizontal section), corresponding to the same configuration as previously, but for these tests, a dummy head was set below the annular jet. A black zone distinguishable above the head is due to its shadow. In Figure 6, an horizontal section round the top of the head is presented. The dummy head is surrounded by a large circle of white smoke, corresponding to the protected area.

Temperature measurements

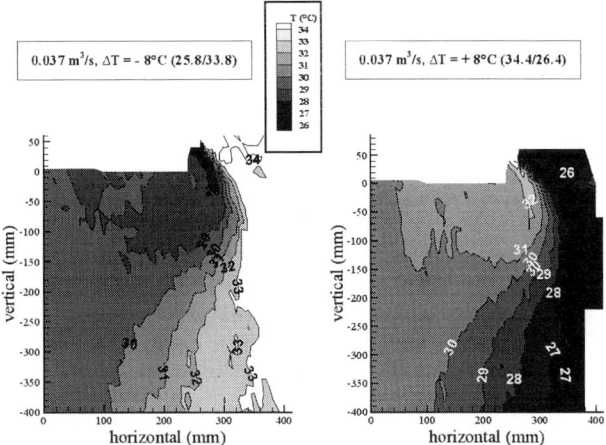

Figure 7 Temperature fields in a half vertical plane under the annular jet.
Cold jet in a warm environment (left). Hot jet in a cold environment (right).

Temperature profiles were made in a vertical section below the annular jet for different situations. Some of them were done with a warm annular jet injected in a cold ambient environment, others were done with a fresh jet injected in a warm ambience. In Figure 7, two cartographies are presented. They correspond to approximately the same flow rate (0.037 m^3/s) and the same absolute temperature difference (8 °C) between the jet and the ambient, but with a negative sign in the left hand side cartography (cold jet) and a positive one in the right hand side profile (warm jet). At first, a large volume at a temperature very close to the medium temperature between the air jet and the ambient can be observed below the device. One can also see that the mixed convection effects due to the local variations of the air temperature do not affect strongly the temperature fields. The isothermal line '30 °C' is the same in the 2 configurations.

Velocity measurements

Velocity measurements were done in the same cases as for the temperature measurements. If we examine now the velocity fields presented in Figure 8, one can see at first, that a big vortex corresponding to the torus flow is located near the jet nozzle. The volume at the medium temperature presented above is also affected by an air stream at a sufficient level to create conditions for comfort. However, one can distinguish some small discrepancies between the aiding mixed convection case and the opposing one. The separation streamlines for the return region and the flowing down region are somewhat the same in the case of a cold jet and for a warm jet. In this last case, local buoyancy forces are pointing upward and its horizontal expansion just downwind the nozzle seems to be reduced compared to the cold jet. Lower down, the horizontal section of the jet seems to be larger that in the other case. This could be due to the fact that the jet becomes turbulent sooner when it is warmed than when it is cold and the entrainment of ambient air is increased, so the descending flow takes up a larger section.

662

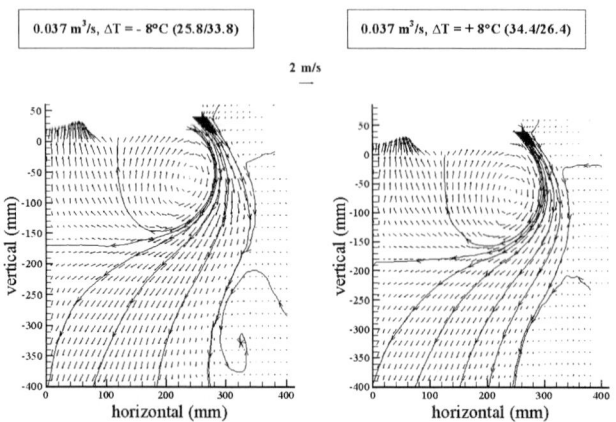

Figure 8 Velocity fields in a half vertical plane under the annular jet.
Cold jet in a warm environment (left). Hot jet in a cold environment (right).

CONCLUSION

A new device, producing a volume of air controlled in temperature and velocity without any wall have been described. Flow visualisations show that this annular jet induces a large vortex at the outlet and, lower down, a flowing down air stream. The recirculation effect is not strongly modified by the presence of an obstacle such as a dummy head.

Temperature fields exhibits also a large isothermal region at a medium temperature between the hottest and the coldest temperatures (jet or ambient). This area corresponds to a protected area in which comfort conditions can be reached.

Velocities measured with a two component LDV system show amplitude of the vortex, a circulatory flow between the annular exhaust and the central air return is quantified. The approximate jet boundary seems to be further from the vertical axis of the flow when the jet is heated. This device can be very useful to achieve a small microclimate in large volumes without any material confinement and a small consumption of energy.

References

ASHRAE (1995), *HVAC Applications I-P edition*, Ashrae handbook, chapter 1-28.

Haines R. W. and Wilson C. L. (1994), *HVAC System Design Handbook,* McGraw-Hill, Inc.

COSTIC (1995), Guide technique de la climatisation individuelle, SEDIT éditeur
Bouteloup J., Le Guay M., Ligen J.(1996), *Climatisation Conditionnement d'air Traitement de l'air,* Les éditions Parisiennes

Rajaratnam (1976), *Turbulent jets*, Elsevier

Rodi W.(1982), *Turbulent buoyant jets and plumes*, Pergamon Press.

© 2000 Elsevier Science Ltd. All rights reserved.
Air Distribution in Rooms, (ROOMVENT 2000)
Editor: H.B. Awbi

AN ISOTHERMAL AIR CURTAIN FOR ISOLATION OF SMOKING AREAS IN RESTAURANTS

J.P. Rydock, T. Hestad, H. Haugen, J.E. Skåret

Building Technology Department
Norwegian Building Research Institute
Oslo, Norway

ABSTRACT

An isothermal air curtain for isolation of smoking areas in restaurants was designed, built and evaluated in a test facility using oil-smoke visualisation and tracer measurements. The test facility was a ventilation test room set up as a small restaurant, with tables, chairs, person simulators (cylindrical heat sources) and balanced mechanical ventilation. Fresh air was supplied in the non-smoking section of the room, exhaust air drawn from the smoking area, and the air curtain was attached to the ceiling between the two sections. The air curtain was a plenum chamber with adjustable slot width and mounting angle fed by a supply fan drawing air from the smoking section of the room. For reasonable room ventilation rates for a restaurant (11 l/s per person supply and exhaust air), the optimised air curtain yielded tracer concentrations in the non-smoking section as low as 5 – 10 % of the values measured at the same time in the smoking section. The limiting factor in the performance of the curtain was found to be the ability to properly supply enough air to the clean-air side of the curtain to prevent recirculation of polluted air from the smoking area into the non-smoking section. This study demonstrates that an isothermal air curtain solution to control contaminant spread need not necessarily require excessive ventilation rates and prohibitive operating costs.

KEYWORDS

Air Curtain, Tracer Testing, Airflow Visualisation

INTRODUCTION

The new amendments to the Norwegian Smoking Law of 1988 require that at least one-half of the tables in a restaurant must be designated as non-smoking and that non-smoking areas are to be kept smoke-free. Erection of physical barriers to achieve compliance with the law is, however, not required. Cigarette smoke in non-smoking areas is considered to be a ventilation problem that requires a ventilation solution.

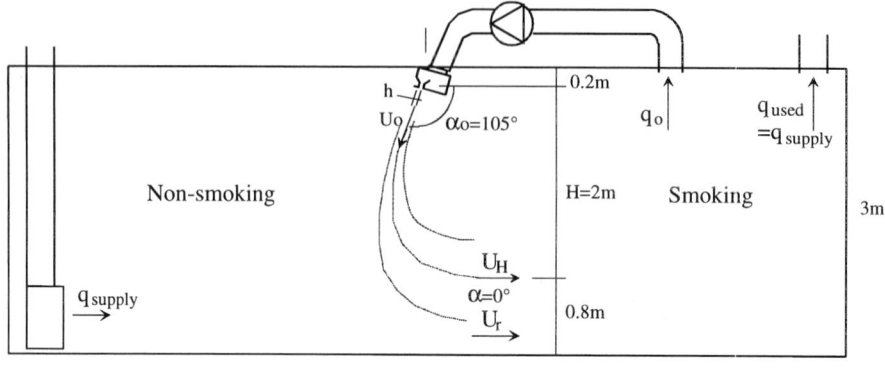

Figure 1 – Schematic of air-curtain in test-room.

One possible approach to keeping one area of a room smoke-free while people are smoking in another area of the same room is with an air curtain. Air curtains have been the subject of a number of research papers since the mid-1970's, primarily as a means of controlling heat loss from entrances to buildings during winter. Less common are applications where isothermal air curtains have been studied for use in contaminant control. Li and Peng (1994), for example, examined a push-pull air curtain in a scale-reduced model that was then implemented to control duck down pollution in a factory in China. As another example, Ho, et. al. (1994) used CFD to design an air curtain to control the spread of maleic anhydride in a polymer manufacturing facility.

This paper presents results of smoke and tracer testing of an isothermal air curtain for controlling the spread of cigarette smoke in restaurants. The study was carried out in the ventilation testing room of the Norwegian Building Research Institute.

DESIGN OF THE AIR CURTAIN

The test facility was a 9m x 6m x 3m ventilation test room set up as a small restaurant, with tables, chairs, person simulators (100 Watt cylindrical heat sources) and balanced mechanical ventilation. Fresh air was supplied in the non-smoking section of the room, exhaust air drawn from the smoking area, and the air curtain was bolted to the ceiling between the two sections (see Figure 1). The air curtain was a plenum chamber (see Figure 2) with adjustable channel width and mounting angle fed by a supply fan drawing air from the smoking section of the room.

The equations below were used for the design of the air curtain. A cross-section of the room is shown in Figure 1, with arrows denoting initial locations for supply and extract air. The isothermal air curtain is supplied with air from the smoking section. The non-smoking section is supplied with as much fresh air as is extracted from the smoking section, yielding balanced mechanical ventilation in the room.

The necessary impulse in the air curtain is determined by the pressure difference between the smoking and non-smoking sections. The air velocity between the zones, U_r is chosen = 0.25 m/s. This velocity corresponds to a dynamic pressure = $0.6 \cdot U_r^2$ =0.04 Pa, which is the pressure difference between the zones.

Assuming the jet centreline has become horizontal approximately 0.8 m above the floor, H is approximately 70% of the height from the floor to the air curtain channel. An analysis of the energy-momentum balance in the jet gives (from Skåret, 1999):

$$y = M (\cos\alpha - \cos\alpha_0) / \Delta p = H \qquad (1)$$

Where M is the momentum flux per unit length of curtain, Δp is the pressure difference, and α, α_0 and H are as defined in Figure 1. Eqn. 1 yields:

$$M = H \cdot \Delta p / (\cos\alpha - \cos\alpha_0) = 2.0 \cdot 0.04 / (1 - (-0.26)) = \underline{0.063} \text{ N/m}$$

We can now use:

$$M = \rho b_0 U_0^2 \ i/\varepsilon \qquad (2)$$

Where U_0 is the start velocity in the curtain, b_0 is the start jet-width, i is the net momentum flux coefficient, taking into account momentum losses, ε is the contraction coefficient for a nozzle, and ρ is the density of air.

We can with good approximation set the channel width $h = b_0 \ i/\varepsilon$ to determine the necessary start flowrate in the air curtain. If we choose h = 0.01 m as a starting value, then:

$$U_0^2 = M/(h*\rho) = 0.063 / (0.01*1.2) \qquad (2a)$$

$U_0 = \underline{2.29 \text{ m/s}}$
$q_0 = 2.29 \cdot 0.01 = 0.0229 \text{ m}^3/\text{s,m} = 23 \text{ l/s,m}$ (For 6 m channel = $\underline{138 \text{ l/s}}$)
                                                                                       ($q_0 = 497 \text{ m}^3/\text{h}$)

The velocity in the jet centre 0.8 m above the floor, U_H, is found approximately from the general equation for a free jet from a channel where the jet length is set to approximately 3m and the vent constant $\cong 7.0$:

$$U_H = U_0 \cdot \sqrt{K_2 h / x} = 2.29 \cdot \sqrt{7 \cdot 0.01/3} = \underline{0.35 \text{ m/s}}$$

The total air volume in the jet at U_H (after 3 m) = q_H

$$q_H / q_0 = \sqrt{2} \cdot U_0 / U_H = \sqrt{2} \cdot 2.29 / 0.35 = 9.25 \qquad \text{or } q_H = 9.25 \cdot 23 = 213 \text{ l/s,m}$$

The entrained airflow into the jet at 3 m = $q_H - q_0$ = 213 – 23 = 190 l/s,m
The entrained airflow from the non-smoking section = 190 / 2 = 95 l/s,m   (For 6 m channel = <u>570 l/s</u>)
To prevent recirculation from the curtain, The fresh air supply to the non-smoking section should be greater than or equal to the entrained airflow from the non-smoking zone:

$$Q_{supply} = 570 \text{ l/s } (=2052 \text{ m}^3/\text{h})$$

This air volume passes under the air curtain with velocity $U_r = 0.25$ m/s. The height of the column passing under the curtain is then $= 0.095 / 0.25 = 0.38$ m, which corresponds reasonably with the available height under the jet.

A necessary condition is that the supply air does not short-circuit to the smoking section before the curtain entrains it.

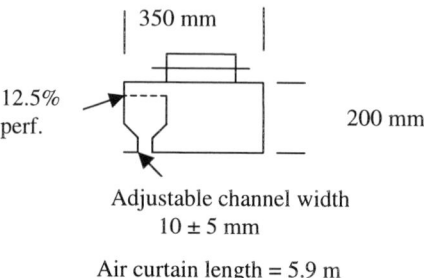

Air curtain length = 5.9 m

Figure 2 – Cross section of the air curtain

TESTING OF THE AIR CURTAIN

The air curtain was evaluated and optimised using smoke and tracer releases. Tracer concentrations were measured with a Bruel & Kjær Multi-gas Monitor Type 1302 and Multipoint Sampler and Doser Type 1303. Concentrations were measured at 5 locations in the test-room: At the exhaust vent in the smoking section, at a height of 1.5 m in the middle of the smoking section, at two locations at a height of 1.5 m in the non-smoking section, and in one of the supply vents in the non-smoking section. Tracer was released at three locations in the smoking section and dispersed with small fans to ensure well-mixed conditions.

Tests were done with air curtain slot widths of 10 mm, 5 mm and 3 mm, with varying mounting angles of the curtain with respect to the non-smoking zone (+15, 0 and –15), with varying supply and extract airflow rates and with several different non-smoking section supply air configurations.

RESULTS AND DISCUSSION

Smoke tests suggested that the curtain was effective at holding smoke out of the non-smoking section of the test room. However, the air curtain could not, in any of the tests, maintain a *completely* smoke-free environment in the non-smoking section. There was always at least a small but noticeable degree of transfer of smoke across the curtain. This was confirmed in the tracer tests. Figure 3 shows an example of a tracer test in the test room. The tracer release was started at approximately 17:00 and shut off at around 17:30. Concentrations rose quickly at the extract air and smoking section measurement locations, and levelled off at approximately

45 parts-per-million (ppm). The total tracer release rate from the three release points was 9 ml/s = .032 m³/h with a supply air flow of 1500 m³/h. Tracer concentrations in the non-smoking section were dramatically lower during the release, demonstrating the air curtains' effectiveness at containing the contaminant. Measurable concentrations of tracer in the supply air were evidence that there was a small degree of short-circuiting between the exhausts and supply ducts, but not enough to account for the tracer concentrations in the non-smoking section of the room.

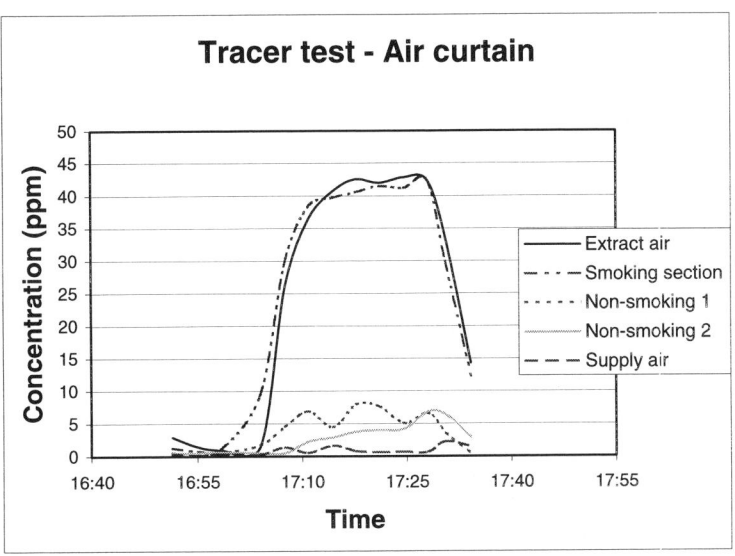

Figure 3 – Tracer test of air curtain.

A 5-mm slot width was clearly superior to both 10-mm and 3-mm slot widths in both smoke and tracer tests. The air curtain also performed optimally when directed at an angle toward the smoking section and with as much source air as possible (in this case the maximum output of the curtain supply fan was 600 m³/h, or 100 m³/h per meter of air curtain). It is interesting to note that the air curtain flowrate in this study was a small fraction of the flowrates in the air curtains discussed in Li and Peng (1994) and Ho and Goodfellow (1994). Li and Peng used a curtain with a 70-mm slot width with a velocity of 3.9 m/s, which corresponds to 983 m³/h per meter of air curtain. Ho and Goodfellow do not give the curtain length or slot width in their paper, but the velocity was 10.2 m/s and the air curtain flowrate was 13,600 m³/h. The more modest flowrate in the air curtain in this study implies better comfort for patrons and a more reasonable operating cost for the proprietor.

Equally good performance was achieved with a non-smoking section supply airflow of 2000 m³/h down to about 1500 m³/h (3/4 of the design value). A fresh-air supply of 1500 m³/h (416 l/s) for a restaurant with a floor area of 54 m² yields a quite reasonable per-person fresh-air supply of 11 l/s (using the recommended design value of 1,4 m²/person in the Norwegian Building Code).

A noticeable decline in performance occurred for a room supply rate below 1500 m³/h. The air curtain also performed relatively poorly with displacement ventilation from floor-based supply air terminals, yielding tracer concentrations in the non-smoking section 30-60% of the well-mixed value in the smoking section. The bulk of the non-smoking section supply air short-circuited across the floor under the curtain and was therefore not available to provide the induced flow necessary to prevent eddies of polluted curtain air from recirculating into the non-smoking section. Considerably better performance (tracer concentrations in non-smoking section down to 5-10% of concentrations in smoking section) was achieved using ceiling-mounted low-impulse supply air terminals.

CONCLUSIONS

This study demonstrates the feasibility of using an air curtain to provide improved air quality to non-smokers in a space to be occupied by both smokers and non-smokers. Both smoke and tracer tests demonstrate that the air curtain is effective at containing pollutants in the smoking zone of a room partitioned with the air curtain. The results suggest, however, that a completely smoke free environment is not attainable in a section of a single room if smoking occurs in another part of the same room and they are partitioned with an isothermal air curtain. A small flux of contaminant across the partition appears to be unavoidable.

The air curtain performed equally well with a room-supply airflow down to 3/4 of the design value (11 l/s per person), provided that the supply air was well mixed in the non-smoking section before entrainment into the air curtain. In addition, the air curtain start flow (100 m³/h per meter of air curtain) was substantially smaller than in previous related studies, demonstrating that an air curtain solution need not necessarily require excessive ventilation rates. This is important for a restaurant application both in terms of comfort for patrons and operating costs for the owner.

ACKNOWLEDGEMENTS

This project received financial support from the Confederation of Norwegian Business and Industry.

REFERENCES

Ho F.C.M. and Goodfellow (1994). The application of computational fluid dynamics to predict contaminant concentration in a polymer facility. *Proceedings of the 4th International Symposium on Ventilation for Contaminant Control,* held in Stockholm, September 5-9, 1994.

Li Q. and Peng S. (1994). Duck down pollution control with recycle air curtain. *Proceedings of the 4th International Symposium on Ventilation for Contaminant Control,* held in Stockholm, September 5-9, 1994.

Skåret J.E. (1999). *Ventilasjon Kompendium* (in Norwegian), Norges byggforskningsinstitutt, Oslo, Norway.

Air Distribution in Rooms, (ROOMVENT 2000)
Editor: H.B. Awbi

JET ASSISTED LOCAL EXHAUST VENTILATION

- THE EFFECT OF JET DIRECTION

G.R. Hunt[1], X. Wen[2] & D.B. Ingham[3]

[1]Department of Applied Mathematics and Theoretical Physics, University of Cambridge,
Silver Street, Cambridge, CB3 9EW, UK.
[2]Environment Centre, University of Leeds, Leeds, LS2 9JT, UK.
[3]Department of Applied Mathematics, University of Leeds, Leeds, LS2 9JT, UK.

ABSTRACT

The effectiveness of flanged exhaust hoods that rely purely on suction for the removal of airborne contaminants and fumes from the workplace is known to be significantly enhanced by the appropriate use of a turbulent blowing jet of air. Systems employing both suction and blowing are commonly referred to as jet-reinforced hoods, and the airflows these hoods induce have been measured and predicted when the blowing jet of air is directed *perpendicular* to the direction of the flow into the exhaust, *e.g.* for a bench mounted Aaberg slot exhaust hood, the air jet is directed vertically upwards from a narrow rectangular slot and the suction into the exhaust is directed horizontally along the surface of the bench.

In this paper we examine the effect which the direction of the blowing jet has on both the air speeds induced by a slot exhaust hood and the shape and size of the region from which contaminant capture is possible. Our new theoretical results show that the performance of jet-assisted hoods may be enhanced considerably if the angle between the initial jet direction and the suction direction is less than 90°. The implications are lower operating costs at no reduction in performance.

KEYWORDS

Local extract ventilation, jet-assisted flows, jet direction, capture region, enhanced airflow.

INTRODUCTION TO JET-ENHANCED EXHAUSTION - THE AABERG PRINCIPLE

'Traditional' local exhaust ventilation systems: Local exhaust ventilation systems (LEVS) remove airborne pollutants from the industrial workplace by creating a low pressure region in front of an exhaust hood that is nominally positioned close to the source of the contaminant generation. Traditionally, the LEVS have relied purely on the suction produced by fans to create the low pressure region. The suction is not focused in any particular direction and, consequently, the hood draws air from all directions and induced air speeds decrease rapidly with increasing distance from the hood. In

ideal circumstances (*i.e.* in a still environment and for neutrally-buoyant and non-diffusive contaminants), contaminant released from any region of the workplace is successfully captured. In practice, capture is successful only from regions where the air speeds induced toward the suction opening of the hood are greater than the level of the background air disturbance and the buoyancy-induced velocity of the contaminant. For effective removal of contaminant, the hood therefore needs to be placed close to the source of generation, which may prove impractical, or a further reduction in pressure is required in front of the hood so as to facilitate capture from larger distances. This further decrease in pressure is normally achieved simply by increasing the suction. The energy requirements of the fans producing the suction may be high and this energy expenditure is very inefficient as a significant fraction of the exhausted air is drawn from typically uncontaminated regions of the workplace, *e.g.* behind the hood. The effectiveness of these hoods is enhanced, to some extent, by attaching a flange to the exhaust inlet thereby limiting the directions from which the exhaust can draw air, although the air speeds still decrease rapidly in front of the hood.

Jet-assisted LEVS: The effectiveness of 'traditional' flanged exhaust hoods is known to be significantly enhanced by appropriately combining a turbulent blowing jet of air with the existing suction. Systems employing both suction and an air jet are referred to as Aaberg, or jet-reinforced, exhaust hoods, see figure 1. When operating with a suitable ratio of blowing and suction strengths, the flow induced by the jet leads clean air away from the region in front of the hood and thereby focuses the suction directly towards the source of the contaminant. This results in quite different and enhanced velocity profiles in front of the hood, which now selects air from a well-defined region. The concentrated directional suction of the Aaberg exhaust hood can then be 'aimed' directly at the source of the contaminant and thus facilitates capture from great distances and with a higher concentration of contaminant in the exhaust than is possible when using conventional hoods.

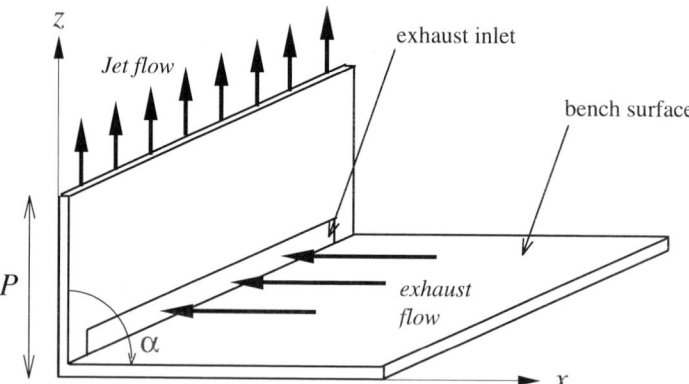

Figure 1: Schematic of a jet-assisted slot exhaust hood. For a conventional jet-assisted hood $\alpha = 90°$.

The jet-assisted airflows induced by these hoods have been studied both by experiment and theoretical analysis when the air jet is directed *perpendicular* to the flow into the exhaust. Air speeds and capture efficiencies produced by prototype Aaberg exhaust hoods have been measured under laboratory conditions by a number of authors, including Pedersen & Nielsen (1991) and Fletcher & Saunders (1993). In these studies the hood comprised a circular exhaust opening surrounded by a disc from the perimeter of which a radial jet was ejected to produce an approximately axisymmetric airflow pattern. Flow visualisations showed that the induced airflow consists of two distinct regions: an *efficient* region, in which air is drawn successfully towards the hood and may be exhausted, and a *labile* region, from which air initially moving toward the hood will be not be sampled but blown back into the workplace via the jet flow. The size and shape of these regions is characterised by the ratio of the momentum flux of the jet to the momentum flux of the exhaust flow. Increasing this ratio increases the range of the hood but reduces the width of the efficient region. Fletcher & Saunders (1993) obtained qualitatively similar findings for a jet-reinforced slot exhaust hood (figure 1) which produces an approximately two-dimensional airflow. By considering a vertical section through the flow then the position of a particular streamline, which we refer to as the *dividing streamline*, traces the 'boundary' between the efficient and labile regions.

A theoretical analysis of the Aaberg exhaust hood by Hunt & Ingham (1992), for the two-dimensional Aaberg slot exhaust hood, and Hunt (1994) and Hunt & Ingham (1996), for the axisymmetric hood, revealed that the shape of the capture region depends upon the geometry of the hood, via the geometric parameter S, and the relative strengths of the exhaust and jet flows, via the operating parameter G, where

$$S = \frac{a}{P}, \quad G = \frac{A}{m}, \quad \text{where} \quad A = cP^{1/2}K^{1/2}.$$

In the above expressions, P denotes the distance between the centre of the exhaust inlet and the orifice of the jet, a is the width of the exhaust inlet, K is the initial kinematic momentum flux of the jet (*i.e.* a measure of the jet strength), m is the volume flow rate of the exhaust, and c is an empirical constant which depends upon the entrainment into the turbulent jet ($c = 3^{1/2}/2\sigma$, where $\sigma = 7.67$ is related to the spreading rate of a two-dimensional jet). The component of the overall airflow produced by the jet-induced flow (*i.e.* in the absence of suction) is controlled by the parameter A.

In its conventional form, the jet flow of the Aaberg is directed perpendicular ($\alpha = 90°$, see figure 1) to the direction of the suction. In this case, the air speeds deduced from the potential flow analysis of Hunt & Ingham cited above show good agreement with the available experimental measurements over a broad range of operating conditions. As a result of this simple analysis, the basic airflow pattern is well understood when $\alpha = 90°$. This close agreement between measurement and predictions gave us confidence to consider a fundamental design modification, as yet unconsidered experimentally, namely, the effect of tilting the flange and jet forwards so that $\alpha < 90°$.

We now present results of a new theoretical model for the assumed two-dimensional airflow pattern generated by an Aaberg slot exhaust hood which examines this effect, and which indicates that the overall flow may be significantly improved if $\alpha < 90°$.

NEW MODEL TO INCLUDE VARIABLE JET DIRECTION

The two-dimensional flow field generated by an Aaberg slot exhaust hood with $\alpha = 90°$ has been determined analytically (Hunt & Ingham, 1992). For $\alpha \neq 90°$ an analytical solution was not sought and a boundary-integral technique used to determine the airflow. Details of this procedure are given by Wen, Hunt & Ingham (1999). The turbulent jet was modelled as a vertical line sink of fluid, the strength of the sink was determined following Schlichting (1968). The jet-induced flow was modelled as inviscid potential flow. For simplicity, the suction inlet was modelled as a horizontal line sink at the base of the exhaust flange; the flow is then characterised by the single parameter G.

Our primary objective is to determine how the airflow and, in particular, the capture region of the hood, responds to changes in the initial jet direction. The basic airflow is of paramount interest, and so sources of contaminant are not modelled explicitly, other than to assume that a neutrally-buoyant contaminant is distributed throughout the flow and that diffusion effects are negligible. Under these assumptions the contaminant simply follows the streamlines. Typical operating conditions which produce a balanced airflow with a prototype Aaberg slot exhaust hood with $\alpha = 90$ result in $G \approx 0.2$

(Pedersen 1991). For illustrative purposes we shall take $G = 0.2$ and examine the flows induced by hoods with $\alpha = 90°$, $75°$ and $60°$.

RESULTS & DISCUSSION

At this stage it is convenient to return to the notion of the dividing streamline. This streamline separates the flow being drawn towards the jet (*i.e.* in the labile region above $\psi = 1$, figure 2) from that travelling towards the suction inlet (*i.e.* the efficient region below $\psi = 1$). Under our assumptions, contaminant released below the dividing streamline will be successfully exhausted while contaminant released above the dividing streamline will be entrained by the jet and 'blown' back into the room.

The effect of tilting the exhaust flange and jet forwards, on the size and shape of the region from which contaminant capture is likely to be successful, is now examined. Figure 2 shows predictions of the induced airflow pattern in front of the hood for $\alpha = 90°$, $75°$ and $60°$ in a region which extends $5P$ and $4P$ in the horizontal and vertical directions, respectively.

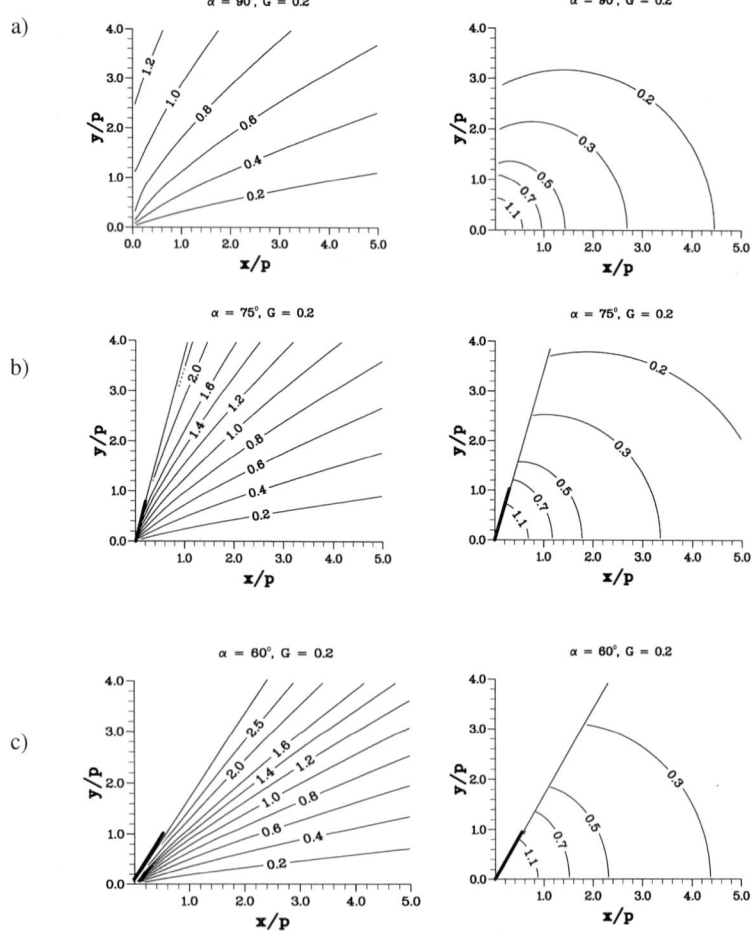

Figure 2: Streamlines (left-hand column) and lines of constant air speed (right-hand column) for a) $\alpha = 90°$, b) $75°$ and c) $60°$, $G = 0.2$. $\psi = 1$ is the dividing streamline.

Note that as α decreases, the dividing streamline $\psi = 1.0$ is displaced towards the bench surface. The width of the capture region is thereby reduced implying increased air speeds in the efficient region. This is confirmed in the plots of constant air speed which predict a significant increase in the air speed in the region in front of the hood with decreasing α, *e.g.* with $\alpha = 90°$ a non-dimensional air speed of 0.3 (the air speeds plotted in figure 2 are scaled on the characteristic air speed m/P) is achieved at a distance of approximately $2.6P$ from the suction inlet; this increases to almost $4.4P$ with $\alpha = 60°$.

This increase in air speed is due in part to an enhanced jet-induced flow and in part to an enhanced suction flow. When $\alpha = 90°$, the suction is focused in a $90°$ sector bounded by the exhaust flange and the work surface of the bench. The angle of this sector decreases as the flange and jet are tilted forwards, and thus the flange limits the directions from which the exhaust can draw air. The suction is thereby confined in an increasingly small sector in front of the hood and thus becomes effectively stronger as α decreases. This effect occurs irrespective of the jet direction.

In order to examine precisely the effect of the jet direction on the overall air flow, the contribution to the total air speed towards the inlet that results solely from the jet-induced flow has been determined. Figure 3 illustrates how this purely jet-induced component $U_\alpha(x,0)$ varies with distance from the exhaust inlet. Air speeds are non-dimensionalised with respect to the jet-induced component $U_{90}(x,0)$ at $\alpha = 90°$. From figure 3 it is evident that for $X/P \geq 1.5$, tilting the flange forwards from $\alpha = 75°$ to $\alpha = 60°$ results in a significant increase in the jet-induced air speeds along the bench surface towards the exhaust inlet. Thus for $\alpha < 90°$, the enhanced air velocities toward the exhaust inlet are not due entirely to the enhanced suction flow but may be strongly attributed to the enhanced jet-induced flow produced by the inclined jet. This effect is also predicted for further decreases in α. In practice, there will be a critical angle at which the jet-induced velocity component attains a maximum value, with any further reduction in the angle then producing a detrimental effect to the flow. Beyond the critical angle, the jet will tend to blow fluid away from the hood. In the current model the jet is assumed to have zero thickness and a critical angle of α is not predicted. The model is therefore inappropriate for small α.

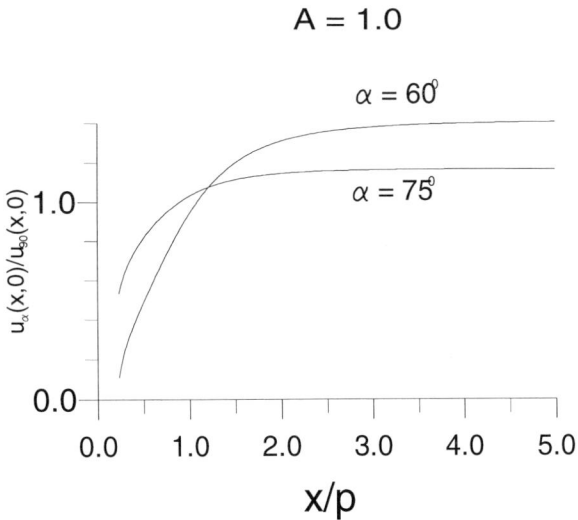

Figure 3: Air speeds induced along the bench surface
solely by the jet with $A = 1.0$ for $\alpha = 75°$ and $60°$.

CONCLUSIONS

Conventional jet-assisted local exhaust ventilation systems, such as the Aaberg exhaust hood, discharge a turbulent jet in a direction perpendicular to the flow into the exhaust inlet (*i.e.* $\alpha = 90°$). However, our results indicate that the airflow produced by these conventional hoods is significantly enhanced by tilting the exhaust flange and jet forwards so that $\alpha < 90°$.

Predictions of the two-dimensional airflow produced by a jet-assisted slot exhaust hood demonstrate that the air speeds towards the hood increase as the exhaust flange and jet are tilted forwards, although the vertical extent of the region from which contaminant capture is possible, decreases. The increase in air speeds towards the hood is due both to the suction and the jet-induced flow.

Inclining the jet forward increases the range over which contaminant capture is possible without the need to increase suction velocities or the strength of the jet, *i.e.* with no increase in operating costs. The implications are for a quieter and more energy efficient exhaust with no loss in performance.

The model described is appropriate for small angles of tilt and further research is required before reliable guidelines for predicting optimum tilt angles to suit a particular application and contaminant can be produced. Currently, mathematical analysis provides the only means of considering the airflows produced by different initial jet directions as prototype hoods for test under laboratory conditions exist for $\alpha = 105°$ and $\alpha = 90°$.

ACKNOWLEDGEMENTS

GRH and XW would like to thank the ESPRC for their financial support.

REFERENCES

Fletcher, B. & Saunders, C.J. (1993) Jet enhanced local exhaust ventilation. *Annals of Occupational Hygiene*, **37**, 15-24.

Høgstead, P. (1987) Air movements controlled by means of exhaustion. ROOMVENT '87. *International Conference on Air Distribution in Ventilated Spaces,* Stockholm.

Hunt, G.R. & Ingham, D.B. (1992) The fluid mechanics of a two-dimensional Aaberg exhaust hood. *Annals of Occupational Hygiene*, **36**, 455-476.

Hunt, G.R. & Ingham, D.B. (1993) A mathematical model of the fluid flow of a slot exhaust hood reinforced by a two-dimensional jet flow. *J. Mathematical Engineering in Industry*, **4**, 227-247.

Hunt, G.R. (1994) The fluid mechanics of the Aaberg exhaust hood. *Ph.D. thesis*, Department of Applied Mathematical Studies, the University of Leeds.

Hunt, G.R. & Ingham, D.B. (1996) Long-range exhaustion - a mathematical model for the axisymmetric air flow of a local exhaust ventilation hood assisted by a turbulent radial jet. Annals of Occupational Hygiene, 40, 171-196.

Pedersen, L.G. & Nielsen, P.V. (1991) Exhaust system reinforced by jet flow. Ventilation '91. *International Conference on Ventilation*, Cincinnati.

Schlichting, H. (1968) Boundary-layer theory. 6th Edition, McGraw Hill Book Company, New York.

Wen, X., Hunt, G.R. & Ingham, D.B. (1999) A boundary integral technique for the numerical modelling of the air flow for the Aaberg exhaust system. Submitted to *Engineering Analysis with Boundary Elements*.

Air Distribution in Rooms, (ROOMVENT 2000)
Editor: H.B. Awbi

NUMERICAL MODELLING OF CLOSELY SPACED SUPPLY DIFFUSERS IN A HIGH AIR CHANGE RATE APPLICATION

T. Abbas[1], M. J. Seymour[2],W. B. Booth[1] & P. M. Rose[2]

[1]BSRIA, Old Bracknell Lane West, Bracknell, Berkshire, RG12 7AH
[2]Flomerics Limited, 81 Bridge Road, Hampton Court, Surrey, KT8 9HH, UK

ABSTRACT

A full size physical mock-up of a room in which temperature was required to be controlled to very tight tolerances ($\pm$ 0.1°C) was constructed in BSRIA's laboratory on behalf of a client. Part of this work involved using numerical modelling techniques to predict the patterns of room air movement. This was carried out for specific configurations of the physical mock-up to provide validation. In addition, several scenarios were investigated using predictions only. The air change rates were all 60 or above per hour.

The key issues with regard to the numerical modelling were the accurate representation of a single (isolated) diffuser and of multiple diffusers in a closely spaced array.

Surveys over an appropriate grid were carried out close to the diffuser(s) using a pole-mounted array of six anemometers with integral temperature sensors. Analysis of the measurements enabled the numerical representation of the diffuser(s) to be optimised. Data visualisation techniques were employed to enable the project team to readily identify the fit of predicted and measured data.

The paper describes the faithfulness of the representation of the specific supply diffuser – on its own and in combination. Further, the prediction of air temperatures and velocities in the close temperature control room is reported for a range of extract locations and furniture layouts. The influence of heat loads (simulated equipment) on room performance is characterised. In some cases, comparison is made with the data collected in the physical mock-up.

The detailed description of the numerical modelling of the supply diffusers is the prime focus of the paper.

KEYWORDS

Empirical, Mock-up, CFD, Airflow, Modelling, Control, Diffuser

INTRODUCTION

Modelling ventilation systems using simulation techniques depends heavily on the accurate specification of boundary conditions. Whilst clearly the heat loads and fabric heat transfer are important, characterising the diffuser is normally critical and often represents a significant challenge. The reason is simple – the airflow through a single diffuser alone is extremely complex and in itself would represent a significant modelling task. In this instance, the diffuser is unusually sophisticated being made up of nine miniature swirl diffusers, each comprising a series of blades. The aim is to be able to represent these diffusers so as to achieve the mixing air velocities typical of the real device without resorting to modelling the detail of each diffuser. The benefit of this approach is that the computation time is significantly reduced for an ordinary desktop PC. This paper describes briefly the physical model of a laboratory facility from which the empirical data is derived for comparison. The facility has very high air change rates with the intention that the proposed design can achieve very uniform conditions. Swirl diffusers are used to create good mixing, and to dissipate the high velocity / momentum associated with the large air change rate providing very tight temperature control in the occupied zone and but with low bulk velocity.

THE PHYSICAL MODEL

The physical model was made up of a twin skin test chamber inside a controlled environmental chamber. The inner chamber represents a close temperature controlled laboratory. The construction, as installed on site, was double skin plasterboard on both sides of 150mm of insulated low density glass fibre mats. The mock up provided an insulated sealed raised floor void to prevent heat loss through the laboratory concrete floor to a depth of 0.2m and a suspended ceiling where the diffusers were sited.

The installed air handling unit (AHU) has a cooling capacity significantly in excess of the 1kW cooling load and provides the air to the room through twelve ceiling mounted multi-outlet swirl diffusers.

The general configuration is described with a diagram in the section entitled Airflow Modelling.

The temperatures were measured using a combination of PRT (Platinum Resistance Thermometers) to within ±0.015°C. The flow analyser was also used to measure local RMS air velocity at ±0.01m/s ±5%.

Further details of the physical modelling and data analysis can be found in Booth et al (2000).

AIRFLOW MODELLING

The test facility was modelled using FLOVENT© which is a purpose designed computational fluid dynamics program for airflow simulations in the built environment. Figure 1 shows the schematic representation of the computer model.

Physical Construction

The computer model only represents the inner chamber, with the controlled environment for the outer chamber represented as a fixed temperature. The surfaces of the inner chamber allow heat transfer based on any local temperature difference and the properties of the surfaces. This is relatively limited because the chamber is designed to have a relatively low heat transfer coefficient. This model includes the access doors and any slight imbalance in airflow was allowed to escape around them.

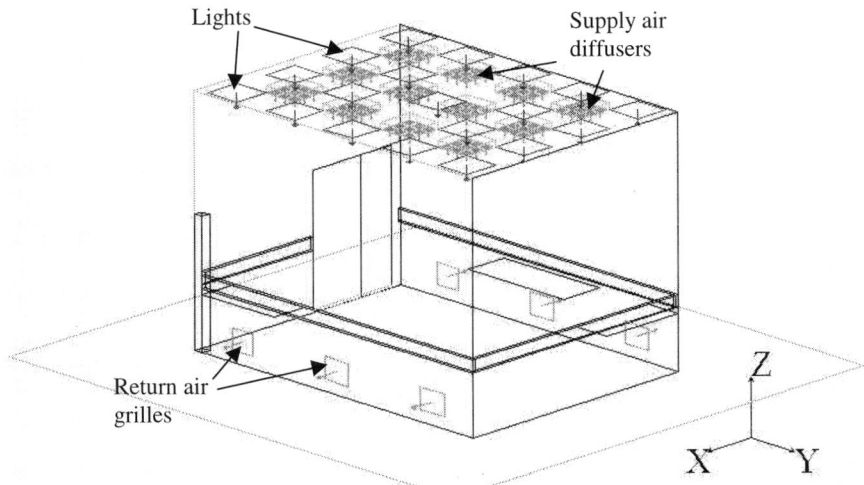

Figure 1 : Schematic representation of computer model

Furniture

Three tables were placed in the chamber as workbenches although no equipment was present in both the physical and computational model. Heat transfer through the tables was assumed to be negligible and so only their physical presence was considered.

Heat Gains

The only heat gains present in this scenario were those due to lighting. However this is considerable since high luminance levels were required, resulting in a heat load of 0.99 kW from 19 fittings. The heat was assumed to be emitted solely by convection since the high air velocities near the ceiling result in low surface temperatures and the lights are physically screened minimising any long-wave radiation. It is expected that some of the heat would be lost to the ceiling void but since the quantity was unknown it was assumed that the full heat gain entered the inner chamber.

Diffusers

The facility is ventilated by twelve 600mm x 600mm units each comprised of 9 miniature swirl diffusers. The air is supplied through a boot containing a perforated plate intended to achieve uniform flow through the diffusers. Each diffuser measures 120mm in diameter with a 30mm hub and has 8 blades. Compared with the chamber dimensions of 4.3m x 5.8m x 4m, these diffusers may be regarded as being very small. It would be impractical to model the detailed geometry and flow through each individual diffuser and so some simplification is needed.

Numerical tests were undertaken to simplify the representation by considering the swirl angle and flowrate superimposed on the basic geometrical information, e.g. diameter etc. From the empirical tests it could be observed that the central diffuser of each tile exhibits a different flow pattern compared with the outer diffusers. The outer diffusers tend to Coanda across the ceiling while the air from the central diffuser cannot penetrate sideways and is forced to project vertically downwards. As a result, the modelling is different for the central diffuser.

To mimic the attached jet, the angle of the jet leaving the swirl diffuser relative to the diffuser face was varied between 10 and 30 degrees in a test model. The jet attached to the ceiling only at lower angles so an angle of 10 degrees was applied to the outer diffusers for the full chamber model. To simulate the downward flow from the central diffuser, the air was projected vertically from the central diffuser of each active tile. None of the specific details, such as the blades, were represented.

COMPARISON

The six diagrams in Figure 2 show that temperature uniformity is easily achieved with only lighting loads; only a few points can be seen to be outside of ± 0.1°C. These are shown as lighter shaded areas at the six heights.

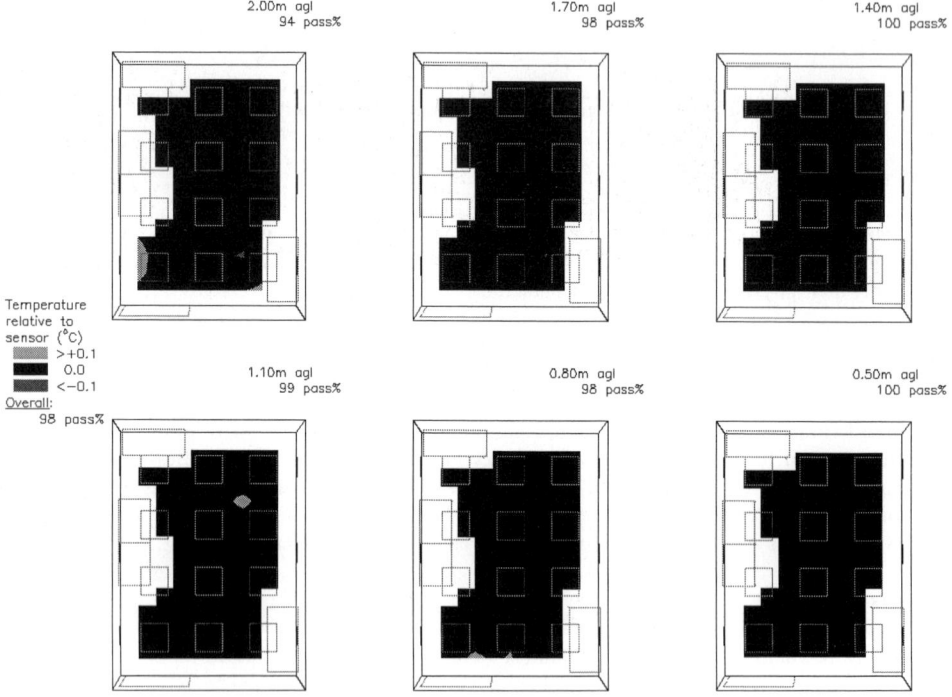

Figure 2 : Physically measured temperature distribution

Figure 3 shows a wider temperature variation, only 85% falling within the ±0.1°C criterion compared with 98% for experiment. The majority (96.8%) of the temperatures however, fall within ±0.15°C so the spread is still small. The probable cause of these larger temperature variations is the over-estimation of the lighting heat source as indicated in the section on heat gains above. This is confirmed by the fact that the mean temperature predicted by the numerical model is higher than the experimental mean temperature. The results are summarised in Table 1.

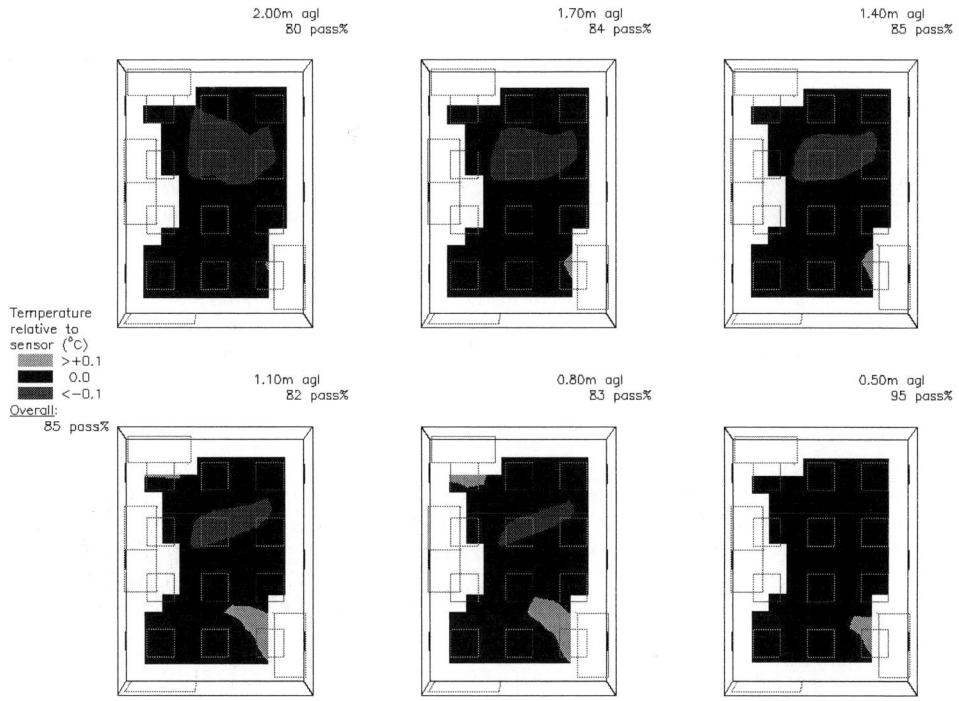

Figure 3 : Numerically predicted temperature distribution

TABLE 1

PASS RATE FOR EXPERIMENTAL AND NUMERICAL TEMPERATURE DATA

Height (m)	Experimental % Pass ±0.1°C	Numerical % Pass ±0.1°C	Experimental % Pass ±0.15°C	Numerical % Pass ±0.15°C
2.0	94.1	80.4	99.0	95.1
1.7	98.0	84.3	100.0	96.1
1.4	100.0	85.3	100.0	96.1
1.1	99.0	82.4	99.0	96.1
0.8	98.0	83.3	100.0	97.1
0.5	100.0	95.1	100.0	100.0

The results were found to be similar with air speed comparisons, some 84% falling below 0.2 ms^{-1} where all the experimental data was in this range. The main difference was due to localised high velocities of the order of 0.3 ms^{-1} which was not seen in the experimental data. The primary causes are thought to be the increased heat gain within the numerical model and the simplification in modelling the supply air diffusers. When considered in the light of the high velocities (several metres per second) at the face of the diffusers, the mixing out of the supply momentum has been successfully achieved by the time the air reaches the occupied zone.

SUMMARY & CONCLUSIONS

Airflow modelling has clearly predicted that this facility will have low air velocities with good temperature uniformity as indicated by experiment. This has been achieved for conditions where the flow could be anticipated to be very unstable due to the low air velocities in the body of the room. Further comparisons for scenarios with more realistic use and heat gains represent an opportunity for further study. The most challenging aspect of the computer modelling, simulating the diffusers, has been achieved with a relatively simple approach although the results indicate that some refinement would still be of benefit.

References

Booth W.B. & Topaltziki A., Application of Physical Modelling Techniques to Optimise a Close Temperature Control Room, Roomvent 2000.

Air Distribution in Rooms, (ROOMVENT 2000)
Editor: H.B. Awbi

DETAILED MEASUREMENTS OF AIRFLOW CHARACTERISTICS OF CIRCULAR CEILING DIFFUSERS

Y. K. Chuah[1], S. C. Hu[1], and J. M. Barber[2]

[1] Dept of Air-Conditioning and Refrigeration National Taipei University of Technology
1, Section 3, Chung-Hsiao E. Rd, Taipei 106, TAIWAN
FAX: 886-2-27314919   Email: f10870@ntut.edu.tw
[2] School of Architecture and Building Engineering, University of Liverpool
Liverpool PO Box 169, UK   Email: JMB1@Liverpool.ac.uk

ABSTRACT

This paper presents a series of detailed measurements of airflow characteristics for a vortex diffuser and two multi-cone circular diffusers. The flow patterns and turbulence characteristics in the diffuser outlet region and in the room were measured by using a three dimensional ultrasonic anemometer. In the region adjacent to the ceiling (0.08m from the ceiling), the air velocity was measured by using a hot-wire anemometer. The results show that the flow patterns in the vicinity of outlet region of the vortex diffuser are three dimensional and highly turbulent. Also, inappropriate design of the multi-cone circular diffuser results in "dumping" of the supply air to the occupied region of the room. The K values of the vortex diffuser investigated are in the range of 2.1 to 2.3. For the multi-cone circular diffusers, the K values cover a large range and are related to the design of the geometry of the diffuser. These detailed measurements are useful for evaluating numerical simulation models as well as for understanding the behavior of room air motion created by different diffusers.

INTRODUCTION

Circular ceiling diffusers are probably the most common types of air terminal devices used for air distribution in commercial buildings. Throw data are commonly provided in the manufacturers' product catalogues. However, the detailed airflow information including airflow patterns in the outlet region, jet centerline velocity decay coefficient (K value) and jet entrainment are not usually available.

Small differences in the design of circular ceiling diffusers can greatly affect the diffuser's air distribution performance. For example, the flow pattern is very much related to the magnitude of the angle between the jet outlet and the ceiling, ϕ. Little of this information is reported in literature. Hirano

(1998) reported on detailed isothermal measurement of flow pattern 0.1 m under the exit plane of a three-cone circular ceiling diffuser and observed notable "dumping" flow patterns. Hu et al. (1999) compared cold air distribution (with a temperature of 8 •) in the occupied zone of a room equipped with various diffusers including a nozzle type diffuser, a multi-cone circular ceiling diffuser and a vortex diffuser. The flow pattern produced by the multi-cone circular diffuser is influenced by the angle, φ. Becher (1966) presented a correlation for jet centerline velocity along the ceiling for a single cone diffuser with angle φ < 30. By using a hot-wire anemometer, Jackman (1973) conducted series of measurement for air movement in rooms with ceiling mounted diffusers, including multi-cone circular diffusers and multi-cone square diffusers. Nielsen (1985) measured the three-dimensional wall jet from grilles and registers. Detailed velocity information close to the diffuser is not reported. The K values reported by Jackman and Nielsen were based on the diffuser neck velocity and diffuser neck area. The K values indicated in the ASHRAE Handbook- Fundamentals (1997) are mostly based on research results published more than four decades ago (Koestal 1955). In the family of circular ceiling diffusers, the vortex diffuser, as shown in Figure 1, has a relatively high entrainment due to the swirl of the exit air, which creates a more thorough mixing of supply air and room air than with most other diffusers. The swirl effect is caused by the stationary radial guide-vanes which increase the turbulence level by imparting a spiral twist or swirl to the supply air. The control disk (Figure 1) of the vortex diffuser can be adjusted, to the lowest level resulting in a horizontal flow, and to the highest level resulting in a downward flow. So the vortex diffuser is a variable outlet area diffuser. Although currently on the market, the authors have been unable to obtain detailed technical performance data for it. Because of the lack of the detailed technical information for either multi-cone circular diffusers or vortex diffusers, this paper focus on detailed measurements of airflow characteristics specifically for these two types of diffuser. Therefore, the objectives of this study were: (1) to obtain detailed flow characteristics of the isothermal jets created in the vicinity of the outlet by a vortex diffuser and two multi-cone circular diffusers (shown in Figure 2 and Figure 3), (2) to study the influence of various geometry parameters on the diffusion performance for the multi-cone circular diffusers, and (3) to examine whether the induction due to twist vanes of a vortex diffuser is sufficiently high to have a significant effect on room air motion.

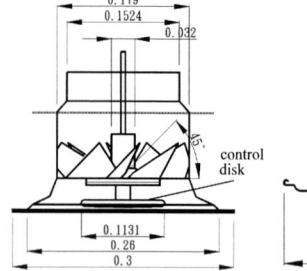 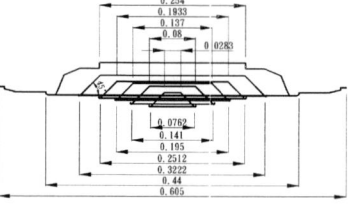

Figure 1. Vortex diffuser used in this study.   Figure 2. Three cone-circular ceiling diffuser.   Figure 3. Six-cone circular ceiling diffuser.

DESCRIPTION OF EXPERIMENTS

A full-scale environmental chamber (length (L) x width (W) x height (H) = 3.8 m x 3.6 m x 2.95 m) was constructed following the ANSI/ASHRAE 130-1996 standard, as shown in Figure 4. The layout and type of air diffusers could be changed according to the various experimental needs. The air volume flow rate was measured by standard nozzles which were located upstream of the diffusers. The flow patterns and turbulence characteristics in the diffuser outlet region and in the room were measured by using a three dimensional ultrasonic anemometer. The experimental setup is shown in Figure 5. In the region adjacent to the ceiling (0.08m from the ceiling), the velocity magnitude measurements were conducted by using a hot-wire anemometer. The three-dimensional ultrasonic anemometer and the hot-wire were mounted individually at the extension end of an XY mechanical traverse table which was driven by personal computer controlled step motors, and can move in orthogonal directions with 0.01-mm position precision. During the measurement, the personal computer and other instruments were placed outside the environment chamber to avoid interference with the airflow. The sampling frequency of 20Hz was sufficient as the frequency response of the measured velocity fluctuations distributed in a range were below 10Hz (at about -100dB). A period of 30 seconds was used in this study. The principles of three-dimensional ultrasonic anemometry lie in the change of sonic traveling time relative to the air velocity. The sensors of a three-dimensional ultrasonic anemometer consist of three pairs of transmitter and receiver probe heads spaced 5-cm apart (Figure 5).

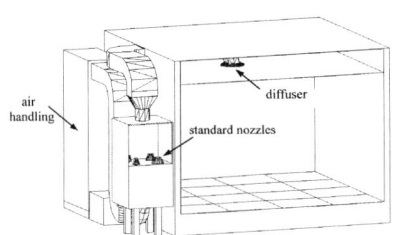

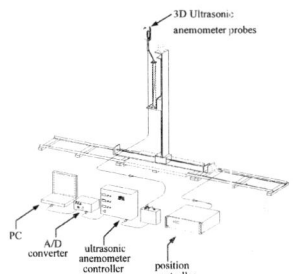

Figure 4. Schematic diagram of the environment chamber

Figure 5. Measurement system setup

RESULTS AND DISCUSSION

The different cases of experimental investigation are shown in Table 1.

Table 1. Cases tested

Case	Diffuser type	Supply flow rate Q
1	A	**47.2 L/s (100cfm)**
2	B	**47.2 L/s (100cfm)**
3	B	**23.6 L/s (50cfm)**
4	C	**47.2 L/s (100cfm)**
5	C	**35.4 L/s (75cfm)**
6	C	**23.6 L/s (50cfm)**

A: three-cone circular ceiling diffuser, B; six-cone circular ceiling diffuser; C: vortex diffuser

Cases of three-cone circular diffuser

Figure 6a shows the flow patterns and Figure 6b shows the turbulence intensity in the vicinity of the diffuser outlet region for case 1. An apparent recirculation zone with center located at 0.13 m under the outer cone was observed under the diffuser. This is very similar to the measured results of Hirano (1998) in which a three-cone diffuser was tested. The dimensions of currently tested diffuser are close to that used by Hirano, except there was a 0.016m diameter hole in the inner cone of the diffuser tested by

Hirano. For the three-cone diffuser, as expected, there was an upward air current in the inner-cone, which may result in surface condensation if a cold air is supplied to a hot/humid room where the induced room air touches the inner-cone's surface. It is noted that the supply air is eventually "dumped" under the diffuser. In such a case, it is impossible to access the jet's centerline velocity decay coefficient (K value). In the present study the resultant turbulence intensity (TI) at a location is defined as $TI = ((\sigma_x^2 + \sigma_x^2 + \sigma_x^2)/3)^{1/2}/V$, where σ_x, σ_y and σ_z are the standard deviation of velocity fluctuation along X, Y and Z directions, respectively, and V is the local mean air velocity at a measured point.

Cases of six-cone circular diffuser

Flow patterns in the vicinity of diffuser outlet region and the jet expansion region for case 2 (six-cone circular ceiling diffuser with Q=47.2 L/s(100cfm)) is depicted in Figure 7. Unlike the "downward projection" in case 1, a horizontal jet was produced due to the horizontal velocity component and the Coanda Effect. Air entrainment induced by the horizontal jet along the jet direction is very apparent, resulting in upward flow pattern throughout the measuring domain. The flow patterns in the vicinity of diffuser outlet region and the jet expansion region of case 3 exhibit the same shape as those of case 2 and are not presented due to the length limitation of the paper. It may be concluded that the geometry design of the multi-cone diffuser affects the flow pattern greatly, where more cones produce a better horizontal jet.

Cases of vortex diffuser

The symmetric test was performed at the diffuser outlet level (the plane section 0.05 m under the diffuser). As indicated in Figure 8, a surprising octagonal shape uneven velocity profile (Vmax= 3.3m/s and Vmin=0.2 m/s) along the radial direction of the diffuser was observed. This means that the air distributions just outside the vortex diffuser are not even. This uneven velocity distribution is mainly due to the spin caused by the twist vanes. A vertical plane section with maximum velocity was selected for further flow distribution measurements. Figure 9 shows the flow patterns in the vicinity of diffuser outlet region and the jet expansion region for cases 4 (vortex diffuser with Q=47.2 L/s (100cfm)). Note that under the diffuser outlet region, the airflows are in upward directions. It was also observed that away from the diffuser the air jet traveled horizontally, attaching to the ceiling and extending to the opposite wall for case 4. No "dumping" was observed in the occupied zone. The flow patterns in the vicinity of diffuser outlet region and the jet expansion region of case 5 show the same trend as those of case 4 and are not presented due to the length limitation of the paper. For a terminal jet velocity of 0.25m/s, the throw distances of case 4 and case 5 were 1.5m and 1.1 m, respectively. The corresponding drops were 0.1m and 0.08 m for case 4 and case 5, respectively. For the same supply airflow rate (case 2 and case 4), the air velocity magnitude of case 4 was measured to be generally lower than that of case 2 at the outlet, and which indicates a larger effective outlet flow area for case 4.

K values

The K values for the six-cone circular diffuser and vortex diffuser are 1.9 and 2.4, respectively. The larger K values obtained for a vortex diffuser is probably due to the rotational effect created by the twist vanes.

CONCLUSIONS

1. It may be concluded that the geometry of the multi-cone diffuser affects the flow pattern greatly, with more cones produce a better horizontal jet.

2. The air distribution patterns in a room produced by the vortex diffuser are asymmetrical. This inlet

velocity asymmetry is mainly due to an asymmetrical spin provided by the twist vanes.

3. The horizontal jet at the diffuser outlet region produced by the twist vanes is very significant. The induction due to twist vanes is sufficiently high to have a significant effect on room air motion.

4. Detailed measurements in this study are useful for evaluating numerical simulation models as well as for understanding the behavior of room air motion created by different types of diffusers.

ACKNOWLEDGEMENT

The support from the National Science Commission of Taiwan is acknowledged.

REFERENCE

ANSI/ASHRAE Standard 130-1996, Method of testing for rating the performance of air outlets and inlets, 1996, American Society of Heating, Refrigerating and Air-Conditioning Engineers, Inc.

ASHRAE Handbook-Fundamentals, Chapter 32, 1997, Atlanta: American Society of Heating Refrigerating and Air-Conditioning Engineers, Inc.

Becher, P., 1966, Air distribution in ventilated rooms. JIHVE, 34,219-227.

Hirano, T. et al.,1997, Measurement of the airflow velocity and turbulence energy around circular ceiling type diffuser to verify CFD results, Proceedings of Technical Meeting, SHASE-Japan,(B-26):510-513 (in Japanese).

Hu, S. C., J. M. Barber and H. Chiang, Full scale thermal performance tests of alternative diffusers when operating with cold air, ASHRAE Transactions, 1999, Part 1, CH-99-6-5.

Jackman, P. J., 1973, Air movement in rooms with ceiling mounted diffusers (including supplements A and B). BSRIA Rep. No. 81, Building Services Research and Information Association, Bracknell, UK.

Koestel, A.(1955) Path of horizontally projected heated and chilled air jets. Trans. ASHAE, 61, pp. 213-232.

Nielsen, P. V., 1985,, Measurement of the three-dimensional wall jet from different types of air diffusers, Proc. CLIMA 2000, Copenhagen Denmark,pp.383-387.

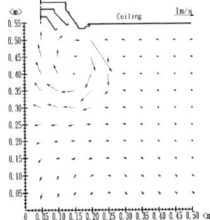

Figure 6a. Flow patterns in the vicinity of diffuser outlet region for case 1.

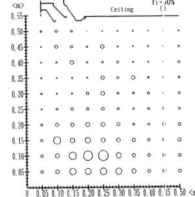

Figure 6b. Turbulence intensity in the vicinity of diffuser outlet region for case 1.

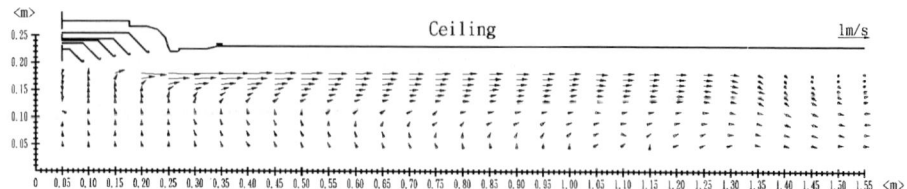

Figure 7. Flow patterns in the vicinity of diffuser outlet region and the jet expansion region for case 2.

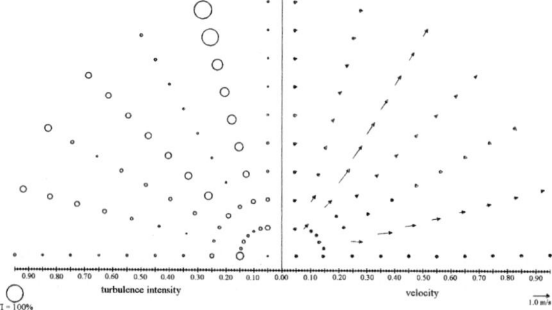

Figure 8. Flow patterns (right hand side of the figure) and turbulence intensity contours (right hand side of the figure) at the outlet plane for case 4 (top view).

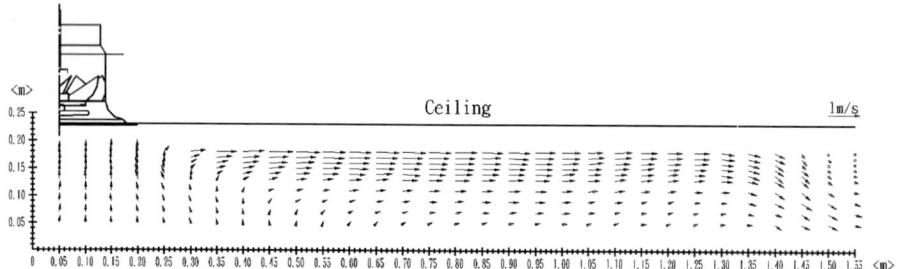

Figure 9. Flow patterns in the vicinity of diffuser outlet region and the jet expansion region for case 4.

Air Distribution in Rooms, (ROOMVENT 2000)
Editor: H.B. Awbi

ANALYSIS OF AIR VELOCITY AND TEMPERATURE DISTRIBUTION IN FURNISHED ROOMS WITH UNDERFLOOR AIR DIFFUSERS

M. De Carli[1], F. Peron[2], P. Romagnoni[2], R. Zecchin[1]

[1] Dipartimento di Fisica Tecnica, Università di Padova, Via Venezia 1, 35131 Padova, Italy
[2] Dipartimento di Costruzione dell'Architettura, IUAV, S. Croce 191, 31135 Venezia, Italy

ABSTRACT

HVAC systems with underfloor air distribution are frequently used in office buildings with raised floor. In this case particular attention must be paid to the design of the air inlet and outlet layout; the air velocity and temperature fields must be very carefully controlled to avoid uncomfortable conditions.

CFD models can be very useful in the design process to analyse the influence of different parameters and different solutions for the air supply inside rooms. Numerical models must always be assessed in order to correctly predict the temperature and air velocity fields. In this paper a comparison between experimental and numerical results is presented in the case of floor air diffusers in a typical furnished office.

KEYWORDS

CFD, ventilation, comfort.

INTRODUCTION

The Underfloor Air Conditioning systems (UFAC) were developed to provide a rational solution of the cooling problems in computer rooms. In fact the high cooling load due to heat dissipated by electronic equipments can be efficiently neutralised using this system. In the last years, owing to its flexibility, much attention has been paid to this type of plant and some works have been carried out on this topic, Han et al. (1999). UFAC systems require the adoption of a false floor in order to allow the insertion of the terminal units (air inlet and outlet) and the connection with the air treatment unit. However the underfloor space can be used also for cables, electrical connections, piping, and data transmission network. In this context, some practical lay-outs have been developed, which allow to extend UFAC to business or administrative offices in which the human comfort is a main matter.

Technical international Standards, as ISO 7730 (1992) and ASHRAE 55-1992 (1995), impose well defined values for air temperature, vertical air temperature difference, air velocity, and surface floor temperature, in order to obtain comfortable conditions. The parameters influencing thermal comfort in the indoor environment have been widely studied and evaluation criteria such as PMV ("predicted mean vote") and PPD ("percentage people dissatisfied") are nowadays of common use, ISO (1992), ASHRAE (1995), CEN (1997), Fanger (1972). Fanger's approach easily allows to predict users reactions (PMV and PPD), as a function of activity level (expressed by the metabolic rate M), clothing thermal resistance (R_c), mean radiant temperature of the surrounding surfaces (T_r), as well as of dry-bulb temperature, relative velocity and humidity ratio of the air (t_a, v_a and W_a respectively). Moreover if the air velocity is appreciable, the parameter DR ("Draft Risk", expressed as percent of people dissatisfied due to draught) should be also evaluated, ISO (1992), ASHRAE (1995), as a function of local air temperature (t_a), local mean air velocity (v_a), and local turbulence intensity (Tu), this last being defined as the ratio between standard deviation and mean value of the local air velocity. Therefore, for given activity level and clothing thermal resistance R_c, the knowledge of air temperature, humidity ratio, air velocity, and mean radiant temperature allow to determine global comfort and discomfort indices; local values of air streams temperature, velocity, and turbulence intensity allow to evaluate DR.

A detailed study of temperature and air velocity patterns in a room is not a simple task: it requires the set up of a climatic chamber, which is not always available. It is also possible to carry out field measurements in order to investigate comfort conditions at a few given locations, as in the case here considered. In the design process, however, it can be necessary to study the air distribution in rooms for different layout of inlet and outlet, different dimensions of the room, different thermal loads. Such study can be achieved by using the CFD technique, e.g. Fletcher (1990), which requires nevertheless a first assessment based on measurements. In this paper a CFD numerical model has been developed and assessed on the basis of experimental data pertaining to a UFAC system in a furnished office, Baggio et al. (1995).

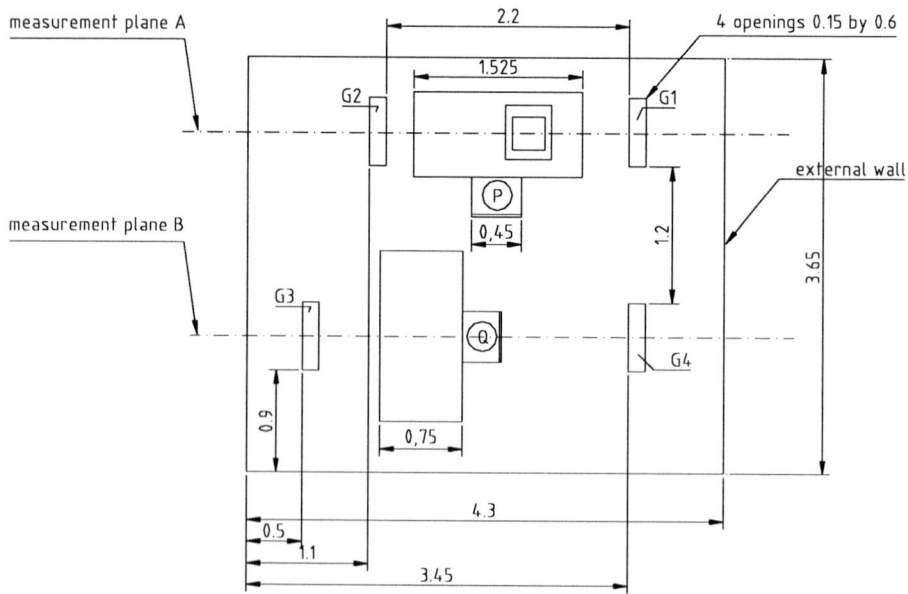

Figure 1: Lay out of the test room (lengths expressed in meters)

EXPERIMENTAL TESTS

A test room (4.3 m length, 3.65 m width, 2.7 m height) equipped with an UFAC system and furnished with two desks, two chairs and one computer, was set up (fig. 1). Only one wall is external (N-W exposure) with a window (3.25 m by 1.5 m) shaded from direct solar radiation by means of an external venetian blind. The other walls, the ceiling, and the floor are adjacent to other internal spaces.

A couple of supply grilles located near the window (G1 and G4) and a couple of return grilles near the door (G2 and G3) provide air supply and exhaust. This typical lay-out on floor surface has the purpose of eliminating stagnation zones. As a reference for air velocity and temperature distribution in the test room, the middle planes A and B (figs. 2 and 3) of the air jets have been investigated. Experimental data have been collected on a grid spaced 0.25 m in vertical direction and 0.20 m in horizontal direction.

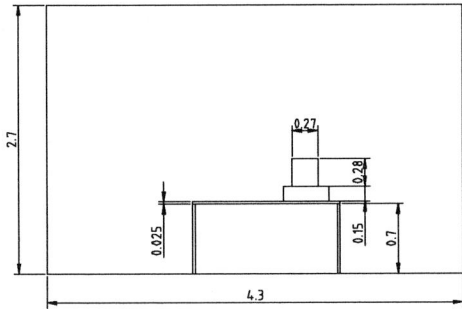 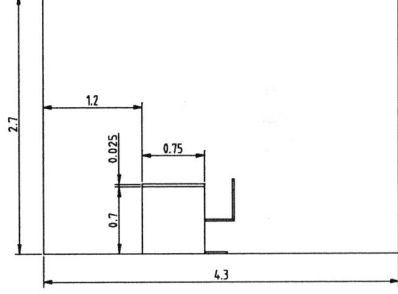

Figure 2: Measurement plane A   Figure 3: Measurement plane B

The air velocity has been measured by means of hot-wire anemometers. The velocity device measures in the ranges 0 to 1 or 0 to 5 m/s with ±2% accuracy fsd at 20°C. An additional precision thermal sensor in the probe allows the continuous measurement of the temperature over a range $0 \pm 30°C$ with the 1,5% fsd accuracy. T-type thermocouples, shielded from radiation effects, have been used for measuring temperature of air and wall surfaces. A data acquisition system HP 75 000, connected to a PC, directly collected the measured values.

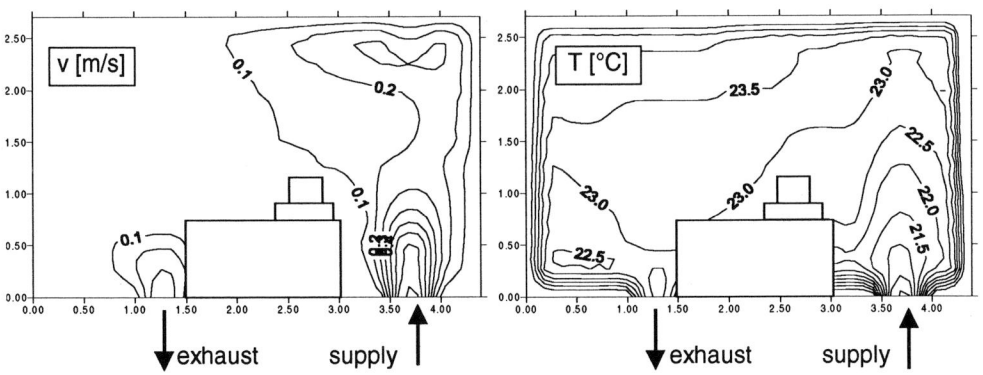

Figure 4: Sample pattern of air velocity
(measured on plane A)

Figure 5: Sample pattern of air temperature
(measured on plane A)

In figs. 4 and 5, patterns of temperature and air velocity on the plane A are reported. It can be noticed that the zone where the maximum vertical temperature difference occurs near the inlet air section and the minimum vertical temperature difference takes place near the return section; the measured temperature difference above the air inlet, between 0.5 m and 1.5 m, is about 1.5°C.

NUMERICAL MODEL

The same test room described in the previous paragraph was simulated by resorting to a commercial software, namely PHOENICS, which allows to evaluate temperature and air velocity fields, Rosten & Spalding (1987). The numerical model was set up by taking advantage from previous experiences (e.g. De Carli et al. (1999), Onishi et al. (1990), Chen & Jiang (1992)) and its assessment was carried out by comparing numerical simulation results with experimental data. Specifically during the assessment, a sensitivity analysis on the values of turbulence intensity of air at the inlet, has been conducted in order to investigate its influence on the patterns of air temperature in the room and the values of comfort indices.

The system has been considered in steady state conditions and the air has been assumed as an ideal gas. Briefly the differential conservation equations governing the considered phenomena, solved by the software, can be written in this general form:

$$\frac{\partial}{\partial t}(\rho\varphi_i) + div(\rho\bar{v}\varphi_i - \Gamma_{\varphi i} grad\varphi_i) = S_{\varphi i} \tag{1}$$

where:

φ_i	= conserved physical quantity;
ρ	= density;
$\bar{v}$	= velocity vector;
$\Gamma_{\varphi i}$	= diffusion coefficient of φ_i ;
$S_{\varphi i}$	= production rate of φ_i ;

From this, by substituting φ_i with the components of velocity vector, u, v, w, the momentum balance can be obtained, while, by substituting φ_i with the enthalpy (h), the equation expresses the conservation of energy. At last, assuming $\varphi_i = 1$ the mass conservation equation can be obtained.

For large values of the Rayleigh number, Ra ($Ra = \dfrac{l^3 \beta g \Delta T}{va}$, where l [m] is the characteristic dimension of the room, β is the thermal isobaric expansion coefficient [K⁻¹], g is the gravity acceleration [m/s²], ΔT is the relevant temperature difference [K], v is the kinematics viscosity [m²/s], a is the thermal diffusivity [m²/s]), the flow is turbulent and a specific model must be considered. The most used formulation, in engineering calculations, is the k-ε model, proposed in Rosten & Spalding (1987), which consists of two differential equations for the transport of turbulent kinetic energy k and of the rate of turbulent energy dissipation ε.

Simulations have been lead under boundary conditions specified in Table 1. For the inlet conditions the equation proposed in Versteeg & Malalasekera (1996) has been used.

The space domain has been discretised in 53 x 57 x 24 cells. In the assessment of the model, it has been verified that a more detailed grid did not modify the results.

Comfort conditions for persons seated on the chairs of the test room (points P and Q) have been investigated. Different inlet turbulence intensity have been investigated (min 0.2, max 0.8), but negligible influence was observed.

TABLE 1
BOUNDARY CONDITIONS OF NUMERICAL SIMULATIONS

Surface temperatures [°C]			Inlet conditions			Internal heat gain [W]
Internal walls	External wall	Window	v [m/s]	T [°C]	Tu [%]	
26	27	28	0.8	20	40	150

RESULTS

The patterns of temperature (fig. 6) and air velocity (fig. 7) on the plane A are reported and can be compared with measurements in the same plane (figs. 4 and 5). It can be seen that the respective zones of low air velocity are in good agreement and the temperature difference (between simulations and measurements) in the same zone is within 1°C. For the two seated persons (points P and Q in fig. 1) the difference in temperature is about 0,5°C, so comfort conditions can be predicted from the CFD results with good precision. For example, the comfort conditions in locations P and Q, both from measured and calculated values, are reported in Table 2, referring to a clothing resistance of 0.5 clo and an activity level of 1.2 met. It can be noticed that PPD's from measured values and from calculated values differ less than 1% between each other. This accuracy can be considered reasonable in comparison with a limit value of 10% PPD suggested by the above mentioned Standards ISO (1992), ASHRAE (1995). No measured data were available for turbulence, in the case considered, so the accuracy of DR from calculated values has not yet been verified and work on this topic is being carried on.

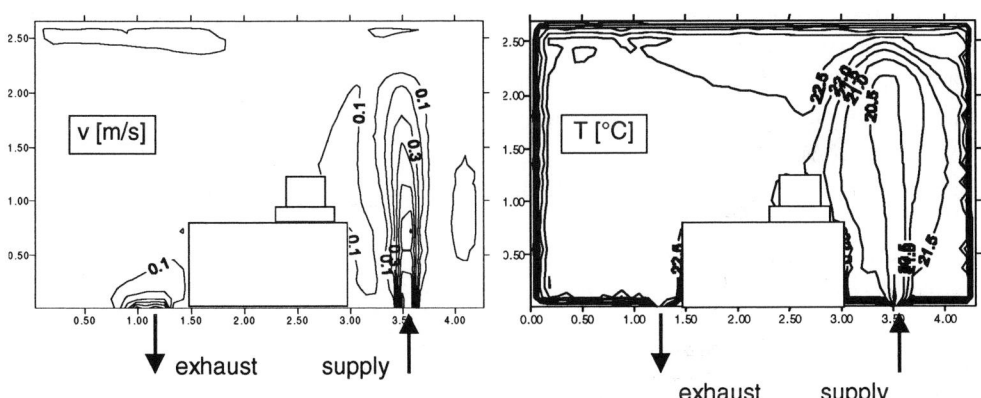

Figure 6: Calculated air velocity pattern on plane A

Figure 7: Calculated temperature pattern on plane A

TABLE 2
EVALUATION OF COMFORT CONDITIONS

Position (fig. 1)	From measurements		From simulations		
	PMV	PPD [%]	PMV	PPD [%]	DR [%]
P	0.02	5.0	-0.21	5.9	9.4
Q	0.02	5.0	-0.16	5.5	0.0

CONCLUSIONS

In this work the tuning of a numerical model on the specific physical system (UFAC) has been carried out, by comparing measurements and calculations. Some results about the suitability of a computer code to correctly predict comfort indices, have been discussed. The model can be used for design purpose, to determine the characteristics of the air flow field within internal spaces, in order to obtain useful indications in different cases, with various values of air flow rate, room dimensions, positions of air supply and exhaust, and different thermal gains. Further results of a parametric study on this topic are reported in De Carli et al. (in printing).

REFERENCES

ASHRAE (1995). ANSI/ASHRAE Standard 55-1992 and Addendum 55a, Thermal environmental conditions for human occupancy. American Society of Heating Refrigerating and Air-conditioning Engineers, Inc., Atlanta, GA.

Baggio P., Romagnoni P., Varutti R. (1995). Thermal and velocity field in a room with an underfloor air conditioning system. Proceedings of 19th International Congress of Refrigeration. **Vol. III b**, 759-766, The Hague, NL.

CEN (1997). prENV 1752, Ventilation for buildings: design criteria for indoor environment.

Chen Q., Jiang Z. (1992). Significant questions in predicting room air motion. ASHRAE Transactions. **AN-92-9-1**.

De Carli M., Peron F., Zecchin R. (1999). Experimental and numerical investigation on temperature and air velocity distribution in a room equipped with split-system air conditioner. Proceedings of Indoor Air, **Vol. 2**, 42-47, Edinburgh, UK.

De Carli M., Peron F., Romagnoni P., Zecchin R. (in printing). A parametric study on comfort conditions with underfloor air-conditioning systems, Quaderno del DCA, IUAV, Venezia, I.

Fanger P. O. (1972). Thermal Comfort: Analysis and Applications in Environmental Engineering. McGraw Hill, New York, N.Y.

Fletcher C. A. J. (1990). Computational Techniques for Fluid Dynamics. Springer-Verlag, Sydney, AUS.

Han H., Chung K.S., Jang K.J. (1999). Thermal and ventilation characteristics in a room with underfloor air-conditioning systems. *Proceedings of Indoor Air*, **Vol. 2**, 344-349, Edinburgh, UK.

ISO (1992). ISO 7730. Moderate thermal environments - Determination of the PMV and PPD indices and specification of the conditions for thermal comfort.

Onishi J., Kurimura M., Tanaka S. (1990). Applicability and limitations of numerical method in Thermal environment analysis for air-conditioned rooms. Proceedings of Roomvent, Oslo, S.

Rosten H. I., Spalding D. B. (1987). The PHOENICS Reference Manual. CHAM TR 200, London, UK.
Versteeg H.K., Malalasekera W. (1996). An introduction to computational fluid dynamics, Longman Ed., Edinburgh, UK.

Air Distribution in Rooms, (ROOMVENT 2000)
Editor: H.B. Awbi

INDIVIDUAL SYSTEM OF AIR INTAKE IN ROOMS WITH TIGHT PARTITIONS

J. Z. Piotrowski[1]

[1] Department of Civil Engineering, Technical University of Kielce,
Kielce, 25-336, POLAND

ABSTRACT

Creating suitable ventilation and microclimatic conditions in buildings with tight partitions is a problem which urgently requires technical solutions. The individual system of air intake has been introduced to solve it. The system enables to use unexplored natural ventilating shafts to set off the air intake. The air will be partly warmed up in the basement.

Preparatory research have shown that the system of air intake stabilises the air exchange on required level and eliminates any disturbances in the ventilation functioning. At the same time other microclimatic and acoustic condition do not get worse.

KEYWORDS

Ventilation system, Natural ventilation, Air intake, Microclimate, Air change rate, Tight partitions, Air handling.

INTRODUCTION

Ensuring suitable ventilation and microclimatic conditions in energy conserving housing and in buildings with tight partitions and window woodwork poses a serious problem. It requires urgent technical solutions (Etheridge & Sandber 1996). The issue is confirmed by numerous complaints on the part of the residents of flats, which have tight, lined timbering (partitions, contacts and window woodwork) are being ventilated with air sucked in via ventilation and combustion gas ducts (Piotrowski 1996). Not only does the latter deteriorate microclimatic standards but it also constitutes a threat to residents (flame blowing out, carbon monoxide blowing in).

In case of a tight partition, ventilation takes place in negative pressure, which is connected with the surplus of air going out. Thus, it is not a normal working environment of a gravitation ventilating duct, where pressure above atmospheric is assumed to be created first, due to the surplus air intake through a leaky partition. Then the pressure is equalised through ducts. Actually, with tight timbering, one or a few ventilation or combustion gas ducts become intake type ducts. Such cases are not only limited to low, 1-2 storey buildings. Numerous instances of air intake onto the first storey in 3-4 storey buildings were noted.

In the Department of Civil Engineering Technical University of Kielce research began, which aimed at designing the system to solve the above mentioned problems, both in the existing buildings and those being constructed. It was assumed at the same time that the system should be simple and demanding relatively small outlay and modernisation alterations on the part of the user. The paper sums up the first stage of work, during which the system assumptions were provided and preliminary investigations were carried out.

INDIVIDUAL AIR INTAKE SYSTEM (IAIS) OPERATION PRINCIPLES

The operation of Individual Air Intake System (IAIS) consists in providing the rooms with fresh air from outside through the existing individual ventilation shafts. Most individual shafts possess unblocked lower section, which is not used. That refers particularly to prefabricated shafts, but it also applies to those made in the traditional way. Through ventilation ducts sections situated on the extension of offtake ducts downward, the air is driven into the rooms. The air supply to the duct is provided via intake ducts installed in cellar or ground floor. The duct outlet should be located just below the ventilation opening, with blowing into another room than that in which outlet is situated. The diagram of IAIS operation is presented in Figure 1 in projection and section of 2- storey building.

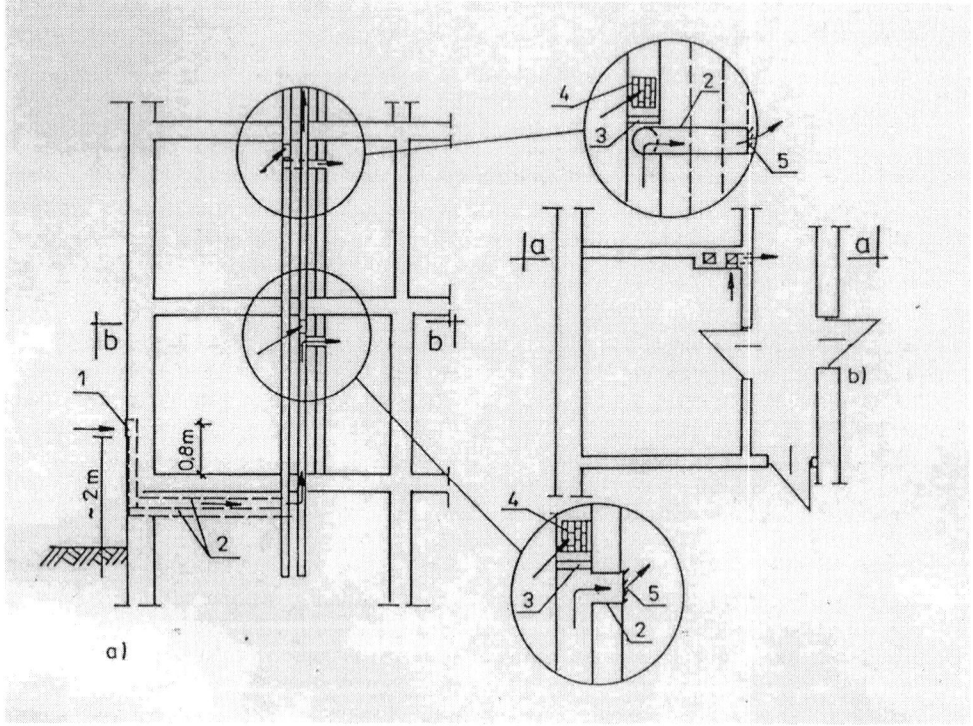

Figure 1: Individual air intake system
a - vertical section b - cross-section
1- outside air intake opening 2 - supply conduit 3 - tight partition 4 - outlet 5 - inlet

Preliminarny analyses demonstrate that the adaptation of not used duct sections is possible in 70% of buildings constructed previously and 90% of buildings constructed at present, especially those of prefabricated blocks. In case of buildings undergoing modernisation, the adaptation of inactive smoke flues creates an additional option.

THE SYSTEM TECHNICAL CONDITIONS

In case inactive smoke flues or combustion gas shafts are used, it becomes necessary to install an insert inside, e.g. in the form of sleeve of aluminium alloy. If leakage is revealed in ventilation ducts, the insert is also indispensable so as to avoid flow disturbances.

The air supply to the lower duct section should be provided by a conduit of circular or rectangular section, made of sheet or PVC. The conduit is to connect a ventilation duct in the cellar or on the ground floor with the outdoor air. Outside air intake opening is supposed to be located approx. 2 m above the ground (Fig. 1). The flow into rooms should proceed directly in case the duct is adjacent to the room, into which blowing in takes place. If it is not possible, the blow-in supply from the duct into the room will be also carried through a PVC or sheet conduit.

The blow-in supply conduits and the number of ducts were specified in the course of preliminary investigations on the basis of indispensable volume of intake air. In this case it is necessary to keep the balance between the air taken in, also via the outer timbering and the air carried away mainly through offtake ducts (1).

$$V_1 + V_2 + V_3 = V_4 \tag{1}$$

where:
V_1, V_3 - air volume flowing into the room from outdoors through leaks in woodwork and outer partitions (V_1) and through the leaks in the door separating staircase and the flat (V_3), dependent on "a" coefficient and pressure difference Δp,
V_2 – air volume flowing in through intake ducts, functionally dependent on Δp at pre-set boundary conditions (local resistance, sections, among others),
V_4 - air volume removed via ventilation network, at specified boundary conditions, functionally dependent on pressure difference resulting from stack effect Δh and wind pressure force w.

Due to the merely initial stage of investigations, the conduit section and the number of air intake ducts were assumed on the basis of air flow investigations, which had already been completed (Piotrowski 1998). The section of the conduit supplying air to the intake ducts was specified to be equivalent to ducts section of ϕ 120 mm whereas the number of intake ducts amounted to 2/3 of the existing air offtakes. Taking into account the size of the supply conduit section, it constitutes approx. 30% of the sum of the total section field of offtake ducts.

Figure 2 presents some options concerning air intake ducts connections and air carrying away through the existing offtake ducts. There were shown the solutions most typically found in practice.

Lengthening the air way thanks to the passage through supply conduits, which cross heated cellars initially heats the intake air. Lengthening the intake air way is also recommended so as to maintain acoustic insulating power.

For the correct air intake system operation two conditions have to be satisfied:
 - room into which the air is blown cannot be the room with the same duct operating as offtake,

- intake section of the duct changing into the offtake one has to be separated with a tight horizontal partition.

The first condition will guarantee the proper circulation of the intake air within the flat, the other will prevent the disturbances of the offtake part operation, thus ensuring its proper functioning. Furthermore, it would be advisable to fix air direction grates in inlet for the residents to enjoy microclimatic comfort.

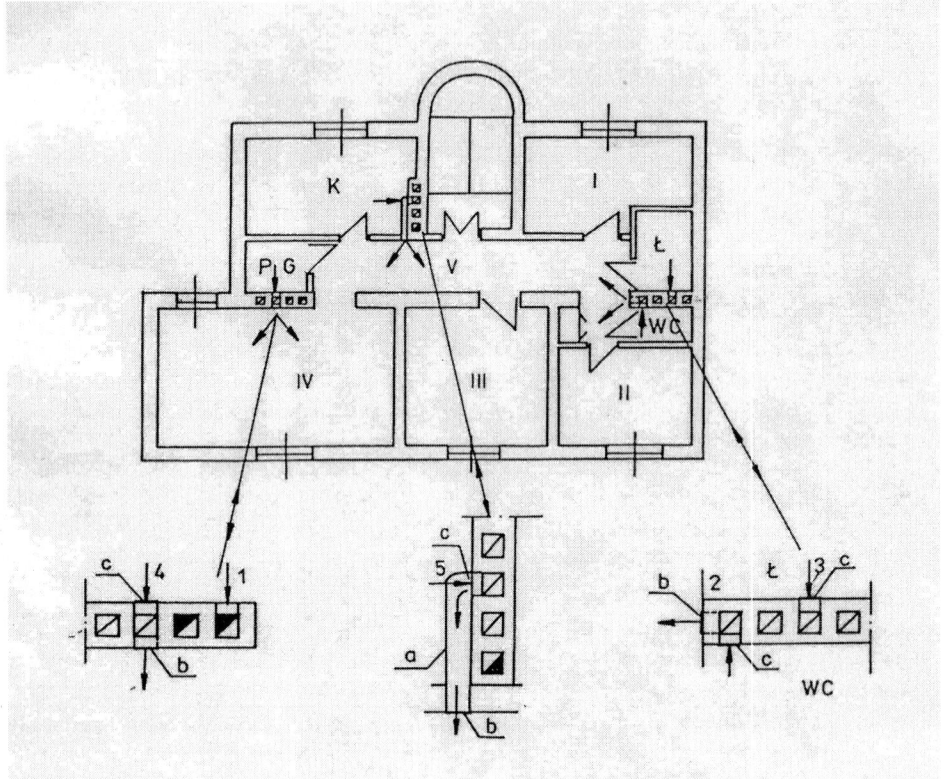

Figure 2: Some options concerning air intake ducts connections
a - intake air supply conduit b – inlet c – outlet
I-IV-rooms /L-bathroom / WC-lavatory / K-kitchen / PG- gas water heater

INVESTIGATIONS INTO AIR EXCHANGE WITH IAIS IN OPERATION

Investigations into air exchange were carried out in a 2-storey building with a usable attic (duplex). The building had tight window woodwork $a = 0.5$-0.7 $m^3(m \cdot h \cdot daPa^{0.7})$. Before modification with IAIS, there occurred functional disturbances in two shafts (1,2 - Fig. 2) cold air was taken in as the result of operation under the conditions of negative pressure. That was both noxious (shaft 2 – offtake from lavatory) and dangerous (shaft 1 – combustion gases of gas water heater). All the remaining shafts functioned properly.

The investigations demonstrated (Fig. 3) that the mean air exchange in the rooms I, II, III (Fig. 2) amounted to $n_1 = 0.3$ h^{-1}. After the elimination of the "disturbing" shafts the air exchange remained at the level $n_2 = 0.15$ h^{-1}. In order to guarantee air exchange level $n_i = 0.6$-0.7 h^{-1}, the intake duct capacity should be $n_i' = 0.55$ h^{-1}, which could be provided by three IAIS ducts. Air intake was initiated in shafts 2, 4 and 5 (Fig. 2) with the air supply into the hall V and spare room IV. As it is shown in Fig. 3, after the introduction of IAIS, the air exchange stabilised at $n_3 = 0.6$ h^{-1}, on average. At the same time, the disturbing activities in shafts 1 and 2 ceased to operate.

Air inlet into the outside air intake opening was located at 1.5 m above the ground level (development with garden space). Weather conditions during the investigations period were determined by temperature parameters $\Delta t = 15$-30°K, and those of wind velocity $w = 0$-8 m/s. Air exchange investigations were carried out with marker method (Piotrowski 1996).

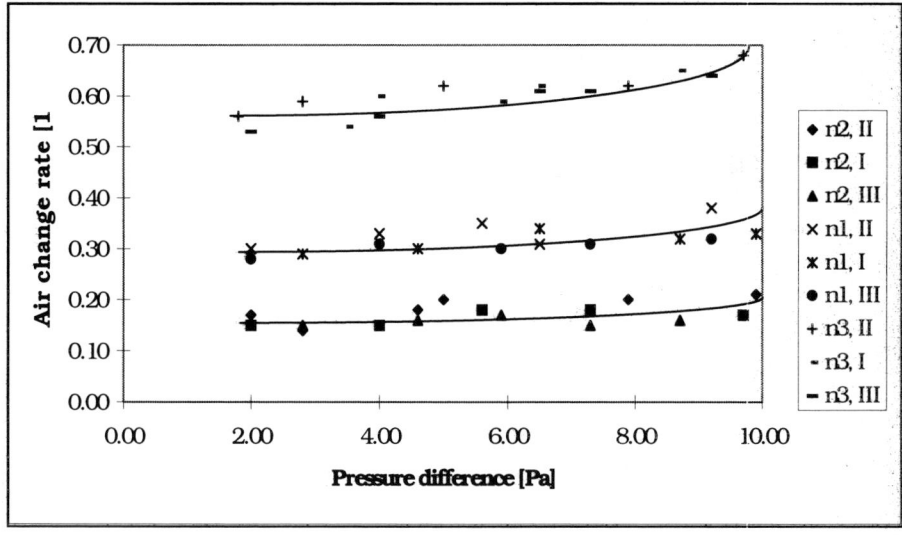

Figure 3: Air change rate in rooms I, II, III

SUMMING UP THE INVESTIGATIONS

The initial stage of investigations demonstrated that IAIS application produced desirable effects: disturbances in the operation of ventilation and combustion gas shafts were got rid of, satisfying air exchange was achieved at tight timbering, microclimatic comfort was maintained. Due to intake ducts location, offtake grate profile and preliminary air heating in supply conduits no discomfort concerning the changes in air flux velocity was experienced, nor was the so-called "aeration" observed. That refers mainly to rooms with air intake. The acoustic conditions did not deteriorate, either.

The reliability of IAIS operation, however, has to be confirmed by thorough investigations, which would account for a wide range of random (weather) and controllable conditions (room arrangement, shaft location). Both air exchange investigations in all rooms and on different storeys as well as those into shaft location impact on microclimatic conditions have to be complemented. In order to make the system generally applicable it is also necessary to verify the calculations of supply sections being coupled with the operation of gravitational ventilation and combustion gas shafts.

CONCLUSIONS

- IAIS is an alternative to air intake control devices or actions aimed at untightening the external timbering.
- As the preliminary investigations showed the system designed improved ventilation without deteriorating other microclimatic conditions.
- The system is applicable in various architectonic arrangements (storey, functional arrangement of rooms).
- IAIS does not require significant changes in construction, technologies or sophisticated equipment. Neither are extensive modernisation works necessary in buildings existing already.
- It is desirable to carry out complementary investigations under various random and controllable conditions as well as verification of assumed supply sections and their impact on microclimatic and acoustic conditions.

REFERENCES

Etheridge D. and Sandberg M. (1996). *Building Ventilation: Teory and Measurment,* John Wiley & Sons, Chichester, UK

Piotrowski J.Z. (1996). Air Exchange through Elements of Building Partition. *Proceeding of the 7th International Conference on Indoor Air Quality and Climate - Indoor Air '96,* Vol. 3, pp 827 - 832.

Piotrowski J.Z. (1998). Neural Network Application for Air Exchange, *Proceeding of the 6th International Conference on Air Distribution in Rooms - Roomvent '98,* Vol. 1, pp 311 - 315.

Air Distribution in Rooms, (ROOMVENT 2000)
Editor: H.B. Awbi

CHARACTERISTICS OF THE ENERGY ECONOMY OF AIR CURTAINS

Qiangmin Li

Department of Thermal Engineering, Tongji University,
Shanghai, 200092, P.R. China

ABSTRACT

This paper calculates plane air curtain's heat transfer, the energy consumption of air curtain in a year as well as it's energy conservation effect. Using Shanghai's climate data, the heat insulation efficiency of air curtain is obtained with dynamics calculation for a year. The air curtain's flow pattern can be shown using flow visualization. Air curtains have been widely used in China's HVAC systems, and have good energy conservation effect.

KEYWORDS

Air curtain, flow visualization, heat insulation efficiency, energy conservation

INTRODUCTION

When an air curtain operates under steady-state, its energy conservation effect can be described by the thermal insulation efficient. The thermal insulation efficiency of air curtain is $\eta = 1 - \dfrac{Q_1}{Q_2}$, with Q_1 representing the energy which is brought in by outdoor, Q_2 represents the heat transfer capability of air curtain due to the temperature difference between indoor air and outdoor air. When the parameters in air-conditioned room are at design conditions, Q_1 and Q_2 are vary with outdoor conditions. This paper calculates the thermal insulation efficient and energy conservation effect of the air curtain by a dynamic method.

THE CALCULATION OF INVADING HEAT Q_2

The stack effect caused by the temperature difference between indoor and outdoor air at the entrance to the air–conditioned room, result in convection between the indoor and outdoor cold or hot air. The amount of heat (Q_2) which is brought in by the outdoor air when it comes indoors is determined by the enthalpy difference between outdoor and indoor air, the flow rate of the convection, and the convection heat transfer coefficient.

$$Q_2 = G^{'} \varepsilon \left(i_w - i_n\right) \quad (\text{kW}) \tag{1}$$

$G^{'}$: the amount of convection between the outdoor and indoor air

ε : the convection heat transfer coefficient $\quad \varepsilon = \dfrac{\gamma_n}{\gamma_w}$

i_w: the enthalpy of outdoor air(kJ/kg dry air)

i_N: the enthalpy of indoor air (kJ/kg dry air)

The amount of the convection flow rate through the gate (G') can be got form the following formula that is recommended by the Design Institute of the Commercial Department (P.R. China).

$$G' = 2.95uB \sqrt{\frac{\gamma_w(\gamma_N - \gamma_w)H^3}{\left(1 + \sqrt[3]{\gamma_w/\gamma_N}\right)^3}} \quad \text{(kg/s)} \tag{2}$$

In this formula:

u: the flowrate coefficient of the gate;

B: the width of the gate (m);

H: the height of the gate(m);

γ_N: the density of indoor air(kg/m^3);

γ_w: the density of outdoor air(kg/m^3)

The indoor air-conditioned design parameters in shops in Shanghai are: the indoor air design temperature $t_c=28°C$, relative humidity $\phi_c=60\%$, outdoor air design temperature $t_w=34C$, relative humidity $\phi_w=65\%$. The height of the gate H is equal to 2.5m, the width of the gate B is equal to 0.9m, the flow rate coefficient of the gate u =0.64.We can calculate the invading heat in summer.

The enthalpy of air and the corresponding time τ at which it occurs are listed in the table below for the climate of Shanghai.

The outdoor enthalpy (kJ/kg dry air)	66.53	70.35	74.38	78.58	82.99	87.57	91.39
Appearing time τ_i (hour)	266	283	371	465	384	173	57

The outdoor relative humidity in Shanghai is 65%, according to the above table, each value of i_{wi} can be obtained corresponding to t_{wi} on i-d diagram, we can work out the density of the air (γ_{wi}), the value of , ε , G_i and Q_{2i}. Finally, we can calculate the invading heat in summer by formula 6. These results are listed in Table 1.

TABLE1

THE CALCULATIVE RESULT OF THE INVADING HEAT IN THE SUMMER

i_{wi}	t_{wi}	Υ_{wi}	G_i	G_i	τ_i	i_{wi}-i_N	ε	Q_2	ΣQ_{2i}
66.53	27.5	1.175	0.000	0.000	266	1.89	1.002	0	0
7035	28.8	1.170	0.281	1012	283	5.71	0.997	1.63×10^6	1.63×10^6
74.38	29.9	1.166	0.429	1546	371	9.74	0.994	5.54×10^6	7.15×10^6
78.58	31.0	1.161	0.561	2020	465	13.94	0.990	1.29×10^6	2.01×10^6
82.99	32.2	1.157	0.647	2330	384	19.35	0.986	1.61×10^6	3.62×10^6
87.57	33.2	1.153	0.722	2601	173	22.93	0.983	1.01×10^6	4.63×10^6
91.39	34.2	1.149	0.790	2846	57	27.85	0.980	4.39×10^6	5.07×10^6

From the Table 1, the total invading heat in summer: $Q_2 = \sum Q_{2i} = 5.07\times10^7$ (kJ)

CALCULATION OF THE AMOUNT OF HEAT TRANSFER OF AN AIR CURTAIN (Q₁)

When the air curtain is in stable operation, the sensible heat due to the temperature difference between indoor and outdoor air can be calculated by the below formula:

$$Q_1 = k(t_w - t_N)HW\tau \quad (kJ) \tag{3}$$

In the formula:

K: heat transfer coefficient of the air curtain (w/m^2.$^\circ$C);

t_w: the dry-ball temperature of outdoor air ($^\circ$C);

t_N: the design dry-ball temperature of indoor air($^\circ$C);

H: the height of the air curtain (m);

W: the width of the air curtain (m);

τ : the appearing time of t_w (hour).

Because the outdoor temperature t_w varies with time, the correspondingly value of τ should be decided according to the climatic data. So, Q_1 should be calculated by a dynamic method.

The American scholar Hayes (1969) has done a lot of research on the heat transfer of air curtains. The transfer heat of the air curtain relates to its minimum bent modules D_{min} and its relative length H/b_0. When the air curtain is in a stable operation, it's heat transfer coefficient K is only related to the relative length of the air curtain, as shown Fig. 1. This result is obtained under the conditions that the height (H) is equal to 2.1m, the indoor temperature (t_N) is equal to 23.3 $^\circ$C, the outdoor temperature (t_w) is equal to 23.9°C. Under other conditions, the heat transfer coefficient must be revised.

$$K=K_1C_A \tag{4}$$

In the fomula:

K_1 can be calculated using Fig. 1;

C_A revised coefficient;

$$C_A = \sqrt{2.95H\left(1 - \frac{273+t_N}{273+t_w}\right)}$$

Besides sensible heat transfer, there is latent heat transfer in the air curtain. The ratio of the total heat transfer to the sensible heat transfer of the air curtain is expressed by the value of m.

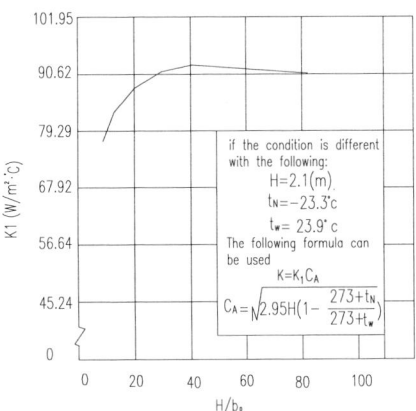

Fig. 1:   Heat transfer coefficient of air curtain

$$m = \frac{sensible heat + latent heat}{sensible heat} = \frac{i_w - i_N}{0.24(t_w - t_N)} \tag{5}$$

So the total heat transfer of the air curtain is:

$$Q_{total} = mQ_1 \tag{6}$$

If the height of the air curtain (H) is equal to 2.5m, width (W) is equal to 0.9m, the outlet thickness (b_0) is equal to 0.07m. We can calculate the total heat transfer of the air curtain in stable operation.

We have known $H/b_0=36$, $t_N = 28°C$, $i_N = 64.33$(kJ/kg dry air). From the Fig.1, we can obtain $K_1=90.62$(w/m^2), then put the parameters into the formula (4),(5),(6), and find the corresponding value of τ for the climatic data of Shanghai. The result is shown in Table 2.

TABLE2
THE CACULATIVE RESULT OF THE TOTAL HEAT TRANSFERRED BY THE AIR
CURTAIN

t_w	t_w-t_N	C_A	K	τ	Q_1	i_w	m	Q_1	ΣQ_1
28.8	0.8	0.140	12.69	283	23228	70.02	7.08	1.64×10^6	
29.9	1.9	0.215	19.49	371	111113	74.03	5.09	5.64×10^6	7.32×10^6
31.0	3.0	0.270	24.45	465	275784	78.21	4.61	1.27×10^6	2.0×10^6
32.2	4.2	0.319	28.87	384	376480	82.60	4.34	1.63×10^6	3.64×10^6
33.3	5.3	0.357	32.38	173	240078	87.15	4.29	1.03×10^6	4.68×10^6
34.2	6.2	0.386	34.96	57	99910	90.96	4.46	4.47×10^6	5.10×10^6

From the Table 2, the total heat transfer through the air curtain in summer is below:
$$Q_1=5.10 \times 10^6 (kJ)$$

ENERGY CONSERVATION EFFECT OF AN AIR CURTAIN

According to the analysis of the climatic data, we can see that the difference between the enthalpy of indoor and outdoor air (i_{wi}-i_N) present a distribution status that the middle is big and the two ends are small. So is the appearing time. Now, in Fig. 2, the time frequency $f(\tau)=\dfrac{\tau_i}{\sum \tau_i}$ is plotted on the ordinate axis, and the difference between the enthalpy of indoor and outdoor air is plotted on the abscissa axis, so the enthalpy frequency distribution in Shanghai is shown in Fig. 2. In this figure, the symbol o is a statistical value which can be obtained from the local climatic data. By doing curve analysis with these statistical values, we can get an empirical formula for the frequency distribution of the difference in the enthalpy.
$$f(\tau)=0.233e^{-0.171(\Delta i-3.32)^2} \tag{7}$$
The error analysis of the formula is shown in the Table 3. The mean square error of the formula $\sqrt{M} = 0.080704573$. So the frequency distribution of enthalpy for the difference between indoor and outdoor represents normal distribution.

TABLE 3
ERROR ANALYSIS OF THE ENTHALPY DIFFERENCE FREQUENCY

Δi	$f(\tau)$statistic value	$f'(\tau)$calculated value	$\Delta = f(\tau)-f'(\tau)$	Δ^2	$M=\Sigma \Delta^2$	$\sqrt{M}$
0.45	0.133	0.0569	-0.0760	0.005780528	0.006513228	0.080704573
1.36	0.142	0.1208	-0.0212	0.000449498		
2.32	0.186	0.1964	0.0103	0.00010769		
3.32	0.233	0.233	0.0000	0.0000000		
4.37	0.194	0.1929	0.00096	0.000000932		
5.46	0.087	0.1065	0.01134	0.000128534		
6.63	0.029	0.0358	0.00679	0.000046044		

The heat invading indoors in the summer (Q_{2i}) is in directly proportional to the enthalpy difference between indoor and outdoor enthalpy and the appearing time τ_i.. Let us take the dimensionless value of the invading heat $Q_a = \dfrac{Q_{2i}}{\sum Q_{2i}}$ and the dimensionless value of enthalpy difference

between indoor and outdoor $i = \dfrac{\Delta i}{i_N}$ as the parameters, then using data in Table 1, the distribution

of the dimensionless invading heat in the summer can be calculated. In the same way, we can take

the dimensionless value of the total heat transfer $Q_b = \dfrac{Q_{1i}}{\sum Q_{2i}}$ and the dimensionless value of

enthalpy difference between indoor and outdoor $i = \dfrac{\Delta i}{i_N}$ as the parameters, then using the data in

the Table 2, the distribution of the dimensionless heat transferred through the air curtain in the
summer can be worked out. The two sets of dimensionless data are listed in the Table 4.

TABLE 4
THE DISTRIBUTION OF THE DIMENSIONLESS ENTHALPY DIFFERENCE AND
DIMENSIONLESS HEAT

The dimensionless enthalpy difference I	0.088	0.151	0.216	0.284	0.366	0.431
The dimensionless invading heat Q_a	0.032	0.109	0.255	0.318	0.200	0.087
The dimensionless transferred heat Q_b	0.00323	0.0111	0.0251	0.0322	0.0203	0.00882

The data in Table 4 can be changed into the distribution curve of dimensionless heat. This is
shown in Fig. 3. In this figure, the dotted line represents the distribution curve of Q_a, the area
embraced by the dotted line and I axis represents the invading heat in the summer. The solid line
represents the distribution curve of the Q_b. The area embraced by the solid line and the i-axis
represents the total heat transferred by the air curtain. The irregular area shown by the oblique line
represents the saved dimensionless heat. The following are known:

$$Q_2 = \sum Q_i \cdot \sum Q_{1i} \qquad (8)$$
$$Q_2 = \sum Q_i \cdot \sum Q_{2i} \qquad (9)$$

We can get some information from Table 4;

$$\sum Q_a = 0.032 + 0.109 + 0.255 + 0.318 + 0.200 + 0.37 = 1$$
$$\sum Q_b = 0.00323 + 0.0111 + 0.025 + 0.0322 + 0.0203 + 0.00882 = 0.10075$$

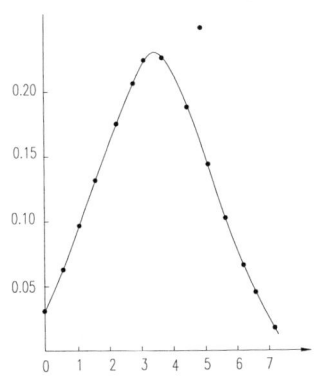
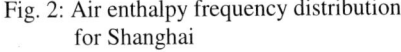
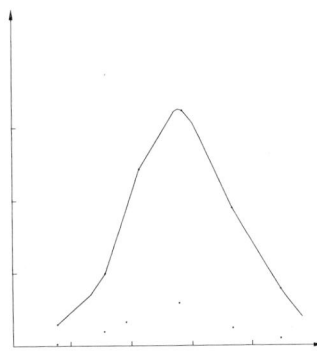

Fig. 2: Air enthalpy frequency distribution   Fig. 3: The distribution curve of dimensionless heat
for Shanghai

From the Table 1:

$$\sum Q_{1i} = 5.07 \times 10^7 \text{ (kJ)}$$

Putting the above data separately into the formula (8),(9), we can get:

$$Q_2 = 1.00 \times 5.07 \times 10^7 = 5.07 \times 10^7 \quad \text{kJ}$$

$$Q_1 = 0.10075 \times 5.07 \times 10^7 = 5.11 \times 10^6 \quad \text{kJ}$$

So, the efficiency of the air curtain for the whole summer is:

$$\eta = 1 - \frac{Q_1}{Q_2} = 1 - \frac{5.11}{50.7} = 0.899$$

The saved cooling capacity of the air curtain for the whole summer is

$$Q_{saved} = Q_1 - Q_2 = 5.07 \times 10^7 - 5.11 \times 10^6 = 4.56 \times 10^7 \quad \text{kJ}$$

According to calculations for a real engineering project, the unit cooling capacity power of the air-conditioning system of a civilian building in Shanghai is given by:

$$p = 1.35 \times 10^5 \, (\frac{kwh}{kJ})$$

So the saved power of the air curtain :

$$p = Q_{saved} \cdot p = 1091 \times 5.65 = 6164 (kW \cdot h)$$

In this example, the plane air curtain is employed. The power of this air curtain is:

$$N = 0.114 kW \cdot h$$

The operation time in the summer is: $\sum \tau = 1733h$

The consumption of power of the air curtain is: $p_{air-curtain} = 0.114 \times 1733 = 198 (kW \cdot h)$

So the really saved power in the summer of each plane air curtain is:

$$P_{reral} = p - p_{air-curtain} = 6164 - 198 = 5966 (kW \cdot h)$$

CONCLUSIONS

1) It is a reasonable method to calculate the energy conservation effect of the air curtain by the dynamic method. It is more practical than the stedy-state method.

2) It is effective to install air curtains in air-conditioned buildings because the plane air curtain has a reasonable dimensions that is suitable for most buildings. This air curtain cannot only be used in new built buildings, but also in the refurbished buildings.

3) This paper calculates the cooling insulation efficiency of a plane air curtain by the dynamic method to be $\eta = 89.9\%$

REFERENCES

F.C.Hayes and W.F.Stoecker.(1969). Heat transfer characteristics of the air curtain ASHRAE Transactions, Vol.75, Part II.

F.C.Hayes and W.F.Stoecker. (1969). Design data for air curtain, ASHRAE Transactions ,Vol.75, Part II.

Li Qiangmin, Peng Xiahui and Folke Peterson. (1994). Duck down pollution control with recycle air curtain, Proceedings of the 4th International Symposium on Ventilation for Contaminant Control, Stockholm, Sweden.

AUTHOR INDEX

Volumes I and II

KEYWORD INDEX

Volumes I and II